SPACE NUCLEAR POWER SYSTEMS 1986

Edited by

Mohamed S. El-Genk
Institute for Space Nuclear
Power Studies
University of New Mexico

Mark D. Hoover
Lovelace Inhalation Toxicology
Research Institute
Albuquerque, New Mexico

VOLUME 5

OF THE SERIES *SPACE NUCLEAR POWER SYSTEMS*

ORBIT
A FOUNDATION SERIES

ORBIT BOOK COMPANY, INC.
MALABAR, FLORIDA

Original Edition 1987

Printed and Published by
ORBIT BOOK COMPANY, INC
2005 Township Road
Malabar, FL 32950

Printed in the United States of America

This material is listed in the Department of Energy Technical Information Center as CONF-860102.

Library of Congress Cataloging-in-Publication Data

Symposium on Space Nuclear Power Systems (3rd: 1986 :
Albuquerque, N.M.)
Space nuclear power systems 1986.

(Space nuclear power systems ; v. 5)

Includes index.
1. Space vehicles—Nuclear power plants—Congresses.
I. El-Genk, Mohamed S. II. Hoover, Mark D. III. Title. IV. Series.
TL1102.N8S96 1987 629.47'443 87–18551
ISBN 0–89464–017–8

10 9 8 7 6 5 4 3 2

PREFACE

The group of researchers that got together back in early 1983 recognized the importance of expanding the development of safe and reliable nuclear power systems for use in the exploration and utilization of space. They visualized the need for direct communication among those involved and interested in this field. Thus, they came together from the University of New Mexico, the Air Force Weapons Laboratory, Sandia National Laboratories, and Los Alamos National Laboratory and formulated the plans for a yearly symposium on the development of nuclear power for space use. Starting in January 1984 the University of New Mexico has sponsored the very successful yearly Symposium on Space Nuclear Power Systems.

This, the 3rd symposium, was organized by the University of New Mexico's Institute for Space Nuclear Power Studies. It was held in Albuquerque, New Mexico on January 13–16, 1986. There were 400 registered attendees representing universities, national laboratories, private industry, and government agencies. They came from the United States and several foreign countries. There were scientists, engineers, managers, and students present. They represented developers as well as potential users of space nuclear power. All sessions were well attended at this very successful symposium. Many new ideas and much recent data was exchanged and discussed.

Many organizations as well as individuals worked hard and contributed generously to sponsor, plan, organize, and manage this symposium. They deserve the credit for this third yearly success.

The Symposium was co-sponsored by the American Nuclear Society, the ANS Trinity Section, Los Alamos National Laboratory, Sandia National Laboratories, and the U.S. Air Force Space Technology Center. There were over 22 participating organizations and exhibitors. Financial support for the symposium came from a number of industrial and federal sponsors including the BDM corporation, GA Technologies, the General Electric Company, and the U.S. Department of Energy.

The Steering Committee of the Symposium consisted of a number of respected members of the technical community from universities, government, and industry. The Committee included: Dr. Jay Boudreau, Los Alamos National Laboratory; Tom J. Farer, President of the University of New Mexico; Colonel John Friel, Air Force Space Technology Center; Goetz K. Oertel, U.S. Department of Energy; Dr. Nester R. Ortiz, Sandia National Laboratories; Ray Romatowski, U.S. Department of Energy; Dr. Duane C. Sewell, Consultant; Dr. Giles Sinkewicz, General Electric Company; and Dr. Earl Van Landingham, NASA Headquarters. The Committee provided general guidance to the local organizers with respect to the direction and technical content of the symposium.

The Organizing Committee of the Symposium was responsible for local planning and execution of the Symposium. The Honorary Chairman was Congressman Manuel Lujan, Jr. (R, NM) to whom is owed a great debt of gratitude for his support and encouragement. Professor David M. Woodall of the University of New Mexico was the General Chairman. The Technical Program Committee was chaired by Professor Mohamed S. El-Genk of the University of New Mexico and co-chaired by David Buden of Science Applications International Corporation. Those individuals were responsible for coordination and planning of the important technical aspects of the Symposium. Mary Lujan and Carlota Klimas of the University of New Mexico served as the Administrative Chairwomen. Their expertise in conference organization contributed to the smooth operation of the Symposium and the competence with which local details were handled. Dr. Mark D. Hoover of the Lovelace Inhalation Toxicology Research Institute, ably handled the publication interface for the Symposium. Richard Jimenez of the Aerospace Corporation performed the duties of Finance Chairman for the Symposium.

Many outstanding speakers were attracted to the symposium which contributed a large measure to its success. A special thanks is due to Symposium Steering Committee member Ernest C. Hardin, Jr. for serving as master of ceremonies at both the luncheon and the banquet. The banquet speaker was Dr. Linda Cordell, Chairwoman of the Department of Anthropology at the University Of New Mexico. Her talk on the Indian cultures of New Mexico was exceptionally enlightening. The luncheon speaker was Dr. Joseph Angelo, Jr. of the Florida Institute of Technology, who gave an informative and entertaining talk entitled, ''Beyond the Space Station or Is There a Martian in Your Future?''

The Plenary Session presented an outstanding opening to the Symposium, emphasizing the importance of technology developments in space. The session was chaired by Dr. Everet Beckner of Sandia National Laboratories. The speakers included: the Honorable James Vaughan, Jr., Acting Assistant Secretary for Nuclear Energy of the U.S. Department of Energy; Colonel Edwin W. A. Peura, Director of the Office of Plans, U.S. Air Force Space Division; and the Honorable Manuel Lujan, Jr. (R, NM).

The Symposium was preceeded by a two day short course on High Temperature Heat Pipes. The featured lecturers included Michael Merrigan and Will Ranken, both of Los Alamos National Laboratory. The material presented included an excellent overview of the field.

Among the short course and Symposium attendees were a large number of undergraduate and graduate engineering students. The most valuable service provided by the symposium to the technical community was probably the interest which it generated among faculty and students of the engineering educational institutions across the country. That impact will be felt in years to come as more trained professionals join the ranks of those who are developing this exciting and important technology.

Duane C. Sewell
Member
Symposium Steering Committee

INTRODUCTION

This volume contains the peer reviewed and edited versions of many of the papers presented at the 3rd Symposium on Space Nuclear Power Systems held in Albuquerque, New Mexico, January 13–16, 1986. It is entitled *Space Nuclear Power Systems 1986* and is the fifth volume in the series *Space Nuclear Power Systems*. Volumes one and two are entitled *Space Nuclear Power Systems 1984* and cover the 1st Symposium, held January 11–13, 1984. Volumes three and four are entitled *Space Nuclear Power Systems 1985* and cover the 2nd Symposium, held January 14–16, 1985. The objective of the symposia, and hence these volumes, is to summarize the state of knowledge in the area of space nuclear power systems and to provide a forum at which the most recent findings and important new developments can be presented. The Symposia have succeeded largely because they attract leading experts in the field at a time when there is a renewed national interest in nuclear power systems for applications in space.

Space Nuclear Power Systems 1986 includes a total of 45 papers prepared by 119 authors from more than 30 organizations. We are grateful to the authors for their cooperation, enthusiasm, and sincere desire to contribute to a clear, useful, and technically sound publication.

For the purpose of organizing the material, the Symposium papers are divided into 14 parts. Part 1 provides a congressional perspective on the U.S. space nuclear power program, addresses some of the promises and challenges of both civilian and military applications of space nuclear power, and discusses some of our national prospects for opening new vistas in space. Part 2 goes into more detail on the power requirements for future missions in space, with special emphasis on utilizing the natural resources of space. Part 3 deals with the critical issue of safety, describing the inherent safety features of current designs, the probabilities of accidents, and the special concerns related to the reentry or ultimate disposal of reactor systems.

Part 4 gives the neutronics characteristics of space nuclear power reactors and describes models for simulating the performance of reactor systems. Part 5 describes progress in designing, simulating, and testing systems for thermal management in space. Part 6 provides an update on progress in the development of energy conversion systems, including the analysis of transient behavior and the development of fuel cells for burst power operation. Part 7 deals with the problems of power conditioning in space, and describes advances in materials technology for superconductors and insulators for use in space reactor environments. Part 8 describes an expert system for developing space power supplies and details user interface and operational issues for thermionic systems. Part 9 gives an update on the suitability and performance of special refractory alloy materials for use in space.

Part 10 analyzes the radiation damage mechanisms for electronic and microelectronic components in space. Part 11 describes recent tests on a ground-based reactor with characteristics relevant to a space nuclear system. Part 12 covers the technologies, feasibilities, and potential costs of using nuclear systems for propulsion in space. Part 13 describes and compares a number of potential options for obtaining multimegawatt quantities of power in space, and also describes the design based limitations of certain systems for providing such power. Part 14 deals with the performance and hardware status of radioisotope thermoelectric generators and describes progress in the design of dynamic systems involving the Rankine or Brayton cycles.

The preparation of this volume has been accomplished through the encouragement, support, and diligence of many individuals. The chairpersons of the technical sessions at the Symposium were a fundamental part of the network we established for conducting the peer review process. We gratefully acknowledge their contributions in nominating appropriate technical experts to review manuscripts and in providing us with their composite reviews. The chairpersons of the Symposium sessions were: **Plenary**, Everet H. Beckner, Sandia National Laboratories; **Space Nuclear Missions and Applications**, Gordon R. Woodcock, Boeing Aerospace Company, and Terry Wallace, Los Alamos National Laboratory; **Radiation Effects**, James Gover, Sandia National Laboratories, and Arvind Kumar, University of Missouri-Rolla; **Thermal Management**, Patrick Magee, General Electric Company, and Michael Merrigan, Los Alamos National Laboratory; **Systems Integration**, Robert V. Anderson, Rockwell International, and John Carson, Air Force Space Technology Center; **Space Reactor Fuels**, R. Bruce Matthews, Los Alamos National Laboratory, and Dale Dutt, Hanford Engineering Development Laboratory; **Nuclear Electric and Nuclear Propulsion**, Robert J. Vondra, Air Force Rocket Propulsion Laboratory, and Joe Wetch, Space Power, Inc.; **Radioisotope Generators**, Jerry R. Peterson, General Electric Company, and James J. Lombardo, U.S. Department of Energy; **Nuclear Safety**, Robert C. Nelson, Air Force Directorate of Nuclear Surety, and Richard B. Harty, Rockwell International; **Energy Conversion**, Gary O. Fitzpatrick, GA Technologies, Inc., and Gerry Stapfer, Jet Propulsion Laboratory; **Reactors and Shielding**, William J. Barattino, United States Air Force, and Sterling Bailey, General Electric Company; **Testing**, Richard Rice, Idaho National Engineering Laboratory, and Wendy Dixon, U.S. Department of Energy; **Refractory Alloys**, Roy Cooper, Oak Ridge National Laboratory, R. William Buckman, Jr., Westinghouse Electric Corporation, and William O. Harms, Oak Ridge National Laboratory; **Control, Power Conditioning, and Distribution**, Arthur D. Schoenfeld, TRW, Inc., and Alfred O. Britting, Jr., Aerospace Corporation; **Advanced Multimegawatt Systems Concepts**, R. J. Sovie, NASA Lewis Research Center, and James H. Lee, Jr., Defense Nuclear Agency; and **Technical Exchange**, Patrick J. McDaniel, Sandia National Laboratories.

We wish to acknowledge the individual technical reviewers who unselfishly contributed their time and expertise to our extensive peer review process. Reliable scientific literature cannot be produced without this kind of dedicated effort. The names and affiliations of reviewers provided to us through the completed manuscript peer review report forms are recorded in the *List of Peer Reviewers* at the end of this volume. We also express our gratitude to the many other individuals, whose names we do not have, who contributed to the peer review process at various stages of the preparation and revision of the Symposium papers.

In the preparation of these volumes for publication, we gratefully acknowledge the skilled support of Mr. Robert E. Krieger, President, The Orbit Book Company, and of Marie Bowles and the other enthusiastic and competent members of his organization. We are also grateful to the management and staff of J. Jarrett Engineering for their excellent typesetting skill. We thank Dr. Stephen Seiffert and his colleagues at The BDM Corporation for valuable editorial assistance. For staff support we gratefully acknowledge the competent help of Carlota Klimas, Mary Bragg, and Michael Vigil of the Institute for Space Nuclear Power Studies, University of New Mexico. For the encouragement to tackle and complete this task, we are grateful to Prof. Frank Williams, Chairman, Department of Chemical and Nuclear Engineering, University of New Mexico, to Prof. David M. Woodall, General Chairman of the 3rd Symposium, and to Drs. Hsu-Chi Yeh, Bruce B. Boecker, and Roger O. McClellan of the Lovelace Inhalation Toxicology Research Institute.

Our final and most heartfelt thanks go to our wives Martha Hoover and Irene El-Genk, and to our children Alison Hoover and Ryan Hoover and Hussein El-Genk for their continued understanding, patience, and encouragement during the preparation of this volume.

Mohamed S. El-Genk
Technical Chairman

Mark D. Hoover
Publications Chairman

Mohamed S. El-Genk is an associate professor of Chemical and Nuclear Engineering and is the director of the Institute for Space Nuclear Power Studies at the University of New Mexico. Dr. El-Genk holds BS (1968) and MS (1975) degrees from the University of Alexandria, Egypt, and a Ph.D. degree (1978) from the University of New Mexico, all in nuclear engineering. From 1978 to 1981 he was an engineering specialist doing light water reactor fuel behavior research at the Idaho National Engineering Laboratory. In 1981 he joined the Faculty at the University of New Mexico. Dr. El-Genk has produced numerous technical articles and reports on light water reactor and liquid metal fast breeder reactor safety, heat transfer and thermal hydraulics, on nuclear fuel irradiation and swelling behavior, on fission gas release, and on the design, modeling, safety, and reliability of space nuclear power systems. He is a Distinguished Faculty Member of the nuclear science and engineering honor society Alpha Nu Sigma and a member of Phi Kappa Phi, the American Nuclear Society, the American Society of Mechanical Engineers, and the American Institute of Chemical Engineers.

Mark D. Hoover is a research scientist at the Lovelace Inhalation Toxicology Research Institute in Albuquerque, New Mexico. Dr. Hoover holds a BS degree (1970) in mathematics and English from Carnegie-Mellon University and MS (1975) and Ph.D. (1980) degrees from the University of New Mexico in nuclear engineering. His research focuses on the inhalation hazards associated with nuclear fuel cycles. Dr. Hoover has published in the open literature on aerosols from fusion energy systems, aerosols from fabrication of mixed-oxide reactor fuels, instrumentation for characterizing airborne particles, generation and characterization of aerosols by laser vaporization, and on the inhalation hazards of plutonium recycle, use of thorium-uranium fuels, handling and use of lithium, use of beryllium in reactor systems, and decommissioning of radioactively contaminated facilities. He is a member of the Honor Societies of Phi Kappa Phi and Tau Beta Pi, and the American Association for Aerosol Research. He is Chairman of the New Mexico Section of American Nuclear Society (ANS) and Program Chairman of the ANS Environmental Sciences Division.

ABOUT THE SYMPOSIUM LOGO

The four corners of the logo represent the points of a compass and symbolize the role space nuclear power can play in the exploration of new frontiers; the large and small circles represent the integrated space system with a nuclear power plant at its center; the radiating spirals, also resembling a stylized atomic nucleus, symbolize the transfer of thermal and electric energy to support the space mission; and the logo's overall appearance of simplicity reflects the emphasis on space nuclear power systems that can operate without failure and provide reliable power for the duration of their mission.

The Series *Space Nuclear Power Systems*
Vols. 1 and 2 — *Space Nuclear Power Systems 1984*
Vols. 3 and 4 — *Space Nuclear Power Systems 1985*
Vol. 5 — *Space Nuclear Power Systems 1986*

TABLE OF CONTENTS

VOLUME 5

Part 1

PLENARY

Part 2

MISSIONS AND APPLICATIONS

Part 3

NUCLEAR SAFETY

Part 4

REACTORS AND SHIELDING

Part 5

THERMAL MANAGEMENT

Part 6

ENERGY CONVERSION

Part 7

CONTROL, POWER CONDITIONING, AND DISTRIBUTION

Part 8

SYSTEMS INTEGRATION

Part 9

REFRACTORY ALLOYS

Part 10

RADIATION EFFECTS

Part 11

TESTING

Part 12

NUCLEAR ELECTRIC AND NUCLEAR PROPULSION

Part 13

MULTIMEGAWATT SYSTEMS

Part 14

RADIOISOTOPE GENERATORS

Part 1

PLENARY

Space Nuclear Power Systems 1986
Edited by M. S. El-Genk and M. D. Hoover

Chapter 1

Congressional Perspectives on Space Nuclear Power

The Honorable Manuel Lujan, Jr.

U. S. House of Representatives (R, NM)
(1st District—New Mexico)
1323 Longworth Office Building
Washington, DC 20515

Introduction

I am both pleased and honored to speak to you here today at this third symposium on Space Nuclear Power Systems. As you know, the Symposia held during the last two years have been a tremendous success—so successful, in fact, that I hope this forum is now an annual tradition and a permanent institution.

I want to take this opportunity to express my thanks and appreciation to some of the people and institutions who have worked hard to make this symposium happen. General Chairman Dr. David Woodall and Technical Chairman Dr. Mohamed El-Genk from the University of New Mexico have done an outstanding job in putting together a highly informative and sophisticated program. And with the help of more than 60 men and women on the committees alone, and the participation of more than 50 organizations, this symposium cannot fail to be another success. I want to take a moment and extend my special thanks to the New Mexico technical community for their continued involvement and co-sponsorship: the Institute for Space Nuclear Power Studies of the University of New Mexico, the Trinity Section of the American Nuclear Society, the Sandia and Los Alamos National Laboratories of the Department of Energy, and the Air Force Space Technology Center. New Mexico's scientific community has a long and distinguished history of nuclear research and technology development, and I know that Albuquerque is delighted to serve again as host city for this symposium.

I believe that the space nuclear power community needs this forum. Here, we share information on the latest developments in research and technology development, as well as program management and resource issues. If we are to use nuclear power in the years ahead to meet our space power needs, and I feel we must, then this symposium serves a vital rôle in achieving that goal. Over the next three days, not only will there be an opportunity to discuss a wide range of technology issues, but there also will be an opportunity to discuss options for future management, resource requirements, and possible missions for space nuclear power, and methods to generate increased support for the programs. These are all critically important issues and ones which we need to share an understanding of, and then to work closely to form a consensus on, during the coming years.

The Need for Adequate Energy Sources

In large part, this country's ability to conduct future programs of both space exploration and space exploitation will be enabled—or limited—by our ability to provide adequate energy supplies. The fact is that we cannot fully explore space without sustained, sufficient and reliable long-term sources of power; we simply cannot exploit space for commercial or defense purposes without energy sources considerably more capable than those we have in our inventory today. All studies conclude that nuclear power is a key component of our future space infrastructure.

As you all well know, in space there is one almost uninterrupted natural source of available power—solar energy. To date, electrical power in space has been largely provided by solar cells, and additionally chemical systems and some nuclear sources—radioisotope thermoelectric generators—have been used for specific applications. While these are all technologies which we need to continue to develop and use, it is clear that they alone will not be adequate to meet our space power needs of the future. As the human race probes deeper and deeper into the universe and farther and farther from Planet Earth, nuclear power sources will ultimately become necessary. Already the Pioneer, Viking and Voyager missions to Jupiter, Mars, Saturn and beyond have pointed to the feasibility and necessity of using nuclear sources to power such spacecraft. And the strategic defense of the U.S. and our allies must depend on nuclear power sources, for nuclear power is the only answer to SDI demands for reliable, survivable, high-wattage sources of power. Success of nearly all elements of the SDI is dependent on major advances in space power.

Civilian Requirements

Let me briefly talk about civilian requirements for space nuclear power. In the future, there are two basic classes of civilian missions and activities in space for which nuclear power is enabling or enhancing: those where great quantities of electrical power are required over extended periods of time and those which operate far from the sun. In the field of space exploration, we will need electrical power supplies for bases established on planetary bodies like Mars or the Moon, and nuclear electric propulsion for manned or unmanned probes into deep space. Closer to the earth, large quantities of power will be required for earth orbital applications as well: electrical power for an expanded, permanently-manned space station, geosynchronous communication platforms, long-lived remote sensing satellites, and air/ocean traffic control systems.

Defense Applications

But potential defense applications seem more probable for nearer term nuclear power applications. One possible near term application is space-based radars for surveillance, acquisition, tracking, and target discrimination. In fact, a hybrid bistatic radar system being developed by General Dynamics could be used for both SDI and for surveillance in air defense systems. Such a system would only require about 35 kilowatts of power, much less than other radar concepts. In addition, battle management systems, weapons platform housekeeping and prime power requirements, and orbit transfer via electric propulsion are all possibilities for nuclear power. And I think it has become clear to those who follow the space nuclear power issue that we must aggressively develop a technology base to provide power to future missions with even higher energy requirements in the multimegawatts scale. With the rapid development of concepts for the President's Strategic Defense Initiative, there is a whole new realm of possible missions and power requirements for nuclear power in the space military arena. While many of the potential mission requirements are, of necessity, classified, I am sure that some of you will touch on the enabling technologies in the last day's sessions. In many cases, the core technologies are the same.

National Initiatives

With all these potential mission applications, the question is: what are we as a nation doing to meet this challenge for the future? The answer, while more heartening than two years ago when the first symposium was held, still gives pause for thought.

The SP–100 Program

First the good news. There have been two recent major positive developments in the space nuclear power field, both related to the SP–100 program, the tri-agency program to develop a thermoelectric reactor system in the 300 kilowatt range. First, as you may know, a Phase II Memorandum of Agreement between the Department of Energy, NASA and the Strategic Defense Initiative Organization was recently signed concerning the conduct of the SP–100 Program. This memorandum defines program scope, management structure, and respective agency responsibilities, as well as providing estimates of resource requirements. This is a clear indication that the SP–100 program is moving ahead and that the Administration and all three agencies are strongly behind this important program.

Second, the Ground Engineering System Test Facility for the SP–100 program has finally been selected—the Hanford Laboratory in Washington state. I have to pause here to comment that Hanford was not my personal first choice. I'm sure you can all appreciate that I would have preferred to have this segment of the program located right here in Albuquerque at Sandia Labs, but the important thing is that this part of the program can now move ahead. With selection of the Ground Engineering System Test Facility completed, the space nuclear power program has taken a significant step toward having the capability of flight-testing a reactor system in the early 1990s.

Increased University Involvement

And I would like to briefly comment on one other aspect of the growing space nuclear power program which I find particularly gratifying and consider to be part of the good news; that is, the increasing interest and involvement of many universities in this research. This university involvement will not diminish the role of industry and the national laboratories at all, but rather will enhance the technical capabilities of the entire program. I would particularly like to commend

SDI's Innovative Science and Technology Office for their work in initiating the "Non-Nuclear Space Power Consortium" under the leadership of Dr. Frank Rose of Auburn University and the "Innovative Nuclear Space Power Institute" under the leadership of Dr. Nils Diaz of California State University (Long Beach). The involvement of these groups and a number of associated universities in the research and development of space nuclear power broadens the program's technical base, and also the political base, for the entire effort. Of equal importance, the university participation will result in development of curricula specifically aimed at space nuclear power applications and the education and training of the science and engineering personnel necessary in the decades ahead to make it all a reality.

Economic Obstacles

Now let me regrettably turn to what I regard as the bad news; and that is that it's going to take an enormous amount of additional work—both political and technological—to continue building on this foundation and to actually make space nuclear power a reality. Let me expand on the trials we all still have to face.

As a nation, we face an economic crisis of frightening proportions. In an effort to dramatically reduce spending, and curb the ever-escalating deficit, the Congress recently passed legislation known as the Gramm-Rudman-Hollings Deficit Reduction Act of 1985. As you probably know, the major mission of this law is to cut growth in the federal deficit to zero by 1991. While the full programmatic ramifications of this bill are not yet well understood, one thing has become perfectly clear: the impact of Gramm-Rudman on new and emerging programs in particular will be serious in all of the federal agencies. In 1986, most federal programs face cuts estimated at up to 5 percent. While no program may be eliminated in 1986 under the terms of this bill, some programs may receive obviously crippling cuts. In 1987, the bite of Gramm-Rudman will deepen and some new programs may be deferred for a period of years. While the details of the specific impact of Gramm-Rudman on selected programs is not clear, what is clear is that the next few years will be a very critical period for the space nuclear power program and its potential mission applications. To continue to develop and demonstrate the emerging technologies, it will first be necessary to ensure that the program's priorities remain relatively high.

This is *our* task—those of us in Congress and on the Science and Technology committee who support the development of space nuclear power and recognize its importance—but this is also *your* task. I recommend that each of you take on the additional responsibility to communicate the importance of the space nuclear power program on an individual basis to your elected representatives and to scientific colleagues not directly involved with the program. We also need to work together to build the general public's understanding of, and support for, space nuclear power.

A National Commission on Space

As you may know, in an effort to anticipate some of the problems the future may hold for our nation's space program, last year the Congress created a National Commission on Space. We directed the Space Commission to study the future of our country's space program and to report back to the Congress and to the President with specific goals and recommendations. The Space Commission is due to submit its report to Congress and the President on March 17th of this year.

As an ex-officio congressional member of the Space Commission, I have taken the opportunity to provide the Commission and its members with suggestions on long-term space policy issues. One of my top priorities has been the question of how best to meet our long-term space power needs. I am happy to report that a variety of very sound recommendations have been made to the Space Commission by a number of individuals concerning our nation's development of space nuclear power. It is my hope that the Space Commission will accept these recommendations from some of the country's most respected experts on space nuclear power.

I personally have urged the Space Commission to accept the following recommendations:

1. The conversion of the SP–100 nuclear reactor from a ground demonstration test in 1992 to a timely flight demonstration test in space;
2. Expansion of the research and development effort in the multi-megawatt power systems area; and
3. The continued enhancement of the technology base for the radioisotope thermoelectric generator and dynamic isotope power system programs.

A concentrated, coordinated and adequately funded national plan for the development of space nuclear power is also necessary if we are to have a robust space program in the future. Let me reiterate and amplify on the point. I cannot emphasize enough my feelings that the Department of Defense, the Department of Energy, and NASA must now identify a specific near-term space mission requiring a nuclear power system if we are to ensure the ultimate success of this program. The commitment to ground testing of the SP–100 reference mission system is not the ideal way to assure timely development of flight-qualified system in these emerging budget times and in fact could result in endless delays.

Cooperation and Commitment

The challenge is clearly before us. Clearly, implementation of these recommendations will require more than a report to Congress and the President. Every participant in this symposium must accept additional responsiblility to assure the success of future space nuclear power programs. I urge you to communicate the importance of this program to your elected representatives, as well as continuing to work with your coinvestigators. It is clear that it is only by working together that we can make sure that the space nuclear power program moves ahead rapidly in the years ahead when the mood of Congress and the President is to cut the budget rather than expand it. I hope that in the months ahead, I can count on your support in that effort. You can certainly count on mine.

Acknowledgments

This paper is a written version of the remarks made by Rep. Lujan at the opening session of the 3rd Symposium on Space Nuclear Power Systems. Rep. Lujan is the ranking Republican member of the House of Representatives Committee on Science, Space, and Technology.

Space Nuclear Power Systems 1986
Edited by M. S. El-Genk and M. D. Hoover

Chapter 2

Nuclear Power—Opening New Vistas in Space

James W. Vaughan, Jr.

U.S. Department of Energy
Washington, DC 20585

Introduction

It is an honor and a pleasure to be with you again this year representing the Department of Energy as you begin this Third Symposium on Space Nuclear Power Systems.

Our country is on the threshold of a new and exciting era—an era marked by virtually unlimited but very achievable vistas in space. The time is right for building on the capabilities and successes of the space shuttle and for exploiting new opportunities that will be afforded by space stations and the many specialized activities related to the President's Strategic Defense Initiative (SDI). Such endeavors will, of course, demand major technological advances on many fronts, not the least of which is power—one of the key aspects of the SDI effort that has initially received less emphasis than some of the more glamorous aspects such as x-ray lasers.

Enabling power systems will have to be capable of operation in the demanding environment of space for extended periods of time and for varying modes and power levels. Many of the power requirements are beyond the capabilities of currently available sources and thus pose serious challenges to our ingenuity and resolve. They will surely call for innovation in power systems technology. Systems incorporating nuclear energy as a prime source of this enabling power are destined to play a key role.

The American public must come to understand the important role nuclear power has played in our space program and the even more vital role it must play in the years to come. The space nuclear power projects we are formulating today are not something brand new; indeed, they are logical extensions of decades of quiet but effective hard work that have allowed the United States to achieve world leadership in space. Future missions will undoubtedly require significantly larger power sources—requirements which can most effectively be met through continued development of space nuclear power sources.

The Department of Energy continues to welcome the new opportunities of the space age and is determined to remain on its present course—one that will continue to provide the means to take us further and give us more flexibility for our platforms in space. We are fully committed to meeting the engineering and management challenges required to develop nuclear power systems for space, and to adapt our development activities to the mission needs of the user agencies.

Research and development of safe and reliable nuclear space power sources is a priority responsibility among our nuclear energy activities. The programs we now have in progress will significantly advance the civilian and military applications of space technology for use in the 1990s and beyond. These programs are in varying stages of conceptual design, development, demonstration, production, or application and they involve systems with distinctive but complementary performance characteristics.

Cost and Credibility

In the midst of the enthusiasm to move forward with these challenging technological improvements for space nuclear power applications, I must also add two notes of caution.

First, all of us are aware of the importance being placed by this Administration on deficit reduction and—with recent enactment of the Gramm-Rudman-Hollings Act—moving aggressively towards a balanced budget. This makes Federal funds for research and development scarcer than ever before and underscores the importance of the entire technology community—government, national laboratories, universities, and industry—utilizing wisely and prudently those funds which are available. Development of nuclear power for space applications is *not* a new program that represents a "free ride" for every innovative project that scientists and engineers have had buried in their file cabinets for years. Each and every technological development which is proposed and pursued must have merit. Even those projects typically identified as high-risk R&D with potentially high payoff will be carefully scrutinized to assure that there is an appropriate "mission pull" that warrants a technology push. I am counting on the professional integrity of everyone in this room and in the technical community you represent to act as both responsible citizens and responsible professionals to help the Federal Government in our effort to assure that we gain the maximum practical benefit for R&D dollars invested.

Second, the technical assurance of safety and reliability in space nuclear power systems is important but not enough. An equal challenge lies in establishing our credibility with the American people and dispelling their misconceptions about nuclear power. An informed and educated public is essential to our ability to continue to safely deploy nuclear energy sources in space vehicles. Let us not lose sight of our responsibility to articulate to the public our many accomplishments and plans for the future—and about our concern for the same things they are concerned about when it comes to nuclear power—costs, reliability, and, most importantly of all, safety. I challenge each of you in your various roles to do your share in accomplishing this objective.

Against this background, let me review the key space nuclear power programs currently underway and highlight some of the recent accomplishments and future directions.

Role of Radioisotope Thermoelectric Generators

The use of radioisotope thermoelectric generators—or RTGs—as an electric power source for space missions has been a key element in some of our nation's most impressive space exploration projects. Eleven days from now—on January 24th—NASA's Voyager II spacecraft will fly past Uranus, the third largest planet in the solar system and the seventh most distant from the sun. This spacecraft is powered by three RTGs that have produced more than 30,000 kilowatt hours of energy.

Nuclear power is the only option for exploring a planet so distant from our sun that it receives 1/400th as much sunlight as Earth. The RTGs for Voyager II and earlier space missions have proven to be safe, reliable, and long-lived sources of electric power. They have exceeded mission requirements by providing power above that required and beyond planned mission lifetimes.

RTGs have employed a variety of thermoelectric conversion and radioisotope heat source technologies spanning a power range of a few to approximately 160 watts. The next two interplanetary spacecraft scheduled to be launched by NASA—Galileo and Ulysses, in May of this year—will be powered by new General Purpose Heat Source RTGs, each rated at 285 watts-electrical.

We believe RTGs will remain a major space power source option. They may even be mandatory for space missions that require power of up to about 1 kilowatt, high reliability, and very long mission life. The remarkable advancements to date in RTG performance would not have been possible without a dedicated commitment to the continued development of thermoelectric materials technology, a sensitivity to the safety issues involved in the development and exploitation of that technology, and a thorough appreciation of future mission requirements. To support future space missions, such as the NASA Mariner Mark II series in the early 1990s, we plan to continue RTG development on several fronts, including:

1. An improvement of the thermoelectric materials,
2. The development and demonstration of a modular RTG, and
3. Further enhancement of the already impressive safety margin for the General Purpose Heat Source that could be used for the modular RTG or a Dynamic Isotope Power System.

The immediate outlook for productive use of these new RTGs is very promising. The units for the Galileo and Ulysses missions have been assembled and tested, and last week were transported to the Kennedy Space Center for those launches this spring. The qualification unit is approaching 10,000 hours in its life test with no unusual occurrences.

The Next Step for Power from Radioisotopes

To cover the power range from 1 to about 10 kilowatts, DOE is beginning to explore with the U.S. Air Force a Dynamic Isotope Power System, often referred to as DIPS.

A dynamic system is conceptually similar to smaller, static RTGs, but is more efficient. In both systems, the heat from the decay of plutonium-238 is converted into electricity. For RTGs, this conversion is accomplished without moving parts, using thermoelectric materials with efficiencies on the order of 6 to 7 percent. In a dynamic isotope power system, moving parts—involving a turbine/alternator set—can convert the same heat into electricity at potential efficiencies of 15 to 20 percent. For power requirements above 1 kilowatt, this improved efficiency translates into both increased power per pound and lower per-unit power costs.

By virtue of its compact size, higher power, and general robustness in the face of threats, DIPs may be suitable for the planned Boost Surveillance and Tracking System—one of the proposed surveillance platforms for the Strategic Defense Initiative. A dynamic isotope or space reactor power system could also be the energy source for the future Space Surveillance and Tracking System, another proposed platform for the SDI program.

Over the years, both Brayton and Rankine cycle dynamic power conversion technologies have been studied, so there exists an experimental data base to support the DIPS technology. Thousands of hours of test time have been logged by prototypic equipment. A reasonable data base already exists for both the Brayton and Rankine power conversion systems, so the DIPS effort will be based on a sound technological foundation. The Brayton system, for example, has seen extensive service in connection with state-of-the-art aircraft engine hardware—over 36,000 hours of test time have been accrued. Conceptual design work has been completed on integrating a Rankine power conversion system with the spacecraft, and similar work is in the offing for the Brayton system. An organic Rankine conversion system has been tested for over 10,000 hours. All of this work has convinced us that DIPS is viable and that the efficiency and reliability objectives can be achieved. We believe that DIPS can span the 1 to 10 kilowatt power range and provide the flexibility and survivability needed to meet Air Force and SDI requirements.

SP-100: The Next Reactor System for Space Power

With the inception of the joint DOD/NASA/DOE SP-100 Space Nuclear Reactor Power System program in February 1983, the United States embarked on a concerted effort to evaluate and ultimately demonstrate the feasibility and readiness of reactor power systems for space application in the range of several hundred kilowatts-electric, beginning in the 1990s.

The SP-100 development program has a high probability of uncovering key technical solutions to practical problems at a time when the pressures of launch schedules are not predominant.

We achieved significant progress in the SP-100 program in 1985. At the conclusion of Phase I of the program, the technical and safety feasibility of such a power system was established and a concept was selected for further engineering development and ground demonstration testing during Phase II. The selected concept employs a compact, liquid metal-cooled, fast-spectrum reactor with out-of-core thermoelectric power conversion. It will be designed, built, and tested at a power of 300 kilowatts-electrical. Last November, we selected the Hanford, Washington, complex as the preferred site at which to conduct the planned reactor ground demonstration tests.

Early results from the SP-100 program are very impressive. The first phase of the program saw a number of specific, technical achievements:

1. SP-100 safety and technical feasibility was established. Guidelines were established to ensure protection of the public and the environment under both normal and accident conditions.
2. Refractory metal fuel pin and uranium nitride fuel fabrication technologies were successfully demonstrated and irradiation tests were conducted, with encouraging results.
3. Thermionic-converter fabrication technology was successfully demonstrated and testing was conducted.
4. A 3 kilowatt-electrical Stirling engine unit was tested for more than 5000 hours and a 25 kilowatt-electrical free-piston Stirling engine was designed, built, and tested.
5. A compact thermoelectric module—which promises a 20-fold increase in output power density and a 20-fold decrease in cost—was designed and built.

More recently, we have issued a Request for Proposals to secure a system development contractor from industry. That contractor will be responsible for designing, fabricating, and directing the testing of a ground prototype reactor and other key subsystems to demonstrate system performance, safety, dependability and flight readiness. As an option, this effort includes the provision for the production of flight units to expedite a flight program when the initial user commitment is finalized. To encourage user commitment, the Ground Engineering System hardware will be designed to meet the requirements of a recently established reference mission—a 300 kilowatt-electrical nuclear electric propulsion flight demonstration for the Strategic Defense Initiative Organization in the early 1990s, for which planning has been initiated.

An interagency agreement to manage and implement the second or Technology Flight Readiness phase has been signed by DOE, DOD, and NASA. Arrangements similar to those used successfully in Phase I are involved, but during Phase II each participating agency has distinct functions for which it is responsible. Specifically, DOE is responsible for ground engineering system development. The Strategic Defense Initiative Organization is responsible for overall program direction and, with the Air Force as its executing agent, for military mission analysis and requirements definition. NASA is responsible for civil mission analysis and requirements definition, as well as for advanced aerospace technology that could ultimately enhance an SP-100 system.

A wide range of emerging civilian and military mission applications would be enabled or greatly enhanced by SP-100. Such applications include surveillance, communications, electric propulsion, materials production, growth space station power, and weapons platform quiescent power. Many of these applications require power levels and other attributes that are not currently available but which are fully consistent with SP-100.

Moving into the Multimegawatt Power Range

Complementing the SP-100 program, we have begun to look toward other potential reactor power sources in the multimegawatt power regime.

This follow-on program, also structured as a joint DOD/DOE endeavor, is aimed at developing multimegawatt reactor concepts with a wide range of applications for the Strategic Defense Initiative. It is now in its initial exploratory phase. In concert with the Strategic Defense Initiative Organization, we are determining technology and safety feasibility. Innovative, evolutionary technological research needed to support the extension of SP-100 technology to even greater power levels and modes of operation are being pursued. We are exploring an effort to couple fuels, materials, and power conversion technologies with reactor concepts to meet a near-term goal of selecting a reference concept within the next 5 years as additional research, development, and test results become available and further concept definition and reviews are completed.

Our multimegawatt activities have focused primarily on identifying reactor and power system concepts that have the potential to meet the expected, demanding requirements of the Strategic Defense Initiative. In addition, extensive planning has been conducted on technology development activities needed to support the most promising multimegawatt reactor concepts.

Building on Past Technology to Assure Safety and Reliability

The United States has been active in developing and applying nuclear technology to the needs of space missions for many years. Efforts in support of space nuclear power and propulsion have been under way since 1951. Since then, 38 nuclear systems have been employed on 23 DOD and NASA missions. All but one of these power systems used RTGs, and that one used a small 500 watt-electrical reactor. Other reactor power systems were studied and developed to varying degrees, but the work fell victim to shifts in budget or program priorities. In fact, space nuclear reactor programs began a period of retrenchment in the late 1960s and many were terminated in 1973. Other programs were sustained at levels barely permitting their survival. The climate, however, has changed dramatically in recent years. The user community—NASA, DOD, and the separate military services—now recognizes the need for development and ultimate demonstration of this technology. Without timely development, demonstration, and a proper focus, projected missions will be forced to rely on more limited, currently available power sources, which would not permit certain missions to be fully accomplished.

Both solar power systems and RTGs will continue to play a key role in supplying our space power needs; each has its own distinct utility and has been successfully proven in space on many occasions. Given our current thrust, dynamic isotopic systems and space reactors will soon be added to the list of space power options. To realize these goals, we must continue to emphasize safety and reliability and tailor our technology development to user and mission requirements.

There is today international consensus on the use of nuclear power sources in outer space. The United States has participated in the United Nations Working Group on the use of Nuclear Power Sources in Outer Space since its inception in 1979. The purpose of this working group is to develop, review, and endorse functional safety criteria for the use of nuclear power sources in space and to determine if new or additional criteria are required. That body has concluded that nuclear power sources can be used safely in outer space, provided that all necessary safety requirements are met.

The safety of nuclear power sources and protection of both the public and our environment, particularly during potential launch accidents, have been and will continue to be of prime importance in the Department's space nuclear power program. Beginning with mission requirements formulated by the user agency, safety criteria are carefully formulated and design safety compliance is confirmed by test and analysis. A key safety factor with regard to space reactors is our ability to launch them in a radioactively cold, clean state and to keep them subcritical until startup in a stable, planned orbit. Moreover, an independent, interagency nuclear safety review panel is chartered to review and evaluate all such safety aspects prior to each launch—a function similar to that performed by the Nuclear Regulatory Commission for commercial reactors.

Past missions, as well as their ground development activities, have an impressive safety record. With the ability to keep reactors subcritical until startup in a stable, planned orbit, there is no doubt that we can and must continue to preserve this record for missions using reactor power sources.

Where do We go from Here?

We have long recognized the need for a cooperative and mutually beneficial working arrangement between the development and user communities. Through close cooperation, the user community can obtain an appreciation for the problems involved in developing space power systems while we gain greater insight into their mission needs and problems. Consequently, both groups benefit and the technological focus of the mission develops logically, hand- in-hand with the development effort.

Despite the hiatus in space reactor development during the 1970s, we understand the range of existing, emerging, and future spacecraft power and user requirements and have incorporated this knowledge into our programs. By adding the DIPS, SP-100, and multimegawatt capabilities to our already proven RTG technology, we will greatly expand the range of civil and military space missions that can be supported.

Conclusion

Before concluding, let me briefly reiterate the challenges we face of ensuring

1. The maximum practical benefit for R&D from the limited funds available, and
2. That the public is made aware of the importance placed on the safety and reliability of space nuclear power sources.

Finally, I would be remiss if I did not take this opportunity to acknowledge the importance of ensuring the continued availability of a capable and enthusiastic supply of human resources. Time is moving on for those of us with years of space nuclear power experience, and this expertise will disappear if we fail to bring new talent into this challenging arena to participate in developing the potential of this technology.

I hope that my remarks today have left no doubt as to the Department's sense of purpose and direction with respect to space nuclear power. As we see it, the determination of mission applications and the evolution of systems technology are natural extensions of our past efforts. Our destination is clear and we feel we have chosen the best route to opening new vistas in space.

Thank you for this opportunity to be with you today. This annual symposium can continue to serve us well in helping to accomplish the common understanding among the technical community, the mission planners, and the public that is essential to our continued progress.

Acknowledgments

This paper is a written version of the remarks made by Mr. Vaughan at the opening session of the 3rd Symposium on Space Nuclear Power Systems. At that time Mr. Vaughan was the DOE Acting Assistant Secretary for Nuclear Energy.

Part 2

MISSIONS AND APPLICATIONS

Space Nuclear Power Systems 1986
Edited by M. S. El-Genk and M. D. Hoover

Chapter 3

Space Power Needs and Forecasted Technologies for the 1990s and Beyond

David Buden and Thomas Albert

Science Applications International Corporation
2109 Air Park Road, S.E.
Albuquerque, NM 87106

A new generation of reactors for electric power will be available for space missions to satisfy military and civilian needs in the 1990s and beyond. To ensure a useful product, nuclear power plant development must be cognizant of other space power technologies. Major advances in solar and chemical technologies need to be considered in establishing the goals of future nuclear power plants. In addition, the mission needs are evolving into new regimes. Civilian and military power needs are forecasted to exceed anything used in space to date. Technology trend forecasts have been mapped as a function of time for solar, nuclear, chemical, and storage systems to illustrate areas where each technology provides minimum mass. Other system characteristics may dominate the usefulness of a technology on a given mission. This paper will discuss some of these factors, as well as forecast future military and civilian power needs and the status of technologies for the 1990s and 2000s.

Introduction

Our purpose is to evaluate space power needs of the 1990s and beyond to forecast possible gains in space power technologies and to identify regimes where certain technologies have unique attributes. We are entering a new era in space exploration—one in which space power needs for both civilian and military applications will change significantly. For many expanded missions, space power will be an enabling technology, the feasibility of the mission will be determined by the available power technology. Alternative power technologies are making great strides. Therefore, the goals for nuclear power need to be reexamined considering the changing needs and roles of these alternative technologies.

Potential Mission

We will discuss both civilian and military missions, but will concentrate on the former. Man's orderly development of an extraterrestrial civilization could follow the scenario shown in Figure 1 (Buden 1985).

The Space Station

The Space Station, scheduled for initial operation in 1992, will serve as a science base, construction and assembly platform, and space processing demonstration facility. It will start at 75 kWe in 1992, five times the size of the largest power system flown in space, and rise to 300 kWe or more by the turn of the century. The Space Station will expand until it becomes a zero-gravity manufacturing facility and a center for assembling the vehicles that will carry man deeper into space. Multimegawatt power plants will be needed to achieve this.

Figure 1 shows a materials Depository, Reclamation, and Resource Center on the line above the Space Station. This is a new item introduced to provide a depot in space for upper stages of launch vehicles, obsolete or nonusable satellites, junk from the Space Station, debris collected in space, parts from refurbished satellites, and so on. Some materials will be used directly as parts for other applications; other materials will be reprocessed and used as a source of raw materials for such purposes as space manufacturing in the Space Station or as fuel for orbital transfer vehicles (OTVs). Power levels of tens of kilowatts will be needed for operations such as melting materials, operating cutting tools, and moving materials.

Geosynchronous Earth Orbit

Geosynchronous Earth orbit (GEO) uses are expected to include large communications satellites and a construction platform to maintain and repair satelites. It could also be the final checkout and repair center for expeditions to Mars. Large communications satellites might require hundreds of kilowatts of power, while the depot might require 25 to 50 kWe.

Lunar Resource Bases

The next stage in extraterrestrial development could be the orderly establishment of lunar resource bases (Angelo and Buden 1985a). A polar orbiter is needed to determine the existence of frozen volatiles at permanently frigid areas of the Earth's moon and to complete the mapping of the lunar surface. Lunar rovers or explorers can be used to further explore the surface and subsurface features. Envisioned here is a more sophisticated rover called a Maneuverable Geoscience Laboratory. This robotic laboratory will traverse the moon while its drilling machinery extracts subsurface rocks for analysis. Man's return to the moon will probably start with "lunar day" excursions (lasting 14 Earth days). These excursions will be followed by semipermanent or permanent manned bases, a permanent base camp for scientific experiments and initial lunar material processing, extensive material processing and mining, and cislunar trading and full self-sufficiency.

Table 1 provides projected levels for each stage of lunar developments (Angelo and Buden 1985a), as well as candidate power supplies. The first stage, where activities are concentrated on detailed surface and possibly subsurface exploration and site preparation by robotic vehicles, will require a few kilowatts of power. Electrochemical power systems can be used with a solar array for recharging; however, the lunar night will cause a 14-day period shutdown in each 28-day cycle. Continuous activity is possible with a dynamic radioisotope unit. An initial manned lunar base of 6 to 12 persons for scientific studies, expanded resource exploration, experiments on material processing, and acclimation to another planetary body will require hundreds of kilowatts of power. One concept suggested is the use of innovative solar technology concepts, wherein lunar satellite relays provide continuous power from one part of the lunar surface to another during the day/night cycle. Another concept

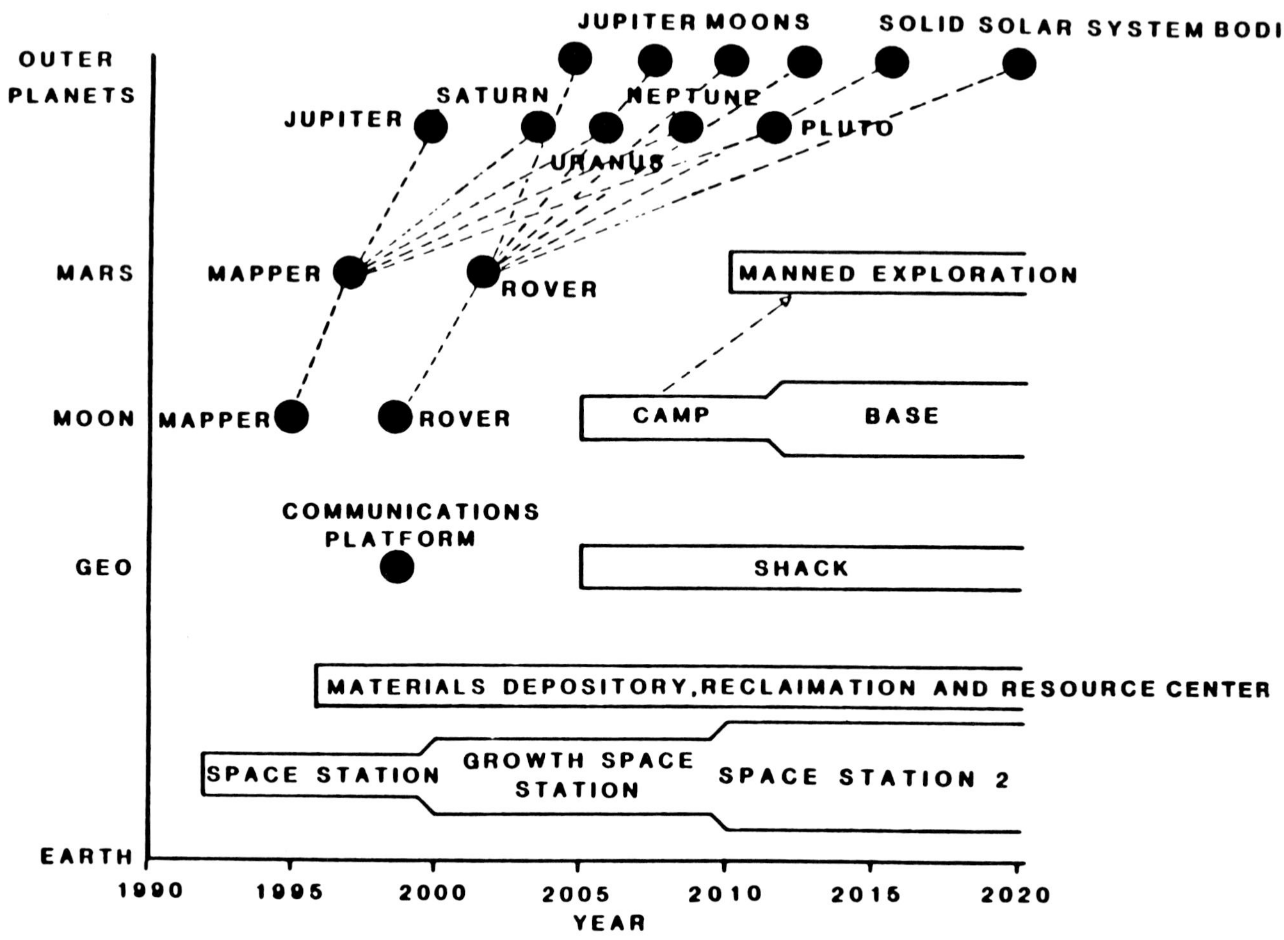

Figure 1 *Scenario for Long-Range Space Plan.*

embodies a modified SP-100 space nuclear power plant to provide continuous power. The plant could be located in a crater, for natural shielding, and augmented by site shielding using lunar regolite materials. The evolution toward early lunar settlements might see expanded research activities, prototyped processing of lunar materials, development of fabrication techniques, and the start of lunar agriculture. Megawatt power levels will be needed, which might use advanced versions of the SP-100 power plant or similar solid-core technology. The next evolutionary stage, mature settlement of 1000 to 10,000 persons, might see cislunar trading, development and export of raw materials (such as oxygen), manufactured products for space stations, support resources for exploration of Mars and beyond, and a thriving agricultural sector. This will require hundreds of megawatts of power. Advanced multimegawatt nuclear power plants can be refueled and readily maintained will be needed. This could evolve into autonomous lunar colonization with advanced nuclear power plants, and a self-sufficient fuel cycle.

Mars Mission Scenario

Mars will play a key role in Man's exploration and conquest of the solar system (Angelo and Buden 1985b and Angelo and Buden 1984). In the Martian scenario described here, orbiting spacecraft and sophisticated surface rovers (the same as those used on the moon) will establish the location for the first manned expedition. The rovers will probably be powered by dynamic radioisotope power sources. The Mars space vehicle can be assembled in low Earth orbit (LEO) and then move to GEO unmanned. There, the crew can board the vehicle and depart for Mars. After several months orbiting Mars and exploring the surface, the crew can return to Earth. Nuclear electric propulsion (NEP) can be used for a manned mission to Mars. Figure 2 shows a NEP vehicle (Atkins 1983). In this Mars mission scenario, a single spacecraft with a Mars Lander could carry a crew of five on a 2.6-year mission to the Red Planet. The NEP system is powered by a megawatt-electric nuclear reactor. Closed air and water life support systems and artificial gravity can sustain the crew throughout this venture.

The NEP Mars vehicle can be assembled at a space station in LEO. An unmanned OTV boosts the Mars spacecraft from the space station to a nuclear safe orbit (typically 700 km). The Mars expedition vehicle, still unmanned, then experience an NEP spiral from the nuclear safe orbit to GEO. With its human payload on-board, the Mars vehicle executes an Earth departure spiral from GEO, an operation that takes 510 days. This is followed by a 39-day duration Mars cap-

Table 1 *Stages of Lunar Base Development and Associated Power Requirements (1992–2092).*

Stage	Activity	Power Level	Probable Power Supply	Remarks
1	Automated Surface Exploration and Site Preparation	Few kWe	– Electrochemical Radioisotopic (RTGs; Possibly Dynamic Conversion)	– Detailed Surface Exploration and Site Preparation by Robot Vehicles
2	Initial Lunar Base (6–12 Persons)	~100 kWe	– Solar Technology – Modified 100 kWe Class Space Reactor	– Initial Scientific Base – Expanded Resource Exploration – Extra Terrestrial Material Experiments
3	Early Lunar Settlements (100–1000 persons)	~1 MWe	– Expanded 100 kWe	– Expanded Research Activities – Prototype Processing Operations – Start of Lunar Agriculture
4	Mature Lunar Settlements (1,000–10,000 Persons)	100 MWe	– Advanced Multimegawatt Reactor Designs	– Cislunar Trading – Lunar Raw Materials – Foodstuffs – Manufactured Products
5	Autonomous Lunar Civilization (More than 100,000 persons)	100s MWe	– Advanced MW Designs	– Full Self-sufficiency – Lunar Nuclear Fuel Cycle – RTG and Reactor Fuel – Waste Repository

ture spiral, which leaves the vehicle in a 3000-km altitude nuclear safe orbit around the Red Planet. The planetary reconnaissance portion of the mission lasts 100 days, including 30 days on the surface by a 3-member crew. A 23-day Mars departure spiral starts the vehicle back to Earth. Mars-to-Earth transfer takes 229 days with optimum coast. This is followed by a 15-day capture spiral to GEO. From GEO, an OTV transfers the crew to the space station for quarantine (if necessary), debriefing, and eventual return to Earth (Atkins 1983).

Also, the same NEP unit for Mars can be used to take mappers and rovers to the outer planets. For example, Neptune mission times can be reduced from 10 to 15 years to 3 to 5 years (see Figure 3) using multi-megawatt NEP rather than chemical propulsion, solar electric propulsion, or 100-kW NEP (Atkins et al. 1983).

Military Missions

Military missions will probably be the technological driver for near-term nuclear space power plant development. As shown in Figure 4, the military needs cover a range of power and energy levels, encompassing communications, surveillance, and weapon systems. Power needs tend to fall into two regimes: (1) a long-duration, continuous regime of many years; and (2) a burst-mode, short-duration regime.

Space Power Subsystem Issues

Many issues should be considered in selecting a power subsystem for a given application. Mass or specific mass is usually the primary figure of merit (FOM) used for judging different candidates because of transportation cost (launch and orbit transfer). Launch cost along is approximately \$1000/kg. Typical of past spacecraft, the power subsystem is 30 percent of the spacecraft mass. For a future spacecraft where large pulses of burst power are needed, the power subsystem could represent 80 percent of the spacecraft mass.

Volume is another significant parameter related to launch costs. For the Space Transportation System, the ability to package the spacecraft, including the power subsystem and OTV, is often limited by volume, without exceeding the mass limit of the launch vehicle.

In addition, many other key technology FOMs should be considered in the selection of power subsystems. These relate not only to each power subsystem's unique characteristics, such as status of technology, size, orientation requirements, costs, start-up times, reliability, and lifetimes, but also to the interrelationships between the power subsystem and the balance of the spacecraft. Consideration must be given to such parameters as shielding for nuclear power sources; platform instability from effluent in open-loop chemical or nuclear systems; effluent contamination of surfaces and sensors; and vibration effects on precision alignment of the platform. Also significant in power subsystem selection is the type of environments (natural, self-induced, or hostile). Table 2 lists user requirement categories and possible power subsystem parameters. A characterization of these parameters for each application provides a basis for power subsystem selection.

The balance of this paper will concentrate on forecasting mass and specific mass, usually the starting points in assessing power subsystems for various applications.

Technology Forecasts

Forecasts of mass and specific mass become increasingly uncertain because of new processes, new

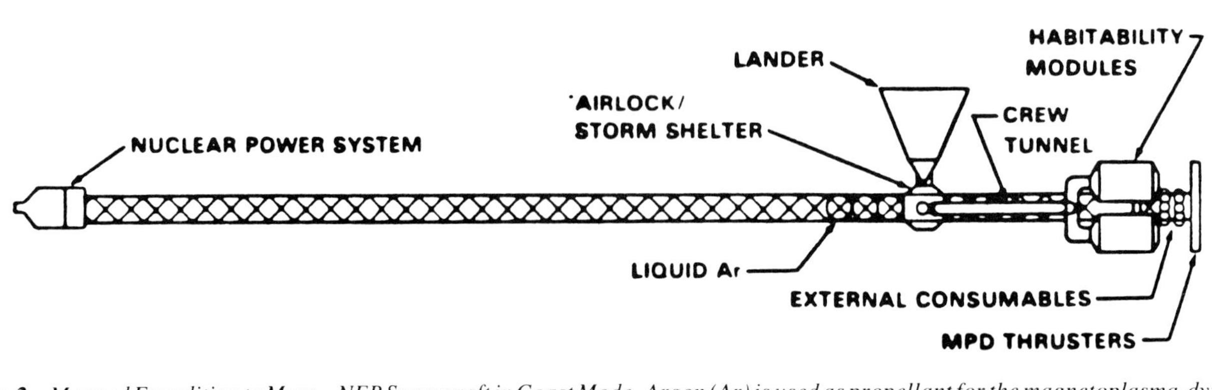

Figure 2 *Manned Expedition to Mars—NEP Spacecraft in Coast Mode. Argon (Ar) is used as propellant for the magnetoplasma-dynamic (MPD) thrusters (Courtesy of Atkins, et al. 1983).*

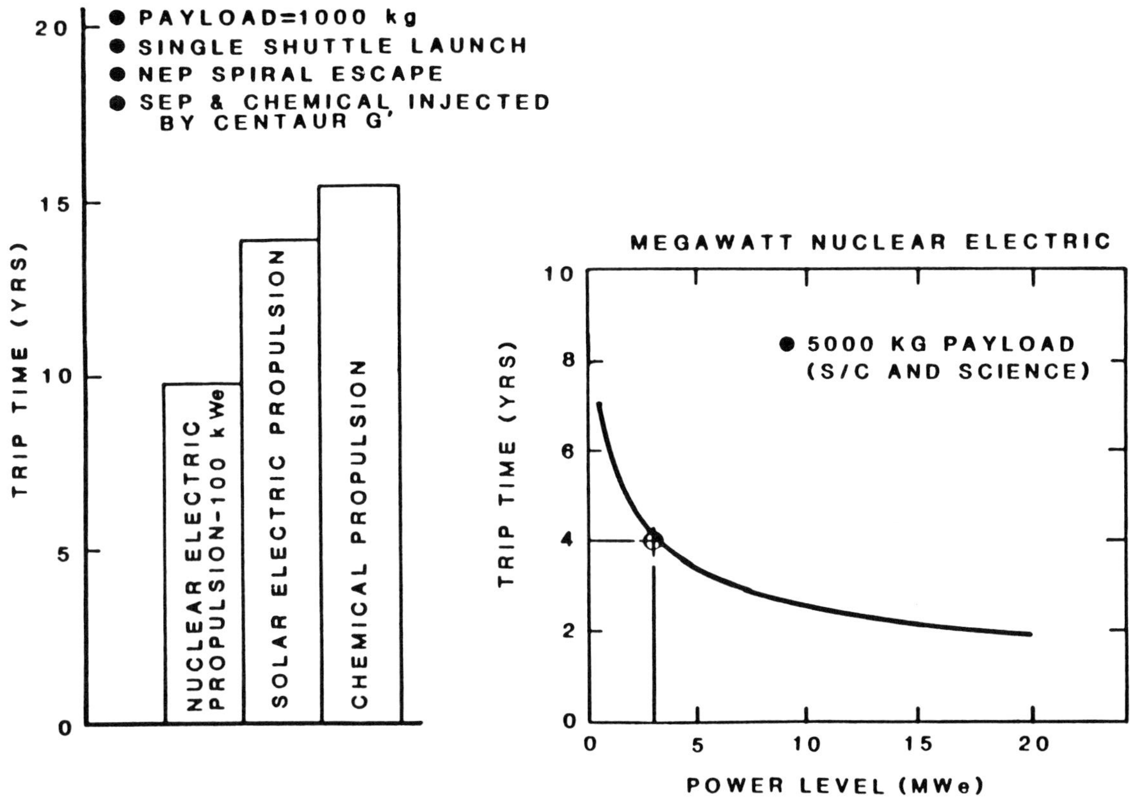

Figure 3 *Neptune Mission Duration.*

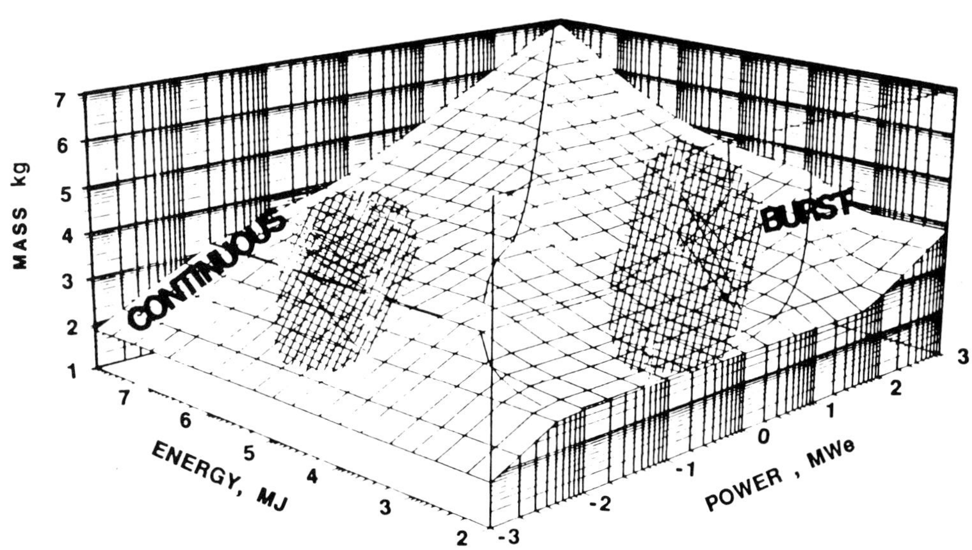

Figure 4 *Forecasted Military Power Needs Regimes.*

material needs, and uncertain funding. We will address the principle power alternatives next.

Solar Photovoltaics

Solar photovoltaics, space proven to 15 kW, have been the ''work horse'' for space power. Current technology is about 11 W/kg, 33 W/m^2, and cost \$1500/W, using silicon cells and NiCd batteries. Significant advances are being made in solar technology, such as a greatly-increased solar cell efficiency, whether used in a panel or a concentrator assembly (see Figure 5 for state-of-the-art (SOTA) forecasts). Silicon conversion efficiency is being improved through the use of narrow grid lines, shallow diffusion, back-surface fields, back-surface reflectors, and front-surface sculpturing. Gallium arsenide is becoming available with an initial efficiency of 16 percent and the possibility of 21 percent. Multi-band gap solar cells involving two or three semi-conductors in a vertical stack, such that the top cells are transparent to that part of the spectrum that they do not convert, can increase efficiency to 25 to 40 percent level.

High Density Storage Systems

The need for higher density storage systems is as important as more efficient solar arrays. This activity is also well underway, with the possibility of reducing energy storage masses by factors of 5 to 10. Technological advancements include NiH_2, Li/Cl and NaS batteries and advanced H_2-0_2 fuel cells. Cost reductions of 50 percent might be possible. The main issues concerning solar photovoltaics with energy storage are that size and mass limit the useful range for civilian and military missions. Furthermore, survivability issues lead to additional limits for military systems.

Radioisotope Generators

Radioisotope generators have proven to be very robust in space applications (see Figure 6 for SOTA forecasts). Units with thermoelectrics have operated successfully for more than 15 years in space. The units are very rugged and compact, but their cost is approximately \$12,500/W. For reducing cost: (1) more efficient electric conversion systems using dynamic converters can increase the electric conversion efficiency of a factor of 3 (effectively reducing fuel cost by one-third); and (2) the heat source can be recovered and the fuel reprocessed after use in space. Radioisotope generators, with their compactness and high survivability ratings, are prime candidates to meet long-duration needs to about 10 kWe.

Solar Thermal Systems

Solar thermal systems use concentrators to focus the sun's energy on a thermal storage media (see Figure 7 for SOTA forecast). This stored energy is converted to electricity using dynamic machinery. Because of their relatively lower mass, size, and drag compared to solar photovoltaic systems, solar thermal systems are serious candidates for space station

Table 2 *User Requirement Impacts on Power Subsystem Options.*

Category	Major Impacts on Power Subsystem
Propulsion	Structural design, packaging, assembly
Structures	Structural design, configuration
Telemetry, Communications, Information Processing	Housekeeping power, reliability, reserve power
Sensors	Effluent control, outgassing
Power Subsystem Specific Requirements	Mass, volume, number of power modes, power conditioning regulation, propellant/coolant usage
Platform Environment	Power transmission, induced effects on applications
Weapons	Characteristics of fire mode, size and mass of subsystem, power conditioning
Platform Stability	Effluent control, sloshing of fluids, dynamic machinery noise
Cryogenics/thermal control	Configuration, mass and volume, effluent control
Consumables	Mass and size, operations, serviceability
Servicieability	Vibration, shielding, packaging
Survivability	Mass, coatings, materials, arrangement
Reliability	Inherent redundancy, simplicity, servicing
Schedule	Development time, risk
Safety	Mass, deployment
Platform Risk	Configuration
Deployment	Configuration
Attitude Control	Effluent, configuration
Disposal	Configuration, mass

power. Also, their relative robustness makes solar thermal a candidate for 100-kWe military missions as an alternative to solar photovoltaics and nuclear reactors. To date, there has been relatively limited component development. The major technology issues to be resolved are the thermal storage receiver and use of dynamic machinery. The specific area is projected to reach 250 W/m^2. The major long-term limitation is size and mass (depending on the solar flux).

Nuclear Reactors

Nuclear reactors for space were extensively developed from 1960 to 1973 (see Figure 8 for SOTA forecasts). One reactor was flight-tested in 1965 (500 W output) and a reactor and 35 kW power subsystem was ground tested for one year. Also, the nuclear rocket program (Rover) demonstrated 2450 K, 4200 MWt, and duration of 1 h (not in the same test). Renewed interest in the late 1970s and early 1980s led to the SP-100 program. This is now entering the ground engineering phase, where a fast spectrum, pin fuel reactor with liquid metal Li coolant and thermoelectrics as the power conversion elements has been selected for development. Forecasted are improved converters such as thermionics and Stirling, refuelable and maintainable reactors, and higher-temperature systems throughout the power plant. Also, feasible are reactors that operate in several modes depending on the various power profiles. The major issues relate to materials, radiological safety acceptance, and final dis-

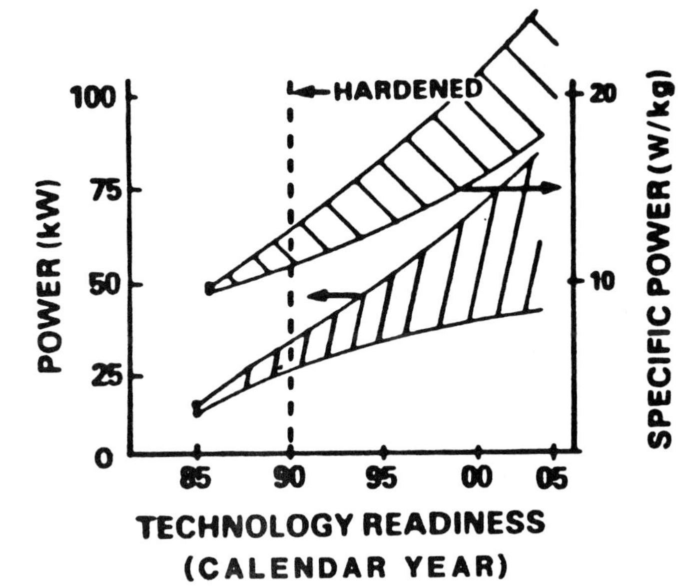

Figure 5 Solar Photovoltaic and Batteries/Fuel Cells State-of-the-Art Forecasts.

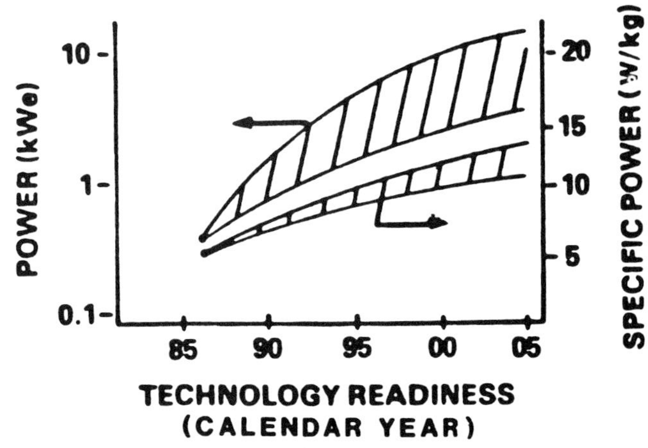

Figure 6 Radioisotope Generators State-of-the-Art Forecast.

posal. The developing infrastructure in space should help end society's concerns for radiological safety. Long-term disposal of spent reactors to higher orbits or escape should alleviate concerns of a reentering reactor.

Burst-Power Candidates

Figure 9 provides a forecast for burst-power candidates. Chemical sources are proceeding along an orderly development path, having demonstrated a 4-MW generator. A 20-MW turbine will be tested in 1986 and a 40-MW turbogenerator, based on F-1 engine technology, is under development. Chemical technology should be available in the 1990s to support demonstration tests requiring 10s to 100s of megawatts and operational platforms beyond. Nuclear open-cycle configurations can also meet the high burst-power needs, but development is down the

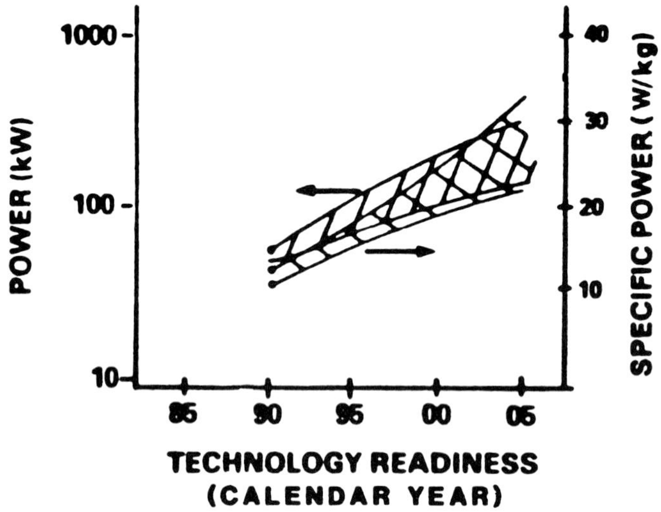

Figure 7 *Solar Thermal Power Systems State-of-the-Art Forecast.*

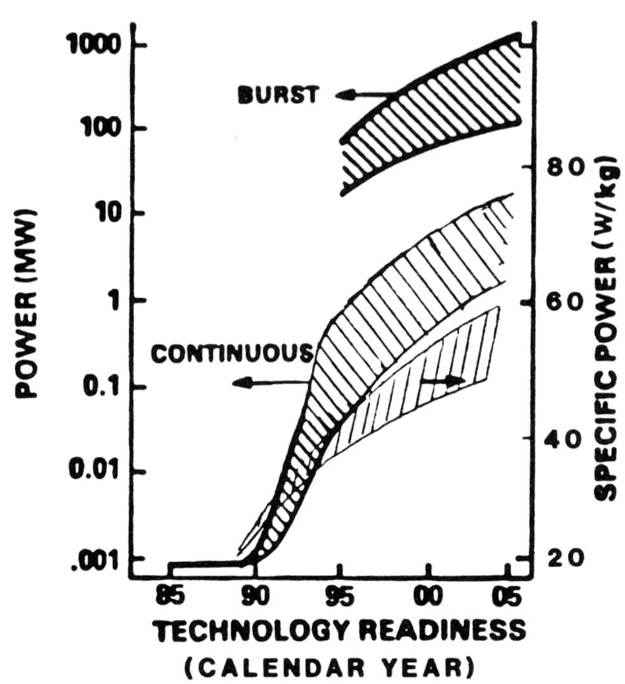

Figure 8 *Nuclear Reactor Power Systems State-of-the-Art Forecasts.*

road. Significant mass savings are possible with nuclear power systems as the need for operating times for weapon platforms increase.

Both chemical and nuclear open-cycle systems have the disadvantages of using fluids for periodic testing, introducing vibrations into the platform, generating thrust from the effluent, and possibly contaminating the platform. Electrochemical sources and flywheels do not have these problems, but tend to be very massive for the time durations of interest. Closed-loop nuclear systems can resolve the problems, but will require significant technological advancements in heat rejection to be mass competitive.

Minimum Mass Forecast

The regimes of minimum mass have been forecasted for technology readiness in the year 2000. This forecast, shown in Figure 10, indicates that advanced solar technology could be the lightest weight to a few hundred kilowatts. Closed-loop nuclear, then, is the lightest weight, long-duration system. For short durations, batteries and flywheels are lightest weight. Higher power and long operating time lead to open-loop chemical systems. Very high power favors open-loop nuclear systems.

Summary

The growth in space power needs have been forecasted as follows:

- Continuous power needs will increase by a factor of 10 per decade;
- Burst-power needs may rise to several hundred megawatts; and
- Duration will increase from 3 to 5 years to 7 to 10 years this century, and continue to increase to 30 years.

Significant gains are being made in technologies alternative to nuclear. In solar technology, this includes the development of solar concentrators and higher ef-

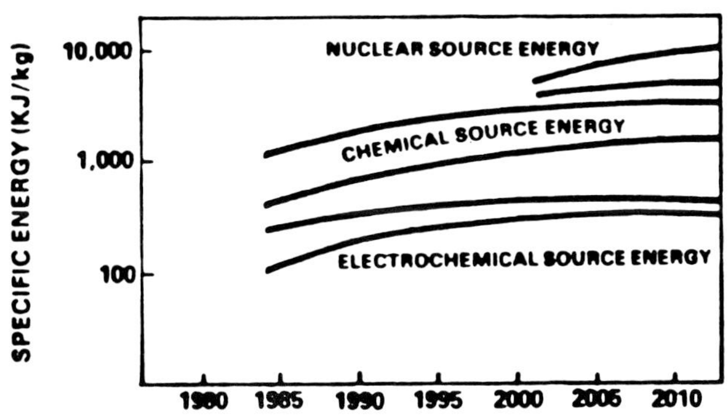

Figure 9 *Burst-Power Sources State-of-the-Art Forecast.*

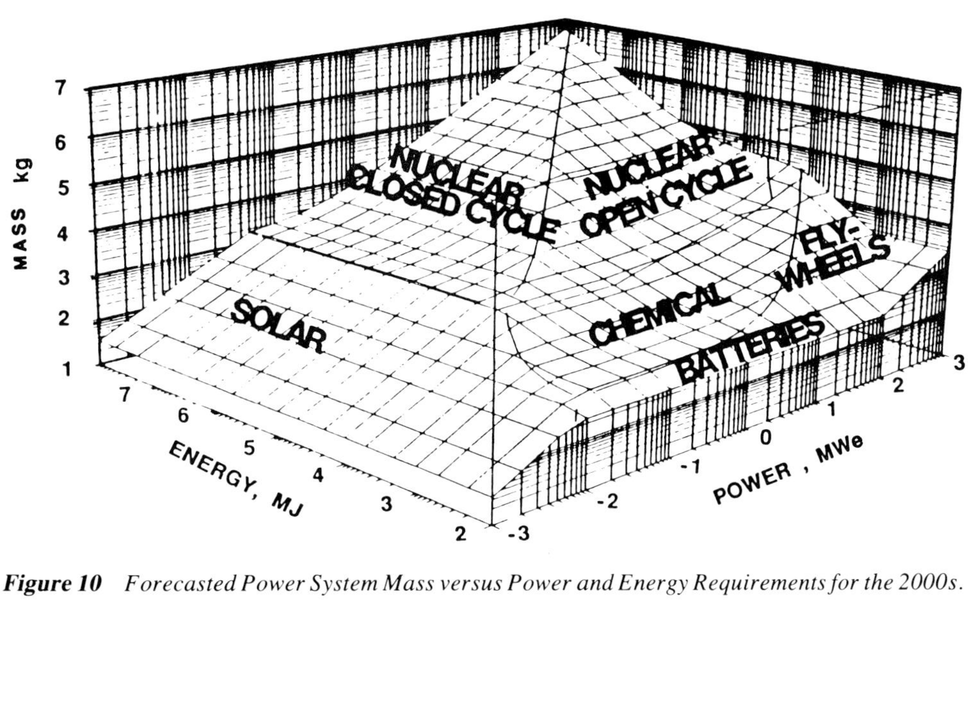

Figure 10 *Forecasted Power System Mass versus Power and Energy Requirements for the 2000s.*

ficient solar cells, and dynamic machinery. These could push the mass advantage of solar from 25 kWe to 50 kWe.

High burst-power issues are formidable and will take much development to resolve the technical issues regardless of the selected technology.

Acknowledgments

This work was sponsored by the Science Applications International Corporation (SAIC).

References

Buden, D. (1985) ''We Are The Future!,'' in *National Commission On Space*, Albuquerque, NM, 4 October 1985.

Angelo, J.A. and D. Buden (1985a) ''Nuclear Energy For Lunar Base Applications,'' in *36th International Astronautical Congress*, Paper No. IAIF–85–180, Stockholm, Sweden, 7–12 October 1985.

Angelo, J.A. and D. Buden (1985b) ''Power Requirements For The Conquest of Mars,'' in *The Case For Mars II*, American Astronautical Society, Vol. 62, Science and Technology Series, San Diego, California, 1985.

Angelo, J.A. and D. Buden (1984) ''Mars—Pathway To The Stars,'' in *21st Space Congress Proceedings*, Cocoa Beach, FL, 26–28 April 1984.

Atkins, K.L. et al. (1983) *Expedition to Mars*, NASA/JPL Report, Jet Propulsion Laboratory, Pasadena, CA.

NASA (1984) *Civil Applications of Space Nuclear Power—Final Report of the Civil Missions Advisory Group*, NASA Headquarters, Washington, DC.

Space Nuclear Power Systems 1986
Edited by M. S. El-Genk and M. D. Hoover

Chapter 4

Power Demands for Space Resource Utilization

Andrew H. Cutler

California Space Institute
U. C. San Diego
La Jolla, CA 92093

Space Resource Utilization (SRU) is a logical step in the conquest of space. SRU will require megawatt-scale power sources for which nuclear power is well suited. Power is needed for transportation as well as processing, and power for propulsion may exceed processing power requirements. Indirect electric propulsion from an Earth (0.5 to 2.0 MWe) or lunar (800 kWe) orbiting transport node will greatly enhance the economics of lunar resource utilization, but indirect electric propulsion is unlikely to play a significant role in asteroidal or Martian resource utilization. Indirect nuclear thermal propulsion ($>>$ 1.7 MWt) is found to be useful for asteroidal resource utilization or initial lunar or martian factory emplacement. Nuclear powered vehicles are not practical for lunar resource utilization. Electrically propelled vehicles with I_{sp}'s well over 5000 s would be useful for Martian resource utilization. Almost any advanced propulsion system would be suitable for asteroidal resource utilization. Several methods of producing lunar propellants are described in detail. Lunar hydrogen production (70 tonne/yr) has the highest power requirements at tens of MW, and various oxygen production schemes (1000 tonne/yr) will require 1 to 4 MW. Each process has unique power conditioning requirements. Thus, considering energy requirements to be a simple output power capacity is misleading and not sufficiently descriptive. Optimal plant design is only possible when power plant and factory designs are integrated.

Introduction

Space nuclear power systems currently are not linked with prospective mission needs. Therefore reactor development programs are hard to start and easy to end. The common perception is that the first mission to actually require a reactor has to bear the entire development cost (even though 20 more benefit from the new technology). This results in the cancellation of missions that actually require a reactor and hence, the cancellation of reactor development. The following conundrum occurs: no one will propose a mission that requires nuclear power, knowing that it entails developing the technology as well as their mission hardware; but who will fund reactor development in the absence of a mission need for it? This cycle can be broken only by showing how reactor development will contribute to a variety of near- and far- term missions. This would allow development to be pursued for its overall contribution to the space program, rather than its benefit to one specific mission.

While reactor design and development are relatively well understood, reasons for developing any given space reactor or power system are not. Appropriate justification is needed before development can begin. The ability to design and build an elegant space nuclear power system is not sufficient; a prospective mission need, such as space resource utilization (SRU), must be shown to benefit from nuclear power. Analysis of SRU power requirements is a step towards making a case for actual reactor development.

This paper discusses the power demands expected for space resource utilization. These power demands could be met by either solar or nuclear power, but the scale and type of power demands seem to favor nuclear power. The prospective use of nuclear power must be justified in each case by demonstrating that the overall nuclear powered system is more economical than other competitive systems. Preliminary design data showing that a reactor could meet these power demands more economically than a solar power system would demonstrate that SRU has a mission need for nuclear power.

The costs, benefits, and motivations for SRU are poorly defined. Thus technical details are necessarily vague. This can be considered a source of frustration, or an opportunity to create an optimal design before some parameters become fixed for programmatic reasons. Any reactor designer who wishes to work on SRU must perform more background work than usual, since design specifications for the SRU plant do not exist. Systems level tradeoffs must be made before sensible design specifications for both the factory and the reactor can be developed. Care must also be taken to understand the impact of uncertain operating parameters, such as feed composition and capacity, on the overall design.

SRU is not a near term mission, and thus is not a motivation for building any reactors in the next 10 years. It is, however, an inevitable activity worth some thought. As there are many SRU schemes discussed in the literature, the designer must be careful to choose those that serve some realistic mission need or market. There is no value in demonstrating that nuclear power can efficiently serve a nonexistent market.

SRU power demands can be met in a variety of ways. Solar cells are one option. It is reasonable to believe that the specific power of solar cell arrays will exceed 200 W/kg by the time SRU occurs. This or other appropriate figures should be used for comparison purposes. While solar cells suffer radiation degradation, it is not dramatic. On the other hand, nuclear reactor cores burn up and must be replaced. Factory life is likely to be 10 years or longer, so core burnup may be a limiting factor. In addition, solar dynamic power systems are immune to radiation, and could easily exceed 200 We/kg or 2000 Wt/kg. Two major factors affecting the trade between solar and nuclear energy are the effects of night periods (which reduce the duty cycle and cause reliability problems) on the factory and powerplant mass required to provide a given average throughput, and the cost and mass of power management and distribution systems which must collect power from an extended solar collector field.

Total cost is the most basic and most often neglected factor in SRU. Any SRU scheme is supporting a potential mission, program or market because SRU costs less than the value of the benefits derived from this mission and less than the alternative methods of meeting the same needs. There is a limit to the value of any program, so there is a limit to the value (and hence allowable cost) of SRU. It is extremely difficult to justify spending more than a few billion dollars per year on SRU. (Since SRU is lowering the Earth launch rate required to achieve a given end result, its maximum value is the marginal cost of the launches that must occur in order to perform the task in question without SRU. Some fraction of NASA's launch operations budget is a good first approximation.) A few billion dollars is surprisingly little in the context of space operations.

Commodity Markets

There are three principal markets for space resource derived products. These are for orbital transfer vehicle (OTV) propellant, propellant and structural material for local use at a lunar or Martian base and structural material in Earth orbit. The first market is reasonably likely to exist, the second is dependent upon program decisions for non economic reasons and the third is quite speculative.

Propellant Market in LEO

The first and most important market is for propellant in low Earth orbit (LEO). Hydrogen/Oxygen (H/O) is a high performance propellant for orbit raising. There are existing stages and engines to use it. It would be programmatically and operationally desirable to make an H/O-propelled reusable orbit transfer vehicle (OTV). By mass, more than 84% of H/O propellant is oxygen, so any market for H/O is also a market for oxygen alone. Since the 1960s, about half of the payload mass launched to LEO has been propellant for orbit raising. Shuttle and Ariane manifests and mission models indicate this will continue for quite some time. This propellant demand is driven by a variety of different needs. No single new technology or program decision is likely to cause a dramatic decrease; in fact, most prospective new programs would cause a substantial increase.

Projected demand (Cutler 1984 and Cooper 1984) for propellant in LEO is in excess of 300 tonnes per year. Obtaining this propellant from an extraterrestrial source would lower the demand for terrestrial propellant and expendables to 50 to 150 tonnes per year, thus lowering the total launch costs from Earth (counting the payloads to be placed in high orbits) between 40% and 65%. A free source of liquid oxygen (LOX) would lower effective launch costs by 68%. By contrast, LOX demand in geosynchronous Earth orbit (GEO) for OTV return would be about 30 tonnes per year. Free LOX delivery would lower the overall demand for terrestrially derived propellant by about 16%, and would lower the total amount of material (propellant + payload) that would have to be launched form earth by $< 12\%$. All space resources are propulsively closer to LEO than to GEO, making the larger, higher leverage market in LEO easier to serve.

Commodity Markets in Space Exploration

A lunar or Martian exploration program would be greatly enhanced by the local availability of propellant or structural materials. This is another possible market for SRU. For example, the total mass that would have to be launched to LEO for simple lunar or martian sample return missions could be cut by ⅓ to ½ by the manufacture of return oxidizer at the destination (Lawton and Frisbee 1985). On-site propellant manufacture would allow missions with a fixed mass budget to carry about twice as much experimental equipment as would otherwise be the case, and would allow the return of far more sample material. These results should apply to manned exploration as well. These markets are easy to serve, since the manufacturing facility need only generate a few times its own mass in product over its life. The scale of these markets is probably a few hundred tonnes over the life of the program.

Structural Material Market in LEO

Finally, large space structures (Billingham et al. 1979), radiation shielding for manned orbiting facilities or shielding for ballistic missile defense satellites (Nozette 1984) are often cited as markets for space resource derived manufactures. The first and second are not realistic markets since they are not economically justified. The third is a very speculative market. Most of the shielding ballistic missile defense (BMD) satellites may need will be required in LEO or slightly higher orbits. Therefore consideration of propellant delivery to LEO can usually be generalized to cover the delivery of shielding materials as well.

Space Resources

Space resources are sources of material in space that may be used to achieve a desired end more economically than could be done using terrestrially derived materials. The moon, the moons of Mars , Mars itself and certain near-Earth asteroids are potential space resources that could lessen the cost of orbit raising from LEO or the propulsion costs of certain interesting manned missions. In addition, almost any body to which it is desirable to send a spacecraft is a potential resource for use by that spacecraft (Ash et al. 1981). Using these resources to supply propellant to LEO is interesting because they may be used to serve a market worth billions of dollars per year which is ongoing and is a major part of our space activities.

The Moon

The moon is an interesting resource because it is energetically closer to LEO than is the Earth's surface and round trip flight times are short. The resources available are highland and mare soils, which are pulverized igneous rock and glass with about 0.2% metallic iron (McKay and Williams 1979). Small concentrations of solar wind implanted hydrogen in these soils may also be a resource. A lunar base program would be of considerable scientific value. There would be significant synergism between a lunar base program and lunar resource utilization.

The Mars System

Phobos and Deimos are interesting resources because they are energetically somewhat closer to LEO than is the Earth's moon, they are believed to be composed of more readily useful minerals (specifically organic and water containing materials) than is the moon, (although they have not been well characterized or sampled) and their exploitation could greatly enhance a manned Mars exploration program, which would be of considerable scientific value. Mars itself is also a resource. Its atmosphere is a source of carbon dioxide and water vapor, from which oxidizer and fuel can be made by simple, reliable chemical processes. Since Mars is in a relatively deep gravity well, use of the Martian atmosphere would be useful in support of an extended Martian exploration program, but probably would not be competitive with Phobos and Deimos resources for export to LEO. Compared to the moon, round trip flight times to Mars are long and round trip opportunities are infrequent.

Near Earth Asteroids

Certain near earth asteroids are energetically extremely close to LEO (Lau and Hulkower 1985). There is no strong scientific rationale for extensive manned exploration of these bodies. Flight times and the waiting times between mission opportunities are somewhat longer than for Mars. These are energetically very accessible, but propellant consumption is the major determiner of cost, not the theoretical energy for each leg of the trip. Like Phobos and Deimos, these asteroids have not been characterized. It is statistically certain that some have useful compositions and that many do not. (Spectral characterization of near earth asteroids is reviewed in McFadden 1984). These asteroids are so accessible that they might be exploited by returning raw ore to LEO (Lewis and Meinel 1983). Recent work indicates that this will not be significantly more economical than lunar or Martian manufacturing (Cutler and Hughes 1985). The advantages of asteroidal ore recovery are that a small, presumably automated, device could go to the asteroid in question, extract some ore, and return it to LEO without any chemical processing. The major argument for this scheme is that the size, cost, and complexity of this asteroid miner would be low, and the manufacturing facility could be in LEO, which is its most convenient and economical location. Factors that militate against this are the costs of carrying the return propellant out to the asteroid, investing money several years before any product is delivered, and ensuring reliability over a long period with no manned maintenance or repair.

Power Requirements for Transportation

Transportation is a major cost in SRU. The lower bound of product cost is determined by transportation costs, and it is seldom particularly low. There are very few propulsion technologies currently available. It is worth discussing the impact various forms of nuclear propulsion could have on SRU transportation economics. It may turn out that substantially more power will be required in the transportation loop than in manufacturing.

Transportation in Lunar Resource Utilization

Schemes for the utilization of lunar resources must involve the transport of needed terrestrial materials to the lunar surface, the actual manufacture of a product on the lunar surface, and the delivery of this product to its destination (usually LEO). This loop has been studied in some detail (Frisbee and Jones 1983, Cutler and Hughes 1985, Davis 1983, Stump 1985, and Salkeld 1966). A reasonable scheme for lunar oxygen delivery to LEO involves manufacture of oxygen on the moon and the import of propellant hydrogen from Earth. Cryogenically propelled vehicles fly between LEO and some intermediate point, such as L1 (neutral point in the gravitational potential between Earth and the moon) or lunar orbit, and between this intermediate point and the lunar surface. There is a transportation node (a facility to receive, service, supply, and launch the vehicles) in LEO and on the lunar surface. There may or may not be one at the intermediate point.

There are few economically viable uses for nuclear propulsion in lunar resource utilization. Replacing cryogenically propelled vehicles with nuclear propelled vehicles would seem promising. Radiation exposure from the reactor core and from passage through the Van Allen belts is not a problem, since liquid oxygen is not a particularly radiation-sensitive material. However, there are several drawbacks that rule this out. A high thrust system is needed to operate to and from the lunar surface, but terrestrial hydrogen is too valuable to use as the sole propellant despite the high performance of a NERVA type rocket. Lunar oxygen does not perform well enough as a thermal propellant to be of use. It causes severe materials problems in the core as well. Nuclear electric propulsion cannot serve the lunar surface due to its low specific thrust, and nuclear thermal propulsion appears uneconomical.

In the LEO - L1 or LEO - lunar orbit leg, high thrust thermal propulsion is ruled out again due to the cost of hydrogen. In cislunar space, low thrust propulsion systems have much longer trip times than high thrust systems. A large fleet of low thrust vehicles would be required to replace each cryogenically propelled vehicle. In addition, the propellant for a low thrust engine would be entirely terrestrially derived. A low thrust transport vehicle that braked propulsively in LEO would require a specific impulse (I_{sp}) in excess of 5000 s to have economic performance equivalent to the available RL - 10 II engine on the basis of terrestrial material consumed. An aerobraking low thrust vehicle would require an I_{sp} of more than 1300 s.

Tethers could be used to transfer momentum from the earth orbiting transport node (probably the space station) to departing spacecraft (Carroll 1984). Since most of the spacecraft departing for the moon would have the same mass and trajectory requirements and spacecraft would fly frequently, this is a simple tether application. A source for the momentum transferred to the spacecraft is required. Momentum could be acquired through tethered deboost of reentering spacecraft or by capture of partially aerobraked returning spacecraft. This is operationally complex, and properly scheduling arrivals and departures would be difficult. It is more likely that momentum would be acquired through the use of thrusters on the transport node.

Some type of electric propulsion would be used due to the high I_{sp}. For example, a typical lunar oxygen delivery loop may involve boosting 150 to 200 tonnes of payload per year with a 1-km/s tether. Since the propellant would be terrestrially derived, I_{sp} would have to be $>$ 1500 s for reasonable economy. Indirect boosting requires a net average thrust of 4.76 to 6.34 N at the transport node. An electrodynamic tether thruster might provide this thrust for 50 to 100 kWe. Power requirements are higher with more traditional electric thrusters. Assuming 50% efficiency, a thruster with an I_{sp} of 2000 s requires 90 to 125 kWe, while one with an I_{sp} of 5000 s requires 225 to 310 kWe. These figures are for continuous operation. Using a more reasonable 60% duty cycle they become 155 to 205 kWe and 390 to 520 kWe, respectively.

Note that if tethered boost of lunar payloads is routine, tethered boost of OTVs and perhaps tethered capture of earth launched payloads will be routine. After all, the purpose of lunar oxygen delivery is to lessen the cost of space activities. Each part of the system must be inexpensive, and thus may be directly exploitable as well as useful in supplying inexpensive propellant. Momentum makeup requirements for capture of Earth launched payloads are 1.5 to 3 times as high as requirements for lunar mission assist, while momentum makeup for OTV boost will be 3 to 4 times that for lunar mission assist.

Tethered boost and deboost from a lunar orbiting transport node also offer significant savings. Due to the lower gravity gradient field and operational constraints in cislunar space, a boost of about 500 m/s would be appropriate. There would be about 600 tonnes of net upward mass flow per year, requiring an average momentum makeup of about 8.62 N. At this point in the loop, terrestrially derived mass such as electric thruster propellant is costly. Conserving expensive propellant requires operating at the highest practical I_{sp}, leading to substantial power demands. At an I_{sp} of 5000 s and 50% efficiency, power demands are 470 kWe at 100% duty cycle, and 780 kWe at 60% duty cycle. (The duty cycle will be controlled by operational constraints and thruster maintenance, not reactor availability).

A lunar orbiting transport node using tether mediated momentum transfer could be an important part of initial factory installation or of a lunar base program. In these cases it is desirable to move the equipment in question to its destination as rapidly as possible. Thrust and power limits determine how quickly material could be sent to the lunar surface. Multimegawatt (tens of newtons) electric propulsion systems for momentum makeup could be useful and justifiable for lunar base or factory infrastructure delivery. All propellant mass will be terrestrially derived until the first lunar factory starts operating. Thus, for initial lunar base or factory emplacement hydrogen thermal thrusters or relatively low I_{sp} electric propulsion systems would be useful, since they outperform cryogenic engines burning entirely terrestrially derived propellants. These high I_{sp} engines would become uneconomical as lunar oxygen becomes available, but could pay for themselves many times over at startup.

Transportation in Martian Resource Utilization

The Martian resource utilization transport loop involves a transport node (probably the space station) in LEO; a transport node in Martian orbit, perhaps on Phobos or Deimos; and some manufacturing facility. The manufacturing facility may be on Phobos, Deimos or the Martian surface, depending on the ores available and the product mix desired.

Indirect electric propulsion from the Earth orbiting transport node is more difficult than in the lunar case because mission opportunities only occur about every two years. Either the transport node would have to be extremely massive (in excess of 3000 tonnes), or the electric propulsion system would have to have very high thrust to acquire momentum rapidly (in excess of 100 N).

Indirect electric propulsion at Mars would suffer from the same disadvantage that it would at Earth—most of the payloads leaving for Earth leave in a short time period every two years or so. In some cases, tethered boosts could react against Phobos, which can be considered an infinite momentum store for near term activities. Gravity losses reduce the efficiency of propellant burned by vehicles boosted from Phobos. In many cases a tethered deboost from Phobos followed by a periapsis burn near Mars uses less propellant than tethered boost followed by a high thrust maneuver. In addition, very long tethers are required to utilize the mild gravity gradient near Phobos, and operations are slow due to this low gradient. During almost all operations of practical interest, the tether would be cut by micrometeoroids several times before the operation was complete. Tethered boosting of an electrically propelled vehicle always makes sense since they are subject to gravity losses anyway, and indirect propulsion can help reduce their flight times.

Flight times are always long in the Earth - Mars system, so electrically propelled freighters may be feasible. They must consume less terrestrially derived propellant than the mass of plant, consumables and delivery system (amortized over a reasonable period of time) imported from Earth to produce sufficient Martian propellant to deliver an equivalent amount of product to LEO.

An approximate relationship between the specific mass of the plant, S_m (the ratio of its mass to its annual output), and the required electric thruster I_{sp} for economical use of electric propulsion can be derived. Using reasonable chemical propellants and engines, the mass ratio to return propellant from Mars orbit to LEO is about 2, while the mass ratio to emplace a factory in Mars orbit is about 3. Thus, the initial mass in LEO to deliver a unit mass of Mars-derived propellant *to* LEO using chemical propulsion is $3 * 2 * S_m$, where S_m is the specific mass of the factory. An electrically propelled product delivery system requires propellant manufacture at Mars, requiring a mass of about S_m in LEO to deliver the factory using the electric freighters. Thus, amortizing factory mass cost over 10 years, assuming aerocapture at both Earth and Mars, and taking the low thrust ΔVs as 7.8 km/s and 9.8 km/s to return from and go to Mars,

$$0.5\, S_m > (\exp(800/I_{sp}) - 1) * (\exp(1000/I_{sp}))$$

For electric propulsion using terrestrially derived propellant to be more economical than Martian propellant for product delivery. Making some simple approximations:

$$0.5\, S_m > \exp(1000/I_{sp}) * 800/I_{sp}$$
$$0.5\, S_m + 1 > 1 + 800/I_{sp} + 800000/I_{sp}^2$$
$$0.5\, S_m + 1 > \exp(800/I_{sp})$$
$$\ln(0.5\, S_m + 1) > 800/I_{sp}$$
$$0.5\, S_m > 800/I_{sp}$$
$$I_{sp} > 1600/S_m$$

Reasonable specific masses are in the range of 0.1 to 0.3, implying that I_{sp} must be greater than 5300 to 16,000 s. Propellant manufacture on the Martian surface leads to a relationship of $I_{sp} > 500/S_m$, due to the large amount of Martian derived chemical propellant consumed in reaching Martian orbit from the surface. Charging interest (using a discount rate) favors electric propulsion. At 10% per year, the minimum required I_{sp}'s are 4200 to 13,000 s to exploit the Martian moons. These approximations require $I_{sp} >> 800$ and $S_m << 1$ to be valid.

Transportation in Asteroidal Resource Utilization

Certain near-Earth asteroids are potential space resources. There are two options for asteroidal resource utilization: (1) have a factory manufacturing products close to finished form at the asteroid, or (2) perform minimal processing at the asteroid and return ore to LEO for conversion to final form. Almost any form of high I_{sp} propulsion is beneficial. In this case power requirements vary substantially depending on the extent of processing carried out at the asteroid.

The transportation loop for asteroidal resource utilization involves a transport node in LEO and one at the asteroid itself. The transportation system is relatively simple, and, unlike the lunar or Martian case, some of the propulsion requirements are for deep space maneuvers (Lau and Hulkower 1985) during which there is no gravity loss penalty for using low thrust propulsion.

To be economical, asteroidal ore recovery depends on aerocapturing the bulk of the returned material in LEO. A nuclear electric propulsion system might be returned propulsively or discarded for safety reasons (to prevent inadvertent reentry during aerocapture). An $I_{sp} >> 10{,}000$ s would be required for any economic possibility of propulsive ore return.

Indirect electric propulsion from the Earth orbiting transport node has the same constraints as for Martian resource utilization. Launch windows are narrow and infrequent. Thus, effective indirect electric propulsion requires an incredibly massive transport node or an extremely powerful thruster.

When asteroidal material is being recovered for processing in LEO, all of the propellant ultimately consumed originates in LEO. Thus, the use of hydrogen thermal thrusters in a high thrust, high I_{sp} vehicle may be economical. On a mass basis, a high thrust system operating at an I_{sp} of 900 s gives performance equivalent to a low thrust system operating at a specific impulse of 2000 to 3000 s, due to gravity losses. Since gravity losses can only be offset if most or all of the hyperbolic excess energy is acquired by the spacecraft near Earth, indirect propulsion does not reduce them.

In the case of asteroidal ore recovery, cost per unit mass is essentially controlled by the amount of propellant required. Any high I_{sp} thruster significantly lowers costs. The only real restraint on system performance is that the system have a lower total mass than other reasonable competing systems. Due to gravity losses, if I_{sp} is < 1400 s, it will be more economical to boost the spacecraft to earth escape with chemical propellants. If I_{sp} is < 2500 s, a NERVA type system would be more economical for Earth escape. This does not necessarily mean that development of a NERVA system would be justifiable, only that use of an existing system would be beneficial. Actual development would depend on the relationship of development and production costs to the overall cost of alternative methods.

Transportation Summary

Tethers allow the use of various kinds of nuclear propulsion in lunar factory emplacement and product delivery. During lunar factory emplacement, powerful nuclear thermal or electric propulsion systems in LEO and low lunar orbit (LLO) are desirable. Once product delivery begins, economical delivery requires I_{sp}'s > 1500 s in LEO, or I_{sp} > 3000 s in LLO. Such thrusters could substantially lower product delivery costs as well as assist in a variety of other space activities. Nuclear electric freighters are not economical, since many are required to do the work of one chemically propelled vehicle.

High I_{sp} nuclear electric freighters may be useful for Martian product delivery. Tether mediated propulsion schemes are probably impractical, severely limiting the role of nuclear propulsion in Martian resource utilization.

Both lunar and Martian factory emplacement could benefit from nuclear thermal propulsion, as one or a few large payloads must undergo substantial ΔVs using entirely terrestrial propellants.

Asteroidal ore recovery can benefit directly from most forms of nuclear propulsion without the use of tethers (which is impractical due to launch window constraints). There are some relatively low performance constraints below which a mixed chemical/nuclear propulsion system is more economical than a nuclear propulsion system alone. Some form of nuclear propulsion is beneficial in almost any asteroidal resource recovery scheme, since propellant usage (and hence cost) is inversely proportional to I_{sp} in deep space.

Power Demands for Extraterrestrial Materials Processing

The detailed power demands for several methods of processing extraterrestrial ores follow. While these numbers are not particularly accurate or precise, they do illustrate the kinds of power demands that can be expected for SRU, and can serve as a starting point for design considerations. Most of the following discussion will focus on the production of oxygen from lunar materials. Some discussion of hydrogen production from lunar regolith is also included.

Lunar Materials Processing

Any system to deliver lunar oxygen to LEO depends on some technique for actually producing oxygen on the moon. Such techniques have been addressed (Cutler 1985, Cutler and Krag 1985, Waldron 1985, Steurer 1985, Haskin and Lindstrom 1979, Oldson et al. 1981, and Gibson and Knudsen 1985). There is little consensus on how lunar oxygen will be produced, or which technologies are most promising (Cutler, manuscript in preparation). However, one thing is certain, power demands will be in the megawatt range. Detailed discussion of power demands for lunar oxygen and hydrogen production are given here.

The moon is composed of igneous materials that are typically 40% to 50% oxygen as oxides. There are no significant amounts of air or water on the moon, so these rocks are the only source of oxygen. The rocks have been precrushed by meteoritic bombardment. The major minerals present are anorthite, olivine and ilmenite, which may be considered ores (McKay and Williams 1979). It should be possible to obtain relatively pure mineral separates from mined regolith (Williams et al. 1979).

Oxygen is chemically combined with metals in the lunar minerals, so extracting oxygen requires coproduction of a metal. Extracting oxygen from metals is very similar to extracting metals from their oxides, so a large body of knowledge on extractive metallurgy is applicable to our problem. Most methods of oxygen production employ a reductant to reduce a metal from the oxide mineral in the first step, extracting oxygen in a more chemically convenient form than metal oxide. Subsequent processing steps extract elemental oxygen and regenerate the reductant.

There are a variety of possible processes for lunar oxygen production. Hydrogen reduction of solid ilmenite (Gibson and Knudsen 1985, Oldson et al. 1981 and Williams 1983), carbon reduction of molten ilmenite (Cutler and Krag 1986) and electrolysis of ferruginous slags (Haskin and Lindstrom 1979) have been compared and discussed (Cutler 1985). Power requirements for HF acid leach (Waldron 1985), vapor phase pyrolysis (Steurer 1985) and carbothermal processing (Cutler and Krag 1986) are given in the literature. Table 1 provides a summary of power demands to produce 1000 tonnes of lunar oxygen per year.

There are significant opportunities to thermally integrate the plant and power plant used for oxygen production via hydrogen reduction of ilmenite. Such integration would doubtless lead to lower cost and mass than for a comparable power unit and factory with a pair of busbars between them. For example, hydrogen could be preheated by direct passage through the reactor core.

Carbothermal processing has been described in some detail (Cutler and Krag 1985 and references cited therein). In this process, regolith is mined and an iron bearing mineral concentrate extracted. This mineral concentrate is smelted with carbon to give slag, iron, and carbon monoxide. The slag is discarded, the iron is decarburized with water or oxygen (produced later in the process), the resultant steel being dis-

Table 1 *Power Requirements for Space Resource Utilization Processes.*

	Power[a]		
	Electric (MWe)	Thermal (MWt)	Temperature (K)
Flouroacid Leaching[b]	2.51	0.952	600
Vapor Phase Pyrolysis[c]	1.24	3.17	3000
Carbothermal Processing	2.35	0.549	363 and 873
Slag Electrolysis	1.99	0.450	363
Hydrogen Reduction:			
with condensation	1.19	1.58	363 and 1445
without condensation	1.27	1.04	363
Hydrogen Recovery from Regolith:			
High hydrogen	3.34	6.12	773
Average hydrogen	5.04	14.3	773
Low hydrogen	11.6	29.2	773

[a] All plants are assumed to operate at 90% duty cycle. The assumptions used to calculate these power requirements are given in the body of the paper. Due to the limited consideration given to loss mechanisms, actual power requirements may be somewhat higher.
[b] From Waldron (1985).
[c] From Steurer (1985).

carded (or perhaps used subsequently), and the product carbon monoxide joining the smelter off gas stream. The off gas is reformed with hydrogen to form hydrocarbons and water. The hydrocarbons are cracked to hydrogen and carbon for recycle. The water is electrolyzed or thermolyzed to make hydrogen for recycle and product oxygen. The oxygen is liquefied, stored, and ultimately delivered to LEO. Simple slag and steel manufactures may be of some use, for example as shielding bricks for the reactor. A block flow diagram with material and energy fluxes is shown in Figure 1. The energy fluxes are either derived by analogy to similar terrestrial processes or by consideration of theoretical heat requirements and the theoretical losses expected from the major obvious loss mechanisms. Power demand is calculated on the basis of 90% duty cycle. Mining (Gertsch 1985) and beneficiation (Agosto 1985) power requirements were obtained from workers in the field. Cracking power is calculated on the basis of turning $(CH_2)_n$ at 573 K into n C + n H_2 at 973 K. Oxygen liquefaction efficiency is taken from handbook data (as 0.86 kWe/kg). Reformer heat rejection requirements are calculated on the basis of 2 H_2 (at 648 K) + CO (at 673 K) forming H_2O + (CH_2) at 673 K. Other power requirements are as described in Cutler and Krag (1985).

A somewhat more direct process can be based on the hydrogen reduction of lunar ilmenite (Gibson and Knudsen 1985, Oldson et al. 1981, and Williams 1983). In this process, regolith is mined and an ilmenite concentrate produced. This concentrate is charged to a fluidized bed reactor, where it is reduced with hydrogen and the spent charge is discarded. The reacted hydrogen contains 5% to 10% water vapor. This effluent could be electrolyzed using a high temperature ceramic electrolyte. Alternatively, the effluent could be cooled and dewatered, and the water electrolyzed or thermolyzed by standard techniques. Block flow diagrams with theoretical energy and materials fluxes are shown in Figures 2 and 3. The energy fluxes are underestimates, since they are based on theoretical heat capacities and theoretical losses from the most obvious major loss mechanisms. Power demand is calculated on the basis of 90% duty cycle. A more closely sized fraction of ilmenite is taken than is the case for carbothermal processing to ensure proper fluidization. The ilmenite charge is assumed to enter the reduction reactor at 363 K, and the spent ilmenite is discharged at 1173 K. Radiative losses from the reduction reactor are calculated by assuming it is a cylinder with height equal to diameter with an emissivity of 0.7. It is surrounded by a radiation shield with an emissivity of 0.15. The reduction reactor is sized so it can hold two hours' throughput with a (fluidized) density of 1.25. Heat losses from the countercurrent heat exchanger and the preheater are assumed to be negligible. High temperature electrolysis is assumed to be 48.4% energy efficient in order to make the heat balance work out. The electrolysis cell is kept hot enough to obviate the need for a hydrogen preheat. Radiation loss from the electrolyzer is neglected. The expected

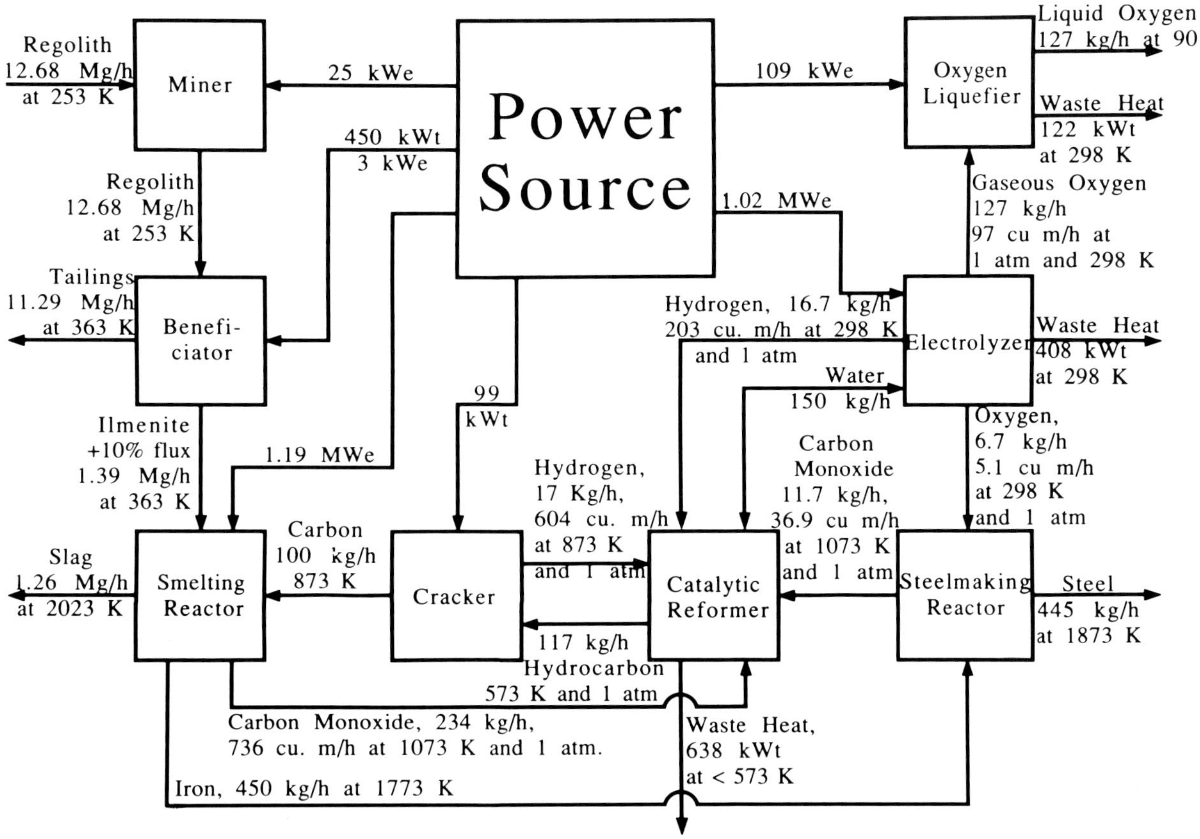

Figure 1 *Power and Materials Fluxes for Carbothermal Oxygen Production.*

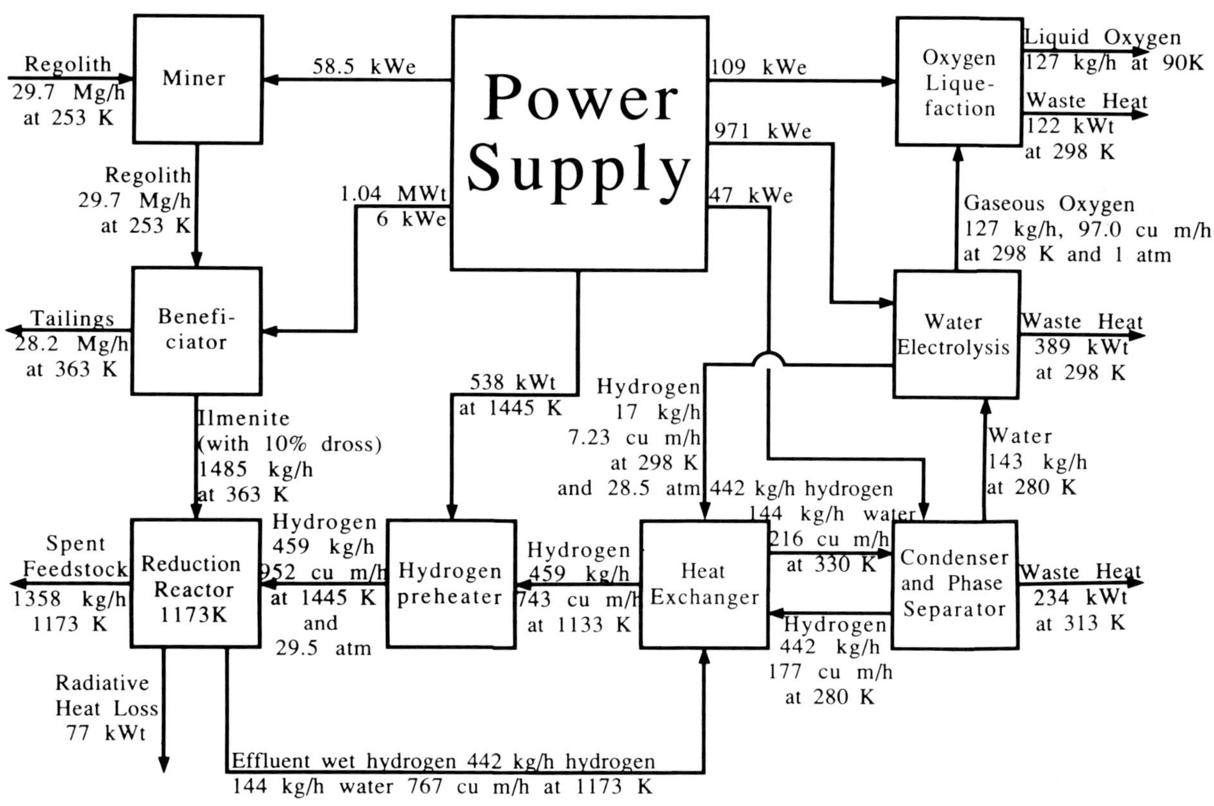

Figure 2 *Power and Materials Fluxes for Oxygen Production via Ilmenite Reduction with Condensation.*

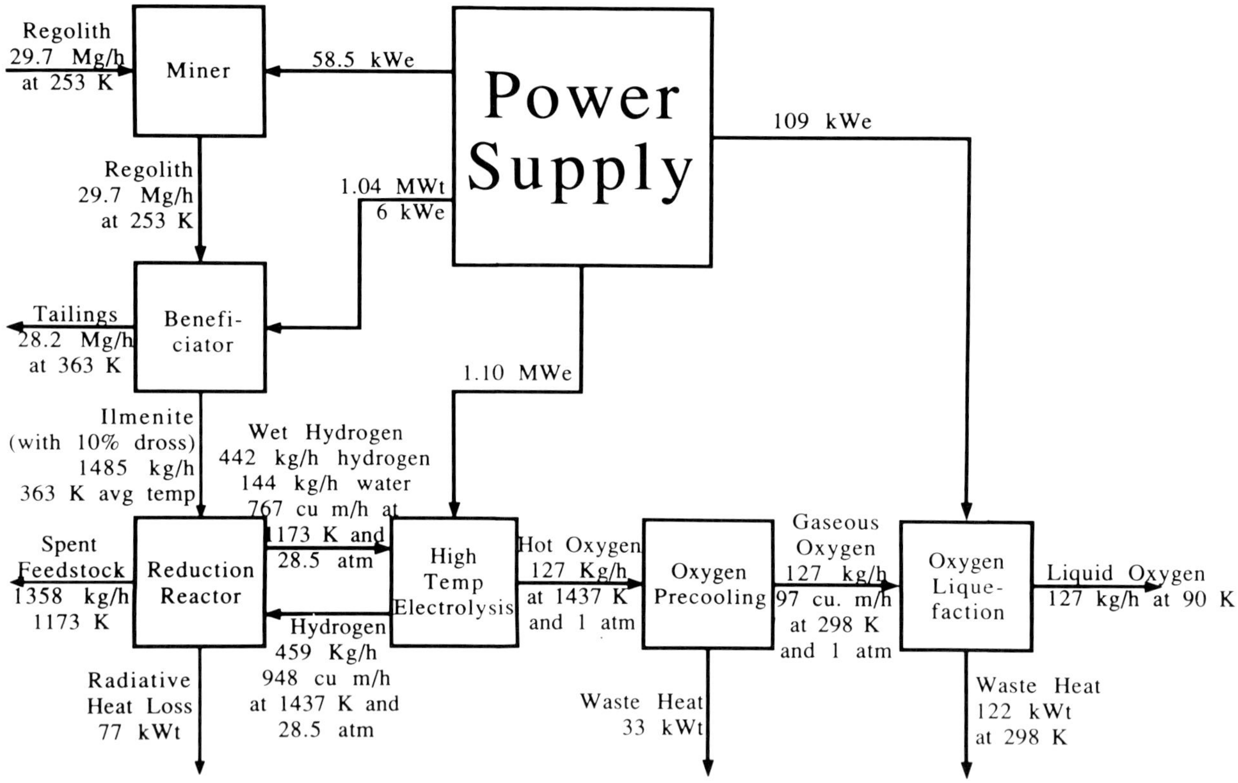

Figure 3 *Power and Materials Fluxes for Oxygen Production via Ilmenite Reduction with High Temperature Electrolysis.*

losses from the electrolyzer would have to be made up by increasing the electric power input.

Oxygen might also be produced by electrolyzing a molten slag (Haskin and Lindstrom 1979). It is probably best to electrolyze some iron containing mineral concentrate, because the separation of oxygen from iron has a lower theoretical energy demand than the separation of oxygen from silicon, aluminum, titanium, calcium, or magnesium. In engineering practice the separation of iron and oxygen is easier and closer to theoretical efficiency than is separation of oxygen from any of the other major lunar elements. Thus, it seems likely that iron will be a coproduct of lunar oxygen production. Each mass unit of oxygen produced corresponds to the production of 3.50 units of iron. In the cases of aluminum, silicon, titanium, calcium and magnesium, the corresponding figures are 1.12, 1.76, 1.50, 2.51 and 1.52, respectively. A block flow diagram with theoretical energy and materials fluxes is shown in Figure 4. This process is quite simple and direct, but it imposes severe materials demands on the anode and crucible, and it rejects a substantial amount of waste heat at high temperature. Rejecting this waste heat by using it to superheat the working fluid above the reactor outlet temperature could lead to significant economy. The actual efficiency of this process will vary substantially with electrolyzer design and feed composition. For design purposes it would be worth considering power unit designs based on efficiencies of between 20% and 50%. The information in Figure 4 is based on an assumed 30% efficiency, in which case an additional 30% of the energy is used to melt the incoming feed and 40% is available as waste heat. Oxygen liquefaction is from handbook data as previously described.

Ultimately, it would be desirable to require no terrestrial fuel for the delivery of lunar oxygen to LEO. Hydrogen is known to exist in the regolith at low concentration and can be partially extracted by heating to temperatures well below the sintering temperature. There is considerable uncertainty about the hydrogen distribution in lunar soils. Thermal desorption of hydrogen from lunar soils has been extensively studied, and compilations of data are available (Gibson 1974). Plant design and power requirements are strongly dependent on how much hydrogen is in bulk minable soil and to what extent it is concentrated in the fines. Recent work indicates substantial concentration (and variability) can be expected in the fines (Bustin et al. 1986). A block flow diagram for a hydrogen extraction plant is shown for three different assumptions regarding hydrogen distribution in lunar soils (with

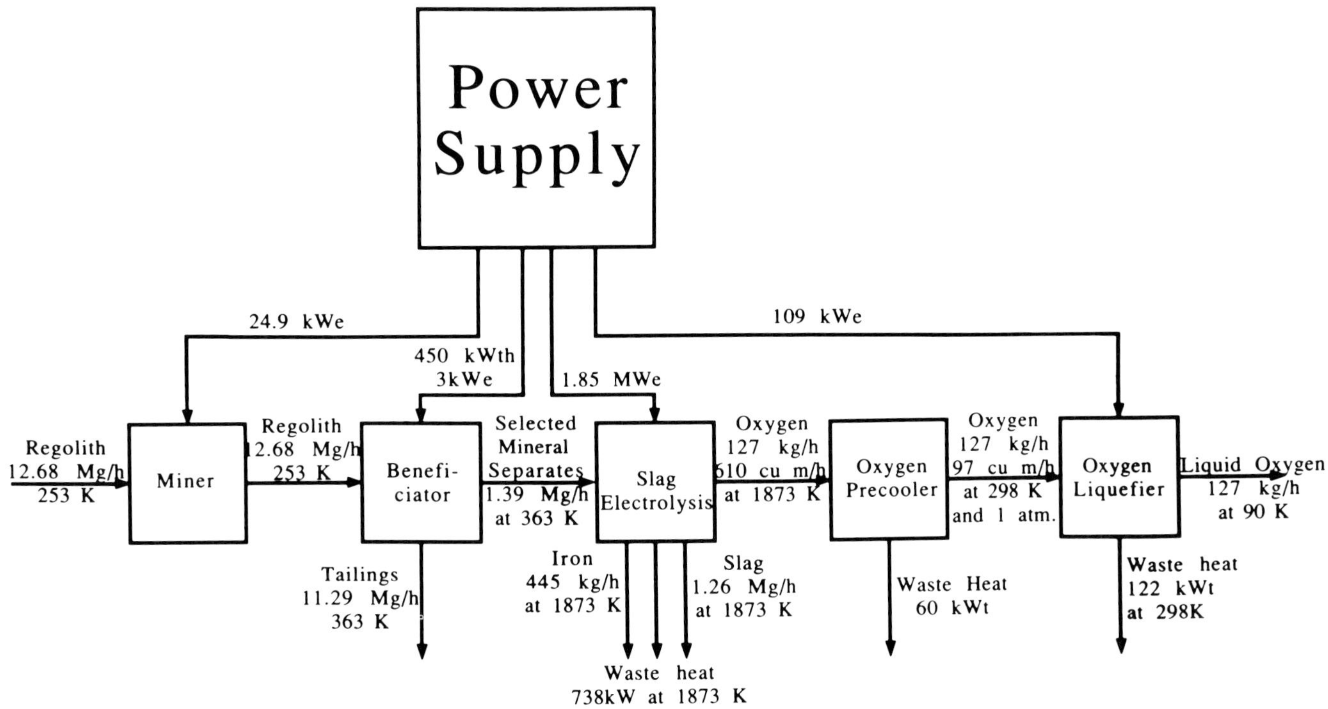

Figure 4 *Power and Materials Fluxes for Oxygen Production via Slag Electrolysis.*

theoretical energy and materials fluxes) in Figure 5 for a plant sized to produce 70 tonnes of hydrogen per year.

The optimistic assumptions in Figure 5 are that soil is mined in which 5% of the material is $< 10\ \mu m$ in size. No power requirements for sieving regolith are readily available. The fine fraction is assumed to have an overall hydrogen content of 350 ppm (slightly higher than the maximum observed value—assuming some additional exploration), of which 60% is recoverable on heating to 873 K and holding for two hours. The retort is assumed to be a spherical vessel that holds two hours throughput of material with an effective bulk density of 1.5. The vessel has an emissivity of 0.7 and a single radiation shield with an emissivity of 0.15. The feedstock is preheated from 253 K to 773 K with thermal energy, and negligible hydrogen loss is assumed. The charge is heated from 773 K to 873 K by induction heating. Power conditioning, as well as the coil efficiency, are assumed to be 96%. Mining is as described by Gertsch (1983 and 1986). Beneficiation power requirements are neglected, as they have not yet been determined for sieving lunar soil.

The average assumptions in Figure 5 are that the feedstock is $< 20\ \mu m$ material, which is 25% of the bulk soil. The feed is assumed to contain 180 ppm of hydrogen, of which 50% is recoverable on heating to 898 K.

The pessimistic assumptions in Figure 5 are that the feedstock is $< 20\ \mu m$ material, which is 20% of the bulk soil. The feed is assumed to contain 110 ppm of hydrogen, of which 40% is recoverable on heating to 923 K.

Solar versus Nuclear Power for Lunar Manufacturing

Insolation at the lunar equator is the same as it is just above Earth's atmosphere—1370 W/m^2. Light and dark periods on the moon are about 14 days long. Continuous sunlight may be available at the lunar poles (Burke 1985). Continuously illuminated locations have not been determined from existing photogeology and thus may or may not exist. The structural problems of erecting vertical megawatt scale solar arrays or collectors have not been addressed, but are expected to be significant.

Martian and Asteroidal Materials Processing

Detailed power demands for the use of Martian and asteroidal resources will not be given here (but see Lawton and Frisbee 1985, Lewis and Meinel 1983, and Leonard et al. 1986). Several general points can be made about Martian and asteroidal materials pro-

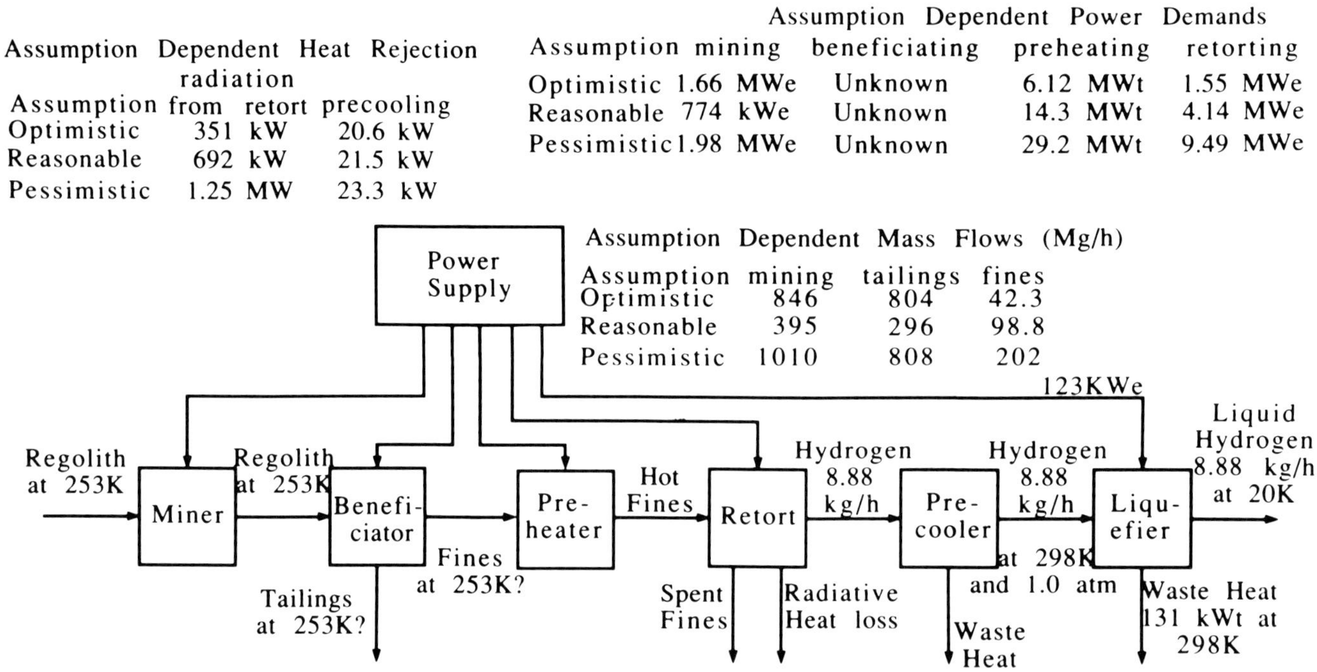

Figure 5 *Power and Materials Fluxes for Hydrogen Production via Retorting of Fines Using Optimistic, Reasonable and Pessimistic Assumptions about Hydrogen Availability.*

cessing. First, in terrestrial industrial practice, all thermal processes and most other processes use significantly more energy than their theoretical requirements. Typical energy efficiencies for metals melting are 60%. Since thermal efficiency increases with increasing size and throughput, one can expect that there will be significant inefficiency in any space processing plant. Calculating power demands on the basis of theoretical requirements will lead to substantial underestimates. Second, extracting water and volatile organics from carbonaceous chondrites or the moons of Mars will bear a lot of similarity to oil shale refining. Thus there is a significant body of terrestrial engineering literature applicable to this process, and some indication that it will not be trivial to do. Third, while one can imagine transporting some plant (such as one of the lunar plants described above) to an asteroid or Mars and operating it with minor modifications, this is certain to be much less efficient than designing a plant specifically for the resource it is to exploit. Such comparisons are not an appropriate means of comparing the exploitation of different resources. Fourth, in a very rough sense, we can expect the power demands for propellant production to be about the same (within a factor of two or three) regardless of which resource we exploit. By mass, the bulk of hydrogen—oxygen or hydrocarbon—oxygen propellants is oxygen. All of the oxygen accessible to us as an ore at the various space resources is available as an oxide—of hydrogen, iron or carbon, and these oxides all have similar heats of formation per oxygen atom. Thus the theoretical energy requirements to obtain the bulk of our propellant are of similar magnitude no matter which resource we exploit. While this is not a strong enough statement to draw any economic conclusions, it does imply that power demands for propellant production will typically be on the close order of a few MW's per 1000 tonnes per year—a useful number to keep in mind.

Solar versus Nuclear Power for Martian and Asteroidal Manufacturing

Solar collectors in the Martian vicinity receive about 43% of the energy density they receive in Earth's vicinity. Beyond this, there are specific factors affecting the trade at each of the three locations of interest—Phobos, Deimos and the Martian surface. There is an economic tradeoff between solar and nuclear power in all space manufacturing schemes. The details of ore processing and solar power availability at each location will determine the most economical power source.

Phobos receives about 1 ¼ h of shadow and 6 ½ hours of sunlight throughout most of the Martian year. There are two periods of approximately one month in which Phobos is in continuous sunlight. Solar powered factories located at the poles of Phobos could operate with this type of duty cycle. Locations elsewhere on Phobos have 3 ½ hour days and nights. Since Phobos is phase locked to Mars, locations from which Mars is visible are eclipsed for up to 1 ¼ hour per orbit (about 1 day) during much of the Martian year. Nuclear power would allow continuous operation. The frequent need to start up and shut down a solar powered factory on Phobos strongly favors nuclear power.

Deimos is in continuous sunlight for most of the Martian year. The north and south poles receive continuous sunlight for about 7 months at a time. There are also periods of about 3 months when Deimos spends about 1 hour per day in the shadow of Mars and 29 ¼ hours in sunlight. Thus, continuous solar powered operation for long periods is feasible. Due to Deimos' low gravity, the entire factory could be moved from pole to pole to take advantage of the continuous sunlight. While this would not be trivial, this is believable and should be considered. The factory would be similar in size to an 18 wheel truck and could be lifted by a few astronauts.

A number of factors favor nuclear power over solar power on the Martian surface. Frequent long nights, atmospheric dispersion, dust storms and the possibility of dust settling on solar collectors all favor the use of nuclear power. In addition, the availability of an atmosphere to aid in heat rejection favors the use of dynamic power conversion technologies.

It is relatively easy for an asteroidal processing facility to remain in constant sunlight. Insolation at the asteroid varies greatly. Accessible asteroids are typically in elliptical orbits with perihelia near 1 AU and apohelia between 2 and 3 AU. Mission opportunities usually involve arriving at the asteroid near perihelion and departing near apohelion. Thus, insolation starts at about 100% of that at Earth and rapidly declines to 10% to 25% of the initial value.

Power Plant Design for Materials Processing

A number of factors must be considered in SRU power plant design. The power supply is expected to be a major fraction of the mass and dollar cost of any SRU factory. The power plant must be integrated with each type of factory design for maximum economy. It is not particularly useful to use design parameters that do not account for power conditioning or that compare factories and power plants that are not integrated with one another. Specifically, it is seldom practical to compare solar and nuclear powering of the same plant (plants of equivalent average output optimized for each power source should be compared) or to compare thermoelectric and dynamic power conversion technologies for meeting identical power demands.

Methods of integrating the power plant and factory include heating of reactant streams, combined heat rejection and direct provision of thermal energy. Power conditioning requirements can also be minimized by careful integration. Careful matching of heat fluxes and temperatures can make materials requirements less stringent and can minimize heat rejection radiator mass and cost. Direct generation of the required frequencies and voltages (often low voltages and DC or $<$ 1 kHz AC) can substantially reduce the amount of power conditioning equipment required. All of these factors favor locating the power plant close to the factory.

Colocation of the power plant and factory imposes stringent safety requirements. Radiation control is essential as the factory requires both scheduled and unscheduled servicing and refurbishment. These tasks cannot be fully automated. Power plant components will occasionally be damaged by accident, mishandling or failure of factory components. The factory would almost certainly be located near a lunar or Martian base and its launch and landing facilities. Thus, there is a significant risk of accident. Proper design is required to keep accidental radiation release to an acceptable level and to facilitate the repair or replacement of damaged components.

Conclusions

SRU is inevitable, but it is not a near term mission. SRU is market driven. The most likely markets for space resource derived products are for OTV propellant or for propellant to support a lunar base or Mars mission program. Nuclear power can play a significant role in both propulsion and materials processing. Tether mediated indirect electric propulsion is a likely part of lunar resource utilization, though, once installed, it will be more heavily used for other purposes. Nuclear electric freight vehicles can significantly improve the economy of asteroidal ore recovery. They are less likely to be useful for Martian resource utilization and are quite unlikely to be useful for lunar resource utilization.

Nuclear power for materials processing (and for propulsion) must always be traded off against solar powered systems. Process power demands are likely to be in the megawatt range, and cannot now be accurately specified. Power plant mass and cost are expected to be significant fractions of overall plant mass and cost. Integrating the power plant and materials processing equipment may lead to significant economies.

Acknowledgments

I would like to thank Peter Krag, Andrew Wolff and one anonymous referee for their substantial efforts reviewing this manuscript and making helpful comments and suggestions, and Terry Wallace for inviting me to write this paper. As with many papers it took on a life of its own and turned out to be a very interesting and useful exercise when it was finally done (many months after it was supposed to be). This work was performed while I was a postdoctoral fellow at the California Space Institute at U. C. San Diego. My present address is Energy Sciences Laboratories, POB 85608, San Diego, CA 92138–5608.

References

Agosto, W. N. (1985) "Electrostatic Concentration of Lunar Soil Ilmenite in Vacuum Ambient," in *Lunar Bases and Space Activities in the 21st Century*, W. Mendell, ed., Lunar and Planetary Institute, Houston, TX.

Ash, R. L., V. Cuda, M. L. Stancati, and J. C. Niehoff (1981) "Outer Planet Satellite Return Missions Using In Situ Propellant Production," *Acta Astronautica*, 8: 511–526.

Billingham, J., W. Gilbreath, and B. O'Leary, eds. (1979) *Space Resources and Space Settlements*, NASA SP-428, U. S. Government Printing Office, Washington, DC.

Leonard, R. F., J. D. Blacic, and D. T. Vaniman (1986) "Application of Nuclear Power to the Mining of the Martian Moons," in *Trans. Third Symposium on Space Nuclear Power*, Conf-860102—SUMMS, held in Albuquerque, NM, 13–16 January 1986.

Burke, J. D. (1985) "Merits of a Lunar Polar Base Location," in *Lunar Bases and Space Activities in the 21st Century*, W. Mendell, ed., Lunar and Planetary Institute, Houston, TX.

Bustin, R., R. H. Carr, and E. K. Gibson (1986) "The Hydrogen Content of Lunar Soils," in *Abstracts of the 17th Lunar and Planetary Science Conference*, held in Houston, TX, 17–21 March 1986.

Carroll, J. A. (1984) "Tether Applications in Space Transportation," IAF paper 84–438, presented at the 35th International Astronautical Federation Congress, held in Lausanne, Switzerland, 5–12 October 1984.

Carroll, J. A. (1985) *Guidebook for Analysis of Tether Applications*, Final report on contract RH4-394049 with the Martin Marietta Corporation, available from Marshall Space Flight Center, Mail Code PS-01, Huntsville, AL.

Cooper, L. P. (1984) *Propulsion Issues for Advanced Orbit Transfer Vehicles*, NASA Technical Memorandum 83624, U. S. Government Printing Office, Washington, DC.

Cutler, A. H. (1985) "Slag - Metal Equilibrium in Lunar Smelting and Arc Electrowinning," Paper presented at the SpaceTech Conference, Anaheim, CA 23–25 September 1985. Available from the Society of Manufacturing Engineers, One SME Drive, Dearborne, MI.

Cutler, A. H. (1984) "$H_2/O_2/Al$ Engines and Their Applications to OTV's," Paper number 84–314, presented at the 35th IAF congress, held in Lausanne, Switzerland, 5–12 October 1984.

Cutler A. H. and M. L. Hughes (1985) "Transportation Economics of Extraterrestrial Resource Utilization," in *Space Manufacturing 5: Engineering with Lunar and Asteroidal Materials*, B. Faughnan and G. Maryniak, eds., American Institute of Aeronautics and Astronautics, New York, pp. 233–244.

Cutler A. H. and P. Krag (1985) "A Carbothermal Scheme for Lunar Oxygen Production," in *Lunar Bases and Space Activities in the 21st Century*, W. W. Mendell, ed., Lunar and Planetary Institute, Houston, TX, and references cited therein (Specifically the multitudinous references to Rosenberg *et al.*).

Davis, H. P. (1983) "Lunar Oxygen Impact Upon STS Effectiveness," Eagle Engineering report 8363, Eagle Engineering, Houston, TX.

Frisbee R. H. and R. M. Jones (1983) "Transportation," section 7 of *Research on the Use of Space Resources*, W. F. Carroll, ed., JPL Publication 83–36, Jet Propulsion Laboratory, Pasadena, CA.

Gertsch, R. E. (1983) "A Method for Mining Lunar Soil," in *Proc. 5th Princeton Space Manufacturing Conference*, which appeared as *Advances in the Astronautical Sciences*, 53: 337–346.

Gertsch, R. E. (1985) Personal Communication. Unpublished analysis indicates that approximate power requirements for lunar mining are 75 to 112 kWe to operate a slusher (as described in Gertsch 1983) which mines 450,000 tonnes/yr of regolith; Earth Mechanics Institute, Colorado School of Mines, Golden, CO.

Gibson, E. K., G. W. Moore, and S. M. Johnson (1974) *Summary of Analytical Data from Gas Release Investigations, Volatilization Experiments, Elemental Abundance Measurements on Lunar Samples, Meteorites, Minerals, Volcanic Ashes and Basalts*, available from Johnson Space Center, Houston, TX.

Gibson M. A. and C. W. Knudsen (1985) "Lunar Oxygen Production from Ilmenite," in *Lunar Bases and Space Activities in the 21st Century*, W. W. Mendell, ed., Lunar and Planetary Institute, Houston, TX.

Haskin L. A. and D. J. Lindstrom (1979) "Electrochemical Preparation of Useful Materials from Ordinary Silicate Rocks" in *Space Manufacturing 3*, Proceedings of the 4th Princeton/ AIAA Space Manufacturing Conference, American Institute of Aeronautics and Astronautics, New York, p. 129.

Helin, E. F., N. D. Hulkower, and D. F. Bender (1984) "The Discovery of 1982DB, the Most Accessible Asteroid Known," *Icarus*, 57: 42–47.

Lau, C. O. and N. D. Hulkower (1985) "On the Accessibility of Near Earth Asteroids," American Astronautical Society Paper 85–352, presented at the AAS/AIAA Astrodynamics Specialist Conference, held in Vail, CO, 12–15 August 1985.

Lawton, E. H. and R. H. Frisbee (1985) *A New Look at Oxygen Production on Mars*, JPL publication D-2661, Jet Propulsion Laboratory, Pasadena, CA.

Lewis, J. S. and C. Meinel (1983) "Asteroid Mining and Space Bunkers" *Defense Science 2OO1 +*, 2: 33.

McFadden, L. A. (1984) "Mineralogical-Petrological Characterization of Near-Earth Asteroids," *Icarus*, 59: 25–40.

McKay, D. S. and R. J. Williams (1979) "A Geologic Assessment of Potential Lunar Ores," in *Space Resources and Space Settlements*, NASA SP-428, J. Billingham, W. Gilbreath and B. O'Leary, eds., U. S. Government Printing Office, Washington, DC, pp 243–255.

Nozette, S. D. ed. (1984) *Defense Applications of Near Earth Resources*, California Space Institute, La Jolla, CA.

Oldson, J., J. Livingston, K. Brown, and R. Edelson (1981) "Design of a Lunar Liquid Oxygen Plant," in *Abstracts of the Cal Space Conference*, held at the University of California, Berkeley, CA, 24–25 March 1981.

Salkeld, R. J. (1966) "Economic Implications of Extracting Propellants from the Moon," *J. Spacecr. Rocket*, 3: 254.

Steurer, W. F. (1985) "Lunar Oxygen Production by Vapor Phase Pyrolysis," in *Space Manufacturing 5: Engineering with Lunar and Asteroidal Materials*, B. Faughnan and G. Maryniak, eds., American Institute of Aeronautics and Astronautics, New York, NY, pp. 123–131.

Stump, W. R. (1985) "Analysis of Lunar Propellant Production," Eagle Engineering Report 85–103B, Eagle Engineering, Houston, TX.

Waldron, R. D. (1985) "Total Separation and Refinement of Lunar Soils by the HF Acid Leach Process," in *Space Manufacturing 5: Engineering with Lunar and Asteroidal Materials*, B. Faughnan and G. Maryniak, eds., American Institute of Aeronautics and Astronautics, New York, NY, pp. 132–149.

Williams, R. J. (1983) "Enhanced Production of Water from Ilmenite: an Experimental test of a Concept for Producing Lunar Oxygen," in *Abstracts of the 14th Lunar Science Conference*, Lunar and Planetary Institute, Houston, TX.

Williams, R. J., D. S. McKay, D. Giles, and T. E. Bunch (1979) "Mining and Beneficiation of Lunar Ores," in *Space Resources and Space Settlements*, NASA SP-428, J. Billingham, W. Gilbreath, and B. O'Leary, eds., U. S. Government Printing Office, Washington, DC, pp. 275–288.

APPENDIX A

Basic Celestial Mechanics and Space Transportation Concepts

Transportation costs, a major factor in SRU economics, are mainly determined by propellant requirements and celestial mechanics. Since there are many opportunities to use nuclear power in the transportation loop, and nuclear engineers are seldom familiar with the (surprisingly simple) basics of space flight, some introductory background on celestial mechanics, propulsion basics and reasonable propulsion technology follows.

Specific impulse, I_{sp}, is a measure of any rocket's efficiency of mass use. I_{sp} has units of time. It is the length of time a kg (mass) of propellant will provide a kg (force) of thrust (a kg of force is a bastard unit equal to 9.81 newtons - the weight of a kilogram on the earth's surface). It is equal to the average exhaust stream velocity, V_e, divided by the earth's gravitational acceleration, g, 9.81 m/s^2:

$$I_{sp} = V_e / g \; . \tag{A1}$$

Electric thrusters turn electrical power into exhaust energy. The exhaust momentum provides thrust. As kinetic energy scales with the square of exhaust stream velocity, at constant mass flow, power requirements vary as the square of I_{sp}. At constant thrust (force, N), power requirements vary linearly with I_{sp}, since, in this case, the mass flow decreases as I_{sp} increases. In either case, the total energy required for a spacecraft to execute a given maneuver varies linearly with I_{sp} while the propellant mass varies inversely with I_{sp}. Some typical I_{sp}'s are shown in Table A1.

Electric propulsion systems are typically "low thrust" ($a << 1$ cm/s^2) due to the achievable specific power of the electrical power source. Chemical and thermal engines are typically "high thrust" ($a >> 10$ cm/s^2) due to the higher achievable specific powers. Low thrust propulsion systems typically have much higher I_{sp}'s, and perform more efficiently than high thrust systems. Many factors lessen or negate the advantages of low thrust propulsion, as will be shown.

Tsiolkovsky's equation (Equation (A2)) relates a reactive vehicle's initial and final masses to its required change in velocity (referred to as ΔV). The required propellant mass can easily be calculated from this (note that propellant tanks are not massless). In most practical cases, M_{final} is substantially larger than the payload mass.

$$\frac{M_{\text{initial}}}{M_{\text{final}}} = exp \left| \frac{\Delta V}{I^*_{sp}\, g} \right| \tag{A2}$$

Thus, for ΔVs substantially less than $I^*_{sp}\, g$, this reduces to $M_{\text{initial}}/M_{\text{final}} = 1 + \Delta V / (I^*_{sp}\, g)$, and the propellant used for a given ΔV is inversely proportional to I_{sp},

$$\frac{M_{\text{propellant}}}{M_{\text{final}}} = \Delta V / (I^*_{sp}\, g) \; . \tag{A3}$$

From this we see that the energy, E, required per unit final mass (for small velocity changes) is proportional to I_{sp}:

$$E = \frac{\text{Propellant}}{2\, M_{\text{final}}} I^*_{sp}\, g \; . \tag{A4}$$

Higher ΔV's favor higher I_{sp} propulsion more than this proportionality would indicate.

Moving a spacecraft to a higher orbit or to escape is called orbit raising. Certain aspects of orbit raising are quite simple, and are important in the following discussions. Equation (A5) gives the speed V of a spacecraft at some point r in an orbit with semimajor axis a:

$$V = \sqrt{(2\mu/r - \mu/a)}, \tag{A5}$$

where the distance r is measured from the center of the body the orbit is around, and the constant μ describes the gravitational attraction of the central body. The value of μ (in m^3/s^2) is 3.991×10^{14} for Earth, 1.329×10^{20} for the sun, 4.909×10^{12} for the moon and 4.284×10^{13} for Mars. The respective average radii (periapsis, the point on the orbit closest to the central body, must be somewhat greater than this) are 6371315 m for Earth, 1738300 m for the moon, and

3380000 m for Mars. The semimajor axes of various orbits of interest are: Earth about the sun, 1.496×10^{11} m; the Moon about Earth, 3.844×10^8 m; Mars about the sun, 2.278×10^{11} m; Phobos about Mars, 9.408×10^6 m; Deimos about Mars, 2.35×10^7 m; and GEO, 4.206×10^7 m.

Equation (A5) can be used to calculate the required velocity change to go from one orbit to another as follows. An orbit is selected that intersects (usually tangentially) both the origin and destination orbit. This is a transfer orbit. The ΔV is the scalar sum of the velocity change required to get from the origin orbit to the transfer orbit, and to get from the transfer orbit to the destination orbit. Velocities must be treated as vectors when the origin, transfer, and destination orbits are not all coplanar. For most orbits of interest, the lowest ΔV will usually be obtained when the transfer orbit is tangent to the periapsis of the destination orbit and is approximately tangential to the origin orbit about ½ an orbital revolution earlier.

Equation (A6) gives the velocity at any point on a hyperbolic (escape) orbit.

$$V = \sqrt{(2\mu/r + C_3)}\ . \tag{A6}$$

This type of orbit is used as a transfer orbit from low Earth orbit (LEO) to a sun centered Earth to Mars trajectory. C_3 is a measure of the spacecraft's energy per unit mass far from the body it is escaping. The velocity very far from Earth ($\sqrt{C_3}$) is the velocity required to embark on the transfer orbit from the orbit of the Earth around the sun to the orbit of Mars around the sun if the Earth's gravitational field were not present. This procedure is reversed at Mars, with the square of the ΔV required to go from the sun centered transfer orbit to the orbit of Mars about the sun (in the absence of Mars) being C_3. The ΔV required to go from this hyperbolic orbit to an elliptical or circular orbit around Mars can then be calculated.

The reason for gravity losses (increased propulsive requirements to go from one orbit to another using thrust in high orbits) can be seen from Equations (A5) and (A6). A low thrust propulsion system must escape, then provide a ΔV of $\sqrt{C_3}$. A high thrust system can provide escape velocity plus a small additional ΔV while in a low orbit and attain the same C_3. Increasing energy is easier at high velocity, since $\Delta E = F * d = F * V * dt$. Further gravity losses are incurred by a low thrust system as it spirals out. This is equivalent to escaping by transferring from one orbit to another slightly higher orbit until escape is achieved. When the ratio of origin to destination orbit semimajor axis is large, transferring via intermediate orbits requires more total ΔV than transfer (using a high thrust system) via a single orbit, as can be seen from Equation (A5).

Momentum can be gradually accumulated at a transport nodes and then be impulsively transferred to a vehicle (with a tether, for example). This technique,

Table A1 *Representative Specific Impulses.*

Engine	Vehicle	Propellant	Specific Impulse (I_{sp}, vacuum)
SRB[a]	Shuttle	NH_4ClO_4/Rubber	270 (sea level)
SSME[b]	Shuttle	LH_2/LO_2	454
OMS[c]	Shuttle	NTO/MMH[d]	313
RCS[e], continuous	Shuttle	NTO/MMH	289
RCS, pulsed	Shuttle	NTO/MMH	224
RL-10	Centaur G and G′	LH_2/LO_2	446
J-2[f]	Saturn	LH_2/LO_2	426
ASE[g]	OTV	LH_2/LO_2	480–490
NERVA[h]	???	LH_2	800–900
Arcjet[i]	???	LH_2	1000–2000
MPD thruster[i]	???	Varies	2000–5000

[a] Solid Rocket Booster.
[b] Space Shuttle Main Engine.
[c] Orbital Maneuvering System.
[d] Nitrogen tetroxide/ monomethyl hydrazine.
[e] Reaction Control System.
[f] Upper stage engine for the Saturn vehicles.
[g] Advanced Space Engine, which is hypothetical at this time.
[h] Nuclear thermal engine with hydrogen propellant.
[i] Electric propulsion.

indirect electric propulsion, avoids some of the penalties of low thrust propulsion, while reaping some fraction of the benefits.

Third body encounters with the moon can be used to increase the C_3 of spacecraft in hyperbolic orbits without using propellant. The maximum possible C_3 increase is 4 km^2/s^2. Launch window constraints will almost always limit the increase to somewhat less than this. Since typical C_3's to go to accessible near Earth asteroids are 10 to 50 km^2/s^2, or 5 to 10 km^2/s^2 to go to Mars, this does not substantially reduce the impulsive ΔV required in LEO nor the low thrust ΔV required after escape. It is not possible to embark on a transfer orbit to any known near earth asteroid from a highly elliptical Earth orbit by an encounter with the moon without substantial propellant expenditure.

Aerobraking is a technique wherein a spacecraft uses a planetary atmosphere to slow down, rather than expending propellant for an equivalent ΔV. An aerobraking maneuver is performed by placing the spacecraft in an orbit with a periapsis inside the central planet's outer atmosphere. The periapsis is chosen so the spacecraft slows down substantially, but not enough to fall out of orbit and land on the planet. The periapsis is later raised above the top of the atmosphere by a small propulsive ΔV. Aerobraking is possible at Earth and Mars. Most schemes for the delivery of lunar or Martian manufactures to LEO depend on aerobraking to be economically viable.

Table A2 provides typical ΔV's, flight times and flight frequencies for impulsive and low thrust systems. These do not account for maneuvering and course correction.

Tethers provide a means of using high I_{sp} thrust in circumstances where this would not otherwise be possible. They also allow the use of one individual propulsion system by many vehicles. The use of tethers would greatly increase the demand for and benefits from nuclear electric propulsion.

Tethers are long tensile structures in orbit (typi-

Table A2 *Representative Flight Times, Frequencies and ΔVs.*

			Flight	
From[a]	To	ΔV (m/s)	Time (d)	Frequency (d^{-1})
LEO	L1	3300	2	.11
L1	LEO[b]	3400	2	16
L1	LEO[c]	100	2	16
LEO	LLO	4360	3	.11
LLO	LEO[b]	4205	3	.07
LLO	LEO[c]	940	3	.07
LLO	LS	2100	.04	.07
LS	LLO	1925	.04	.07
L1	LS	2700	.5	∞
LS	L1	2525	.5	∞
LEO	LMO[b]	5831	260	.0013
LEO	LMO[c]	3715	260	.0013
LEO	LMO[d]	10272	260	.0013
LEO	PO[b]	5605	260	.0013
LEO	PO[c]	4290	260	.0013
LEO	PO[d]	9750	260	.0013
LMO	LEO[b]	5843	260	.0013
LMO	LEO[c]	2250	260	.0013
LMO	LEO[d]	10320	260	.0013
PO	LEO[b]	5100	260	.0013
PO	LEO[c]	1450	260	.0013
PO	LEO[d]	9210	260	.0013
LEO	AST[e]	4440	700	.0024
LEO	AST[d,e]	8690	700	.0024
AST	LEO[c,e]	356	660	.0024

[a] The acronyms are: LEO - Low Earth Orbit; LLO - Low Lunar Orbit; LMO - Low Mars Orbit; L1 - first Earth Moon Lagrange Point; LS - the Lunar Surface; PO - Phobos Orbit; and AST - Asteroid.

[b] Without aerobraking.

[c] With aerobraking, including 100 m/s for circularization.

[d] Low thrust propulsion with minimum achievable gravity losses (escape velocity + C_3) and without aerobraking.

[e] The 1 September 1989 mission opportunity to 1982 DB (Helin et al. 1984) is used in this example.

cally >> 1 km, discussed in detail by Carroll (1984 and 1985)). They allow spacecraft to exchange momentum without the use of propellant. Thus, one spacecraft can accumulate momentum and transfer it to another to reduce the ΔV requirements of performing some maneuver. This can only be done with reasonable ease when near (within a diameter or two) a planetary body such as Earth, the moon or Mars. Tethers can be used for indirect propulsion, where thrust from some engine gradually accumulates as orbital energy and is later impulsively transferred to another vehicle.

Around Earth, electrically conducting tethers can be used as the armature of an electric motor where Earth itself is the magnet. (This works because Earth has a substantial magnetic field and is surrounded with an electrically conducting plasma at low altitudes. The plasma is used to close the circuit from the top to the bottom of the wire.) Such an electrodynamic tether consumes no propellant and can provide thrust more efficiently than most electric thrusters (2 to 8 kWe/N, with the low power in polar orbits and the high in equatorial orbits).

Part 3

NUCLEAR SAFETY

Space Nuclear Power Systems 1986
Edited by M. S. El-Genk and M. D. Hoover

Chapter 5

Accident Probabilities of a NERVA Derivative Reactor for Space Power Applications

R. B. Simonelli, J. A. Peoples, and J. W. H. Chi

Westinghouse Electric Corporation
Advanced Energy Systems Division
P.O. Box 158
Madison, PA 15663

This paper presents a preliminary assessment of the accident probabilities of a NERVA Derivative Reactor used for multi-megawatt space power applications. The analysis was based on a reactor design capable of passive reentry breakup and fuel dispersal in the upper atmosphere. The worst-case accident inferred was intact reentry and intact impact, followed by ejection of the control rods from the core and either water immersion or core compaction.

Introduction

Gas cooled, solid core graphite reactors based on the Nuclear Engine Rocket Vehicle Application (NERVA) technology have been shown to be attractive power sources for Strategic Defense Initiative (SDI) multimegawatt space power applications (Morgan et al. 1984 and Chi et al. 1986). A reactor of this type is called a NERVA Derivative Reactor (NDR). A simplified accident-probability analysis of a NDR reactor for a space power application was performed. The objective was to infer the worst case accident and determine critical reactor design areas where the incorporation of appropriate design features may reduce the inherent risks.

Description of the Power Source

Figure 1 is a schematic of a power source system that utilizes a NDR. The power source system produces baseload power, alert mode power and burst power using a single, multi-mode NDR. The key features of the multi-mode reactor are illustrated in Figures 2 and 3. The reactor core, shown in Figure 2, consists of bundled fuel modules. Each fuel module, shown in Figure 3, consists of 6 hexagonal fuel elements that surround a central hexagonal support element. The fuel elements consist of fuel beads (UC + ZrC coatings) in a graphite matrix co-extruded with the coolant channels. The support elements contain a central coolant channel and an annular flow channel. Two independent flow circuits provide the three levels of power. The reactor power for baseload and alert modes of power is removed by coolant flow through the tie tubes. The reactor power for the burst mode of operation is removed by coolant flow through the main coolant channels in the fuel elements.

Baseload and alert modes of power are produced by closed-cycle Brayton power conversion using He or He-Xe as the thermodynamic working fluid. Burst power is produced either by using hydrogen in an open cycle turbo-generator system or through an energy storage system such as flywheels. Figure 4 illustrates the system with open cycle Brayton for burst power production.

The reactor is controlled by control drums located in the radial reflector. Control rods/poison wires are located in the core to maintain subcriticality under all normal and postulated credible accident conditions during prelaunch, launch, and post-mission operations. The use of control rods is assumed in the analysis.

Accident Probability Analysis

The accident probability analysis focused on the radiological safety of the multi-mode nuclear reactor. It addresses each of the safety critical aspects and acci-

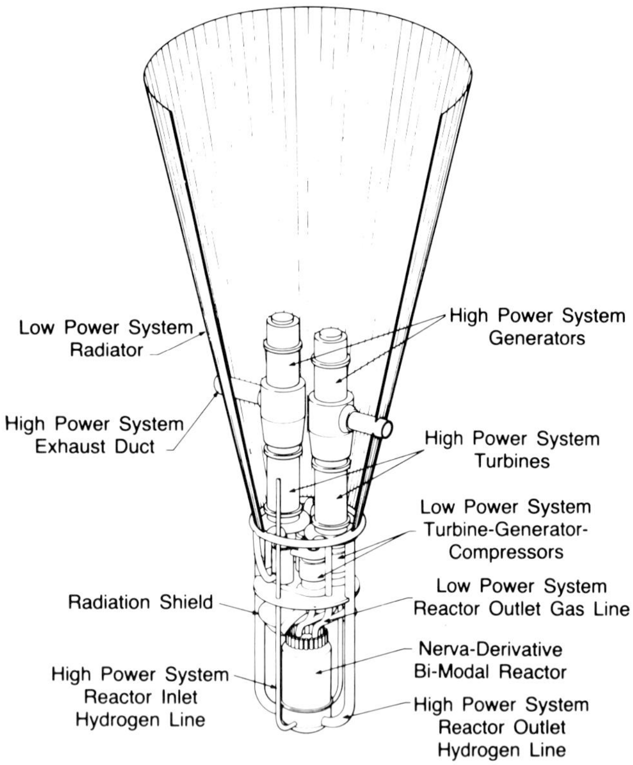

Figure 1 *Trimetric of a Multi-mode Power Source System that utilizes the NERVA Derivative Reactor.*

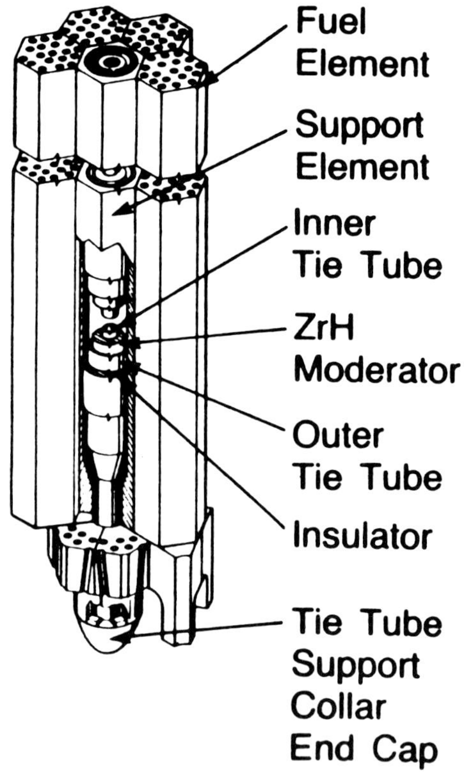

Figure 3 *Schematic of a Multi-mode Power Source System using a NERVA Derivative Reactor.*

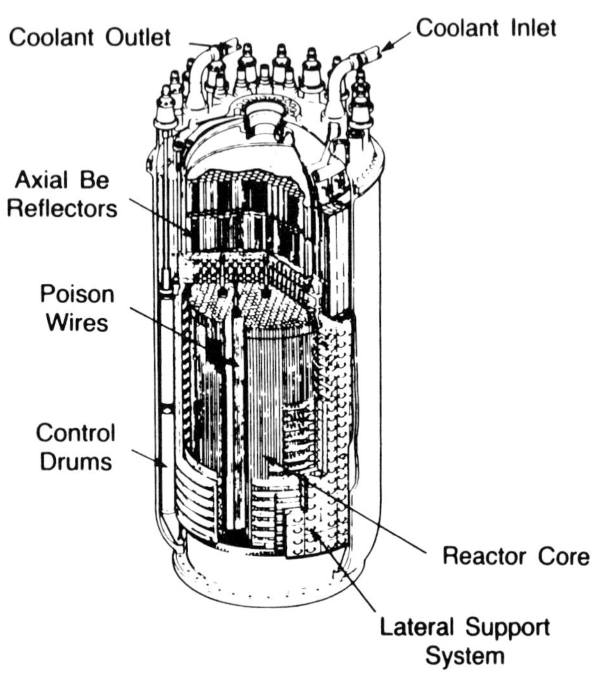

Figure 2 *Illustration of a NERVA Derivative Reactor.*

dent factors according to MIL-STD-1574A (1984). The nuclear reactor safety design is based on the assumption that the nuclear system will be deployed in a nuclear safe orbit; the reactor will not be started up until it is deployed in the intended orbit. Consequently, the principal nuclear safety design objective is to minimize the potential interaction of the radioactive materials with the Earth's population and environment so that exposure levels are within the limits established by international standards. The principal means of achieving this objective is to maintain subcriticality under all normal and postulated credible accident conditions (prelaunch and launch).

For a multimegawatt application, the design of a nuclear reactor for intact reentry and intact impact is expected to incur a large and probably unacceptable mass penalty. Consequently, it was assumed that the reactor is designed to ensure passive reactor breakup and fuel dispersal in the upper atmosphere (as in NERVA) in the event of launch abort and reentry. A conceptual design of a NDR for space electric power production has not yet been made. Therefore, for the purpose of the accident probability analysis, some of the basic nuclear safety features of the NERVA reactor were assumed. Table 1 presents a summary of the

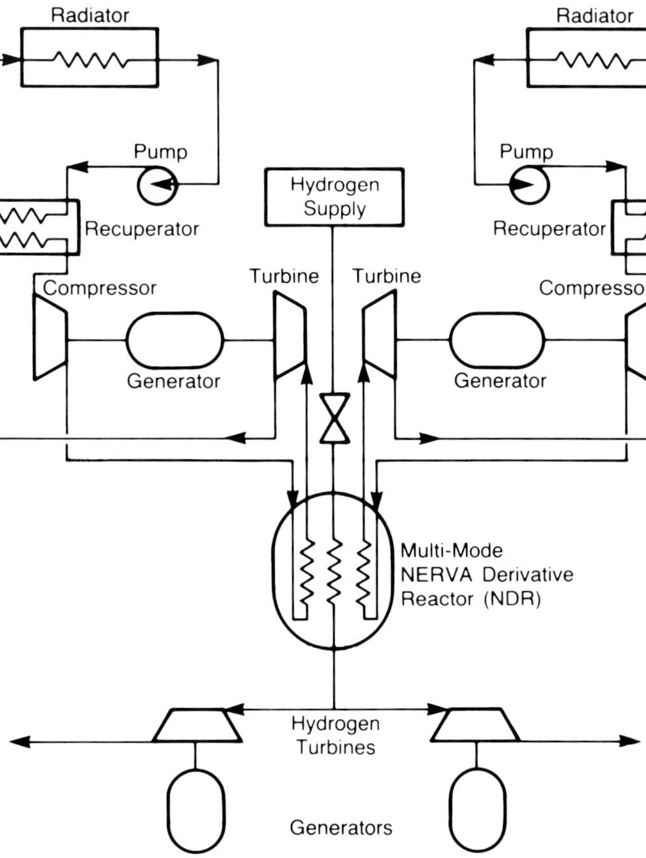

Figure 4 *The Fuel Module of a NERVA Derivative Reactor.*

Table 1 *Summary of Nuclear System Design Features to Meet the Principal Safety Objective: Maintain Subcriticality.*

Phase	Environments	Methods to Meet the Objectives
Ground Handling and Transportation	• Water immersion	• Control rods in core maintain subcriticality
	• Nearness of human	• Control rods in core maintain subcriticality
	• Fire	• No hazard
	• Dropping	• Control rods in core maintain subcriticality
	• Puncture	• Control rods in core maintain subcriticality
Prelaunch	• Explosion-overpressure	• Control rods in core maintain subcriticality
	• Fireball	• Control rods in core even if the reflectors are damaged and fall off
	• Shrapnel impact	• Control rods in core even if the reflectors are damaged and fall off
	• Water impact	• Control rods in core maintain subcriticality
	• Land impact	• Control rods in core maintain subcriticality
	• Liquid propellant immersion	• Control rods in core maintain subcriticality
	• Liquid propellant fire	• Control rods in core maintain subcriticality
	• Solid propellant fire	• Control rods in core maintain subcriticality
	• Thermochemical reactions	• Control rods in core maintain subcriticality
	• Post-impact water immersion	• Control rods in core maintain subcriticality
Ascent	• Explosion-overpressure	• Control rods in core maintain subcriticality
	• Fire	• Control rods in core maintain subcriticality
	• Shrapnel	• Control rods in core maintain subcriticality
Reentry		• Reactor disassembles passively and fuel disperses in the upper atmosphere
Water Impact	• Water immersion	• Control rods remain in intact core
Land Impact	• Core compaction	• Control rods remain in intact core
Post Impact	• Water immersion	• Control rods remain in intact core

nuclear system features designed to meet the various safety objectives. For the NDR, control rods or poison wires are assumed to be inserted in the core, where they remain until a stable orbit is attained.

The NERVA reactor had been shown (analytically) to be capable of passive reentry breakup and fuel dispersal in the upper atmosphere. It is assumed that the NDR, which is a smaller reactor in the range of power anticipated, also will have this capability. This will be achieved through the use of a metallurgical fuse attached to the reactor vessel and activated by reentry heating. Once the reactor vessel breaks up, the graphite fuel elements will disperse and burn up in the upper atmosphere. Then, fuel beads will burn up, and the fuel will be aerosolized and dispersed. Table 2 compares the NERVA reactor with a NDR. While both reactors are capable of passive reentry and breakup, the higher melting temperatures of the reactor vessel of the NDR require the use of a metallurgical fuse to aid reentry breakup.

Operational Sequences and Possible Accidents

As a part of the accident probability analysis, the operational sequences, events and possible accidents of the nuclear system were delineated. These are

- Factory assembly,
- Transport to launch site by land-truck,
- Receipt, inspection and storage,
- Assembly checkout,
- Transport to the launch pad,
- Mating of the reactor system with the booster,
- Combined prelaunch checkout,
- Launch liftoff, pitchover, and stage separation,
- Orbital startup and mission operation,
- Reactor shutdown and mission termination, and
- Orbital decay, reentry, and disposal.

For the multi-mode reactor power source, no launch facility exists today that can permit its deployment in a single launch. The operational sequences and potential accidents are based on a typical launch vehicle similar to the Saturn C5, planned for use in the NERVA Reactor-In-Flight (RIFT) program. Such a system is assumed to have a first stage that uses more than 4.5 million pounds of LOX and RP-1, and a second stage with a total LOX and LH_2 capacity of 1 million pounds. In addition is the hydrogen inventory for the burst power and weapon systems.

Accident Probabilities

The operational sequence and accident probabilities for the multi-mode reactor are summarized in Figure 5. The probability of truck accidents is based on data provided by the U.S. Department of Transportation (Federal Highway Administration 1982 and 1983). Other data on accident probabilities were published by General Electric (1978).

The worst case accident, that which would result in the maximum release of radioactivity, is expected to be the case when the reactor lands intact on Earth, as a result of either launch abort or orbital decay reentry, and becomes supercritical as a result of one of the following:

- Ejection of the control rods followed by core compaction, or
- Ejection of the control rods followed by immersion in water.

In either case, the core becomes supercritical, heats up rapidly and disassembles as a result of graphite-air reactions and vaporization, leading to the release of fission products, UC fuel, and radioactivity. The probabilities of such accidents are extremely low: 1×10^{-14} to 4×10^{-12} for core compaction and 5×10^{-15} to 2×10^{-12} for water immersion. Figure 5 shows that the specific probability of occurrence depends on when and where the accident occurs (impacts on land or on water).

Table 2 *Comparison of Differences in Reactor Designs.*

	NERVA	NDR
Reactor coolant	Hydrogen	He-Xe/He/H_2
Reactor vessel	Aluminum	Superalloy/Refractory Metal Alloys
Maintain subcriticality under all accident conditions	Poison wires in core	Poison wires/control rods in the core
Permanent post-mission reactor shutdown	Poison wires in core	Poison wires/control rods in the core
Reactor, during reentry	Passive breakup and fuel dispersal	Passive breakup and fuel dispersal
Fuel, during reentry	Burnup and dispersal	Burnup and dispersal

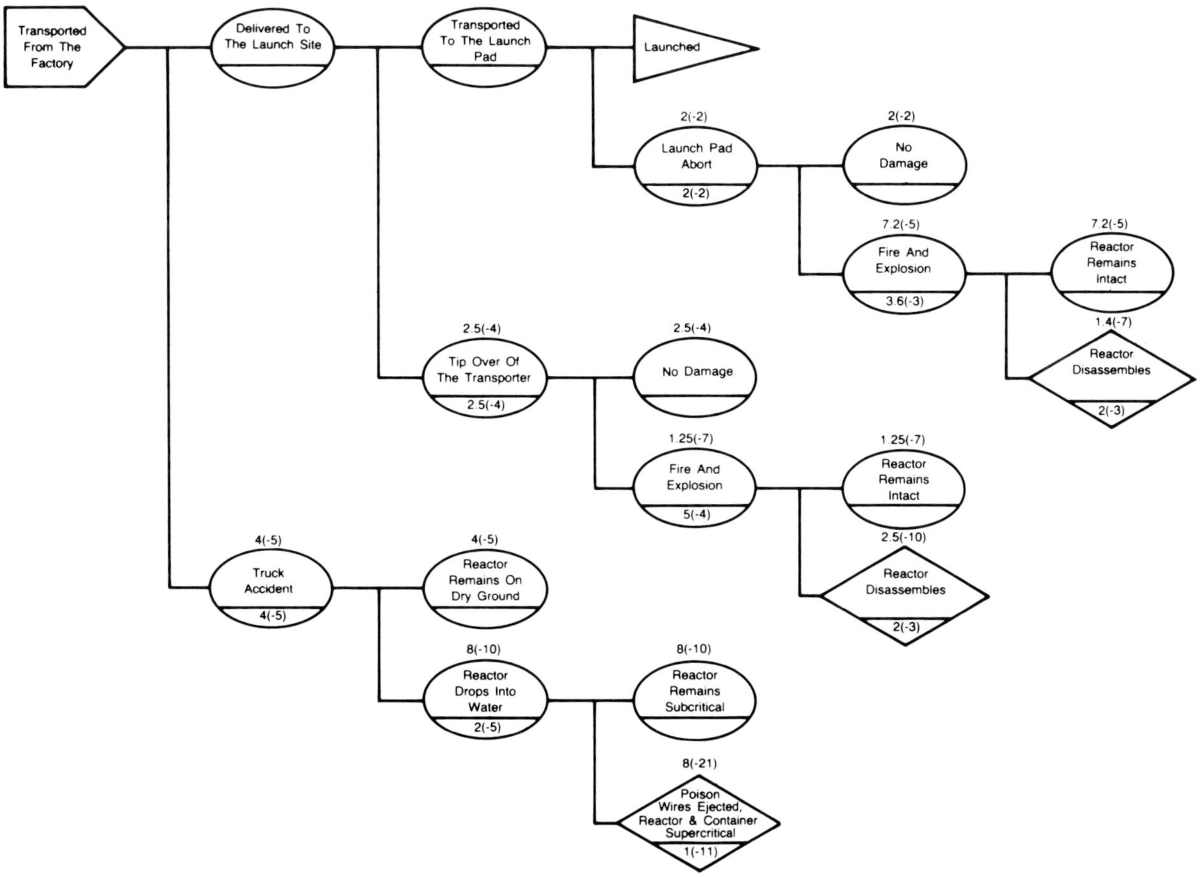

Figure 5a *Operational Sequence Tree and Accident Probabilities of the Multi-mode NERVA Derivative Reactor.*

Conclusions and Recommendations

This simplified accident probability analysis shows that the probability for criticality accidents is very low when the NDR is designed for reentry breakup and fuel dispersal in the upper atmosphere. However, analysis on the corresponding risks should be performed to assess the probability-risks involved under the complete operational sequences and accidents. Such an analysis can be implemented only in conjunction with a reactor conceptual design for a specific power source system and a specific mission. A similar probability-risk analysis should be carried out for a reactor designed for intact reentry and intact impact to compare the probability-risks.

Since the worst case accident that leads to reactor criticality is associated with intact reentry and intact impact, it may be surmised that a reactor designed for intact reentry and intact impact may have higher probabilities for criticality accidents. Clearly, the complete probability-risk analyses recommended are needed to resolve this issue.

Acknowledgment

This work was performed as a part of the Integrated Spacecraft Total Energy System Analysis—Nuclear Power Sources Study for TRW under Air Force Contract M62494GQ35.

References

Chi, J. W. H., K. Kesavan, R. E. Morgan, and J. E. Schmidt (1986) "Comparison of Multi Megawatt Space Nuclear Power Systems," in *Space Nuclear Power Systems 1986*, M. S. El-Genk and M. D. Hoover, eds., Orbit Book Co., Malabar, FL.

GE (1978) "Reactor Systems Preliminary Nuclear Safety Analysis," in *Manned Space Flight Nuclear Systems Safety*, Vol. III, GE-725D4201-3-2, General Electric Company, Valley Forge, PA.

Morgan, R. E., J. W. H. Chi and B. L. Pierce (1984) "Comparison of High Power Nuclear Systems for Space Applications," Proceedings of the IECEC.

US DOT (1982) *Accidents of Motor Carriers of Property 1980-81*, Federal Highway Administration, Bureau of Motor Carrier Safety, Washington, DC.

US DOT (1983) *Accidents of Motor Carriers of Property 1982*, Federal Highway Administration, Bureau of Motor Carrier Safety, Washington, DC.

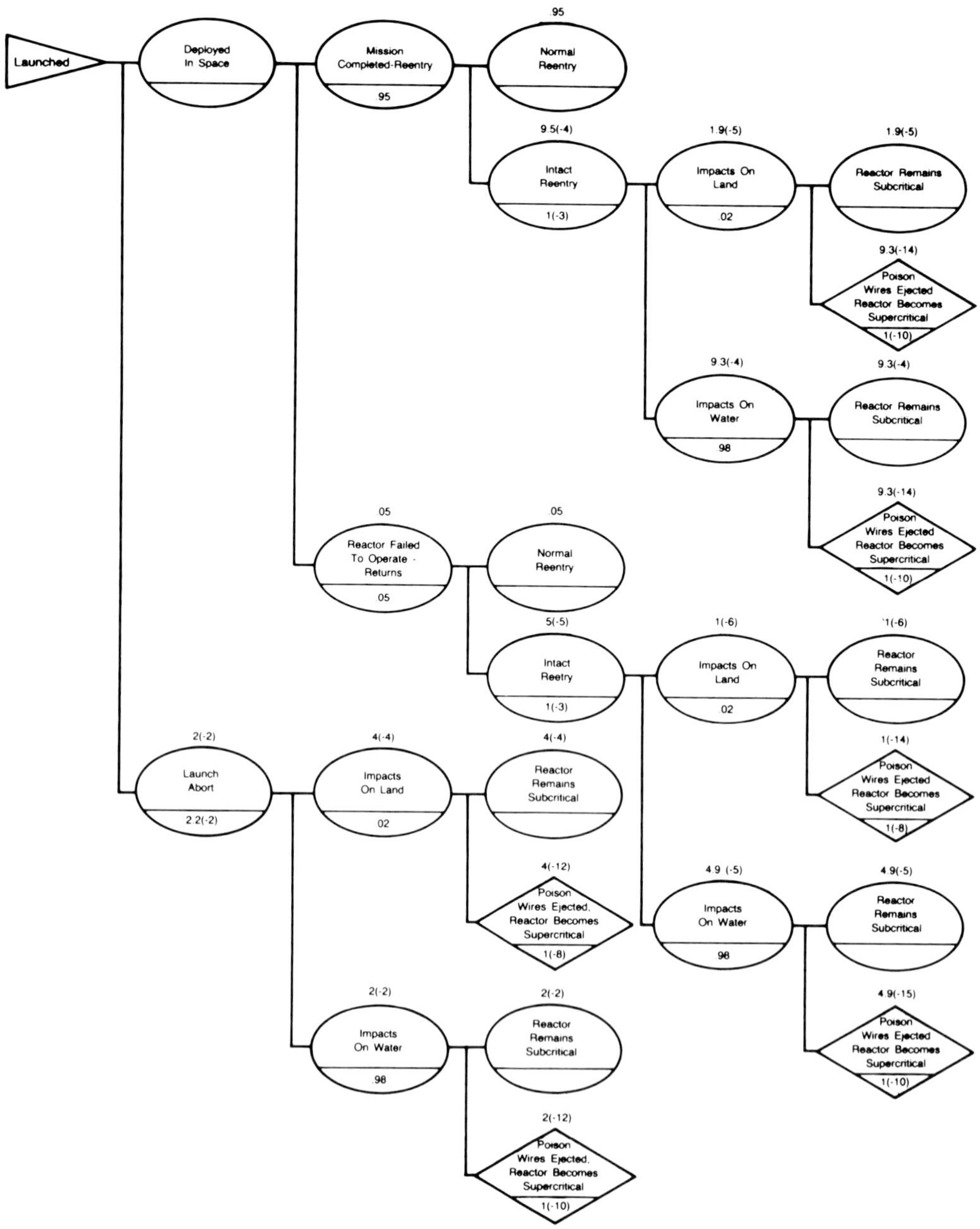

Figure 5b *Operational Sequence Tree and Accident Probabilities of the Multi-mode NERVA Derivative Reactor (continued).*

Space Nuclear Power Systems 1986
Edited by M. S. El-Genk and M. D. Hoover

Chapter 6

Inherent Safety of the Thermionic Space Reactor

M. G. Stamatelatos, A. W. Barsell, and C. R. Fisher

GA Technologies Inc.
P. 0. Box 85608
San Diego, CA 92138

The safety performance of the GA SP-100 in-core thermionic space nuclear reactor was evaluated throughout its mission phases from prelaunch to end-of-life disposal. The main issues addressed here are safety design features, the potential for accidental criticality, overheating and undercooling accidents during operation, launch pad accidents, and health hazards related to those accidents. Criticality was analyzed for compaction (with and without outer reflector), water immersion, and fire conditions. Due to the negative reactivity of the in-core refractory metals, subcriticality margins were sufficient to encompass uncertainties. The inherent safety of the design was demonstrated for mission operation accidents of loss of coolant main circulation pump and sudden loss of coolant. The thermal and mechanical environments of a launch pad abort fireball were analyzed with respect to survivability of the radioactive fuel and toxic chemical material, particularly beryllium. While survival of the tungsten emitter around the fuel is predicted, neither radioactive nor toxic chemical hazards would exceed safety standards beyond the site even if all such material were released.

Introduction

Surety has been an important concern in the SP-100 space power program. Surety is defined as the integration of safety, safeguards, environmental protection, reliability, and quality assurance. The relative importance of the various surety concerns depends largely on the mission phase. For the purpose of the present discussion, the mission was divided into five phases: prelaunch, launch, ascent, operations, and end-of-life disposition. The major surety concerns addressed in this portion of the SP-100 program are shown in Table 1 in relation to the selected mission phases.

This ''surety matrix'' shows that, even in a simplified approach, surety issues are complex and often affect different mission phases in different ways. In some cases, having addressed a surety issue for one phase of the mission automatically solves the problem for any other mission phase where that issue is meaningful. In other cases, the solution of a surety issue for one phase has little to do with the necessary solution for another mission phase.

This paper focuses on safety issues; in particular, safety design features, subcriticality during accidents, overheating and undercooling accidents during operation, launch pad accidents, and health hazards related to those accidents.

Safety Criteria and Compliance

The safety requirements are based on the ''Nuclear Safety Criteria and Specifications for Space Nuclear Reactors,'' OSNP-1 (Rev. 0), issued by the Office of Space Nuclear Projects of the U.S. Department of Energy in 1982. These are supplemented by requirements issued by the SP-100 Project Office to reflect up-to-date programmatic and environmental concerns. OSNP-1 requirements essentially cover three categories:

1. Safety design features,
2. Subcriticality under credible accident conditions, and
3. Radiation protection criteria for normal and off-normal conditions throughout the mission.

The safety design features are consistent with normal safety practices that have been applied to terrestrial power reactors for many years. These include items such as no single-point failure in safety-related systems; redundant and diverse shutdown and residual heat removal systems; and specification of the necessary safety functions for the control, protection, and heat removal systems. In addition, the SP-100 Project Office has specified a requirement for incorporating an additional passive permanent shutdown system into the design for use at the end of life of the reactor.

The SP-100 requirements on subcriticality cover credible accident conditions involving core compaction, flooding, and expansion to a fuel-to-moderator ratio that is optimum for criticality.

The radiation protection criteria in OSNP-1 are implicitly based on regulatory limits used for terrestrial reactors. GA has further interpreted the overall health protection requirements to be explicitly consistent with quantitative limits stipulated by 10CFR100 for radiologic hazards, and by occupational toxicity limits from industrial standards for circumstances when these standards are applicable to SP-100 systems.

One safety concern for SP-100 that has no counterpart in terrestrial reactors is the planned (end-of-life) or unplanned (accidental) reentry of the reactor into the atmosphere. This may present a threat of toxic or radioactive material release and dispersal under aerothermal loads. Land, water, or air contamination and population exposure to the radioactive and/or toxic materials (such as beryllium) consequently may occur. Insofar as this is a probabilistic concern, it must be dealt with accordingly. In the interim, the SP-100 Project Office has requested that, for end-of-life reentry, the design should provide either for a fine dispersal of the radioactive and toxic materials into the atmosphere or for an "essentially intact" reentry. The thermionic SP-100 system contains high-temperature refractory materials that are not expected to easily ablate and disperse under the reentry heat loads. Consequently, the thermionic SP-100 system was designed for essentially intact reentry.

Safety Features of the Thermionic SP-100

The thermionic SP-100 has a number of important features that contribute to the excellent surety performance as documented in this report. Many of these

Table 1 *The SP-100 Surety Matrix.*

MISSION PHASES \ CONCERNS	SAFETY				SAFEGUARDS
	EXPOSURE TO RADIATION		EXPOSURE TO HAZARDOUS (NON-NUCLEAR) MATERIALS	FAILURE TO OPERATE AS REQUIRED	READILY ACCESSIBLE SPECIAL NUCLEAR MATERIAL (SNM)
	UNPLANNED CRITICALITY	NON-CRITICALITY EXPOSURE			
PRE-LAUNCH	• UNPLANNED STARTUP	• STS REQUIREMENTS	• STS REQUIREMENTS		
LAUNCH AND ASCENT	• UNPLANNED STARTUP • DRY COMPACTION • IMMERSION/FLOODING • COMPACTION & SOIL IMMERSION • EXPANSION & WATER IMMERSION • FIRE (MELTING)	• FIRE • EXPLOSION	• FIRE • EXPLOSION		• SNM LOCATION
LOW EARTH ORBIT (LEO) OPERATION		• ORBITAL DECAY REENTRY & EARTH IMPACT	• ORBITAL DECAY REENTRY & EARTH IMPACT	• SAFETY SYSTEMS RELIABILITY • ACCIDENTS • EOL SHUTDOWN	• ORBITAL DECAY REENTRY
BOOST: TO ORBIT (LEO; NSO)	• UNPLANNED STARTUP	• BOOST FAILURE REENTRY	• BOOST FAILURE REENTRY		• BOOST FAILURE REENTRY (FOOTPRINT)
BOOST: FROM LEO		• BOOST FAILURE REENTRY	• BOOST FAILURE REENTRY		• BOOST FAILURE REENTRY
NUCLEAR SAFE ORBIT (NSO) OPERATION				• SAFETY SYSTEMS RELIABILITY • ACCIDENTS • EOL SHUTDOWN	
END OF LIFE (EOL) DISPOSAL	• WATER IMMERSION • DRY COMPACTION • COMPACTION & SOIL IMMERSION • EXPANSION & WATER IMMERSION	• ORBITAL DECAY REENTRY • EARTH IMPACT	• ORBITAL DECAY REENTRY		• ORBITAL DECAY REENTRY

features are inherent in the concept and some are engineered features, tailored to the particular mission requirements.

Multiple Barrier Fuel Containment

As illustrated in Figure 1, the core of the reactor comprises 176 fuel elements, each of which in turn comprises six thermionic converters or cells. Each thermionic converter contains a cylinder of nuclear fuel enclosed in a closed tungsten cup. The tungsten cups then are enclosed in a trilayer structure made of refractory metal (niobium). All of the above components are enclosed further in a continuous sheath tube that is closed in a continuous, welded structure. The result of the above construction is essentially triple containment of the nuclear fuel and any included fission products, making release of the contents unlikely for a variety of postulated event scenarios. These events include reentry, fire, explosion, and transportation accidents. The multiple containment feature not only contributes to safety, but also safeguards protection of the nuclear fuel.

Refractory Metal Construction

The SP-100 structure that encloses the reactor coolant system and reactor core is virtually all refractory metal alloy, Nb-1%Zr. This material is strong and tough and has a very high melting point relative to that of alloys of the iron-nickel family. The high melting point is particularly important when analyzing event sequences posing potential challenges to surety requirements. The material is able to maintain an intact structural integrity despite the existence of postulated events that heat the material to high temperatures. These events include fire, reentry, and loss of coolant. The ability of the reactor to maintain its structural integrity keeps the core in a predictable configuration, enhancing recovery and preventing unintended criticality.

The large volume fraction of refractory metal in the core (tungsten, niobium, and tantalum) provides strong absorption of neutrons of epithermal energy. This feature limits the reactivity increase, due to flooding or immersion, to relatively small amounts. Therefore, design measures to prevent accidental criticality are easy to accomplish and do not significantly perturb the design as optimized for performance.

Low Power Density Reactor Core

Because the thermionic reactor core density is set by the power conversion process rather than by coolant heat transfer limits, the core is relatively large and normally operates with a low power density (power per unit volume), especially relative to terrestrial liquid metal-cooled fast reactor designs. Another important feature of the core, relative to heat transfer, is that the fuel element sheath structure is very thick and made of niobium alloy, which has a high thermal conductivity. The above features combine to produce a core with very small temperature gradients even when the fuel elements are arranged on a unity pitch-to-diameter ratio. This results in a core design not subject to the concerns about fuel rod bowing, flow maldistribution, and local overheating, which can damage the core of the typical liquid metal-cooled fast reactor.

The ability to space the fuel rods closely together has another payoff in the safety evaluation. Closely spaced rods result in a small coolant volume fraction in the core (typically, about 10%). This sharply limits the amount of water that can enter the core for accident scenarios involving immersion of the reactor. Therefore, measures to deal with this concern are relatively easy to provide without significant design perturbations.

Low Temperature Coolant System

In the thermionic reactor concept, the reactor coolant normally operates at the heat rejection temperature of the power conversion process. This permits the use of low melting point liquid metals with relatively low vapor pressures in the operating temperature range and for postulated accidents. The SP-100 design uses eutectic sodium-potassium alloy having a melting point of 261 K. This coolant is liquid over the entire operating temperature range. The result of this feature is that there is no concern for overstressing the reactor coolant system due to freezing with subsequent thawing. This provides abundant margin against exceeding allowable stresses in piping, reactor vessel, and other equipment in the coolant loop.

Redundant and Diverse Decay Heat Removal

The SP-100 design employs redundant and diverse decay heat removal pumps. The main coolant pump is an alternating current induction pump, requiring an external power supply for operation. The decay heat removal pump is a self-contained, conduction pump whose power is provided by thermoelectric elements that are an integral part of the pump. Heat is supplied to the thermoelectric elements by the coolant being pumped. This ensures that, as long as there is decay heat to be removed, the pump will continue to operate with no external power supply or controls. The coolant loop design employs a flow diode, or jet pump, to limit backflow in the parallel pumps with no mechanical check valves. This feature contributes to the high reliability of decay heat removal for the system.

Redundant and Diverse Reactivity Control

The control drums and their controls ensure reactor shutdown for all postulated conditions where they are available. In addition, the design includes redundant

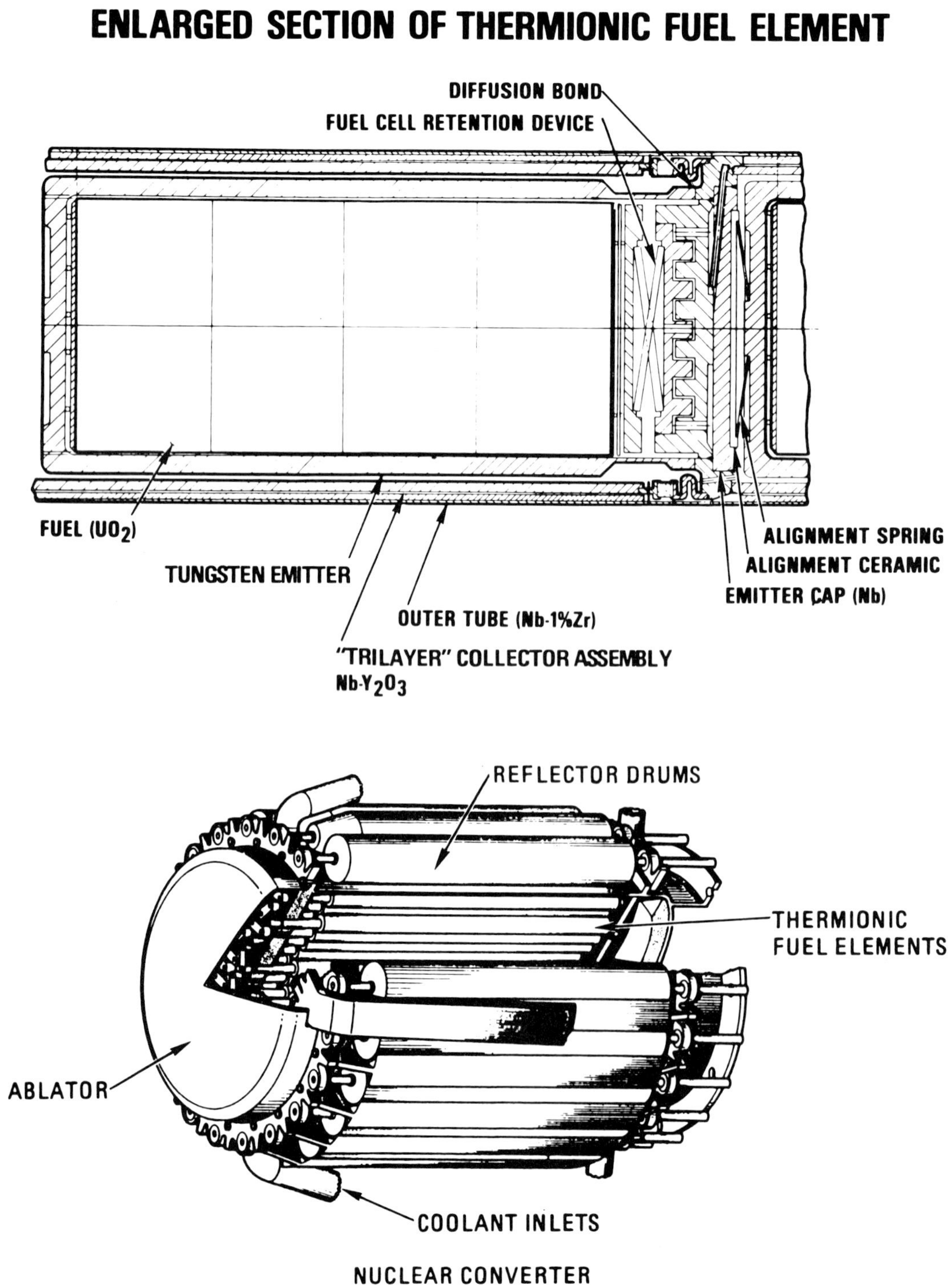

Figure 1 *Illustration of the SP-100 Thermionic Reactor and Fuel Cells.*

and diverse shutdown rods with independent controls. These rods are inserted, by linear motion, into the reactor when shutdown is needed. The reactor is brought to cold shutdown by either the drums or rods even when the reactor is flooded and immersed, or the core is compacted by an accident environment. Both of these shutdown devices are inserted at the end of reactor life (by the controls and by an independent, end-of-life shutdown device) to ensure a long decay time of radioactive materials to limit radiological hazard to persons following return of the reactor to Earth after orbital decay.

Subcriticality During Accidents

This section deals with the potential for unplanned criticality during the launch, ascent, boost, and deorbit phases of the mission through accidents involving changes in reactor geometry and material compositions. These include reactor compaction, water flooding, water immersion, liquid propellant immersion, reactor melting due to liquid and solid propellant fires, and credible combinations thereof. Scenarios during which such events may take place include reactor:

- Fall on launch pad and potential immersion in liquid propellant;
- Impact with ground, concrete, or water due to failure to launch or failures during ascent;
- Impact with ground, concrete, and water following accidental reentry due to boost failure;
- Impact with ground, concrete, and water following planned reentry due to orbital decay at the end of life; and
- Launch pad fires and explosions.

These scenarios were investigated by the Los Alamos National Laboratory (LANL) and by GA and were reduced to a set of benchmark accidents by the SP-100 Project Office. These accidents were considered to be worst-case limiting accidents. The criticality analyses discussed in this section evaluate the potential for criticality of the SP-100 thermionic reactor under these limiting conditions, which, because of conservative assumptions, involve geometry and composition configurations that are worst-case extrapolations of credible scenarios. In some cases the model extrapolations are such that the resulting configurations are beyond those physically attainable under the given accident scenarios.

The accidental criticality benchmarks can be divided into two categories: (1) ''baseline'' accidents and (2) ''more severe'' accidents. The first category contains four accidents identified by the SP-100 Project Office to lead to reactor configurations that must remain subcritical without any modifications of the current baseline design. The second category contains three accidents that are more severe and less probable than those in the first category. Should the present SP-100 thermionic baseline reactor not maintain subcriticality under second category accident conditions, modifications of the baseline design must be provided to ensure that subcriticality is maintained.

The criticality calculations were performed by GA with a two-dimensional discrete ordinate computer program called TWOTRAN (Lathrop and Brinkley 1970). Subsequently, they were verified by LANL (Sapir and Brandon 1984a and 1984b) using the two-dimensional transport program called TWODANT (Alcouffe 1984) and the Monte Carlo program called MCNP (LAMC Group 1981). The Monte Carlo analyses were benchmarked against data from LANL critical experiments (Hansen et al. 1975). The results reported here represent a composite of the results from GA and from LANL analyses.

The seven criticality cases are summarized in Tables 2 and 3. The first table shows the four baseline criticality accidents. The first case is the result of a complete axial and radial compaction uniformly eliminating all void in the core while retaining the control drums and the shutdown rods.

The second case is a variation of the first case in which drums are removed from the compacted core and replaced with an infinite soil reflector that has the following composition: 40% mineral, 35% water, and 25% air.

The third case represents an intact geometry in which drums and reflector have been replaced by an infinite water reflector and water has replaced voids and coolant (NaK-78).

The fourth case is the state of the reactor after being subjected to a 2433 K (2160°C) fire for 630 s. The control drums and the reflector are assumed to have been removed and the shutdown rods are inserted. A separate thermal analysis has shown that no reactor melting occurs when subjected to such a fire.

The effective multiplication factors for all cases show subcriticality with sufficient margins to allow for computational uncertainties.

Table 3 summarizes the results of the remaining three criticality calculations, which represent potentially more severe accidents than those in Table 2.

The first case in Table 3 is a geometrically maximum possible compaction of the reactor into a sphere. This case is obviously not attainable physically.

Table 2 *Los Alamos Selected Accidents—Baseline Accident Conditions.*

Case No.	Case Description	K_{eff}[a]
1.	Axial/radial compaction; dry core; drums, reflector, rods in	0.927
2.	Axial/radial compaction; no drums; no reflector; rods in; soil reflection	0.973
3.	Intact geometry; water immersion (reflector); water replaces NaK; water replaces voids; no drums; no Be reflector; rods in	0.928
4.	2433 K, 630 s fire; no drums; no reflector; rods in	0.772[b]

[a]Maximum uncertainty in K_{eff} is ±0.02.
[b]No melting or slumping of the core is expected.

The second case represents a uniform geometry change (expansion) coupled with water immersion to yield the worst criticality configuration. Since the baseline reactor lattice is undermoderated, the optimum fuel-to-moderator ratio is attained when the core becomes uniformly splayed to 1.4 times its original pitch. Such a uniform expansion is obviously very conservative and not attainable physically to the degree represented by the analytic model.

Finally, the third case represents the extreme result from a prolonged exposure of the reactor (1000 s) to a large solid- and liquid-propellant fire at the maximum possible temperature of 3023 K (2750°C). Since melting of the vessel is expected to a large degree, it is assumed that everything outside the tungsten layer of the thermionic fuel cells (TFC) has melted. (Tungsten melts at temperatures well above 3023 K (2750°C).) The configuration for criticality analysis is therefore a pile of TFCs laying on an infinite soil reflector.

The results in Table 3 indicate that subcriticality is maintained under a set of very severe conditions *with significant margins and without any necessary modifications of the baseline design*.

In conclusion, accidental SP-100 thermionic criticality is not possible under the series of SP-100 Project Office prescribed conservative accident scenarios ranging in likelihood from improbable to virtually impossible to attain physically.

Overheating and Undercooling Accidents in Orbit

The thermionic SP-100 reactor design is equipped with both inherent and engineered safety features to completely eliminate or to reduce the occurrence probability of accidents.

Potential accidents leading to unplanned criticality were reviewed earlier in connection with mission phases where such accidents pose the greatest threat to public safety.

In this section, the system consequences of credible accidents causing excess heating or insufficient cooling are examined. Two of these accidents were singled out because they represent extreme conditions whose potential severity is of concern. If these accidents could be mitigated by the inherent safety features of the SP-100 reactor and they could be shown not to lead to core melt or even to core damage, then the concern about these and other less consequential accidents would be diminished considerably.

The first accident involves loss of most (90%) of the circulation by failure of the main system pump (ALIP) following reactor trip. The remaining circulation occurs as a result of the TEMP pump, which is assumed to work.

The thermal transient analyses was performed with SPTRAN (Pfremmer 1984), a GA general system

Table 3 *Los Alamos Selected Accidents—Severe Accident Conditions for Margins of Baseline Design.*

Case No.	Case Description	K_{eff}[a]
5.	Complete spherical compaction; dry sore; drums in; drums in; reflector in; rods in (spherical shell)	0.967
6.	Water immersion/flooding; no drums; no reflector; rods in; geometry changed to optimum moderation (1.4 times original pitch)	0.952
7.	Geometry affected by 2750°C 1000 s fire; no drums; no reflector	0.851[b]

[a]Maximum uncertainty in K_{eff} is ±0.02.
[b]Conical pile of TFEs, soil reflection at base.

transient response computer program. The results are shown in Figure 2, which shows the highest temperature component in the system, the emitter of a thermionic fuel cell, and the NaK temperature at the reactor exit.

The emitter temperature drops from 1740 K to thermal equilibrium with the collector, which is initially at 1100 K. This temperature drop is so rapid that it offsets (by far) the temperature rise due to the fission-product decay-heat source.

The coolant temperature rises, reaching a maximum at 1260 K within approximately 30 s after main pump trip. This temperature is 100 K above the NaK boiling point, but the resulting system pressure is well within the design strength limits of the cooling system so that no loss of coolant is expected.

Nevertheless, if it is assumed that the coolant is somehow lost following reactor trip, then the second and more severe accident, total loss of coolant, occurs.

This accident was analyzed with a two-dimensional heat transfer model developed for the GA version of the TAP (Finkelstein 1968) computer program. In this analysis, the initial temperature of the thermionic fuel element at the instant of the loss of coolant was assumed to be the equilibrium temperature between the emitter and the collector. Within the reactor vessel, the sole radial heat transfer mechanism was by radiation between fuel elements; axial heat transfer was by conduction along the fuel elements and through the end reflectors in series with radiation across the end plenum. The results of the analysis (Figure 3) show that the fuel element temperature at the core center rises by only 125 K after total loss of coolant. This temperature rise is small enough that no ensuing core damage is expected. After the temperature peaks, all generated heat is rejected to space passively by radiation.

Thus it appears that a complete loss of coolant accident in the thermionic SP-100 reactor following reactor trip is completely inconsequential from the safety standpoint. This benign behavior of the core under such an important thermal transient can be attributed to its very large heat capacity, namely, 1.2×10^6 J/K, and to its ability to withstand the high temperatures for passive heat rejection.

In view of this result, it appears that the active decay heat removal system in the thermionic SP-100 re-

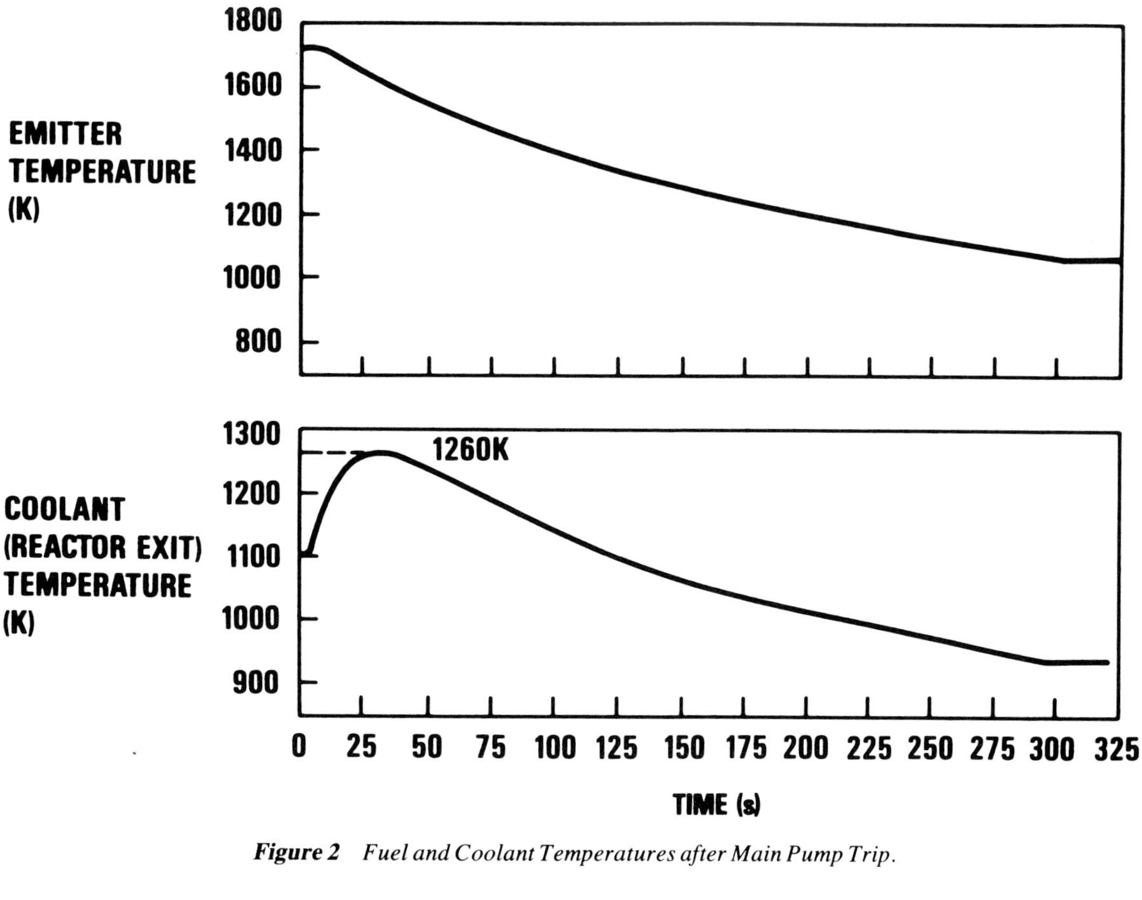

Figure 2 *Fuel and Coolant Temperatures after Main Pump Trip.*

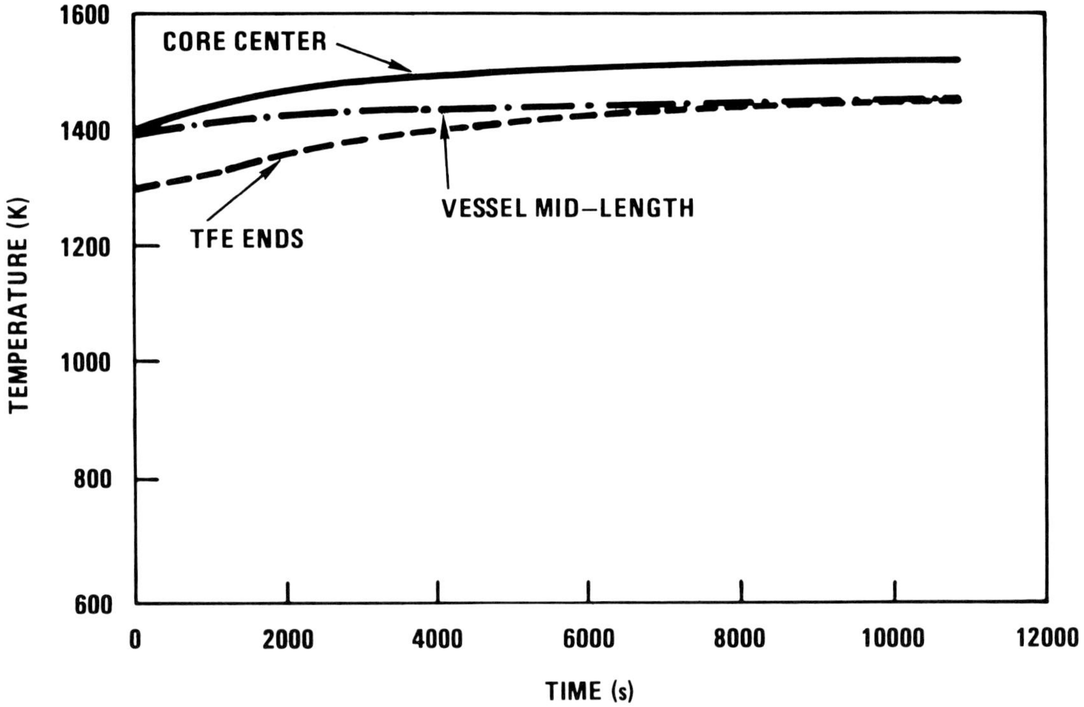

Figure 3 *Core Temperature during a Loss of Coolant Accident.*

actor is not a truly "safety" system, since its failure may only have an effect on the availability of the SP-100 but not on safety.

Launch Pad Accidents

Safety concerns regarding hazards due to accidents during the prelaunch and launch phases focus on aerosolization and dispersion of toxic materials, mainly Be and BeO. Since the reactor has not operated yet, there are no significant fission products or activation products. Less than 0.5 Ci of ^{235}U is present as radioactive material.

During the prelaunch phase, fires and explosions can occur during transportation accidents and during storage building conflagrations. These accidents typically are not severe enough either to melt through the refractory containment layers around the hazardous materials or to mechanically rupture these layers. Thus, no release is expected.

During the launch phase, a postulated launch pad abort explosion potentially can be more severe than prelaunch accidents. A launch pad abort pertains to the short time period of ignition and liftoff of the booster rockets from the launch pad. Potential accident scenarios can lead to explosive combustion of the liquid propellant and/or burning of the solid propellant. Therefore, consequences of this event are analyzed in this section.

The potential hazards of SP-100 materials and their antecedent conditions or causes during postulated launch pad aborts are summarized in Table 4. There are three types of hazards to the people in the vicinity of the launch pad: inhalation of released small aerosols, ingestion of fallout, and external exposure. The first two pertain to both radioactive and toxic materials. The third pertains to radioactive material only.

The hazards due to inhalation of particles in the passing cloud are of primary concern because:

- The potential magnitude and time scale of the accident create a cloud containing hazardous materials,
- Beryllium and ^{235}U are much more biologically harmful on inhalation than on ingestion,
- ^{235}U is a rather weak external gamma radiation source, and
- Particles large enough to fall out within a short enough distance to cause a concentrated ground deposition are too large to be resuspended to any significant degree.

Therefore, the analysis focuses on cloud inhalation rather than on ingestion or external hazards.

The analysis described below shows that mechanical forces of the explosion aerosolize a portion of the BeO, B_4C, and UO_2 materials. These aerosols will vaporize quickly in the fire. The Be metal will vaporize slowly during the longer-term solid propellant fire, if exposed directly to the fire.

Table 4 *Potential Hazards and Causes for Launch Pad Abort.*

Hazard/Pathway	Antecedent Conditions/Causes
1. Inhalation of Toxic and Radioactive Materials	1. Combustion of Propellant on Launch Pad Liftoff
– Fragmentation and aerosolization	– Mechanical force of debris/shock waves
– Vaporization	– Liquid and solid propellant fires
– Interaction of cloud with inhabitants	– Fireball liftoff, meteorology
2. Ingestion of Toxic and Radioactive Fallout	2. Atmospheric Dispersion and Fallout over Farmland/Reservoirs
– Food/milk pathway	– Vegetable and soil contamination
– Fresh water pathway	– Fresh water contamination
3. External Exposure to Radioactive Fallout	3. Atmospheric Dispersion and Fallout over Inhabited Areas

Regardless of the fraction of Be, BeO, B_4C, and UO_2 released, the hazards are calculated to be below accident limits at distances where onsite personnel or members of the public are likely to be. This applies to any combination of release to the elevated fireball or, subsequently, to the ground.

The SP-100 reactor is planned to be launched inside the Space Shuttle and then deployed upon reaching parking orbit. Major potential combustion sources on the launch pad include about 3300 tonnes (1.5 million pounds) of LOX/LH_2 (equivalent volumes) in the external tank (ET) and the two solid rocket boosters on the sides of the ET, each booster with 1000 s of burn time at sea level.

An analysis of potential launch abort accident scenarios has been performed as described by Bell (1984). There were 14 accidents or initiating events identified for the launch/ascent phases that could potentially result in an explosion. Seven failure mode categories were identified, four of which apply to the launch pad phase and six of which apply to the ascent phase. For the four failure modes applicable to the launch phase, Bell (1984) defined four generic explosion environments as follows:

1. Belly blast (planar wave front impinging on the side of the shuttle),
2. Nose blast (explosion forward with enveloping wave on shuttle sides),
3. Tail blast (aft of shuttle), and
4. Cargo bay axial blast (within shuttle aft of SP-100).

These are listed approximately in descending order of severity. Cases 1 and 2 are bounding worst-case conditions and so are specified by Bell (1984). Since conditions for cases 3 and 4 are not yet specified, only cases 1 and 2 are analyzed in the current work.

Table 5 summarizes the dynamic parameter specifications. The debris impacting the SP-100 has a lower velocity for the nose blast case. However, the mass per unit area of the debris is much greater for the nose blast case, because the shuttle fuselage is imploded into the SP-100 by the enveloping wave (compared to a planar wave for the belly blast case). Thus the debris impulse is worst for the nose blast case. Also, the shock wave impulses are less severe than the debris wave impulses.

The thermal environment is specified in Figure 4. It consists of a short-term (approximately 10 s) liquid propellant burn and a long-term (to 1000 s) solid propellant burn. For the former, the peak environment temperature is 2900 K. For the latter it is just over 3000 K, according to the conservative theoretical assessment at LANL.

For a specified launch abort environment, a deterministic model was used to evaluate the consequences. This model is schematically outlined in Figure 5, where the ovals represent inputs, the dia-

Table 5 *Dynamic Loads on SP-100 during Postulated Launch Pad Explosions.*

	Belly Blast (Planar)	Nose Blast (Enveloping)
Debris Wave		
Velocity (m/s)	1,600	1,110
Mass/area (kg/m^2)	58	380
Impulse load (N-s)	5,600	25,430
Shock Wave		
Peak pressure (MPa)	21	44
Duration of shock (10^{-3} s)	0.3	0.75
Impulse load (N)	342	1,717

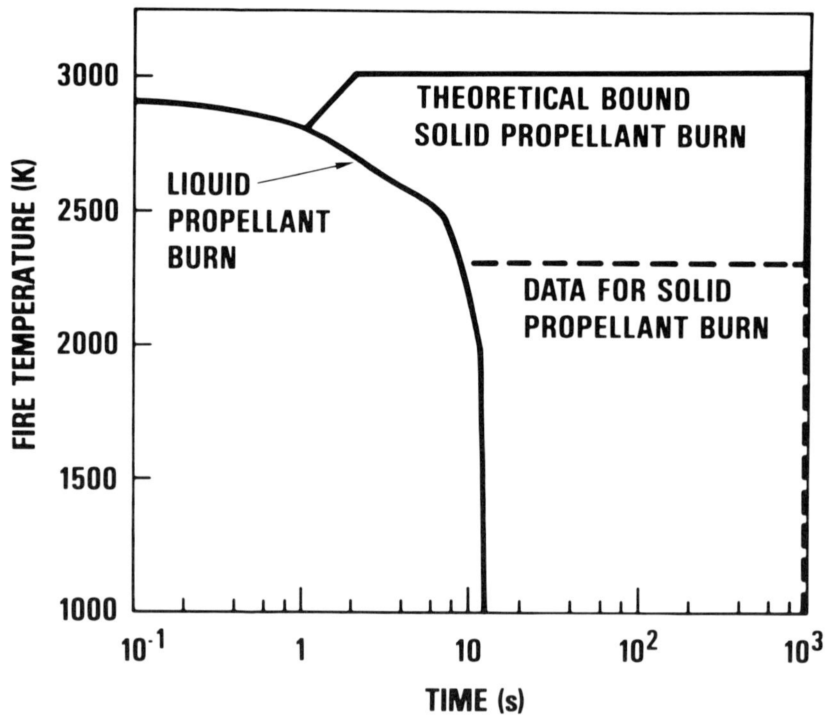

Figure 4 *Thermal Environment for Launch Pad Abort Accidents.*

mond indicates the final output, and the rectangular boxes represent intermediate steps and results. A specification of the debris and air shock wave impulses produced by the conditions (liquid propellant combustion) is the starting point of the analysis.

Essentially, all BeO becomes pulverized by the mechanical forces of the explosion, specifically by the debris wave. This is particularly true for the enveloping debris wave of the nose blast case, where the kinetic energy transfer exceeds by 2 orders of magnitude the predicted energy required to produce particles smaller than 10 μm throughout the BeO and to rupture the container. For the belly blast case the planar debris wave produces a kinetic energy transfer that is marginally equivalent to the predicted energy absorbed by the metal container upon deformation (strain energy) and by the ceramic during the compression and fragmentation process. The shock wave effects are relatively smaller in both cases and cause little or no aerosolization.

For the Be reflector drums, the beryllium metal does not aerosolize. The B_4C material inside the drum is likely to aerosolize based on the same considerations as above for the BeO.

A one-dimensional thermal model in cylindrical coordinates, called VESSEL 1-D, was developed (Harris 1984) as depicted in Figure 6 to represent the SP-100 reactor vessel and internals. The heating from the fire is applied at the circumferential surface and it gradually penetrates radially into the concentric annular rings, through the outer reflector, then through the vessel wall, and then into the fuel element and the control rod bundle. The core structure was simplified to allow a one-dimensional analysis.

Example spatial temperature distributions at different times are illustrated in Figure 7. This case is for a bare reactor (without outer reflector) exposed to the short-term liquid propellant fire. Other cases included the outer reflector.

The VESSEL 1-D thermal analysis shows that no melting of the Be outer reflector or of the reactor vessel occurs as a result of the liquid propellant fire. If the outer reflector remains on the vessel after the debris and shock waves, the maximum Be temperature is 900 K, and the vessel does not heat up. Even with the outer reflector removed, the reactor internals heat up very little, as shown in Figure 7.

For the combined liquid and solid propellant fire, using the maximum fire temperature of the solid propellant burn from Figure 4, some melting and vaporization of SP-100 hazardous materials is predicted by the thermal model. The Be reflector melts within 152 s. After 182 s, the Nb-1%Zr vessel sidewall melts. Up to that time, the fuel element/shutdown rod bundle,

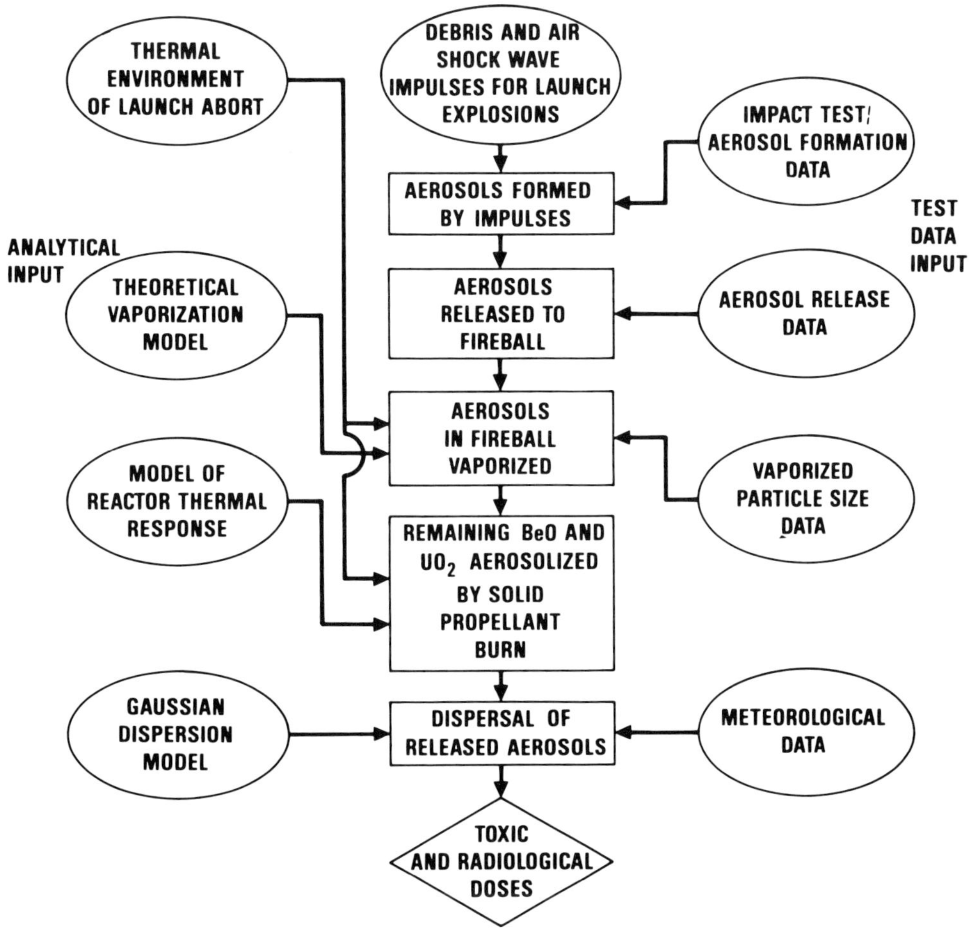

Figure 5 *Methodology for Launch Pad Abort Hazards.*

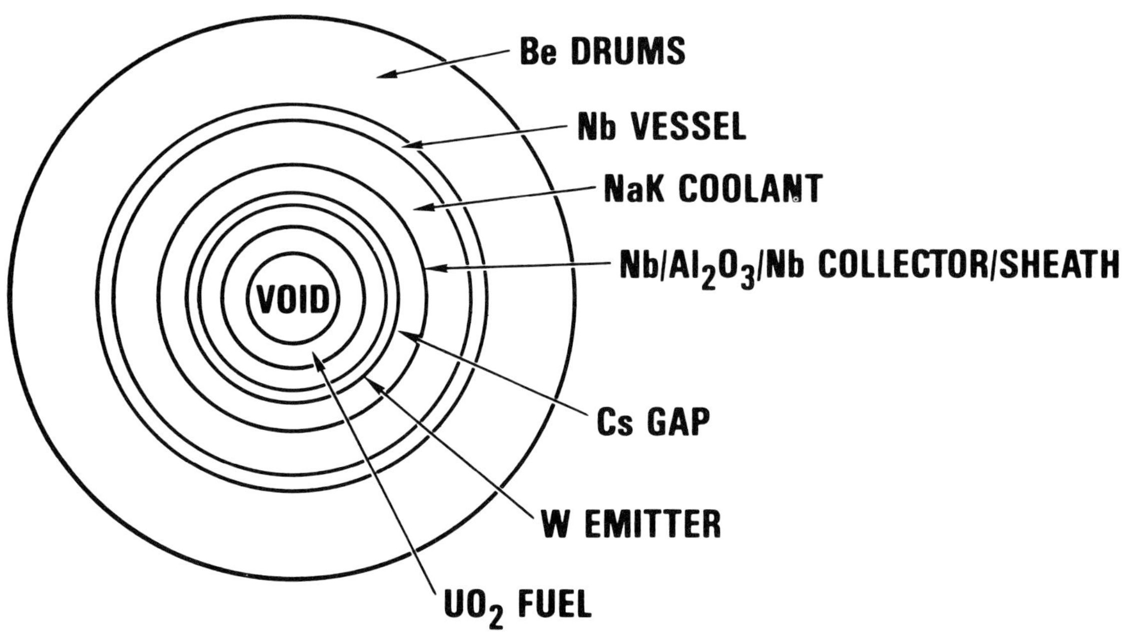

Figure 6 *VESSEL-1D Thermal Model Geometry.*

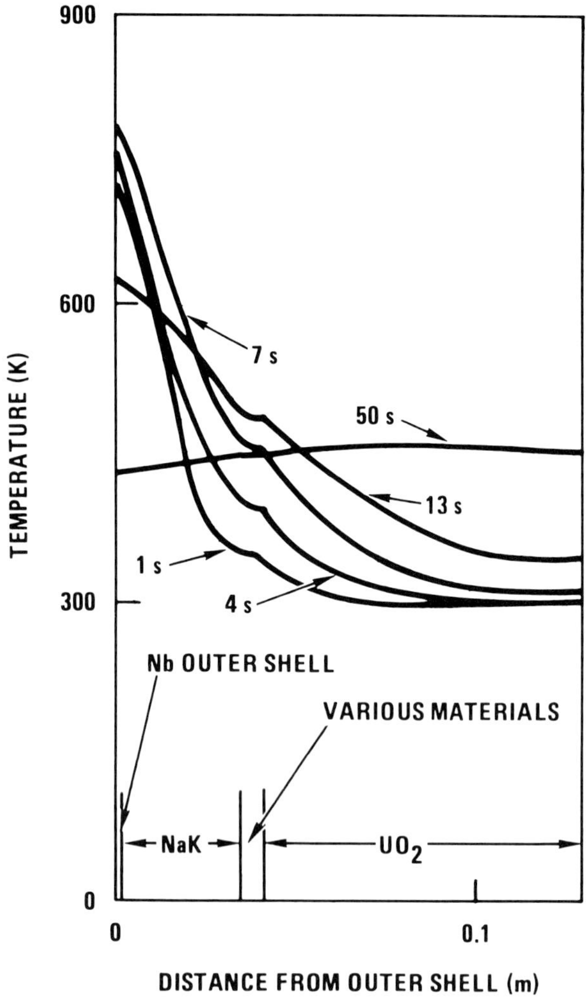

Figure 7 *Spatial Temperature Distributions in the Reactor at Selected Times.*

protected by the radial BeO reflector, remains significantly below its melting point.

Events after vessel sidewall melting are difficult to predict without a detailed two-dimensional analysis. If the fuel elements separate from the grid plates, they would collapse into a randomly oriented pile. In this case, the fuel element outer layers exposed to the fire will melt at 500 s. However, the tungsten emitter and the UO_2 do not melt and survive intact. If the fuel elements do not separate from the grid plates, only the outer fuel elements in the bundle melt in their outer layers.

Hazards for Launch Pad Accidents

For the purpose of consequence analysis, the elevated release of Be or BeO and UO_2 in the liquid propellant explosion fireball as it lifts off the ground was considered. To scope the magnitude of the consequences, separate dispersion calculations were performed assuming 100% release. If the results are acceptable for this conservative unrealistic case, they will be acceptable for any fractional release. For any estimated release fraction, the hazards can be scaled down proportionally.

The toxic or radioactive material released prior to about 10 s will be carried aloft by thermally driven air currents and then transported downwind. The downwind exposure to such material will depend on the effective rise height and diameter of the fireball cloud as well as on the quantity of material release, gravitational settling, and meteorological conditions. Table 6 illustrates estimates, taken from the OSM (1975), of the cloud height of rise and diameter for various quantities of liquid propellant combusting (charge). For the full contents of the ET, the cloud rise is over 2500 m and the cloud diameter is over 255 m.

The following cloud characteristics are assumed for the hazards calculations:

- Cloud diameter = 250 m, and
- Cloud height of rise = 2000 m.

The value of the height of rise is conservative relative to the OSM value. This is prudent in view of the possibility of an inversion layer acting to restrict the cloud rise.

A scoping calculation of the inhalation dose was performed in which all of the toxic or radioactive material was assumed to be released to the fireball in the form of particles small enough that fallout or gravitational settling may be neglected. In this case, the initial concentrations in the cloud are:

- Be or BeO: 40,600 $\mu g/m^3$, and
- ^{235}U: 5.2×10^{-8} Ci/m^3.

For an elevated release, the maximum ground concentration occurs for an unstable weather category (that is, Pasquill category A), since it minimizes the time for the cloud to reach the ground. There are standard curves for the atmospheric (Gaussian) dispersion characteristics of a point or a volume puff source (the latter corresponding to this case) for each Pasquill category. For example, Appendix D, Volume 2 of the OSM (1975) gives the equations and parametric dispersion results in graphical form for elevated release. The parameters are mean wind speed, u, height of release, h_o and inversion layer height, h_i. The DISPERS computer program (Barsell 1984), which uses curve

Table 6 *Estimated Stabilized Explosion Cloud Height and Diameter.*

Quantity of Propellant (lb)	Height of Rise (m)	Diameter (m)
3.9×10^1	200	10
1.5×10^3	500	31
2.4×10^4	1,000	77
1.2×10^5	2,000	131
3.9×10^5	2,000	190
9.5×10^5	2,500	255
2.0×10^6	3,000	323
6.2×10^6	4,000	479

fits to these dispersion curves, was used for this analysis.

Scoping calculations of the toxic and radiological inhalation doses versus distance were made based on the following parameters:

- Release height h_o is 2000 m,
- Inversion layer height h_i is 2000 m (case 1),
- No inversion layer (case 2),
- Stability is Pasquill category A,
- Wind speed is 4 m/s (annual average for Cape Canaveral),
- Initial cloud concentrations of Be and ^{235}U are as above,
- Initial cloud diameter is 250 m, and
- Breathing rate is 3.47×10^{-4} m^3/s (NRC Regulatory Guides 1.3 and 1.4).

Results are depicted in Figures 8 and 9. It was found that the peak ground exposure (total quantity inhaled

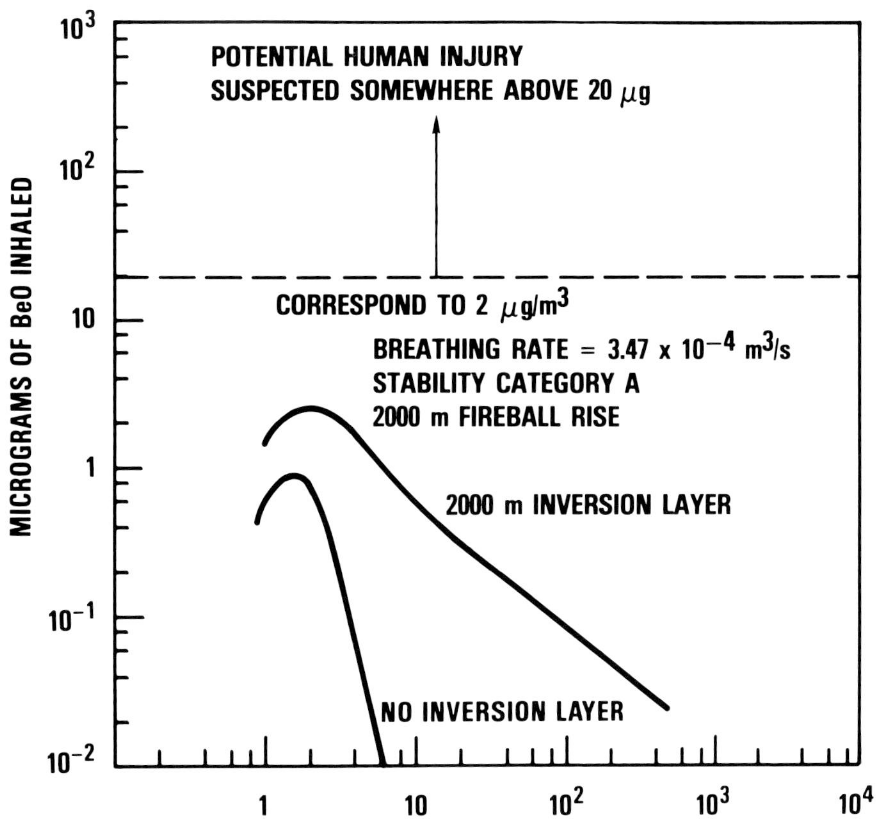

Figure 8 *Hazards Resulting when all Beryllium is released to the Fireball.*

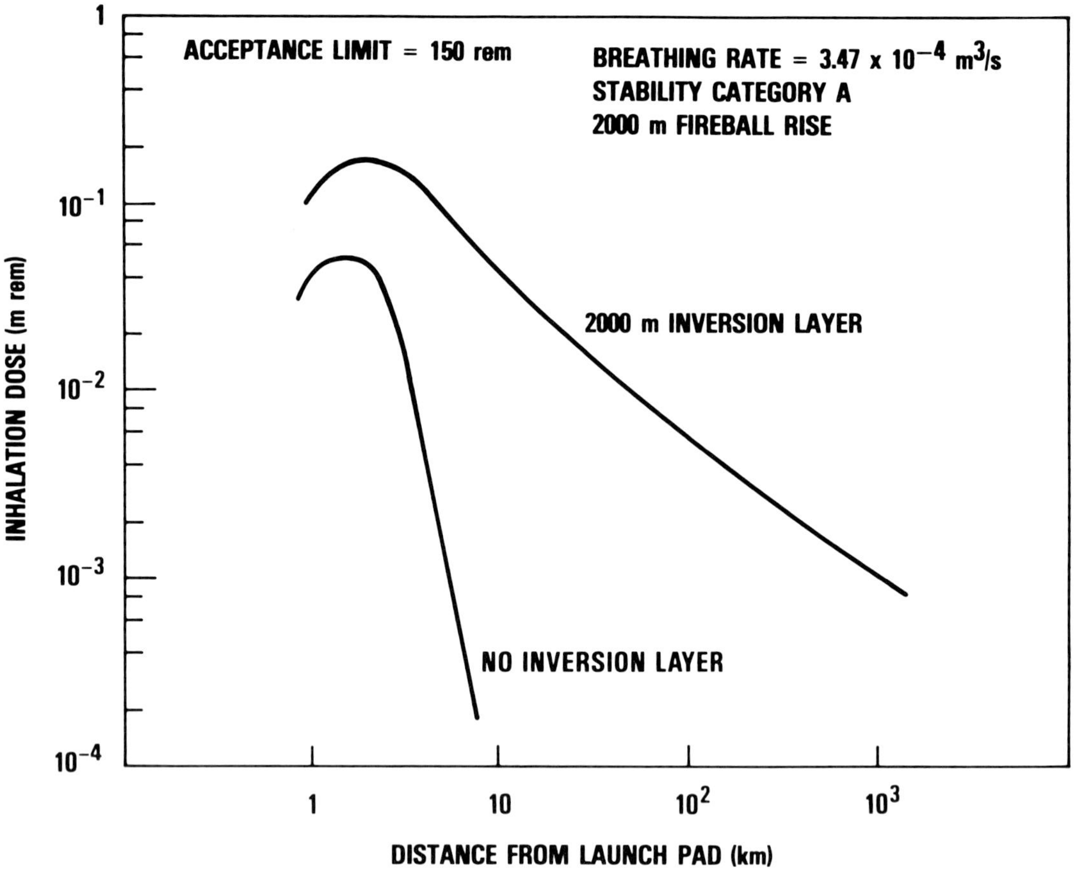

Figure 9 *Hazards from the Release of All ^{235}U to the Fireball.*

from the passing cloud if a person were standing outdoors there) occurred at about 2 km downwind from the launch pad. This means that the cloud passes overhead at nearer distances. Peak values of chemical dose were 2.8 μg for a 2000 m inversion layer height and 0.9 μg for no inversion layer. Peak radiological doses due to ^{235}U were below 0.2 mrem, which is insignificant.

For the Cape Canaveral site, the nearest occupational personnel would be about 3 km away and their doses would be somewhat lower than the above maximum values. Only in the SW and SSW directions are there personnel within 5 km from launch site. The offsite population essentially begins at 20 km away, a distance at which the dose is much reduced.

The meteorological data for the Cape Canaveral site (Vol. 3 of the OSM, (1975)) indicate that Pasquill category A occurs only about 1% of the time. For Pasquill category B (occurring 3% of the time), the peak chemical doses are 0.5 μg and 0.2 μg for a 2000 m inversion layer height and for no inversion layer, respectively. These maximum doses occur at distances of 5 km and 4.5 km, respectively. For Pasquill category C (occurring 15% of the time), the peak doses and distances are 0.2 μg at 70 km for a 2000 m inversion layer height, and 0.1 μg at 50 km for no inversion layer.

These results show that the maximum dose decreases and that the distance to maximum dose increases for more stable weather conditions. The dose versus distance curves cross, however, and the doses at greater distances may be relatively higher for stable weather categories. For example, the dose at 70 km is relatively higher for Pasquill category C than for category A. However, both doses are exceedingly low compared to limits of acceptable human exposure and are unimportant. For all weather categories, the inhalation doses, both chemical and radiological, are acceptable.

In this study, the limit of acceptable human exposure to Be via the inhalation pathway was taken to be 20 μg. This value was derived from the industrial standard for occupational exposure of 2 $\mu g/m^3$ for an 8-h working day, assuming a breathing rate of

3.47×10^{-4} m^3/s (NRC Regulatory Guides 1.3 and 1.4), 100% retention of inhaled particles, and 8 h of exposure. Doubtlessly, this limit is below levels resulting in human injury, but the margin is unknown at present.

The limit of acceptability for ^{235}U radiological inhalation doses is taken to be 150 rem to the critical organ (lung), in conformance with accident dose regulations. This limit is about 6 orders of magnitude above the calculated maximum doses.

Conclusions

The analysis showed the GA thermionic reactor to be inherently subcritical during accidents resulting in compaction of the core or immersion of the reactor in water or a fire. In large part, this is attributed to the negative reactivity of the in-core refractory materials. During operation the design withstands loss of the main NaK coolant circulation pump without losing the coolant. Should coolant be lost for any reason, the design possesses sufficient high temperature capability that decay heat is rejected passively to space (by radiation) without material melting or degradation.

For severe launch pad explosions, the bulk of the unirradiated fuel should remain encased in the tungsten emitter. Even for complete release of all fuel and beryllium, radioactive and toxic chemical offsite doses are within limits of acceptability.

Acknowledgments

The work reported here was sponsored by the Department of Energy, San Francisco Operations Office, under contract number DE-AT03-84SF-11906. DISCLAIMER: This study is preliminary in nature.

References

Alcouffe, R. E. et al. (1984) *User's Guide for TWODANT: A Code Package for Two-Dimensional, Diffusion-Accelerated, Neutral Particle Transport*, Report LA-10049-M, Los Alamos National Laboratory, Los Alamos, NM, March 1984.

Barsell, A. W. (1984) Unpublished Report, GA Technologies Inc., San Diego, CA.

Bell, C. R. (1984) ''Launch Phase: Compaction of the Core by Explosion Overpressure. Task 1 - Problem Specification,'' draft memo dated 23 May 1984, Los Alamos National Laboratory, Los Alamos, NM.

Finkelstein, T. (1968) *Manual for the Thermal Analyzer Program*, Report SE ARP 8889-5000, North American Aviation - Rocketdyne, Canoga Park, CA, January 1968.

Harris, P. A. (1984) Unpublished Report, GA Technologies Inc., San Diego, CA.

Hansen, G. E. et al. (1975) *Critical Masses of Oralloy Lattices Immersed in Water*, Report LA-2026, Los Alamos National Laboratory, Los Alamos, NM.

Lathrop K. D. and F. W. Brinkley (1970) *Theory and Use of the General Geometry TWOTRAN Program*, Report LA-4432, Los Alamos National Laboratory, Los Alamos, NM, May 1970.

Los Alamos Monte Carlo (LAMC) Group (1981) *MCNP-A, General Monte Carlo Code for Neutron and Photon Transport, Version 2B*, Report LA-7396-M, Los Alamos National Laboratory, Los Alamos, NM, revised April 1981.

NUS Corp. (1975), *Overall Safety Manual*, Volume 2, Technical Models, and Volume 3, Reference Data, prepared for the US Department of Energy, Washington, DC.

Pfremmer, R. D. (1984) Unpublished Report, GA Technologies Inc., San Diego, CA.

Sapir, J. L. and D. I. Brandon (1984a) *MCNP Criticality Calculation for Compaction of the GA SP-100 Reactor Design*, Report SP-TN-105, Los Alamos National Laboratory, Los Alamos, NM.

Sapir, J. L. and D. I. Brandon (1984b) *Neutronic Calculations for the Revised GA SP-100 Reactor*, Report SP-TN-106, Los Alamos National Laboratory, Los Alamos, NM.

Space Nuclear Power Systems 1986
Edited by M. S. El-Genk and M. D. Hoover

Chapter 7

Reentry Safety of the Thermionic Space Reactor

A. W. Barsell, V. Hwa, and M. G. Stamatelatos

GA Technologies Inc.
P.O. Box 85608
San Diego, CA 92138

A reentry trajectory, heating, and ablation analysis has been performed for the thermionic SP-100 reactor design during Earth orbital decay and severe steep angle reentries. The reactor assembly, including reactor, reflector drums and radiator, was predicted to survive the heating and oxidation environments of reentry without breakup or ablation. Even for separate reentry of the reactor components, no significant ablation of fuel or toxic materials was calculated. Upon ground impact, the reactor was predicted to become buried and inaccessible. The high temperature and high strength capability of the materials in the thermionic reactor design lead to an essentially intact reentry. Nevertheless, the hazards associated with dispersion of the radioactive fuel and toxic materials in the atmosphere or after impact were assessed and shown to comply with safety criteria. This conclusion holds even for reentry soon after reactor operation in space.

Introduction

This paper deals with reentry safety during preoperative phases, ascent to initial orbit, and boost to mission orbit, as well as post-operation phases including boost from low Earth orbit (LEO) and orbital decay from nuclear safe orbit (NSO). Before reactor operation, the concerns focus on the nonradioactive toxic materials, such as Be, LiH, and B_4C, and on the unirradiated UO_2 fuel containing ^{235}U. After operation, the concerns extend to include long-lived fission products, actinides, and activation products such as ^{94}Nb. This work is a part of the overall surety assessment of the SP-100 thermionic reactor (Stamatelatos and Barsell 1985).

The SP-100 Project Office has established two reentry safety criteria options. One is the assumed dispersal of the hazardous (particularly radioactive) material into the atmosphere. The other is "essentially intact" reentry of the reactor to the point of impact, meaning in one contiguous piece without breakup of the reactor core. This analysis demonstrates that the thermionic reactor is inherently compatible with the latter criterion option. Ths hazards of post-impact external radiation and postulated toxic material release are analyzed as a function of time of reentry. In addition, the hazards of atmospheric dispersion, either as an accidental release or as a planned alternate reentry option, are analyzed.

Trajectory Analysis

Two sets of reentry conditions are analyzed. One is Earth reentry from orbital decay. This case covers all unperturbed, gradually decaying Earth orbits, including both LEO and NSO. It also covers failures to ignite boosters in a temporary parking orbit or LEO. In all cases, the computational significant portion (last orbit) of the reentry begins at about 116 km (380 kft). The flight conditions at this altitude are:

- Circular velocity (V = 7.84 km/s (25,700 ft/s)); and
- Near horizontal flight direction, that is, small flight angle ($\lambda < 1°$).

Orbital decay reentry may occur over a range of mission phases. The earliest possibility would be accidental decay from parking orbit before the reactor has operated. Alternately, it may be caused by reentry from LEO after the reactor has operated (deliberately or via failure of higher orbit boosters), or from NSO long after reactor operation (up to hundreds of years decay time).

The other set of reentry conditions represents those generally called "steep angle" reentries. The SP-100 Project Office agreed on a bounding set of steep angle reentries for Phase 1. This set is considered to represent severe steep angle heating conditions for potential reentries associated with ascent guidance, boost failures, or orbital transfer misalignment failures. Steep angle reentry can occur from before operation for ascent or boost failures to just after operation for attempted EOL boost from LEO to a higher decay orbit.

The specified bounding set of steep angle conditions is based on control angle misalignment upon booster ignition from a parking orbit into an attempted geosynchronous Earth orbit (GEO) as the mission orbit. For a mission of this type, the SP-100 assembly (stowed configuration) and the Centaur G booster vehicle will be transported inside the Space Shuttle to an initial parking orbit, which can range typically from 240 to 390 km (130 to 210 NMi). The present analysis assumes a 278 km (150 NMi) parking orbit; the parking orbital altitude is not expected to affect the final heating transients significantly.

The Centaur G booster is assumed to be deployed and fired. For the GEO orbit, the boost velocity increment is 2.4 km/s. The control alignment angle then is assumed to incorrectly boost the vehicle in a vector direction that is within the orbital plane but not along the intended flight path. As a result of this misalignment, the SP-100 subsequently can reenter the Earth's atmosphere at a steep angle and at a velocity different from the circular velocity (depending on the booster vector angle).

Figures 1 and 2 show the calculated entry velocity and angle as a function of misalignment angle α. These are polar coordinate (r, θ) graphs, with circumferential dimension θ being the misalignment angle α and the radial dimension r being the entry velocity or entry angle.

It was found that the steepest entry angle(19°) occurred for $\alpha = \pm 120°$. Steeper entry angles are not possible for the conditions analyzed. The entry angle declines rapidly towards zero as α drops below 60°. At about $\alpha = \pm 32°$, the entry angle is zero, meaning that the orbital perigee is 116 km. At smaller angles, the assembly is sent into an elliptical orbit with a perigee higher than 116 km (which was the intent of the boost). In this elliptical orbit, there would be multiple opportunities for a successful second boost into GEO. Table 1 summarizes the entry conditions at 116 km altitude as a function of the misalignment angle α.

There is a narrow range of misalignment angles (32° to 38°) that corresponds to multi-orbit, grazing reentry, that is, perigee below 116 km but insufficient penetration into the atmosphere to be captured on the first pass.

The entry velocity increases as α decreases from 180° (retrofire). At 90°, the entry velocity is just above

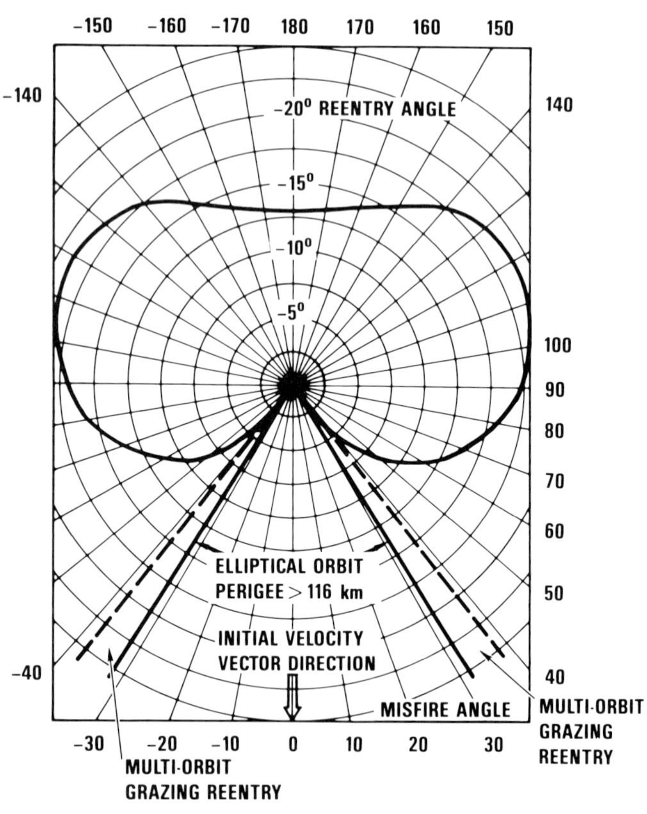

Figure 1 *Reentry Angle as a function of Misfire Angle on GEO Boost.*

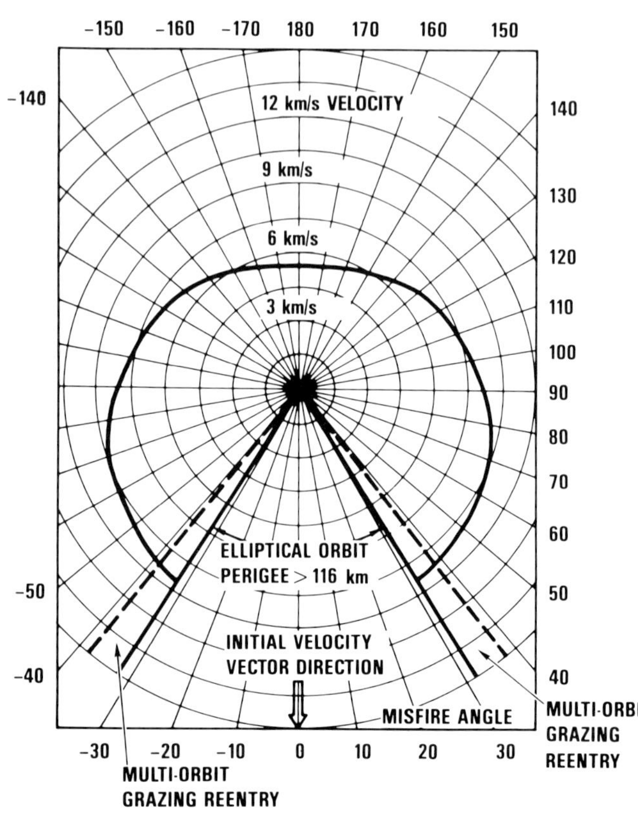

Figure 2 *Reentry Velocity as a function of Misfire Angle on GEO Boost.*

Table 1 *Reentry Conditions for SP-100 Assembly after GEO Boost Control Failure or Orbital Decay.*

Misfire Angle (deg)	Conditions at 116 km Velocity (km/s)	Flight Angle (deg)
120	7.08	-18.7
60	9.34	-11.2
45	9.75	- 6.9
Orbital Decay	7.84	- 0.2

orbital decay velocity (7.85 km/s). At the maximum prompt reentry condition ($\alpha = 38°$), the entry velocity is just over 9.75 km/s (32,000 ft/s).

Reentry of the reactor assembly attached to the spacecraft via the extended boom was not one of the configurations subjected to detailed aerodynamic and thermal analysis. This was because a simple, hand-calculated structural analysis indicated that the boom would not withstand reentry drag forces. The dynamic structural analysis was based on reentry trajectory data and calculated kinematic parameters, (such as velocity, acceleration, and angle of attack). From these parameters, the state of instantaneous momentum conservation was computed.

From the instantaneous force equilibrium and the constituitive relationships, the SP-100 boom's structural response was determined as a function of altitude, using standard stress equations.

The stress calculation indicated that the tensile force created by the drag moments about the center of gravity reaches and exceeds the ultimate strength of the boom at an altitude between 75 and 90 km. At this point, the boom fails and the reactor assembly follows an independent trajectory. This altitude precedes the major portion of the reentry heating. Thus, the boom could be assumed to have been severed before reentry.

While a major focus was on the SP-100 assembly in the stowed configuration, free flight reentry of the assembly components (fuel elements, reflector drums, reactor, and radiator and radiation shield) were also analyzed. The approach was not to perform detailed stability analyses of the configurations and not to analyze only the predicted flight attitude. Rather, the range of flight attitudes (spinning, tumbling, stable end-on, and so on) was analyzed.

The SP-100 assembly stowed configuration is illustrated in Figure 3. For the assembly reentering end-on (reactor first), the configuration resembles a cylinder/cone combination. Assembly components analyzed include the reactor and radiation shield, the reactor alone, an individual reflector drum, and an individual fuel element. Drag coefficients for these components utilized data given by Cropp (1965) for cylinders.

Figure 4 depicts the ballistic coefficients of the configurations in hypersonic continuum flow. The worst case (highest ballistic coefficient) occurred for the detached reactor. Thus, the detached reactor would receive heating much greater than that for the reactor attached to the assembly. The parachute effect of the shield and radiator in slowing down the reactor is evident.

Terminal velocities of the configurations at sea level were calculated based on Reynolds number dependent drag coefficients. They range up to just over 200 m/s with the worst case being the reactor detached from the shield and radiator assembly.

The trajectories, aerodynamic heating and oxidation heating along the reentry flight paths were calculated with the GA Technologies' version of the RESTORE computer program (Elliott 1984). The results agreed well with benchmark calculation results from Los Alamos National Laboratory (Biehl 1984).

Thermal Analysis

The basic assumptions of the aerodynamic heating rate (Elliott 1965) follow:

- Velocity-cubed dependent heating rate in free molecule flow,
- Modified Fay and Riddell correlation for stagnation point heating in continuum flow,
- Transition bridging between the above two flow regimes,
- Hot wall correction factor, and
- Nonequilibrium flow effects neglected.

Oxidation heating was computed in continuum flow along the trajectory and added to the aerodynamic heating. This was found to increase the surface heating by about 25%. The rate of oxidation was based on the following assumptions:

- Infinite reaction rates of the metal or graphite with oxygen; and
- Reynolds analogy for boundary layer flow (mass flow proportional to heat flow).

There are at least three possible reactions of the niobium with oxygen. The predominant reaction in a normal air environment produces Nb_2O_5. In an oxygen-sparse environment it is possible to form suboxides, namely NbO_2 or NbO. All these reactions have essentially the same heat of reaction, per unit mass of oxygen; namely, about 24 kJ/g of O_2. For the graphite ablator, the heat of oxidation is about 8 kJ/g of O_2, a factor of 3 lower than for niobium oxidation.

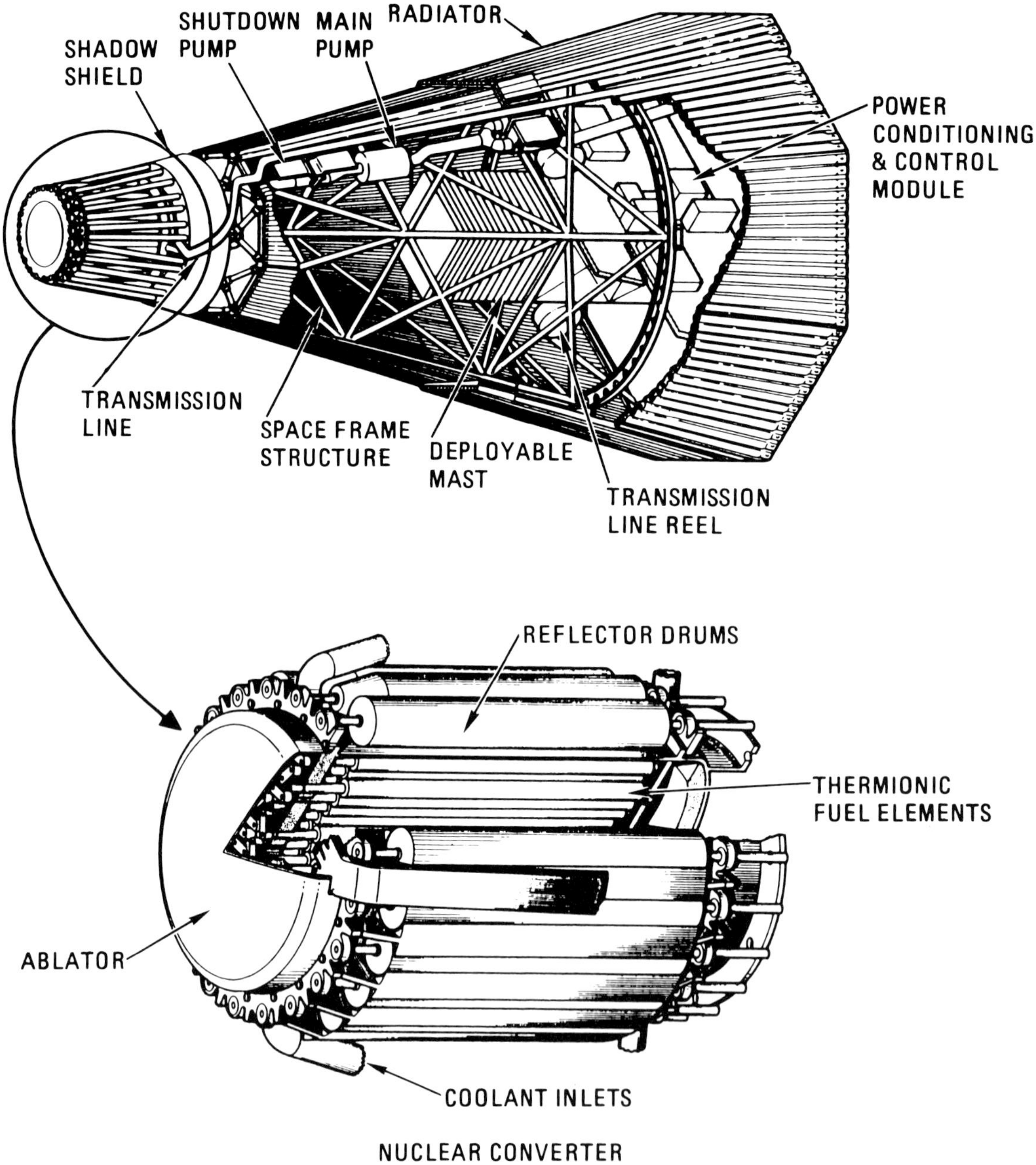

Figure 3 *Thermionic Reactor Assembly in the Stowed Configuration showing Details of the Reactor.*

The thickness of niobium or graphite oxidized was calculated assuming that all oxidized material at the surface would be lost instantaneously. This could occur for niobium by melting its oxides, which have a low melting temperature of around 1500 K, or by spalling the oxide and shearing it off by the aerodynamic flow forces. For graphite, the oxide formed is gaseous. Typical values of thickness of Nb oxidized range from tenths of a mm to several mm, depending on the configuration and trajectory as well as the reaction assumed. The depth of graphite oxidized is similar to that for Nb_2O_5. This degree of oxidation does not cause breakup or degradation of the reactor assembly.

The RESTORE trajectory heating transients for the assembly reentering end-on are summarized in Figures 5 and 6 and Table 2. For the steep angle cases analyzed, the highest integrated reentry heating to the SP-100 was found to occur on booster misalignment at the shallowest angle ($\sim 40°$), which results in a prompt capture by the Earth's atmosphere, that is, reentry on the first pass. At that condition, the integrated heating was nearly equal to that for orbital de-

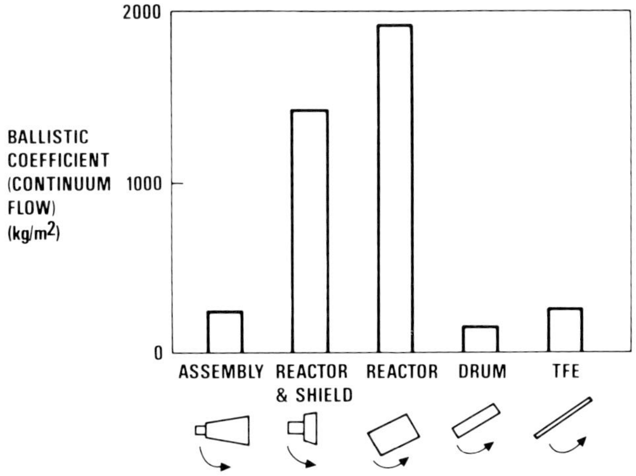

Figure 4 *Tumbling Ballistic Coefficient.*

cay. The highest reentry peak heat rate was found to occur at a booster misalignment angle of about 60°.

There is a narrow corridor of misalignment angles (32° to 38°) resulting in elliptical multiple pass trajectories that graze the atmosphere. These grazing orbits create a unique series of heating and cooling cycles, with a period of several hours between heating times. Since the RESTORE code is not suited for multiple grazing orbits, a special computer code called SKIP (Barsell 1971) was employed for misalignment angles from 32° to 38°. These grazing orbits were found to represent worst cases for accumulated amounts of oxidation and are used for determining the ablator thickness required. Further analysis is needed to assess cyclic thermal stresses.

From comparisons with the calculated heat of ablation, that is, the integrated heat required to bring the materials to and through melting, it is evident that there is insufficient reentry heating to ablate any significant portion of the reactor. Also, the heat rate is insufficient to melt the outer layers of the radiator because of the reradiation heat loss.

The thermal responses of the reactor assembly and components during orbital decay reentry and steep angle reentry from the specified conditions of orbital boost misalignment were analyzed. Using the GA-TAP computer program (Elliott 1984) thermal models were developed for the radiator and the space frame (Figure 7), for the reactor (Figure 8), for an individual fuel element, and for an individual reflector drum.

Results show that, for orbital decay, no breakup of the stowed assembly configurations occurs, no melting occurs on the radiator space frame or the reactor, and the outer reflector remains attached.

For the steep angle reentry trajectories, the radiator does not melt but the space frame composite reaches its failure point, according to a conservative analysis. If the reactor detaches and tumbles or reenters end-on (head first), no melting or ablation on the reactor occurs, regardless of the detachment altitude. Again, the drums remain attached. The ablator on the reactor head prevents melting of the vessel top head. Without

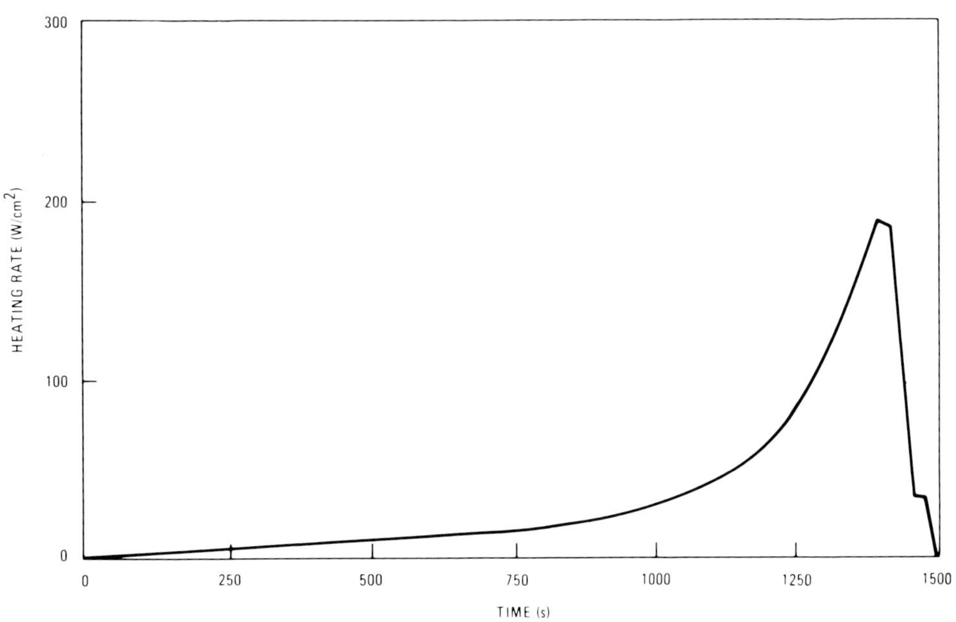

Figure 5 *Orbital Decay Reentry Heating for End-on Assembly.*

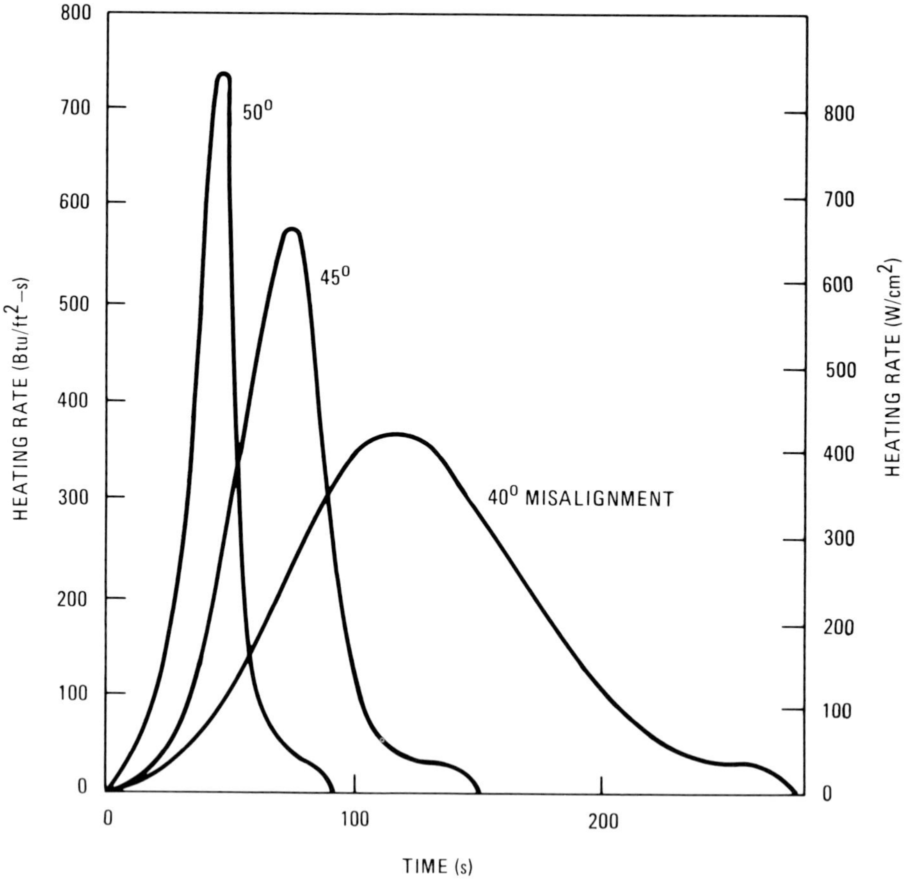

Figure 6 *Heating to the Reactor Head during Steep-Angle Reentry of Assembly.*

the ablator, the reactor head would ablate only down to the top reflector, and the drums would detach.

It is concluded that the reactor reenters "essentially intact" (with sufficient margin) to the point of ground impact. It, thereby, meets and exceeds the reentry safety criterion for intact reentry.

Based on the calculated drum temperatures during orbital decay and steep angle reentries, no melting of the beryllium is predicted. Furthermore, the amount of oxidation on the Nb-1%Zr cladding is slightly less than the cladding thickness. These results demonstrate that detached reflector drums will not ablate and will reenter intact.

The calculated thermionic fuel element (TFE) temperatures for the orbital decay and steep angle reentries show that, for the orbital decay, the surface ablates down to the inner Nb-1%Zr layer (collector) but the fuel remains contained. For steep angle reentry, the ablation proceeds through the collector trilayer, but the fuel and the tungsten emitter do not

Table 2 *Reentry of SP-100 Assembly End-on (RESTORE Results).*

GEO Boost Misfire Angle (deg)	Peak Heat Rate (W/cm^2)	Integrated Heating (J/cm^2)	Thickness of Nb Oxidized (mm)
120	517	8,740	0.5
60	829	19,300	0.5
45	653	30,340	0.9
40	417	48,800	1.4
32–38	Grazing multi-orbital reentry		
0–32	Long-life elliptical orbit		
Orbital decay	191	52,070	1.5

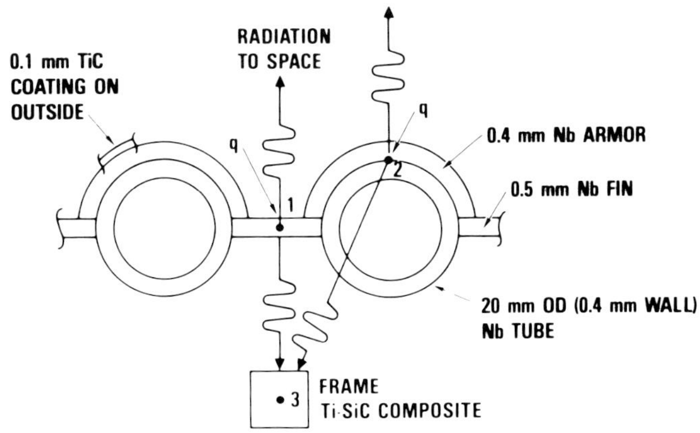

Figure 7 *Thermal Model of the Radiator and the Space Frame.*

ablate. These results demonstrate the infeasibility of reentry dispersal for the thermionic reactor fuel. On the other hand, they demonstrate the feasibility of ''essentially intact'' reentry.

Impact Analysis

A simplified analysis was conducted, based on the procedures developed by Stevenson (1980) and Riera (1968), to assess the dynamic-structural behavior of the SP-100 reactor during reentry impact. The target was assumed to be a rectangular reinforced concrete slab 6 m wide, 6 m long, and 0.75 m thick. The shear velocity of the reinforced slab induces, under dynamic behavior, the equivalent of a hard rock surface.

The results of the dynamic impact analysis show that, if one assumes a rigid SP-100 assembly (reactor/drum/shield/radiator), a deep penetration of approximately 1.27 m (50 in.) of the missile into the soft target (rectangular reinforced concrete slab) occurs. If one assumes a flexible SP-100 reactor, the total deformation incurred during impact phase is on the order of 20% of the total length, given a rigid target.

Data from available literature (such as by Otter 1973) indicate that the SP-100 thermionic reactor is more likely to survive impact on a hard surface (for example, 15 cm of concrete) without gross rupture of the vessel, and is likely to penetrate through the concrete and become buried in the ground (inaccessible to people at the surface).

Hazards Analysis

Reentry Hazards

To assess the hazards of LEO missions, the inventories of 43 key radionuclides, including fission products, actinides and activation products, were tracked as a function of reactor operation time and decay time after shutdown (to 1000 yr). These inventories were then weighted by their dose conversion factors (rem to Ci ratios) for inhalation, ingestion, and external radiation pathways. These hazard index factors were summed over all nuclides as a partial measure of the overall hazard for each pathway.

The calculations show that, during reactor operation, the inventory of hazardous fission products builds up quickly (within months). Hazardous activation product and actinide inventories build up more gradually. At six months of operation, the total inhalation hazard index is already one-third of the hazard index after seven yr of operation (end of life - EOL). The same occurs for the ingestion hazard index. The external radiation hazard index at six months is 84% of that at EOL and, thus, is very weakly dependent on operational time.

After reactor shutdown, the external radiation hazard index (Figure 9) declines quickly at first, then gradually according to the half life of ^{137}Cs (30 yr). At about 200 yr, the external radiation hazard index starts to become controlled by ^{94}Nb. Since this radionuclide is very long-lived, the external hazard index is relatively constant after 200 yr.

For the inhalation and ingestion pathways, the amount of rapid drop in the hazard index (Figures 10 and 11) shortly after operation is small because of the dominance of ^{90}Sr at the end of operation. The hazard index factors decline gradually according to the half life of ^{90}Sr (29 yr) until very long-lived nuclides become more important. For inhalation, this occurs at around 160 yr, after which the inhalation hazard index is dominated by ^{239}Pu. For ingestion, this occurs later at about 320 yr, after which ^{94}Nb controls the hazard

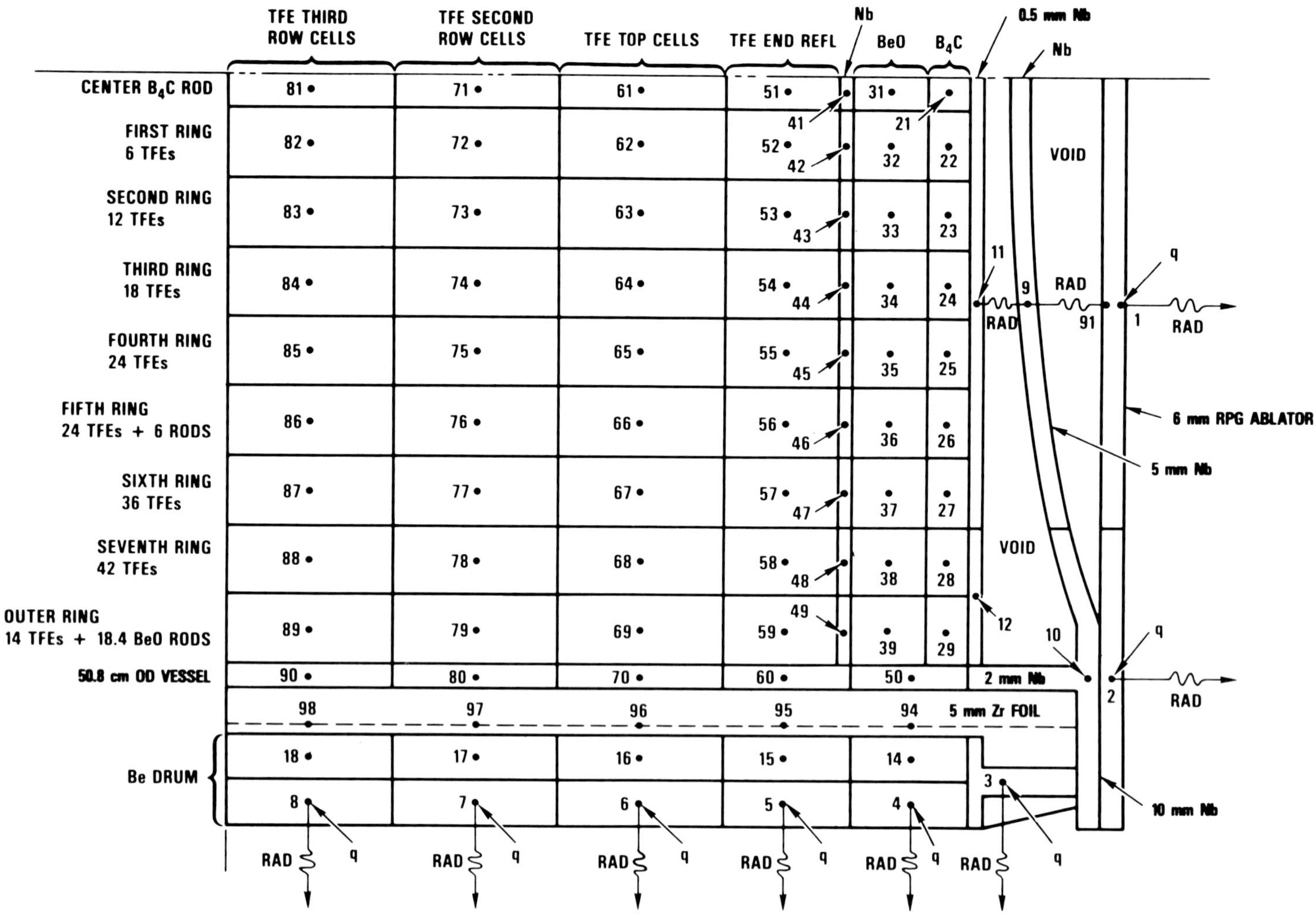

Figure 8 *Two-dimensional Thermal Model of the Reactor Vessel.*

index. Orbital lifetimes beyond these times produce little effect on the hazard index factors.

A specific hazards analysis was performed of the dispersal of all radioactive and nonradioactive toxic material in the atmosphere as bounding condition to assess the feasibility of this reentry disposal option. Two distinct atmospheric dispersion models were applied. One analyzed worldwide dispersion of small particles (less than several micrometers in diameter), for which inhalation was the major exposure pathway. The other analyzed the ground or fresh water deposition of larger particles, for which ingestion and external radiation exposure pathways were also important (in addition to inhalation of resuspended particles).

A simple small particle model developed at the U.S. Weather Bureau (Machta 1962 and 1982) was applied to determine air concentrations of the toxic or radioactive material at ground level as a function of latitude and burnup altitude.

It was found (Table 3) that peak calculated concentrations of toxic and radioactive materials were at least six orders of magnitude below maximum permissible concentration (MPC) levels. Be and ^{239}Pu are relatively the most hazardous. However, the maximum individual doses are negligible and the total radiation population dose is below 10^3 manrem, a level corresponding to low probability of even one somatic health effect. It was concluded that fine particle dispersion of SP-100 materials in the atmosphere at 300 yr would not pose a potential inhalation hazards to the world population.

For the large particle fallout model, the initial source was modeled as a point or a line source con-

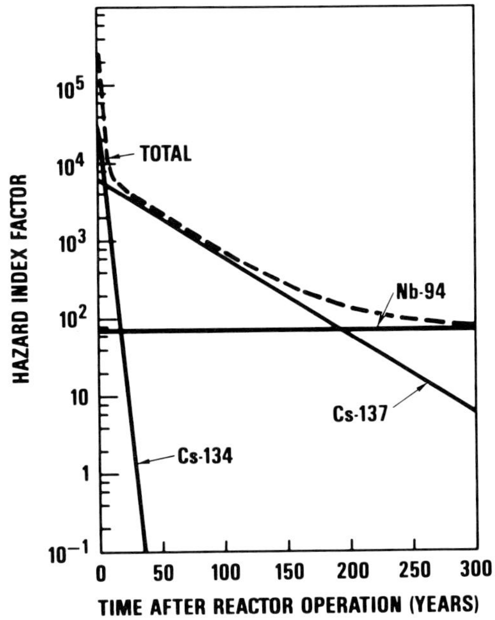

Figure 9 *Hazard Index Factor versus Time for External Radiation.*

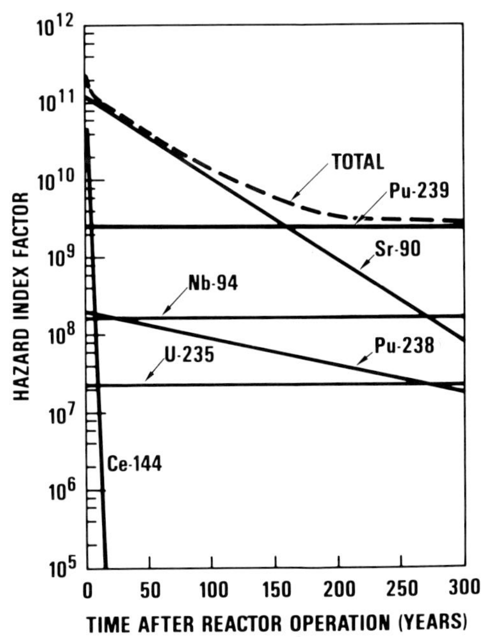

Figure 10 *Hazard Index Factor versus Time for Inhalation.*

centration corresponding to the trajectory segment where burnup occurs. The source turns into a cloud, as the particles are acted upon by gravity (downward), and by wind conditions (toward the downwind direction) and eddy diffusion (lateral expansion). Use was made of standard solutions of integrated dose and ground deposition density as functions of distance.

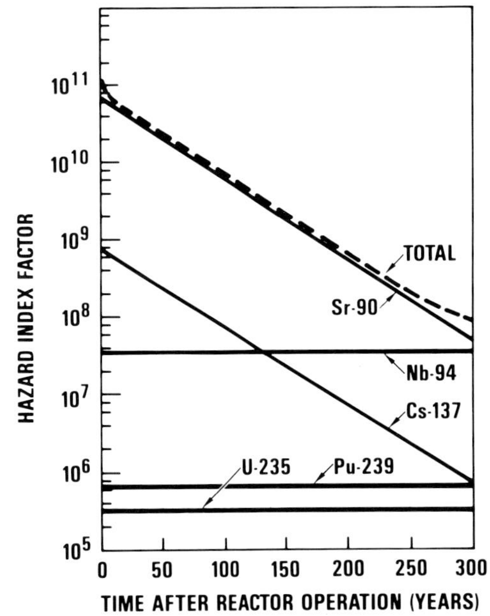

Figure 11 *Hazard Index Factor versus Time for Ingestion.*

These standard solutions are given for each Pasquill stability category in the Overall Safety Manual, Volume 2 (1975) and in Nuclear Regulatory Guidelines. Other parameters are average wind speed, release altitude, and particle size.

The range of particle sizes was estimated from analysis of ablated particle breakup, using the Weber criterion for critical particle size. Critical particle diameters for metals, BeO and UO_2 along the reflector drum and fuel element trajectories were calculated.

The results indicate that the ablated particle sizes, initially around 200 μm to 800 μm in diameter, will be stable at high altitudes (above 70 km). As the particles descend, the UO_2 and BeO particles will become unstable and will likely break up. The post-breakup particle sizes are likely to be distributed, with many fine (micrometer size) droplets and some possible drops of fractional particle size (up to from 100 μm to 240 μm). The metals Nb and Be are not likely to break up.

Based on this particle size distribution, the lowest release (ablation) altitude, and average weather conditions, the ground deposition density distribution was calculated. At the maximum ground deposition density location, the following corresponding radioactive and chemical doses were calculated for a 300-yr time of reentry:

- The 50-yr commitment ingestion dose for farmland contamination, assuming no crop or milk control or impounding;

Table 3 *Summary of Hazards (Reentry at 300 yr).*

Pathways to Humans	Random Impact Probability	Key Materials	Minimum Safety Margin
Small particle inhalation	1.0	^{239}Pu	10^6
		BeO	10^6
Large particle fallout in fresh water	$< 1.5 \times 10^{-3}$	^{94}Nb	10^3
		^{90}Sr	10^4
		BeO	10^5
Large particle fallout on farmland (food-milk)	4×10^{-2}	^{94}Nb	35
		^{90}Sr	300
		BeO	10^5
External radiation	0.3	^{94}Nb	20
		^{137}Cs	10^5
Inhalation of resuspended particles	0.3	BeO	10^6
		^{239}Pu	10^6

- The 50-yr commitment ingestion dose for fresh drinking water, assuming perfect solubility of particles;
- The external radiation dose of ground deposition, based on a smooth ground surface; and
- The 50-yr commitment inhalation dose of resuspended ground deposition, assuming resuspension factors from WASH-1400.

The results are summarized in Table 3.

Post-Impact Hazards

The post-impact hazards were evaluated assuming unlikely aerosolization or respirable particles released to the air or unshielded external radiation exposure. Release fractions were based on data from the Los Alamos National Laboratory for heated UO_2 upon cooling in air. For this calculation, no credit was taken for the multiple refractory metal containment layers in the thermionic reactor design.

The calculated 50-yr commitment inhalation radiation dose is depicted as a function of time of reentry in Figure 12. Major radioisotope contributors also are shown in this figure. At early times after operation, the inhalation dose is dominated by ruthenium radionuclides. After about 5 yr of decay before reentry, ^{90}Sr becomes the highest dose contributor until about 100 yr, after which ^{238}Pu and ^{239}Pu dominate.

Probabilistic calculations indicate that 1000 m seems to be a reasonable downwind distance at which to compare inhalation doses with accident limits. Shorter distances are unlikely to include any people. Thus, the practiced regulatory limit of 150 rem to the critical organ is shown in Figure 12 for comparison with doses at 100 m and 1000 m. This limit is not exceeded for orbital decay times beyond about 1 yr. Doses, due to the ingestion pathway, were calculated to be lower than the inhalation doses.

Post-impact hazards via the external pathway consist of gamma radiation from the intact reactor and ground shine from deposition of released and dispersed activity after impact. Based on the above release data, the latter can be shown to be negligible. A shielding calculation was performed for a core having the approximate size of the SP-100 thermionic reactor. It accounted for self-shielding of the UO_2 fuel but conservatively neglected the shielding effect of the refractory metals in the core. Results for selected times are depicted in Figure 13 as a function of distance. It is seen that the dose rate drops off rapidly with time after operation, then approximately with a 30-yr half life (namely, of ^{137}Cs) until several hundred years.

A probabilistic calculation indicates that people are unlikely to be present in a random land area inside a circular distance of 140 m. At this distance, the dose rate at the end of operation is 0.6 rem/h. Exposure at this rate for a period of 10 h could result in a whole body dose of 6 rem, which is a factor 4 below the accident regulatory limit of 25 rem to the whole body (10CFR100). A 10-h period is a reasonable time for reactor location and mitigation action within a popu-

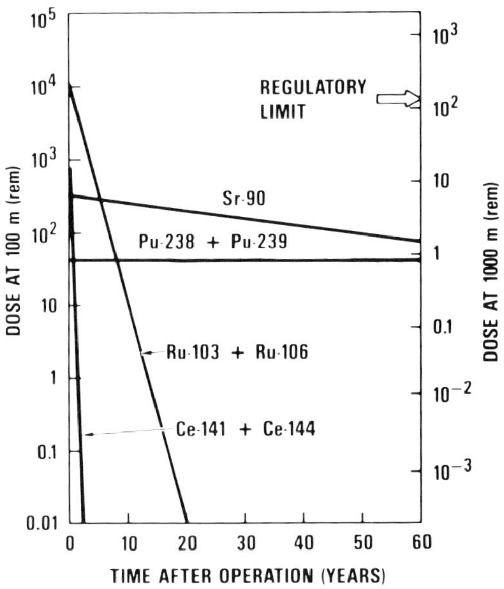

Figure 12 *Inhalation Dose for Post-Impact Release.*

lated area. The regulatory limit would not be exceeded beyond 70 m for 10 h.

For reentry after 1 yr from end of operation, with the same assumptions, the regulatory limit would not be exceeded beyond 20 m. At 10 yr after operation, the corresponding distance would be 10 m.

Summary and Conclusions

As a part of the thermionic SP-100 surety assessment (Stamatelatos and Barsell 1985) the planned (end-of-life) or unplanned (accidental) reentry of the thermionic space reactor and its response to thermal and mechanical loads have been analyzed. Results demonstrate compliance with the "essentially intact" option of the SP-100 Project Surety Criteria. Extensive use of refractory metals such as niobium in the design and the use of a thin nose ablator enable the reactor vessel and internals to withstand melting and be fully intact at Earth impact after reentry from Earth orbit or from ascent or boost abort. This conclusion is independent of the potential ablation or breakup of the radiators/space frame, which is predicted only for a worst-case steep angle reentry (not for orbital decay), or of the radiation shield. Analysis of potential flight modes, trajectories and aerodynamic/oxidation heating, coupled with a two-dimensional thermal analysis, formed the basis of this conclusion.

On ground impact, the thermionic reactor vessel is expected to remain essentially in one piece, based on calculated terminal velocities and a review of impact data. It most likely would become buried in the ground, inaccessible for approach or contact.

An assessment was made of the potential health hazards due to accidents during reentry of the radioactive and nonradioactive toxic materials in the reactor assembly. In relative terms, the external radiation hazard declines rapidly soon after reactor shutdown, then gradually according to the 30-yr half life of the fission product ^{137}Cs. At about 200 yr of orbit lifetime, including 7 yr of continuous operation, the ex-

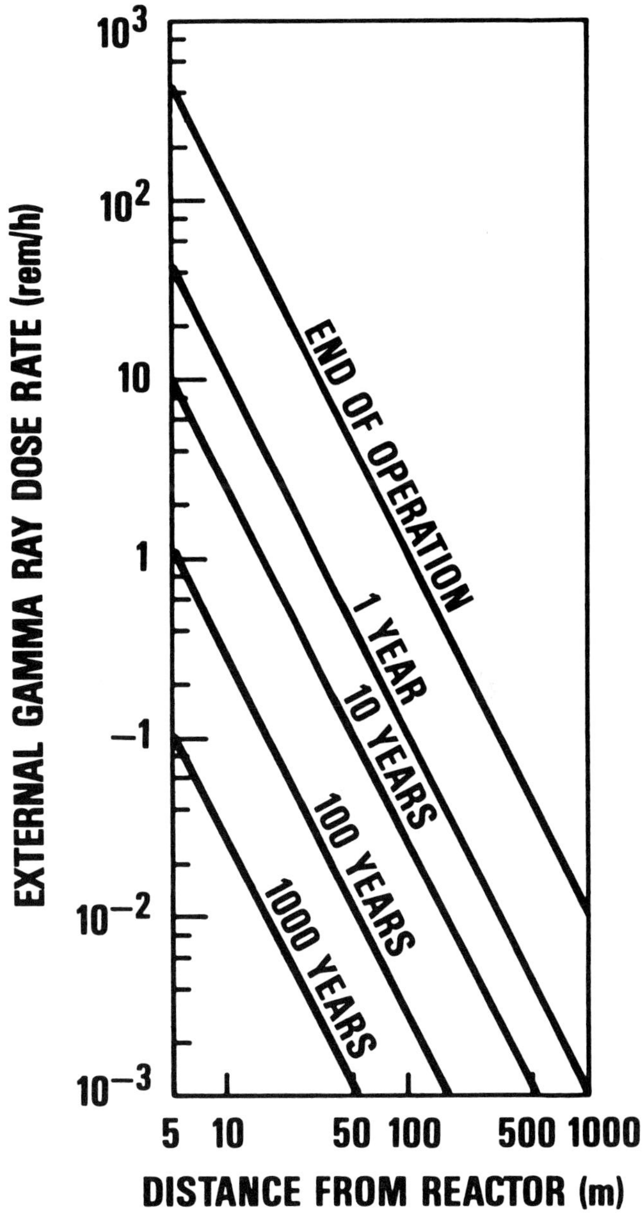

Figure 13 *External Radiation from Intact Reactor as a function of Distance.*

ternal radiation hazard becomes relatively constant, controlled by the long-lived activation product ^{94}Nb. Radioactivity inhalation and ingestion (internal) hazards immediately after shutdown are controlled by the fission product ^{90}Sr, and, thus, decline slowly. At about 160 yr, the inhalation hazards level off, controlled by the long-lived actinide ^{239}Pu. Ingestion hazards level off at about 320 yr, dominated by ^{94}Nb.

In absolute terms, all post-operation reentry times, including those from LEO, are ''safe'' relative to accident limits on human exposure to toxic and radioactive material. For unlikely accidental atmospheric release, worldwide dispersion of any small particles (such as Be or UO_2) would comply with chronic exposure guidelines, and local ground deposition of any large particles would comply with accident standards for terrestrial reactors. An accidental post-impact release or unshielded external exposure would comply with accident standards at distances where people are likely to be, probabilistically. While the combined safety margin, considering all exposure pathways, increases with reentry time, little further benefit is predicted for reentry times beyond about 50 yr after operation, corresponding to a circular earth orbit altitude of 700 km.

Acknowledgments

The work reported here was sponsored by the Department of Energy, San Francisco Operations Office, under contract number DE-AT03-84SF-11906. DISCLAIMER: This study is preliminary in nature.

References

Barsell, A. W. (1971) ''Unperturbed Multiple Elliptical Orbit Reentry of Isotopic Heaters,'' *J. of Spacecraft of Rockets*, 8 (12):1235.

Biehl, F. (1984) *Personal Communication*, Los Alamos National Laboratory, Los Alamos, December 1984.

Cropp, L. O. (1965) *Analytical Methods Used in Predicting the Reentry Ablation of Spherical and Cylindrical Bodies*, Report SC-RR-65-187, Sandia National Laboratories, Albuquerque, NM, September 1965.

Elliott, R. D. (1966) *Methods for Predicting Aerodynamic Heating Rates to High-Velocity Bodies in the Upper Atmosphere*, Report NAA-SR-11704, Atomics International, Canoga Park, CA, January 1966.

Elliott, R. D. (1984) Unpublished Report, GA Technologies Inc., San Diego, CA.

Machta, L. (1962) ''Worldwide Fallout Since 1959—Meteorological Aspects,'' *JCAE Hearings on Radiation Standards*, pp. 141–170, June 1962.

Machta, L. (1982) ''Transfer of Tracer Materials from the Stratosphere to Ground Level,'' Unpublished Weather Bureau Manuscript, March 1982.

Otter, J. M., et al. (1973) *Aerospace Safety Program Summary Report*, Report AI-AEC-13100, Atomics International, Canoga Park, CA, July 1973.

NUS Corp. (1975) ''Overall Safety Manual,'' Volume 2 Technical Models, prepared for the US Department of Energy, Washington, DC, June 1975.

Riera, Jorge D. (1968) ''On the Stress Analysis of Structures Subjected to Aircraft Impact Forces,'' *Nuclear Engineering and Design*, 8:416–426.

Stamatelatos, M. G. and A. W. Barsell (1985) *Thermionic SP-100 Surety Assessment*, USDOE Report GA-A18111, GA Technologies, San Diego, CA, September 1985.

Stevenson, J. D. (1968) *Structural Analysis and Design of Nuclear Power Plant Facilities*, ASCE Manual and Reports on Engineering Practice - No. 58.

Space Nuclear Power Systems 1986
Edited by M. S. El-Genk and M. D. Hoover

Chapter 8

A Feasibility Assessment of Nuclear Reactor Power System Concepts for the NASA Growth Space Station

Harvey S. Bloomfield and Jack A. Heller

National Aeronautics and Space Administration
Lewis Research Center
Cleveland, OH 44135

A preliminary feasibility assessment of the integration of reactor power 8system concepts with a projected growth Space Station architecture was conducted to address a variety of installation, operational, disposition and safety issues. A previous NASA-sponsored study, which showed the advantages of Space Station-attached concepts, served as the basis for this study. A study methodology was defined and implemented to assess compatible combinations of reactor power installation concepts, disposal destinations, and propulsion methods. Three installation concepts that met a set of integration criteria were characterized from a configuration and operational standpoint, with end-of-life disposal mass identified. Disposal destinations that met current aerospace nuclear safety criteria were identified and characterized from an operational and energy requirements standpoint, with delta-V energy requirements as a key parameter. Chemical propulsion methods that met current and near-term application criteria were identified and payload mass and delta-V capabilities were characterized. These capabilities were matched against concept disposal mass and destination delta-V requirements to provide a feasibility of each combination.

Background

A preliminary feasibility assessment of the integration of reactor power system concepts with a projected growth Space Station architecture was conducted to address a variety of potential installation, operational, disposition, and safety issues.

A previous NASA Lewis Research Center-sponsored study (Silverman et al. 1984) evaluated a variety of nuclear power concepts for the Space Station in the multi-hundred kilowatt range and concluded that free-flyer platform reactor power systems were less attractive than Space Station-attached concepts using structural boom and tether mounts. Therefore, the study goals were to identify attractive Space Station-attached reactor power system concepts and assess their installation and disposition feasibility. The scope of the study was limited to normal installation and disposition operations at the beginning and end of mission, respectively, and did not consider abort or severe malfunction scenarios.

The study methodology shown in Figure 1 was selected to provide a logical approach for assessing the feasibility of each installation concept, with respect to a matrix of disposal locations and propulsion methods. The approach used to select installation concepts was based on defining a set of criteria for a projected nuclear powered growth Space Station to screen candidate concepts. Selected concepts were then characterized from a configuration and operational viewpoint, with disposal mass as a key parameter. Disposal destinations that met current aerospace nuclear safety criteria were identified and characterized in terms of operational and energy requirements, with delta-V energy requirements as a key parameter. Propulsion methods that met the near-term time frame criterion were identified and characterized from a payload mass and delta-V energy capability viewpoint. These propulsion capabilities were then matched against installation concept disposal mass and disposal destination delta-V energy requirements to assess the feasibility of each combination. The study matrix consisted of three installation concepts, six disposal destinations, and two chemical propulsion methods, as shown in Table 1. Identification of reactor power system installation locations on the Space Station was based on a projection of growth options for a 300-kWe power level. This included consideration of reactor power system location impact on projected in-place solar power systems, payloads, and occupied station modules, as well as on station and Shuttle operations. A projected "Dual-

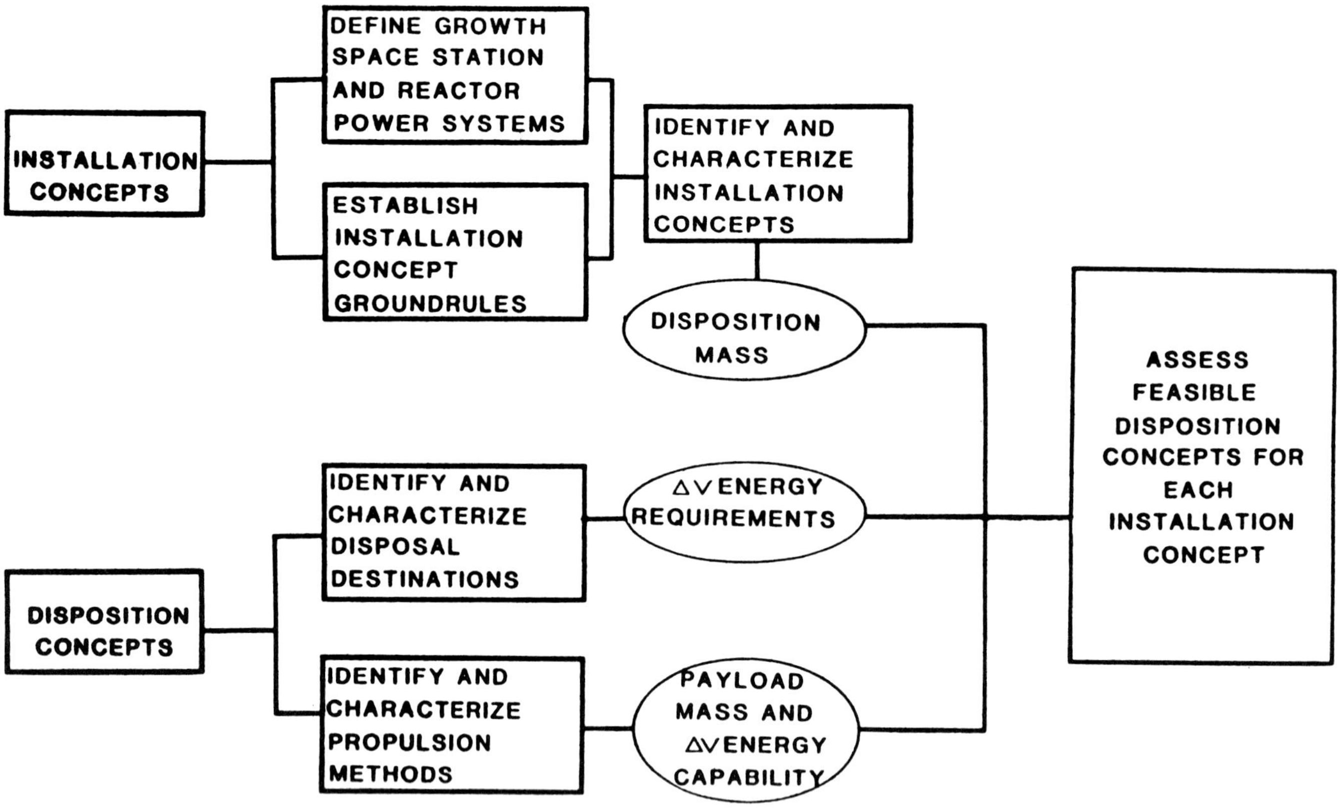

Figure 1 *Study Methodology to Assess System Installation and End-of-Life Disposition.*

Keel'' Space Station configuration, and major components assumed to exist prior to reactor power system installation, is shown schematically in Figure 2.

Three attractive installation concepts were identified for further characterization:

1. A single 300-kWe reactor power system using instrument-rated shadow shielding, mounted on a 2-km-long tether attached to the station upper payload boom;
2. A single 300-kWe reactor power system using shaped 4-pi man-rated shielding, mounted on a 60-m boom extension below the station lower payload boom; and
3. A dual-balanced 150-kWe reactor power system employing two operating power systems, each using shaped 4-pi man-rated shielding, mounted on 100-m boom extensions of the station crossbeam.

The reactor power systems proposed were based on the SP-100 Program Thermoelectric baseline (SiGeGaP) conversion concept. Power level, shielding, and boom configuration were modified as required for the different installation locations. The two SP-100 reactor power system configurations assumed for this study are shown in deployed form in Figure 3. Identification of disposal destinations was generally based on the application of the aerospace nuclear safety philosophy of ''delay and decay.'' This philosophy provides for sufficient isolation time to reduce radioactivity levels and, thus, meet radiation safety standards before exposure to the population. This can be accomplished by placing a shutdown reactor power system in a long-life orbit of more than 300 years. Alternatively, and currently favored, the reactor power system can be directed to solar or lunar destinations that provide a near zero probability of Earth reencounter and, hence, infinite isolation time by precluding exposure to the Earth's populace. In addition, all destinations, except solar impact and solar system escape, allow for future reactor retrieval, if desired.

Six disposal destinations were identified for further characterization. They include four non-Earth encounter destinations, one long-life orbit (1000 km) destination, and one Earth return destination. Although the Earth return destination did not meet the delay and decay safety philosophy, it was included for purposes of comparison. A variety of trajectory options has been characterized for all non-Earth encounter destinations, and a controlled reentry and Shuttle recovery option has been characterized for the Earth

Table 1 *Study Matrix showing Installation Concepts, End-of-Life Reactor Disposal Destinations, and Chemical Propulsion Methods Considered.*

Installation Concepts	Disposal Destinations	Chemical Propulsion Methods
1. Single 300-kWe Tether Mount	1. Long Life Orbit	1. Cryogenic Propellant
2. Single 300-kWe Boom Mount	-1000 km	2. Storable Propellant
3. Dual 150-kWe Balanced Boom Mount	2. Solar Orbit	
	-Earth escape	
	-Circular heliocentric	
	-Planetary swing-by	
	3. Solar System Escape	
	-Hyperbolic trajectory	
	-Jupiter swing-by	
	4. Solar Impact	
	-Direct	
	-Jupiter swing-by	
	5. Lunar Impact	
	6. Earth Return	
	-Controlled re-entry	
	-Shuttle recovery	

return destination. Identification of propulsion methods was based on consideration of near-term, state-of-the-art technologies. Two generic, high thrust chemical propulsion concepts were characterized to cover a range of payload mass and potential destination capabilities. The propulsion concepts included high performance cryogenic propellants and both liquid and solid storable propellants for a range of integrated and nonintegrated system options. Solar and nuclear powered electric propulsion vehicle concepts were not included in this study because of their relatively immature vehicle system technology. However, their significant increase in payload capability to high delta-V disposal destinations warrants further evaluation in future reactor power system disposal studies. Of particular interest are feasibility studies of disposal scenarios that use the Space Station reactor power system to provide electricity for an integrated onboard electric propulsion vehicle. This concept has the potential for simple, low cost disposal, but further study of systems integration issues is required.

Installation Concept Ground Rules

A set of ground rules was initially applied to all installation concepts to provide a consistent basis for comparison. These included:

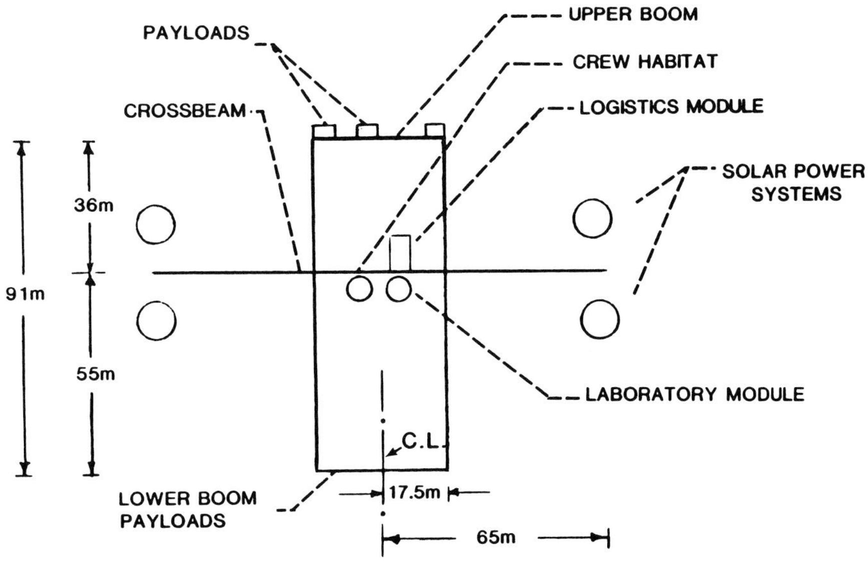

Figure 2 *Projected Dual Keel Space Station Configuration.*

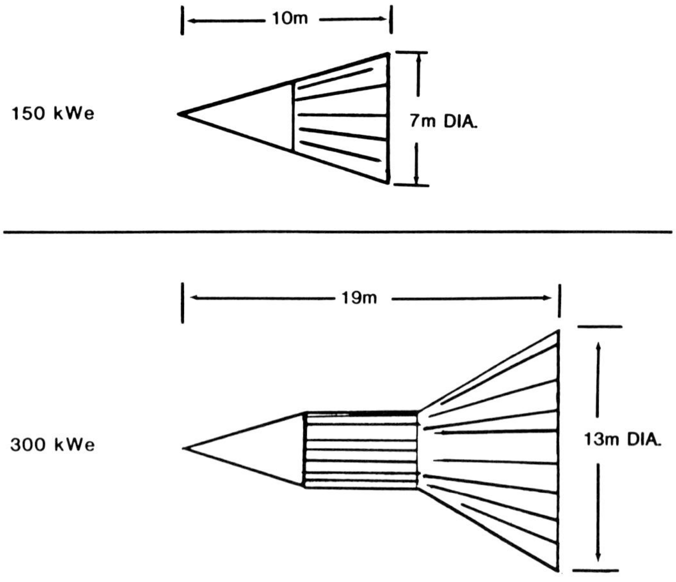

Figure 3 *SP-100 Nuclear Power System Reference Design Envelopes. (From General Electric 1984.)*

1. All concept configurations must maintain the existing Space Station center of gravity location while minimizing impact on station payloads and/or operations;
2. A uniform exposure dose rationale, to be used for all installation locations, shall be based on a maximum total biological radiation exposure dose of 20 rem per quarter year (to the eyes) for all normal Space Station activities;
3. Boom-mount installation concepts shall use man-rated shaped shield configurations employing a shadow cone to shield Space Station activity locations and a potentially removable spherical geometry portion for all other activity locations;
4. Boom-mount installation concepts shall use the 5-m-square booms with a lineal mass of 17 kg/m, and the tether concept shall use a 0.1-m-diameter, 2-km-long tether with a mass of 1400 kg;
5. Single 300-kWe reactor power systems will be based on an 8-MWt reactor power level with reactor dimensions of 88 cm long by 69 cm diameter; and
6. Dual 150-kWe reactor power systems will be based on a 4-MWt reactor power level with reactor dimensions of 74 cm long by 55 cm diameter.

As the study progressed, it became clear that further definition was required for the first two ground rules. Both 300-kWe concepts will require large ballast moments to offset their unbalanced power system locations and masses. A nonoperating spare reactor power system was provided as ballast mass when possible. Thus, a replacement reactor power system could already be onboard the Space Station, and replacement time would not be subject to Shuttle launch schedules.

In the case of the tether-mount concept, the location of the Space Station center of gravity is preserved by mounting a spare deployed nonoperating tether and reactor power system directly opposite, and in-line with, the operating system. However, for the boom-mount, single-reactor power system concept, an equal mass system was not possible because of potential interference with upper boom payload functions. Therefore, a longer moment arm, lower mass approach was used in combination with a spare, undeployed reactor power system located on the Space Station.

The second ground rule, meeting a 20 rem per quarter biological exposure dose rate for all normal Space Station activities, had a number of issues that needed resolution. First, a definition of normal activities was required. Then, for each activity it was necessary to determine location, time spent at location, existing in-place shielding, and finally, the natural background biological dose rate including attenuation by existing shielding. Existing shielding includes attenuation by Space Station and Shuttle skin (wall) thickness for on-station and in-shuttle locations, and astronaut space suit visor thickness for Extra Vehicular Activity (EVA) operations. Conservative dose estimates then were computed. For example, operations in the Space Station were assumed to account for more than 98 percent of each astronaut's time. This includes all activities within the confines of Space Station modules with local shielding of 2.4 g/cm^2 to reduce natural radiation levels to 7.1 mrem/h. EVA operations were subdivided into two distinct types occurring at different locations for the boom-mount installation concepts. Of the 34.2 total EVA hours allocated, 32.2 hours were assumed to occur within the confines of the man-rated shadow shield provided for each boom mount installation concept. From a radiation dose standpoint these are referred to as routine EVA operations since reactor-attributed dose rates are less than natural background levels. For the tether-mount installation concept, all EVA operations are in a low dose rate environment and, therefore, are considered routine.

Emergency EVA operations were defined to occur in the relatively high dose rate regions outside the shadow shield cone angle of the boom-mount installation concepts. A two-hour exposure duration was assigned to this activity, which was assumed to occur in the highest dose rate region 300 m from the Space Station center of gravity. Natural background radiation dose rates for all EVA operations were attenuated to 9.0 mrem/h by the assumed 0.9 g/cm^2 shielding thickness of the astronaut space suit visor. Shuttle orbital operations, including approach, docking, separation, and departure, also required further definition for dose calculations. A two-part scenario for approach and docking was based on calculational data supplied by NASA/Johnson Space Center (Anderson 1985). Shuttle flight profiles during the final 300 m purposely were assumed to traverse the highest dose rate regions for each boom-mount installation concept to provide conservatism in total exposure dose. For the tether concept, all final flight paths are equivalent from a radiation dose standpoint since natural background radiation is the major dose contributor.

The Shuttle fly-by activity was defined to represent, for example, a "wave-off" from an attempted rendezvous. A total of 0.4 hours was assigned to this activity, which is assumed to occur at the momentary Shuttle stop location of 300 m from the Space Station. For purposes of dose rate calculations, a circular fly-by maneuver at a 300-m radius around the Space Station was assumed equivalent to a stationary hold at 300 m.

300-kWe Tether-Mount Installation

Although the tether-mount installation concept relies on a relatively new method of power transmission in space, a recent conceptual design study (Bents 1985) provided the feasibility basis for its consideration. The tether-mount concept has the potential to provide both the lowest launch and disposal mass of all the installation concepts identified. This significant advantage is achieved by trading reactor power system shield mass for separation distance via a long tether.

The concept selected was based on a single, 300-kWe reactor power system attached to the Space Station by a 2-km-long by 0.1-m-diameter tether that is also an electrical transmission line. The 2-km separation distance selected allows the existing SP-100 instrument rated shield design to serve as a man-rated configuration for all locations near the Space Station. However, additional shielding was added for Shuttle approach and departure maneuvers outside of the shield cone. The basic SP-100 shield geometry was modified to include a tapered side shield, or wing. The wing provides a uniform 2.3-mrem/h dose rate along the Space Station orbital plane out to 300 m, which is sufficient to limit all Shuttle operation doses to negligible levels. A total modified shield mass of 2300 kg was estimated of which 800 kg is due to the additional wing. Total integrated dose for all normal activities is estimated at 20.37 rem.

Potential power system interference with payloads on the upper boom is due to their functional requirements. These payloads consist of experimental observatories that require an unobstructed view of deep space; a distant, operating reactor power system within their field of view may prove to be unacceptable. The final configuration selected for the 300-kWe tether-mount concept is shown schematically to scale in Figure 4, and the corresponding mass statement is given in Table 2. It includes the required ballast in the form of an equal and opposite deployed spare (non-operating) reactor system located at the end of a 2-km tether below the Space Station. This ballast configuration was selected over other potential ballast location within the confines of the Space Station framework from a simple mass penalty rationale. Locating the required ballast moment arm of 2.1×10^7 kg·m at the furthest point below the Station's center of gravity on the lower payload boom would have required an unacceptable ballast mass of about 4×10^5 kg. The selected ballast location should not pose any interference penalties for communication payloads located on the lower boom.

300-kWe Boom-Mount Installation

The boom-mount installation concept was identified for consideration because it preserves the Space

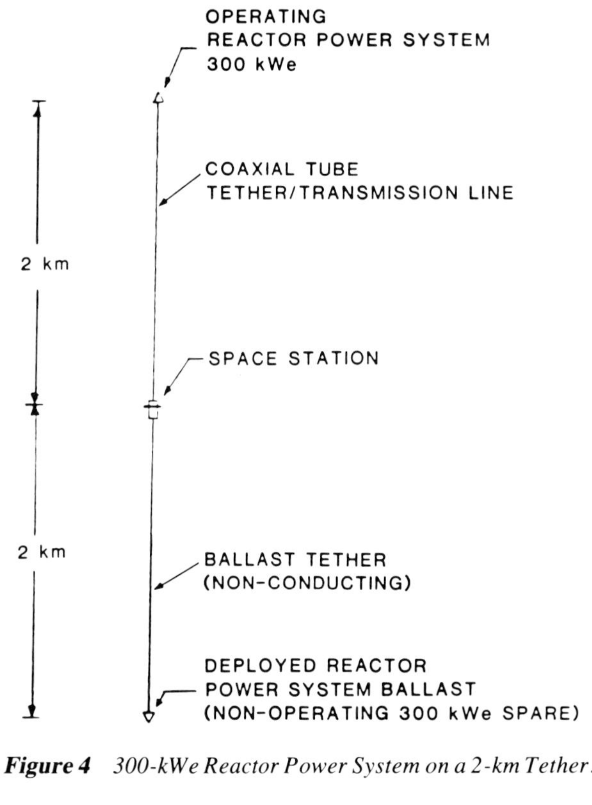

Figure 4 *300-kWe Reactor Power System on a 2-km Tether.*

Table 2 *Estimated Reactor Power System Mass for the 300-kWe Tether Configuration, including a Spare System for Ballast.*

Item	Mass (kg)
300-kWe Reactor Power System without Shield	6,800
Modified Shadow Shield	2,300
2-km Tether	1,400
Total Disposition Mass	10,500
Ballast	
Spare Reactor Power System	9,100
Attached Tether	1,400
Attached and Ballast Subtotal	10,500
Total Launch Mass	21,000

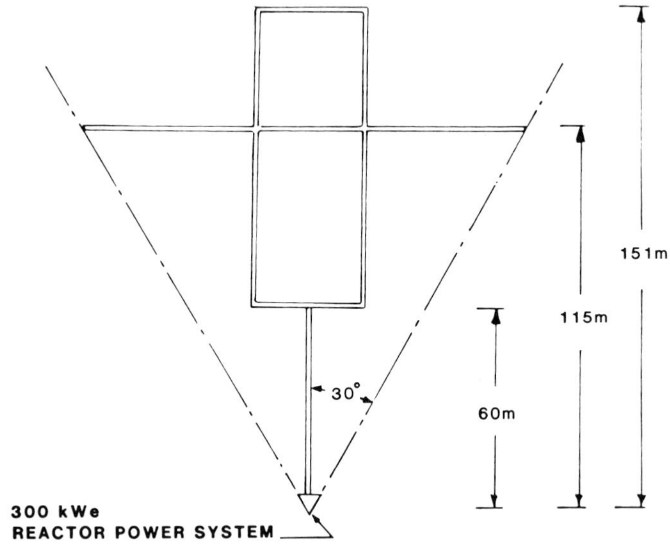

Figure 5 *300-kWe Reactor Power System on a 60-m Boom extended below the Space Station Lower Beam, showing 30° Half-angle Shield Geometry.*

Station's vertical gravity gradient with a single 300-kWe reactor power system attached to a conventional boom. The power system is located on a 60-m boom extending downward from the station's lower payload boom and is line-oriented along the local gravity gradient for minimum interference. The shielding configuration for this concept is based on a man-rated conical/spherical-shaped shield design. It provides a reactor attributable dose rate of 1.5 mrem/h at 115 m inside the shield shadow at the habitat location, and a dose rate of 200 mrem/h at 160 m outside the shield shadow. The reactor power system-Space Station geometry relationship is shown to scale in Figure 5. The shield geometry consisted of a combined 30-degree, half-angle cone shadow shield and a 330-degree removable spherical shield.

Two installation issues for this concept have been identified. The first is the location of ballast mass required to maintain the existing location of the Space Station center of gravity, and the second is potential power system interference with lower boom payloads.

A spare undeployed reactor power system is located under the upper payload boom within the Space Station framework, and a second smaller ballast mass is located above the station at the end of a long tether. A 1-km tether length was selected for the additional ballast location. The proposed use of multiple, thin monofilament line tethers and the low volume (about 2.5 m^3 for water) of the additional ballast should not interfere with upper boom payload functions. The second issue, that of potential interference with lower boom payload functions, cannot be resolved at this time. The final configuration for the 300-kWe boom-mount concept is shown to scale in Figure 6, and the corresponding mass statement is given in Table 3.

Dual 150-kWe Balanced Boom-Mount Installation

The dual 150-kWe balanced boom-mount installation concept differs from the previous single reactor concepts by employing two operating power systems, thereby providing an added measure of reliability. In addition, locating each reactor power system on a 100-m extension of the Space Station crossbeam preserves the Station's center of gravity without the use of ballast or interference with operations. The total reactor-to-habitat separation distance of 165 m required a 20-degree shield cone half-angle to shadow the entire Space Station. The installation geometry and dimensions used for this concept are shown in Figure 7. Dose rates inside the shadow shields of both reactor power systems were specified to provide a total dose rate of 1.5 mrem/h at the intersection of the Space Station crossbeam and vertical centerline, the same specification used for the single 300-kWe boom-mount concept. A dose rate of 115 mrem/h at 160 m was selected outside the shadow shield of each reactor power system to limit total integrated dose to 20 rem per quarter. The final configuration selected for the 150-kWe dual-boom concept is shown to scale in Figure 7. The corresponding mass statement is given in Table 4. A significant feature of this concept is the capability of splitting both the launch and disposition masses into two separate packages, each consisting of a complete 150-kWe reactor power system.

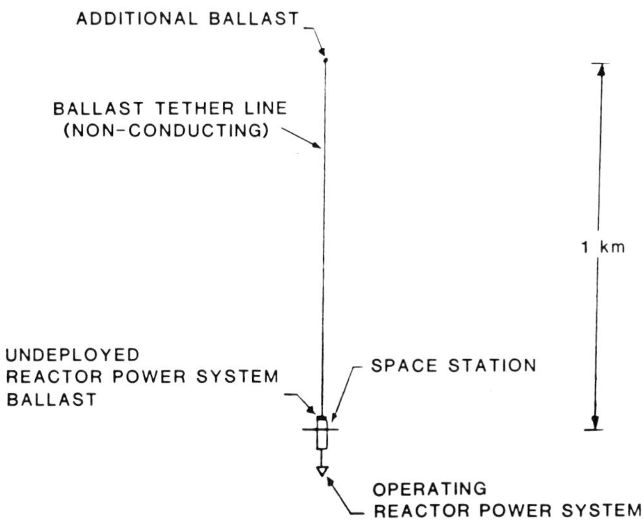

Figure 6 *300-kWe Reactor Power System on a 60-m Boom extended below the Space Station Lower Beam, Location of Ballast to Preserve Station's Original Microgravity Distribution.*

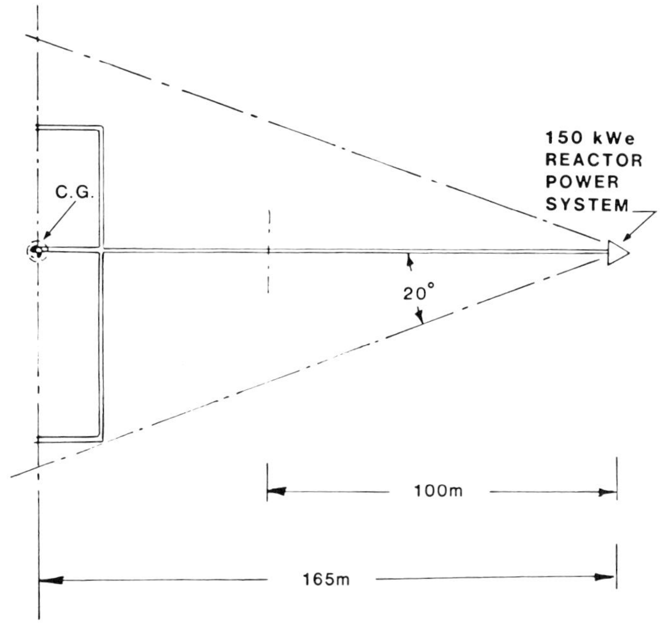

Figure 7 *Dual 150-kWe Reactor Power Systems Located at Each End of the Space Station's Extended Boom, and Geometry of the Shadow Shield.*

Disposal Destinations

The disposal destinations described herein have been selected to provide characterization of a representative group of potential locations; they do not represent the results of a selection process to identify preferred locations. With the exception of the Earth return cases, all destinations studied are acceptable from the generally accepted space nuclear safety philosophy espoused by the U.S. Department of Energy (to be published). The six destinations identified in Table 1 were characterized from an energy (delta-V) and operational requirements viewpoint. The advantages and disadvantages of each destination are described and a summary comparison is provided. The operational characterization of each destination included the number of burns, time interval between burns, specific launch time and guidance requirements, and retrieval availability. A detailed description of the orbital mechanics involved in disposing nuclear waste is given by Thompson et al. (1974). Retrieval availability may be of potential interest from a reactor power system, post-operation degradation evaluation aspect. All other items are of potential interest from a propulsion system operational reliability aspect.

Long-Life Earth Orbit

An Earth-orbit altitude of 1000 km has been selected to conservatively satisfy the requirement of a

Table 3 *Estimated Reactor Power System Mass for the 300-kWe Extended Boom Configuration, including Ballast Provisions.*

Item	Mass (kg)
300-kWe Reactor Power System without Shield	6,800
Shaped Shield	25,400
Total Disposition Mass	32,200
Attachment Boom	1,020
Ballast	
Spare Reactor Power System Ballast	32,200
Attachment Boom	1,020
Additional Mass @ 1 km	2,490
Attachment and Ballast Subtotal	36,730
Total Launch Mass	68,930

Table 4 *Estimated Reactor Power System Mass for the Dual 150-kWe Lateral Boom Configuration, maintaining Original Microgravity Distribution.*

Item	Mass (kg)
Two 150-kWe Reactor Power Systems without Shield	8,000
Two Shaped Shields	33,900
Total Disposition Mass[a]	41,900
Two Attachment Booms	3,400
Ballast	0
Attachment Subtotal	3,400
Total Launch Mass[b]	45,300

[a]May be disposed of in two separate packages of 20,950 kg each.
[b]May be launched in two separate packages of 22,650 kg each.

greater than 300-year orbital lifetime. To achieve an increase in circular orbital altitude from 500 to 1000 km, two propulsion maneuvers are required. After the reactor power system payload has been detached from the Space Station, the propulsion system imparts a delta-V of 130 m/s to the payload, placing it in an elliptical transfer orbit. After about a 1-hour coast time up to the 1000-km altitude, a second burn is initiated with a delta-V of 130 m/s to circularize the orbit. In the event of a second burn propulsion failure, the resulting elliptical orbit would have an orbital decay lifetime exceeding 3 years, which should be sufficient to permit corrective action. The advantages of employing the high Earth parking orbit are

- A very low delta-V requirement of 260 m/s,
- Insertion can be initiated at any calendar time,
- A variety of propulsive methods can be used,
- A simple, onboard event timer constitutes all of the guidance and control required, and
- The reactor is available for future retrieval, if desired.

The only disadvantage identified for this destination is the requirement for two propulsion burns to insert and circularize the orbit.

Solar Orbits

Solar orbits identified for characterization are

- Elliptical Earth escape,
- Circular heliocentric orbits at greater and fewer than 1.0 Astronomical Units (AU) (mean distance Earth to Sun about 150,000,000 km), and
- Planetary swing-bys.

These orbits retain the disposed payload in the solar system for potential future retrieval while providing essentially infinite delay and decay time.

Earth Escape

This highly elliptical orbit around the Sun can be accomplished by a single burn that provides a delta-V of about 3200 m/s. The advantages of this solar orbit approach are

- A moderate delta-V requirement of 3200 m/s,
- Insertion can be initiated at any calendar time,
- Single burn propulsion,
- No spacecraft guidance required, and
- Reactor available for future retrieval, if desired.

Circular Heliocentric Orbits

To provide a clear, positive separation from the Earth's orbit around the Sun, the reactor power system could be placed in essentially circular orbits of greater and fewer than 1.0 AU. For this study the 1.10 AU and 0.90 AU orbits were selected to provide examples of the representative delta-Vs required for circular orbits that do not cross the Earth's actual elliptical heliocentric orbit. For each orbit, two burns are required for a total delta-V of about 4500 m/s with a coast period of about 16 months between burns. The advantages of this solar orbit approach include initiation at any time, power system available for future retrieval if desired, and simple onboard event timer required for guidance.

Disadvantages include the requirement for two propulsive burns and a moderately high delta-V of 4500 m/s.

Planetary Swing-bys

Destinations in highly elliptical orbits that do not intersect Earth's orbit were also investigated. These consist of orbits achieved by using the gravitational attraction of planets such as Mars and Venus. However, the total delta-V required for each case is about the same as that required for the circular heliocentric orbits previously discussed, 4500 m/s. A single initial burn is required; a second burn is eliminated by the use of the respective planetary gravitational attraction to complete the maneuver. In addition, the swing-by cases will require up to a 2-year time interval between optimal initiation of reactor disposal operations to take advantage of the relative location of the Earth with respect to these planets to minimize trip time. The advantages of this solar orbit approach are the same as those for circular heliocentric orbits, except it requires only a single burn. Disadvantages include the same moderately high delta-V requirement of 4500 m/s and up to a 2-year interval between separation and launch.

Solar System Escape

Several solar system escape scenarios were examined. These destinations provide a permanent, nonretrievable disposal location.

Direct

The direct solar system escape can be achieved with only a single propulsive burn. The reactor power system payload is placed in a hyperbolic trajectory in the plane of the ecliptic, and can be launched at any calendar time with no guidance required. The major disadvantage of this permanent disposal scenario is that it requires a large delta-V of about 8760 m/s, which severely limits the choice of a propulsion scheme.

Jupiter Swing-by

A lower energy variation of the direct escape destination can be employed by guiding the package into a trajectory approaching, and then swinging by, Jupiter. This reduces the total required delta-V to 7600 m/s, but requires a mid-course correction burn maneuver to achieve the desired proximity behind Jupiter. Precise launch times to Jupiter are also required and are limited to one specific time per year because of the relative positions of Earth and Jupiter.

Solar Impact

An additional permanent, nonretrievable destination is the Sun, itself. Two variations of solar impact were investigated.

Direct

To achieve direct solar impact, a single burn delta-*V* equal and opposite to that of the Earth's mean orbital velocity of 29 800 m/s is required. This extremely high energy requirement is a major disadvantage; however, the launch can occur at any time and no guidance is required.

Jupiter Swing-by

With the added complexity of spacecraft guidance and mid-course correction, solar impact can be achieved with a greatly reduced delta-V of 7600 m/s. In this concept, the propulsion vehicle is guided to pass near the "front" of Jupiter. Thus, Jupiter's orbital velocity nearly cancels that of the vehicle resulting in a near-zero velocity with respect to the Sun; and, subsequently, the payload impacts the Sun. This method, however, is limited to about a once-a-year launch opportunity when the Earth and Jupiter are in optimum positions relative to each other.

Lunar Landing

A potentially useful destination for the reactor power system would be a remote location on the Moon. In this approach, the payload would be soft-landed at the desired site. A delta-V of 4000 m/s would be required to acquire a lunar orbit, and then a second delta-V of about 1800 m/s would be required for a precision soft landing on the Moon. An onboard event timer would provide sufficient guidance accuracy to place the reactor power system within the desired area.

One potential advantage of this destination is the possibility of reusing portions of the intact reactor power system for future lunar missions. At present, there does not appear to be any policy prohibiting the deposit of nuclear (fissile) material on the Moon; however, an examination of potential accident modes and consequences is needed to assess this option.

Earth Return

In the event that it becomes necessary to return the reactor power system to Earth for scientific, economic, or political reasons, several hypothetical return concepts were examined. These concepts do not maintain the aerospace nuclear safety philosophy of delay and decay; however, they have been evaluated for reasons of general interest and study comparison. Two cases were considered:

- Intact controlled reentry of the reactor power system landing at a preselected remote landing area, and
- A conceptual scenario of a return of the (stowed configuration) reactor power system in the Shuttle.

Controlled Reentry

This Earth destination option requires reentry with a single-burn, expendable propulsion system that includes an onboard event timer guidance mechanism. With the very low delta-V requirement of about 150 m/s, the package is easily returned to Earth. The preselected impact area could be over deep ocean or a remote Free-World land location where final retrieval could be accomplished with parachutes. The transit time to fully de-orbit is not critical, and the fuel requirements of the propulsion vehicle are minimal. However, an examination of potential accident modes and their consequences from a nuclear safeguards aspect is required to assess this option.

Shuttle Return

A second method of achieving a controlled safe Earth destination is through the use of the Shuttle. In this method, the shutdown reactor power system must be disassembled and reconfigured for stowage in the Shuttle payload bay. This appears to be a major issue requiring consideration of at least the following:

- Disassembling the reactor power system to meet Shuttle mass limits,
- Refolding or removing radiator sections,
- Locating the reactor and shield to provide Shuttle crew radiation protection,
- Locating the reactor power system and possible ballast to meet Shuttle center-of-gravity requirements, and

- Removing reactor decay heat in the Shuttle bay.

Overriding these major issues is the question of Shuttle crew safety. For example, failure of the decay heat removal system or a hard landing impact could result in release of liquid metal into the Shuttle bay. Based on these concerns, the Shuttle return option does not appear to be either viable or desirable, and should remain hypothetical.

Summary of Disposal Destinations

The results of the characterization of all of the disposal destinations are shown in Table 5, which summarizes the energy requirements and operational characteristics of each destination. In terms of energy requirements alone, both the long life orbit and Earth return destinations are clearly superior. However, if the operational advantages of a single burn or maneuver are of interest, the Earth escape, solar orbit destination is attractive. It combines moderate energy requirements with a single burn capability, and no requirement for launch time or guidance. Another criterion of potential interest is future retrieval availability. All destinations except solar impact and solar system escape have this capability.

Chemical Propulsion Capability

The assessment of chemical propulsion capability is summarized in terms of requirements and recommendations for each disposal destination. Integration concept issues are shown in Table 6. Performance for a storable propellant (I_{sp} = 350 s) is shown in Figure 8 as a function of propellant mass versus payload mass for a number of disposal destinations. Figure 8 is based on selected solutions of the following equation for the characteristic impulsive velocity increment, delta − V, as a function of payload mass fraction:

$$\text{delta} - V = 9.815\, I_{sp} \,\text{Ln}\,(1 + MF\,(W_L/W_T - 1) \quad (1)$$

where

delta − V = velocity increment, m/s,
I_{sp} = specific impulse, s,
MF = propellant mass fraction,
W_L = reactor power system disposal (payload) mass, kg,
W_T = total initial vehicle mass, kg,

and

$$W_T = W_L + W_P + W_H \quad (2)$$

$$MF = W_P/(W_P + W_H), \text{ assumed to be } 0.85 \quad (3)$$

Table 5 *A Summary of Disposal Destinations showing Energy Requirements and Operational Characteristics for Each Destination.*

Destination	No. of Burns or Manuevers Required	Propulsion System Δ V Requirements (m/s) First Burn	Second Burn	Total	Time Interval Between Burns (h)	Launch Time Requirements	Requirement for Guidance	Retrieval Availability
Long Life Earth Orbit								
- 1000 km	2	130	130	260	1	None	On-Board Event Timer	Yes
Solar Orbits								
- Earth Escape	1	3200	—	3200	—	None	None	Yes
- Circular Heliocentric	2	TBD	TBD	4500	11500	None	On-Board Event Timer	Yes
- Planetary Swing-By	1	4500	—	4500	—	Once Every 2 yrs For Alignment	On-Board Event Timer	Yes
Solar System Escape								
- Direct	1	8760	—	8760	—	None	None	No
- Jupiter Swing-By	1	7600	—	7600	—	Once Every Year For Alignment	On-Board Event Timer	No
Solar Impact								
- Direct	1	29800	—	29800	—	None	None	No
- Jupiter Swing-By	1	7600	—	7600	—	Once Every Year For Alignment	None	No
Lunar Landing	2	4000	1800	5800	Several	None	On-Board Event Timer	Yes
Earth Return								
- Controlled Re-entry	1	150	—	150	None	None	On-Board Event Timer	Yes
- Shuttle Return	—	—	—	—	—	None	—	Yes

Table 6 *A Summary of Chemical Propulsion Disposal Scenarios in terms of Requirements for Each Destination and Concept Integration Issues.*

Concept Integration Level	Propellant Requirements	Payload/Propulsion Elements: Integrated In Reactor Shutdown Radiation Environment	Payload/Propulsion Elements: Integrated In Space	Payload/Propulsion Elements: Exposed To Long-Term Reactor Radiation	Payload/Propulsion Elements: Exposed To Long-Term Space Environment	Capability For High Energy Destinations
Full Integration	Storable Only	None	Propellant And Upper Stage To Payload	Propellant And Upper Stage	Propellant And Upper Stage	No
Partial Integration	Storable Or Cryogenic	Propellant To Upper Stage And Payload	(1) Upper Stage To Payload (2) Propellant To Upper Stage And Payload	Upper Stage	Upper Stage	Yes
Not Integrated	A Storable Only	Propellant And Upper Stage To Payload	Propellant And Upper Stage To Payload	None	Propellant And Upper Stage	No
	B Storable Only	Propellant And Upper Stage To Payload	(1) Propellant To Upper Stage (2) Propellant And Upper Stage To Payload	None	Propellant And Upper Stage	No
	C Storable Or Cryogenic	Propellant And Upper Stage To Payload	Propellant And Upper Stage To Payload	None	None	Yes-For Tether Concept Only
	D Storable Or Cryogenic	Propellant And Upper Stage To Payload	(1) Propellant To Upper Stage (2) Propellant And Upper Stage To Payload	None	None	Yes

where

W_P = propellant mass, kg,

and

W_H = vehicle hardware or burnout mass, kg.

For the low energy, 1000-km Earth orbit destination, the use of a Shuttle-compatible upper stage, such as the solid rocket motor Payload Assist Module (PAM) series, is recommended. Two methods of propulsion system/payload integration are equally feasible. The fully integrated concept provides a single Shuttle launch of the integrated propellant/upper stage to low Earth orbit, where it is attached in a single operation to the reactor power system payload in a radiation-free environment prior to reactor startup. The major integration issue identified for the concept is the unknown effect of long-term exposure of the propellant and the upper stage to the operating reactor radiation and the space environment. An alternate method of propulsion system/payload integration is Option C of the nonintegrated concept. This option also provides single Shuttle launch of the integrated propellant/upper stage to low Earth orbit, where it is integrated with the reactor power system payload after reactor shutdown. This eliminates the long-term exposure issue of the fully integrated method but requires propulsion system/payload integration in the radiation environment of the shutdown reactor. As shown, this requirement is not considered a major feasibility issue for EVA radiation exposure.

For higher energy destinations, chemical propulsion capability is limited only by the number of Shuttle launches required to place the propulsion system into low Earth orbit. From a cost standpoint, the combination of tether-mount installation and Earth escape destination is clearly the most attractive because the propulsion system can be placed in low Earth orbit with a single launch of a Shuttle-compatible upper stage vehicle (nonintegrated concept Option C). However, all installation concept payloads can reach Earth escape destination with the partially integrated and nonintegrated Option D concepts. As shown in

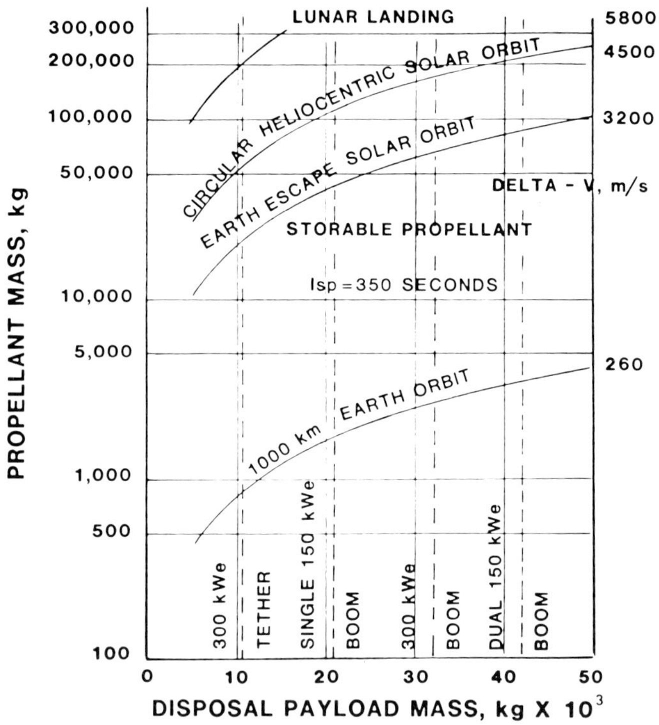

Figure 8 *Performance of a Storable Propellant ($I_{sp} = 350$) as a function of Propellant Mass versus Payload Mass for a Number of Disposal Destinations. The Disposal Masses of the Principal Reactor Configurations are shown along the Abscissa.*

Table 6, comparison issues for these concepts reveal that both concepts require one integration operation in the shutdown reactor radiation environment and two in-space integration operations. The characteristic that distinguishes the concepts is the difference in the type of propulsion components to be integrated in the shutdown radiation environment. The partially integrated concept requires only propellant integration, while the nonintegrated concept requires integration of both the propellant and the upper stage. From a component mass aspect, both are approximately equivalent since propellant is about 85 percent of the total propulsion vehicle mass. However, from the operationally more important aspect of component volume, the nonintegrated concept is less attractive because it includes the additional volume of the upper stage. In addition, we defined a propellant integration option for the partially integrated concept, which was based on transferring propellant by pumping it into empty propellant tanks that had been integrated with the upper stage and payload prior to reactor startup. This option is particularly attractive from an operational standpoint because the integration requires only the attachment of fluid couplings. Furthermore, from a structural standpoint, it is advisable to hard mount the propulsion system and payload inside the framework of the Space Station 5-m square boom. This could easily be accomplished before reactor startup when the payload, upper stage and empty propellant tanks are integrated. Therefore, it is concluded that the partially integrated concept, using pumpable liquid propellant, is the most attractive scenario for high energy destinations.

The liquid propellant to be used can be either storable or cryogenic. The only criterion for selection examined in this study was that of propellant mass. The higher specific impulse of cryogenic propellant yields a significant mass savings of about 35 percent for Earth escape destination. However, other selection criteria may be equally important. For example, the potential liquid propellants include cryogenic hydrogen/oxygen, storable bipropellant NTO/MMH (nitrogen tetroxide/monomethylhydrazine), storable monopropellant hydrazine (N_2H_4) and storable water electrolyzed to gaseous hydrogen/oxygen. The use of water in Space Station operations may have particular significance in propellant selection. The state-of-the-art of high specific impulse (370 s) water electrolysis propulsion is proven in low thrust rocket engines, and electrolysis for space and terrestrial life support applications is well developed. In addition, the nontoxicity and high density properties of water make it an excellent candidate for mass makeup on any volume-limited Shuttle payloads required for Space Station operations. Therefore, the issue of propellant selection for high energy destinations should remain unresolved until further definition of water availability and synergism with Space Station operations is determined.

Study Conclusions

The results of this study have confirmed the feasibility of installation and disposal of the three nuclear reactor power system concepts assessed for Space Station application. From the viewpoint of launch and disposal mass, and therefore operational cost, the tether-mount is the preferred concept. However, in terms of potential Space Station interference issues and required technology development the boom-mount concepts are more attractive, with the 150-kWe dual-boom concept preferred over the 300-kWe single-boom installation.

All three concepts were found to be easily boosted to a safe 1000-km Earth orbit with current and near-term, chemical propulsion Shuttle-compatible upper-stages. Should a higher energy destination be required, Earth-escape elliptical solar orbit is preferred. Propellant mass requirements for this destination strongly favor the tether-mount concept, which needs only one-third to one-fourth the propellant for the single- and dual-boom mount concepts, respectively. Although the mass advantage of the tether-mount installation permits boost to Earth escape with a single-launch, fully integrated Shuttle/Centaur G-Prime upper stage, this propulsion concept was less attractive than the partially integrated propulsion scenario from an operational propellant transfer viewpoint. Furthermore, although an additional 35 percent reduction in propellant mass requirements for Earth escape can be realized by use of cryogenic propellant, the synergistic operational benefits associated with the use of water in Space Station operations may significantly affect this selection. In general, no insurmountable technical barriers to feasibility have been identified and only modest technology advancements and/or requirements definition are required to verify concepts and resolve open issues.

Acknowledgments

The study effort reported here was performed at the Lewis Research Center of the National Aeronautics and Space Administration. Assistance and contributions by Lewis Research Center personnel are gratefully acknowledged, including members of the Space Station Systems Directorate.

References

Anderson, C. (1985) *Personal Communication*, NASA Lyndon B. Johnson Space Center, Houston, TX.

Bents, D. J. (1985) *Tethered Nuclear Power for Space Station*, NASA TM 87023, NASA Lewis Research Center, Cleveland, OH.

General Electric (1984) *SP-100 Program: Ground Engineering System (GES) Baseline Design Study*, Report No. 84DS4272, General Electric Company, Philadelphia, PA, November 1984

Silverman, S. W., H. J. Willenberg, and C. Robertson (1984) *Applicability of 100 kWe-Class of Space Reactor Power Systems to NASA Manned Space Station Missions*, Final Report, NASA CR-174696, Report prepared for the National Aeronautics and Space Administration by Boeing Aerospace Company, Seattle, WA.

Thompson, R. L. et al. (1974) *Study of Extra-terrestrial Disposal of Radioactive Wastes*, NASA TMX-71557, NASA Lewis Research Center, Cleveland, OH.

U.S. DOE (to be published) *Nuclear Safety Guidelines for the Use of Fission Reactors in Space*, to be published by the U.S. Department of Energy, Washington, DC.

Part 4

REACTORS AND SHIELDING

Space Nuclear Power Systems 1986
Edited by M. S. El-Genk and M. D. Hoover

Chapter 9

Neutronics Characteristics of Space Power Reactors

Winston W. Little and John O. Barner

Pacific Northwest Laboratory
P. O. Box 999
Richland, WA 99352

This report presents a brief description of the neutronics characteristics of selected conceptual designs for space reactors. Parametric studies also are included to delineate the effects of possible design changes. Although a wide range of power densities and sizes is considered, the focus is on reactors that provide roughly 1 Mwt to 50 Mwt. Both ceramic (UO_2 and UN in clad pins) fueled reactors and cermet (Mo, W, and Ta matrix) fueled reactors are examined. Although ceramic fuels have become the standard for power reactors, cermet fuels exhibit several desirable attributes for space applications. These attributes include greater structural strength and a more reliable reactivity expansion coefficient.

Introduction

The general objective of this report is to describe the key neutronics characteristics of compact, fast reactors for space applications. A more specific objective is to compare the characteristics of ceramic fueled space reactors with cermet (ceramic dispersed in a metal matrix) fueled space reactors. All neutronics calculations are based on the Evaluated Nuclear Data File Version V (ENDF/V), as processed by the Los Alamos cross section codes. Effective multigroup cross sections are generated using the Battelle 1DX code (Hardie and Little 1969).

This report is divided into the following six topics, which will be discussed in sequence:

- Critical Mass,
- Alternative Fissile Fuels,
- Reflector Worth and Control Considerations,
- Reactivity Coefficients,
- Reactivity Lifetime, and
- Calculational Model.

Critical Mass

The critical mass for reactors containing ceramic and cermet fuel materials was computed as a function of core volume. The fuel materials selected for consideration are those generally regarded as appropriate for long life operation at elevated temperatures. They include

- Ceramics: UO_2 and UN, and
- Cermets: Mo, W, and Ta (with UN as the embedded fuel).

For the ceramics, the fuel form was assumed to be cylindrical pins with Nb clad. The core composition was assumed to be 25 v/o coolant, 25 v/o clad, 40 v/o fuel, and 10 v/o void (this could alternatively be considered as 50 v/o fuel at 80% theoretical density to accommodate swelling).

For the cermets, the fuel form was assumed to be a solid matrix containing coolant channels and embedded ceramic (UN) fuel. The cermet reactor composition was assumed to be 25 v/o coolant, 65 v/o cermet (matrix plus embedded ceramic fuel), and 10 v/o void.

The ^{235}U critical mass as a function of core volume is shown in Figure 1 for the UO_2 and UN ceramic fueled reactors previously discussed. All reactors were assumed to be cooled by natural lithium, and the reflector was assumed to be Be with a thickness of 10 cm and a volume fraction of 0.80. As the core volume was varied, criticality was maintained in the ceramic fueled reactors by adjusting the uranium enrichment, with an upper enrichment limit of 93% ^{235}U. The marks overlaid on the curves indicate specific enrichments, beginning with 93% at the lower left and regressing to 75%, 50%, 40%, and 35%.

Figure 2 gives the ^{235}U critical mass as a function of core volume for the Mo, W, and Ta cermet fueled reactors. The volume fraction of fuel in the cermet was adjusted to maintain criticality, and the enrichment was held at 93%. In this case, the marks overlaid on

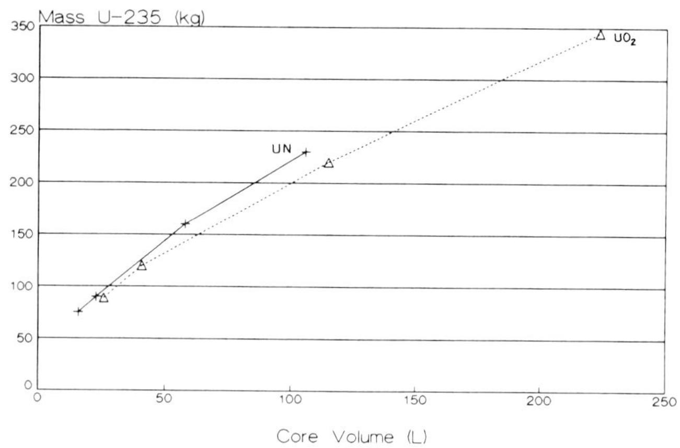

Figure 1 Critical Mass in Ceramic Reactors.

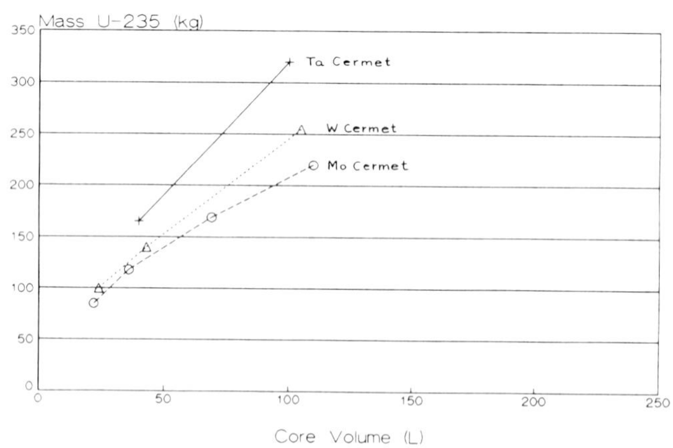

Figure 2 Critical Mass for Cermet Fueled Reactors.

the curves indicate specific ceramic volume fractions in the cermet itself, beginning with 50% at the lower left and regressing to 40%, 30%, and 25%.

As shown, the UN fueled reactors have a smaller minimum critical mass than the UO_2 fueled reactors due to the UN's higher density (14.3 versus 10.96 g/cm^3). However, for a given volume, the critical mass of a UN fueled reactor is larger due to the relatively high absorption cross section of N.

For the cermet reactors, Mo is seen to offer the lowest critical mass, followed by W and Ta. Indeed, Ta is a rather strong poison, and the effect of its high absorption cross section becomes increasingly pronounced in the larger, softer spectrum reactors. Table 1 provides a breakdown of total mass by component for a 35-liter Mo cermet reactor.

Alternative Fissile Fuels

For small to moderate powers, space reactor size is often determined by criticality limitations. For these circumstances, use of a more reactive isotope, either ^{233}U or ^{239}Pu, could be considered as an appropriate replacement for ^{235}U. Figure 3 shows the effect of using either ^{233}U or ^{239}Pu in place of ^{235}U in both the 93% enriched UN ceramic reactor and the 40 v/o Mo cermet reactor. As shown, both ^{233}U and ^{239}Pu offer savings in volume and critical mass by about a factor of three. Also, the cermet reactor fueled with either ^{233}U or ^{239}Pu has an appreciably lower critical mass than the ceramic reactor fueled with 93% ^{235}U.

Reflector Worth and Control Considerations

Small space reactors have high neutron leakage, which indicates that appreciable reactivity control can be achieved by varying the reflector configuration. Control systems can be designed to move reflector materials either vertically or horizontally away from the immediate vicinity of the core. Rotating rods, with poison on one side and reflector on the other, also can be designed, but often have less worth and add the complication of absorber cooling.

Table 1 Inventory of 35-Liter Mo Cermet Reactor.

Element	Mass (kg)
^{6}Li	0.29
^{7}Li	4.1
N	7.3
Mo	131.0
^{235}U	114.0
^{238}U	8.6
Be	120.0
TOTAL	385.3

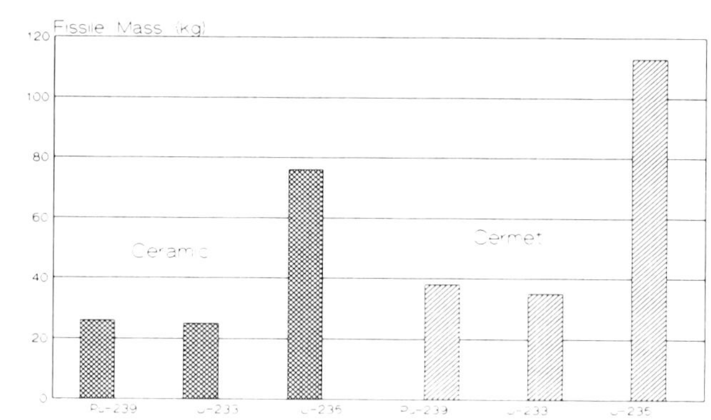

Figure 3 Effect of ^{233}U and ^{239}Pu on Critical Mass.

Rough estimates of the magnitude of reflector control available are given in Figures 4 and 5, where k_{eff} is a function of reflector (Be) volume fraction and thickness, respectively. The nominal reference reactor for these calculations is a 35-liter Mo cermet core surrounded with a 10-cm beryllium reflector at 80% volume fraction. Note that the reflector worth saturates significantly after about 10 cm to 15 cm.

Another less common technique for reactivity control—but one that is particularly suitable for space reactors—is to split the reactor core into halves (vertically), and move the halves apart to reduce reactivity. This technique is widely used in critical assemblies, but is inappropriate for large power reactors because of their low leakage and high fuel mass. However, this technique is attractive for high leakage, low mass space systems operating in a weightless environment. For ^{235}U-fueled reactors, the control magnitude is given by the approximation

$$dk = -.5\, dL/L \tag{1}$$

where dL is the gap width and L is the core height.

Reactivity Coefficients

The reactivity worths of selected elements were computed for a 35-liter Mo cermet reactor. The reactivity coefficients were computed by homogeneously adding a small amount to the core, and taking the difference in Keff. The results for the light elements and the heavy elements are shown in Figures 6 and 7, respectively.

The reactivity coefficients are positive for all elements shown except ^{10}B, ^{6}Li, and Ta. This indicates that the positive leakage reduction component dominates the negative absorption component.

The hydrogen worth in the ^{235}U-fueled reactor previously described was computed to be 0.188 dk/kg, which corresponds to an H_2O worth of about 0.021

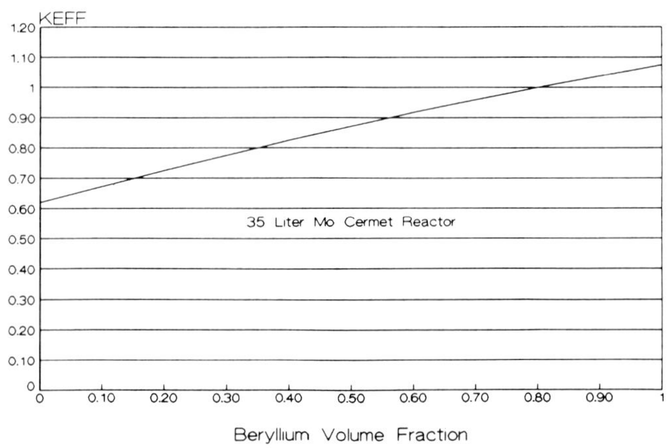

Figure 4 *Effect of Reflector Density on k_{eff}.*

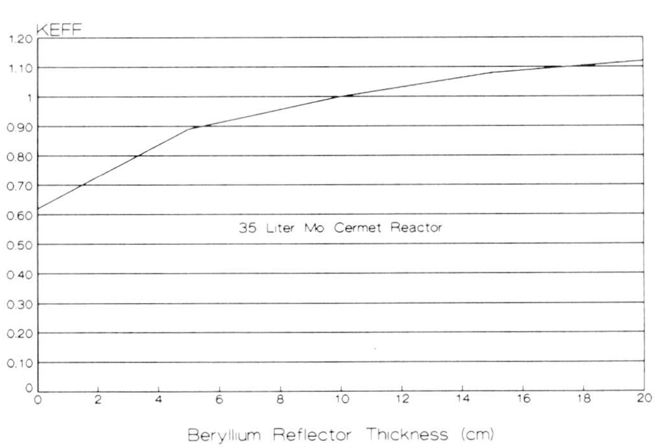

Figure 5 *Effect of Reflector Thickness on k_{eff}.*

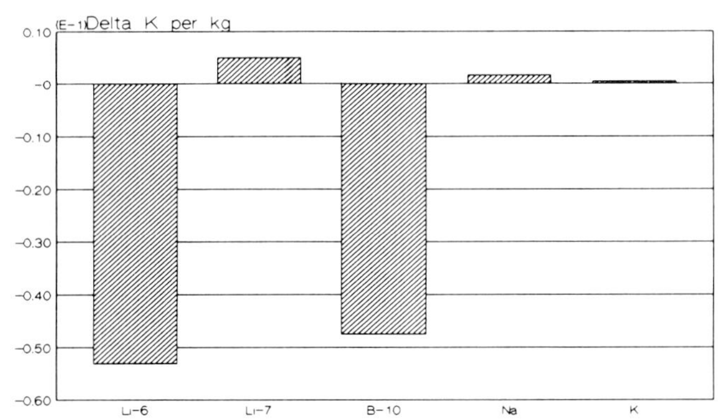

Figure 6 *Reactivity Coefficients for Light Elements.*

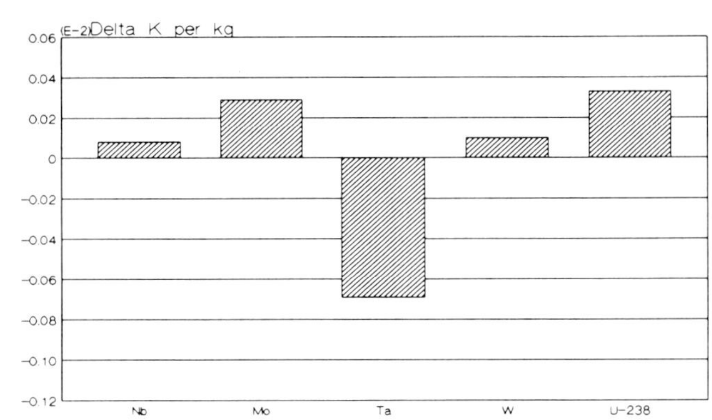

Figure 7 *Reactivity Coefficients for Heavy Elements.*

dk/kg. This coefficient is rather high, and could lead to concern over potential flooding accidents. It should be noted that one way to substantially reduce the hydrogen worth is to use ^{239}Pu rather than ^{235}U as the fissile fuel. For the same size Mo cermet reactor, with ^{239}Pu in place of ^{235}U, the hydrogen worth is reduced to 0.038 dk/kg, or about 0.004 dk/kg H_2O.

Finally, Table 2 presents the spectrum averaged absorption and fission cross sections, averaged over the core, for a more detailed understanding of the reactivity coefficients. These numerical values, together with appropriate perturbation expressions, can be used to estimate reactivity changes for other conditions.

Reactivity Lifetime

The operational lifetime of a space reactor can be limited by numerous factors. This section addresses the restrictions imposed by reactivity loss. The reactivity loss during operation of a fast spectrum space reactor can be approximated by the expression

$$dk = -C(1 - BR)\, t\, P / M \tag{2}$$

where:
C = Constant (about 0.25 for ^{235}U),
dk = Reactivity loss,
BR = Breeding ratio,
P = Thermal power (MWt),
M = Fissile mass (kg), and
t = Time (years).

From Equation (2), it can be seen that the reactivity loss can be reduced by increasing the breeding ratio (for example, by adding fertile fuel) or by increasing the fissile inventory. For a given reactor size, substitution of ^{239}Pu for ^{235}U would tend to reduce the reactivity loss during burnup by increasing the breeding ratio, but increase the reactivity loss by decreasing the critical mass. As shown in Table 3, and graphically in Figure 8, the net effect is that the ^{235}U reactor has the lowest reactivity loss during burnup.

Table 2 *Spectrum Averaged Cross Sections for a 35-Liter Mo Cermet Reactor.*

Element	Cross Section (Barns)
H, absorption	.13 − 3 (i.e., 0.00013)
Be	.11 − 1
N	.45 − 1
^{10}B	.17 + 1
^{6}Li	.80 + 0
^{7}Li	.24 − 4
K	.36 − 1
Na	.18 − 2
Nb	.10 + 0
Mo	.70 − 1
W	.14 + 0
Ta	.40 + 0
^{235}U	.19 + 1
^{238}U	.30 + 0
^{239}Pu	.21 + 1
^{235}U, fission	.16 + 1
^{238}U, fission	.94 − 1
^{239}Pu, fission	.18 + 1

Calculational Model

The results presented in this paper were based on a multigroup, one-dimensional (spherical), diffusion theory model. It is reasonable to ask what the effect would be of using a more sophisticated calculational model.

To this end, Keff calculations were performed using a one-dimensional transport (S8) model and a two-dimensional diffusion model (R-Z geometry, with L/D = 1). Figure 9 shows the results of these calculations for three Mo cermet reactors (corresponding to 50 v/o, 40 v/o, and 25 v/o UN in cermet). As shown, not only are the results from the more sophisticated models comparable to the results from the one-dimensional diffusion model (within about 2% dk), but the approximations tend to be canceling. Thus, it

Table 3 *Burnup Parameters for a 35-Liter Mo Cermet Reactor.*

Parameter	^{235}U Fuel	^{239}Pu Fuel
^{235}U (kg)	114	0
^{238}U (kg)	9	54
^{239}Pu (kg)	0	70
Breeding Ratio	0.009	0.09
dk/(50 MWd/kg)	−0.033	−0.06

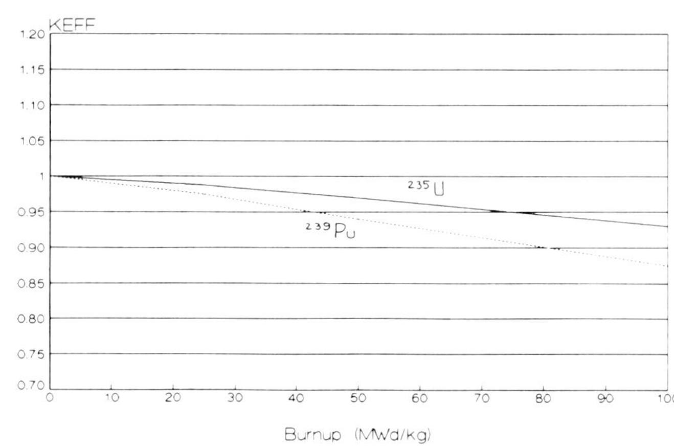

Figure 8 *Burnup in 35-Liter Mo Cermet Reactors.*

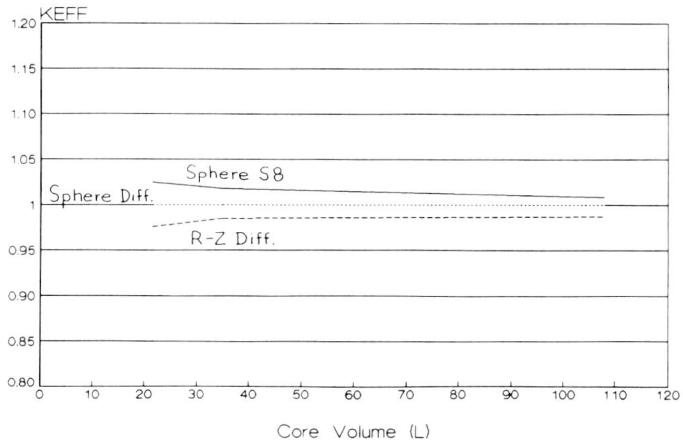

Figure 9 *Effect of Model on Eigenvalue Estimates.*

is concluded that the one-dimensional diffusion model is adequate for the survey calculations described in this paper.

Conclusions

There are no appreciable differences in neutronic characteristics between UN and UO_2 ceramic fueled space reactors. UN has a somewhat higher mass density, which permits a smaller critical mass.

The differences between ceramic and cermet fueled reactors, however, are significant. Ceramic fueled reactors can achieve a smaller minimum critical mass due to their high fuel volume fraction. And the larger parasitic absorption in W and Ta cermets reactors causes these reactors to have a substantially higher critical mass than comparable ceramic fueled reactors.

For either ceramic or cermet fueled reactors, reactivity can be varied by various neutron leakage control devices. Moving fuel would appear to be one of the most attractive techniques for leakage and reactivity control in space reactors.

Use of either ^{233}U or ^{239}Pu in place of ^{235}U has the potential of reducing both critical mass and core volume by about a factor of three. Use of ^{239}Pu can also appreciably reduce the effect of water flooding on reactivity.

Acknowledgments

This work was performed at the Pacific Northwest Laboratory under a Department of Energy contract.

References

Hardie, R. W. and W. W. Little (1969) *IDX, A One-Dimensional Diffusion Code for Generating Effective Nuclear Cross Sections*, BNWL-954, Pacific Northwest Laboratory, Richland, WA, March 1969.

Space Nuclear Power Systems 1986
Edited by M. S. El-Genk and M. D. Hoover

Chapter 10

Simulation of the Preliminary General Electric SP-100 Space Reactor Concept using the ATHENA Computer Code

C.D. Fletcher

Idaho National Engineering Laboratory
EG&G Idaho Inc.
P.O. Box 1625
Idaho Falls, ID 83415

The capability to perform thermal-hydraulic analyses of a space reactor using the Advanced Thermal-Hydraulic Energy Network Analyzer (ATHENA) computer code is demonstrated. The fast reactor, liquid-lithium coolant loops, and lithium-filled heat pipes of the preliminary General Electric SP-100 design were modeled with ATHENA. Two demonstration transient calculations were performed simulating accident conditions. Calculated results are available for display using the Nuclear Plant Analyzer (NPA) color graphics analysis tool in addition to traditional plots. ATHENA-calculated results appear reasonable for both steady state full power conditions, and the two transients. This analysis represents the first known transient thermal-hydraulic simulation using an integral space reactor system model incorporating heat pipes.

Introduction

The capability to evaluate the performance of advanced reactor concepts is being developed at the Idaho National Engineering Laboratory (INEL) as a part of the Exploratory Research and Development Program. Advanced reactor designs under consideration are the Training Research and Investigation Gulf Atomic (TRIGA) reactor, the Modular High Temperature Gas Reactor (MHTGR), the compact Liquid Metal Reactor (CLMR), the Process Inherent Ultimate Safety (PIUS) reactor, and the SP-100 and Multimegawatt space reactors.

INEL capabilities to model a TRIGA reactor using the ATHENA computer code have been reported (Davis 1985). The purpose of this paper is to demonstrate the INEL capability to perform thermal-hydraulic analysis for a space reactor. A preliminary General Electric SP-100 design concept was selected for this demonstration. The design incorporates a fast reactor with liquid-lithium primary coolant loops that transfer heat to lithium-filled, high temperature heat pipes. These pipes are radiatively coupled to thermoelectric generators on the evaporators of potassium-filled, low temperature heat pipes that, in turn, radiate to space.

The unique features of the SP-100 reactor preclude modeling it with commonly used light water reactor thermal-hydraulic systems codes, such as RELAP5/MOD2 and TRAC-PF1. Specifically, the liquid metal fluids and heat pipe components of SP-100 are beyond the capabilities of these codes. However, the ATHENA code, with capabilities similar to RELAP5/MOD2, can model liquid metal systems and has been upgraded through the inclusion of heat pipe and zero gravity capabilities.

This paper describes the reactor concept, the ATHENA computer code, the ATHENA SP-100 model, the Nuclear Plant Analyzer (NPA) output display, and the results of ATHENA transient demonstration calculations. A more detailed discussion of this work is reported elsewhere (Fletcher 1985).

Reactor Description

Figure 1 presents a schematic of the preliminary General Electric SP-100 space reactor concept. A detailed description of the reactor is provided by General Electric (1983). The concept consists of a reflector-controlled fast reactor, operating at 2 MWt, that is directly cooled by three liquid-lithium coolant loops. Forced loop flow is maintained by one or more thermoelectric electromagnetic (TEM) pumps in each loop. To power the pumps, a portion of the heat in each of the primary coolant loops is radiated to space through thermoelectric converters. Thus, the pumps are reliable, because no moving parts are involved, and inherently safe, because the pumped flow rates vary directly with coolant temperature.

Heat is transferred from the primary coolant system via a heat exchanger on each loop. Each heat exchanger interfaces with 40 high-temperature, lithium-filled heat pipes. The condenser ends of these pipes are radiatively coupled to thermoelectric converters with a rated power output of 100 kWe. Heat passing through the converters is routed through 240 low-temperature, potassium-filled heat pipes to radiator panels for dissipation to space.

One or more accumulators on each loop provide primary coolant system pressure control. The accumulators are spring-and-bellows devices that accommodate fluctuations in coolant volume with changing temperatures.

ATHENA Code Description

The ATHENA computer code (Roth 1985) is based on the RELAP5/MOD2 computer code (Ransom 1983), and has similar but expanded capabilities. The RELAP5/MOD2 code has been used extensively for best estimate analysis of light water reactors and related experimental facilities. Both codes are based on a two-fluid hydrodynamic model that allows unequal phasic velocities and temperatures in a calculational cell. Both codes have models to represent the physical processes of critical flow, vertical and horizontal stratified flow, and heat transfer from structures. Both codes have models to represent point reactor kinetics, control systems, and components such as valves, pumps, and turbines. The major difference between ATHENA and RELAP5/MOD2 is ATHENA's ability to simulate working fluids other than water. A description of ATHENA and its general capabilities is the subject of another paper in these proceedings (Carlson 1986) and the reader is referred there for a more complete description of the code.

To perform the analysis presented in this report, ATHENA was upgraded to allow user specification of the acceleration due to gravity. Furthermore, ATHENA was upgraded to provide the capability for modeling heat pipes. The ATHENA heat pipe model assumes a complete separation of the liquid and vapor phases inside the pipe. Interphase drag forces are removed and a separate pressure solution is generated within each phase. The pressure solutions are consistent axially within each phase and radially between the phases with the liquid and vapor pressures differing by a capillary head calculated from

$$P_v - P_l = (2s/r)\cos a$$

where
P_v = vapor pressure,
P_l = liquid pressure,
s = surface tension of liquid,
r = effective wick pore radius, and
a = angle between the heat pipe radial direction and wick meniscus tangential contact direction.

While the ATHENA heat pipe model simulates the basic thermal-hydraulic phenomena of heat pipe operation, it is still considered a preliminary model. The

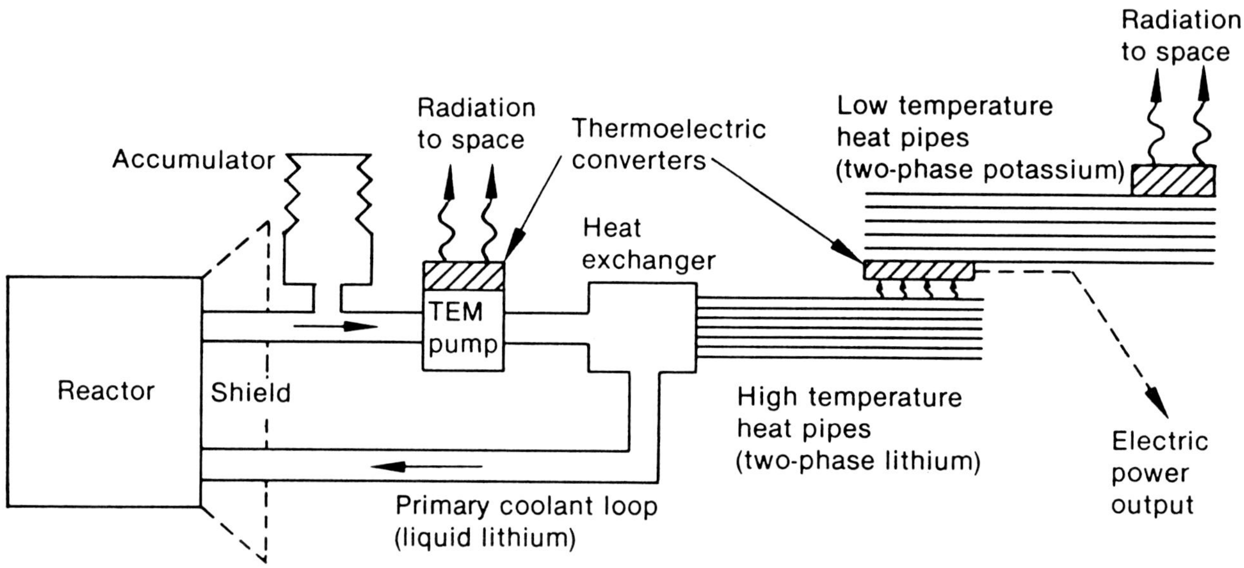

Figure 1 *Schematic of Preliminary General Electric SP-100 Concept.*

demonstration calculations in this paper are the first applications of the ATHENA heat pipe model.

ATHENA SP-100 Model Description

Figure 2 is a nodalization diagram for the ATHENA SP-100 model. The diagram shows the layout of calculational cells, connecting junctions, and heat structures for the model. Each of the three primary coolant loops was modeled independently. Note that the 40 high temperature heat pipes on each loop were combined into an ATHENA heat pipe component model. Heat structures were used to represent the most important metal structures in the reactor: core rods, TEM pump radiators, and high temperature heat pipe evaporator and condenser section walls. Design parameters for the SP-100 space reactor were obtained from the conceptual design study (General Electric 1983).

For this demonstration analysis, the ATHENA model excluded the low temperature heat pipes. These pipes are filled with potassium, a working fluid not currently available in ATHENA. The ATHENA capabilities to simulate heat pipe behavior will be demonstrated with the high-temperature, lithium-filled heat pipes. The capability with other heat pipe working fluids is expected to be comparable. In lieu of the low temperature heat pipes, the ATHENA model placed temperature boundary conditions on the outer wall surfaces of the model's heat pipe condenser sections. These temperature boundary conditions were calculated independently for each loop during each time step based on the heat transfer rates through the pipes and the resistances of the two radiation heat transfer paths in series between the condensers and space. A 255 K space background temperature was assumed. It is expected that use of the temperature boundary conditions, instead of modeling the low-temperature heat pipes, will moderately alter the transient thermal response.

The core power was calculated by the ATHENA point reactor kinetics model. Reactor kinetics feedback effects due to Doppler, and moderator and structure expansion effects were simulated. No movement of the reflector control drums was assumed.

Behavior of the TEM pumps was simulated with ATHENA time-dependent junction components. These junctions (between components 210 and 220, 310 and 320, and 410 and 420 in Figure 2) forced the loop flow as a function of coolant temperature in each of the loops. Given a coolant temperature, TEM pump radiator heat loss rates were calculated. From these, and the efficiencies involved with the thermoelectric generators and the electromagnetic pumps, the pumped mass flow rates were calculated and used to

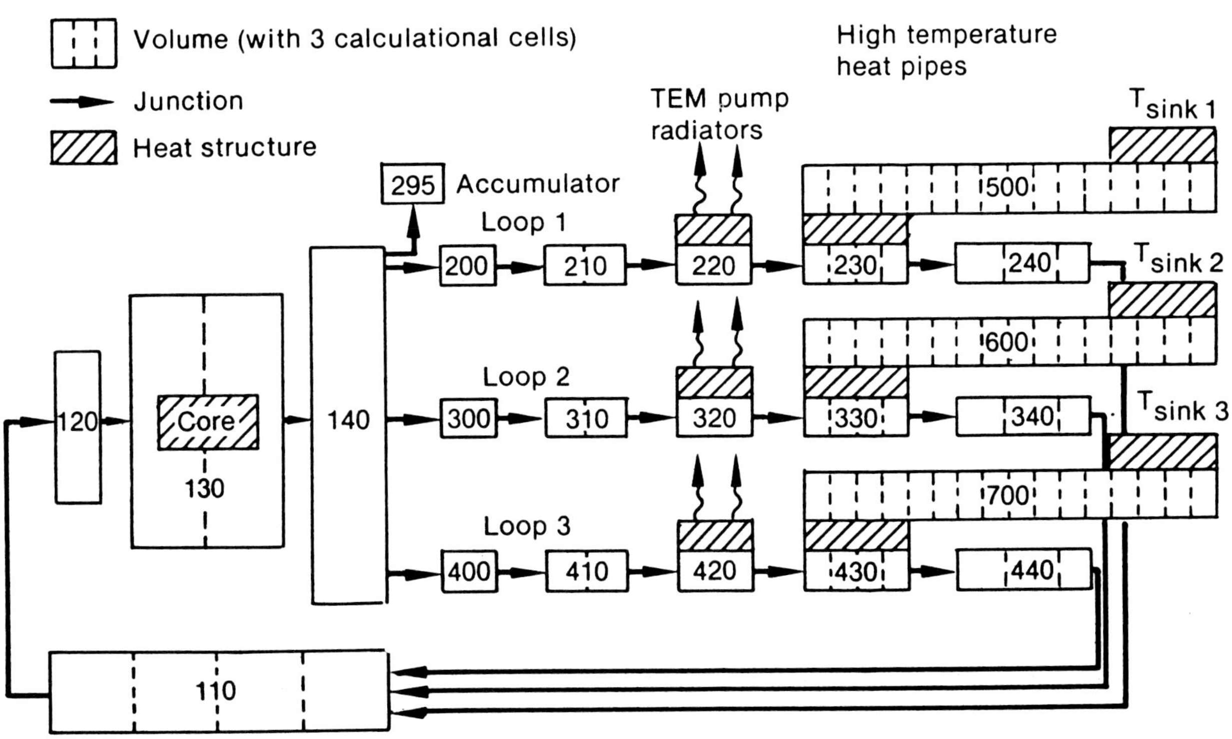

Figure 2 *SP-100 ATHENA Nodalization.*

drive the junctions. The TEM radiator heat loss rates also were used as cooling boundary conditions for the heat structures modeling these radiators. Constant efficiencies of 3% for the electric generator, and 20% for the pump were based on the conceptual design study.

The primary system coolant pressure control of the accumulators was simulated with an ATHENA time-dependent volume component. ATHENA calculated the deviation in coolant system volume as a result of changes in temperature, and the effect of that deviation on accumulator spring deflection, spring force, and system pressure. The calculated pressure was then used to control the ATHENA model primary coolant pressure by means of the time-dependent volume.

The SP-100 electric power output was calculated by an ATHENA control variable that summed the heat transfer rates from the three heat pipe condenser sections and then multiplied by a constant 5.4% thermoelectric generator efficiency.

Table 1 presents a comparison of full-power steady-state operating conditions, as calculated by ATHENA, and desired conditions from the General Electric (1983) conceptual design study. All comparisons are in excellent agreement, indicating that the ATHENA model well represents the SP-100 design.

Nuclear Plant Analyzer Display

The Nuclear Plant Analyzer (NPA) is a color graphics analysis tool at INEL. The NPA presents a visual simulation of thermal-hydraulic phenomena occurring during a transient in a reactor system.

Coupling an ATHENA calculation with the NPA requires the development of a display mask and interface program. The format of the visual display is determined by the mask, while the content of the display is determined by the interface.

Figure 3 shows the NPA mask for the SP-100 model. The mask represents the primary and heat pipe systems of the SP-100 reactor and has a format similar to the ATHENA model nodalization shown in Figure 2. Generally there is a one-to-one correspondence between the nodes in the mask and calculational cells of the ATHENA model. The major exception is that, for completeness, the NPA mask includes representations of the low-temperature heat pipes that are not directly modeled with ATHENA. The temperature behavior displayed for these pipes was calculated by ATHENA based on the heat transfer rates of the adjacent high-temperature heat pipes and the space background temperature. The void fraction behavior displayed for the low-temperature heat pipes was the same as for the adjacent high-temperature heat pipes.

The color of each node in the NPA mask is driven by the thermodynamic conditions in the corresponding region of the ATHENA model. Temperatures are indicated by color, ranging from blue (coldest) to red (hottest). Void fractions are indicated by the intensity of the color.

In addition to temperature and void fraction behavior, the NPA display includes digital output of primary system pressure, accumulator level, core power, TEM radiator powers, heat pipe powers, electric power output, primary loop mass flow rates, and fluid temperatures in the cold legs and heat pipes.

Results

Partial Loss of Loop Flow (Transient 1)

The first demonstration calculation performed with the ATHENA SP-100 model involved the partial loss of Loop 1 flow. This transient simulates the response of the SP-100 system to an accident where, due to meteorite impingement, one of the two TEM pump radiators on Loop 1 is damaged. As a result, one of the two Loop 1 TEM pumps could be expected to stop functioning. Because the two Loop 1 pumps are in

Table 1 *Desired and ATHENA-Calculated Full Power Conditions.*

Parameter	Desired	Calculated
Core Power (MWt)	2.000	2.000
Mass flow Rate per Loop (kg/s)	4.4	4.396
Loop Differential Pressure (Pa)	24440	31058
Primary System Pressure (Pa)	100000	99432
Hot Leg Temperature (K)	1546	1538.6
Cold Leg Temperature (K)	1506	1502.7
TEM Radiator Power per Loop (Wt)	45670	45334
Heat Pipe Power per Loop (Wt)	620997	621332
Electric Power Output (We)	100000	100937
Heat Pipe Temperature (K)	1500	1480.9

parallel, the surviving pump could be expected to continue pumping at its full flow. As a boundary condition for this calculation, the Loop 1 pumped flow is assumed to be one half the full flow. The remainder of this section discusses the ATHENA calculation results for this transient.

As the Loop 1 flow slowed down, the reactor outlet temperature increased as shown in Figure 4. The Loop 1 cold leg temperature decreased while the Loop 2 and 3 cold leg temperatures increased. Thus, little change in reactor inlet temperature was noticed. With the reactor inlet temperature stable and the outlet temperature rising, the average primary system temperature rose moderately. The thermal expansion of primary coolant caused the accumulator level to increase as shown in Figure 5. Furthermore, the rising primary system average temperature caused an initial negative reactivity insertion, thereby reducing core power as shown in Figure 6.

Figure 7 shows the heat transfer rates at the condensers of the high temperature heat pipes. The lower Loop 1 primary side flow resulted in decreasing power though the heat pipes on that loop. However, the decreasing power caused the Loop 1 low-temperature heat pipe temperature to fall, thereby reducing the Loop 1 high temperature heat pipe temperature. The Loop 1 heat pipe power decreased until the reduction in its temperature stabilized its power. The Loop 2 and 3 high temperature heat pipe powers initially rose, in response to the lower Loop 1 heat removal, then fell in response to the declining core power. Calculated electric power output for this transient is shown in Figure 8. The early rise in electric power was an effect caused by simulating the low-temperature heat pipes with temperature boundary conditions. This response is not expected in the prototype system.

Figure 9 presents the primary system and high-temperature heat pipe pressure responses. Only minor changes in primary system pressure are observed. The changes are in response to the varying accumulator level shown in Figure 5. The Loop 1 high-temperature heat pipe pressure fell in response to the decreasing Loop 1 heat pipe power and sink temperature. Similarly, Loop 2 and 3 high temperature heat pipe pres-

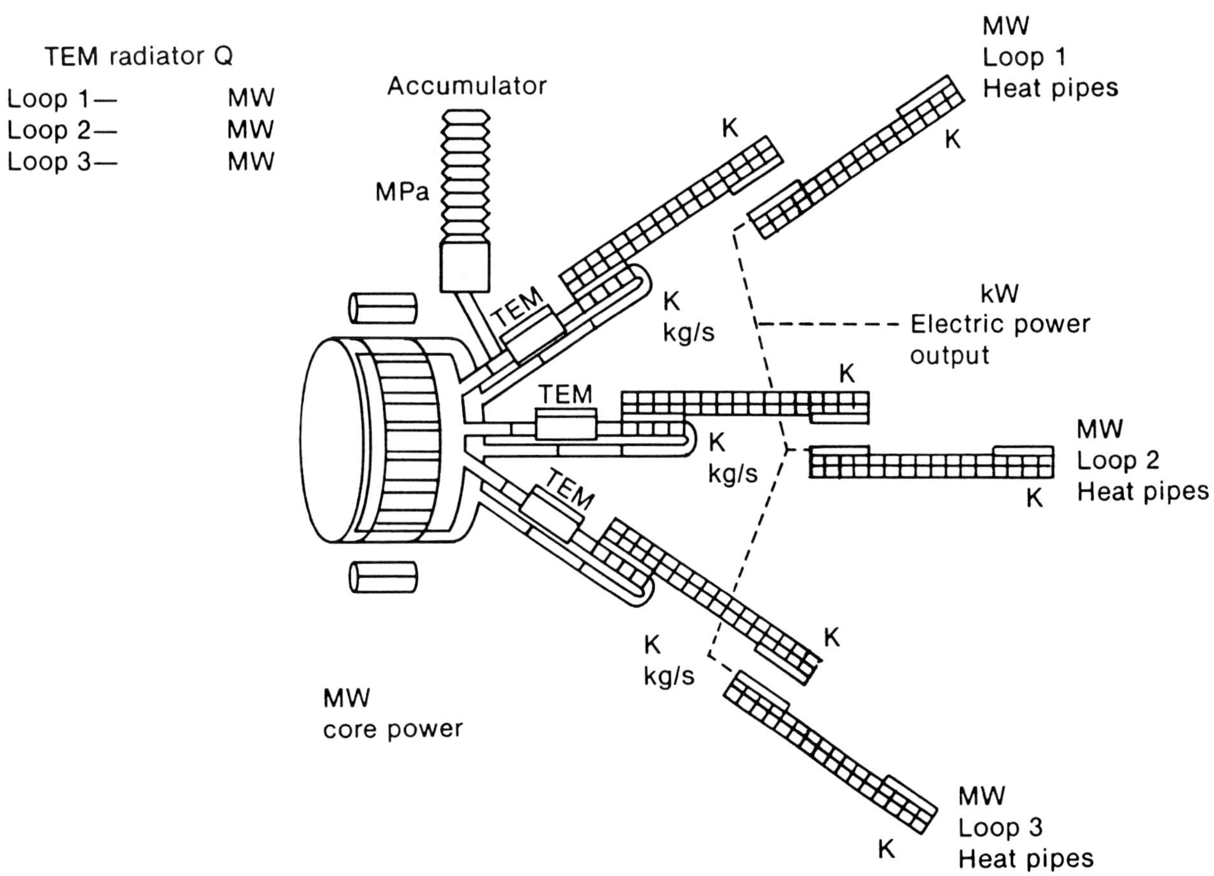

Figure 3 *SP-100 Nuclear Plant Analyzer Mask.*

sures rose in response to increasing heat pipe powers and sink temperatures.

Figure 10 shows vapor velocities at the centers of the high-temperature heat pipes. The Loop 1 velocity fell and the Loop 2 and 3 velocities rose over the first 7 *s* of the transient in response to the diverging heat pipe powers shown in Figure 7. Interestingly, the velocity trends reversed after about 7 *s*. This was due to the changing heat pipe pressure responses shown in Figure 9. As the Loop 2 and 3 heat pipe pressures increased, the vapor densities and mass flow rates increased. Thus, the Loop 2 and 3 heat pipe vapor velocities actually were lower than their initial values even though the heat transfer rates through these pipes had increased.

The transient calculation was continued for 200 *s*, whereupon reactor response had essentially stabilized. The core power fell until the decline in primary system average temperature caused a positive reactivity insertion and a new steady operating condition was approached. The new operating condition included a lower-than-original core power and flow, asymmetric loop flows, asymmetric heat pipe powers, and lower electric power output.

Loss of Loop 1 Low Temperature Heat Pipes (Transient 2)

Damage to the SP-100 reactor due to meteorite impingement is a potential operating concern. Since the low-temperature heat pipe radiator panels are on the exterior of the reactor, perforation of these pipes is a likely consequence of meteorite impingement. A transient calculation was performed using the ATHENA SP-100 model to demonstrate its response in this type of sequence. The transient assumed that all of the 80 low-temperature heat pipes carrying heat from Loop 1 (one-third of the total heat sink) are penetrated by meteorites, allowing their working fluids to

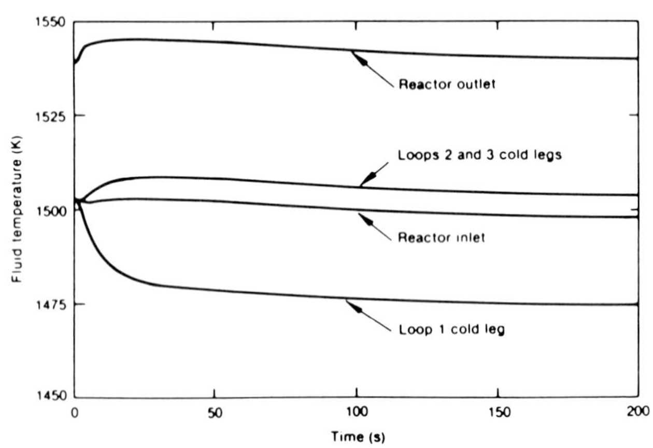

Figure 4 *Transient 1 Primary System Temperature Responses.*

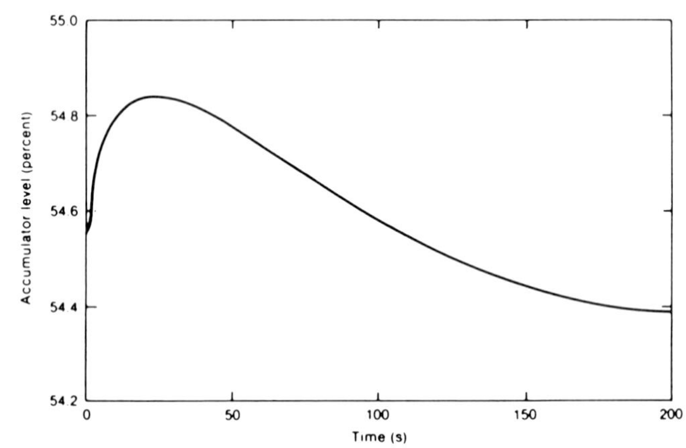

Figure 5 *Transient 1 Accumulator Level Response.*

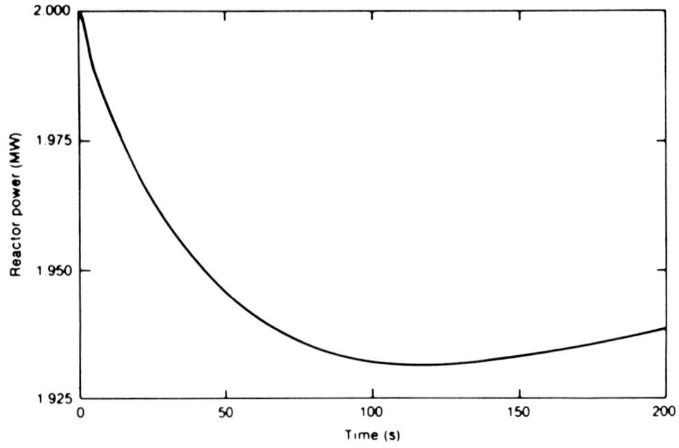

Figure 6 *Transient 1 Core Power Response.*

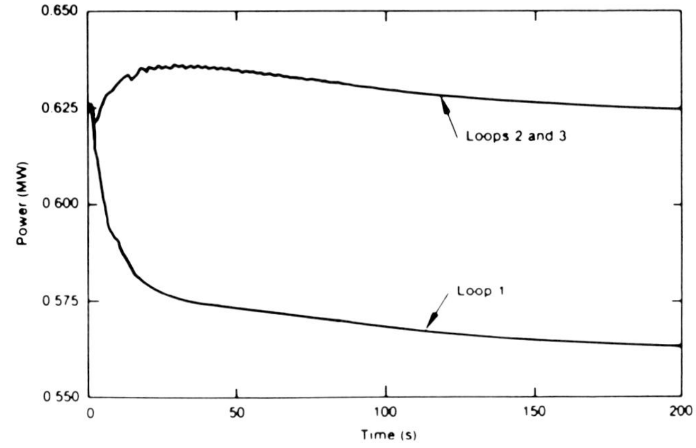

Figure 7 *Transient 1 High Temperature Heat Pipe Powers.*

escape to space. In the model, these effects were simulated by removing all of the condenser heat structures on the Loop 1 high temperature heat pipes. The remainder of this section discusses the ATHENA calculation results for this transient.

When the Loop 1 low temperature heat pipe effectiveness was lost, a power mismatch existed, with the core power far exceeding the total of the heat removal rates through Loops 2 and 3. As a result, primary system temperatures rose as shown in Figure 11. Note the Loop 1 cold leg temperature diverged from those of Loops 2 and 3 and approached the hot leg temperature. The reactor inlet temperature exceeded the Loop 2 and 3 cold leg temperatures because Loop 1 continued to flow even though the heat sink on that loop was lost. With the rising average primary system temperature, the primary coolant expanded, causing the accumulator level to increase as shown in Figure 12. Furthermore, as a result of the increasing temperatures, a negative reactivity insertion reduced the core power as shown in Figure 13.

The high temperature heat pipe power responses are shown in Figure 14. The Loop 1 heat pipe power vanished at time zero as a result of the transient initiator. The Loop 2 and 3 heat pipe powers initially increased to carry the excess core power, then decreased in response to declining core power.

Figure 15 shows electric power output response. This power was calculated by summing the high temperature heat pipe powers from Figure 14 and then multiplying by the constant 5.4% thermoelectric generator efficiency.

Figure 16 presents the primary system and high-temperature heat pipe pressure responses. Primary system pressure changes are in response to the varying accumulator level and the associated compression

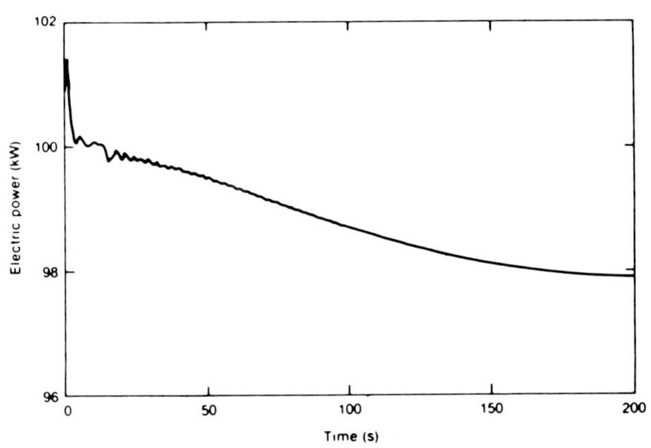

Figure 8 *Transient 1 Electric Power Output.*

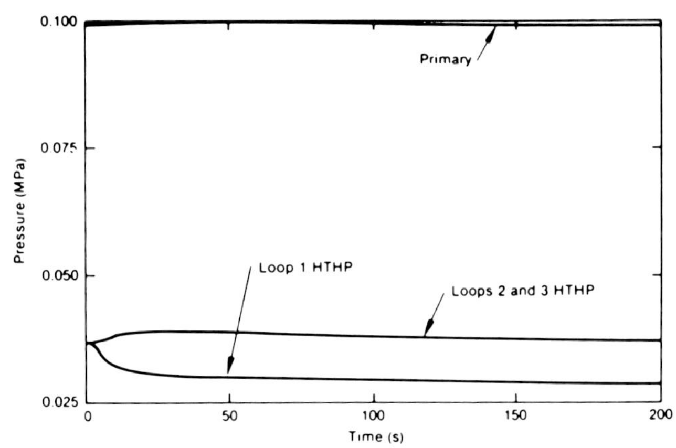

Figure 9 *Transient 1 Pressure Responses.*

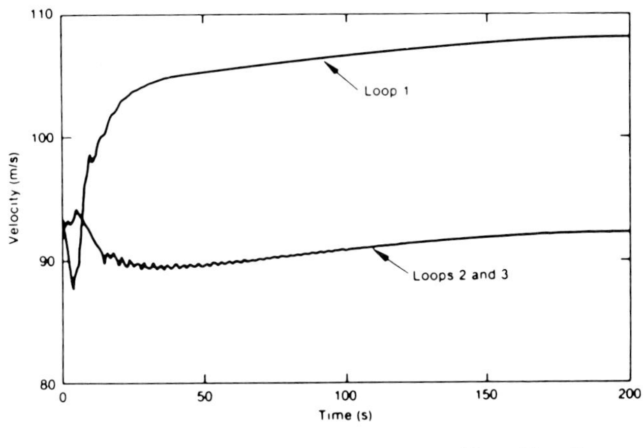

Figure 10 *Transient 1 High Temperature Heat Pipe Vapor Velocities.*

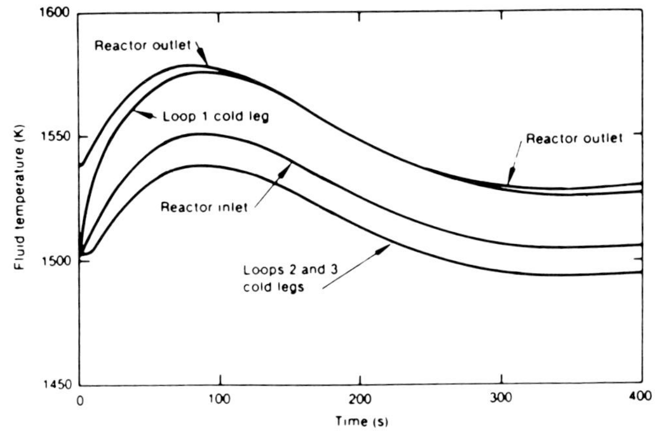

Figure 11 *Transient 2 Primary System Temperature Responses.*

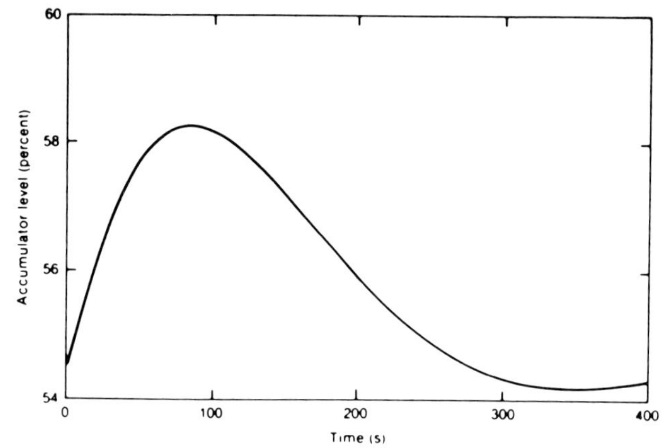

Figure 12 Transient 2 Accumulator Level Response.

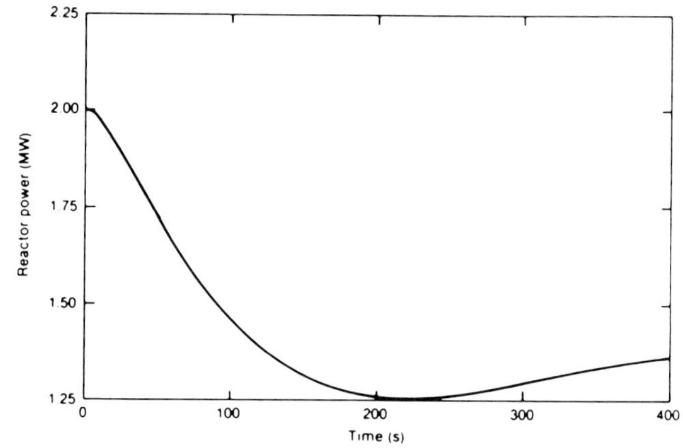

Figure 13 Transient 2 Core Power Response.

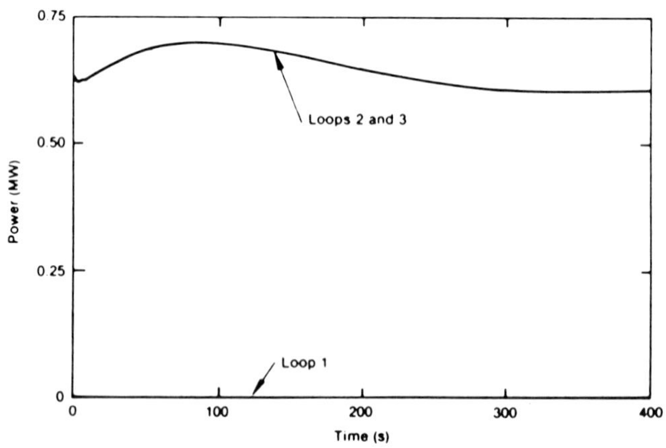

Figure 14 Transient 2 High Temperature Heat Pipe Powers.

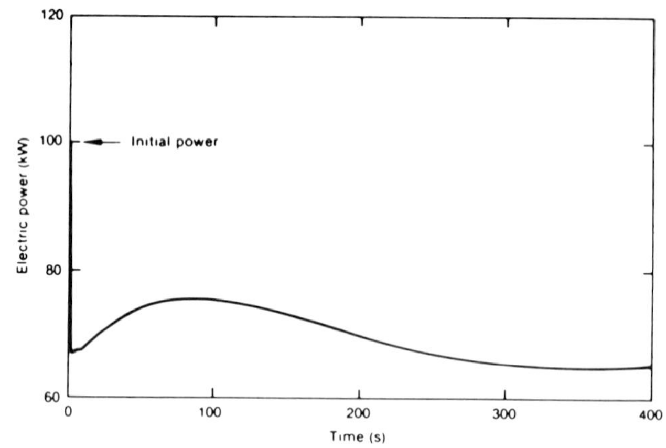

Figure 15 Transient 2 Electric Power Output.

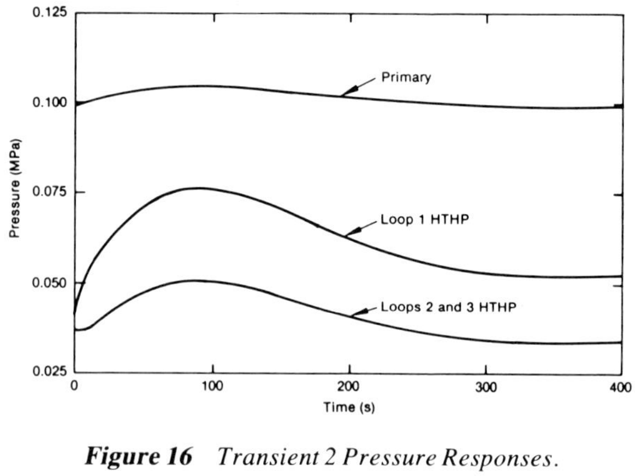

Figure 16 Transient 2 Pressure Responses.

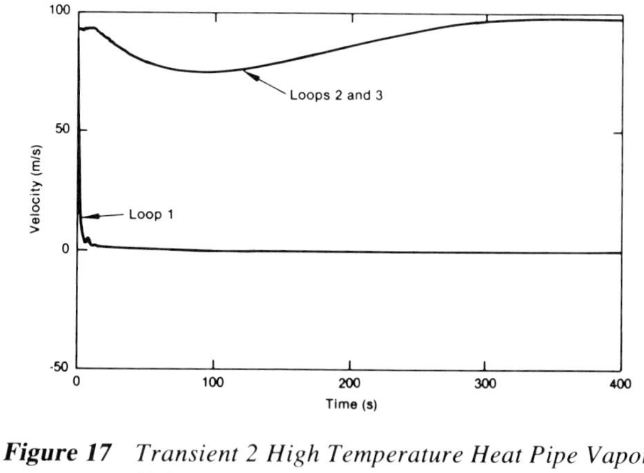

Figure 17 Transient 2 High Temperature Heat Pipe Vapor Velocities.

and expansion effects of the accumulator spring. Even though its heat transfer rate vanished, the Loop 1 heat pipe pressure increased as its fluid was heated gradually to the hot leg temperature. The Loop 2 and 3 heat pipe pressures also increased, but not as much as the Loop 1 pressure. This is because their condensers continued to function while the Loop 1 heat pipe condenser did not.

Figure 17 shows the high temperature heat pipe vapor velocities at mid-pipe. The Loop 1 heat pipe vapor flow stopped after about 90 *s*. Loop 2 and 3 vapor velocities followed the same behavior experienced for Transient 1: vapor velocity primarily tended to vary inversely with pressure, rather than directly with heat pipe power.

The transient was continued for 400 *s*, whereupon reactor response had essentially stabilized. The core power stabilized at a new lower level, with power dissipation evenly split between Loops 2 and 3. Reactor temperatures at the new steady condition are near their original values.

Conclusions

The capability of the ATHENA computer code to simulate a SP-100 space reactor has been demonstrated. This demonstration included the simulation of transient heat pipe behavior in a zero gravity environment. The ATHENA model represents the first known space reactor transient integral system model incorporating heat pipes.

ATHENA-calculated results appear reasonable for both steady state full power conditions, and the two demonstration transient sequences. The results of the transient calculations suggest the preliminary General Electric SP-100 reactor design is capable of withstanding the effects of severe transients, such as the loss of one-third of the total heat sink.

INEL is extending its work in this area. The ATHENA heat pipe model is being improved to include performance limiters such as sonic, wicking, and boiling limits. The ATHENA SP-100 model is being upgraded to simulate the current General Electric SP-100 concept that has been selected for development. To support this upgrade, the capability to model a potassium working fluid is being added to ATHENA.

Acknowledgments

This research was performed at the Idaho National Engineering Laboratory for the Idaho Operations Office of the U.S. Department of Energy under Contract DE-AC07-76D01570.

References

Carlson, K. E. (1986) "A Simulation Tool for Advanced Space Reactor Design and Analysis," in *Trans. Third Symposium on Space Nuclear Power Systems*, CONF-860102-Summs, held in Albuquerque, NM, 13–16 January 1986.

Davis C. B. (1985) *Development of Capability to Model a TRIGA Reactor using ATHENA*, Report EGG-RST-6878, EG&G Idaho Inc., Idaho Falls, ID, May 1985.

Fletcher, C. D. (1985) *Capabilities of the ATHENA Computer Code for Modeling the SP-100 Space Reactor Concept*, Report EGG-RST-7025, EG&G Idaho Inc., Idaho Falls, ID, September 1985.

General Electric Co. (1983) *SP-100 Conceptual Design Study Final Report*, 2 Volumes, Report 83SDS4262, Philadelphia, PA, 7 December 1983.

Ransom, V. H. et al. (1983) *RELAP5/MOD2 Code Manual*, 2 Volumes, Draft Report EGG-SAAM-6377, EG&G Idaho Inc., Idaho Falls, ID, September 1983.

Roth P. A. et al. (1985) *ATHENA Code Manual, Volume 1: Code Structure, System Models, Solution Methods, and Input Data Requirements*, Draft Report EGG-RST-7034, EG&G Idaho Inc., Idaho Falls, ID, September 1985.

Space Nuclear Power Systems 1986
Edited by M. S. El-Genk and M. D. Hoover

Chapter 11

SNPSAM-Space Nuclear Power System Analysis Model

Mohamed S. El-Genk and Jong T. Seo

The Institute for Space Nuclear Power Studies
Department of Chemical and Nuclear Engineering
The University of New Mexico
Albuquerque, NM 87131

This research developed a Space Nuclear Power System Analysis Model (SNPSAM) for analyzing integrated thermoelectric (TE) SP-100 systems. SNPSAM is a modular computer model that takes into account the interactions and feedback processes among the system's major components. These components are the nuclear reactor, the primary coolant loop, the primary heat exchanger, the thermoelectric-electromagnetic (TE-EM) pump, and the TE energy converter. Each component was modeled and verified individually before being coupled to the other models in the system. The users of SNPSAM can select between SiGe/GaP and SiGe for the TE generators both in the main TE converter and in the TE-EM pumps, and between liquid Lithium (Li) and NaK-78 for the primary coolant. SNPSAM was used to study the performance of the integrated TE SP-100 system during steady state and transient operation. The system's parameters during steady state operation were calculated as functions of the reactor thermal power and coolant type (Li or NaK-78). These operating parameters include the reactor core and coolant temperatures, the coolant mass flow rate and specific pumping power, and the overall thermal-to-electric conversion efficiency as well as the electric power output. The results showed that the TE SP-100 system is a nonlinear system and that the system's conversion efficiency and electric power output strongly depend on the reactor thermal power. While the limiting factor for the NaK-78 cooled system was the boiling point of the coolant, that for the Li cooled system was the melting point of the cladding temperature. The specific pumping power for the system decreased as the reactor thermal power increased; it was about 7 percent higher for NaK-78 coolant than for the Li coolant. This research also investigated the system's transient behavior following step and ramp external reactivity insertions into the reactor core.

Introduction

A major challenge in the design of a space nuclear power system is attaining a high level of reliability while keeping the system's mass and volume as low as possible (Vaughan 1985 and El-Genk and Seo 1986a). Achieving this goal requires selective choices of the various system's components, as well as a good understanding of the operating characteristics of the integrated system. These issues are particularly important because an integrated power system that consists of optimized components might not be an optimum system. In an attempt to investigate these issues, a modular simulation computer model, SNPSAM (Space Nuclear Power System Analysis Model), was developed at the University of New Mexico's Institute for Space Nuclear Power Studies (ISNPS) for analyzing integrated thermoelectric (TE) SP-100 systems (Bailey et al. 1985 and GE 1983). This interactively coupled model is capable of predicting the performance of the integrated system during both steady state and transient operations, the system's load following characteristics, and the integration of the power system with payload. Additionally, SNPSAM can be used for optimizing the design and assessing the reliability of the TE SP-100 system with an electric power output up to 200 kWe. In recent research, for example, the present version of SNPSAM (mode 0) was used to study the effects of a partial loss of the radiator's surface area caused by the impact of meteoroids on the system's overall performance and on the decay heat thermal management (El-Genk et al. 1986a and 1986b); it was also used to predict the system's load following characteristics without a shunt regulator of the electric load (El-Genk et al. 1986b).

This paper describes SNPSAM-mode 0 and discusses the results of the integrated TE SP-100 system's performance studies using SNPSAM. The system's electric power output, conversion efficiency, coolant temperatures, and specific pumping power were calculated as functions of the reactor thermal power and the liquid metal coolant type (Li or NaK-78) during steady state operation. This paper

NOMENCLATURE

English

- A: Cross sectional area (m^2)
- $\bar{A}$: Flow area without wire wrap (m^2), Equation (17)
- a: Cross sectional area of channels in heat exchanger (m^2)
- B: Magnetic flux density (gauss)
- b: Width of coolant duct in EM pump (m)
- C: Specific heat (J/kg·K)
- c: Conversion factor, Equation (11)
- D: Inside diameter (m)
- D_e: Equivalent hydraulic diameter (m)
- d_e: Equivalent hydraulic diameter of heat exchanger channel (m)
- $\dot{F}$: Fission product decay power (W)
- f: Friction factor
- G: Constant, Equation (17)
- g: Fraction of decay power generated in the coolant, Equation (13)
- H: Wire wrap lead (m)
- h: Heat transfer cooefficient (W/m^2·K)
- I: Electric current (amp)
- J: Electric current density (amp/m^2)
- k: Thermal conductivity (W/m·K)
- L: Length (m)
- l: Length of channels in heat exchanger (m)
- M: Mass (kg)
- $\dot{M}$: Mass flow rate in the heat exchanger header (kg/s)
- $\dot{m}$: Mass flow rate (kg/s)
- N: Number of channels in heat exchanger, Equation (23)
- N: Number of components in the primary loop, Equation (37)
- P: Power (W)
- ΔP: Pressure loss (N/m^2)
- Pe: Peclet number
- p: Fuel rod pitch (m)
- Q: Coolant volumetric flow rate (m^3/s)
- q: Heat flux (W/m^2)
- R: Electrical resistance (ohm)
- Re: Reynolds number
- T: Temperature (K)
- t: Time (s)
- U: Thermal conductance (W/K)
- V: Electric voltage (volt)
- w: Width of channels in heat exahanger (m)
- x: Coordinate
- Z: Delayed neutron precursor latent power (W)

Greek

- α: Temperature coefficient of reactivity (K^{-1}), Equation (7)
- α: Seebeck coefficient (volt/K), Equation (32)
- β: Delay neutron fraction
- $\bar{\beta}$: Average delay neutron fraction
- η: Efficiency
- γ: Correction factor, Equation (30)
- χ: Fission yield of fission product, Equation (12)
- Λ: Neutron life time (s)
- λ: Decay constant (s^{-1})
- ρ: Reactivity in the core, Equation (1)
- ρ: Electrical resistivity (ohm·m), Equation (31)
- τ: Thomson coefficient (volt/K)
- ω: Density (kg/m^3)

Subscripts

- a: Annulus
- av: Average
- C: Cold header of heat exchanger
- CN: Control drum
- c: Coolant, Equation (15)
- c: Reactor Core, Equation (17)
- c: Cold shoe of TE converter, Equation (32)
- cav: Coolant average
- cl: Cladding
- $clav$: Cladding average
- D: Decay power
- d: Doppler effect, Equation (7)
- EM: Electromagnetic pump
- e: Electric power
- ex: Exit
- FB: Feedback
- f: Fuel, Equation (8)
- f: Fringe current, Equation (19)
- fav: Fuel average
- H: Hot header of heat exchanger
- HX: Heat exchanger
- h: Hot shoe of TE converter
- i: Index
- in: Inlet
- J: Joule effect
- L: External load
- n: Fission power, Equation (1)
- n: N leg of TE converter, Equation (32)
- o: Initial condition
- P: Peltier effect
- p: Pipe, Equation (22)
- p: P leg of TE converter, Equation (32)
- pn: P and N legs of TE converter
- $pump$: TE-EM pump
- r: Projected area for one wire wrap over one lead, Equation (17)
- s: System
- T: Thomson effect
- TE: Thermoelectric generator
- tot: Total
- w: Wall
- wav: Wall average

Superscripts

- a: Annulus
- C: Coolant channel in heat exchanger
- c: Core
- H: Coolant header in heat exchanger
- (n): Iteration number

also discusses the system's transient behavior following both a step and a ramp external reactivity insertion as well as the effects of the initial steady state power, the mode of power change on the maximum electric power output attainable, and the system's settling time during the transient. A full description and design parameters in SNPSAM are available elsewhere (Seo 1986).

Description of SNPSAM

Table 1 summarizes the design parameters and nominal operating conditions of the TE SP-100 system analyzed in this paper. The nuclear reactor core and the primary heat exchanger designs are similar to those of the early General Electric (GE) SP-100 design concept (see Figure 1) (Bailey et al. 1985).

Components Models

The major components of the TE SP-100 system modeled in SNPSAM are the nuclear reactor core, the primary coolant loop, the TE energy converter, the TE-EM pumps, the primary heat exchangers, and the main radiator (see Table 2). Each model was verified individually before being coupled to the system's other components models. This modular approach enables the user to employ the system's model for optimizing the system's performance and for investigating the effects on the integrated system's performance of the different component designs.

Nuclear Reactor Model

The nuclear reactor model, based on the DSNP (Dynamic Simulator for the Nuclear Power-Plant) concept (ANL 1983), employs a simple lumped parameter approach and consists of five coupled submodels:

- A neutron kinetics and reactivity control model,
- A reactivity feedback model,
- A decay heat model,
- A reactor thermal model, and
- A reactor hydraulic model.

These submodels are described in the following subsections.

Neutron Kinetics and Reactivity Control Model

The neutron kinetics model solves the neutron point kinetics equations using six delayed neutron groups. It calculates the time derivative of the reactor thermal power generated by fission, $DP_n(t)/dt$, and the time de-

Table 1 *SP-100 System Characteristics in SNPSAM.*

Parameter	Value
Operating Parameters:	
Nominal thermal power (kWt)	2000
Nominal electric power (kWe)	112
System efficiency (%)	5.6
Core inlet temperature (K)	1254
Core outlet temperature (K)	1284
Fuel average temperature (K)	1376
Cladding average temperature (K)	1288
TE hot shoe temperature (K)	1237
TE cold shoe temperature (K)	857
Coolant mass flow rate (kg/s)	14.1
Design Parameters:	
Fuel material	UN
Cladding material	Nb-1%Zr
Coolant material	Li or Nak-78
TE material	SiGe/GaP or SiGe
Pump magnet	Hiperco-27
Nominal magnetic flux density (gauss)	1000
Number of TE couples in main converter	172800
Number of heat pipes in main radiator	240
Number of heat pipes in TE-EM pump radiator	6
Number of primary coolant loops	3
Number of TE-EM pumps	3
Number of heat exchangers	12

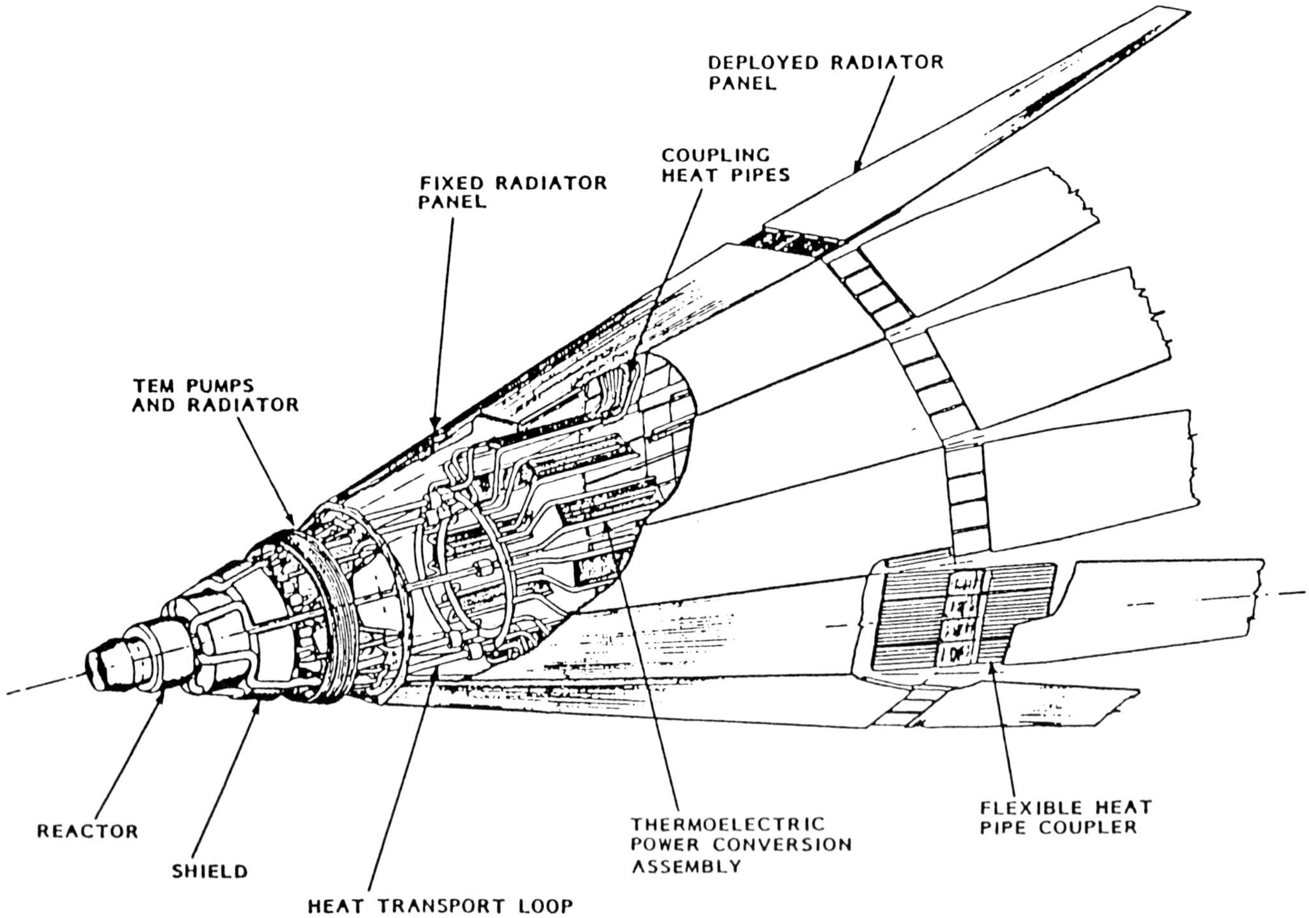

Figure 1 *Thermoelectric (TE) SP-100 System.*

rivative of the delayed neutron precursors, $dZ_i(t)/dt$, using the following two coupled equations:

$$\frac{dP_n(t)}{dt} = \left(\frac{\rho(t) - \overline{\beta}}{\Lambda}\right) P_n(t) + \sum_{i=1}^{\beta} \lambda_i Z_i(t) , \quad (1)$$

$$\frac{dZ_i(t)}{dt} = \left(\frac{\beta_i}{\Lambda}\right) P_n(t) - \lambda_i Z_i(t) , \quad (2)$$

where

$$\rho(t) = \rho_{CN} + \rho_{FB} . \quad (3)$$

In Equation (3), $\rho(t)$ is the total reactivity input, ρ_{FB} is the net feedback reactivity, and ρ_{CN} is the external reactivity insertion into the reactor core. During steady state operation, ρ_{CN} is constant and equal to ρ_0 however, during transient operation, it could take any of the following forms:

- Step Function

$$\begin{aligned} \rho_{CN}(t) &= \rho_0 , && \text{for } t < t_1 , \\ \rho_{CN}(t) &= \rho_0 + \Delta\rho , && \text{for } t \geq t_1 , \end{aligned} \quad (4)$$

- Ramp Function

$$\begin{aligned} \rho_{CN}(t) &= \rho_0 , && \text{for } t < t_1 , \\ \rho_{CN}(t) &= \rho_0 + \Delta\rho(t - t_1) , && \text{for } t_1 \leq t \leq t_2 , \\ \rho_{CN}(t) &= \rho_0 + \Delta\rho(t_2 - t_1) , && \text{for } t > t_2 , \end{aligned} \quad (5)$$

- Arbitrary Function

$$\begin{aligned} \rho_{CN}(t) &= \rho_0 , && \text{for } t < t_1 , \\ \rho_{CN}(t) &= f(t) , && \text{for } t_1 \leq t \leq t_2 , \\ \rho_{CN}(t_2) &= f(t_2) , && \text{for } t > t_2 , \end{aligned} \quad (6)$$

where t_1 is the initiation time of the transient, and t_2 is an arbitrary time after the initiation of the transient.

Reactivity Feed Back Model

This model calculates the reactivity feedback caused by the changes in the densities of the fuel and the coolant with temperature, as well as by the Doppler effect in the fuel. The reactivities induced by Doppler effect, core expansion, and coolant expansion are calculated as follows:

- Doppler feedback effect:

$$\rho_d = \alpha_d \ln\left(\frac{T_{fav}(t)}{T_{fav}(0)}\right) , \tag{7}$$

- Fuel temperature and density feedback effect:

$$\rho_f = \alpha_f \left(T_{fav}(t) - T_{fav}(0)\right) \text{ , and} \tag{8}$$

- Coolant temperature and density feedback effect:

$$\rho_c = \alpha_c \left(T_{cav}(t) - T_{cav}(0)\right) . \tag{9}$$

- The total feedback reactivity is then given by

$$\rho_{FB} = \rho_d + \rho_f + \rho_c . \tag{10}$$

Decay Heat Model

Although the decay heat model in SNPSAM cannot be used for calculating the decay heat in the reactor core after shutdown, it can calculate the power generated by the radioactive decay of the fission products during the reactor operation. Since the model assumes that the decay heat is primarily produced by gamma

Table 2 *Description of the Components Models in SNPSAM (mode 0).*

Components	Model	Model Description
Nuclear reactor	Neutron Kinetics and Reactivity Control Model	• Point kinetics model with six delayed neutron groups. • Calculates the thermal power generated in the core due to fission for a step, ramp, or any arbitrary reactivity insertion function.
	Reactivity Feedback Model	• Calculates reactivity feedback due to Doppler effect, fuel expansion, and coolant expansion.
	Decay Heat Model	• Calculates the power generated in the core during operation due to the gamma decay of fission products. • It divides the fission products into four groups where each group has an effective decay half life.
	Reactor Thermal Model	• A lumped parameter model. • Calculates the fuel and cladding average temperatures. • Calculates the core coolant average and exit temperatures.
	Reactor Hydraulic Model	• Calculates the pressure drop in the reactor core region and in the annulus between the core baffle and the reactor vessel.
Primary Coolant Loop	TE-EM Pump Model	• Couples the EM pump model to both the TE model and a 2-D thermal analysis model for the pump magnet. • Calculates the pumping head and the overall pump efficiency.
	Piping Hydraulic Model	• Calculates the pressure drop in the primary coolant loop pipings.
	Heat Exchanger Hydraulic Model	• Calculates the flow distribution in the coolant channels and the total pressure drop in the heat exchanger.
	Heat Exchanger Thermal Model	• The energy balance model of the HX coolant channels is coupled to the main TE converter model. • Calculates the temperatures of the coolant and the wall in the HX coolant channels. • Calculates the total thermal energy extracted from the heat exchanger and the exit coolant temperature from the heat exchanger.
Energy Conversion	TE Model	• A Transient 1-D finite element model. • Uses temperature dependent material properties. • Handles any combination of boundary conditions. • Calculates the electric power output, conversion efficiency, and the waste heat rejection.
Radiator	Waste Heat Rejection Model	• Direct radiation model from the cold shoes of the TE generators (both in the main TE converters and in the TE-EM pumps) into space. • Takes into account the effect of the sonic limit of the radiator's potassium heat pipes.

radiation, the fission products are divided into four fission product groups according to their half life (see Table 3). Thus, the decay power is directly proportional to both the concentration and the production rate of the fission products in the core,

$$P_D = \sum_{i=1}^{4} c_i \int_{\text{time step}} \dot{F}_i dt , \tag{11}$$

where

$$\dot{F}_i = \chi_i P_n - F_i \lambda_i . \tag{12}$$

Reactor Thermal Model

The reactor core thermal model calculates the fuel and the cladding average temperatures, and the core coolant average and exit temperatures using a lumped parameter approach. (At the time of publication, a two dimensional fuel rod model was developed and incorporated in SNPSAM. This model enables the user of SNPSAM to calculate the radial and axial temperature distribution in the fuel pin for both open and closed gap conditions, as well as the location and temperature of the hot spot in the cladding.) The equations for calculating the fuel and cladding average temperatures are

$$T_{fav} = \int_{\text{time step}} ((P_n + P_D(1-g) - (T_{fav} - T_{clav})U_f) / M_f C_f) dt , \text{ and} \tag{13}$$

$$T_{clav} = \int_{\text{time step}} ((T_{fav} - T_{clav})U_f - (T_{clav} - T_{cav})U_{cl} / M_{cl} C_{cl}) dt . \tag{14}$$

The coolant temperature at the exit of the reactor core and the average coolant temperature in the core are calculated, respectively, as:

$$T_{ex} = \int_{\text{time step}} ((P_D g + (T_{clav} - T_{cav})U_{cl} - \dot{m} C_c (T_{ex} - T_{in})) / M_c C_c) dt , \tag{15}$$

$$T_{cav} = \frac{(T_{in} + T_{ex})}{2} . \tag{16}$$

Reactor Hydraulic Model

Figure 2 shows a cross sectional view of the reactor core and a schematic of a single flow channel in the core. The coolant inlet and outlet nozzles of the reactor core are connected at the bottom of the reactor vessel. The coolant enters the vessel and flows up through the annular gap between the reactor vessel and the core baffle until it reaches the upper plenum. Then, the coolant reverses its direction and flows downward through the core to the lower plenum. A wire wrap separates and keeps a uniform spacing between the fuel rods in the core. The coolant leaves the reactor vessel through the outlet nozzle to the hot leg of the primary coolant loop. The reactor hydraulic model calculates the pressure drop in the reactor core as the sum of the pressure losses in the annulus between reactor vessel and core baffle, ΔP_a, and in the flow channels in the reactor core, ΔP_c, (see Figure 2). In this research, the Chiu-Rohsenow-Todreas (CRT) model is used to calculate the pressure losses in the triangular fuel channels in the core with wire wrapped fuel rods (Walter and Reynolds 1981):

$$\Delta P_c = \left(\frac{L_c}{D_e^c}\right) f_c \left(\frac{\dot{m}^2}{2\omega_c A_c^2}\right) \left(1 + G \frac{A_r}{A} \frac{D_e^c}{H} \frac{p^2}{((\pi p)^2 + H^2)}\right) . \tag{17}$$

In the annular region between the core baffle and the reactor vessel, the pressure loss is calculated using the equivalent hydraulic diameter concept:

$$\Delta P_a = \left(\frac{L_a}{D_e^a}\right) f_a \left(\frac{\dot{m}^2}{2\omega_c A_a^2}\right) . \tag{18}$$

Primary Coolant Loop

As shown in Figure 3, the primary coolant loop of the integrated SP-100 system consists of three flow

Table 3 *Fission Product Groups for Decay Heat Calculations.*

Group	Decay Constant, $\lambda_i (s^{-1})$	Fission yield, χ_i
1	1.0×10^{-5}	0.27
2	3.0×10^{-3}	0.21
3	1.0×10^{-2}	0.16
4	4.0×10^{-2}	0.36

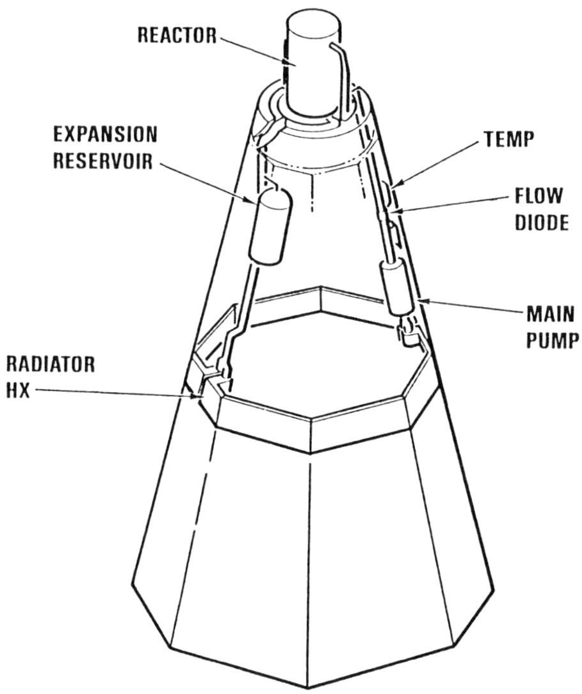

Figure 2 *Cross Sectional View of the Reactor Core and a Flow Channel.*

paths and twelve heat exchangers with common inlet and exit headers. The coolant flows out of the core through three coolant flow paths, each equipped by a TE-EM pump. The coolant flowing out of the pumps is combined in the hot header and fed to the primary heat exchangers; the flow exiting the heat exchangers then combines in the cold header and returns to the reactor core. The major components of the primary coolant loop include the TE-EM pumps, the pipings, and the heat exchangers. In the TE-EM pump, the electric current is supplied by a separate TE converter which is placed between coolant ducts in the pump and the pump's radiator. It should be noted that the TE-EM pumps are located in the hot leg of the primary coolant loop to maximize the TE converter's efficiency and, consequently, the overall pump efficiency (El-Genk and Seo 1986b).

The primary coolant heat exchangers transfer the thermal energy from the primary coolant by conduction through the walls of heat exchangers to the TE generators in the system's main energy converter (GE 1983). The primary coolant loop model consists of four coupled models:

- A TE-EM pump model,
- A piping hydraulic model,
- A heat exchanger hydraulic model, and
- A heat exchanger thermal model.

These models are described in the following subsections.

TE-EM Pump Model

The TE-EM pump is a self-induced, direct current electromagnetic pump equipped with a hiperco-27

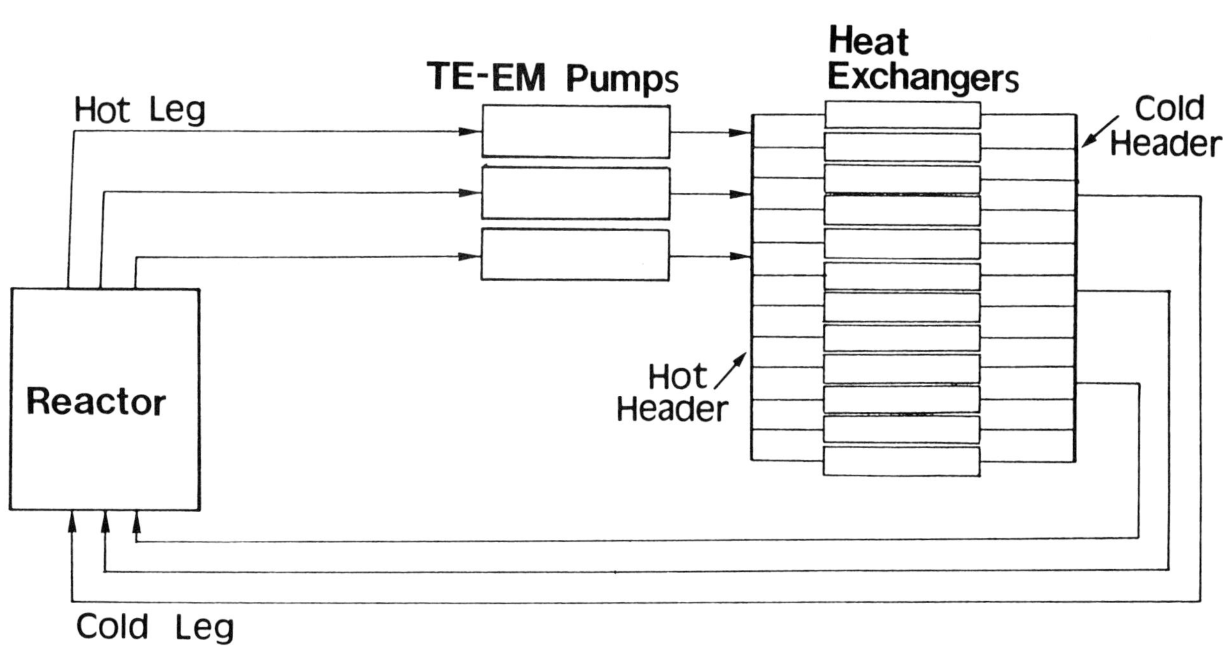

Figure 3 *A Line Diagram of the Primary Coolant Loop.*

magnet and SiGe/GaP TE converter. Each pump has two coolant ducts; the coolant enters the pump through one duct, reverses its direction 180 degrees, and exits through the other duct. Both ducts are surrounded by an S-shaped self-induced magnet which provides the magnetic flux necessary for operating the pump. The magnet is thermally insulated from the walls of the coolant ducts by multifoil insulation which keeps the temperature of the magnet well below its Curie point (the temperature at which the ferromagnetic substance becomes para-magnetic). The magnet is conductively coupled to the pump's radiator for cooling. The average temperature of the magnet, as well as the optimum number of insulation foils, can be calculated using a two-dimensional thermal analysis model (El-Genk et al. 1986a and Buksa 1987). Subsequently, the magnetic flux strength can be determined as a function of the magnet's average temperature.

The electric current provided by the pump's TE converter flows through the coolant ducts in a perpendicular direction to both the magnetic field and the coolant flow. The TE generators, themselves, are conductively coupled at their hot and cold shoes to the coolant flowing through the pump's ducts and to the pump's heat pipe radiator, respectively. Six potassium heat pipes transmit the waste heat from the cold shoe of the pump's TE converter to the pump's radiator. Within the converter, the TE generators are connected in parallel to maximize the electric current supplied to the pump. Figure 4 contains a cross sectional view of the TE-EM pump and Figure 5 shows a line diagram describing the pump's operation. Because the pump's TE converter continues to operate as long as there is a temperature differential between the coolant and pump's radiator, the TE-EM pump provides a passive and extremely reliable means for circulating the primary coolant.

The TE-EM pump model assumes a constant coolant flow area and a uniform magnetic flux density through the coolant ducts. The model also assumes the TE converter's cold shoe temperature to be equal to the pump's radiator surface temperature. Thus, the model neglects the temperature drop between the evaporator and the condenser of the radiator's heat pipes.

El-Wakil (1971) and Barnes (1953) give the pressure rise across the pump duct as

$$\Delta P_{\text{pump}} = \left(\frac{B}{10^4 b}\right)\left(\frac{R_w R_f}{\bar{R}}\right) I - \left(\frac{B^2}{10^8 b^2}\right)\left(\frac{R_w + R_f}{\bar{R}}\right) Q \,, \quad (19)$$

where $\bar{R} = R_e R_w + R_w R_f + R_f R_e$, and the EM pump efficiency can be calculated by:

$$\eta_{EM} = \frac{\Delta P_{\text{pump}} Q}{I\left(\frac{BQ}{10^4 b} + \frac{10^4 b R_e \Delta P_{\text{pump}}}{B}\right)} \,. \quad (20)$$

However, the overall efficiency of the TE-EM pump is calculated as the product of the EM pump efficiency and the TE conversion efficiency:

$$\eta_{\text{overall}} = \eta_{EM} \cdot \eta_{TE} \,. \quad (21)$$

More details on the TE-EM pump model can be found elsewhere (El-Genk et al. 1986a and Buksa 1987).

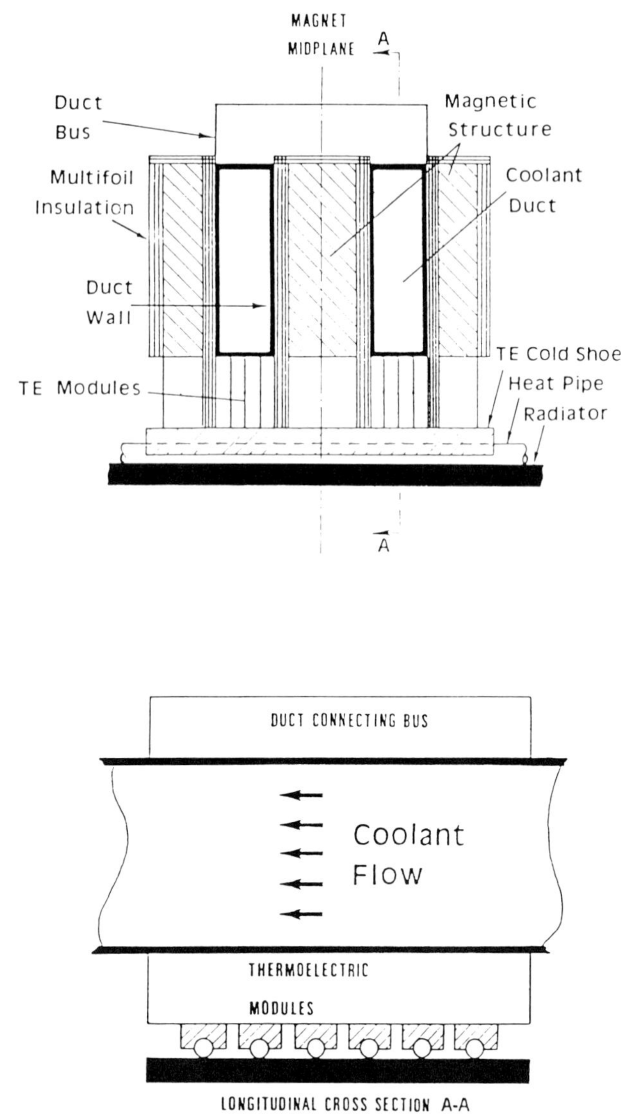

Figure 4 *Cross Sectional Views of the Self-Induced Thermoelectric-Electromagnetic (TE-EM) Pump.*

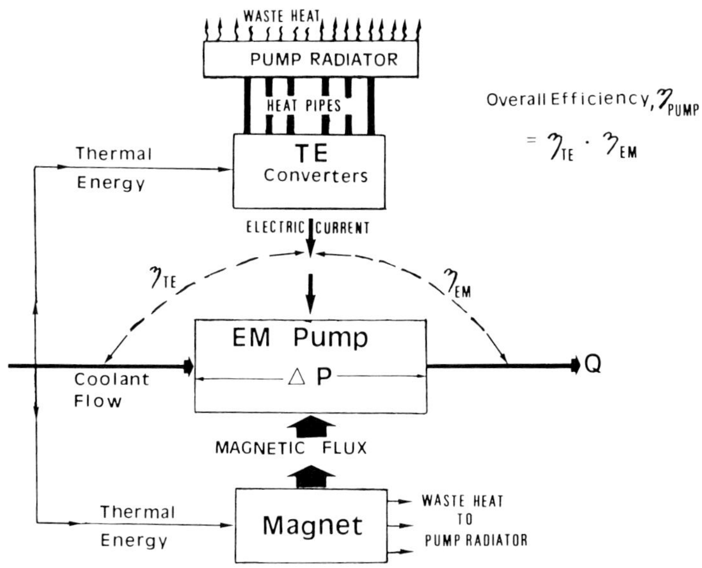

Figure 5 *A Schematic Representation of the TE-EM Pump Model.*

Piping Hydraulic Model

Because the primary coolant piping system is thermally insulated by insulation multifoils, the heat losses from the pipes were assumed to be negligible. Consequently, the coolant temperature at the core exit can be assumed to equal that of the coolant inlet temperature to the TE-EM pumps. Additionally, the coolant inlet and outlet temperatures of the heat exchangers can also be considered equal those at the exit of the pump and at the inlet of the reactor core, respectively. Thus, the pressure drop in the primary coolant loop piping system can be calculated as

$$\Delta P = \left(\frac{L_p}{D_p}\right) f_p \left(\frac{\dot{m}^2}{2\omega_c A_p^2}\right) . \quad (22)$$

Heat Exchanger Hydraulic Model

As mentioned previously, the primary coolant loop employs a total of 12 heat exchangers connected in parallel with common hot and cold headers. Each heat exchanger consists of 2 header tubes, a hot and a cold header, connected by 30 rectangular heat transfer channels (see Figure 6a). The coolant flow area of the headers is sloped in a similar fashion so that the coolant flow rate in each rectangular channel is the same. This uniform coolant flow results in the same temperature gradient in the coolant channels in the direction of the flow. From the mass conservation,

$$\dot{M}_{tot} = \sum_{i=1}^{N} \dot{m}_i . \quad (23)$$

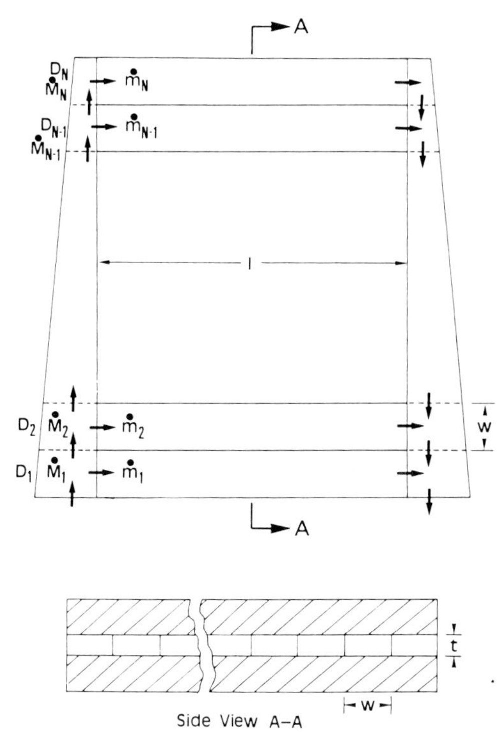

(a) Hydraulic Model

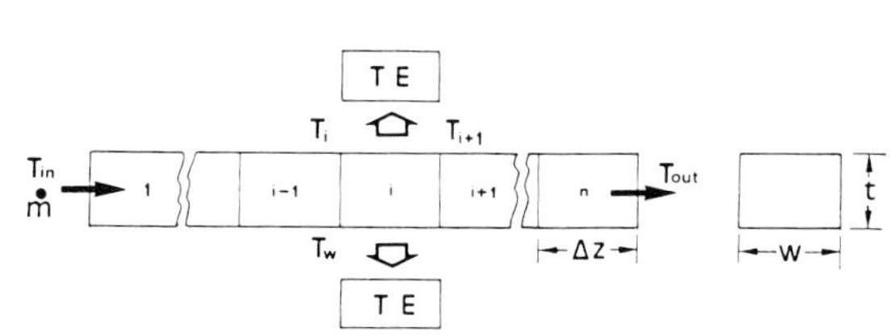

(b) Thermal Model

Figure 6 *Diagrams of the Primary Heat Exchanger Model.*

The TE generator segments are placed on both sides of the heat exchanger perpendicular to the direction of the flow. Although the hot shoe temperatures within each segment are the same, they are different from those for the next segment. The TE generators in each segment are connected in series while the segments are connected in parallel.

The heat exchanger hydraulic model calculates the coolant flow rate in each coolant channel and the total pressure drop in the heat exchanger. To ensure a proper coolant flow in the heat exchanger, the pressure should be balanced in all the coolant channels. For example, the pressure drop in the first channel should equal the sum of the pressure losses in the second channel and in the segments of the hot and cold headers measured between the centers of these two channels (see Figure 6a). Therefore, the difference in friction pressure loss between any two subsequent channels should equal the sum of the pressure losses in the hot and cold headers between the centers of these two channels; thus,

(a) For the 1st and 2nd channels:

$$\Delta P_1^c - \Delta P_2^c = \frac{1}{2}(\Delta P_{1H}^H + \Delta P_{2H}^H + \Delta P_{1C}^H + \Delta P_{2C}^H) \quad (24a)$$

(b) For the i-th and (i + 1)-th channels:

$$\Delta P_i^c - \Delta P_{i+1}^c = \frac{1}{2}(\Delta P_{iH}^H + \Delta P_{(i+1)H}^H + \Delta P_{iC}^H + \Delta P_{(i+1)C}^H) \quad (24b)$$

(c) For the (N-1)-th and N-th channels:

$$\Delta P_{N-1}^C - \Delta P_N^c = \frac{1}{2}(\Delta P_{(N-1)H}^H + \Delta P_{NH}^H + \Delta P_{(N-1)C}^H + \Delta P_{NC}^H) \,. \quad (24c)$$

The pressure loss in the i-th segment heat exchanger header is calculated by

$$\Delta P_i^H = \left(\frac{w}{D_i}\right) f_i^H \left(\frac{\dot{M}_i^2}{2\omega_c A_i^2}\right) , \quad (25)$$

and the pressure loss in the i-th coolant channel is calculated by

$$\Delta P_i^c = \left(\frac{l}{d_e}\right) f_i^c \left(\frac{\dot{m}_i^2}{2\omega_c a_i^2}\right) . \quad (26)$$

Substituting Equations (25) and (26) into Equations (24a,b,c) results in a set of n-1 coupled equations. These equations and Equation (23) are solved simultaneously to calculate the mass flow rates as well as pressure losses in the coolant channels and in the headers of the heat exchanger.

The friction factor in Equations (17), (18), (22), (25), and (26) was evaluated as a function of the flow regime:

(a) Laminar flow ($Re \leq 2100$):

$$f = \frac{64}{Re} \quad (27a)$$

(b) Transition flow ($2100 \leq Re \leq 3000$):

$$f = 0.03048 + 1.522 \times 10^{-5} Re \quad (27b)$$

(c) Turbulent flow ($3000 \leq Re \leq 3 \times 10^6$):

$$f = 0.0056 + \frac{0.5}{Re^{0.32}} . \quad (27c)$$

For the transition flow regime (Equation (27b)), the value of the friction factor was calculated using a linear interpolation between the laminar and turbulent flow regimes. This interpolation was necessary to obtain a continuous function for f as a function of Re. Such a function enables the integrated system model to respond to any changes in the reactor power and, consequently, to the changes in the coolant flow condition from laminar to turbulent or vice versa. The continuous curve for the friction factor as a function of Re is shown in Figure 7.

Heat Exchanger Thermal Model

The thermal model of the heat exchanger calculates the coolant and wall temperature distributions in the coolant channels in the direction of the coolant flow. These temperatures are calculated through the coupling of the heat exchanger thermal model to the main TE converter model.

As shown in Figure (6b), the energy balance equation for i-th segment in a coolant channel is given by

$$P_i = \dot{m}C_c(T_i - T_{i+1}) = h_i A_w (T_{av} - T_{wav}) . \quad (28)$$

In Equation (28), T_{i+1} was calculated using an iteration scheme with the assumption that $T_{i+1}^{(o)} = \beta T_i$, where β is an arbitrary constant less than 1.0:

$$T_{i+1}^{(n+1)} = T_i - \frac{h^{(n)}A_w}{\dot{m}C_c}(T_{av}^{(n)} - T_{wav}^{(n)}) . \quad (29)$$

For the liquid metal coolant, the heat transfer coefficient between the coolant and the walls of the heat exchanger (El-Wakil 1971) is

$$h = \frac{k}{d_e \gamma}(5.8 + 0.02\mathrm{Pe}^{0.8}) , \quad (30)$$

where γ is a correction factor that accounts for the fact that heat removal from the coolant channel occurs at two walls instead of a single wall (see Figure 8). As Equation (30) shows, the heat transfer coefficient for liquid metal coolant is constant for all practical applications, where Reynolds number is less than 2×10^4, and independent of the coolant mass flow rate in the loop.

TE Converter Model

The TE converter model allows the TE material properties (the Seebeck coefficient, α, the thermal conductivity, k, and the electrical resistivity, ρ) to vary with temperature (El-Genk and Seo 1986b). The TE model is capable of handling any combinations of boundary conditions at the hot and cold shoes of the TE converter and of performing both steady state and transient calculations. The TE generators both in the TE-EM pump and in the system's main energy converter are heated by the primary coolant and radiatively cooled into space.

The transient governing equation for the temperature distributions in the P and N legs of a TE generator is given by

$$\frac{\partial}{\partial x}\left(k\frac{\partial T}{\partial x}\right) + \rho J^2 + \tau J\frac{\partial T}{\partial x} = \omega C\frac{\partial T}{\partial t} . \quad (31)$$

To solve Equation (31), with the prescribed initial and boundary conditions, an additional equation is needed to calculate the electric current, I.

$$I = \frac{\alpha_{pn}(T_h - T_c)}{R_L + R_n + R_p} . \quad (32)$$

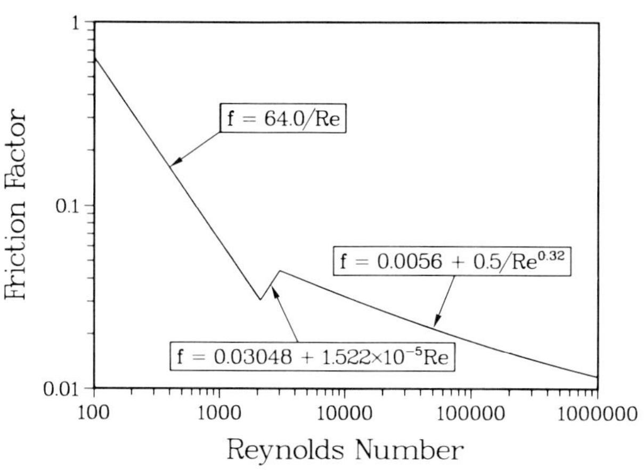

Figure 7 *Friction Factor as a function of Reynolds Number.*

By definition, the electric voltage generated across the external load, R_L, is given as (Angrist 1965 and Soo 1968):

$$V = R_L I = \alpha_{pn}(T_h - T_c) - (R_n + R_p)I , \quad (33)$$

where $R_n = \rho_n L/A$ and $R_p = \rho_p L/A$. The conversion efficiency of the TE generator is calculated by

$$\eta_{TE} = \frac{P_e}{P_h} , \quad (34)$$

where the thermal input at the hot shoe, P_h, and the electric power output, P_e, are given, respectively, as:

$$P_h = P_p + q_h A - \frac{1}{2}P_J - \frac{1}{2}P_T , \quad (35)$$

$$P_e = I^2 R_L . \quad (36)$$

Because of the nonlinear, nonhomogeneous nature of Equation (31), as well as the time dependent and/or nonlinear boundary conditions, a closed form solution is not attainable. However, a solution was obtained using the finite element method coupled to an implicit time integration algorithm (Bathe 1978). More details on the method of solution and model verification are given elsewhere (El-Genk and Seo 1986b).

Integrated System Model

The integrated TE SP-100 system model (see Figure 9) was assembled through coupling the models for the nuclear reactor, for the primary coolant loop, including the primary heat exchangers and the TE-EM pumps, and for the main TE converter. To use this

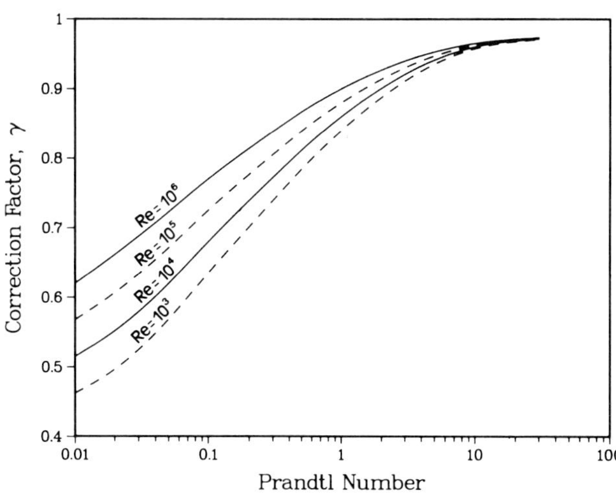

Figure 8 *Correction Factor for the Heat Transfer Correlation in the Primary Heat Exchanger.*

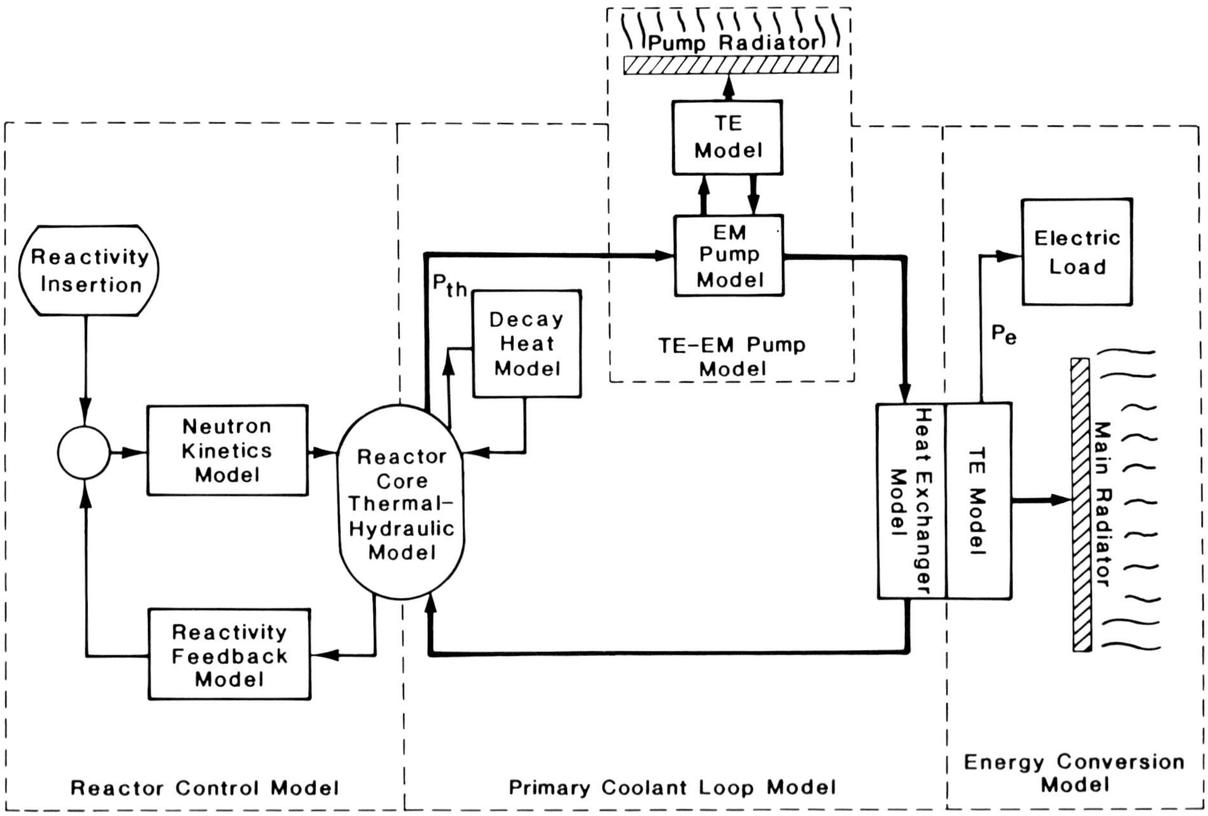

Figure 9 *Block Diagram of the SNPSAM (Space Nuclear Power System Analysis Model).*

model, once the system design parameters are given, the input to the integrated system model must provide the amount of external reactivity or the reactor thermal power, the mode of operation (steady state or transient), and the ambient temperature.

The input to the reactor model is the sum of the external reactivity insertion by the control drums and the reactivity feedback generated by the change in the core temperatures and densities. The total reactivity input is fed to the neutron kinetics model which calculates the fission power and produces the input for the decay heat model. These results are then fed to the core thermal model which calculates the average temperatures of the fuel, the cladding, and the coolant as well as the coolant temperature at the exit of the reactor core. The output of the core thermal model is coupled to the reactivity feedback model to provide a continuous feedback.

In the integrated system model, the thermal coupling of the reactor with the primary coolant loop components is achieved via the primary coolant. Thus, the inlet coolant temperature to the TE-EM pumps can be assumed to be equal to that of the outlet temperature from the reactor core. In the TE-EM pumps, the conversion of thermal energy into electricity is accomplished by the TE generators attached onto the pump's ducts. These generators receive heat from the coolant and partially convert it into electrical power which is then fed back to the coolant passing through the ducts. The waste heat from the pump's TE converters is radiated into space through the pump's radiator. In the TE model, therefore, the hot side temperature of the TE converter is assumed to be the same as the coolant temperature. Additionally, the model also neglects the temperature drop across the thermal boundary layer of the coolant at the duct wall as well as across the duct wall because of the high thermal conductivity of the liquid metal coolant and the small thickness of the duct wall.

The inlet temperature to the heat exchanger is taken to be the same as the outlet coolant temperature from the TE-EM pump, and the outlet coolant temperature from the heat exchangers is the same as the coolant inlet temperature to the reactor core. The amount of energy transmitted from the coolant flowing through the heat exchanger channel to the TE converter depends on both the coolant bulk temperature and the temperature of the heat exchanger channel's walls that are in contact with the hot shoes of the TE generators. Therefore, an iteration scheme is used to calculate the temperatures at the hot and cold shoes of the TE generators as functions of the coolant bulk temperature and the space ambient temperature. Subsequently, the system's overall energy and momentum equations are solved simultaneously to calculate the temperatures and mass flow rate of the coolant in the primary coolant loop.

From the overall energy balance of the integrated system, the difference between the energy transferred to the coolant in the reactor and the energy extracted from the coolant in both the TE-EM pumps and the heat exchangers should equal the change in the sensible heat of the primary coolant loop components including the coolant itself, that is,

$$\begin{pmatrix}\text{reactor}\\ \text{thermal}\\ \text{power}\end{pmatrix} - \begin{pmatrix}\text{heat removal}\\ \text{in primary}\\ \text{heat exhanger}\\ \text{and TE-EM pump}\end{pmatrix} = \begin{pmatrix}\text{change in sensible}\\ \text{heat of primary}\\ \text{loop components}\end{pmatrix} \quad (37)$$

$$P_{\text{reactor}} - (P_{HX} + P_{\text{TE-EM}}) = \sum_i^N M_i C_i \frac{dT_i}{dt}$$

From the overall momentum balance of the integrated system, the total pressure losses in the primary loop including the reactoı, a heat exchanger (heat exchangers are connected in parallel), and pipings should equal the pressure rise in a single TE-EM pump (pumps are connected in parallel):

$$(\text{total pressure loss}) = (\text{total pressure gain}) \quad (38)$$
$$\Delta P_{\text{total}} = \Delta P_{\text{pump}}$$

An estimate of the total pressure losses in the primary coolant loop can be obtained by multiplying the sum of the pressure losses in the reactor core, pipings, and a heat exchanger by a weighting factor of 1.9 in order to account for the pressure losses in the bendings, couplings, and headers (Walter and Reynolds 1981). The coupled system's overall momentum balance and energy balance equations (Equations (37) and (38)) are then solved to calculate the coolant mass flow rates and temperatures in the various components of the primary coolant loop.

The total electric power output of the system is calculated as

$$P_e = \eta_{TE} \cdot P_{HX} , \quad (39)$$

where η_{TE} is the calculated conversion efficiency of the TE converter (El-Genk and Seo 1986b). The overall conversion efficiency, η_s, of the integrated power system is given as

$$\eta_s = \frac{P_e}{P_{\text{reactor}}} = \eta_{TE}\left(\frac{P_{HX}}{P_{\text{reactor}}}\right) . \quad (40)$$

The design parameters of the TE SP-100 system analyzed by using this model are listed in Table 4. Although these parameters could be different from those of the concept selected by SP-100 program, the results of the analysis presented later in this paper are still valid. As

Table 4 *Design Parameters of the TE SP-100 System in SNPSAM Input.*

Parameter	Value
Nuclear Reactor:	
Number of Fuel Rods	1296
Fuel rod lattice	triangular
Fuel rod diameter (cm)	0.74
Fuel rod pitch (cm)	0.7981
Core height (cm)	45.4
Core baffle diameter (cm)	32.8
Reactor vessel diameter (cm)	109.0
Total fuel mass (kg)	25.5
TE-EM Pump:	
Number of pumps	3
Number of ducts/pump	2
Duct height (cm)	5.8
Duct width (cm)	1.53
Duct length (cm)	22.8
Duct wall thickness (cm)	0.064
Nominal magnetic flux density (gauss)	1000
Nominal magnet temperature (K)	950
Number of TE couples	204
TE element length (cm)	0.64
TE element cross-sectional area (cm^2)	0.174
Number of heat pipes	6
Heat pipe cross sectional area (cm^2)	10.75
Number of multifoils	30
Primary Heat Exchanger:	
Number of heat exchangers	12
Number of coolant channels/heat exchanger	30
Header length (cm)	80.0
Coolant channel length (cm)	25.4
Coolant channel width (cm)	2.54
Coolant channel height (cm)	0.35
Coolant channel wall thickness (cm)	0.01
Header inlet diameter (cm)	4.13
Header end diameter (cm)	1.9
Number of TE couples/channel	480
Primary Coolant Loop:	
Number of loops	3
Hot and cold leg length (cm)	120.0
Hot and cold leg diameter (cm)	8.26
Heat exchanger leading pipe length (cm)	30.0
Heat exchanger leading pipe diameter (cm)	4.13
TE Converter and Radiator:	
Number of TE couples	172800
Number of TE couples connected in Series	1440
Number of TE couples connected in parallel	120
TE element length (cm)	0.38
TE element cross-sectional area (cm^2)	0.14
Number of heat pipes	240
Number of TE couples/Heat Pipe	720
Radiator area/TE couple (cm^2)	4.2
Heat pipe cross-sectional area (cm^2)	10.75

shown in Figure 10, the user of the SNPSAM can select the operating parameters from a menu that includes the operating conditions (steady state or transient), coolant type (Li or NaK-78), and TE material (SiGe/GaP or SiGe). Thus, the integrated system model and the analysis presented here have more universal applications.

Results and Discussion

The integrated system model, SNPSAM, was used to predict the system's behavior during steady state and transient operation. The only inputs to the integrated model (SNPSAM) for steady state calculations are the reactor thermal power and space ambient temperature. However, for the transient calculations, the input parameters include the external reactivity insertion, initial reactor thermal power before the transient, and the space ambient temperature. SNPSAM was used to study the effects of the reactor thermal power and the liquid metal coolant type on the system's performance during steady state operation. In addition, the effects of sonic limit of the radiator's heat pipes on the waste heat rejection and on the system's operating characteristics were studied. Finally, the effects of the initial steady state power, and the reactivity insertion mode (a step or a linear function) on the system's maximum electric power attainable and the settling time were investigated.

Limitation on Waste Heat Rejection

The waste heat from the TE SP-100 system's main energy converter is radiated into space from the surface of the main radiator. Potassium heat pipes are used to thermally couple the cold side of the TE converter to the radiator's panels. Because the heat pipes are good thermal conductors, the temperature drop across the heat pipe can be neglected during steady state operation. Figure 11 compares the sonic limit (Prenger 1979) for a potassium heat pipe (10.75 cm^2 cross sectional area) with the radiator's heat rejection capability per heat pipe for an ambient temperature of 250 K. As this figure shows, the radiator operation curve intersects the heat pipe sonic limit curve at a temperature of 680 K (or reactor thermal power of 750 kWt). Below this temperature, the system's heat rejection is restricted by the heat pipe's sonic limit which is significantly lower than the radiator's heat rejection capability.

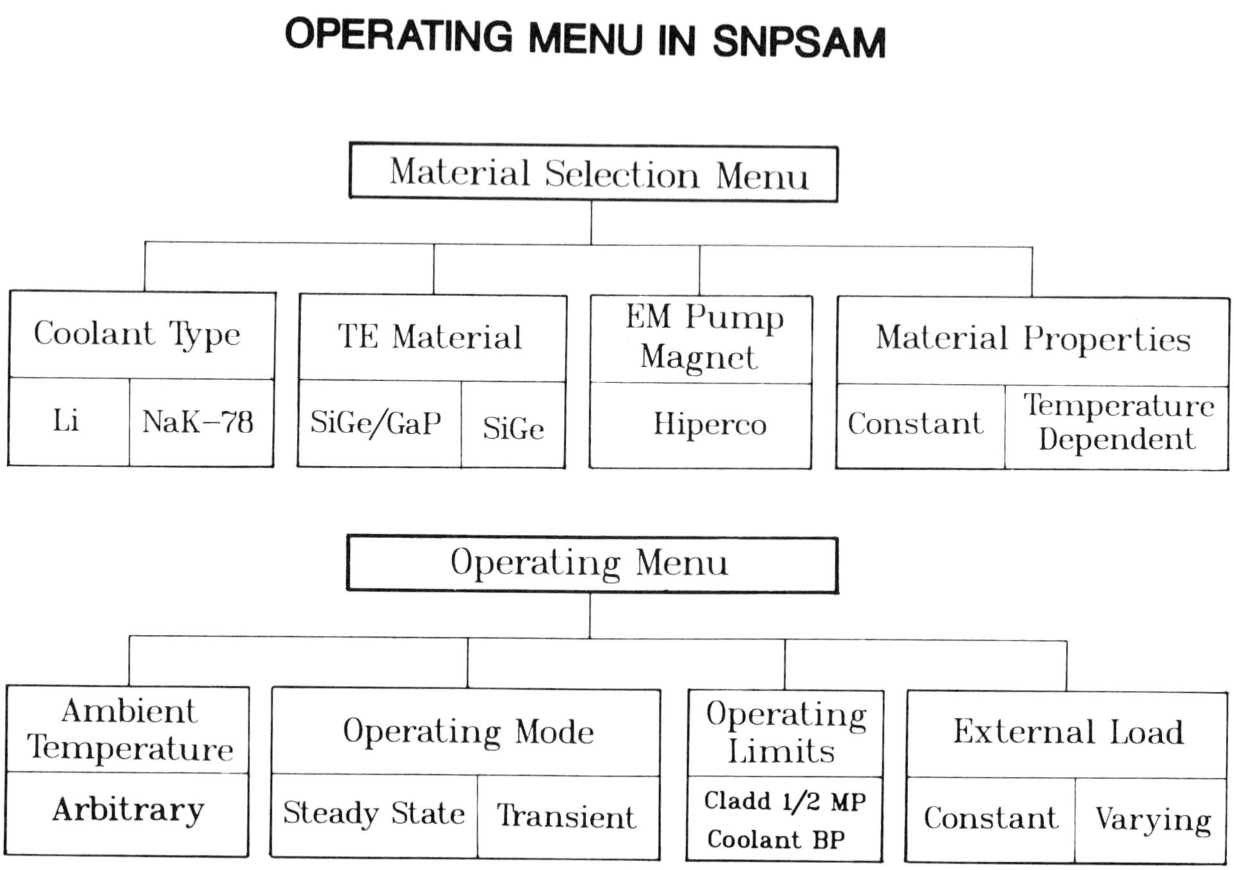

Figure 10 *Calculational Menu in the SNPSAM.*

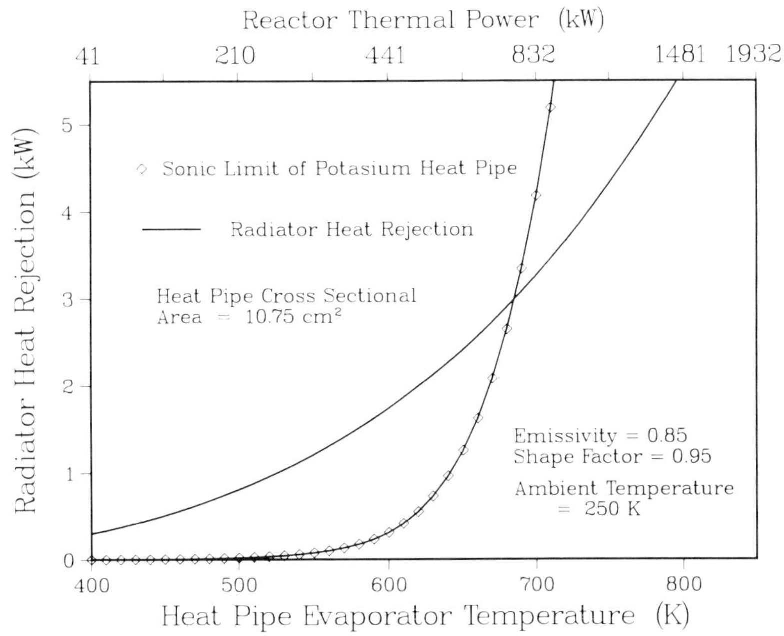

Figure 11 *Effect of the Sonic Limit on the Radiator's Heat Rejection per Heat Pipe.*

These results indicate that the radiator's heat pipe will not only influence the system's operation at low power levels but will also affect the decay heat thermal management following a reactor shutdown. As shown in Figure 11, the sonic limit for the radiator's potassium heat pipes controls the system's heat rejection for the reactor thermal power less than 750 kWt for Li cooled system; this corresponds to a thermal power of 800 kWt for NaK-78 cooled system. At higher power levels, the system's heat rejection is limited by the radiator's capability rather than by the sonic limit of the radiator's heat pipes. It should be noted, however, that the reactor thermal power at which the system's heat rejection becomes controlled by the radiator's heat pipes depends on several design parameters. These parameters are the heat pipe working fluid and the cross sectional flow area of the heat pipe, the radiator's surface area available per heat pipe, and the radiator's surface emissivity and view factor. The effect on the system's overall performance of such a restriction on the waste heat rejections due to the heat pipe's sonic limit at lower reactor thermal powers will be discussed in the following sections.

Steady State Results

Table 5 compares the steady state operating parameters calculated using the integrated system model (SNPSAM) for a reactor thermal power of 2.0 MW with those reported for the earlier version of the General Electric (GE) Company's SP-100 baseline design (GE 1983). As indicated in Table 5, the predicted operating parameters of the integrated TE SP-100 system were generally in good agreement with those reported for the GE baseline design. It should be noted, however, that the present predictions were based on the system's design parameters listed in Table 4, which could be different from those of the GE's baseline design.

In the following steady state analysis, the reactor thermal power ranged from as low as 100 kWt to the maximum value at which one of the system's operation limit is reached. In SNPSAM, the system's operation limits are specified as the coolant boiling point and/or half the melting temperature of the cladding material.

Core and Coolant Temperatures

Figure 12 shows the calculated fuel and cladding temperatures as functions of the reactor thermal power. As this figure indicates, at any given reactor thermal power, the fuel and cladding temperatures are almost independent of the coolant type. However, the maximum reactor thermal power strongly depends on the coolant type. In this figure, the maximum power obtainable is represented with bullets for the Li coolant and with blocks for the NaK-78 coolant. The re-

Table 5 *Comparison of Operating Parameters of SNPSAM and GE Baseline Design.*[a]

Parameters	SNPSAM[b]	GE[c]
Nominal thermal power (kWt)	2000	2000
Electric power output (kWe)	112	110
System efficiency (%)	5.6	5.5
Temperatures (K)		
Core inlet	1254	1310
Core outlet	1284	1350
Fuel average	1376	1445
Cladding average	1288	1350
TE hot shoe	1237	1273
TE cold shoe	857	823
Coolant mass flow rate (kg/s)	14.1	13.2

[a]Both systems used Li coolant.
[b]SNPSAM used SiGe/GaP TE material and neglected heat losses in the primary loop piping.
[c]GE used LaS_x TE material and considered the heat losses from the primary loop.

sults show that the operation limit for the Li cooled system is half the cladding (Nb-1%Zr) melting temperature (1370 K) and the coolant boiling temperature (1057 K) for the NaK-78 cooled system (see Figure 13). The calculated maximum steady state reactor thermal power achievable is 2340 kWt and 1150 kWt for the Li and NaK-78 cooled system, respectively. This suggests that, unlike the NaK-78 cooled systems, the growth potential of the Li cooled system is very good. Depending on the exact value of the elastic-ductile transition temperature of the cladding material, the maximum power of the SP-100 reactor core could be increased as long as this limit is kept safely below the boiling point of the Li coolant (1590 K at 0.1 MPa) at the system's design pressure. As shown in Table 6, the maximum electric power level of the TE SP-100 system that is cooled by the liquid Li strongly depends on the maximum cladding temperature allowed during steady state operation. For example, increasing the maximum operating temperature of the cladding from half to only 55% of its melting point increases the electric power output from 143 kWe to approximately 207 kWe. This increase in the system's electric power output is caused by both the increase in the reactor thermal power and conversion efficiency of the TE converter.

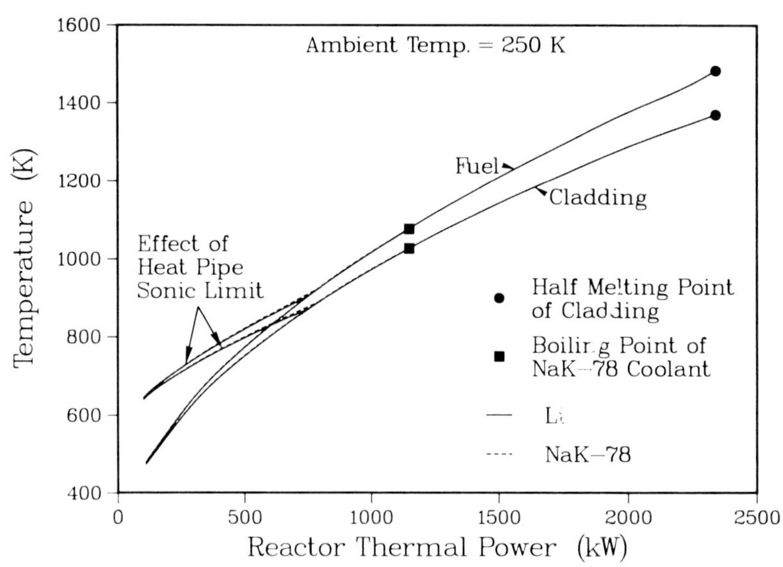

Figure 12 *Steady State Behavior of the Fuel and Cladding Temperatures as a function of the Reactor Thermal Power.*

As showm in Figures 12 and 13, the temperatures of the reactor core and of the coolant increase nonlinearly as the reactor thermal power increases. Also, for a given reactor thermal power, the coolant temperature at the core exit for the NaK-78 cooled system is greater than that for the Li cooled system because the specific heat of the NaK-78 coolant is about ¼ that of the Li coolant. At lower reactor thermal powers, the constraint on the system's waste heat rejection due to the sonic limit of the radiator's heat pipe (see Figure 11) causes the core and coolant temperatures to be much higher than those obtained when the effect of the radiator's heat pipes sonic limit was neglected.

Energy Conversion and Electric Power Output

As shown in Figures 14 and 15, the electric power output and the energy conversion efficiency of the system increase nonlinearly as the reactor thermal power increases. These results suggest that extrapolating the system's performances to different operating power level assuming a linear system (that is, constant conversion efficiency) is inappropriate. The results also suggest that operating the SP-100 system at a high power level has two interrelated advantages: higher conversion efficiency and, consequently, higher electrical power output.

As Figure 15 shows, the system's maximum electric power achievable with NaK-78 coolant is less than half that with Li coolant based on the aforementioned operating limits of the system. Achieving a higher electric power with NaK-78 coolant requires employing an energy converter with higher conversion efficiency and/or pressurizing the primary coolant loop to increase the boiling temperature of the NaK-78 coolant. Although, for a given reactor thermal power, the coolant mass flow rate for NaK-78 cooled system is much higher than that for Li cooled system, the heat transfer coefficient in the heat exchanger is almost independent of the coolant mass flow rate (see Equation (30)). This constant heat transfer coefficient results in the same TE conversion efficiency and electric power output at the same reactor thermal power regardless of the coolant type. Figures 14 and 15 demonstrate that the radiator's heat pipes sonic limit only negligibly affects both the conversion efficiency and the electric power output at lower reactor thermal power values. These negligible effects occur because, at the same reactor thermal power, the temperature differential across the TE converter remains constant regardless of the coolant temperature.

Mass Flow Rate and Specific Pumping Power

The mass flow rate in the primary coolant loop depends on both the total friction pressure losses in the primary coolant loop and the overall efficiency of the TE-EM pumps. These quantities, in turn, depend on the reactor thermal power because changing the reactor thermal power also changes the coolant temperature. Figures 16 and 17 present the predicted coolant mass flow rates and specific pumping powers as a function of the reactor thermal power. The specific pumping power is defined as the thermal power consumed by the pump's TE converter to remove 1 kWt of the reactor thermal power. The coolant mass flow rate was determined from the system's coupled energy and momentum balance in the primary coolant loop (Equations (37) and (38)). As Figure 16 shows, at a given reactor thermal power, the NaK-78 coolant mass flow rate was signigicantly higher than that for Li coolant. Such higher mass flow rates for NaK-78 cooled system occur because the specific heat of the NaK-78 coolant is much lower than that of the Li coolant; thus, a higher circulation rate is required for the former to remove the same reactor thermal power.

As delineated in Figure 17, the specific pumping power for the NaK-78 cooled system is higher than that for the Li cooled system because of the higher

Table 6 *Effect of Cladding Temperature on the Li Cooled TE SP-100 System's Electric Power Output.*

Limiting Factors	Maximum Reactor Thermal Power (kWt)	Maximum Electric Power Output (kWe)
50% Cladding melting point (1370 K)	2340	143
55% Cladding melting point (1507 K)	3035	207
Coolant boiling point (1590 K) (58.7% cladding melting point)	3500	260

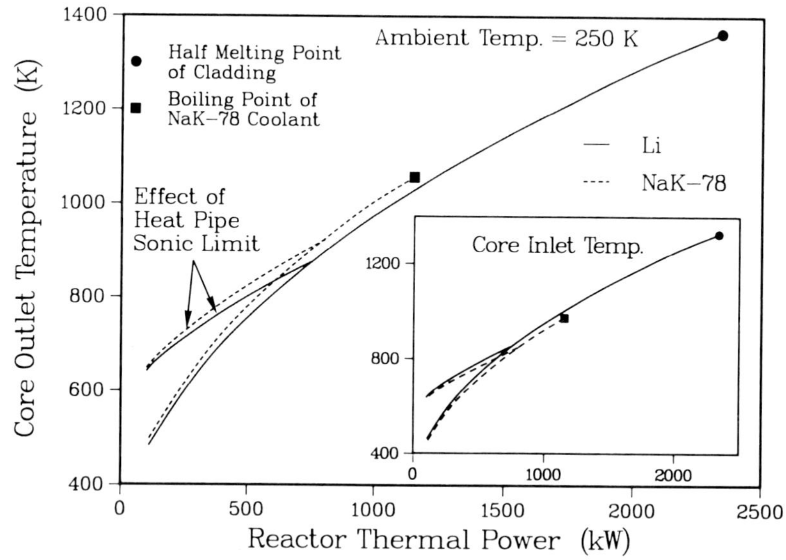

Figure 13 *Steady State Behavior of the Coolant Temperature in the Core as a function of the Reactor Thermal Power.*

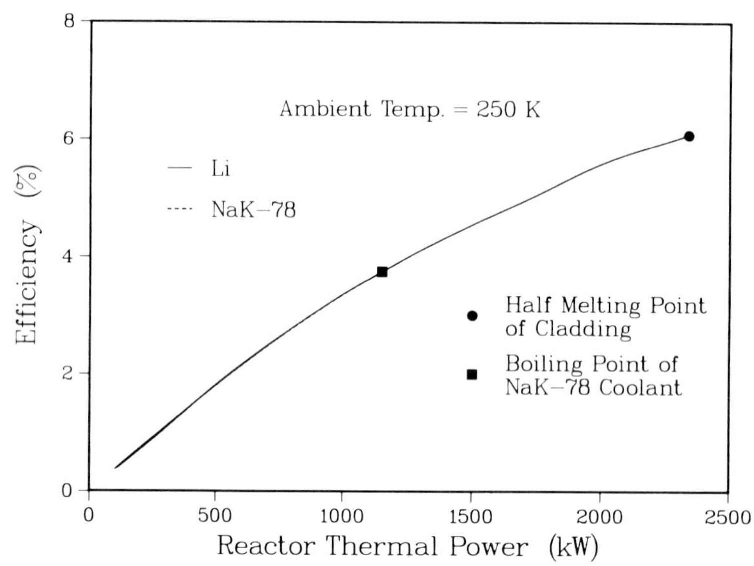

Figure 14 *Steady State Behavior of the System's Conversion Efficiency as a function of the Reactor Thermal Power.*

mass flow rate requirements for the former at the same reactor thermal power. For both coolants, however, the specific pumping power decreases as the reactor thermal power increases. The increase in the reactor thermal power increases the coolant temperature and, consequently, the overall efficiency of the TE-EM pumps. At the same time, increasing the reactor thermal power increases the coolant flow rate and, consequently, the friction pressure losses in the primary loop. However, for reactor thermal powers less than 1500 kWt, the increase in the TE-EM pumps' efficiency outweighs the increase in the pressure losses in the primary coolant loop. These effects cause the specific pumping power of the system to decrease as the reactor thermal power increases until it reaches a minimum at the reactor thermal power of 1500 kWt for the Li cooled system. For the NaK-78 cooled system, the coolant boils before reaching the point of minimum specific pumping power of the system. At higher reactor thermal powers, however, the pressure losses continue to increase as the reactor thermal power increases (that is, coolant mass flow rate increases) while the pump's efficiency approaches an asymptote. Consequently, the specific pumping

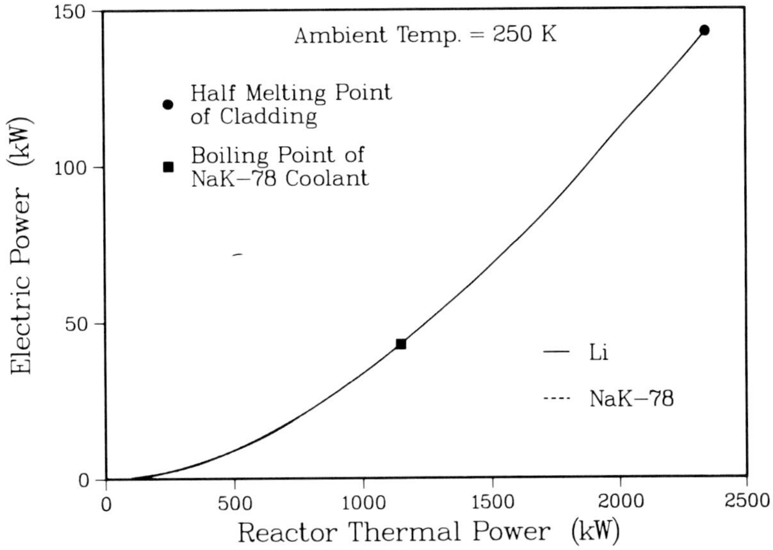

Figure 15 *Steady State Behavior of the System's Electric Power Output as a function of the Reactor Thermal Power.*

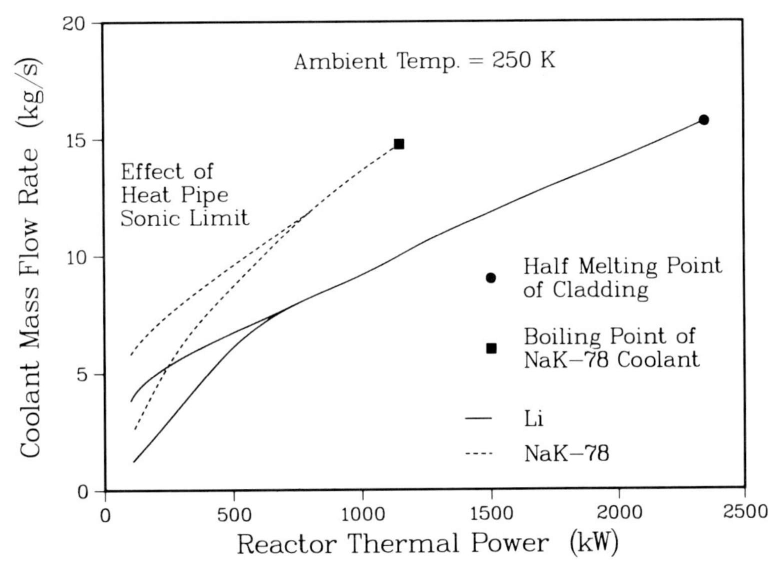

Figure 16 *Steady State Behavior of the System's Coolant Mass Flow Rate as a function of the Reactor Thermal Power.*

power begins to increase as the reactor thermal power increases beyond 1500 kWt for Li cooled system.

At lower reactor thermal powers where the system's waste heat rejection is limited by the sonic limit of the radiator's potassium heat pipes, the specific pumping power increases rapidly as the reactor thermal power decreases (see Figure 17). At these low reactor thermal powers, the coolant temperature in the primary coolant loop is much higher than that calculated by neglecting the effect of the radiator's heat pipe sonic limit (see Figure 13). Although this high coolant temperature improves the efficiency of the TE-EM pumps, it requires a higher coolant circulation rate (see Figure 16), resulting in a higher pressure losses and, consequently, higher specific pumping power for the integrated system. However, it should be noted that, at these low reactor thermal power levels, the specific pumping power for the Li cooled system is significantly lower than that for the NaK-78 cooled system. These results suggest that there is a

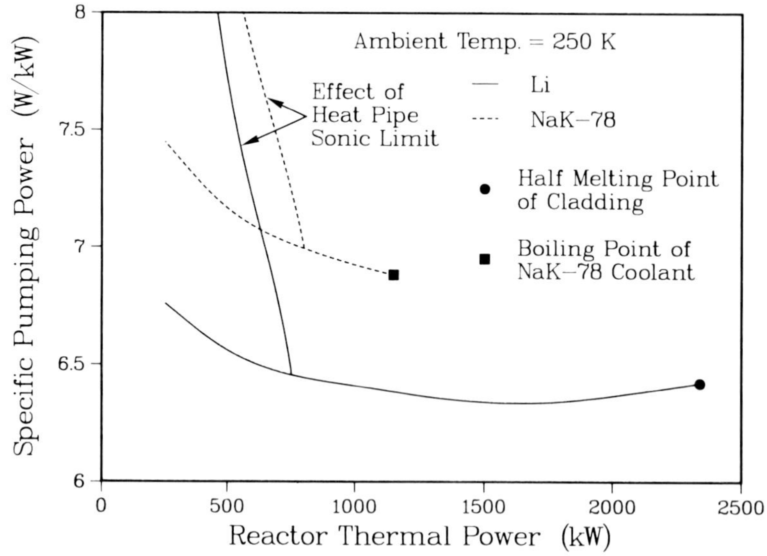

Figure 17 *Steady State Behavior of the System's Specific Pumping Power as a function of the Reactor Thermal Power.*

high penalty for operating an SP-100 system at a thermal power below approximately 750 kWt for the same radiator's heat pipe design mentioned earlier.

Transient Analysis

The transient operation of the integrated TE SP-100 system was initiated by a reactivity insertion into the nuclear reactor core, which had been operating at a specific steady state power level. The external reactivity coupled to the feedback reactivities from the core changes the reactor thermal power and, hence, the reactor and coolant temperatures. Only the Li cooled system was used in this investigation because it can operate at a higher steady state thermal power than the NaK-78 cooled system (see Figure 15).

To investigate the transient behavior of the integrated system, a step reactivity insertion of 0.178 dollar was introduced into the reactor core, which had been operated at a steady state thermal power of 2.0 MWt. This step reactivity insertion is the maximum reactivity insertion that can be introduced into the reactor without violating the system's operating limit; that is, the cladding temperature dose not exceed half its melting point. The reactivity insertions into the core are plotted in Figure 18a as functions of time after the initiation of the transient. As delineated in this figure, the negative temperature coefficient of reactivity caused the effective reactivity insertion in the core to decrease to zero, go negative, and finally reapproach zero indicating a steady state, approximately 400 seconds after the initiation of the transient.

As Figure 18b indicates, the reactor thermal power increased rapidly during the first stage of the transient, reaching a peak of 2950 kWt, 50 seconds after the initiation of the transient. Subsequently, the reactor thermal power decreased and eventually reached another steady state thermal power of 2316 kWt, approximately 400 seconds after the initiation of the transient. As shown in Figure 18c, although the coolant temperatures increased as the reactor thermal power increased, after reaching their peak values, they decreased more slowly than the reactor thermal power. The coolant temperature peaks were slightly higher than their final steady state values but were significantly higher than their initial steady state values at 2.0 MWt reactor thermal power. The slow decrease of the coolant temperatures after reaching their peak values was caused mainly by the stored sensible energy in both the fuel and the cladding during the transient.

Figure 18d shows the system's electric power output as a function of time after the initiation of the transient. As this figure shows, the electric power output increased slowly and reached its new steady state power of 140 kWe, approximately 400 seconds after the initiation of the transient. This implies that the large peak in the reactor thermal power is only negligibly transferred to the electric power output because of the large energy storage capacity of the system's components and the coolant. Although the electric power did not peak as sharply as the reactor thermal power, it began to increase at the same time as the

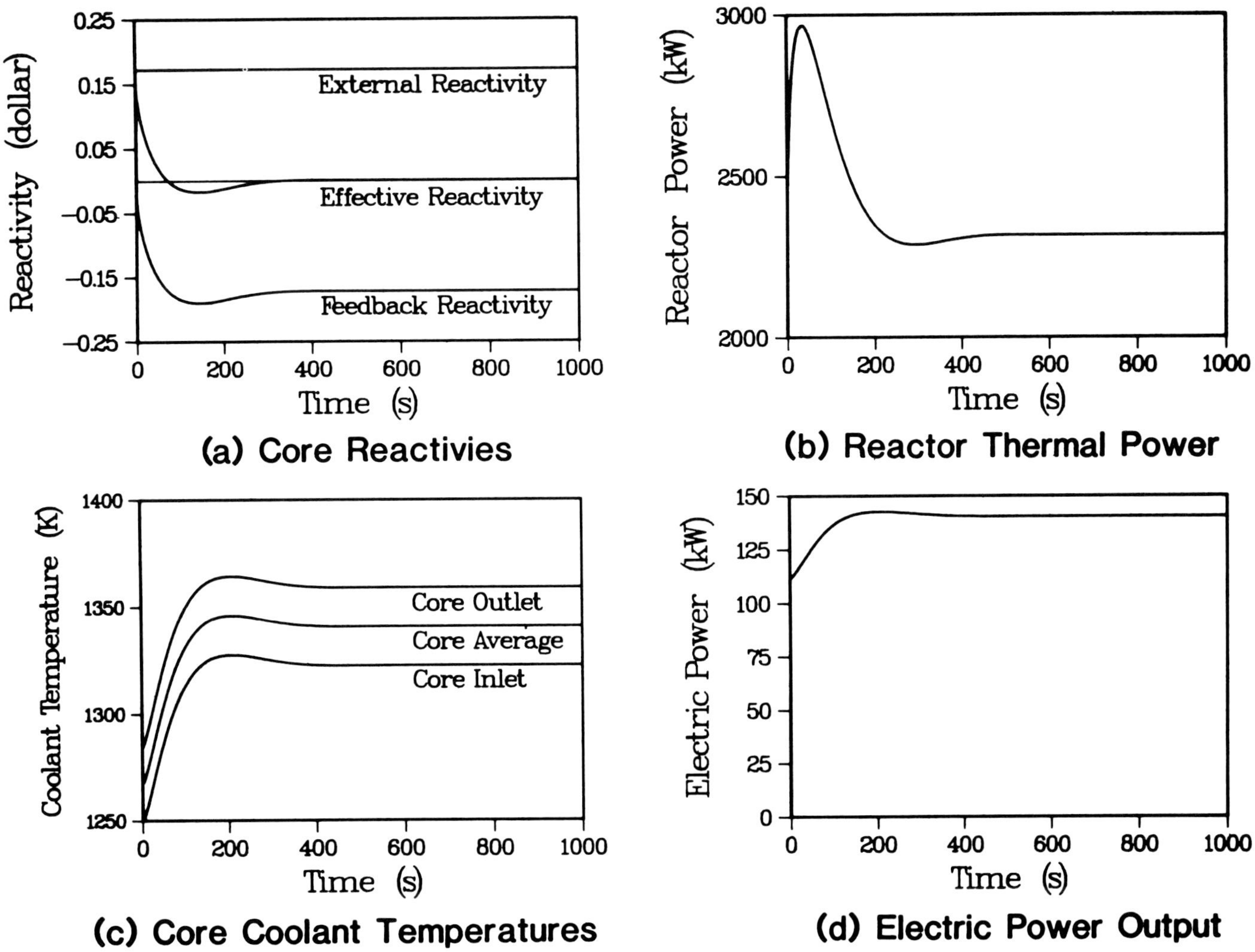

Figure 18 *Transient Behavior of the System's Operating Parameters after a Step External Reactivity Insertion.*

coolant temperature. This immediate response of the TE converter is due to the Peltier effect of the TE generator rather than to the transient heat conduction (Raag 1969).

The maximum steady state thermal power achievable, without exceeding half the melting temperature of the cladding, was calculated as a function of the steady state power level before the transient and the rate of reactivity insertion. Table 7 lists the values of the input parameters and the results of the system's transient response. In Table 7, a reactivity insertion time zero indicates a step reactivity insertion, while 300, 600, and 900 seconds insertion times indicate ramp reactivity insertions. The maximum reactivity insertion is the value of external reactivity which could be inserted into the core without causing the cladding temperature to exceed half its melting point during the transient. The maximum thermal power and the maximum electric power are those achieved with the maximum reactivity insertion at steady state following the transient. The settling time is defined as the time required for the electric power output of the system to reach and remain within 98 percent of its final steady state value.

As shown in Table 7, with the same initial system's steady state thermal power, the maximum reactor thermal power achievable after the transient is almost independent of the maximum value and the mode of reactivity insertion. However, the maximum external reactivity inserted into the core increased as the reactivity insertion time increased. This increase in the maximum reactivity insertion resulted in a slight increase in the maximum steady state thermal power following the transient. The results presented in Table 7 also show that the settling time increased as the reactivity insertion time increased; however, the time required to reach steady state after the termination of the reactivity insertion process decreased as the reactivity insertion time increased. As Table 7 also shows, lowering the initial steady state power causes both the maximum thermal and electric power outputs after the transient to be lower than those obtained with a high initial steady state power. In summary, initiating the transient at a low steady state thermal power has two major effects: increasing the settling time and slightly lowering the maximum power outputs. On the other hand, increasing the reactivity insertion time results in a slightly higher maximum power output and longer settling time.

Summary and Conclusions

This research developed a simulation computer model SNPSAM (Space Nuclear Power System Analysis Model) for analyzing the TE SP-100 systems. The results of the steady state operation demonstrated that the TE SP-100 system is a nonlinear system and that the system conversion efficiency and the electric power output strongly depend on the reactor thermal power. As the reactor thermal power increased, the coolant temperature increased resulting in a high conversion efficiency. In the present version of SNPSAM, the maximum temperatures in the reactor core are not allowed to exceed the boiling point of the coolant and/or half the melting point of the cladding material. Because the boiling point for NaK-78 is lower than that for Li, the limiting factor for the NaK-78 cooled system was the coolant boiling point, and the limiting factor for the Li cooled system was the temperature of the cladding. The maximum power attainable with the Li cooled system was approximately twice as much as that with the NaK-78 cooled system. These results suggest that, unlike NaK-78 cooled system, the growth potential of Li cooled TE SP-100 to higher power levels is very good. Results also showed that the electric power output of the system could be increased from 143 kWe to 207 kWe by only allowing the maximum cladding temperature to be as high as 55 percent of its melting temperature.

The specific pumping power for both the NaK-78 and the Li cooled system decreased as the reactor thermal power increased; however it was, on average, 7% lower for the latter. Although increasing the reactor thermal power requires a faster coolant circulation (that is higher pressure losses), it results in a higher conversion efficiency for the TE-EM pump's TE generators and, consequently, lower specific pumping power. However, for a reactor thermal power greater than 1500 kWt, the TE conversion efficiency approaches an asymptote, thus causing the system's specific pumping power to increase as the reactor thermal power increases.

At a reactor thermal power less than approximately 750 kWt with the Li coolant, the waste heat rejection will be limited by the sonic limit of the potassium heat pipes (with 10.75 cm^2 heat pipe cross sectional area) in the main radiator; this power level will vary with the design of the heat pipes. Such a limitation causes a

Table 7 *Transient Operation Limits.*

Initial Power (kWt)	2000				250			
Reactivity Insertion Time (s) / Parameters	0	300	600	900	0	300	600	900
Maximum reactivity insertion[a] ($)	0.1778	0.1835	0.1871	0.1900	1.6577[b]	1.6921	1.7052	1.7114
Peak thermal power (kWt)	2967	2562	2450	2415	34.3 × 10^6	4390	3460	3110
Power pulse half width (s)	145	–	–	–	3.3 × 10^{-4}	–	–	–
Maximum thermal power (kWt)	2316	2330	2333	2337	2212	2274	2288	2311
Maximum electric power (kWe)	140.3	141.5	141.9	14.2	130.7	136.4	138.6	139.7
Settling time (s)	273	440	740	1040	371	611	895	1158

[a]Delay neutron fraction, β = 0.00766; maximum reactivity without causing cladding temperature to exceed half its melting point.
[b]Prompt Critical.

precipitous increase in the specific pumping power. These results suggest that operating a TE SP-100 system at a lower power level may impose higher pumping penalty on the system. The system operating temperatures such as fuel, cladding, and coolant temperatures, strongly depend on the reactor thermal power. However, unlike the coolant temperatures, the fuel and cladding average temperatures were almost independent of the coolant type.

The transient behavior of a Li cooled SP-100 system was investigated by varying the magnitude and the insertion time of external reactivity into the reactor core for different initial steady state thermal powers. The maximum thermal and electric powers attainable without exceeding half melting point of the cladding were 2337 kWt and 142.2 kWe, respectively. While changing the mode of reactivity insertion only slightly affected the maximum power attainable, the settling time after the transient increased proportionally to the reactivity insertion time. Furthermore, the initial steady state power before the transient only negligibly affected the maximum power attainable and the system settling time after the transient.

Acknowledgments

This work was supported by the Nuclear Technology office of the Air Force Weapons Laboratory, under contract No. F29601-84-C-0080 to the University of New Mexico.

References

Ambrus, J. H. (1985) ''Mission Applications Overview,'' in *Space Nuclear Power Systems 1984*, M. S. El-Genk and M. D. Hoover, eds., vol. 1, pp. 43–52, Orbit Book Co., Malabar, FL.

Angrist, S. W. (1966) *Direct Energy Conversion*, Allyn and Bacon, Inc., Boston, MA.

ANL (1983) *The Simulation Language of DSNP: Dynamic Simulator for Nuclear Power-Plants*, ANL-CT-77-20, Argonne National Laboratory, Idaho Fall, ID.

Bailey, H. S., et al. (1986) ''Space Power Reactor Designs for Broad Mission Applicability,'' in *Space Nuclear Power Systems 1985*, M. S. El-Genk and M. D. Hoover, eds., Orbit Book Co., Malabar, FL.

Barnes, A. H. (1953) ''Direct-Current Electromagnetic Pumps,'' *Nucleonics*, 11(1):16–21.

Bathe, K. J. (1982) *Finite Element Procedures in Engineering Analysis*, Prentice-Hall, Inc., Englewood Cliffs, NJ.

Buksa, J. J. (1987) *SP-100 System Shutdown Thermal Analysis Using SNPSAM and a Thermally Coupled, Self-Induced Thermoelectric-Electromagnetic Pump Model*, M.S. Thesis, Chemical and Nuclear Engineering Department, The University of New Mexico, Albuquerque, NM.

Cobble M. H. (1982) ''Transient Analysis of a Thermoelectric Generator,'' in *Proceedings of the 4th International Conference on Thermoelectric Energy Conversion*, held in Arlington, TX, pp. 52–55, March 1982.

El-Genk, M. S., J. J. Buksa, and J. T. Seo (1986a) ''A Self-Induced Thermoelectric-Electromagnetic Pump Model for SP-100 Systems,'' in *Trans. 3rd Symposium on Space Nuclear Power Systems*, CONF-860102-Summs, held in Albuquerque, NM, 13–16 January 1986.

El-Genk, M. S., J. T. Seo, and J. J. Buksa (1986b) ''Load Following and Reliability Studies of Thermoelectric SP-100 Systems,'' in *Proc. 21st Intersociety Energy Conversion Engineering Conference*, held in San Diego, CA, 25–29 August 1986.

El-Genk, M. S. and J. T. Seo (1986a) ''Trends and Limits in the Upgrading of SP-100 Nuclear Powered Space System'', in *Space Nuclear Power Systems 1985*, M. S. El-Genk and M. D. Hoover, eds., Orbit Book Co., Malabar, FL.

El-Genk, M. S. and J. T. Seo (1986b) ''Analysis of the Transient Behavior of Thermoelectric Generators,'' in *Trans. 3rd Symposium on Space Nuclear Power Systems*, CONF-860102-Summs, held in Albuquerque, NM, 13–16 January 1986.

El-Wakil, M. M. (1971) *Nuclear Heat Transport*, International Textbook Company, Scranton, PA.

General Electric Company (1983) *SP-100 Conceptual Design Study, Final Report, Volume II - Reactor and Shield Design*, 83SDS4261, General Electric Company, Philadelphia, PA.

Prenger, F. C. Jr. (1979) *Heat Pipe Computer Program (HTPIPE) User's Manual*, LA-8101-M, Los Alamos National Laboratory, Los Alamos, NM.

Raag, V. (1969) *Optimization of Thermoelectric Generators for Fixed Temperature and Fixed Heat Input Operation*, Memorandum #4, Jet Propulsion Laboratory, Pasadena, CA.

Seo, J. T. (1986) *SNPSAM—Space Nuclear Power System Analysis Model*, M.S. Thesis, Chemical and Nuclear Engineering Department, The University of New Mexico, Albuquerque, NM.

Soo, S. L. (1986) *Direct Energy Conversion*, Prentice-Hall, Inc., Englewood Cliffs, NJ.

Vaughan, J. W., Jr. (1986) ''Nuclear Power Applications in Future Space Activities'', in *Space Nuclear Power Systems 1985*, M. S. El-Genk and M. D. Hoover, eds., Orbit Book Co., Malabar, FL.

Walter, A. E. and A. B. Reynolds (1981) *Fast Breeder Reactors*, Pergamon Press, Elmsford, NY.

Part 5

THERMAL MANAGEMENT

Space Nuclear Power Systems 1986
Edited by M. S. El-Genk and M. D. Hoover

Chapter 12

Experimental Test of Liquid Droplet Radiator Performance

A.T. Mattick and M.A. Simon

Aerospace and Energetics Research Program FL-10
University of Washington
Seattle, WA 98195

The liquid droplet radiator (LDR) is a heat rejection system for space power systems wherein an array of heated liquid droplets radiates energy directly to space. The use of submillimeter droplets provides large radiating area-to-mass ratio, resulting in radiator systems which are several times lighter than conventional solid surface radiators. An experiment is described in which the power radiated by an array of 2300 streams of silicone oil droplets is measured to test a previously developed theory of the LDR radiation process. This system would be capable of rejecting several kWt of heat in space. Furthermore, it would be suitable as a modular unit of an LDR designed for 100-kWt to 1-MWt power levels. The experiment provided confirmation of the theoretical dependence of droplet array emissivity on optical depth. It also demonstrated the ability to create an array of more than 1000 droplet streams having a divergence less than 1 degree.

Introduction

The liquid droplet radiator (LDR) is evolving rapidly as a lightweight system for heat rejection in space power systems. By using recirculating free streams of submillimeter droplets to radiate waste energy directly to space, the LDR potentially can be an order of magnitude lighter than conventional radiator systems that radiate from solid surfaces. The LDR also is less vulnerable to micrometeoroid damage than are conventional radiators, and it has a low transport volume. Three major development issues of this new heat rejection system are 1) aiming the droplet streams with sufficient precision to avoid fluid loss, 2) radiative performance of the array of droplet streams that comprises the radiating element of the LDR, and 3) the efficacy of the droplet stream collector, again with respect to fluid loss. This paper reports experimental results bearing on the first two issues—droplet aiming in a multikilowatt-sized system, and radiated power from a large droplet array.

Figure 1 illustrates the operational principles of the LDR. Waste heat from a power system is transferred to the radiator working fluid in a heat exchanger, and the fluid is sprayed into space as a collimated array of droplet streams aimed toward a collector. In contrast to conventional radiators wherein heat is transferred from the fluid to solid radiating surfaces, the LDR fluid droplets radiate directly to space. This greatly reduces vulnerability to single point failure due to puncture of manifolds by micrometeoroids. The cooled fluid is captured by the droplet collector, repressurized, and returned to the heat exchanger.

Using this radiator configuration to achieve low mass and reduced micrometeoroid vulnerability imposes constraints on fluid selection and fluid handling that are not shared by solid-surface radiators. The fluid must exhibit vapor pressures of $< 10^{-7}$ Torr so that evaporative losses do not negate the intrinsic mass advantage of the LDR (Mattick and Hertzberg 1981). Several liquid metals have been identified as suitable candidates at high temperatures, and Dow 705 silicone fluid is appropriate near 300 K (Mattick and Hertzberg 1981 and 1985a). The generation of $\simeq 10^6$ collimated droplet streams (for a 1-MWt heat rejection) and their efficient collection poses a unique fluid-handling challenge for space applications. Figure 2 illustrates an approach for droplet generation in which orifice plates, each having 10^3 to 10^4 orifices, are mounted on a manifold. Our tests of a 2300-hole orifice plate described below indicate that a stream parallelism of $\simeq 1$ degree is possible, meeting the collimation requirements for a space-deployed LDR. Several research groups are involved in developing and testing droplet collection concepts. Previous tests of a rotating drum design (Figure 3a) and a linear collector design (Figure 3b) have indicated that extremely low splashing rates are possible for

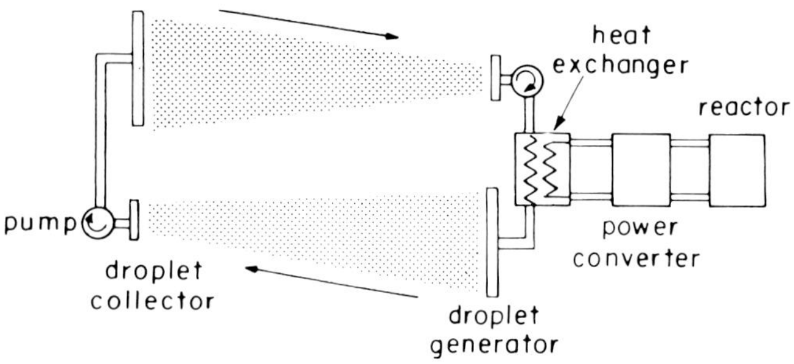

Figure 1 *Schematic of a Dual-Pass Liquid Droplet Radiator for a Space Nuclear Power System.*

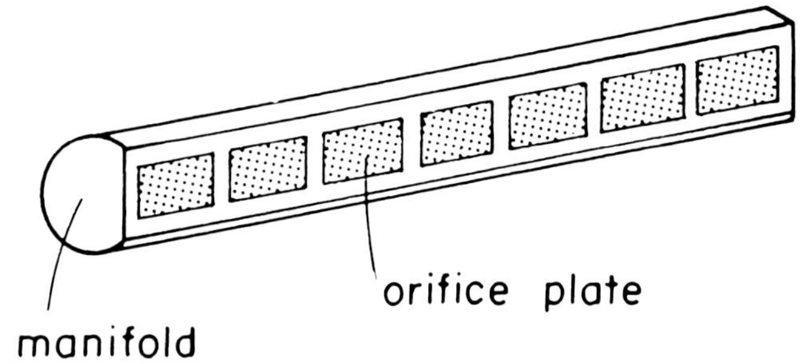

Figure 2 *Droplet Generator System using an Array of Orifice Plates.*

appropriate flow regimes and that efficient repressurization also is possible. Thus, the unique fluid-handling requirements of the LDR appear to be achievable without compromising the effectiveness of the LDR concept; however, substantial engineering development is needed to realize a practical radiator system.

The experiment reported in this article was primarily intended to test the effectiveness of an array of heated droplets as a radiating element for the liquid droplet radiator. Previous trade studies on LDR systems (Taussig and Mattick 1984 and AFRPL 1985) have shown that the in-flight droplets comprise 10 to 25 percent of the LDR system mass. This fraction is relatively constant for systems having various heat-rejection powers and rejection temperatures because the ''wet'' masses of fluid-handling components increase nearly in proportion to the projected area of the radiating droplet sheet. Therefore, the power/mass ratio of an LDR system critically depends on the power/mass ratio of the radiating droplets. In this sense, our measurements of the power radiated by a droplet array serve as an experimental test of LDR performance.

The measurements were made using a cylindrical column of 2300 droplet streams of Dow 704 silicone fluid, which provides a range of diametrical optical depths between 2.0 and 3.0. Although Dow 705 is superior to Dow 704 for space LDR applications (due to its lower ultimate vapor pressure), Dow 704 has virtually identical thermal and spectroscopic properties to Dow 705, and was used in the current experiments because of its lower cost. The orifice plate used to generate this array of droplet streams is comparable (in number of orifices, orifice diameter and lateral dimensions) to a plate that would be used on the generator design of Figure 2. The droplet generation system would be appropriate for an LDR module capable of rejecting several kilowatts of thermal energy. Moreover, a comparison of the measured radiative properties of the cylindrical array of droplets with theoretical predictions required that the spatial distribution of droplets be reasonably uniform and the streams be collimated to within a few degrees. An operational LDR will require comparable droplet stream collimation to ensure all droplets reach the collector. Thus, a test of LDR radiative performance necessitated a demonstration of droplet generation performance meeting specifications typical of a practical LDR system.

Radiation by an Array of Droplets

The power radiated by both sides of a planar solid-surface radiator of projected area A and temperature T is $P = 2\epsilon A\sigma T^4$, where σ is the Stephan-Boltzmann constant and ϵ is the hemispherical emissivity of the surface. For an LDR with a planar array of droplets (as shown in Figure 1), it is convenient to express radiated power in a similar form, by ascribing a hemispherical sheet emissivity ϵ_s to the boundary surfaces of the droplet sheet. Radiation emerging from the droplet sheet arises from thermal emission by droplets and from multiple scattering of this emission as it is transported from the interior regions of the sheet to the boundaries. A given droplet radiates a power $P_d = \epsilon_o A_d \sigma T^4$ where A_d is the droplet surface area, T_d is the temperature, and $\epsilon\sigma$ is the droplet emittance. Since the droplet temperature varies in the droplet propagation direction z, as well as in the direction normal to the sheet (into the plane of Figure 1), we define a local sheet emissivity to be

$$\epsilon_s(z) = I(z)/\sigma\bar{T}_d^4(z), \tag{1}$$

where I is the hemispherical intensity at position z downstream of the droplet generator, and $\bar{T}_d$ is the average droplet temperature at position z. The droplet

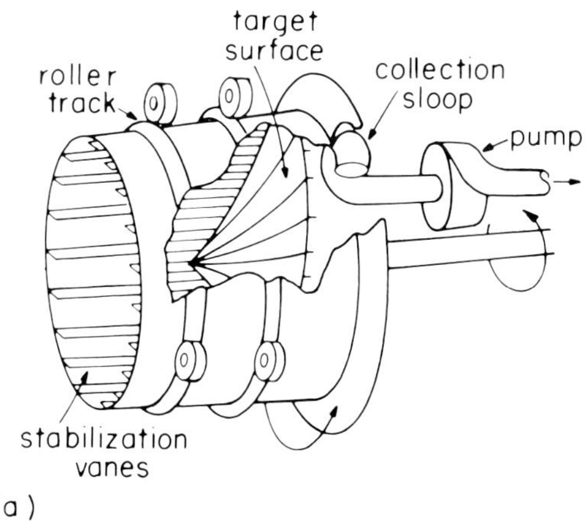

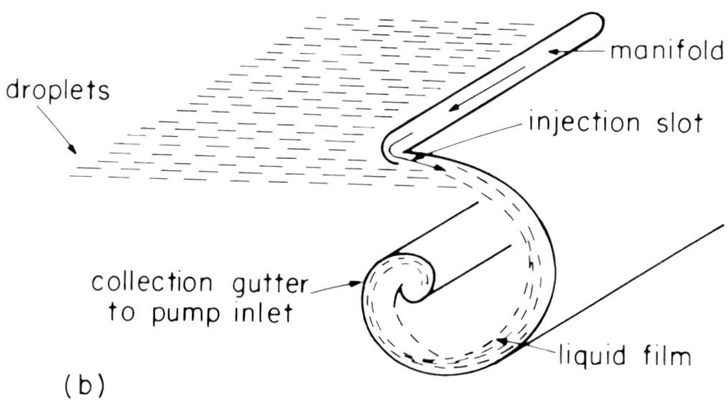

Figure 3 *Droplet Stream Collection Concepts: a) Rotating Drum Collector; b) Linear Collector.*

sheet emissivity is obtained from the solution of an integro-differential equation of radiative transfer in the sheet, taking into account a spatially varying droplet temperature T_d. For droplet diameters large compared with radiation wavelengths, as is the case with typical LDR designs, the scattering is isotropic (Van de Hulst 1957) and the local sheet emissivity is only a function of ϵ_o and the local optical depth, defined as $\tau \equiv n \pi r_d^2 S$, where n is the number density of droplets, r_d is the droplet radius, and S is the thickness of the sheet.

We previously reported our solutions for the sheet emissivity as a function of τ and ϵ_o (Mattick and Nelson 1986). It was found that the development of a nonuniform temperature profile in the sheet thickness direction has only a small effect on sheet emissivity for practical LDR designs. Moreover, the power radiated by the entire sheet can be approximated with an accuracy of a few percent by

$$P_s \cong 2\sigma\epsilon_s^* T^4_{\text{eff}} A_s, \qquad (2)$$

where ϵ_σ^* is the sheet emissivity at the midpoint of the sheet, and $T_{\text{eff}} = \sqrt{T_o T_1}$, where T_0 is the peak droplet temperature (at the generator) and T_1 is the minimum temperature (at the collector), and A_s is the projected area of the droplet sheet. Figure 4 plots the emissivity ϵ_σ^* as a function of optical depth τ and droplet emissivity ϵ_o for a planar sheet. For rarified droplet sheets (small τ), the sheet emissivity increases linearly with τ. Since the mass/area also increases linearly with τ, the radiator power/mass is constant in this region. At high optical depths the sheet emissivity approaches a constant that depends on droplet emittance, and the power/mass decreases as $1/\tau$. The choice $\tau \simeq 1/2\epsilon_o$ yields a reasonable compromise between minimizing droplet sheet area and maximizing power/mass (Mattick and Hertzberg 1985b).

The dependence of droplet sheet emissivity on optical depth and droplet emittance clearly is of central importance to LDR performance. Our experimental program is intended to test the theoretical predictions of this dependence and to demonstrate the heat-rejection capability of an array of heated droplets. The experiment described here was preceded by radiation measurements using a single droplet stream and using a cylindrical array of 465 droplet streams, both carried out with silicone fluids. In the single-stream experiment, the temperature decrease of droplets was measured as a function of radiation time. From these measurements, a droplet emittance of 0.6 ± 0.1 was determined for 600-μm-diameter droplets of Dow 705 fluid. The 465-stream experiment utilized a cylin-

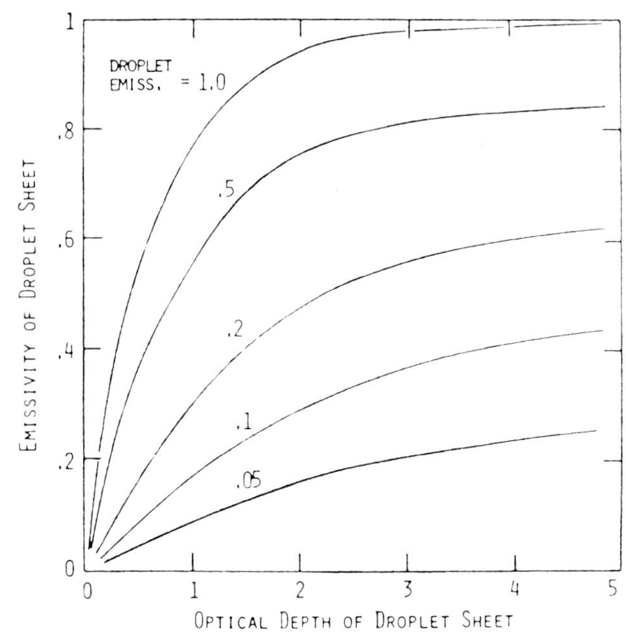

Figure 4 *Variation of the Hemispherical Emissivity of a Uniform Temperature Droplet Sheet with the Optical Depth of the Sheet and Droplet Emittance.*

drical array of 350-μm-diameter droplets of Dow 704 fluid (virtually identical spectroscopically to Dow 705) to achieve optical depths in the range 0.8 to 1.8. The power radiated by the droplets was measured directly to determine a surface emissivity for a cylindrical array, as a function of diametrical optical depth. The theoretical analysis of radiation by a cylindrical array was carried out similar to the analysis of a planar array. Therefore, the experiment provided a test of the analytical approach. Although the variation of droplet array emissivity with diametrical optical depth followed the theoretical curves to within experimental uncertainty, the accuracy of the data was not sufficient to provide a precise test of the theory. The chief experimental uncertainty arose from the nonuniformity and temporal variation of droplet distribution, owing to the marginal performance of the orifice plate used. A fit of the experimental data to the theoretical droplet cylinder emissivity yielded a droplet emittance of 0.6 ± 0.1, in accord with the single-stream experiment. The emittance of a droplet of Dow 704 or Dow 705 fluid is not expected to vary significantly with diameter for diameters $\gtrsim 100\ \mu$m because the optical depth of a droplet for wavelengths of thermal emission at $T \simeq 330$ K is above unity in this diameter range.

The 2300-stream experiment described here was designed to provide a more accurate test of the theory, by incorporating improvements in droplet stream generation to achieve a more uniform and collimated array. A cylindrical array was chosen instead of a rectangular array in order to keep the number of droplet streams manageable. To accurately model a planar LDR droplet sheet, the width of the array would have to be several times the thickness. Thus, a rectangular array having a given optical depth in the thickness dimension would require an order of magnitude more streams than a circular array having the same optical depth (diametrical).

Experiment

The heat transfer experiment was designed to allow a determination of the emissivity of a cylindrical array of droplet streams over an optical depth range of 2.0 to 3.0. To allow for temperature and flow stabilization, the experimental apparatus was a continuous-flow, closed-loop system, as illustrated in Figure 5. The droplets were formed by breakup of liquid jets issuing from an orifice plate mounted on the droplet generator plenum. The jets were directed into a 0.15-m ID evacuated chamber provided with a test section beginning 0.1 m downstream of the orifice plate, approximately the jet breakup length. Thermal radiation measurements and optical depth measurements were carried out via side-arm access ports in the test section. The droplets were collected at the bottom of the chamber and the liquid was fed to a sealless, magnetic drive pump, providing repressurization to between 30 kPa and 300 kPa. The liquid was heated in the droplet generator feed pipe to between 310 K and 325 K.

The most critical component of the experimental apparatus was the orifice plate, because accurate measurements of the emissivity of the droplet array require that the droplet size and spacing be uniform and the streams well collimated. The directional stability of a droplet stream is severely undermined by irregularities in the shape of the orifice, and the previous 465-stream radiation experiment was hampered by this difficulty. The orifice plate for that experiment utilized 1-mm-thick aluminum substrate with 175-μm-diameter straight holes spaced 1 mm apart. The use of a straight, rather than tapered, orifice was dictated by the supplier's fabrication capabilities and the limited strength of the substrate. The straight geom-

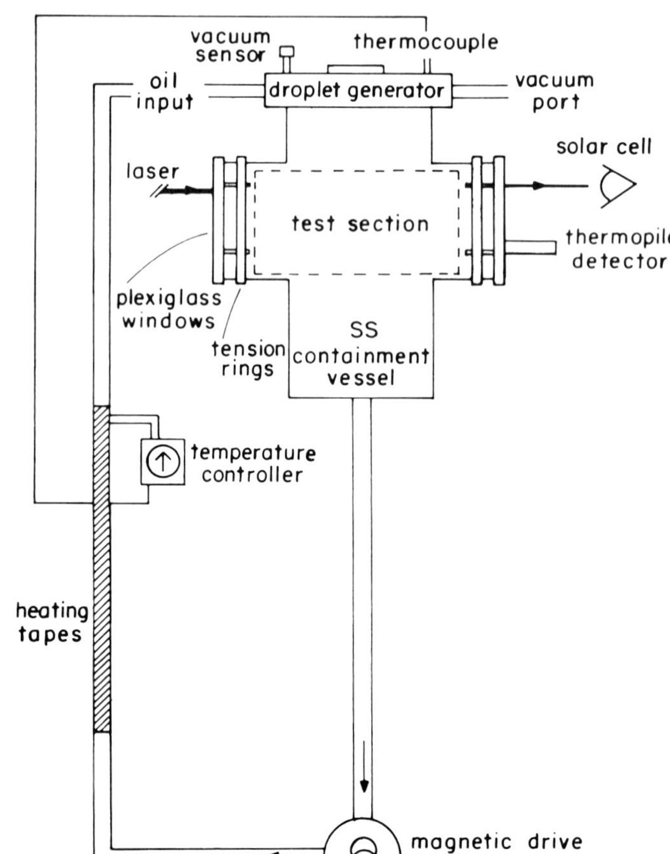

Figure 5 *Experimental Configuration for measuring Power Radiated by a Cylindrical Array of 2300 Droplet Streams.*

etry contributed to directional instabilities as well as large pressure losses. For the current experiment, the Fiber Operations Division of Celanese Corporation (Jet Manufacturing Department, Box 1000, Narrows, VA) was contracted to develop a more efficient and directionally stable orifice plate. This plate was fabricated in stainless steel and utilized tapered entrance regions and a low aspect-ratio straight section. An exit diameter of 175 μm was chosen as a compromise between minimizing the number of orifices required to achieve desired optical depths and producing droplet diameters (approximately twice the orifice exit diameter) representative of high performance LDR systems (100 μm to 500 μm for Dow 705 fluid). The orifices were arranged in a concentric circular pattern with a radial spacing of 1.5 mm and a circumferential spacing of 1 mm, leading to an array diameter of 66 mm.

The orifice plate exhibited small pressure losses, and stable droplet production could be achieved at plenum pressures as low as 30 kPa. At pressures above 150 kPa, flexing of the plate led to a divergence of droplet streams by several degrees. Photographs of the streams indicated a close uniformity of droplet sizes and spacings throughout the array, and at low pressure the streams were collimated to within 1 degree. Although a 35-μm mesh filter was used upstream of the orifice plate, degradation of stream collimation would occur over extended operation $\gtrsim$ 12 hours) due to particulates clogging the orifices. Satisfactory performance was restored after ultrasonic cleaning of the plate. After the experimental program was completed it was found that particulate formation in our system arose from excessive temperatures at the walls of the heater section. The excellent performance of the 2300-orifice plate, under clean fluid conditions, leads us to believe that fabrication of droplet generators for 1-MWt LDR systems is within reach using existing technology.

The diametrical optical depth was determined by measuring the transmission of a He-Ne laser beam through the droplet array. Since the laser wavelength λ = .633 μm is much smaller than the droplet diameter (nominally 350 μm), the optical cross section for laser scattering by a droplet is virtually identical with the geometrical frontal area of a droplet, upon which the droplet array optical depth is based. The laser detector was placed $\simeq$1 m from the test section, ensuring that only transmitted, not scattered, radiation was detected (detector area $\simeq$10 mm^2). The optical depth is given by $\tau = \ln(P_{\text{off}}/P_{\text{on}})$ where Poff and P_{on} are the detected powers with the droplet streams off and on, respectively. The optical depth was observed to exhibit periodic modulations as high as 100 percent. These were correlated with the contacts of gear teeth in the pump. The strength of these pressure fluctuations and the depth of modulation of τ varied with the oil temperature (via viscosity) and mean pressure. The modulation of optical depth could be kept below 10 percent by using an oil temperature of 310 K to 325 K, and pressures below 40 kPa. This effect constituted the principal source of inaccuracy in the determination of droplet array emissivity. Control of the mean optical depth (aside from the pump-caused modulation) was achieved by acoustically driving the droplet generator plenum at frequencies between 2 kHz and 4 kHz using a piezo-electric wafer, and by varying the flow rate.

A thermopile detector, with a flat spectral response over the wavelength range of 0.2 μm to 50 μm and a sensitive area of 1.5 mm $\times$ 1.5 mm, was used to measure thermal radiation from the heated droplets. A cooled circular aperture was placed between the detector and the droplet array to limit the field of view to a cone having a half angle of 6°. A blackened plate was placed behind the droplet array, which overlapped the detector's field of view. The plate temperature T_b was kept near the temperature of the chamber walls and was monitored by a thermocouple during experimental runs.

Only the difference in detected power with the droplet streams on and off was used to calculate the droplet array emissivity. With the streams off, the radiation throughout the test chamber has a blackbody spectral distribution at the background temperature T_b and a power $P_{\text{off}}(T_b)$ reaches the detector. The same radiation distribution would prevail if the streams were on and if the droplet temperature equalled the background temperature. Also, the same power $P_{\text{off}}(T_b)$ would reach the detector. In the latter case, the detected radiation would include radiation emitted by the droplets, $P_d(T_b)$, and radiation from the background, $P_b(T_b)$ which is transmitted through the droplet array or scattered by the droplet array into the detectors acceptance cone. With the droplet streams present and at an arbitrary droplet temperature T_d, the same background power $P_b(T_b)$ will be transmitted or scattered into the detector, along with a contribution $P_d(T_d)$ from thermal emission by the droplets. Thus, the difference in power with the heated streams on and off is

$$P_{\text{on}} - P_{\text{off}} = P_d(T_d) - P_d(T_b) \ . \tag{3}$$

By measuring the power *difference* with the streams on and off, the background power $P_b(T_b)$ is eliminated from the determination of droplet array emissivity. This method will be accurate as long as the test chamber surfaces are highly absorbant and kept at a uniform, constant temperature.

Because of the small acceptance angle of the apertured detector and the small area (5-mm-diameter circle) of the droplet array viewed by the detector, the normal emissivity ϵ_n, rather than hemispherical emissivity, was determined by this experiment. The power emitted by droplets which reaches the detector is given by

$$P_d(T) = \epsilon_n A_d F \sigma T^4, \tag{4}$$

where A_d is the detector area and F is the configuration factor between the detector and the cylindrical boundary of the droplet array (facing the detector). The detector response was calibrated using a second plate, fitted with a heater and thermocouple, which was placed at the location of the droplet cylinder boundary, so that the plate-detector configuration factor was the same as the droplet array-detector configuration factor. The plate was coated with several layers of flat black lacquer having an emissivity $\epsilon_c = 0.96$ over the range of calibration temperatures (300 K to 330 K). The response of the thermopile was found to be linear with power over the calibration temperature range. The normal emissivity of the droplet array is determined from

$$\epsilon_n = \epsilon_c \frac{V_{on} - V_{off}}{T_d^4 - T_b^4} / \frac{dV}{d(T_c^4)}, \tag{5}$$

where V_{on} and V_{off} are the detector voltages with the droplet streams on and off, respectively, and dV/dT_c^4 is the slope of the thermopile calibration curve of detector voltage versus T_c^4, where T_c is calibration plate temperature. In this expression, it is assumed that the droplet array emissivity is not a function of temperature. Any temperature variation of ϵ_n would reflect the temperature dependence of droplet emittance ϵ_o. The optical depth of a 350-μm-diameter droplet is > 3 over most of the blackbody radiation spectrum at 300 K to 330 K. Therefore, little variation of ϵ_o with temperature is expected. This was confirmed by our experimental results, which showed no correlation of ϵ_n with droplet temperature. The droplet temperature T_d was taken as the liquid temperature in the droplet generator plenum (measured with a thermocouple), because over the 0.1-m flight path from generator to test section, the decrease in droplet temperature due to radiation is less than 0.05 K. Since the chamber pressure was kept below 5 mTorr, the conductive and convective heat loss over this transit distance was negligible.

Experimental Results

More than 100 measurements of droplet array radiation were made using droplet temperatures between 310 K and 325 K, and adjusting the flow rate and droplet generator drive frequency to produce diametrical optical depths between 2.0 and 2.8. The background temperature T_b was kept near 295 K. Figure 6 shows the values of normal emittance derived from the experiments, with the data grouped at nine discrete values of optical depth. The error bars correspond to one standard deviation in derived emissivity, based on the data spread. Also plotted in Figure 7 are theoretical curves of normal emissivity of a cylindrical array of uniform temperature droplets as a function of optical depth and droplet emittance ϵ_o. The observed variation of normal emissivity with optical depth is seen to agree with the theoretical variation. The best fit of the data with theory corresponds to a droplet emittance of 0.48 ± 0.05. This value of droplet emittance is somewhat lower than values derived in previous experiments at smaller optical depths, although all the experiments agree to within experimental error.

The droplet emittance derived as a fitting parameter

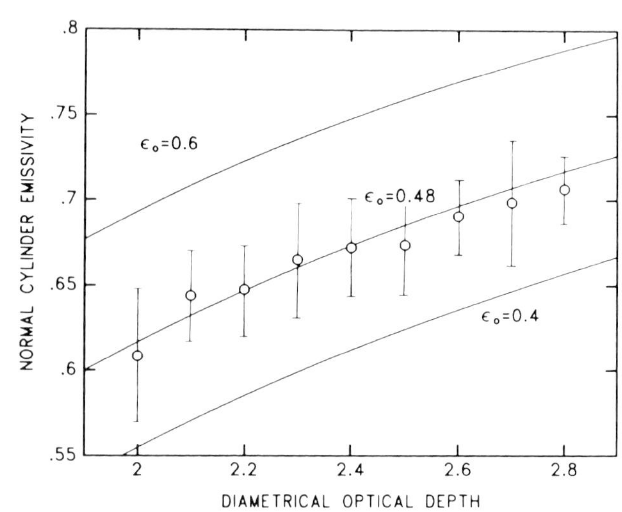

Figure 6 *Experimental (Circles) and Theoretical (Solid Lines) Variation of Normal Emissivity with Diametrical Optical Depth for a Cylindrical Array of Droplets.*

between theory and experiment is surprisingly low. The spectral emittance of a droplet at a given wavelength is a function of the optical depth of the droplet and the index of refraction of the liquid at that wavelength. At large optical depths the emittance is only a function of index of refraction (reflection-limited emittance) as illustrated in Figure 7. Figure 8 shows a transmission spectrum of a 100-μm thickness of Dow 704 silicone fluid for a wavelength range of 5 μm to 50 μm, which encompasses 95% of the blackbody power radiated at 320 K. It is evident that a 350-μm thickness (equal to the droplet diameter) would have an optical depth $\gtrsim 3$ over this entire wavelength range. Thus, we would expect the droplet emittance to be reflection-limited. However, a reflection-limited emittance of 0.48 corresponds to an index of refraction $n = 6$, a remarkably high value. The index of refraction of Dow 704 fluid at visible wavelengths is $n \simeq 1.47$. To resolve this issue, we intend to carry out a spectral reflectance experiment for Dow 704 over a wavelength range of 5 μm to 30 μm.

Summary and Conclusions

An experimental test was made of the performance of a liquid droplet heat-rejection system that would be capable of rejecting several kilowatts of thermal energy as part of a liquid droplet radiator for space power applications. The emphasis of this work was to demonstrate the efficient heat-rejection capability of the liquid droplet radiator concept and to test theoretical predictions of radiation by an array of

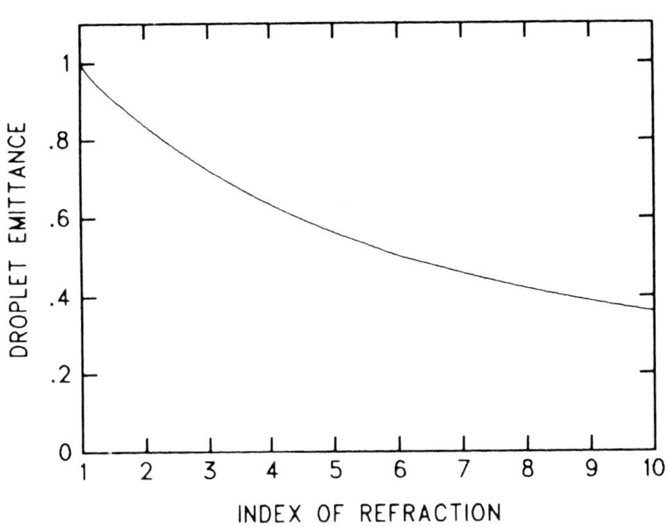

Figure 7 *Emittance of an Optically Thick Droplet as a function of Index of Refraction.*

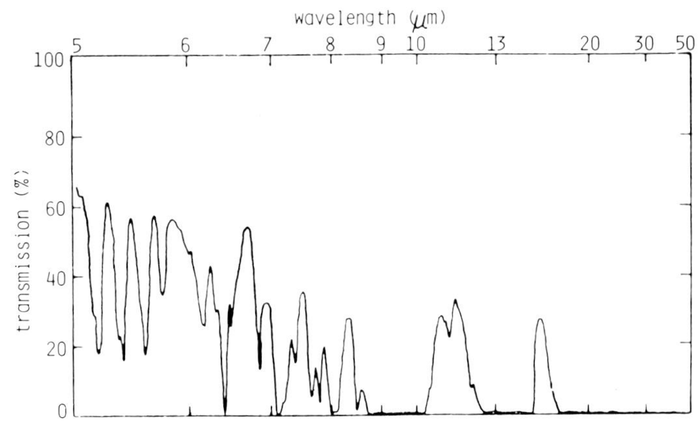

Figure 8 *Transmission Spectrum of Dow 704 Silicone Fluid (100* μm Thickness).

droplets for an optical depth range between 2.0 and 3.0. An accurate test required that the cylindrical array of droplets have the same characteristics as required for a practical LDR system for space. Thus, the experiment also was, by necessity, a demonstration of the ability to generate a uniform, collimated array of droplets.

The dependence on diametrical optical depth of the normal emissivity of a cylindrical cloud of 2300 droplets of Dow 704 silicone fluid was found to agree with theoretical predictions to within the $\pm$ 10% relative accuracy of the measurements. However, the value of droplet emittance derived from the measurements was found to be only 0.48, which implies an anomalously high index of refraction for the silicone fluid at infrared wavelengths. Although this value of droplet emittance would lead to a very low mass for a low-temperature LDR system, it is puzzling from a fundamental standpoint. Therefore, the authors are undertaking an experimental program to determine the droplet emittance in a more direct manner in order to test the theoretical predictions of droplet cloud radiation without using emittance as an adjustable parameter.

The orifice array developed for this experiment was capable of reliable generation of a uniform array of droplet streams, collimated to within 1 degree. This represents an important demonstration in the development of a technology base for the LDR. A droplet generation system for a 1-MWt LDR would utilize between 100 and 1000 similar orifice plates, depending on rejection temperature. A significant, though not unexpected, finding was that the fluid must be kept free of all particulates to achieve long-term, reliable droplet generation.

Acknowledgments

This work was supported by Air Force Office of Scientific Research Grant 83-0367 and National Aeronautics and Space Administration Grant NAG 1-327.

References

AFRPL (1985) *Liquid Droplet Radiator Systems Investigation*, Final Report on Contract F04611-84-C-0025, to be published as TR-86-XXXX, McDonnell Douglas Astronautics Company, Huntington Beach, CA.

Mattick, A.T.and A. Hertzberg (1981) "Liquid Droplet Radiators for Heat Rejection in Space," *J. Energy*, 5:387-393.

Mattick, A.T. and A. Hertzberg (1985a) "Liquid Droplet Radiator Performance Studies," *Acta Astron.*, 12:591-598.

Mattick, A.T. and A. Hertzberg (1985b) "Liquid Droplet Radiator Technology Issues," in *Space Nuclear Power Systems 1984*, M.S. El-Genk and M.D. Hoover eds., Orbit Book Co., Malabar, FL.

Mattick, A.T. and M. Nelson (1986) "Heat Transfer Studies on the Liquid Droplet Radiator," in *Space Nuclear Power Systems 1985*, M.S. El-Genk and M.D. Hoover eds., Orbit Book Co., Malabar, FL.

Taussig, R.T. and A.T. Mattick (1984) "Droplet Radiator Systems for Spacecraft Thermal Control," AIAA paper No. 84-1797, *AIAA 19th Thermophysics Conference*, held in Snowmass, CO, 25-28 June 1984.

Van de Hulst, H.C. (1957) *Light Scattering by Small Particles*, Wiley, New York.

Space Nuclear Power Systems 1986
Edited by M. S. El-Genk and M. D. Hoover

Chapter 13

Simulation of Heat Pipe Rapid Transient Performance Using a Multi-Nodal Implicit Finite Difference Scheme

James S. Peery and Frederick R. Best

Nuclear Engineering Department
Texas A&M University
College Station, TX 77843

A model to simulate heat pipe rapid transients has been developed. This model uses a one-dimensional development of the continuity and momentum equations to solve for the velocity and pressure distributions in both the liquid and vapor regions. A two-dimensional development of the energy equation is used to determine the temperature distributions in the liquid and vapor regions, as well as in the walls of the heat pipe. The vapor and liquid regions are coupled through mass and energy transfer due to evaporation and condensation. The model used for this phenomenon is based on the physical conditions of the vapor and liquid for a given node. However, this model for evaporation and condensation not only causes the energy equation to be nonlinear but also constrains the time step to 10^{-4} seconds for convergence to be reached. The model has been run for small transients up to 2 seconds to produce temperature distributions and demonstrate the convergence difficulties associated with the evaporation/condensation model used.

Introduction

Heat pipes are being considered as a part of the thermal management system of many space applications because they are capable of passively transporting large amounts of thermal energy over considerable distances with essentially no temperature drop (Colwell 1977). A heat pipe consists of liquid and vapor in a closed void-capillary system that comprises an evaporator, an adiabatic section, and a condenser. When one end of the structure is heated, liquid evaporates, increasing the vapor content and causing the vapor-liquid interface to recede into the capillary structure. The vapor moves toward the cooler end of the closed structure and condenses, raising the liquid level in the wick capillary structure. The liquid is then returned to the heated end by capillary forces.

Heat pipe performance is sufficiently well understood in a steady state regime to allow satisfactory heat pipe design. However, conservative design of heat pipes can be the limiting factor in a thermal management system. Such conservation is necessary due to uncertainties in the modeling of heat pipe performance during transients (Best and Wick 1984): startup, shutdown, and rapid changes in power. The ability of a heat pipe to operate through such transients relies on the available capillary forces, which are affected by friction and acceleration pressure losses.

Mathematical modeling of heat pipe performance has been developed for both transient and steady state modes; however, transient heat pipe model development is limited. The modeling of the response of the vapor region in the event of rapid transients has not been fully investigated. The purpose of this study is to model the performance of a screened wick heat pipe in transient modes, and to determine the limiting conditions under which the heat pipe will operate.

Method

A multi-nodal implicit finite difference model has been developed to study the performance of wicked heat pipes under rapid transients. This model includes a nodal scheme in the vapor as well as the liquid-wick regions. The heat pipe being modeled is a rectangular heat pipe that was built and is being tested at Texas A&M University (DeHart and Best 1986). It has an aluminum body, uses water as a working fluid, and can use a multitude of wick materials. Electric heat is used in the evaporator section of the heat pipe, and a water-ice cooling jacket is used in the condenser. The heat pipe is well insulated during operation. Frozen startup is, at present, not being considered. It is assumed that initially the heat pipe is evacuated, then a given volume of liquid is introduced, and finally the heat pipe is allowed to come into equilibrium with the ambient conditions. At this point, the heater and cooling water pump are started and the transient begins.

The conservation equations (continuity, momentum, and energy) are applied to a control volume. The conservation equations are given in Eulerian form as (Slattery 1972):

Continuity:

$$\frac{d}{dt}\int_{\underline{V}} \rho d\underline{V} = 0 \tag{1}$$

Energy:

$$\frac{d}{dt}\int_{\underline{V}} \rho e d\underline{V} = \int_{\underline{V}} \rho [\vec{F} \cdot \vec{U} + \dot{r}] d\underline{V} + \int_{\underline{A}} (\vec{\vec{T}} \cdot \vec{U}) \cdot \vec{n} d\underline{A} + \int_{\underline{A}} q d\underline{A} \tag{2}$$

Momentum:

$$\frac{d}{dt}\int_{\underline{V}} \rho \vec{U} d\underline{V} = \int_{\underline{A}} \vec{t} d\underline{A} + \int_{\underline{V}} \rho \vec{F} d\underline{V} \tag{3}$$

A backwards difference approximation to the spatial derivative is employed to solve these equations. The time derivative is approximated with an implicit scheme. The three conservation equations are solved separately for the liquid and vapor phases, and are coupled through mass and energy transfer due to evaporation and condensation at the interface of the two phases. In addition, the model allows for the variations in the thermal properties of the working fluid. A fourth equation is obtained by assuming saturation conditions for both the vapor and liquid regions. This equation of state is used to find the density of each node (Meyer et al. 1967).

The continuity equation is used to find the velocity distribution in the heat pipe. A one-dimensional model is used for the continuity equation for both regions, with the heat pipe divided into axial nodes. Its final form is given for both the vapor and liquid as:

Vapor:

$$U_j = ((A\rho U)_{j-1} + \psi + V(\rho^n - \rho)/\Delta t)/(\rho A)_j \tag{4}$$

Liquid:

$$U_j = ((A\rho U)_{j+1} - \psi + V(\rho^n - \rho)/\Delta t)/(\rho A)_j \tag{5}$$

where ψ is the evaporation/condensation mass flow rate. Using the boundary conditions $U_l(x = n + 1) = 0$ and $U_v(x < 0) < 0$, Equations (4)

NOMENCLATURE

English

- A: Cross sectional area for flow
- A': Cross sectional area the liquid projects onto the vapor
- e: Internal energy
- $\vec{F}$: Sum of all body force vectors per unit mass of the fluid
- F: Evaporation coefficient
- g: Gravity
- M: Molecular weight of the working fluid
- $\vec{n}$: Unit outward normal vector
- n: Number of axial nodes
- P: Pressure
- q: Contact energy flux into the fluid
- R: Gas constant
- $\dot{r}$: Rate of internal heat generation per unit mass
- $\vec{\vec{T}}$: Tensor of all surface stresses exerted on the fluid
- T: Temperature
- $\vec{t}$: Surface stress vector $\vec{t} = \vec{\vec{T}} \cdot \vec{n}$
- t: Time
- Δt: Time step
- $\vec{U}$: Vector velocity
- U: Axial velocity
- V: Control volume
- x: Axial coordinate position
- Δx: Axial node length

Greek

- θ: Orientation angle measured from horizontal
- ρ: Density
- ψ: Evaporation/condensation mass flow rate

Superscripts

- n: Previous time step

Subscripts

- i: Radial level index
- j: Axial level index
- l: Liquid
- v: Vapor

and (5) can be solved for each node by marching through the grid once ψ is known for each node.

The most difficult item to deal with was modeling the evaporation and condensation from a porous structure. Past transient models use the energy input divided by the latent heat of evaporation as an approximation to the mass flow rate leaving the liquid region as evaporation. This model is valid only for steady state, since some fraction of the energy input must go into stored energy when the heat pipe is undergoing a transient. This is shown in Figure 1 for the aforementioned rectangular heat pipe and a power input of 1.5 kW. The development and use of phenomenological models for evaporation and condensation have been investigated. A model given by Subbotin et al. (1969) is currently being used and is given by

$$\psi = \frac{8\ P_v A'}{\sqrt{\frac{2\pi RT}{M}v}} \left[\frac{1}{F} - \frac{1}{2} - \sqrt{\frac{1}{F^2} - \frac{1}{F} + \frac{P_l}{4P_v}\sqrt{\frac{T^v}{T_l}}}\right] \tag{6}$$

where F is the evaporation coefficient. The model's limitation is the value for the evaporation coefficient. The literature reports values for F that range from 1 to .0005 (Ivanovskii et al. 1982), depending on the amount of impurities in the working fluid. A value for F that results in acceptable performance modeling will have to be determined. In this model, it is assumed that P_l/P_v equals 1.0; therefore, the temperatures of the vapor and liquid regions must be determined to find ψ.

The energy equation is developed to determine the temperatures in the heat pipe. A two-dimensional model is used for the energy equation to account for the energy transport through the walls of the heat pipe. For this model, the heat pipe is separated into radial and axial nodes; however, only one radial node is used for the liquid and vapor regions. The final form of the energy equation derived from equation(2) for any given node is given by

$$\begin{aligned} - [QT]_{j-1,i} - [QT]_{j,i-1} + [QT]_{j,i} \\ - [QT]_{j,i+1} - [QT]_{j+1,i} = [QT]^{n}_{j,i} \end{aligned} \tag{7}$$

where Q can contain conduction, convection, and transient terms depending on the node to which the equation is applied. In the wall, only axial and radial conduction are considered. Energy transfer by axial and radial conduction, axial convection, and evaporation/condensation is considered for both the vapor and liquid regions. The heat pipe is assumed to be perfectly insulated; thus, the gradient of the temperature

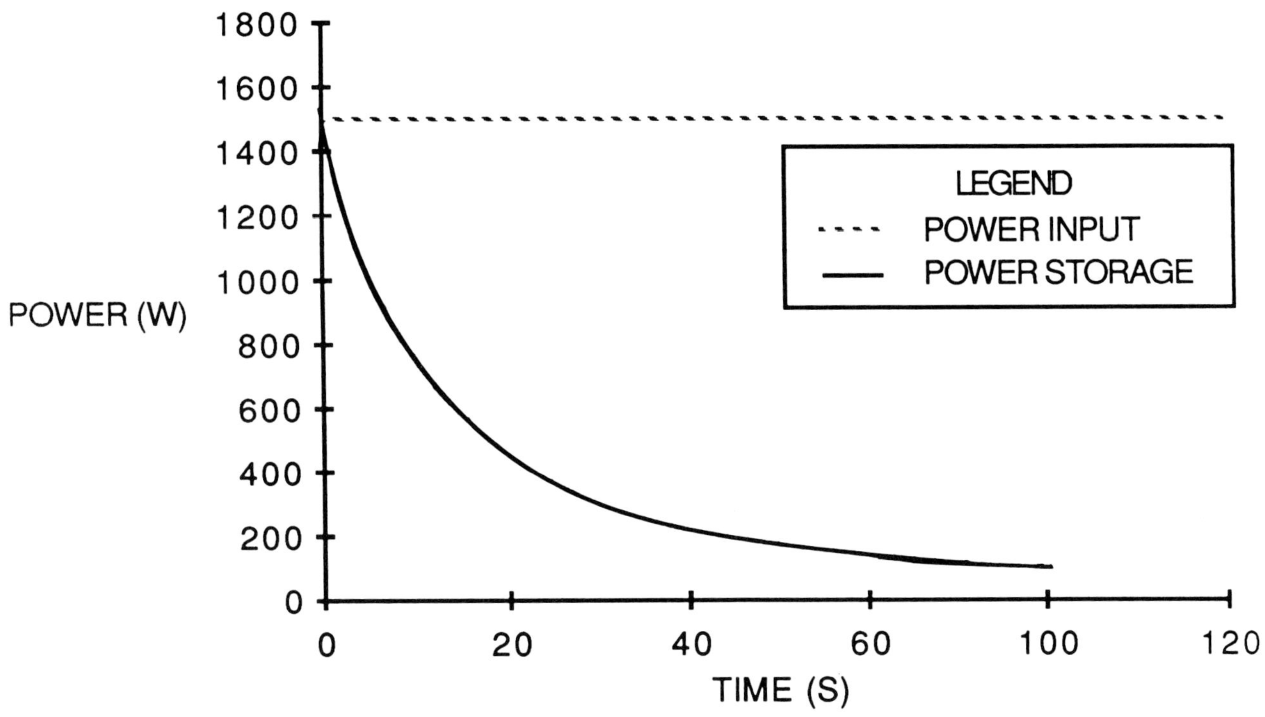

Figure 1 *Power Input and Power Storage versus Time for the TAMU Heat.*

at any boundary is equal to zero. In addition, the power inputs of the electric heater and coolant water temperature serve as boundary conditions. When Equation (6) is substituted into Equation (7) for the liquid and vapor nodes, the energy equation becomes nonlinear. An iterative technique is used to solve the set of equations. The converged solution determines not only the temperatures but also the evaporation or condensation mass flow rates throughout the heat pipe. This information allows Equations (4) and (5) to be solved.

The pressure distribution is determined by solving the momentum equation. Similar to the continuity equation, a one-dimensional model is used for the momentum equation for both regions, with the heat pipe divided into axial nodes. With the velocities known, the pressure gradient between any two nodes can be found. Reducing Equation (3) to finite difference form and solving for the pressure gradient in both the vapor and liquid nodes results in the following equations:

Vapor:

$$\frac{dP}{dx} = \frac{(\rho U)^n - \rho U}{\Delta t} + \frac{(\rho U^2)}{\Delta x}j-1 - \frac{(\rho U^2)}{\Delta x}j - (\rho g \sin\theta)j - \left.\frac{dP}{dx}\right|_f \tag{8}$$

Liquid:

$$\frac{dP}{dx} = \frac{(\rho U)^n - \rho U}{\Delta t} + \frac{(\rho U^2)}{\Delta x}j+1 - \frac{(\rho U^2)}{\Delta x}j - (\rho g \sin\theta)j - \left.\frac{dP}{dx}\right|_f \tag{9}$$

where $dP/dx|_f$ is the pressure drop due to friction forces. In these equations, the same boundary conditions apply as for the continuity equations. Two assumptions are made to find the actual pressures in each node:

1. The last node in the condenser is flooded. ($P_l = P_v$ for that node), and
2. The vapor pressure in the last node of the condenser is equal to the saturation pressure given by the temperature in that node.

These assumptions and the following two equations provide the means to find the pressure in each node.

Vapor:

$$dP/dx = (P_j - P_{j-1})/\Delta x \tag{10}$$

Liquid:

$$dP/dx = (P_j - P_{j+1})/\Delta x \tag{11}$$

With the initial conditions specified, the model solves for the distribution in temperature, pressure, and velocity for the next time step. The model marches through time implicitly, using the previous distributions to calculate the distributions at the new time step. The user is required to specify all physical dimensions of the heat pipe and to input the initial conditions of the heat pipe, including the heating and cooling requirements. Using run time dimensioning, the model allows for the number of nodes to be specified, although this feature is limited by the memory of the computer.

Results

The heat pipe model has been started from an isothermal state of 303 K. However, steady state has not yet been determined, primarily due to the extremely small time steps required to solve both the energy and evaporation/condensation equations iteratively. The initial time step is 10^{-4} seconds. This constraint on the time step was previously reported when using this model for evaporation/condensation (Subbotin et al. 1969). Figure 2 shows the convergence of the liquid and vapor on the first time step for the first evaporator node. The liquid appears to converge after three iterations steps while, it takes approximately 17 iteration steps for the vapor to converge. However, this result is for convergence in the eleventh decimal place; a less strict convergence criterion might allow for convergence on three to four iterations.

Other models have been used for an evaporation/condensation mass flow rate and the results for each time step vary significantly between each model. For the rapid transients studied thus far, the pressure continues to increase inside the pipe as the heat pipe heats up. Moreover, the velocity of the liquid and vapor increase with each time step. The velocity of the liquid and vapor increases from the end of the condenser to a maximum in the adiabatic region and then decreases as it approaches the beginning of the evaporator section. The pressure in the vapor decreases through the evaporator and adiabatic regions due to friction and inertial forces and then experiences some pressure recovery in the condenser. The liquid pressure decreases from the condenser to the evaporator primarily due to friction forces.

Since the temperatures of the liquid and vapor are

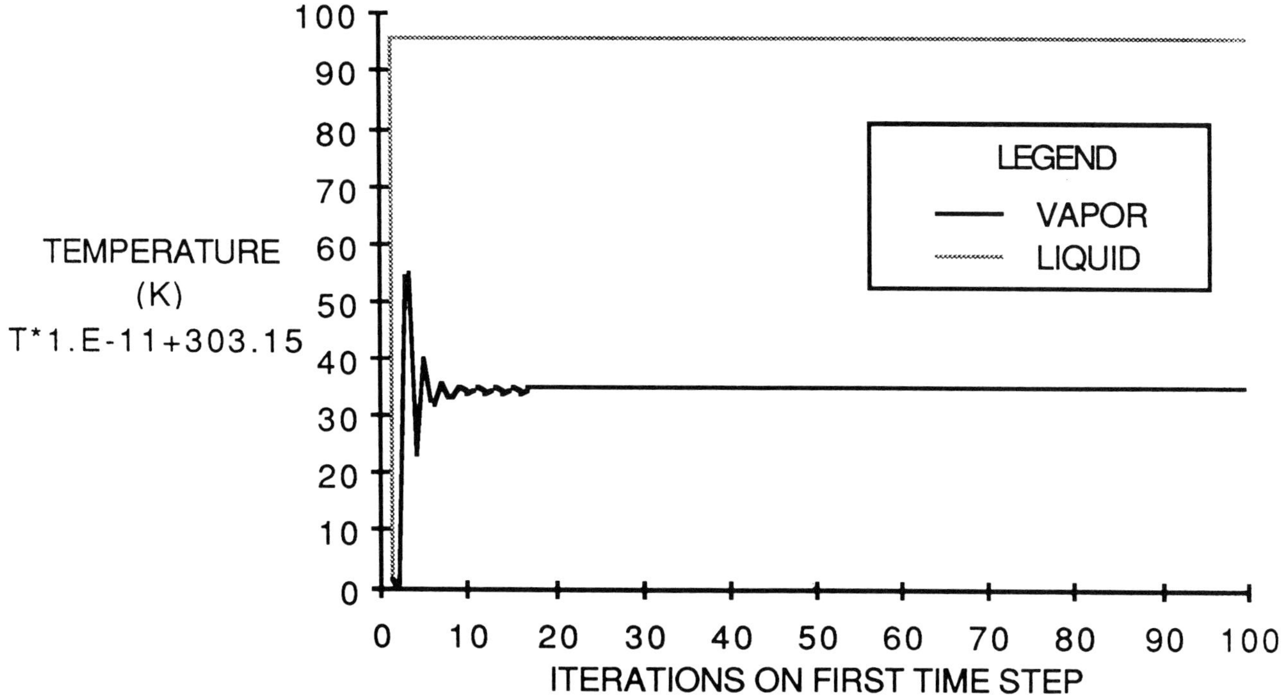

Figure 2 *Convergence of the Liquid and Vapor Temperatures for the First Evaporator Node for a 1.5-kW Power Input.*

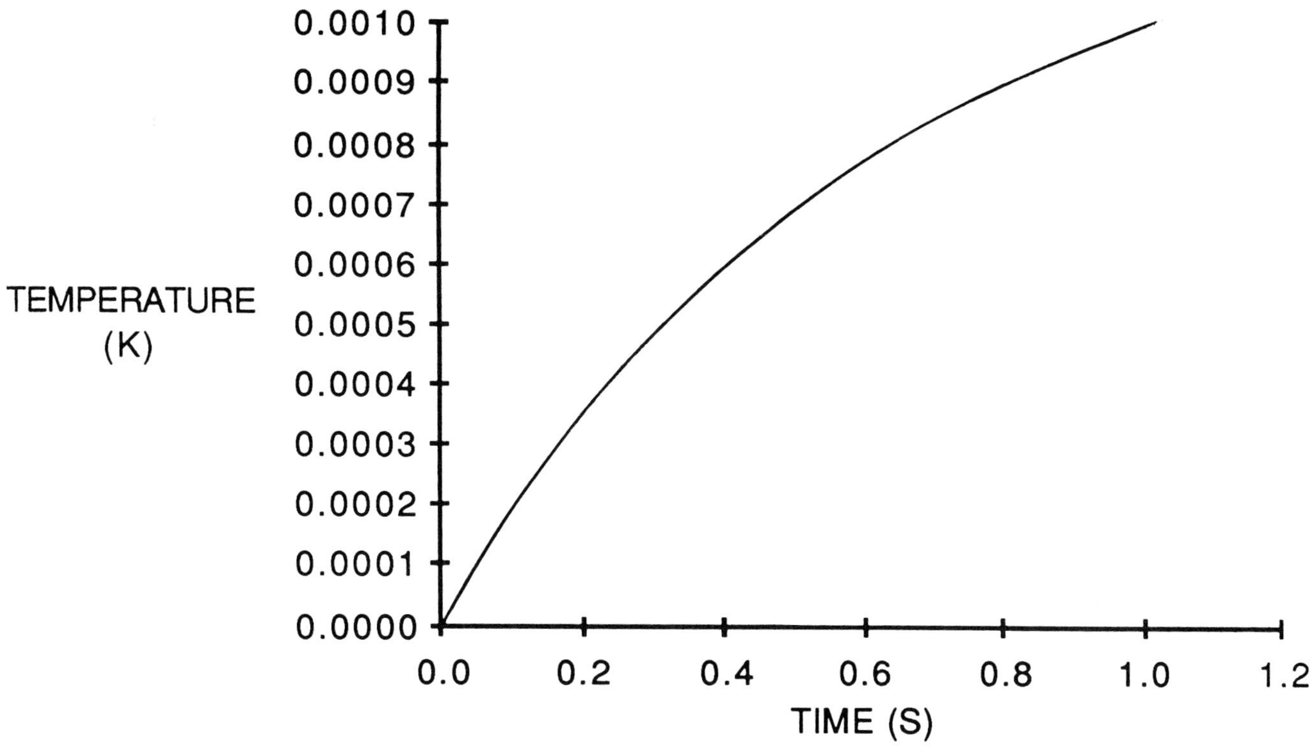

Figure 3 *Temperature Difference between the Liquid and Vapor in the First Evaporator Node for a 1.5 kW Power Input.*

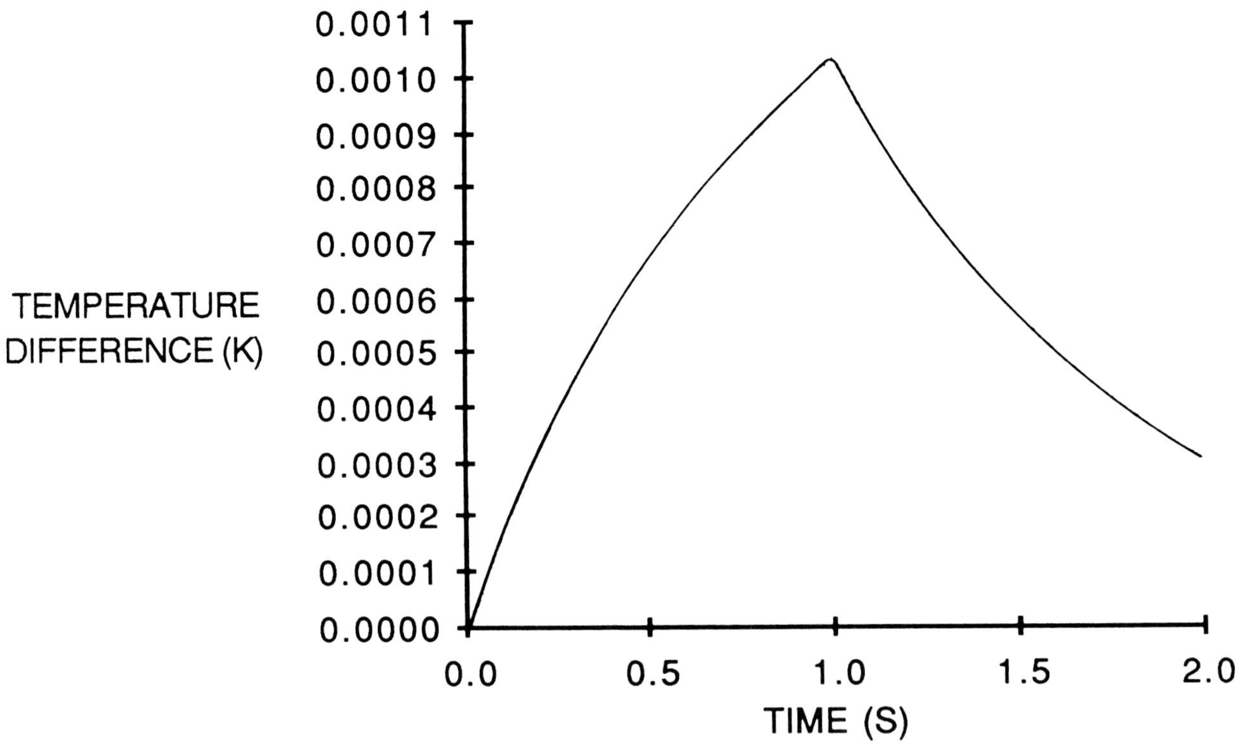

Figure 4 *Temperature Difference between the Liquid and Vapor in the First Evaporator Node for a 1.5-kW Power Input for One Second.*

almost identical for a given node, a plot of the temperature difference between the liquid and vapor during a transient provides the most information. Figure 3 is a plot of the temperature difference between the liquid and vapor for the first node of the evaporator during a 1 second transient. Initially, the heat pipe is at 303.15 K and the power input is 1.5 kW. As expected, the temperature difference increases through the transient but appears to start leveling off. Figure 4 is a plot for the identical conditions as Figure 3, but, the power is turned off at 1 second in the transient. The model is run through to 2 seconds, at which time the temperature difference is seen returning to zero.

Summary and Conclusions

A model to predict the rapid transient response of heat pipes has been developed. This model is unique because it solves for the velocity, temperature, and pressure distributions for the vapor region of the heat pipe. In addition, it uses an evaporation/condensation source that depends on the physical properties of the liquid and vapor. However, this also constrains the model due to the small times steps required to obtain a converged solution. Methods to increase or accelerate the time step must be investigated. Currently, the project's principle effort is directed toward validating the computer model by comparison with experimental work. Future work will include a melting/freezing model for the liquid in the wick, as well as a free molecular flow model for the vapor when it is at very low pressures.

Acknowledgments

This work is being carried out at Texas A&M University with funding supplied by the Air Force Weapons Laboratory, Kirtland AFB, New Mexico.

References

Best F. R. and R. Wick (1984) "Measurement and Analysis of Mission Required Rapid Heat Pipe Transients Using Internal Optical Observations," Unsolicited Proposal to the Air Force, Air Force Weapons Laboratory, Kirtland AFB, NM.

Colwell, G. T. (1977) *Prediction of Cryogenic Heat Pipe Performance*, Final Report, NASA Grant NSG-2054, Goddard Space Flight Center, MD.

DeHart M. D. and F. R. Best (1986) "Heat Pipe Entrainment Analysis Via Optical Methods", in *Trans. Third Symposium On Space Nuclear Power Systems*, CONF-86-0102-Summs, held in Albuquerque, NM, 13–16 January 1986.

Ivanovskii, M., V. Sorokin, and I. Yagodin (1982) *The Physical Principles of Heat Pipes*, Clarendon Press, Oxford.

Meyer C. A. et al. (1967) *ASME Steam Tables*, American Society of Mechanical Engineers, New York.

Slattery, J. C. (1972) *Momentum, Energy and Mass Transfer in Continua*, McGraw-Hill Book Co., New York.

Subbotin, V. I., D. N. Sorokin, D. M. Ovechkin, and A. P. Kudryavtsev (1969) "Heat Exchange on Boiling of Metals under Natural Convection Conditions", Nauka, Moscow.

Space Nuclear Power Systems 1986
Edited by M. S. El-Genk and M. D. Hoover

Chapter 14

Liquid Droplet Heat Exchanger Studies

A.P. Bruckner, D.E. Hedges, and S. Yungster

Aerospace and Energetics Research Program FL-10
University of Washington
Seattle, WA 98195

Recent analytical and experimental investigations of the liquid droplet heat exchanger (LDHX) concept for space power applications are described. The performance of the LDHX is compared to that of a conventional heat exchanger for heat rejection applications in a Brayton cycle, using the mass-specific heat exchanger effectiveness as a figure of merit. It is shown that the LDHX has an order of magnitude advantage over the conventional heat exchanger. Furthermore, significant improvement in cycle efficiency and power to mass ratio is possible. Two-phase flow experiments in a laboratory scale LDHX, using air and water as the two media, show very good agreement with the quasi-one-dimensional model used in the parametric studies.

Introduction

A number of recent studies by Bruckner et al. (1982, 1984, and 1986) and Taussig et al. (1985) have addressed the feasibility of direct contact heat exchanger concepts for space power systems. These advanced concepts offer a potential of low mass and the advantage of reduced risk of single point failure, low pressure drop, compact design, high temperature capabilities and high heat transfer rates. The most promising of the concepts investigated to date is one using a highly dispersed heat transfer medium, such as droplets, in a gas. This class of heat exchanger is referred to here as the liquid droplet heat exchanger (LDHX).

A critical design requirement of the LDHX is complete separation of the heat exchange media following heat transfer. Bruckner and co-workers (1984 and 1986) have developed an approach in which the droplets are centrifuged out of the heat exchange chamber by means of a swirling gas flow. Recently, an alternative configuration, a straight, co-flowing LDHX, was proposed by Taussig et al. (1985). In that approach a sharp 180° turn in the exhaust flow causes the droplets to be separated from the gas by centrifugal force.

In this paper the swirling flow (that is, vortex) LDHX geometry is discussed. Figure 1 shows a schematic of the concept. The working gas is injected tangentially at the outer diameter of the heat exchange chamber, and fine sprays of uniform droplets are injected radially outward from a central manifold. The gas spirals inward to the central manifold, where it exits the chamber. The swirling motion of the gas centrifuges the droplets to the outer wall. As they reach the chamber periphery, the droplets coalesce into a liquid film that is dragged along the wall by the swirling gas and the tangential momentum of the arriving droplets. Several skimmers set into the curved wall remove the liquid film. The spent liquid is continuously pumped back to the heat source or heat sink, depending on the application of the LDHX. The net gas flow is radially inward and the net droplet flow is radially outward; hence, this LDHX configuration operates as a counterflow device with the potential for a very high effectiveness.

LDHX Modeling and Performance

A quasi-one-dimensional, two-phase, steady flow model of the LDHX, developed by Bruckner and Shariatmadar (1986), has been used for parametric studies aimed at developing practical and effective design configurations. The flow model consists of nine governing equations, that is, continuity equations, radial and angular momentum equations and energy equations for both heat transfer media (gas and droplets), and the ideal gas law. The interaction of droplets and gas is included in the model as single-sphere drag terms (Boothroyd 1971) in the radial and angular momentum equations. Heat transfer is assumed to follow the Nusselt number correlation for single spheres (Davidson and Harrison 1971). These standard correlations are used because universally

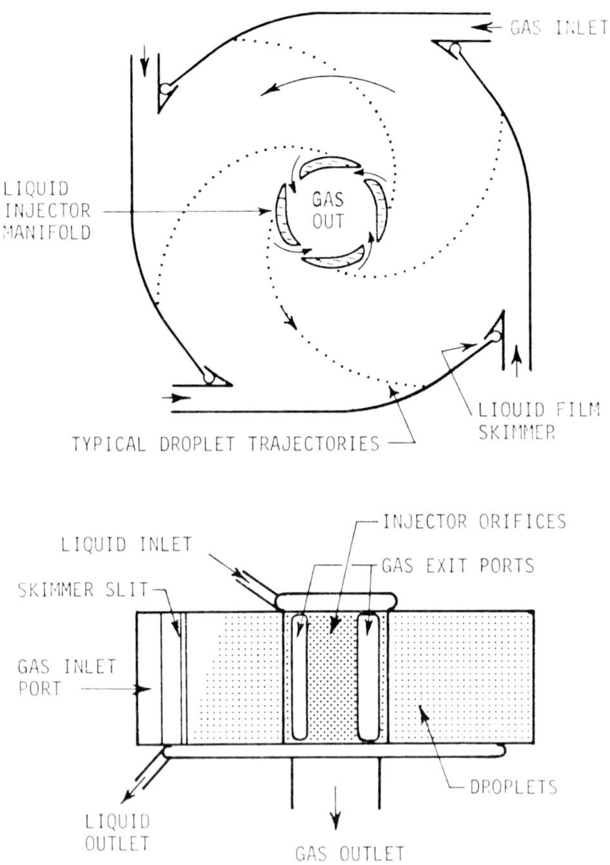

Figure 1 *Schematic of Vortex-Type Liquid Droplet Heat Exchanger (LDHX). For clarity, only four droplet trajectories are shown.*

accepted transport correlations for gas/droplet type two-phase flows are not available. The single particle correlations are adequate at this stage of the investigation for parametric studies and heat exchanger concept comparisons.

To compare different space heat exchanger design concepts at the same heat transfer rate, a new figure of merit, the mass-specific heat exchanger effectiveness has been used (Taussig et al. 1985):

$$Z = \frac{Q}{[C_g \Delta T_i M_{hx}]} = \frac{\epsilon}{M_{hx}} ,$$

where Q is the rate of heat transfer of the heat exchanger, $C_g = \dot{m}_g c_{pg}$ is the heat capacity rate of the gas (where $\dot{m}_g$ is the mass flow rate and c_{pg} is the specific heat at constant pressure), M_{hx} is the heat exchanger mass, ΔT_i is the media inlet temperature difference and ϵ is the heat exchanger effectiveness. A large value of Z is desirable, because it indicates a design with a low mass for a given heat transfer capability.

For the pressure (< 15 atm) and temperature (< 1200 K) range considered in a wide variety of space power systems, the thin shell theory with a safety factor of 3 is reasonably adequate for pressure vessel wall mass modeling. For high temperature applications, such as the heat source heat exchanger in a Brayton cycle, the refractory alloy Nb-1%Zr has been assumed for the construction material (Coombs et al. 1971). For low temperature applications, such as waste heat rejection in a Brayton cycle, the material was assumed to be titanium, due to its high strength-to-mass ratio. Note that although the material selection will affect the absolute value of the heat exchanger performance parameters, it will not affect the relative merit between different concepts.

The heat exchanger mass specific effectiveness, Z, is affected by many parameters, such as the system conditions, droplet diameter, injection velocities of both media, temperature difference between the media, media thermal properties, and ratio of media capacity rates. For the purpose of comparing the present LDHX concept to other heat exchanger designs, the low-temperature heat sink requirements for a 100-kWe regenerative Brayton cycle (Figure 2) were adopted for parametric studies. The working gas assumed in the present case is a mixture of 67% He and 33% Xe (MW = 46). The working liquid is Dow 705 silicone fluid, which has a very low vapor pressure at temperatures up to 400 K and other desirable properties, as described by Bruckner and Shariatmadar (1986). This fluid is also used directly in the liquid droplet radiator (Mattick and Hertzberg 1982) assumed for the present study. After the system conditions (that is, gas pressure, inlet temperature, temperature change, and flow rate) and working media have been specified, the primary influencing parameters left are droplet diameter, media injection velocities, media inlet temperature difference, and media capacity rate ratio. Since the rest of the power system, for example, the size and mass of the radiator, is affected by the media capacity rate ratio, this ratio (C_d/C_g) is fixed and set to be 4 for all subsequent calculations. This value represents a reasonable balance among the heat exchanger efficiency, particle volume fraction (≈1% to 3%), and radiator loading.

The optimum heat exchanger configuration and its corresponding operating conditions have been evaluated by varying the droplet injection velocity, gas inlet (or exit) velocity, droplet sizes, and media inlet temperature difference, independently (Bruckner and Shariatmadar 1986 and Taussig et al. 1985). Since heat transfer is a rate phenomenon, the amount of heat transfer is governed strongly by the droplet flight time but is only mildly and indirectly affected by the velocity difference between the two media due to the

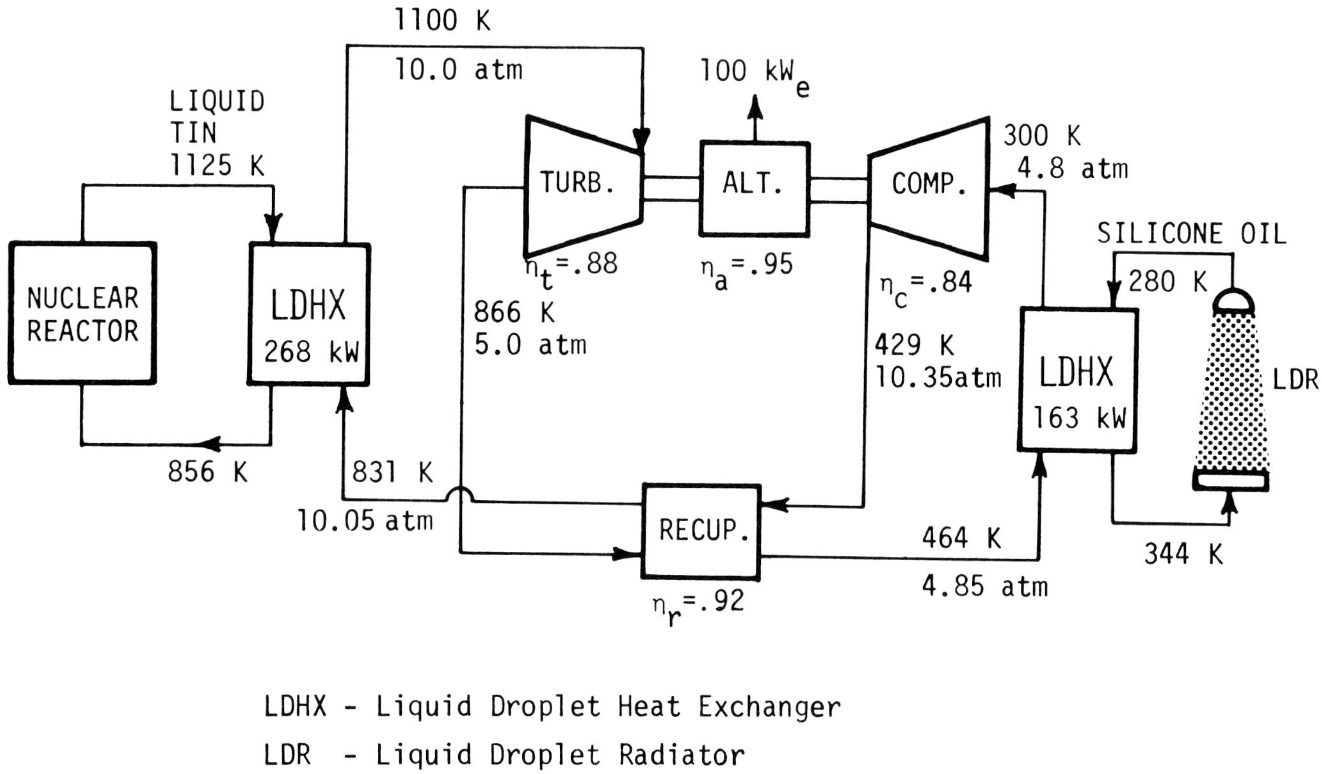

Figure 2 *Brayton Cycle Utilizing Liquid Droplet Heat Exchangers and a Liquid Droplet Radiator (Working Gas is 67% He, 28% Xe).*

Reynolds number dependency. Therefore, considering all the other parameters to remain constant, the LDHX can be made more compact by reducing the droplet injection velocity within practical limits. Furthermore, for a fixed gas flow rate, increasing the gas velocity will also reduce the heat exchanger size and mass (since the flow area required is reduced, the height of the LDHX is reduced). Taussig et al. (1985) have discussed the practical limits of the above parameters.

Figure 3 compares the heat exchanger performance, Z, of different design concepts and different droplet sizes for a particular radiator loading (that is, fixed droplet flow rate, and inlet and outlet temperatures). Due to its cross-counterflow configuration, it is clear that the vortex LDHX is superior to its coflowing counterpart at low values of ΔT_i. When higher values of ΔT_i are available, the two different LDHX design concepts become more competitive in terms of Z. Figure 3 also includes a conventional finned tube heat exchanger sized for the same conditions as the LDHXs. The finned tube heat exchanger is a 7-pass (on the gas side) cross-counterflow device (Taussig et al. 1985). The LDHX concepts have an order of magnitude advantage over the conventional heat exchanger in terms of the Z figure of merit. The vortex LDHX configuration appears to be especially advantageous in this regard.

Application of the LDHX in a Brayton Cycle

The performance, specific mass, and long-term reliability of space Brayton cycles are largely governed by the corresponding parameters of their heat exchangers and radiators. Therefore, this is an application area where the LDHX could have a significant beneficial impact. Accordingly, a recuperated 100-kWe Brayton cycle has been evaluated to determine the impact of the vortex LDHX on the overall power cycle efficiency and system mass. The system is similar to the one illustrated in Figure 2. The system masses were computed assuming a high power density nuclear reactor and an advanced, lightweight radiator having a surface area density of 0.5 kg/m^2 and an emissivity of 0.35 (Taussig et al. 1985). The peak and minimum cycle temperature were assumed to be 1100 K and 282 K respectively.

The results for two power cycles are shown in Figure 4. The dashed lines represent a Brayton power cycle with lightweight but conventional finned tube-and-shell heat source and heat sink heat exchangers similar to the designs of Coombs et al. (1971). Tin is

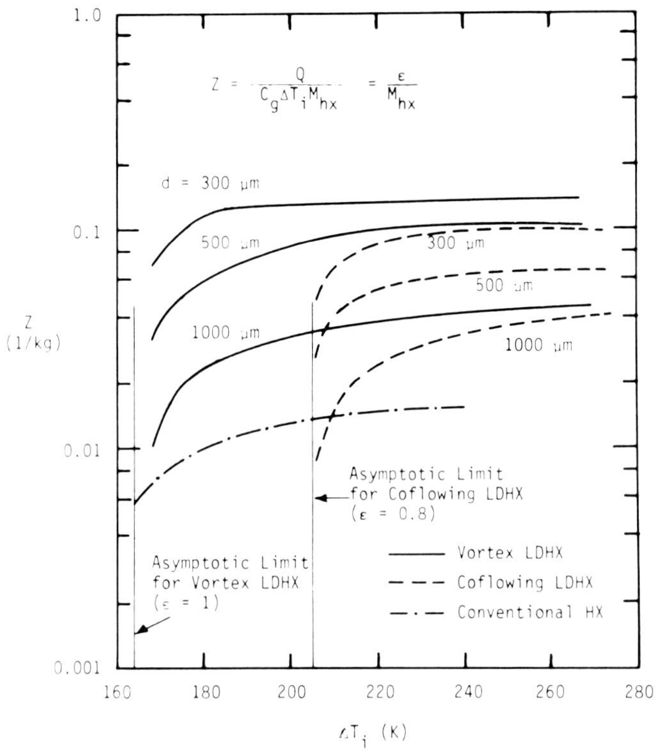

Figure 3 *Optimized LDHX Performance as a function of Media Inlet Temperature Difference for Various Droplet Diameters, d.*

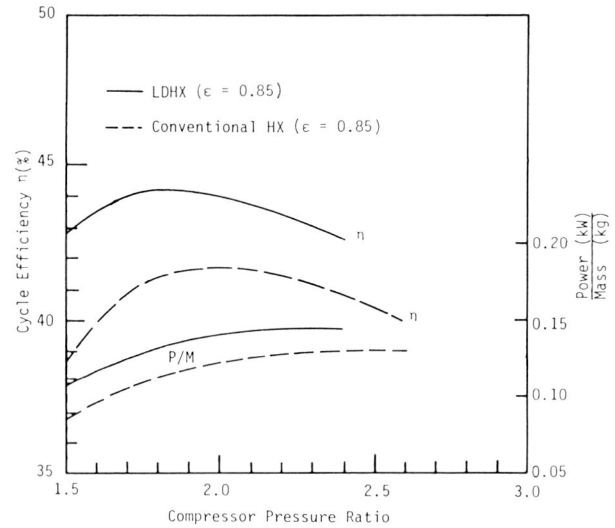

Figure 4 *Cycle Efficiency and Power per Unit Mass for 100-kWe Recuperated Brayton Space Power System with and without Vortex LDHX. Peak Cycle Temperature = 1100 K; Minimum Cycle Temperature = 282 K.*

the heat source liquid and Dow 705 is the heat sink liquid; these media were selected for their extremely low vapor pressures at their corresponding operational temperatures (Bruckner and Mattick 1984). The Helium-Xenon mixture described earlier is used as the gaseous working fluid. The solid lines represent a power cycle with vortex LDHXs for the heat source and heat sink heat exchangers, which use the same heat transfer media as the conventional heat exchangers. The heat exchanger effectiveness is the same in both cases ($\epsilon = 0.85$). A conventional recuperator of corrugated plate design is used in both sets of calculations (Coombs et al. 1971). (The pressure differential between the two flow loops in the recuperator does not permit the use of an LDHX as a recuperator. However, a direct-contact heat wheel type heat exchanger may be suitable for this purpose (Taussig et al. 1985)).

A significant increase in overall cycle efficiency is achieved through the use of LDHXs. This advantage is due to the decrease in pressure losses in LDHXs compared to the conventional heat exchangers and produces a two-fold effect on the total power system mass. First, higher efficiency implies that less heat is transferred into and out of the Brayton cycle for a given power output. Second, as discussed in the previous section, both the heat source and heat sink LDHXs are lighter than their conventional heat exchange counterparts.

The power per unit mass (P/M) continues to increase with increasing pressure ratio beyond the point where the power cycle efficiency peaks. As the pressure ratio increases, the amount of recuperation decreases and the minimum temperature in the recuperator rises. The minimum recuperator temperature is the same as the maximum waste heat temperature in the cycle. Therefore, the effective radiator temperature rises as the pressure ratio increases, leading to a smaller, lighter radiator, because higher temperature radiators reject heat more effectively. Figure 4 shows the P/M peak at a pressure ratio of approximately 2.5. At higher pressure ratios, the cycle efficiency becomes so low that the amount of waste heat overbalances the increase in radiator effectiveness and both the radiator and heat source masses increase, causing a reduction in P/M.

Experimental Studies

The two-phase flow dynamics of the LDHX have been investigated using a small-scale vortex chamber. A schematic of the current facility is shown in Figure 5. Experiments to date have not included studies of heat transfer. Although heat transfer changes

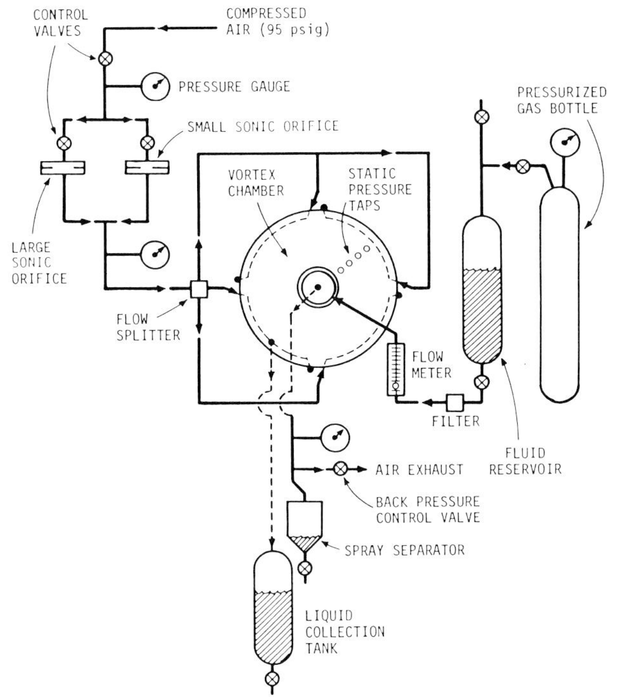

Figure 5 *Schematic of Experimental LDHX Facility.*

Figure 6 *Close-Up of Experimental LDHX Vortex Chamber, showing Gas and Liquid Feed Lines, Transparent End Walls, Pressure Taps and Liquid and Gas Exhaust Tubes.*

the local properties of the gas, these changes do not affect, to first order, the two-phase flow dynamics.

Figure 6 is a photograph of the vortex chamber. The end plates of the chamber are made of Plexiglas to permit flow visualization studies. The upper end plate is equipped with an array of pressure taps to measure the radial static pressure profile in the flow and can be rotated in 30° increments to check for azimuthal non-uniformities. The diameter and height of the chamber are 28 cm and 5 cm, respectively. The working gas is air and is injected into the vortex chamber from a concentric gas plenum through four equally-spaced 0.6 cm × 5.0 cm longitudinal ports at the chamber periphery. The nominal gas pressure in the chamber can be varied from 1 to 2 atm by means of a back pressure control valve in the gas exhaust system. The gas flow rate is controlled by a pair of sonic orifices in the gas supply line. For a given pair of orifices, three different mass flows can be selected at a given line pressure by using one orifice at a time or both orifices simultaneously. With the existing set of orifices, the gas flow rate can be varied from 4 to 30 g/s.

Two different liquid injector/gas exhaust manifold geometries have been tested. In both cases the liquid is injected from 100-μm dia. orifices drilled in the sides of 2.4-mm O.D. brass tubes, each having 15 equally spaced orifices. The first manifold configuration studied consists of a 7.8-cm O.D. aluminum tube having four equally spaced gas exhaust ports (similar to the schematic of Figure 1). The solid areas between these ports are machined with a series of axial grooves to hold the brass liquid injector tubes. There are 24 such tubes in four groups of six, for a total of 360 injector orifices. Figure 7a is a photograph of this geometry. Each injector tube can be rotated ± 45° about its own axis to allow the injected droplets to be imparted any desired tangential velocity component within the range $\pm V_i/\sqrt{2}$, where V_i is the injection velocity.

The second manifold configuration consists of 36 equally spaced, free-standing injector tubes, similar to those described above, arranged on a base circle of the same diameter as in the first configuration (7.8 cm). Figure 7b illustrates this geometry. In this case there are a total of 540 injector orifices. The gas ex-

Figure 7 *Flow Visualization of Droplets in Experimental LDHX Chamber for Two Different Liquid Injector Configurations: a) 24 Injector Tubes set in Solid Manifold Wall having Four Gas Exhaust Ports (Mass Flow of Water = 12.6 g/s; Mass Flow of Air = 24.9 g/s), and b) 36 Free-Standing Injector Tubes, Equally Spaced on Base Circle (Mass Flow of Water = 12.6 g/s; Mass Flow of Air = 25.1 g/s).*

hausts through the spaces between the tubes. This particular geometry offers a more uniform spatial distribution of the injected liquid droplets, permits a wider range in the orientation of the injected liquid streams, and provides a more uniform exhaust configuration for the gas.

In both configurations the injector tubes are fed from a liquid manifold mounted above the vortex chamber. This manifold is in turn supplied with liquid from a 3.8-liter reservoir that is pressurized with compressed air. The liquid flow rate is measured by a rotameter. To prevent clogging of the injector orifices, a 15-μm filter is used to remove any particulate matter present in the liquid. In experiments conducted to date, double-distilled, deionized water has been used as the droplet medium. The liquid jets have been injected purely radially and permitted to break up into droplets naturally, that is, no external excitation has been used to force uniform droplet formation. The average droplet size is ~190 μm. Future experiments will include a piezoelectric transducer in the liquid feed manifold to provide the necessary acoustic excitation to generate uniform droplets.

The gas leaving the vortex chamber is exhausted at the center of the bottom plate of the chamber and is directed to an exhaust plenum chamber that serves as a liquid/gas separator (see Figure 5). Any liquid leaving the vortex chamber with the exhaust gas is trapped in this separator. The fraction of liquid trapped in the exhaust gas can thus be determined. Four liquid skimmer slits, having apertures of 0.8 mm × 50 mm, are provided adjacent to the gas inlet ports in the vortex chamber to remove the liquid film that forms at the periphery. To minimize vertical liquid runoff due to gravity and to guide the liquid to the skimmers, the outer wall of the chamber is lined with 3M Riblet film. The surface of this plastic film is textured with closely spaced parallel micro-grooves, typically 150 μm wide and 150 μm deep, which channel the liquid to the skimmers via capillary action. Although the Riblet film maintains a contiguous film and significantly reduces vertical runoff, some of the liquid nevertheless runs down to the bottom plate, where it forms spiral streams that are dragged out through the gas exhaust aperture by the inward spiraling gas. With the present configuration about 65% of the liquid is removed by the skimmer slits. In future experiments, the outflow of liquid along the bottom plate will be eliminated by machining gutters at the periphery of the plate. In zero-g the runoff problem would not occur.

Flow visualization with still photographs and videotape was performed to allow closer study of the two-phase fluid dynamics, and to help identify areas of improvement in the design. Droplet trajectories at various liquid and gas mass flows were imaged with stroboscopic illumination. The strobe was operated with a flash duration of 1 μs and a frequency of approximately 60 Hz. Single-flash illumination was also used.

Figures 7a and 7b are photographs of the two-phase flow in the chamber for each of the two liquid injector configurations discussed previously. In both cases, the water mass flow rate is 12.6 g/s, corresponding to a droplet injection velocity of 4.5 m/s in Figure 7a (24 injector tubes) and 3.0 m/s in Figure 7b (36 injector tubes). The air mass flow rate is 27.9 g/s in Figure 7a and 25.1 g/s in Figure 7b. The gas rotates clockwise. Chamber pressure is near-atmospheric in each case. The droplets are clearly visible throughout the field of view. Large bright spots are caused by droplets out of focus, reflections from parts of the apparatus, and liquid flowing along the bottom plate.

The videotaped runs allowed slow motion and freeze-frame visualization of droplet formation, transit and collection, as well as the transient start-up and shutdown of liquid phase flow. It has been observed that the crossflow of gas at the injector tubes does not adversely affect the droplet formation process at any of the relative mass flows studied so far. Liquid jet breakup into droplets occurs almost immediately upon exiting the injector tubes. At low to medium gas mass flow rates (up to ~17 g/s), the individual droplet streams retain their identity for a considerable radial distance, gradually fanning out as a result of turbulence. In the case of the liquid injector configuration having the 36 equally spaced injector tubes, the spatial distribution of droplets becomes homogeneous about half-way to the outer wall. At high gas mass flow rates (~25 g/s), the videotape clearly indicates a more homogeneous distribution of droplets throughout the chamber. No evidence of aerodynamic shattering of droplets, nor any evidence of significant droplet breakup or agglomeration resulting from droplet-droplet collisions, has been observed. Furthermore, no entrained droplets have been detected in the gas flow leaving the chamber and no salient secondary flow effects have been observed.

Visualization of droplet collection indicated that the skimmers were basically effective. However, sidewall runoff, due to gravity, overloaded the skimmers at the base of the chamber. This caused the excess fluid to form the spiral streams mentioned previously. Additionally, it was noted that the approximately 5-cm sidewall distance from the skimmers to the gas inlets collected a significant amount of liquid that could not be skimmed off. This fluid encountered the gas inlet, and was carried into the core, some of it splashing onto the end plates. These effects pointed out the need to widen the skimmer openings near the bottom plate and to relocate the skimmers as close as possible to the gas inlet ports in future designs.

Gas flow parameters in the vortex chamber have been measured using a pitot pressure probe and static wall pressure ports connected to a kerosene manometer bank. (At the time of writing, the manometer bank was being replaced by a Scanivalve pressure transducer system connected to a Fluke datalogger. This system permits pressure surveys to be accomplished rapidly, with a considerably larger number of samplings per run.) Stagnation data were measured at one radial station at a time with the pitot probe, while static pressures were measured at all ports simultaneously by photographing the manometer bank. Pressure surveys of the chamber have been carried out at a variety of gas and liquid mass flow rates, including the special case of no liquid injection, and at different azimuthal locations.

Figure 8 shows representative results of radial pressure surveys, one for each of the two liquid injector configurations tested. Theoretical radial pressure distributions based on the quasi-one-dimensional model of Bruckner and Shariatmadar (1985) are plotted for comparison. The liquid flow rate is 12.6 g/s in each case, and the respective gas flow rates are 27.9 g/s in Figure 8a and 25.1 g/s in Figure 8b. In the case of the static pressures the experimental points represent the averaged result of three separate static pressure measurements. Slight pressure fluctuations within the chamber caused the set of experimental data points to shift up and down in unison ~0.03% about the average value. The amount of shift is indicated by the error bars. The source of the pressure fluctuations is thought to be flow turbulence and is presently under investigation. In Figure 8b there are two plotted points very close together at a radius of 9 cm. These were measured 90° apart and show that the flow is azimuthally uniform. Note that to date stagnation pressure data have been obtained only for the liquid injector configuration of Figure 7a (24 injector tubes). The stagnation pressures were obtained one at a time and show considerable scatter, attesting to the difficulty of obtaining accurate pitot pressure measurements in the droplet-laden environment of the vortex chamber. This problem is being addressed in the ongoing studies.

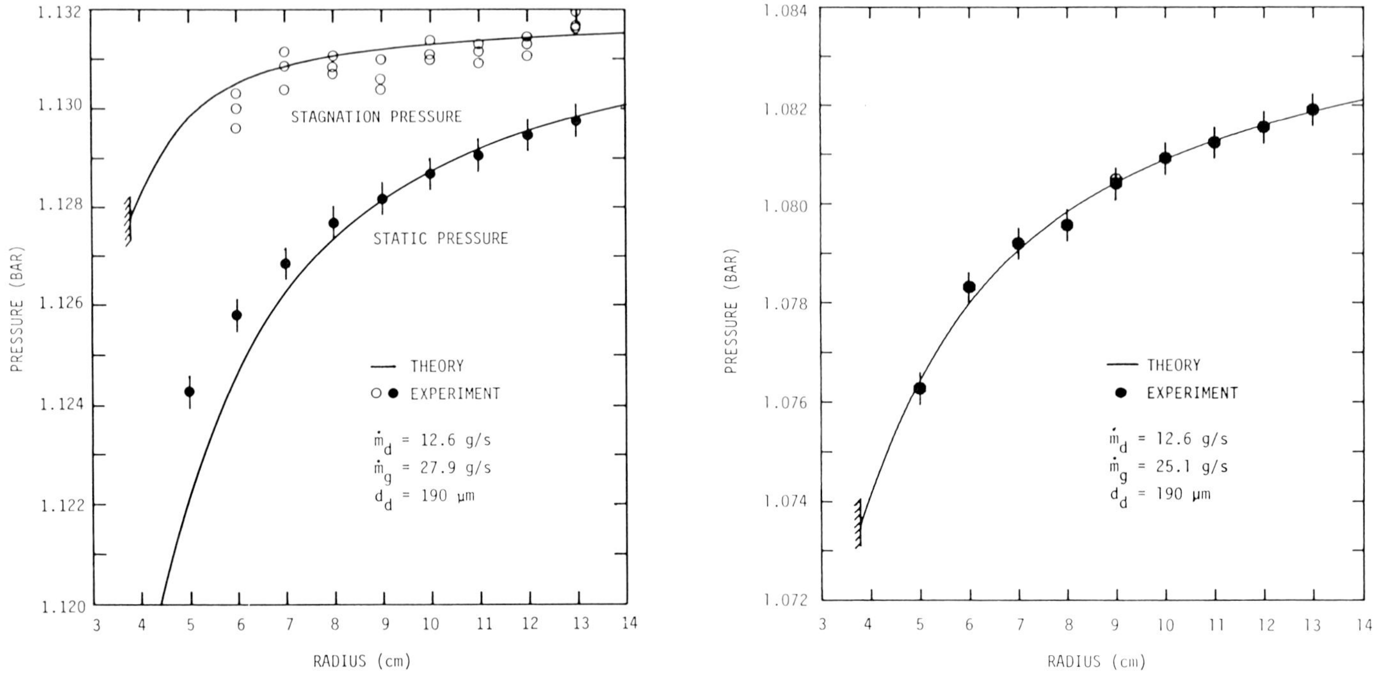

Figure 8 *Radial Profiles of Static and Stagnation Pressures in Experimental LDHX Chamber for Two Different Liquid Injector Configurations: a) 24 Injector Tubes set in Solid Manifold Wall having Four Gas Exhaust Ports, and b) 36 Free-Standing Injector Tubes, Equally Spaced on Base Circle (Dashed Lines indicate Location of Injector Tubes, and* $\dot{m}_d$, $\dot{m}_g$ *and d denote Liquid and Gas Mass Flow Rates and Droplet Diameter, Respectively).*

Agreement between theory and experiment is very good, especially in the case of the static pressures, validating the analytical approach. The static pressure data for the 36 injector tube configuration (Figure 8b) show particularly good corroboration with the theoretical model. In the 24 injector tube case (Figure 8a) the measured static pressures at the smaller radii lie above the theoretical curve, indicating somewhat lower tangential gas velocities than predicted by the model. This may be a result of friction on the solid parts of that configuration's liquid injector manifold. This frictional effect is not included in the analytical model. At lower gas mass flows (hence, lower velocities) the experimental data for the 24 injector tube case match the theoretical predictions more closely (Bruckner and Shariatmadar 1986). The stagnation pressure data in Figure 8a follow the predicted profile reasonably well in spite of the greater scatter of the measurements.

Additional experimental work is required to address some of the problem areas observed to date, such as the vertical liquid runoff at the outer wall, the overloading and positioning of the skimmers, and the stagnation pressure measurements. Furthermore, the heat transfer properties of the LDHX need to be established. This will require a modification of the experimental facility to provide heating and/or cooling of one of the heat exchange media, and will require that the chamber be insulated to reduce heat transfer through the walls. Heat transfer measurements will necessitate the use of a low vapor pressure liquid medium, such as Dowtherm A, to avoid phase change effects. More difficult to implement will be a study of the device characteristics in a microgravity environment. Such an investigation could be carried out aboard one of NASA's zero-g simulation aircraft, for example.

Conclusions

Direct contact heat transfer in the LDHX permits high effectiveness and low pressure drop in a compact, lightweight geometry that avoids many of the limitations of conventional heat exchangers. At least a one order of magnitude decrease in heat exchanger mass for the same heat exchanger effectiveness is possible with the use of LDHXs. Brayton power system mass also is reduced by the increase in thermal cycle efficiency associated with lower pressure losses when using LDHXs. The combined mass savings amounts to a 10% to 15% reduction in overall power system mass. The results of experimental investigations in a laboratory-scale LDHX chamber show very good agreement with the quasi-one-dimensional model used in the parametric studies and indicate that there is no detectable droplet carryover in the exhaust gas, no droplet shattering or agglomeration and no evidence of undesirable secondary flow effects. Additional tests are required to establish the heat transfer properties of the LDHX and its operational characteristics in a zero-g environment.

Acknowledgments

This work was supported by NASA Grant NAG 1–327. The authors are indebted to R.T. Taussig and V.C.H. Lo, Spectra Technology, Inc., Bellevue, WA, for helpful discussions and to A. Shariatmadar for his assistance in the laboratory. Thanks are also due to F. Marentic, 3M Company, St. Paul, MN, for making the Riblet film available for the experimental studies.

References

Boothroyd, R.G. (1971) *Flowing Gas-Solids Suspensions*, Chapman and Hall, London, pp. 24–29.

Bruckner, A.P. and A. Hertzberg (1982) "Direct-Contact Droplet Heat Exchangers for Thermal Management in Space," in *Proc. 17th Intersociety Energy Conversion Engineering Conference*, held in Los Angeles, CA, 8–12 August 1982, pp. 107–112.

Bruckner, A.P. and A.T. Mattick (1984) "High Effectiveness Liquid Droplet/Gas Heat Exchanger for Space Power Applications," *Acta Astronautica* 11: 519–526.

Bruckner, A.P. and A. Shariatmadar (1986) "Heat Transfer and Flow Studies of the Liquid Droplet Heat Exchanger," in *Space Nuclear Power Systems 1985*, M.S. El-Genk and M.D. Hoover, Eds., Orbit Book Co., Malabar, FL.

Bruckner, A.P. (1985) "The Liquid Droplet Heat Exchanger – A Lightweight Approach to Thermal Management in Space Power Systems," Paper No. IAF–85–185, in *36th Congress of the International Astronautical Federation*, held in Stockholm, Sweden, 7–12 October 1985.

Coombs, M.G., C.J. Morse, and C.E. Richard (1971) *Preliminary Design Study of Nuclear Brayton Cycle Heat Exchanger and Duct Assembly*, NASA CR-72816, AiResearch Manufacturing Co., Los Angeles, CA.

Davidson, J.F. and D. Harrison (1971) *Fluidization*, Academic Press, New York, p. 524.

Mattick, A.T. and A. Hertzberg (1982) "The Liquid Droplet Radiator – An Ultralightweight Heat Rejection System for Efficient Energy Conversion in Space," *Acta Astronautica* 9:165–172.

Taussig, T.R., W.J. Thayer, V.C.H. Lo, K.M. Sekins, and A.P. Bruckner (1985) "Direct Contact Heat Exchangers for Space," Paper No. AIAA–85–1077, in *AIAA 20th Thermophysics Conference*, held in Williamsburg, VA, 19–21 June 1985.

Space Nuclear Power Systems 1986
Edited by M. S. El-Genk and M. D. Hoover

Chapter 15

A Self-Induced Thermoelectric-Electromagnetic Pump Model For SP-100 Systems

Mohamed S. El-Genk, John J. Buksa, and Jong T. Seo

Institute for Space Nuclear Power Studies
Department of Chemical and Nuclear Engineering
The University of New Mexico
Albuquerque, NM 87131

An integrated model of a self-induced, thermoelectric-electromagnetic (TE-EM) pump was developed. The model consists of three coupled models: an electromagnetic (EM) model, a two-dimensional thermal analysis model of the magnet, and a steady state/transient thermoelectric (TE) converter model. The integrated pump model incorporates the temperature-dependent magnetic properties of the magnetic material (Hiperco-27), as well as temperature dependent TE material properties (SiGe/GaP). Parametric analyses showed that the pressure head of the pump depends on the number of foils insulating the magnetic structure from the coolant flowing through the pump's ducts up to some asymptotic number of foils. Beyond this point, increasing the number of foils has no effect on the pump pressure head. For a given coolant circulation rate, the pump's pressure head and overall efficiency are strongly dependent on the coolant temperature and the conditions of the pump's radiator. Although a reduction in the heat rejection capability of the pump's radiator and/or a loss of one of the radiator's heat pipes significantly reduces the pump's pressure head and overall efficiency, it results in a slight increase in the magnetic flux density and the TE conversion efficiency.

Introduction

All three space nuclear powered system concepts pursued through the technology assessment phase of the SP-100 program used liquid metal cooled reactors (Anderson et al. 1985, Bailey et al. 1985, and Homeyer et al. 1985). For example, the primary coolant of the thermoelectric (TE) concept and the secondary coolant of both the thermionic and Stirling concepts are circulated by thermoelectric-electromagnetic (TE-EM) pumps. Using liquid metal coolants in space nuclear power system applications makes the electromagnetic (EM) pump very attractive because it has an inherent decay heat removal capability and no moving parts. The pump utilizes the Faraday principle to generate a force field perpendicular in two dimensions to both a magnetic field and an electric current. Figures 1a and 1b present crosssectional views of the General Electric concept of the TE-EM pump. In Figure 1b, the coolant flows into and out of the plane of the figure in the right and left coolant ducts, respectively (or vice versa depending on the magnetic field lines). The TE-EM pump is equipped with a self-induced magnet and TE generators, which, respectively, provide the magnetic field and the DC current-necessary for the operation of the pump. The magnetic structure surrounds the coolant ducts in an S shape (into the plane of Figure 1b). The hot shoe of the pump's TE converter is connected to the pump's ducts and the cold shoe is connected to the potassium heat pipes of the radiator. These heat pipes transport the waste heat of the TE generators to the radiator's panels where it is rejected to space. The amount of electric current produced by the TE generators depends on the temperature differential between the coolant and the pump's radiator; the effective magnetic flux density depends on the temperature of the magnetic structure as well as the current generated by the TE generators. The magnetic field induced by the flow of the electric current from the P-type to the N-type TE elements creates the relatively high magnetic flux strength exhibited by the magnetic structure. However, at elevated temperatures, the magnetic saturation induction of the magnetic material (Hiperco-27) decreases rapidly with temperature; eventually the magnetic material becomes paramagnetic at its Curie point of 1240 K (temperature at which paramagnetization occurs). To keep the magnet temperature sufficiently below its Curie point, the magnet is thermally insulated from the coolant flowing through the pump's ducts by insulation foils and is cooled by conduction to the pump's radiator. The TE-EM pump's design is complex and requires an integrated model to accurately understand its performance.

The objectives of this work were to: (1) develop an integrated model for a self-induced, direct current, TE-EM pump, and (2) use the TE-EM pump model to study the effects on the pump's performance of the number of insulation foils wrapped around the pump magnet, a partial loss of the pump radiator's heat rejection capability, and the failure of one or more of the pump radiator's potassium heat pipes.

Model Description

The TE-EM model, which is based on the early General Electric concept for the SP-100 system, consists of three coupled models: an EM pump model (Barnes 1953 and Blake 1956), a magnet thermal analysis model (Buksa 1987) and a TE converter model (El-Genk and Seo 1986). The EM pump model calculates the pressure head generated by the pump as a function of the coolant mass flow rate, the effective magnetic flux density and the electric current supplied by the pump's TE converter. The effective magnetic flux density is calculated as a function of the magnet's temperature, calculated by the magnet thermal model, and the electric current is calculated by the TE converter model.

The magnet thermal model, a two-dimensional steady state model, calculates the average temperature of the magnet structure as a function of the coolant temperature flowing through the pump's ducts and the number of foils wrapped around the magnet structure. The effective magnetic flux density in the coolant duct depends on the shape of the magnet. Therefore, for simplicity, the thermal model of the magnet specifies a reference value for the nominal magnetic flux density and the corresponding magnet temperature. Given this input, the model then follows the saturation induction curve for the Hiperco-27 material in determining the magnetic flux density as a function of the magnet's temperature (see Figure 2).

The P and N couples, which form the pump TE converter, are sandwiched between the pump's duct walls and the TE cold shoe. These TE elements contact the right and left ducts, respectively. Therefore, the electric current flows upward in the right duct and downward in the left duct. As shown in Figure 1b, the coolant ducts are electrically connected by an electric bus to form a closed electric loop. The magnetic field generated by the circulation of the electric current through the pump structure provides the magnetic flux density needed for developing the deriving pressure head in the pump duct.

The TE converter model can handle any combination of boundary conditions at the hot and cold shoes of the TE generators for both steady state and transient operation. The model also considers the change in the TE material properties with temperature. The equations governing the operation of the TE generators are solved using the finite element method coupled with an implicit time integration algorithm. The results of the TE model have been separately verified before being incorporated in the TE-EM pump model. More details on the development and the verification of the TE converter model are available elsewhere (El-Genk and Seo 1986 and Seo 1986).

NOMENCLATURE

English

- a: Width of insulation foils (m)
- B: Magnetic flux density (Gauss)
- b: Width of pump's coolant ducts or half the magnet width (m)
- E: Electric potential across the pump (volt)
- h: Height of pump magnet (m)
- I: Electric current (A)
- k: Thermal Conductivity ($W/m{\cdot}K$)
- N_{foil}: Number of insulation foils
- P: Power (W)
- Q: Coolant volumetric flow rate (m^3/s)
- R: Electric resistance (ohms)
- T: Temperature (K)
- x: Coordinate
- y: Coordinate

Greek

- ΔP: Pump pressure head (Pa)
- η: Conversion efficiency
- ϵ: Surface emissivity of insulation foils
- σ: Stefan-Boltzmann constant

Subscripts

- c: Coolant
- e: Effective, or electric
- EM: Electromagnetic pump
- f: Fringe or foil region
- i: Foil/magnet interface
- M: Mechanical
- m: Magnet region
- TE: Thermoelectric
- $TE\text{-}EM$: Thermoelectric-Electromagnetic
- r: Radiator

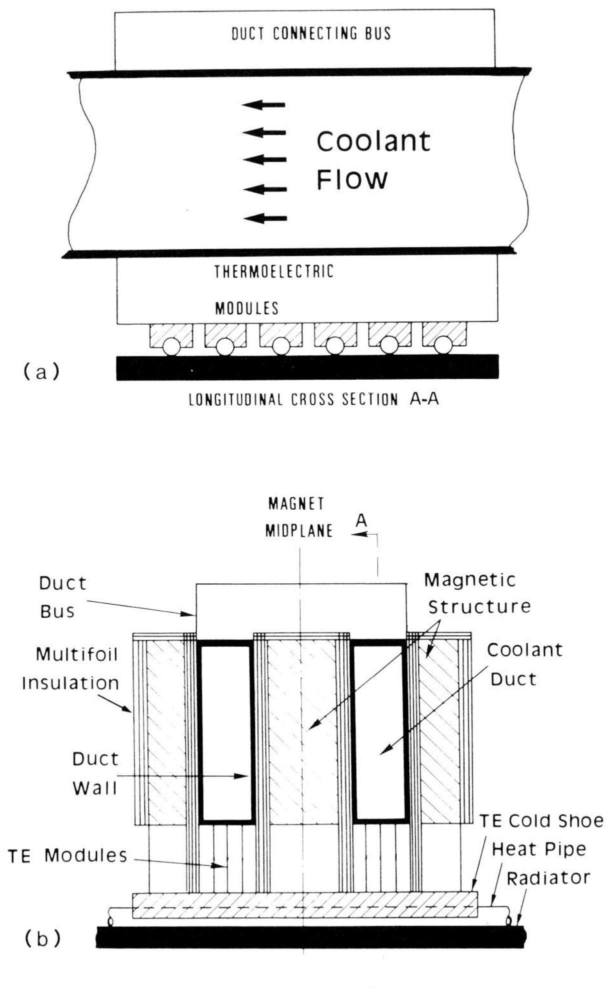

Figure 1 *Cross Sectional Views of the General Electric Concept of Thermoelectric-Electromagnetic (TE-EM) Pump (from GE 1983).*

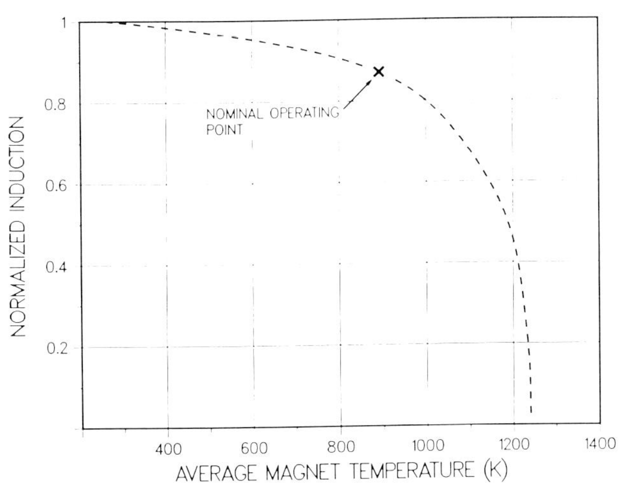

Figure 2 *The Normalized Saturation Induction of the Hiperco-27 Alloy as a function of Temperature (Dokko 1985).*

Figure 3 illustrates the coupled TE-EM pump model. As the coolant flows through the pump ducts, thermal energy is transferred by conduction to the pump's TE converter and to the magnet structure. (However, the insulation foils wrapped around the magnet retards the heat transfer from the coolant to the magnet.) The TE converter partially converts this thermal energy input into electric power, which is supplied to the EM pump; the TE waste heat is then carried out from the TE cold shoe by potassium heat pipes to the pump's radiator where it is rejected into space. The TE generators in the pump's converter are connected in parallel to maximize the electric current supplied to the pump. The TE converter is thermally coupled at its hot shoe to the pump ducts' walls and the TE cold shoe is conductively coupled to the magnetic structure. For simplicity, the temperature drop across the TE cold shoe, the radiator's heat pipes and the radiator panel is neglected. The overall efficiency of the TE-EM pump, $\eta_{TE\text{-}EM}$ is the product of the conversion efficiency of the TE converter, η_{TE}, and that of the EM pump, η_{EM}.

The integrated TE-EM pump model was used to assess the pump's performance as a function of the number of insulation foils placed between the pump's duct walls and the magnetic structure, as well as the coolant volumetric flow rate and temperature. Note that in a TE SP-100 system (Bailey et al. 1984), the coolant's flow rate and temperature are input parameters to the pump; they are determined by the integrated system's overall energy and momentum balances (El-Genk and Seo 1986).

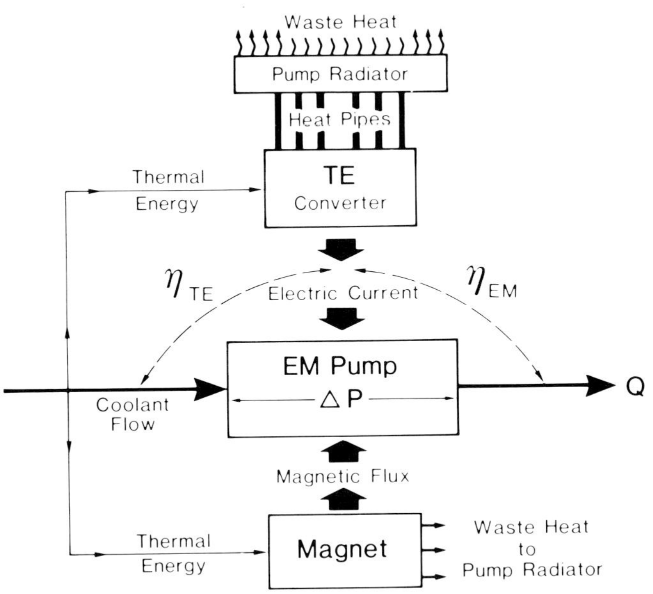

Figure 3 *A Schematic Representation of the Coupled TE-EM Pump Model.*

The following sections fully describe the EM pump model and the magnet thermal model. A brief description of the TE model is also included in this paper; a detailed discussion and analysis of the TE converter model is presented elsewhere (El-Genk and Seo 1986 and Seo 1986).

EM Pump Model

The EM pump model is based on the classic models of Barns (1953) and Blake (1956). The pressure head developed across the pump's ducts can be calculated from

$$\Delta P = \frac{BI_e}{10^4 b} . \tag{1}$$

The total electric current, I, supplied to the pump by the TE converter is

$$I = I_e + I_w + I_f , \tag{2}$$

where I_e is the effective current passing through the coolant flowing through the pump's ducts, and directly contributing to the development of the pumping force. The electric current, I_w, is that which leaks through the duct walls, and I_f is the fringe current. The latter traverses through the coolant in the areas of weak magnetic field at the entrance and exit of the pump's ducts. As a result, I_w is considered a loss since it does not contribute to the pumping force developed in the pump's ducts. Similarly, the fringe current is also considered a loss. Although neglected in this effort, under certain circumstances the fringe current can produce substantial pressure gradients when there exists a magnetic field in the entrance and exit regions of the pump. Using insulation vanes at the pump's entrance and exit has been suggested to increase the fringe resistance, thus minimizing the fringe current losses.

As stated, the TE generators in the pump's TE converter are connected in parallel to maximize the electric current supplied to the pump. As a result, the total electric current supplied to the pump duct, I, is the sum of the currents generated by all the individual TE generators. However, the voltage rise across a single TE generator equals the total potential drop across the pump including the ducts walls, the connecting bus and the TE cold shoe. The electrical resistances of the connecting bus and the TE cold shoe are assumed to be negligible. Thus, the total resistance of the pump's electric circuit is equal to the sum of the effective duct resistance and the equivalent internal resistance of the TE converter.

Barnes (1953) has shown that combining Equations (1) and (2) with a simple circuit analysis (see Figure 4) gives the following expression for the pressure rise in the pump:

$$\Delta P = \frac{IB}{10^4 b} \left(\frac{R_w R_f}{\bar{R}}\right) - \left(\frac{B^2 Q}{10^8 b^2}\right) \left(\frac{R_w + R_f}{\bar{R}}\right) \tag{3}$$

where

$$\bar{R} = R_e R_w + R_w R_f + R_f R_e . \tag{4}$$

Note that the pressure rise across the pump, ΔP, is a linear function of the total electric current, I, and the coolant flow rate, Q, but is a nonlinear function of the effective magnetic flux density, B. In reality, the fringe resistance, R_f, is a complicated expression involving the pump geometry, the magnetic field distribution, and the coolant velocity profile in the pump ducts, and is best determined by an experiment (Barnes 1953). Blake (1956) and Watt (1958) have suggested using a fringing resistance ratio, which is a function of the electrode dimensions, to calculate the wall and fringing resistances. Recently, an analytical expression for the fringe resistance has been forwarded (Hughes and McNab 1983). In the current model, however, the fringe resistance was assumed to be a multiple of the duct walls' resistances, calculated using the resistivity of the duct wall material and the wall area perpendicular to the flow of the electric current. Interface resistances due to brazed or welded joints are neglected here in order to utilize the simplified equivalent circuit of Figure 4.

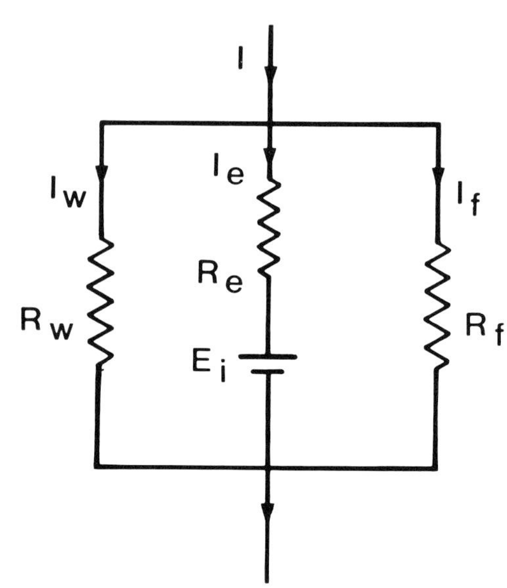

Figure 4 *An Equivalent Electric Circuit of the TE-EM Pump.*

The electric potential across the coolant duct,

$$E = \left(\frac{BQ}{10^4 b}\right) + \left(\frac{10^4 b\, R_e\, \Delta P}{B}\right), \qquad (5)$$

incorporates both the circuit potential drop as well as the "counter-EMF", E_i, produced by the coolant flow in the duct and was used to calculate the electric power input to the pump:

$$P_e = I\, E. \qquad (6)$$

The mechanical power developed by the pump is

$$P_M = \Delta P\, Q. \qquad (7)$$

Subsequently, the EM pump efficiency is

$$\eta_{EM} = \frac{P_M}{P_e} = \frac{\Delta P\, Q}{\left[\left(\frac{BQ}{10^4 b}\right) + \left(\frac{10^4 b\, R_e\, \Delta P}{B}\right)\right] I} \qquad (8a)$$

and the overall efficiency of the pump is

$$\eta_{\text{TE-EM}} = \eta_{\text{TE}} \cdot \eta_{\text{EM}}\,, \qquad (8b)$$

where the conversion efficiency of the TE converter, η_{TE}, is calculated separately by the TE model.

This model assumes that the magnetic field strength and the electric current density along the length of the pump's ducts are uniform; it neglects the effects of the magnet's shape and other effects such as the "armature-reaction" effect on the uniformity of the magnetic field strength in the ducts. It was assumed that design features such as a magnetic field compensation (Blake 1956), duct tapering (Barnes 1953), and/or magnetic pole shaping (Barnes and Cage 1955) would overcome these effects. In this modeling effort, neither the magnetic structure's size and shape nor the self actuating nature of the TE-EM pump was considered.

Magnet Thermal Model

The need to model the thermal behavior of the pump's magnet arose from the knowledge that ferromagnetic materials become paramagnetic at their Curie temperatures. At the Curie temperature, the very high magnetic susceptibility of ferromagnetic materials decreases to a value typical of paramagnetic materials. Even at slightly elevated temperatures, spontaneous magnetization is reduced due to the thermal agitation of the magnetic domains in the ferromagnetic structure. Changing the operating power level of the reactor in an TE SP-100 system will change the coolant temperature. Therefore, the pump's magnetic structure is expected to experience frequent changes in temperature. In order to effectively predict the operation of the TE-EM pump, a two-dimensional, steady state model was developed to calculate the magnetic flux density as a function of the magnet's temperature.

Figure 2 presents the normalized saturation induction versus temperature for the Hiperco-27 alloy. This alloy is the ferromagnetic material chosen for the magnetic structure of the TE-EM pumps in the original GE TE SP-100 concept (Bailey et al. 1985) as well as in the current modeling effort. Because the effective magnetic flux density in the pump's ducts depends on many design variables, such as the magnetic structure's size and shape, the change with temperature of the normalized magnetic flux density was assumed, for simplicity, to be identical to the normalized saturation induction of the magnet material. Given a reference average magnetic flux density and corresponding magnet temperature, Figure 2 was used to determine the magnetic flux density at the calculated average magnet temperature. This temperature is calculated using the magnet thermal model.

Figure 5 presents a schematic of the magnet thermal analysis model. The thermal energy from the coolant is conducted through the duct walls and the insulation foils to the magnet structure, which is cooled solely by conduction to the cold shoe of the TE converter. Thus, the thermal energy transferred from the coolant to the magnet is conducted out of the magnet to the pump's radiator, where it is rejected into space.

The standard high-temperature multilayer insulation used in space applications consists of molybdenum foils with small zirconia particles attached to their surfaces. These particles maintain an open gap between the foils and prevent their compaction. In a space environment, the gaps separating the foils are occupied by vacuum and the heat transfer between the foils occurs only by radiation. The number of foils, their thickness and properties, and the gap size determine the effective thermal conductivity of the insulation:

$$k_f = a\epsilon\sigma(T_i + T_c)(T_i^2 + T_c^2)/(N_{\text{foil}} + 1)\,. \qquad (9)$$

In this model, certain assumptions were made to simplify the thermal analysis of the magnet. For example, the temperature drop across the coolant thermal boundary layer at the duct walls was neglected because of the high thermal conductivity of liquid metals, and because the Nusselt number for liquid

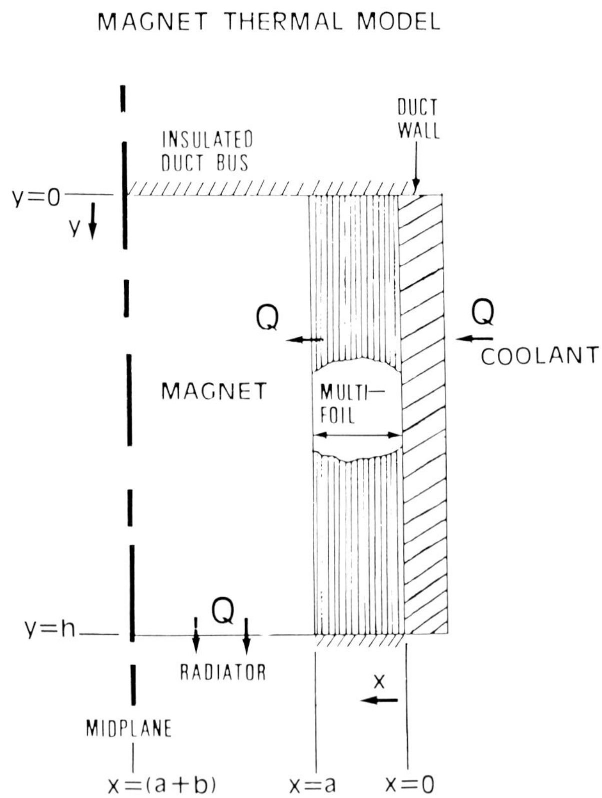

Figure 5 *A Schematic of the Thermal Analysis Model of the Magnet in a TE-EM Pump.*

metals is almost independent of the flow conditions up to a Reynolds number of 3 x 10^4 (El-Genk and Seo 1986 and El-Wakil 1971). As a result, the surface of the foils in contact with the coolant duct walls was considered isothermal at the coolant's temperature. Similarly, the magnet's surface in contact with the TE cold shoe, the face opposite to the duct bus, was considered isothermal at the radiator's surface temperature. The temperature drop in the pump's duct walls was also neglected because their thickness (0.064 cm) is very small compared to the dimensions of the magnet (5.8 cm x 0.7 cm). The magnet's structure surrounding either coolant duct was assumed identical (see Figure 1) because the pump is symmetric at the magnetic structure's midplane. The ducts' surfaces in contact with the electric bus and the magnet at the opposite end of the pump's radiator were assumed adiabatic.

Additionally, the axial heat conduction in the insulation foils was neglected because the thickness of the insulation foils (0.0032 cm) is much smaller than their length (5.8 cm). As a result, a one dimensional heat conduction model was employed to calculate the temperature drop across the insulation foils. However, because the heat from the coolant to the magnet flows in the perpendicular direction to the heat conduction out of the magnet to the radiator, a two-dimensional, steady state model was used to calculate the temperature distribution in the magnet (see Figure 5). The thermal model for the foils and that of the magnet were thermally coupled at the foils/magnet interface ($x = a$).

The temperature distribution in the insulation foils region and in the magnet was found by solving the following sets of governing equations and corresponding boundary conditions.

- *The insulation foil region* ($0 \leq x < a$, *and* $0 \leq y \leq h$)

Governing equation:

$$\frac{d^2T_f}{dx^2} = 0 \tag{10}$$

Boundary conditions:

$$T_f(x = a) = T_i(y) , \tag{11}$$

and

$$T_f(x = 0) = T_c , \tag{12}$$

where T_c is the temperature of the coolant in the pump's ducts and T_i is the temperature at the magnet/multifoils interface.

- *The magnet structure* ($a \leq x \leq (a + b)$, *and* $0 \leq y \leq h$)

Governing equation:

$$\frac{\partial^2 T_m}{\partial x^2} + \frac{\partial^2 T_m}{\partial y^2} = 0 . \tag{13}$$

Boundary conditions:

$$\frac{\partial T_m}{\partial x}(a + b, y) = 0 , \tag{14}$$

$$T_m(a,y) = T_i(y) , \tag{15}$$

$$\frac{\partial T_m}{\partial y}(x,0) = 0_r, \text{ and} \tag{16}$$

$$T_m(x,h) = T_r , \tag{17}$$

where T_r is the radiator's surface temperature, which is determined independently by the TE converter model.

To calculate the interface temperature between the multifoil insulation and the magnet structure, T_i, an additional equation is needed. This equation can be obtained from the heat balance at the interface:

$$k_m \frac{\partial T_m}{\partial x}(a,y) = k_f \frac{dT_f}{dx}(a,y) , \qquad (18)$$

where k_m is the thermal conductivity of the magnet material, and k_f is the effective thermal conductivity of the multifoil insulation as per Equation (9).

Equations (10) and (13) were solved simultaneously for the temperature distributions in the insulation multifoil region and in the pump's magnet as functions of both the coolant and the radiator temperatures. Figure 6 presents the calculated temperature distribution in the magnet and the insulation foils of the pump for coolant and radiator temperatures of 1256 K and 850 K, respectively. An arithmetic average of the hottest and coldest temperatures in the magnet was then calculated. Subsequently this average temperature was used in conjunction with Figure 2 to determine the effective magnetic flux density in the coolant ducts of the TE-EM pump.

Thermoelectric Converter Model

A parallel research effort for the simulation of an integrated TE SP-100 system at the University of New Mexico's Institute for Space Nuclear Power Studies has resulted in the development of a comprehensive steady state/transient TE converter model (El-Genk and Seo 1986 and Seo 1986). The TE model, which uses temperature dependent thermoelectric materials properties, determines the electric voltage and current generated by a single TE couple and, consequently, by the entire TE converter in the pump. In order to minimize duplicate reporting, the reader is referred to the aforementioned articles for in-depth discussions of the complete TE model.

In the TE-EM pump, both isothermal TE hot side and radiative cold side boundary conditions were used to numerically solve the governing equations in the TE converter model. The TE cold side was assumed to be the same as the radiator's temperature. This implies that the temperature drop across the radiator's heat pipes is negligible, which is a good assumption as long as the waste heat rejection of the pump is limited by the radiator rather than the sonic limit of the heat pipes. As Figure 7 indicates, for potassium heat pipes under the conditions delineated in the figure and for a radiator temperature in excess of 540 K, the pump's waste-heat rejection is limited by the radiator (surface temperature, surface area per heat pipe, emissivity and view factor). However, for temperatures less than 540 K, the pump's waste-heat rejection is constrained by the sonic limit of the radiator's heat pipes. This sonic-limited heat rejection, which is lower than the radiator heat rejection capability, causes a reduction in the pump's total waste-heat rejection. Figure 7 also shows that the radiator temperature at which the pump's heat rejection becomes controlled by the heat pipes sonic limit changes along with the total number of heat pipes and/or the radiator's heat rejection capability. Under some circumstances, such as opera-

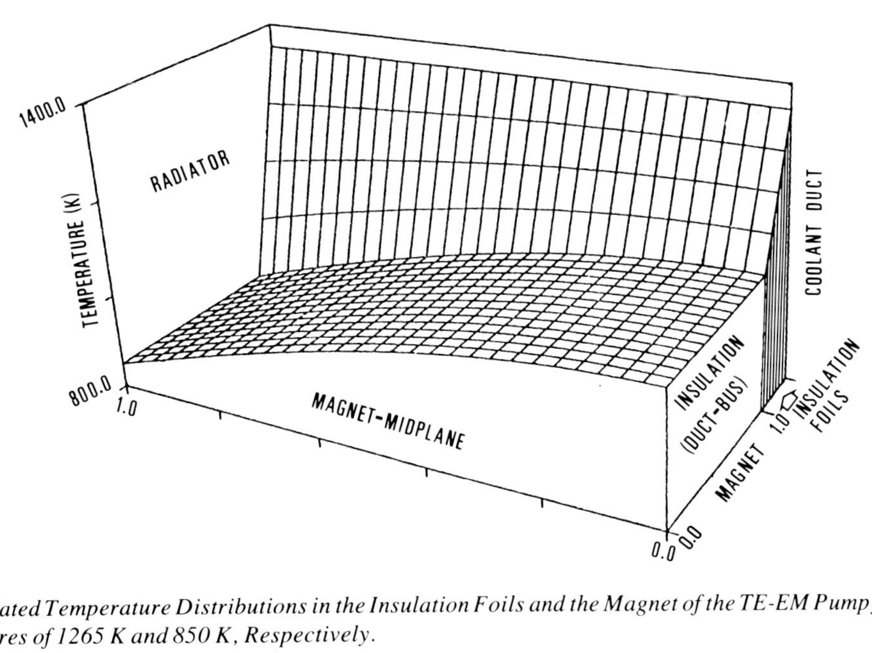

Figure 6 *The Calculated Temperature Distributions in the Insulation Foils and the Magnet of the TE-EM Pump for Coolant and Radiator Temperatures of 1265 K and 850 K, Respectively.*

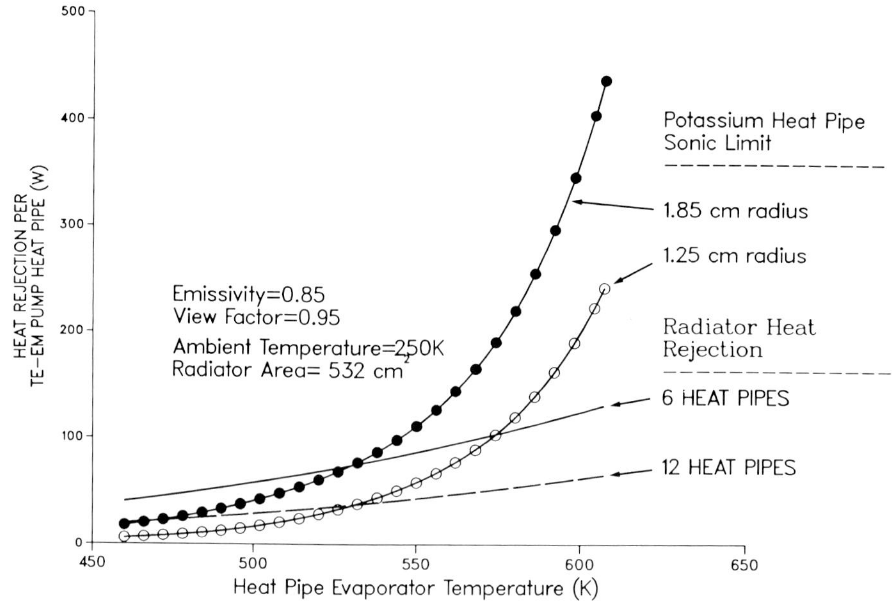

Figure 7 *Limitations on the Waste Heat Rejection in the TE-EM Pump due to the Radiator's Heat Pipes Sonic Limit.*

tion at low thermal power levels (less than 100 kW) or after a reactor shutdown in the TE SP-100 system, the sonic limit of the pump's heat pipes could concomitantly limit the pump's total waste heat rejection, resulting in lower pump efficiency and coolant circulation rate.

Results and Analysis

The development of a thermally coupled TE-EM pump model was required for inclusion in an integrated SP-100 system analysis computer model SNPSAM (Space Nuclear Power System Analysis Model), developed at the University of New Mexico's Institute for Space Nuclear Power Studies (El-Genk and Seo 1986). Other articles in this volume present the results of using SNPSAM to study the steady state and transient behaviors as well as the thermal management of the decay heat in an TE SP-100 system (El-Genk and Seo 1986 and El-Genk et al. 1986). In this paper, the integrated TE-EM pump model was used separately to investigate the effects on the pump's performance of the number of insulation foils wrapped around the magnet, the degradation of the radiator's heat rejection capability, and/or the failure of one of the radiator's potassium heat pipes. Table 1 lists the design and operating parameters of the TE-EM pump analyzed in this section. In this design, the waste heat of the pump's TE converter is transmitted by six potassium heat pipes to the pump's radiator, where it is rejected into space. As shown in Figure 1a, the TE generators in the pump's converter are divided into six identical segments, each conductively coupled at its cold shoe to a separate heat pipe. However, the six TE segments, each having a total of 34 TE generators, were connected in parallel to maximize the electric current and to stabilize the electric voltage across the pump. Thermally coupling the individual TE segments to the radiator's heat pipes can be disadvantageous because the failure of a single heat pipe results in the loss of the electric current generated by the TE generators in the segment attached to that heat pipe. This problem can be solved by connecting the radiator heat pipes by integral fins so that if a heat pipe fails, its heat load can still be carried out by the remaining heat pipes without a measurable decrease in the electric current generated by the pump's TE converter (see Figure 1a).

The performance of the TE-EM pump was studied as a function of the number of foils used to insulate the magnetic structure from the coolant flowing through the pump's ducts. Figures 8 and 9 show the effect of the number of foils used on both the pump's pressure head and overall efficiency. As these figures illustrate, the number of insulation foils has no effect on either the TE efficiency or the current supplied by the TE converter to the pump because they are solely dependent on the temperature differential between the coolant and the TE cold shoe as well as the value of the external load. However, the number of insulation foils directly influences the average temperature of

Table 1 *TE-EM Pump Design and Operating Parameters.*[a]

Parameter	Value
Pump Design	
Duct width	0.0153 m
Duct height	0.058 m
Duct length	0.228 m
Insulation foil thickness	0.00762 cm
Insulation gap width	0.0152 cm
Insulation foil emissivity	0.4
Magnet material	Hiperco-27
Magnet width	0.7 cm
Magnet height	5.8 cm
Magnet thermal conductivity	50 W/mK
TE Converter Design	
TE material	SiGe/GaP
N element length	0.64 cm
P element length	0.64 cm
N element cross-sectional area	0.174 cm^2
P element cross-sectional area	0.174 cm^2
Number of TE generators	204
Pump's Radiator Design	
Number of TE generators/heat pipe	34
Number of radiator's potassium heat pipes	6
Total radiator area	532 cm^2
Radiator emissivity	0.85
Operating Parameters	
Radiator view factor	0.95
Ambient temperature	250K
Nominal coolant temperature	1250K
Nominal mass flow rate	4.7 kg/s
Nominal electric current	2.1 kiloamps
Reference magnet temperature	950 K
Reference magnetic flux density	1000 Guass
Fringe resistance ratio	40

[a]General Electric preliminary design.

the magnet and, consequently, the magnetic flux density in the ducts. As indicated in Figure 8, for any given coolant temperature, increasing the number of insulation foils decreases the magnet's average temperature and raises its effective magnetic flux density. This increase in the magnetic flux density induces a higher pressure head, but has no effect on the overall pump efficiency.

At higher power levels and, hence, higher coolant temperatures, a minimum number of foils is required to keep the magnet's temperature sufficiently below its Curie temperature of 1240 K. As shown in Figure 9, the pump's pressure head increases as the number of insulation foils increases up to some asymptotic number of foils; beyond this point increasing the number of foils has almost no effect on the pump's performance. Note that the asymptotic number of foils will change depending on the exact foil/gap specifications (material thickness and emissivity) and the operating conditions (coolant and radiator temperatures) of the system. Since the coolant temperature increases as the thermal power of the SP-100 system is increased, the asymptotic number of foils will also increase with increasing thermal power (see Figure 8). For example, the results delineated in Figure 8 show that the optimum number of foils increases from two foils to about six foils for system thermal power levels of 1.5 and 2.7 MW, respectively. In Figure 8 and 9, the values of thermal power and the corresponding coolant temperature and flow rate were calculated using SNPSAM (El-Genk and Seo 1986).

As stated, the present pump model incorporates six potassium heat pipes for transfering the waste-heat from the pump's TE converter to the pump's radiator. A degradation of the surface emmisivity of the raditor will result in an increase in both the radiator and the TE cold shoe temperatures. This increase in the TE cold shoe temperature reduces the TE conversion efficiency and the electric current supplied by the TE converter to the pump. Additionally, the higher radiator temperature reduces the amount of waste-heat rejected from the magnet structure to the radiator by conduction, causing the magnet's average temperature to increase and the pump's pressure head to decrease. As demonstrated in Figures 10 and 11, the pump's pressure head and overall efficiency are sensitive to the degradation in the heat rejection capability of the pump's radiator. The pressure head of this TE-EM pump drops rapidly as the radiator's heat rejection capability decreases. For example, a 20% reduction in the radiator's heat rejection capability causes the pump's pressure head to drop by approximately 11% from its nominal value of 17.5 kPa, and the overall pump efficiency to decrease by about 8% from its nominal value of 0.7%.

Investigating the effect of a heat pipe failure on the performance of the TE-EM pump required knowing the type of heat pipe/TE converter coupling. As discussed, the pump design considered here assumes that the loss of a radiator heat pipe results in a loss of the heat sink only for those TE generators in the segment attached to the heat pipe that failed. Consequently, losing a single heat pipe thermally and electrically disconnects one sixth of the TE generators. At any given radiator temperature, the radiator's ability to reject heat remains constant (assuming no area loss or emissivity degradation occurred) and the thermal power transferred by a TE generator to the radiator remains constant (assuming constant coolant and ambient temperatures). Therefore, a reduction in the total number of the active TE generators due to the failure of one or more heat pipes will result in a decreased radiator temperature. In addition, losing a heat pipe will increase the radiator's surface area

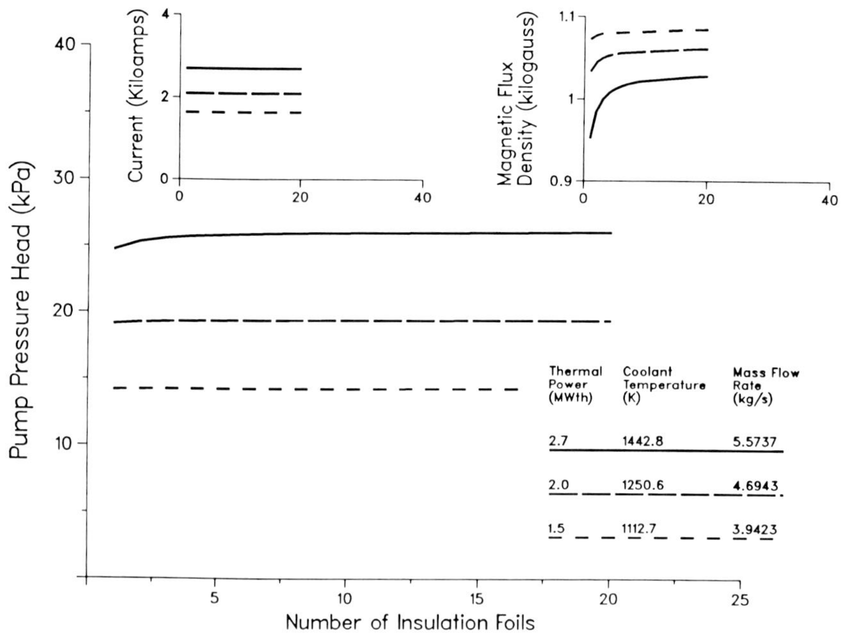

Figure 8 *Effects of the Number of Insulation Foils of the Magnet on the Pump's Pressure Head and the Electric Current Supplied by the TE Converter.*

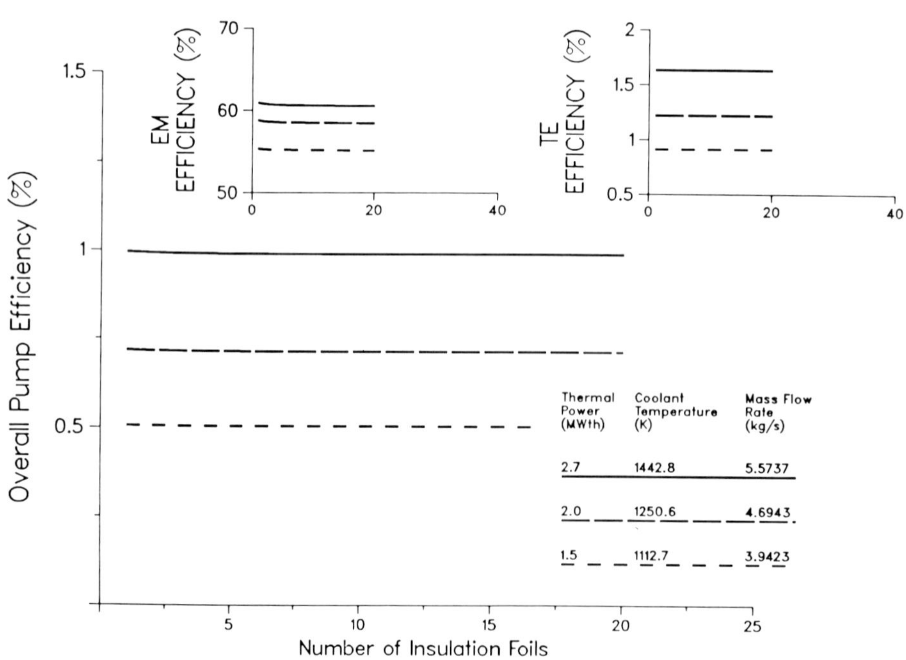

Figure 9 *Effects of the Number of Insulation Foils of the Magnet on the Efficiencies of the TE Converter and the TE-EM Pump.*

available for each of the remaining TE generators and, consequently, lower the radiator's surface temperature. The lower radiator temperature, and resulting TE cold side temperature, increases the magnetic flux density (via lowering the average magnet temperature), the temperature differential across the TE generators, and, consequently, the TE conversion efficiency. As shown in Figures 10 and 11, a loss of one of the radiator's six heat pipes lowers both the TE cold side temperature and the electric current generated by the TE converter, but increases the TE conversion efficiency and the magnetic flux density. The

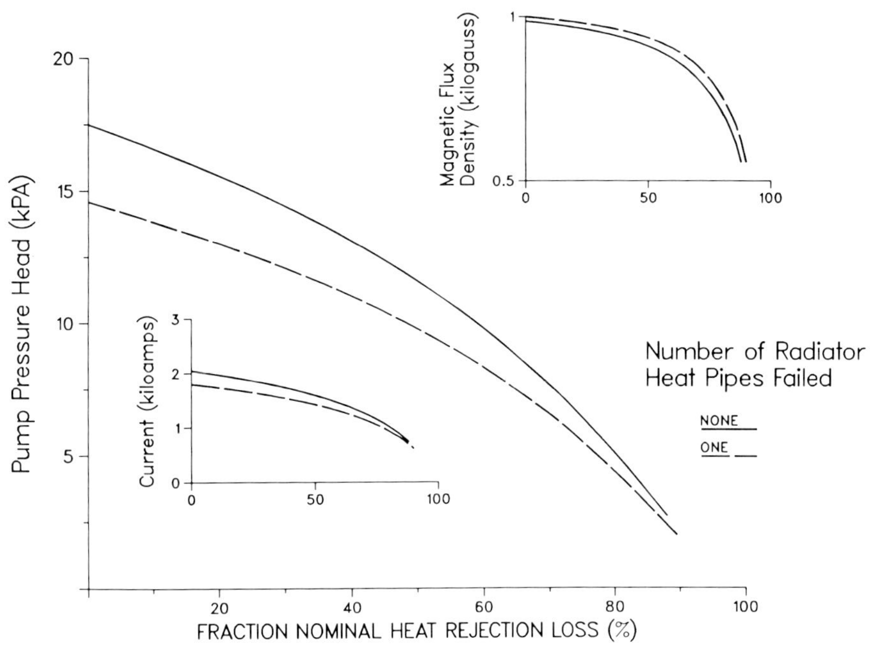

Figure 10 *Effects on the Pump's Pressure Head, Electric Current, and Magnetic Flux Density of the Degradation in the Radiator's Waste Heat Rejection and/or the Failure of a Heat Pipe.*

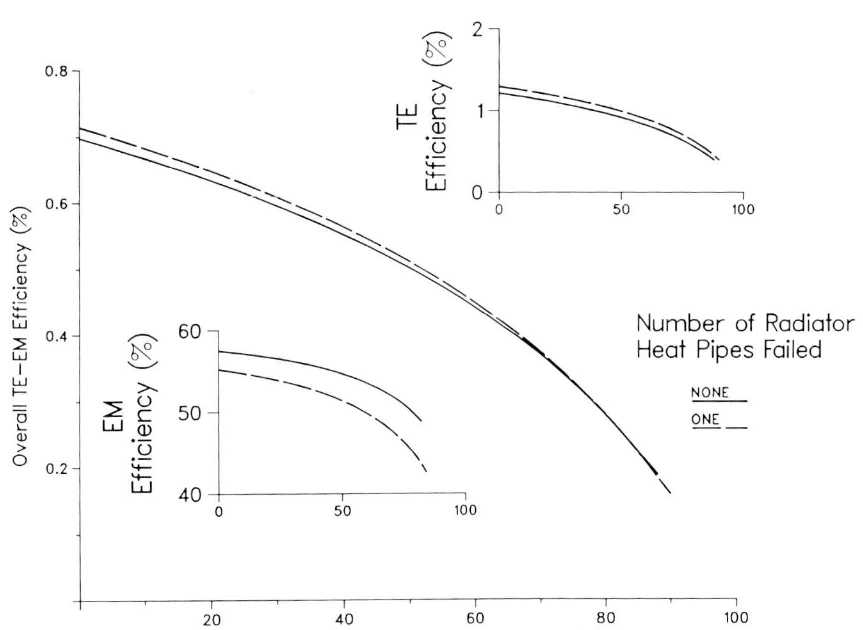

Figure 11 *Effects on the Pump's Overall Efficiency of the Degradation in the Radiator's Waste Heat Rejection and/or the Failure of a Heat Pipe.*

overall effect is a substantial decrease in the pressure head accompanied by a very slight increase in the overall efficiency of the pump (the latter is due to the increased TE efficiency). As the results in Figures 10 and 11 indicate, losing one of the radiator's heat pipes caused less than a one- sixth (approximately 16%) reduction in the pump's pressure head, but slightly increased the pump's overall efficiency.

Conclusions

An integrated TE-EM pump model was developed and used to parametrically study the pump's performance. Results of the parametric analysis suggested that the pump's pressure head and overall efficiency depend on the number of foils insulating the magnet from the coolant flowing through the pump's ducts up to some asymptotic number of foils, beyond which increasing the number of foils has no effect on the pump performance. This asymptotic number of insulation foils increases as the operating power (or coolant temperature) increases. The pump pressure head and overall efficiency are highly dependent on the coolant temperature and flow rate, which are determined by the reactor thermal power in the context of the integrated power system. From a reliability viewpoint, a reduction in the heat rejection capability of the pump's radiator increases the average magnet temperature and, subsequently, reduces the pump's pressure head and electric current while only slightly affecting the pump's overall efficiency. At a given coolant temperature, a failure of a heat pipe in the pump's radiator, although it reduces the number of active TE generators and the total electric current produced, it lowers the radiator temperature and, consequently, increases the pump's efficiency. These results suggest that it is imperative to (1) armor the pump radiator for prolonged system reliability and performance and (2) design the pump system for a desired end-of-life performance taking into consideration expected radiator failures and degradations.

Acknowledgments

This work was supported by the U.S. Air Force Space Technology Center and the Nuclear Technology Office of the Air Force Weapons Laboratory under Research Contract No. F29601-84-C-0080 to the University of New Mexico.

References

Anderson, R.V., R.B. Harty, and J.L. McNerney (1985) ''Analysis of a Reactor Stirling Engine Space Power System,'' in *Trans. 2nd Symposium on Space Nuclear Power Systems*, Conf-850103—Summs, Institute for Space Nuclear Power Studies, University of New Mexico, Albuquerque, NM.

Bailey, S.S., J.D. Stephan and J.E. Van Hoomissen (1985) ''Liquid Metal Cooled Reactors for Space Power Applications,'' in *Trans. 2nd Symposium on Space Nuclear Power Systems*, Conf-850103—Summs, Institute for Space Nuclear Power Studies, University of New Mexico, Albuquerque, NM.

Barnes, A.H. (1953) ''Direct Current Electromagnetic Pumps,'' *Nucleonics*, 11(1): 16–21.

Barnes, A.H. and J.F. Cage, Jr. (1955) ''Pumps — Electromagnetics,'' in *Liquid-Metals Handbook, Sodium-Nak Supplement*, C.B. Jackson eds., the Bureau of Ships, Department of the Navy, Washington, D.C.

Blake, L.R. (1956) ''Conduction and Induction Pumps for Liquid Metals,'' *Proceedings I.E.E.*, Paper No. 2111 U, 104(A): 49–63

Buksa, J.J. (1987) *The Decay Heat Thermal Management of Liquid Metal Space Nuclear Reactor Systems*, M.S. Thesis, Department of Chemical and Nuclear Engineering, The University of New Mexico, Albuquerque, NM.

Dokko, W. (1985) ''EM Pump Performance Prediction,'' in *Heat Transport Technology Review — Thermoelectric System*, Los Alamos National Laboratory, Los Alamos, NM, June 1985.

El-Genk, M.S. and J.T. Seo (1986) ''SNPSAM - Space Nuclear Power System Analysis Model,'' in *Trans. 3rd Symposium on Space Nuclear Power Systems*, Conf-860102—Summs, Institute for Space Nuclear Power Studies, University of New Mexico, Albuquerque, NM.

El-Genk, M.S. and J.T. Seo (1986) ''Analysis of the Transient Behavior of Thermoelectric Generators,'' in *Trans. 3rd Symposium on Speace Nuclear Power Systems*, Conf-860102—Summs, Institute for Space Nuclear Power Studies, University of New Mexico, Albuquerque, NM

El-Genk, M.S. and J.J. Buksa, and J.T. Seo (1986) ''Decay Heat Thermal Management of an SP-100 System,'' in *Trans. 3rd Symposium on Space Nuclear Power Systems*, Conf-860102—Summs, Institute for Space Nuclear Power Studies, Albuquerque, NM.

El-Wakil, M.M. (1971) *Nuclear Heat Transport*, ANS Publications, International Textbook Company, Scranton, PA.

General Electric Company (1983) *SP-100 Conceptual Design Study, Final Report, Vol. II — Reactor and Shield Design*, 83SDS4261, General Electric Company, Philadelphia, PA.

Homeyer, W.G. et al. (1984) ''Thermionic Reactors for Space Nuclear Power,'' *Space Nuclear Power Systems 1984*, M.S. El-Genk and M.D. Hoover, eds., Orbit Book Co., Malabar, FL.

Hughes, W.F. and I.R. McNab (1983) ''A Quasi-One-Dimensional Analysis of an Electromagnetic Pump Including End Effects,'' in *Liquid Metal Flows and Magnetohydrodynamics*, Progress in Astonautics and Aeronautics, 84: 287–312.

Seo, J.T. (1986) *SNPSAM - Space Nuclear Power System Analysis Model*, M.S. Thesis, Department of Chemical and Nuclear Engineering, The University of New Mexico, Albuquerque, NM

Watt, D.A. (1958), ''The Design of Electromagnetic Pumps for Liquid Metals,'' *Proceedings I.E.E.*, Paper No. 2763 U, December 1958, pp. 94–103.

Part 6

ENERGY CONVERSION

Space Nuclear Power Systems 1986
Edited by M. S. El-Genk and M. D. Hoover

Chapter 16

High Temperature Thermoelectric Energy Conversion

Charles Wood

Jet Propulsion Laboratory
California Institute of Technology
4800 Oak Grove Drive
Pasadena, CA 91109

The theory and current status of materials research for high-temperature thermoelectric energy conversion are reviewed. Semiconductors are shown to be the preferred class of materials for this application. Optimization of the figure of merit of both broad- and narrow-band semiconductors is discussed as a function of temperature. Phonon scattering mechanisms are discussed, and basic material guidelines are given for reduction of thermal conductivity. Two general classes of materials show promise for high temperature figure of merit (Z) values, namely the rare earth chalcogenides and the boron-rich borides. The electronic transport properties of the rare earth chalcogenides are explicable on the basis of degenerate or partially degenerate n-type semiconductors. Boron and boron-rich borides exhibit p-type hopping conductivity, with detailed explanations proposed for the transport differing from compound to compound. Some discussion is presented on the reasons for the low thermal conductivities in these materials. Also, ZTs greater than one appear to have been realized at high temperature in many of these compounds.

Introduction

Considerable advances were made in the late 1950s and early 1960s in the theory and development of materials for high-temperature thermoelectric energy conversion. This early work culminated in a variety of materials, spanning a range of temperatures, with the product of the figure of merit, Z, and temperature, T, (that is, the dimensionless figure of merit, ZT, of the order of one). This experimental limitation appeared to be universal and led a number of investigators to explore the possibility that a $ZT \sim 1$ also represents a theoretical limitation. This was found to be incorrect.

Research in the thermoelectric field has languished in the past two decades, with only a small amount of research effort being conducted on high temperature thermoelectric materials and most of that in the Soviet Union. The silicon-germanium alloys represent the current state of the art in high temperature thermoelectric materials. These, like their lower temperature counterparts the bismuth and lead chalcogenides, are exclusively broadband semiconductors with conventional transport mechanisms, that is, they conduct by itinerant motion of charge carriers. Examination of the equation for thermoelectric conversion efficiency (see below) shows that for a real breakthrough to occur (the achievement $ZT >> 1$) much higher hot-junction temperatures are required in addition to high Z values. This involves a search amongst the refractory materials for likely candidates, and has led to exploratory research into new classes of high-temperature semiconducting materials. Recent work on these materials, the boron-rich borides and the rare-earth chalcogenides, will be reviewed after a brief introduction to the theory.

Theory

The efficiency (η) for the conversion of heat to electrical energy using a thermoelectric couple with a hot junction at temperature T_1 and a cold junction at temperature T_0 (see Figure 1) is given by (Joffe 1957):

$$\eta = \frac{T_1 - T_o}{T_1} \times \frac{M-1}{M + (T_o/T_1)}, \qquad (1)$$

where $M = (1 + Z(T_1 + T_0)/2)^{1/2}$,

and Z is a materials parameter, called the figure of merit, which is related to the Seebeck coefficient (often misnamed the thermoelectric ''power''), α_1, α_2; the electrical resistivity, ρ_1, ρ_2, and the total ther-

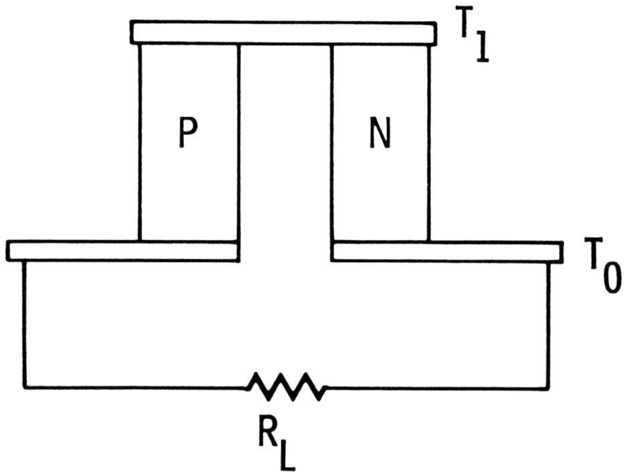

Figure 1 *Semiconductor Thermocouple.*

mal conductivity κ_1, κ_2 of the legs 1 and 2 of the thermocouple as follows:

$$Z = \frac{(\alpha_1 - \alpha_2)^2}{((\rho_1\kappa_1)^{1/2} + (\rho_2\kappa_2)^{1/2})^2} \, . \tag{2}$$

The equation for conversion efficiency (η) shows that the thermodynamic efficiency of an ideal reversible engine (first term) is reduced by the irreversible losses of heat conduction ($\propto\kappa$) and Joule heating ($\propto\rho$) in the thermocouple (second term). Increasing the figure of merit (Z) and the hot junction temperature (T_1) increases the conversion efficiency; in the latter, by increasing both terms in Equation (1). For the purpose of comparing materials, it is convenient to define Z for a single material as

$$Z = \frac{\alpha^2}{\rho\kappa} = \frac{\alpha^2\sigma}{\kappa} \, , \tag{3}$$

where σ is the electrical conductivity. The dimensions of Z are $\deg^{-1}$ and it is common to compare the properties of materials in terms of dimensionless parameter ZT.

Itinerant Semiconductors

Semiconductors are found to have the highest Z values and are, therefore, the preferred class of materials for thermoelectric energy conversion. As illustrated in Figure 2, metals have high electrical conductivity, but the Seebeck coefficients are too low. Insulators can have high Seebeck coefficients, but are too electrically resistive. A good compromise can be made by choosing highly-doped semiconductors with α and ρ values intermediate between the properties of insulators and metals.

Optimization

For a non-degenerate extrinsic semiconductor the Seebeck coefficient and electrical conductivity are related as follows:

$$\alpha = \pm \frac{k}{e}\left(A + \ln\left(\frac{Ne\mu}{\sigma}\right)\right) \, , \tag{4}$$

where A is a constant whose value depends on the type of charge carrier scattering mechanism; μ is the mobility of the carrier of charge, e, and effective mass, m^*; and N is the effective density states:

NOMENCLATURE

English

- a: Lattice constant
- c: Fractional charge carrier concentration
- C_v: Specific heat at constant volume
- e: Electronic charge
- E_g: Energy gap
- E_T: Vibrational energy transported by charge carrier
- G_R: Polaron binding energy
- ΔG: Charge carrier hopping activation energy
- h: Planck's constant
- ΔH_R: Change in energy on removal of charge
- k: Boltzmann's constant
- l: Phonon mean free path
- L: Lorenz factor
- m^*: Charge carrier effective mass
- M: Atomic mass
- N: Effective density of states
- q: Charge of electron or hole
- r: Scattering constant
- ΔS: Entropy
- T: Temperature
- v: Phonon velocity
- Z: Figure of Merit

Greek

- α: Seebeck coefficient
- β: Material parameter (Equation(8))
- Υ: Gruneisen constant
- ϵ: Coefficient proportional to constant lattice force
- η: Conversion efficiency
- θ_D: Debye temperature
- κ: Thermal conductivity
- μ: Charge carrier mobility
- ξ: Expansion coefficient
- ρ: Electrical resistivity
- σ: Electrical conductivity
- τ: Relaxation time

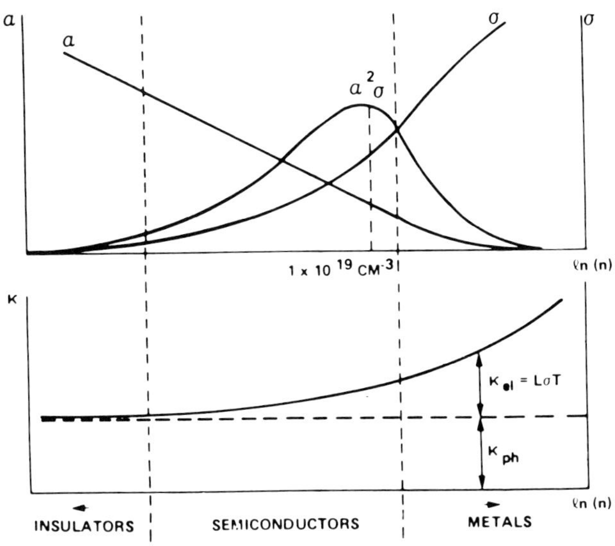

Figure 2 *The Themoelectric Properties of Metals, Semiconductors and Insulators. (From Ioffe 1957.)*

$$N = 2(2\pi m^* kT/h^2)^{3/2} . \quad (5)$$

Obviously as α increases, σ decreases. However, it can be shown (Joffe 1957) that $\alpha^2\sigma$, the numerator of Z, has an optimum value at a semidegenerate concentration of charge carriers of about10^{19}cm^{-3}. This value has to be modified slightly to allow for any significant contribution made by the electronic conduction to thermal conductivity (κ_{el}), since in the denominator of Z, the total thermal conductivity is

$$\kappa = \kappa_{ph} + \kappa_{el} = \kappa_{ph} + \mathrm{L}\sigma\mathrm{T} \quad (6)$$

where κ_{ph} is the lattice or phonon contribution and L is the Lorenz constant. Factoring in the κ_{el} term yields a maximum figure of merit of

$$Z_{\max} = \frac{4}{AT}\left(\frac{\kappa_{el}}{\kappa_{ph}}\right)\left(1 + \frac{\kappa_{el}}{\kappa_{ph}}\right) . \quad (7)$$

At high temperatures, a third term, κ_{rad}, should be included in Equation (6), because, radiative heat transfer may contribute to the total thermal conductivity. Dixon et al. (1975) have calculated the effect of heat transfer by radiation (photons) on lowering $Z_{\max}$. However, other investigators (Keyes and Bauerle 1967 and Kelley and Szego 1964) have shown that, at the high doping levels and effective mass values typical for thermoelectric materials, this effect is negligible up to ~ 2000 K.

Since the optimum doping level corresponds to a semi-degenerate carrier concentration Chasmar and Stratton (1959) and Rittner (1959) used generalized Fermi-Dirac functions to calculate exact Z values for arbitrary degrees of degeneracy. They also included various types of charge-carrier scattering in their analyses. Both the magnitude of $(ZT)_{\max}$ and the depth of the Fermi-level below the conduction band at optimum doping increase with the value of a material parameter:

$$\beta = 8.952 \times 10^{-6} (\mu/\kappa_{ph}) (T/300)^{5/2} m^{*3/2} . \quad (8)$$

A plot, derived from their curves, showing the variation of $(ZT)_{\max}$ with β for various values of scattering parameter, r, is shown in Figure 3.

It is seen from Equation 8 that Z can be related to fundamental material parameters (Chasmer and Stratton 1959, Rittner 1959, and Goldsmid and Douglas 1954), that is

$$Z \propto m^{*3/2} \mu/\kappa_{ph} \quad (9)$$

which suggests that the ideal thermoelectric material should satisfy the somewhat conflicting requirements of high effective mass, high mobility, and low lattice thermal conductivity.

By expressing α, σ, and κ in terms of Fermi-integrals and substituting into Z it can be shown that the value of Z is approximately proportional to the scattering constant r (Figure 3). As proposed by Rittner (1959) and Joffe (1957), this suggests that ionized-impurity scattering with $r = 2$ is the most desirable scattering mode.

Temperature Variation of *Z*

The exact temperature dependence of *ZT* can be determined from Figure 3 by simply selecting the

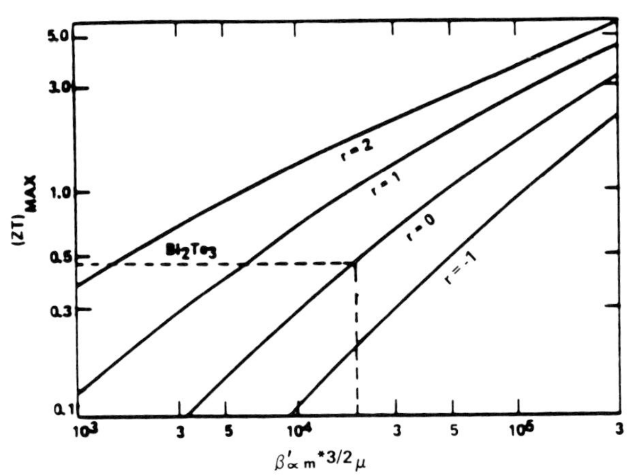

Figure 3 *$(ZT)_{max}$ versus β' for Charge Carrier Scattering by: Impurities ($r = 2$), Ionic Lattice ($r = 1$), Acoustic-Mode Lattice Vibrations ($r = 0$), and Highly Disordered (Amorphous) Lattices ($r = -1$).*

appropriate curve for the predominant scattering mechanism and substituting the temperature dependencies of μ and κ_{ph} in Chasmar and Stratton's (1959) dimensionless parameter β. For example, for acoustic-mode lattice scattering of charge carriers ($r = 0$), and assuming $\kappa_{ph} \propto T^{-1}$, gives $\mu \propto T^{-3/2}$ and, therefore, $\beta \propto T^2$. Whereas, ionized impurity scattering ($r = 2$) gives $\mu \propto T^{+3/2}$ and $\beta \propto T^5$. However, these rapid increases of β and hence, ZT with increasing temperature will be severely limited at high temperatures with the advent of intrinsic conduction. The contribution to the conduction processes from charge of opposite sign will be subtractive in the Seebeck coefficient. Furthermore, the thermal conductivity will rise rapidly at the onset of significant charge-carrier recombination. The temperature at which either of these processes will start to make a significant contribution will depend primarily on the value of the band gap, the effective mass and the mobility ratios of the charge carriers.

An optimum band gap for a particular temperature of operation is apparent. In materials with large gaps, the mobilities are often low with correspondingly low electrical conductivities. If the band gap is too small, as indicated above, intrinsic conduction will set in. Criteria for optimum band gaps have been established by Chasmar and Stratton (1959) assuming transport properties similar to those of Bi_2Te_3. Alternatively, ignoring the presence of minority carriers, Aigrain (1960) states that the temperature at which intrinsic conduction occurs, T_i is of the order of $E_g/2k$, where T_i can be equated to the hot-junction temperature, T_h. It should be noted that the energy gap, E_g, is the value at T_h, not room-temperature.

Thermal Conductivity

A simple method of estimating the thermal conductivity (κ) for any entity is to employ the classical kinetic theory of gases formula:

$$\kappa = 1/3\, C_v l v \tag{10}$$

where C_v is the specific heat at constant volume, v is the velocity and l is the mean free path. Dugdale and MacDonald (1955) expressed the phonon mean free path as $l \sim a/\xi Y T$, where a is the lattice constant, ξ is the expansion coefficient and Y is the Gruneisen constant. Substituting into Equation 10 gives

$$\kappa_{ph} \sim \frac{1}{3} C_v \frac{a}{\xi Y T} v \tag{11}$$

Liebfried and Schlomann (1954) derived the following relationship:

$$\kappa_{ph} \sim \frac{\bar{M} a \Theta_D^3}{T Y^2} \tag{12}$$

where $\bar{M}$ is the average atomic mass and θ_D is the Debye temperature. It would be expected that a material of high mean atomic mass would have high lattice thermal conductivity but the mass dependence of $(a\theta_D{}^3)$ predominates and κ_{ph} falls with increasing mass as shown in Figure 4. The figure also shows a κ_{ph} dependence on mass difference and degree of ionicity in the bonding. Equations 11 and 12 can be shown to be equivalent (Dugdale and Macdonald 1955).

Experimental Results

As mentioned earlier, the highest Z values are found in semiconducting materials. An inherent advantage of semiconductors is that they present a wide choice of elements and compounds for selection of properties suited to a particular application; for example, thermoelectric energy conversion. In par-

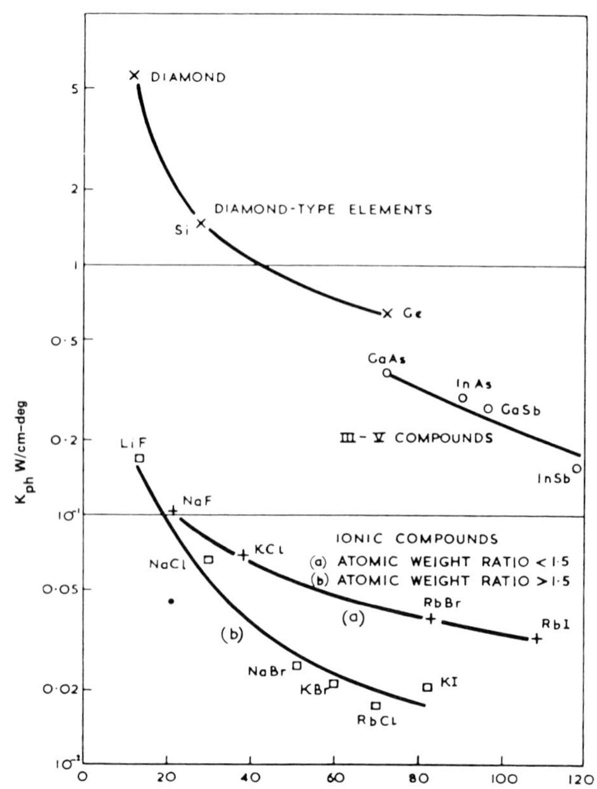

Figure 4 *Dependence of Lattice Thermal Conductivity on Mean Atomic Weight for Covalent and Ionic Crystals. (From Goldsmid 1964.)*

ticular, they offer the flexibility of tailoring their electrical properties to this particular application.

Guided by the theoretical work described above, a number of materials have been developed with respectable Z values for operation in the temperature range from room temperature up to ~ 1000°C. These are shown in Figure 5 where it can be seen that each material has an optimum temperature range of operation. A plot of $ZT = 1$ is seen to envelope the peaks of these curves and led to the belief that $ZT \sim 1$ represents a limit for any material. Numerous attempts to determine a theoretical limit have been unsuccessful (Donahoe 1960, Simon 1962, Rittner and Neumark 1963, Littman and Davidson 1961, and Ure 1972) and it seems safe to say that no theoretical limit on Z exists.

Examination of the equation for conversion efficiency (η) shows that considerable advantage can be obtained by operating the junction at higher temperatures. Hence the thrust of current research into thermoelectric energy conversion is toward high-temperature refractory materials. The question of stability of materials at high temperature involves not only high melting points, but also low dissociation vapor pressures. Furthermore, it is essential that the electrical properties do not change, as reflected by the absence of phase-changes, and precipitation of and changes in defect concentration.

One class of refractory itinerant semiconductors receiving considerable attention for high temperature thermoelectric energy conversion applications is the rare earth chalcogenides.

Rare Earth Chalcogenides

Structure: Rare earth chalcogenides generally form the following binary compounds: RX, $R_3X_4 - R_2X_3$, RX_2, where R represents the rare earth and X the chalcogenide atoms S, Se, or Te. The RX compounds crystallize in a fcc NaCl structure, the $R_3X_4 - R_2X_3$ form solid solutions in bcc (Th_{3P4}) orthorhombic or tetragonal structure, and the RX_2 compounds usually are found in either cubic or tetragonal structure (Flahaut 1978).

Since RX_2 readily dissociates by evolution of the chalcogen at temperatures well below the melting point, and the RX compounds are generally too metallic, only compositions in the homogenous range $R_3X_4 - R_2X_3$ have been extensively investigated for high-temperature thermoelectric applications. There are often two, and sometimes three, polymorphic modifications of the solid solutions $R_3X_4 - R_2X_3$:

1. A low-temperature(< 900°C - 1000°C) orthorhombic α-phase;
2. An intermediate temperature (900°C to 1300°C) tetragonal β-phase; and

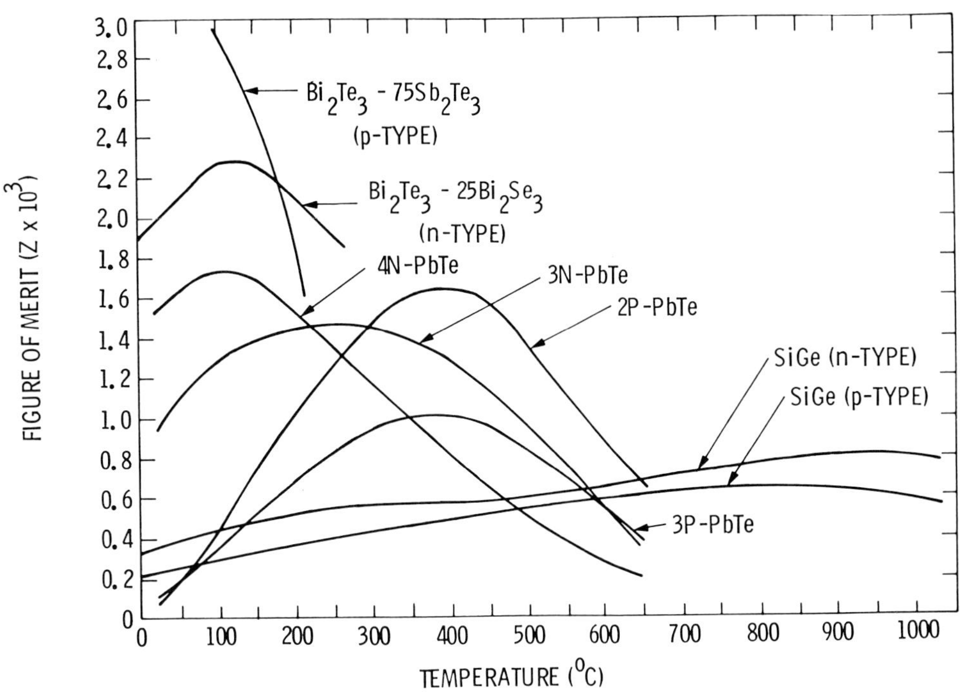

Figure 5 *Figure of Merit (Z) of Selected Thermoelectric Materials as a function of Temperature.*

3. High temperature (> 1300°C) cubic (Th_3P_4) metastable Y-phase (Flahaut 1978).

It is unclear from the literature whether these polymorphic forms occur over the whole compositional range from R_2X_3 to R_3X_4. Furthermore, with the exception of La_2S_3, it is not yet resolved as to whether phase $\beta - R_2X_3$ is truly a polymorph or is representative of a ternary compound involving oxygen (Besancon 1973).

In the cubic Th_3P_4 structure there are 28 lattice sites comprising 12 rare earth atom sites, each having eight chalcogen atom neighbors, and 16 chalcogen atoms, each having six rare earth neighbors. It is convenient to designate the unit cell as $4(R_{3-x}V_xX_4)$ where V is a rare earth vacancy. For R_2X_3 compounds 1 and ⅓ of the 12 rare earth sites are vacant, that is, x = ⅓, and for the R_3X_4 compounds there are no vacant sites, that is, $x = 0$. The ionic character of the lattice can be described by the formula $(R^{3+})_{3-x}V_x(X^{2-})_4\,(e^{-1})_{1-3x}$. Starting with R_2X_3, the two R^{3+} ions contribute six electrons to the chemical bonding which are taken up by the three X^{2-} ions. Thus, there are no excess electrons available for conduction and all R_2X_3 compounds are insulators. As additional rare earth atoms are added, R^{3+} ions are introduced at random vacancy sites in the lattice, thus contributing three electrons per ion.

At the other extreme, in R_3X_4 the three R^{3+} ions contribute nine bonding electrons while four X^{-2} ions accept only eight electrons giving rise to an excess of one conduction electron per formula unit. The concentration of $R_{3-x}X_4$ units is N_Ad/M ($= 6.25 \times 10^{21}$ cm^{-3} for Ce_3S_4) where N_A is Avogadros number, d is the density and M is the molecular weight. Thus, the perfect crystal with no vacancies ($x = 0$) has a free electron concentration of this value (6.25×10^{21}cm^{-3}for Ce_3S_4). The presence of x vacancies requires $(1 - 3x)$ electrons in the conduction band. The electron concentration per cm^3 for $R_{3-x}X_4$ can be written as $4(1 - 3x)/a^3$, where a = lattice constant. Hence, from R_2X_3 to R_3X_4, there is a transition to metallic conduction as the vacancies are filled with rare earth ions.

Since stoichiometric R_2X_3 compounds are insulators, most materials examined for thermoelectric applications are hot-pressed powders or single crystals of $R_{3-x}X_4$ with $x <$ ⅓, that is, varying degrees of excess of rare earth atoms. These stoichoimetries have been found to yield degenerate wide band-gap semiconductors, with electrical conductivities of $\sim 10^3$ ohm^{-1}cm^{-1}, carrier concentrations greater than $\sim 10^{20}$ cm^{-3} and charge carrier mobilities of ~ 1 to 10 cm^2/volt sec. Unfortunately, the polymorphic form being studied and reported upon is not always specified, the samples are generally not known to be single-phase and the degree of oxygen contamination is often unknown, all of which hinder the thorough understanding and utilization of these compounds.

Thermal conductivity: Lattice thermal conductivities (κ_{ph}) of the $R_{3-x}X_4$ compounds generally have low values in the range 0.005 to 0.01 watt/cm-deg. (Smirnov 1972) which are primarily a result of three factors. First, the Debye temperatures (θ_D) are low (~ 200 K to 400 K) and κ_{ph} is proportional to θ_D^3. Second, these are complicated crystal structures with a fairly large number of atoms (28) per unit cell. Since there are three acoustic modes per unit cell the number of optic modes, N = 28 − 3. Examination of the dispersion curve, ω versus k, shows that these optic modes of vibration have very low propagation velocities. Therefore, the majority of phonons are optical and have low velocities. Third, rare earth ion vacancies produce mass fluctuations that, in addition to the mass difference between the rare earth and chalcogen ions, enhance phonon scattering.

Transport Mechanisms: From the above discussion it appears that rare earth chalcogenides behave as highly degenerate n-type semiconductors over most of the composition range R_3X_4 to R_2X_3. For compositions very close to R_2X_3 (low carrier concentrations) or at extremely high temperatures, the degree of degeneracy diminishes to a semi- or non-degenerate condition. At R_2X_3, the compound is an insulator. The transport properties can be described by the equations for almost-free electrons, or degenerate semiconductors, provided x does not approach ⅓. However, the conduction band appears to be narrow from the large values of effective mass and the small values of the mobility. Cutler et al. (1964) have suggested that, from the similarity of behavior of praseodymium sulfide and cerium sulfide, it is the outer 5d and 6s electrons that are involved in the conduction process, not the inner shell 4f electrons. However, group-theoretical calculations by Goryachev et al. (1968) show that the conduction band is more complex, comprising a mixture of 6s, 5d and 4f states. There also appear to be localized levels of d and f character with high density of states. If these states produce a large gradient in the density of states $g(E)$ at the Fermi-surface, then the Seebeck coefficient, which is proportional to $d(1n(\mu g))dE$ (Fritshe 1971), can be large.

Thermoelectric Properties: Since the rare earth chalcogenides are so similar in properties, it is of interest to consider which specific rare earth compound has the highest potential thermoelectric performance. Disregarding thermal conductivity difference (they appear to differ very little from each other between the various rare-earth chalcogenides), examination of Equation (9) facilitates selecting the best compositions. The room-temperature physical properties of a large number of rare earth chalcogenides have been tabulated by Zhuze et al. (1971). Their data have been employed to calculate values for $m^{*3/2}\mu$ listed in Table 1.

It is seen that $La_{3-x}Te_4$ has the greatest potential for high Z values. This is supported by the work of Golikova and Rudnik (1978) who claim that La_3Te_4 has the a value of $0.91 \times 10^{-3}deg^{-1}$ and, hence, a ZT of 1.41 at 1550 K. Zhuze et al. (1971), reporting essentially the same data, quote a Z of 0.53×10^{-3} at 1400 K for the nominal composition La_3Te_4. This and thermoelectric data reported on other rare earth chalcogenides are illustrated by the plot ZT versus T in Figure 6.

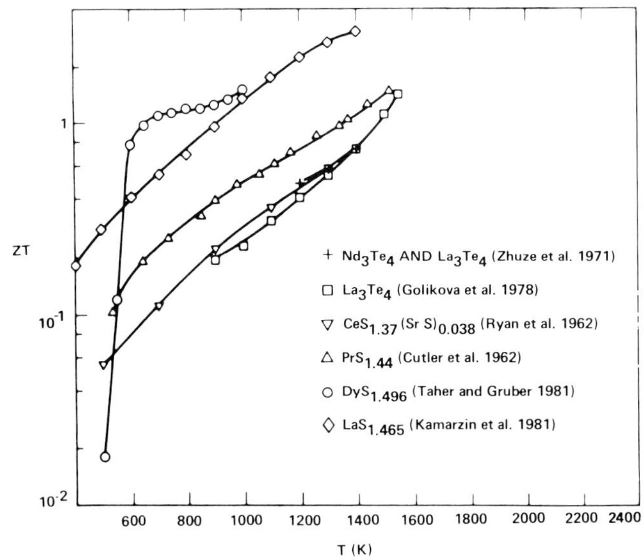

Figure 6 *ZT Values for Rare Earth Chalcogenides.*

Although the transport properties for compositions close to and including R_3X_4 have been elucidated, further work appears necessary to clarify the transport mechanism near the composition R_2X_3. Golikova and Rudnik (1973) state that as $x \rightarrow 1/3$, that is, R_2X_3 compounds, hopping-type conductivity prevails, although no details are given. Earlier, Cutler and Leavy (1964) made a detailed study of the electronic transport properties of high resistivity $Ce_{3-x}S_4$ samples. Hall effect measurements as a function of temperature showed that for samples with $x = 1/3$ the mobility was indeed activated. They interpreted their data as indicating two extreme conduction processes to be operative:

1. A hopping process with electrons in localized states (these states were proposed to arise from fluctuations in potential resulting from random lattice vacancies) and
2. Motion of non-localized electrons.

The $Ce_{3-x}S_4$ compounds exhibit a continuous variation between the two extremes depending upon composition and temperature. The data of Taher et al. (1974) on Y-phase R_2X_3 compounds presumably could be fitted to this model. Thus, small polaron hopping presents a model to explain the data on high resistivity R_2X_3 compounds.

Summarizing, it appears possible to realize values for ZT of greater than one at high temperatures in the rare earth chalcogenide system. However, in nearly every case, the thermal conductivity was either estimated from values for other rare-earth chalcogenides and/or the data were extrapolated from low temperatures. Accurate thermal conductivity data are difficult to obtain, particularly at high temperatures and, therefore, the quoted values of ZT should be viewed with some reservation. Notwithstanding, considerable room for improvement of performance exists at lower temperatures, ~800 to 1300 K, that is, in the temperature range where the Si-Ge alloys are preeminant.

Table 1 *Properties of Rare Earth Chalcogenides (R_3X_4) at Room Temperature.*

Compound	Mobility ($cm^2/V \cdot s$)	Effective Mass, m^*	$m^{*3/2}\mu$
La_3S_4	3.5	3.6	23.9
Ce_3S_4	3.1	2.8	14.5
Nd_3S_4	3.2	2.7	14.2
Pr_3Se_4	2.6	2.6	10.9
La_3Te_4	11.5	1.8	27.8
Ce_3Te_4	4.2	2.1	12.8
Pr_3Te_4	6.2	1.6	12.5
Nd_3Te_4	5.0	2.0	14.1

In addition, although the dissociation vapor pressures of the rare earth chalcogenides at high temperatures are surprisingly low for sulfur, selenium and tellurium compounds, the phase stability is still in question. Ryan et al. (1962) ascribed changes in thermoelectric properties on high-temperature cycling to oxygen contamination. This was discounted by Cutler et al. (1962) whose studies pointed to disproportionation of high-temperature Y-phase of $Ce_{3-x}S_4$ into a sulfur-rich intermediate temperature β-phase in equilibrium with a sulfur-poor Y-phase. Considerable improvement in phase stability was obtained by doping with the alkaline earths, Ba, Ca and Sr. It was conjectured that this improvement resulted from a filling of vacancy sites. Obviously, more phase, mass and lattice defect stability studies are needed before these materials can be safely employed in thermoelectric applications for extended periods at high temperatures.

Hopping-Type Semiconductors

In the discussion so far we have considered materials with conventional semiconducting transport properties, that is, ones which conduct by the itinerant motion of charge carriers. However, there is another broad class of materials that conduct by the charge carriers hopping from site to site.

Ure and Heikes (1961) have derived expression for the transport properties of narrow band or mixed valency semiconductors in which charge carriers are trapped by polaron formation and transport occurs by charge hopping from site to site. The following expression for the electrical conductivity was derived:

$$\sigma = \frac{c(1-c)e^2}{akT}\tau_o^{-1} \exp\left(-\frac{\Delta G}{kT}\right), \qquad (13)$$

where a is the diffusion length or distance between sites, assumed to equal the lattice constant; c is the charge carrier concentration; τ_o^{-1} the a priori transition probability at infinite temperature which is related to the lattice vibrational frequency; the ΔG is the hopping activation energy. The Seebeck coefficient was given by

$$\alpha = \frac{k}{e}\left[-\epsilon\frac{\Delta H_R}{kT} - \ln\left(\frac{c}{1-c}\right)\right]\exp\left(\frac{-\Delta G}{kT}\right), \qquad (14)$$

where ΔH_R is the change in energy associated with the removal of a charged particle and is related to the lattice force constants at a site. Experimentally is of the order of one. Hence,

$$ZT = \frac{k}{\kappa_{\text{ph}}a}\left[\frac{\Delta G_R}{kT} - \epsilon\frac{\Delta H_R}{kT} - \ln\left(\frac{c}{1-c}\right)\right]^2 \qquad (15)$$
$$\tau_o^{-1}c(1-c)\exp\left(\frac{-\Delta G}{kT}\right).$$

Optimization

In order to maximize (ZT) the coefficient and the polaron binding energy G_R were set equal to zero by Ure and Heikes (1961); thus

$$ZT = \frac{k}{\kappa_{\text{ph}}a}\left[\ln\left(\frac{c}{1-c}\right)\right]^2 c(1-c)\,\tau_o^{-1}. \qquad (16)$$

Differentiation of this equation with respect to c yields a maximum at $c \sim 0.1$. However, this value should be accepted with reservation. Ure and Heikes (1961) state that the assumption $\epsilon = 0$ is not born out experimentally. Furthermore, setting $\Delta G = 0$ raises the electrical conductivity but lowers the Seebeck coefficient. Thus, setting the term $\epsilon\Delta H_R/kT \sim 1$ in the Seebeck coefficient and optimizing ZT leads to a c value closer to 0.2.

Temperature Variation of Z

In materials that exhibit hopping-type conductivity, the charge carrier mobility and, hence, the electrical conductivity increase with the increase in temperature. If the conduction mechanism is by small polaron hopping between sites that are equivalent in energy then the Seebeck coefficient should be roughly independent of temperature.

In a disordered small polaron material with charge carriers hopping between inequivalent sites, vibrational energy may be transported with the carrier as it hops. This will give rise to a temperature dependent term in the Seebeck coefficient (Wood and Emin 1984), that is,

$$\alpha = (T\Delta S + E_T)/qT, \qquad (17)$$

where the first (standard) term is proportional to the average change of entropy, ΔS, of the material when a charge carrier is injected into it and the second term is proportional to the average vibrational energy transported with a carrier as it hops, E_T. This additional contribution to the carrier produced heat flow enhances the Seebeck coefficient. Thus, in a small polaron material, in addition to the electrical conductivity rising with temperature the Seebeck coefficient may also increase with temperature. Consequently, Z will increase drastically with temperature and can be considerably greater than the limit predicted by Ure and Heikes (1961) at high temperatures. However, this increase in Z may be tempered somewhat by the carrier produced heat flow now introducing a κ_{el} contribution, which is not present in the absence of vibrational energy transport.

Thermal Conductivity

The discussion of lattice thermal conductivity in broadband semiconductors is, obviously, also applicable to narrowband, hopping-type, semiconductors. However, the electronic contribution to the thermal conductivity will differ in the two classes of materials. In general, for hopping conduction the electronic contribution is zero. This is because equal amounts of work and energy are given up by or returned to the lattice when the charge carrier hops (Ure and Heikes 1961). However, for certain distinctive types of hopping, for example, hopping between inequivalent sites, small amounts of energy can be transported (Wood et al. 1985).

Experimental Results

Examples of materials that fall into the class of hopping-type semiconductors are the boron-rich borides, the alkali-halides and the transition metal oxides. We will consider below the performance of some boron-rich borides as high temperature thermoelectric materials.

Boron and Borides

Structure: Boron forms refractory compounds with large number of elements. Several of these compounds have been investigated for high-temperature thermoelectric conversion. Elemental boron is highly refractory (m.p. ~2800 K) and its thermoelectric properties have also been studied (Golikova 1979). Boron is generally accepted to have at least three and possibly four allotropic forms: α-rhombohedral ($R\bar{3}m$, a = 5.075 A = 58°06′); β-rhombohedral ($R\bar{3}m$, a = 10.145 A = 65°17′); α- or I-tetragonal ($P\bar{4}2m$, a = 8.75 A, c = 5.06 A) which may not correspond to an allotropic form and β- or II- or III- tetragonal ($P4_122$, a = 10.12 A, c = 14.14 A). Boron can also occur in the amorphous form. Many boron-rich borides are structural analogs of one of the four boron crystalline modifications. These structures are characterized by an arrangement of an icosohedral cluster of B_{12} atoms bound either to one another or to isolated atoms by directed bonds. Excellent surveys of the properties of boron and borides have been published by Golikova (1973), Matkovich (1972), and Samsonov et al. (1962).

Thermal Conductivity: The surprisingly low values for the high-temperature thermal conductivity of boron and boron-rich borides has been attributed to the complexity of the crystal structure (Petrov et al. 1964 and Slack et al. 1971). At low temperatures the dominant contribution to the thermal conductivity is due to long-wavelength acoustic phonons. The contributions to the specific heat by the acoustic modes are associated with the vibrations of the unit cell as a whole. Consequently, this contribution should vary inversely proportional to the number of atoms/unit cell (N_c). Thus, the magnitude of κ should vary inversely with N_c supposing the velocity and mean free path changes insignificantly from one boride to another. In addition, the acoustic branches are cut off at smaller values of wavevector while the number of optical vibrations increase as N_c increases. For example, in β-boron with 105 atoms/unit cell there are three acoustic branches and 312 optical branches in the dispersion curve. As the temperature is increased, the relative contribution of the high frequency (optical) modes to the thermal conductivity increases and can exceed the contribution by the acoustic modes. The group velocity of optical vibration is much lower than the sound velocity and, therefore, their contribution to the thermal conductivity is low. Because of the structural complexity of boron and its analogues, groups of atoms in equivalent positions are widely separated. For an optical wave to propogate, a definite phase relationship must exist between these like groups of atoms. Perturbations arising from other groups of atoms in the path of propogation cause a breakdown fo phase relationships, that is, independent local vibrations occur. Thus, the nature of phonon propogation in boron-rich borides closely approaches that in amorphous solids with correspondingly low values. Not all boron-type structures appear to conform to this hypothesis—the thermal conductivity of α-boron (Golikova et al. 1979) being a case in point that warrants further verification.

Transport Mechanisms: Boron and all of the boron-rich borides described above display a notable characteristic behavior: they exhibit a rising α and σ with increasing temperature over a wide temperature range. This generally has been interpreted to be a result of the hopping of charge carriers, a conclusion often supported by measurements showing that the mobility increases as the temperature increases. However, the models proposed to account for this conduction mechanism have differed from material to material or in some instances for the same material but over different temperature ranges.

Despite some differences between interpretations of the experimental data most investigators (Golikova 1979, Werheit and Leis 1970, and Berezin et al. 1977) appear to have arrived at the conclusion that hopping occurs between high densities of localized states that arise naturally in the structure of β-boron. Berezin et al. (1977) and Golikova et al. (1976) have invoked an amorphous concept to explain the origin of these localized states. The complicated structure of β-boron with a large unit cell of 105 atoms results in groups of non-equivalent atoms in the lattice having different coordination numbers, that is, 13% of the atoms have coordination numbers of 8 and 9. These like-atoms are spaced about 10 interatomic distances apart and, thus, to a first approximation, there is no long-range order. It is these atoms that are supposed to give rise to the localized levels.

The previous discussion is also applicable to the boron-rich borides. Close parallels have been drawn between the properties of β-boron and α-AlB_{12} (Golikova et al. 1976) and to a lesser extent $B_{1-x}C_x$ (Werheit et al. 1981), although some strong evidence recently has been accumulated in support of small polaron-hopping in $B_{1-x}C_x$ compounds (Wood and Emin 1984).

An alternative hypothesis has been proposed for the origin of hopping process in $B_{14}Si$ which invokes a medium-range disorder model (Pistoulet et al. 1978). The experiments conducted so far on $B_{14}Si$ have been performed soley on chemically vapor deposited specimens. Cast specimens of B_9Si exhibited an entirely different behavior with much higher mobilities (Djafarov et al. 1980). The question arises as to how much the results on $B_{14}Si$ are influenced by the nature of the deposit or whether the cast specimens of B_9Si were single-phase.

It is of considerable interest and value to speculate on the origin of the hopping-type conduction in boron-rich borides since a thorough understanding of the mechanism will allow the tailoring and optimization of these materials for high-temperature thermoelectric and other applications. However, it may be more a matter of semantics in attempting to distinguish among the various models (localized states, small polaron hopping, medium range disorder, and so on) when the charge carriers are strongly localized.

Thermoelectric Materials: Only four materials have been investigated in any depth for thermoelectric applications: β-boron (105 atoms/unit cell), $B_{14}Si$ (isostructural with β-boron and with 105 atoms/unit cell), $B_{1-x}C_x$ (an analog of α-boron with 15 atoms/unit cell) and α-AlB_{12} (same space group as β- or II-tetragonal boron with from 187 to 208 atoms/unit cell). Thermoelectric data on these materials are shown in Figure 7.

In summary, as with the rare earth chalcogenides, ZT values in excess of one appear realizable at high temperatures. More work is needed to extend the high Z values down to lower temperatures. The achievement of this objective will be greatly aided by a better understanding of both the electronic and thermal transport properties.

Conclusions

The excellent theoretical work carried out two decades ago adequately explained the transport behavior and effectively guided the development of thermoelectric materials of high conversion efficiencies. The more significant contributions involved the estimation of optimum doping concentrations, the reduction of thermal conductivity by solid solution doping and the development of a variety of materials with $ZT \sim 1$ in the temperature range 300 K to 1200 K. It was also shown that $ZT \sim 1$ is not a theoretical limitation although, experimentally, values in excess of one were not achieved.

In the subsequent years, work has continued with emphasis on high-temperature energy conversion. A number or promising materials have been discovered with indications that $ZT > 1$ is realizable. These materials can be divided into two classes: (1) the rare earth chalcogenides with high vacancy concentrations, which behave as conventional highly degenerate n-type semiconductors at room temperature, with the degree of degeneracy decreasing to partial degeneracy at higher operating temperatures; and (2) the boron-rich borides which exhibit hopping p-type conductivity behavior.

It is not clear why the rare earth chalcogenides are generally n-type and the boron-rich borides and β-boron generally p-type semiconductors, but it could be related to large mobility ratios of majority

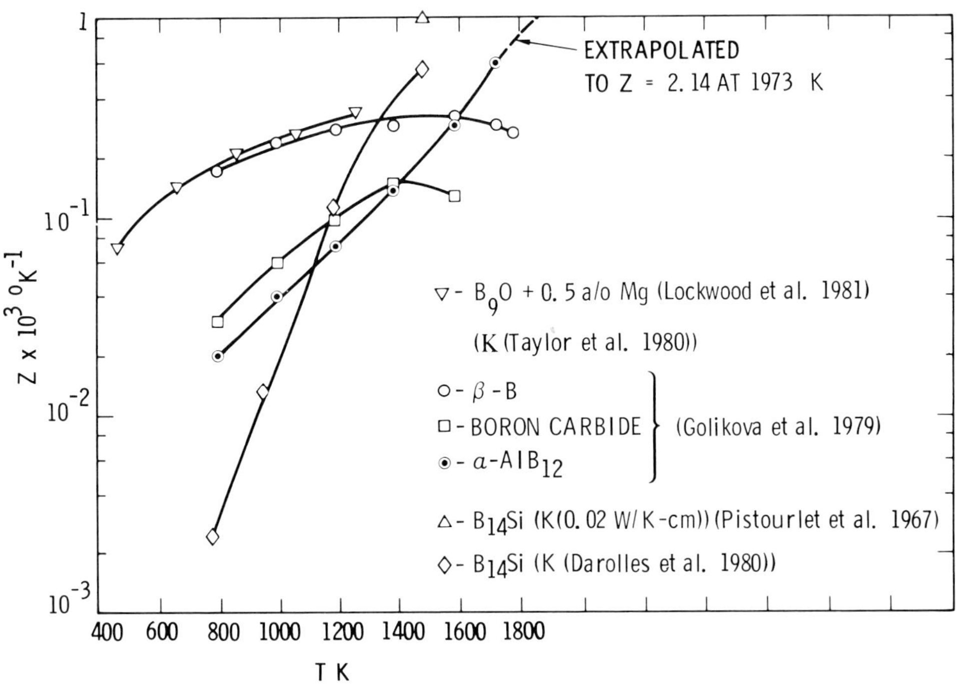

Figure 7 *ZT Values for Boron-Rich Borides.*

and minority charge carriers. It would be desirable, from the view of materials compatibility, if the thermocouple could be fabricated from n- and p-legs of the same compound. However, it appears to be extremely difficult to dope rare earth chalcogenides p-type or boron-rich borides n-type.

A considerable amount of work still needs to be done to relate both the thermal and electronic transport to the composition and crystallographic structure in these classes of materials—particularly the identification of the hopping centers in the case of borides. Current theory is adequate to describe the conditions for optimization of the thermoelectric Figure of Merit in materials that exhibit hopping conductivity and further theoretical work is needed. Since, in this class of materials, the conventional relationship between Seebeck coefficient and electrical conductivity is violated in a manner favorable to the achievement of high Zs, the importance of further basic research into this area should not be overlooked. Hopefully, this report will stir some further interest in this area or more generally into the transport properties of refractory materials.

Acknowledgments

The research described in this paper was carried out by the Jet Propulsion Laboratory, California Institute of Technology, under contract with the National Aeronautics and Space Administration.

References

Aigrain, P. R. (1960) *Thermoelectricity*, P. H. Egli, ed., Wiley, p. 180.

Berezin, A. A. et al. (1977) *Boron and Refractory Borides*, V. I. Matkovich, ed., Springer Verlag, pp. 52–64

Besancon, P. (1973) *J. Solid State Chem.*, 7:232.

Chasmar, R. P. and R. J. Stratton (1959) *J. Electron. Control*, 7:52.

Cutler M. and J. F. Leavy (1964) *Phys. Rev.*, 133:A1153.

Cutler, M., J. F. FLeavy, and R. L. Fitzpatrick (1964) *Phys. Rev.*, 133:A1143.

Cutler, M., J. C. Appel, G. L. Guthrie, and S. W. Kurnick (1962) Final Report, Contract No.: BS-77144. ARPA Order No. 8159, 31 August 1962.

Darolles, J. M., T. Lepetre, and J. M. Dusseau (1980) *Phys. Stat. Sol. (a)*, 58:K71.

Dixon, A. J., M. E. Ertl, and H. J. Goldsmid (1975) *Energy Conv.*, 14:47.

Djafarov, E. O., O. A. Golikova, and M. I. Aliyev (1980) *Akad. Nauk Azerb. SSR Doklady*, 36:523 and 18.

Donahoe, F. J. (1960) *Electl. Engng.*, 79:488.

Dugdale, J. S. and D. K. C. MacDonald (1955) *Phys. Rev.*, 98:1751.

Flahaut, J. M. (1978) *Handbook on the Physics and Chemistry of Rare Earths*, K. A. Geschneidner and L. Eyring, eds., North Holland.

Fritsche, H. (1971) *Sol. State Comm.*, 9:1813.

Goldsmid, H. J. (1964) *Thermoelectric Refrigeration*, Plenum Press.

Goldsmid, H. J. and R. W. Douglas (1954) *Brit. J. Appl. Phys.*, 5:386.

Golikova, O. A. (1979) *Phys. Stat. Sol. (a)*, 51:11.

Golikova, O. A. et al. (1976) J. *Less-Comm. Metals*, 47:129.

Golikova, O. A. and I. M. Rudnik (1978) *Izv. Akad. Nauk, SSSR Neorg. Mater.*, 14:17.

Golikova, O. A., N. E. Solovev, Ya. A. Ugai, and V. A. Feigelman (1979) *Phys. Semicond.*, 13:486.

Goryachev, Yu M., T. G. Kutsenok, and L. I. Zadvornyi (1968) *Izv. Akad. Nauk SSR Neorg. Mat*, 4:1912.

Joffe, A. F. (1957) *Semiconductor Thermoelements and Thermoelectric Cooling*, Info-search Limited, London.

Kamarzin, A. A., K. E. Mirnov, V. V. Sokolov, Yu N. Malovitsky, and I. G. Vasilyera (1981) *J. Cryst. Growth*, 52:619.

Kelley, C. M. and G. C. Szego (1964) *Coll. Energy Sources and Conversion*, p. 651, held in Cannes, 16 March 1964.

Keyes, R. W. and J. E. Bauerle (1961) *Thermoelectricity: Science and Engineering*, R. R. Heikes and R. W. Ure, eds., Interscience, NY, p. 91.

Liebfried, G. and E. Schlomann (1954) *Nachr. Akad. Wiss. Goettingen II Math.-Physik. Kl.*, 2:71.

Littman, H. and B. J. Davidson (1961) *J. Appl. Phys.*, 32:217.

Lockwood, A., C. Wood, B. H. Reynolds, and N. Elsner (1981) Electrochemical Society Meeting, held in Minneapolis, MN., May 1981.

Matkovich, V. I. (1977) *Boron and Refractory Borides*, Springer-Verlag, NY.

Petrov, Q. V., M. S. Germaidze, O. A. Golikova, A. Yu Kiskachi, and V. N. Matreev (1969) *Sov. Phys. Solid State*, 11: 741.

Pistoulet, B., J. L. Robert, J. M. Dusseau, F. Roche, P. Girard, and L. Ensuque (1979) *J. Less Common Metals*, 67:131, and *Physics of Semiconductors* (1978), p. 793.

Pistoulet, B., J. L.. Robert, J. M. Dusseau, J. M. Darolles, B. Armas, and C. Combescure (1967) *Proc. Internat. Solar Electric Conf.*, p. 887, held in Toulouse, France, 1–5 March 1967.

Rittner, E. S. (1959) *J. Appl. Phys.*, 30:702.

Rittner, E. S. and F. G. Neumark (1963) *J. Appl. Phys.*, 34: 2071.

Ryan, F. M., I. N. Greenberg, F. L. Carter, and R. G. Miller (1962) *J. Appl. Phys.*, 33:864.

Samsonov, G. V., L. Ya. Markovskii, A. G. Zhigach, and M. G. Valyashko (1962) *Boron its Compounds and Alloys*. NTIS Publication, AEC-TR-5032, July 1962.

Simon, R. (1962) *J. Appl. Phys.*, 33:1830.

Slack, G. A., C. W. Oliver, and F. H. Horn (1971) *Phys. Rev.*, 4:1714.

Smirnov, I. A. (1972) *Phys. Stat. Sol. (a)*, 14:363.

Taher, S. M. and J. B. Gruber (1981) *Mat. Res. Bull.*, 16:1407.

Taher, S. M. A., J. G. Gruber, and O. C. Olsen (1974) *J. Chem. Phys.*, 60:2050.

Ure, R. W., (1972) *Energy Conv.*, 12:45.

Ure, R. W. and R. R. Heikes (1961) *Thermoelectricity: Science and Engineering*, Interscience NY, p. 339.

Werheit, H. and H. G. Leis (1970) *Phys. Stat. Sol.*, 41:247.

Werheit, H., K. deGroot, and W. Malkemper (1981) *J. Less Common Metals*, 2:153.

Wood, C. and D. Emin (1984) *Phys. Rev. B.*, 29(8):4582.

Wood, C., D. Emin, and P. E. Gray (1985) *Phys. Rev. B.*, 31:6811–6814.

Zhuze, V. P., V. M. Sergeeva, and O.A. Golikova (1971) *Sov. Phys. Solid State*, 11:2071 and 13:6689.

Space Nuclear Power Systems 1986
Edited by M. S. El-Genk and M. D. Hoover

Chapter 17

Analysis of the Transient Behavior of Thermoelectric Generators

Mohamed S. El-Genk and Jong T. Seo

The Institute of Space Nuclear Power Studies
Department of Chemical and Nuclear Engineering
The University of New Mexico
Albuquerque, NM 87131

This research developed and verified a steady state/transient model for thermoelectric (TE) generators. The model takes into account the change in the material properties with temperature and handles any combination of boundary conditions at the hot and cold shoes of the TE generator. The model was verified by comparing its predictions with the available exact solutions for simplified boundary conditions, and by performing an energy balance on the TE generator. Finally, the model was used to investigate the steady state and transient behavior of radiatively cooled SiGe/GaP TE generators.

Introduction

Among the energy conversion systems being considered for space nuclear power applications, the thermoelectric (TE) generators are the most developed energy converter; they have had extensive space flight experience in the Radioisotope Thermoelectric Generators (RTGs) (Johnson 1985 and El-Genk and Seo 1985). Recently, the TE space nuclear power system concept was selected by the SP-100 program to meet the electric power needs in the range of 100 kW to 300 kW. This concept employs a fast flux nuclear reactor as the heat source and TE generators in the energy converter, and as the electric power source in the thermoelectric-electromagnetic (TE-EM) pumps (GE 1983). Although the TE generators operate at a near steady state power in the RTGs, in space nuclear power systems, the generators are subjected to various operating transients, including those that occur during the reactor start-up and shutdown and during a change in the reactor thermal power and/or in the electric load. Therefore, it is important to understand the transient behavior of the TE generators.

This paper presents a steady state/transient model for the TE generators. The model considers changes in the TE material properties with temperature, and can handle any combination of boundary conditions at the hot and cold shoes of the TE generators (Seo 1986); previous studies, in contrast, were limited to constant material properties and isothermal boundary conditions (Angrist 1965 and Cobble 1982). Although assuming constant TE material properties was necessary to obtain closed-form analytical solutions, it could, nonetheless, result in a large error in predicting the performance of the TE generator during both steady state and transient operation. After the present TE model was developed, it was verified by comparing its predictions with the available exact solutions for constant material properties and simplified boundary conditions. For higher order boundary conditions, such as radiative cooling and time-dependent heat flux, the results of the model were verified by performing an energy balance on the TE generator. Finally, the TE model was used to study both the steady state and transient behavior of SiGe/GaP TE generators. In this study, the hot shoe of the TE generators was subjected to a time-dependent heat flux and the cold shoe was radiatively cooled.

Model Descriptions

As shown in Figure 1, a TE generator consists of two semiconductor elements: a P-type material and an N-type material. The two elements are connected thermally in parallel and electrically in series. The heat input at the hot shoe is partially rejected at the cold shoe with the difference converted into electricity. The difference in the electric polarity between the P and N elements causes electric current to pass in the P element in the same direction as the thermal energy flow (that is, from the hot shoe to cold shoe); while in the N element, the current passes in the opposite direction to the flow of thermal energy (that is, from the cold shoe to hot shoe). The electric load is connected between the P and N elements at the cold shoe where the total current flowing through the external load is equal to that generated in both elements. To simplify the analysis, the following assumptions were introduced:

- There is no heat transfer from the heat source to the heat sink except through the TE elements, with no lateral heat transfer since the TE elements are thermally insulated. Therefore, the heat flow through the TE elements can be treated as a one-dimensional problem.
- The cold shoe is directly coupled to the radiator with no interfacial resistance to the heat flow.
- The junction electrical contact resistance is negligible compared with the bulk resistance of the N and P elements.
- The cross-sectional areas of the P and N elements are constant.
- Thermal contact resistances between the TE element and the hot and cold shoes are negligible.

In the present analysis, the TE material properties (the Seebeck coefficient, the thermal conductivity, and the electrical resistivity) were allowed to vary with temperature as shown in Figure 2 (Stapfer 1985). Such a consideration causes the governing equations to be nonlinear; thus, a closed analytical solution is unattainable. Other difficulties are caused by the nonlinear boundary condition of radiative cooling at the cold shoe and by the nonhomogeneous nature of the temperature governing equation due to the internal heat generation induced by Joule and Thomson effects. The temperature distributions in the TE elements are determined by

$$\frac{\partial}{\partial x}\left(k\frac{\partial T}{\partial x}\right) + \rho J^2 + \tau J\frac{\partial T}{\partial x} = wc_p\frac{\partial T}{\partial t}\cdot \qquad (1)$$

The terms on the left hand side of Equation (1) represent the heat conduction, Joule heating, and Thomson heating, respectively. Thomson coefficient, τ, is given by Kelvin's relation as

$$\tau = T\frac{d\alpha}{dT}\,. \qquad (2)$$

The initial condition for Equation (1) could be either a uniform temperature or a spatial temperature distribution:

$$T(x,0) = T_0\,, \qquad (3a)$$

or

$$T(x,0) = T_0(x). \qquad (3b)$$

NOMENCLATURE

English

- A: Cross sectional area (cm^2)
- c_p: Specific heat (J/g)
- F: Radiation viewfactor, Equation (6)
- I: Electric current (a)
- J: Electric current density (A/cm^2)
- k: Thermal conductivity (W/cm·K)
- L: Length of the TE element (cm)
- P: Power (W)
- q: Heat flux (W/cm^2)
- R: Electrical resistance (ohm)
- s: Time step interval (s)
- t: Time (s)
- T: Temperature (k)
- v: Test function, Equation (16)
- V: Voltage (V)
- w: Density (g/cm^3)
- x: Axial coordinate (cm)

Greek

- α: Seebeck coefficient (V/K)
- η: Conversion efficiency
- ϵ: Surface emissivity of the radiator, Equation (6)
- ϕ: Basis function, Equation (15)
- ρ: Electrical resistivity (ohm-cm)
- σ: Stefan-Boltzmann constant (5.669×10^{-12}W/$cm^2 \cdot K^4$), Equation (6)
- τ: Thomson coefficient (V/K), Equation (2)

Subscript

- a: Ambient
- c: Cold shoe
- e: Electric
- h: Hot shoe
- i: Index
- j: Index
- L: External load
- N: Number of elements, N TE element
- 0: Initial condition
- pn: P and N elements
- P: Peltier heat, P TE element

Superscript

- k: Time step index, Equation (23)
- (n): Iteration index, Equation (25)

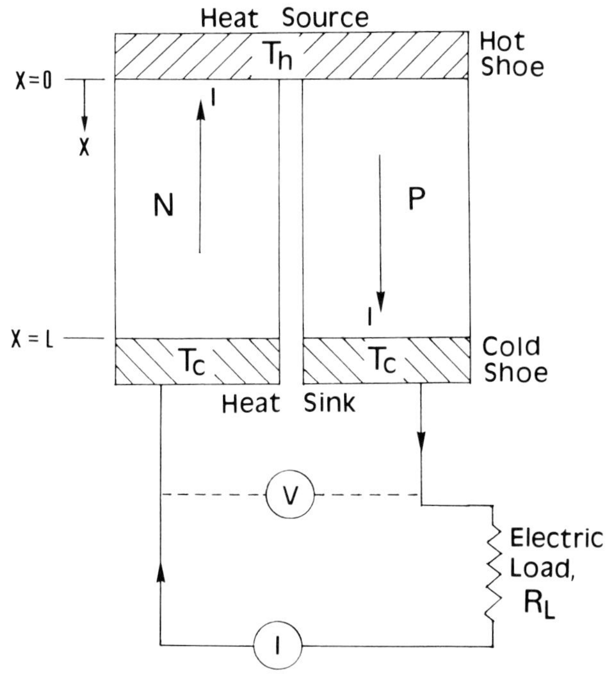

Figure 1 *Schematic Diagram of a Thermoelectric Generator.*

The boundary condition at the hot shoe of the TE elements ($x = 0$) could be either a time-dependent temperature or heat flux:

$$T(0,t) = T_h(t) \ , \tag{4a}$$

or

$$-k \frac{\partial T(0,t)}{\partial x} = q_h(t) \ . \tag{4b}$$

Also, the boundary condition at the cold shoe of the TE elements ($x = L$) could be either a time-dependent temperature or heat flux:

$$T(L,t) = T_c(t) \ , \tag{5a}$$

or

$$-k \frac{\partial T(L,t)}{\partial x} = q_c(t) \ . \tag{5b}$$

For a radiatively cooled cold shoe, Equation (5b) becomes

$$-k \frac{\partial T(L,t)}{\partial x} = q_c(t) = \sigma \ \epsilon \ F \ (T^4(L,t) - T_a^4) \ . \tag{6}$$

For steady state operation, the term on the right hand side and the Thomson heating term of Equation (1), the initial condition (Equation (3)), and the time variable in the boundary conditions (Equations (4) through (6)) are dropped out.

In order to solve Equation (1), with the prescribed initial and boundary conditions, an additional equation is needed to calculate the electric current. By definition, the electric voltage generated across the external load is given as (Angrist 1965 and Soo 1968):

$$V = R_L I = \rho_{pn}(T_h - T_c) - (R_n + R_p)I \ , \tag{7}$$

where

$$R_n = \rho_n \ L/A, \text{ and}$$
$$R_p = \rho_p \ L/A \ .$$

Rearranging Equation (7) gives the following expression for the electric current flowing through the circuit of a TE generator:

$$I = \frac{\alpha_{pn}(T_h - T_c)}{R_L + R_n + R_p} \ . \tag{8}$$

Equation (8) is solved simultaneously with Equation (1) for the temperature distribution in the P and N elements as well as the temperatures at the hot and cold shoes (T_h and T_c), respectively.

After the temperature distributions in the P and N elements of the TE elements are calculated, the electric power output and the conversion efficiency of the TE generator can be given, respectively, as

$$P_e = I^2 R_L \ , \tag{9}$$

and

$$\eta = \frac{P_e}{P_h} \ , \tag{10}$$

where P_h is the total heat flow into the hot shoe of the TE generator and is given by

$$P_h = P_P + \left[q_h - \frac{L}{2}(\rho J^2) - \frac{L}{2}\left(\tau J \frac{\partial T}{\partial x}\right) \right] A \ . \tag{11}$$

In Equation (11), q_h is the heat conduction term at the hot shoe (Equation 4b) and P_P reflects the heat transmitted from the hot shoe to the cold shoe by the Peltier effect

$$P_P = \alpha_{pn} T_h I \ . \tag{12}$$

Method of Solution

A closed-form solution was not possible because of the nonlinear, nonhomogeneous nature of the temperature distribution governing equation as well as the time dependent and/or nonlinear boundary conditions. Therefore, a solution was obtained by using the finite element method coupled to an implicit time integration algorithm (Bathe 1982). The variational form of Equation (1) is

$$\int_0^L \left\{ Cv \frac{\partial T}{\partial t} + k \frac{dv}{dv} \frac{\partial T}{\partial x} - v\, Q(t) - v\, G(t) \frac{\partial T}{\partial x} \right\} dx = 0 \,, \quad (13)$$

where,

$$Q(t) = \rho J^2(t) \,,$$
$$G(t) = \tau J(t) \,, \quad (14)$$
$$C = w\, c_p \,.$$

By using the Galerkin approximation, the solution and test functions of Equation (13) are given as

$$T_N = \sum_{i=1}^{N} T_i(t)\, \phi_i(x) \,, \quad (15)$$

and

$$v_N = \sum_{i=1}^{N} v_i(x)\, \phi_i(x) \,, \quad (16)$$

where $\phi_i(x)$, $i = 1, 2, \ldots, N$ are the basis functions. Substituting Equations (15) and (16) into Equation (13) and then rearranging yields

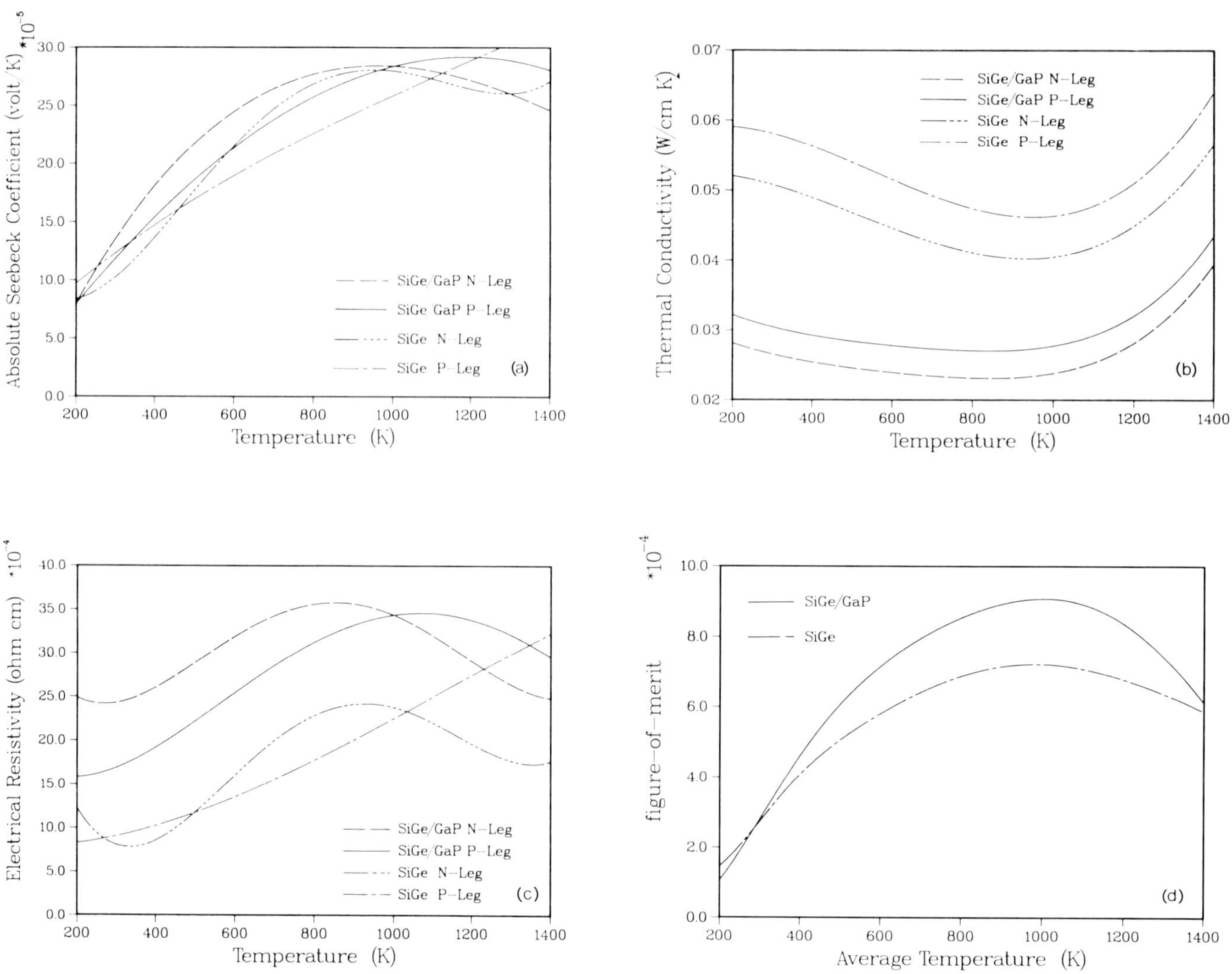

Figure 2 *SiGe/GaP and SiGe TE Material Properties as a function of Temperature: (a) Absolute Seebeck Coefficient, (b) Electrical Resistivity, (c) Thermal Conductivity, and (d) Figure of Merit.*

$$\sum_{i=1}^{N} v_i \left[\sum_{j=1}^{N} \left\{ C_{ij} \dot{T}_{ij}(t) + K_{ij} T_j(t) - G_{ij} T_j(t) \right\} - Q_i \right] = 0 , \quad (17)$$

where

$$C_{ij} = \int_0^L C\phi_i(x)\phi_j(x)dx , \quad (18)$$

$$K_{ij} = \int_0^L k \frac{d\phi_i}{dx} \frac{d\phi_j}{dx} dx , \quad (19)$$

$$Q_i = \int_0^L \phi_i(x) Q(t) dx , \quad (20)$$

$$G_{ij} = \int_0^L \phi_i(x) \frac{d\phi_j}{dx} G(t) dx . \quad (21)$$

Equation (17) can be written in matrix form as

$$[C]\{\dot{T}\} + [H]\{T\} = \{Q\} , \quad (22)$$

where

$$[H] = [K] - [G] .$$

For the time integration, the backward difference algorithm was used and it can be written as

$$\{\dot{T}\} = \frac{\{T\}^k - \{T\}^{k-1}}{s} , \quad (23)$$

where s is the time interval between steps and the superscripts k and k-1 denote steps. Substituting Equation (23) into Equation (22) and rearranging yields the final matrix equation:

$$\Big[[C] + s[H]\Big] \{T\}^{k+1} = s\{Q\}^{k+1} + [C]\{T\}^k . \quad (24)$$

Because the right hand side of Equation (24) consists of known values, the most recent temperature vector, $\{T\}^{k+1}$ was calculated using Gauss elimination. The matrices and vectors were calculated by choosing piecewise linear shape functions.

When the radiative boundary condition at the cold shoe (which is non-linear due to the fourth power of the temperature) was incorporated, several factors had to be considered. If the difference between the cold shoe temperature and the ambient temperature is small, the cold shoe temperature can be found with reasonable accuracy through the linearization of the boundary condition (Carslaw 1973). However, if the temperature of TE cold shoe is several hundred degrees higher than the ambient temperature (which is the case in the SP-100 system), the linearization approximation will result in a significant error in predicting the cold shoe temperature. In the latter case, an iteration scheme was developed to solve for the cold shoe temperature. The scheme began with an initial guess of $T_c^{(n)}$ and solves for a new temperature, $T_c^{(n+1)}$, using the following relation:

$$T_c^{(n+1)} = \left(\frac{q_c^{(n)}(t)}{\sigma \epsilon F} + T_a^4 \right)^{1/4} , \quad (25)$$

where the cold shoe heat flux, $q_c^{(n)}(t)$, is calculated from the temperature gradient, $[-k\, \partial T^{(n)}/\partial x]_{x=L}$, at the cold shoe temperature, T.

The calculated temperature distribution and the hot and cold shoe temperatures were then used to determine the operating parameters of the TE generator. These parameters include the electric current flowing through the TE generator, the voltage across the external load, the electric power output, the amount of heat being rejected from the cold shoe, and the thermal-to-electric conversion efficiency.

Model Verification

In order to determine the accuracy of the finite element solution, the model predictions were compared with the exact solutions for transient heat conduction problems with simplified boundary conditions as well as constant TE material properties. First, the model predictions of the temperature distribution and conversion efficiency for the problem with isothermal hot and cold shoes were compared with the exact solutions reported by Cobble (1982) (see Figure 3). The TE material and the design parameters of the TE generator used in this comparison were

TE Material: $(SiGe)_{0.95}(GaP)_{0.05}$,
N-leg Length: 0.38 cm ,
P-leg Length: 0.38 cm ,
N-leg Area: 0.14 cm^2 ,
P-leg Area: 0.14^2, and
External Resistance: 0.0161 ohm .

As shown in Figure 3, the model predictions of the temperature distribution in the P leg were in excellent agreement with the exact solution, and the predicted TE conversion efficiency was within 0.2 percent of the exact solution.

Next, the numerical solution was compared with the exact solution for the transient heat conduction

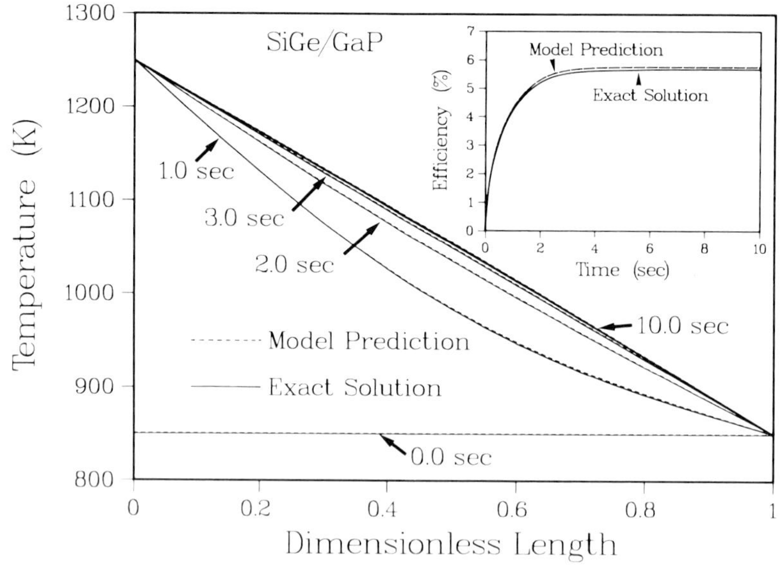

Figure 3 *Comparison of TE Model Predictions with the Exact Solution for the Problem of Transient Heat Conduction with Isothermal Boundaries.*

problem with time-dependent heat flux at the hot shoe and an isothermal cold shoe (Carslaw 1973). Figure 4 compares the model predictions with the exact solution for a SiGe/GaP TE generator subjected to a constant heat flux of 15 W at its hot shoe, while the cold shoe temperature was held constant and equal to the initial temperature of 450 K. As shown in Figure 4, the model predictions of temperature distribution in the *P* leg and the temperature of the hot shoe, T_h were in good agreement with the exact solutions.

An additional verification of the model was carried out using an energy balance on the TE generator. This energy balance requires that the thermal energy input to the TE generator during the time required to reach steady state should equal the sum of the electric energy produced, the thermal energy rejected, and the change in sensible heat of the TE generator. Since the energy balance results were always in excellent agreement with the exact solutions, the energy balance approach was subsequently used to check the model predictions for the cases where an exact analytical solution was not attainable. Such cases include

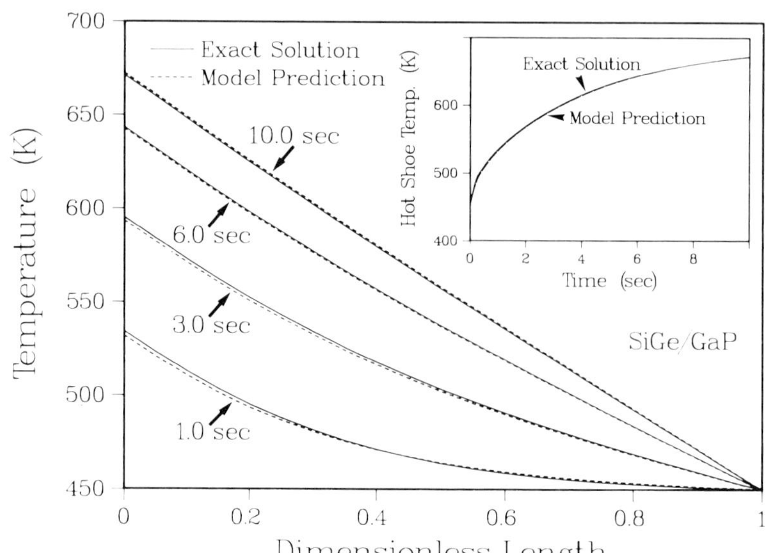

Figure 4 *Comparison of TE Model Predictions with the Exact Solution for the Problem of Transient Heat Conduction with Isoflux Hot Shoe and Isothermal Cold Shoe.*

temperature-dependent material properties and/or time-dependent heat flux or radiative cooling.

These excellent agreements between the model predictions and the exact solutions have not only shown the soundness of the present numerical solution, but also have established a confidence in applying the present model to more complicated problems. One example is that encountered in the TE converter of the SP-100 space nuclear power system (El-Genk and Seo 1986, Seo 1986, and Buksa 1986). For this problem, in addition to the necessity of considering temperature-dependent TE material properties, the cold shoe of the TE generator is radiatively cooled to space and the hot shoe will be subjected to either a constant or a time-dependent heat flux. The steady state and transient solutions to this problem were obtained using the present TE model. The results are discussed in the following section.

Results and Discussion

The TE converter of the SP-100 space nuclear power system may be subjected to a constant or a time-dependent heat flux at the hot shoe during steady state and transient operation, respectively. Examples of the transient operation of the SP-100 system include the reactor start-up and shutdown, and changing of the reactor thermal power and/or the system's external electric load demand. In the present analysis, the nominal input power to a single TE generator was taken to be 11.57 W, which corresponds to a nominal thermal power of 2.0 MW for the SP-100 nuclear reactor (GE 1983). An ambient temperature of 250 K and thermoelectric material of SiGe/GaP were selected as base parameters. The dimensions of the TE elements are the same as those listed in the previous section. For this base case, the steady state behavior of the TE generator was analyzed in order to understand its operating characteristics and to provide the initial conditions for subquent transient analyses.

Steady State Analysis

The results of the steady state operation of a SiGe/GaP TE generator are presented in Figures 5 and 6 where the physical properties of the TE materials were taken to be temperature-dependent. In these figures, the value of 500 K for the ambient temperature, although unrealistic for a space environment, was introduced into the analysis only for the purpose of comparison with the results for the ambient temperature of 250 K. As delineated in Figure 5, the temperature for both the P and N elements decreased nonlinearly as the distance measured from the hot shoe increased. This nonlinearity in the temperature distributions is attributed to the internal heat generation due to the Joule and Thomson effects. However, the heat generation due to these effects represents only a few percent of the total heat transported at the hot shoe. Note that assuming constant material properties not only causes the term for Thomson heating ($\tau\, \partial T/\partial x$) in the temperature distribution governing equation to drop out, but also results in different hot and cold shoe temperatures and, consequently, a different conversion efficiency.

Figure 5 also shows that as the thermal input power to the TE generator increased, the hot shoe temperature increased, but at a continuously decreasing rate. The cold shoe temperature also increased as the thermal input power increased, but at a slower rate than the hot shoe temperature, thus causing the temperature differential across the TE generator to increase. The slow rise in the cold shoe temperature with input power is caused by the increased heat rejection rate by radiation into space as the cold shoe temperature increases; heat rejection increases proportionally to the fourth power of the cold shoe temperature. The increasing temperature differential and hot shoe temperature cause the conversion efficiency to increase as the thermal power input to the TE generator increases (see Figure 6). Note that the TE conversion efficiency increases nonlinearly with the thermal input power. In addition, the ambient temperature slightly affects the TE hot shoe and cold shoe temperatures but, has a negligible effect on the TE conversion efficiency.

The results presented in Figure 6 also show that for a thermal input power less than approximately 8 W per TE generator, the electric power output is highly nonlinear with the thermal input power; it increases slowly, but at an increasing rate, as the thermal input power increases. However, for a thermal input power higher than 8 W per TE generator, the electric power output increases almost linearly with the thermal input power.

Transient Analysis

The calculations for the transient behavior of the TE generators were performed for two types of input power changes: step and ramp input power changes. The initial temperature distribution in the TE generator was the steady state temperature distribution corresponding to a thermal input power of 5.79 W. As in the steady state analysis, the cold shoe of the TE gen-

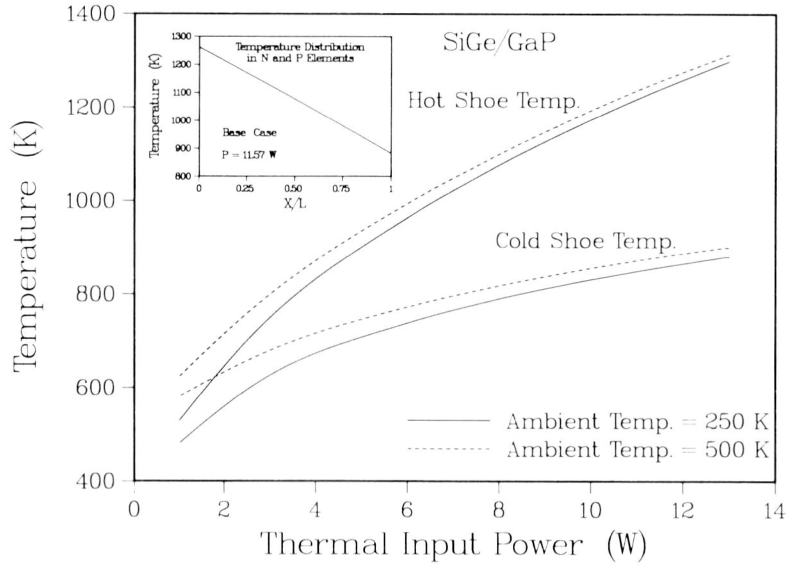

Figure 5 *Predicted Temperatures at the Hot and Cold Shoes of a SiGe/GaP TE Generator as functions of Input Thermal Power.*

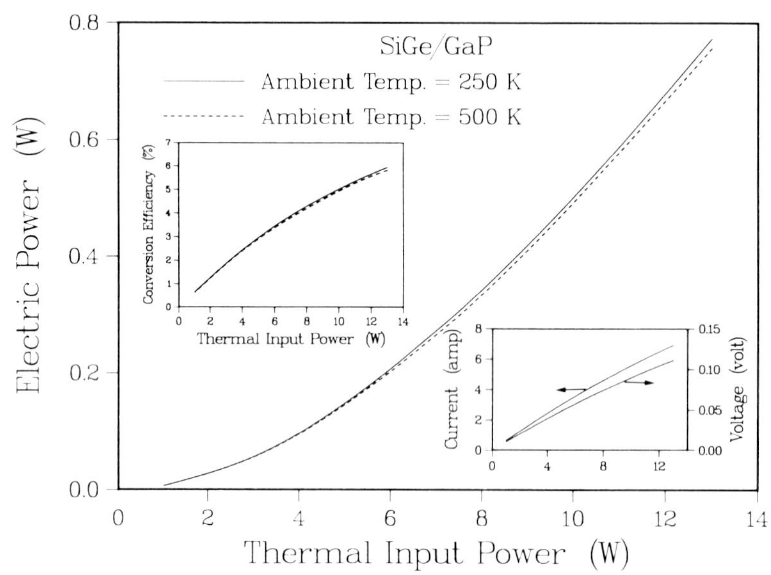

Figure 6 *Electrical Power and Conversion Efficiency of a SiGe/GaP TE Generator as functions of Input Thermal Power.*

erator was cooled by radiation into space with an ambient temperature of 250 K.

Step Input Power Changes

A step input power to the TE generator is unlikely to occur during the operation of a space nuclear power system because of the damping effects of the coolant and of the components in the primary coolant loop between the nuclear reactor and the main TE energy converter. However, to understand the basic transient behavior of the TE generator, a step increase in the input power to 200 percent (11.57 W) of the initial input power was introduced to the TE model. This input power was maintained constant for 100 seconds to allow the TE generator to reach steady state and was followed by a step decrease to its initial value (see Figure 7a). As delineated in Figure 7a, the electric power output increased immediately approaching steady state 32 seconds after the initiation of the transient. However, the temperatures of hot and cold shoes followed the change in the input power with a time lag (Figure 7b). The hot shoe temperature increased sharply due to the immediate effect of the increased input power. Note that only about 30% of the total thermal power input is transferred immediately from the hot shoe to the cold shoe of the TE generator due

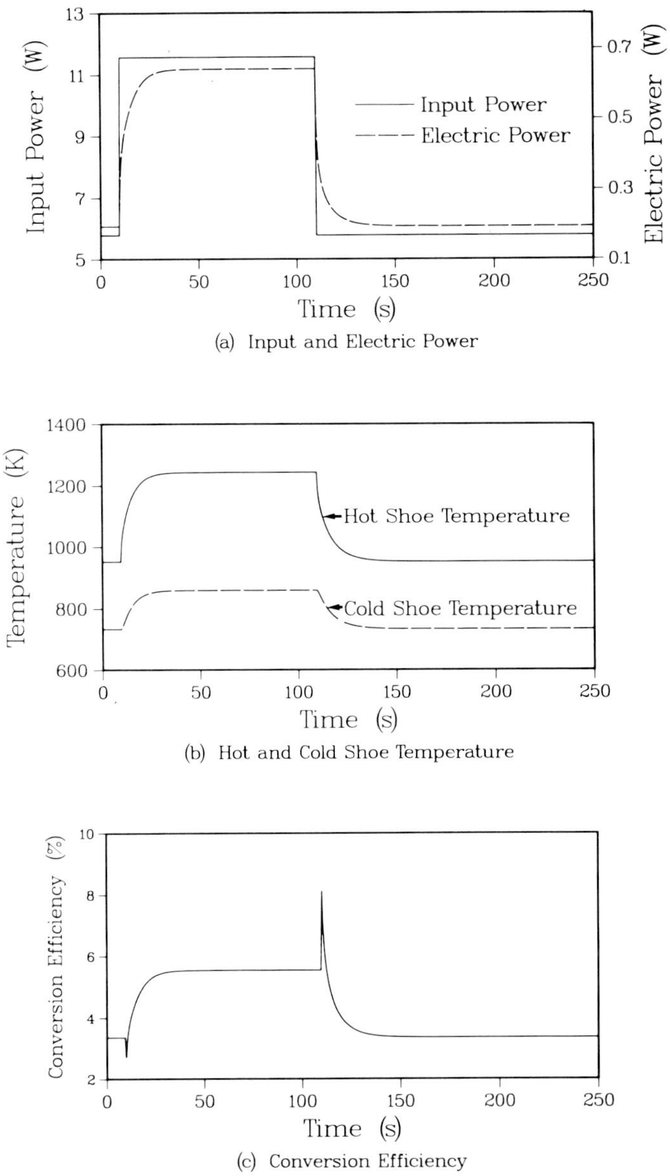

Figure 7 *Transient Response of Operating Parameters of a SiGe/GaP TE Generator following a Step Change in the Input Thermal Power.*

to the Peltier effect (Raag 1969). The other 70% of the thermal input power is controlled by the transient conduction through the *P* and *N* legs of the TE generator. As a result, although the cold shoe temperature started to increase simultaneously with the hot shoe temperature, it increased at a slower rate due to the time delay caused by the accumulation of thermal energy as a sensible heat in the TE generator. The Peltier effect also was responsible for the immediate increase in the electric power output following the step increase in the input power.

As shown in Figure 7c, the TE conversion efficiency dropped sharply immediately after the step increase of the input power, but sharply increased right after the step decrease of the input power. Following the step increase in the input power, the input thermal energy was partially stored in the the TE generator as a sensible heat. This, in turn, reduced the fraction of the input thermal energy being converted into electricity and consequently caused the TE conversion efficiency to initially drop as shown in Figure 7c. As the temperature of TE element increased, the fraction of the input thermal energy being converted into electricity increased as did the TE conversion efficiency. Eventually, the conversion efficiency approached its steady state value of 5.54 percent (initially, the conversion efficiency was only 3.3 percent). A similar, but opposite, response of the TE conversion efficiency occurred following the step decrease in the input power to its initial value.

Ramp Input Power Changes

The TE generators in a space nuclear power system are more likely to be subjected to a ramp input power change, rather than a step input power change. As shown in Figure 8, the input thermal power was increased to 200 percent of its initial steady state value over a period of 25 seconds, followed by a holding period of 50 seconds. Finally, the input power was decreased to its initial value over a period of 25 seconds. As shown in Figure 8, the responses of all output parameters (temperatures, conversion efficiency, and eelectric power) followed closely the change in the input power but at slower rates. The main difference between the step and ramp changes in the input power was the response of the TE conversion efficiency. The change in the energy stored in the TE element during the ramp increase (or decrease) of the input power was smaller than that experienced during the step input power change, resulting in a smaller and smoother drop (or peak) in the conversion efficiency in the former.

Summary and Conclusions

The research developed a dynamic model for analyzing the steady state and transient performance of TE generators. This model takes into account the change in the physical properties of the TE material with temperature and is also capable of handling any combination of boundary conditions at the hot and cold shoes. These capabilities enabled the TE model to be incorporated into the integrated SP-100 system modeling currently being developed at the University of New Mexico's Institute for Space Nuclear Power

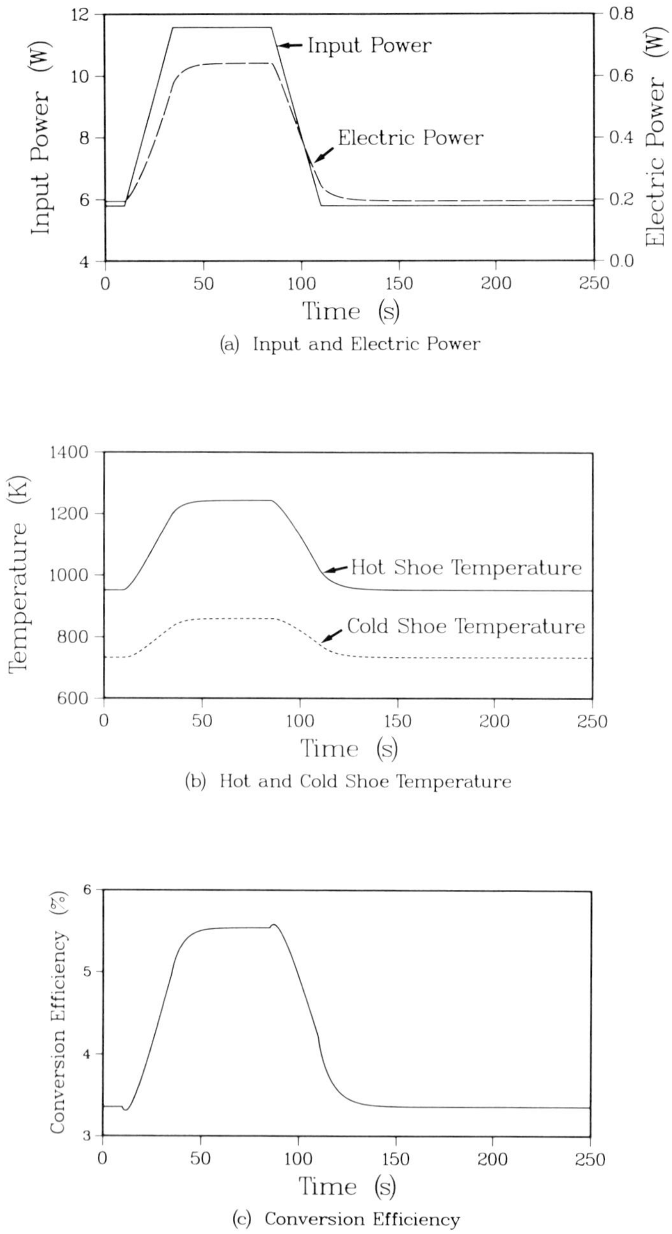

Figure 8 *Transient Response of Operating Parameters of a SiGe/GaP TE Generator following a Ramp Change in the Input Thermal Power.*

Studies. An exact analytical solution is not attainable for higher order boundary conditions with temperature-dependent material properties. Therefore, the model uses the finite element method coupled to a backward difference time integration algorithm to solve for the temperature distribution and the TE energy conversion efficiency.

The accuracy of the model was verified by comparing its predictions with the available exact analytical solutions of problems with constant material properties and simplified boundary conditions. For transient operation with isothermal hot and cold shoes, the model predictions of the temperature distribution and the TE conversion efficiency for a SiGe/GaP TE generator were in excellent agreement with the exact solutions. Model predictions were also in good agreement with the exact solution of the problem of transient operation with an isoflux hot shoe and an isothermal cold shoe. Additional verifications of the model predictions were carried out using an integral energy balance on the TE generator. The results of the energy balance approach were always in excellent agreement with exact solutions. These agreements with the exact solutions established confidence not only in the method of solution, but also in the results for those cases where an exact analytical solution is not attainable. Such cases include transient operation with temperature-dependent TE material properties and time-dependent input heat flux and a radiative cooling boundary conditions.

A steady state analysis of SiGe/GaP TE generators was also conducted. In the analysis, the hot shoe was subjected to a constant heat flux input and the cold shoe was radiatively cooled. The results showed that the TE conversion efficiency and the electric power output increase nonlinearly with increasing the thermal input power. The electric power increased slowly with increasing the thermal input power up to 8 W/TE generator, but increased rapidly at high input power values. These results strongly suggest that an analysis of the TE SP-100 system based on a linear extrapolation of the TE conversion efficiency could result in erroneous predictions of the overall system's performance.

Following a step increase in the input thermal power, the temperatures of the hot and cold shoes increase, but at different rates. Both the cold shoe temperature and the electric power output begin to increase immediately after the initiation of the transient because of the Peltier effect. However, the thermal energy stored in the TE element as sensible heat reduces the fraction of the thermal energy converted into electricity, causing the TE conversion efficiency to drop initially. Subsequently, the conversion efficiency increases approaching a higher asymtotic value. A similar but opposite response occurred following the step decrease of the input thermal power. For a ramp input power change, the response of all output parameters was slower than that experienced with the step input power change to the same peak thermal input power.

Acknowledgments

This work was supported by the Nuclear Technology Office of the Air Force Weapons Laboratory, under Contract No. FI29601-84-C-0080 to the University of New Mexico.

References

Angrist, S.W. (1965) *Direct Energy Conversion*, Allyn and Bacon, Inc., Boston, MA.

Bathe, K.J. (1982) *Finite Element Procedures in Engineering Analysis*, Prentice-Hall, Inc., Englewood Cliffs, NJ.

Buksa, J.J. (1986) *A Self-Induced Thermoelectric-Electromagnetic Pump and Decay Heat Thermal Management in a SP-100 System*, M.S. Thesis, Chemical and Nuclear Engineering Department, The University of New Mexico, Albuquerque, NM, May 1986.

Carslaw, H.S. and J.C. Jaeger (1973) *Conduction of Heat in Solids*, Oxford University Press.

Cobble, M.H. (1982) "Transient Analysis of a Thermoelectric Generator," in *Proceedings of the 4th International Conference on Thermoelectric Energy Conversion*, held in Arlington, TX, March 1982.

El-Genk, M.S. and J.T. Seo (1985) "Trends and Limits in the Upgrading of SP-100 Nuclear Powered Space System," in *Trans. 2nd Symposium on Space Nuclear Power Systems*, CONF-850103-Summs, held in Albuquerque, NM, 14-16 January 1985.

El Genk, M.S. and J.T. Seo (1986) "SNPSAM - Space Nuclear Power System Analysis Model," in *Trans. 3rd Symposium on Space Nuclear Power Systems*, CONF-860102-Summs, held in Albuquerque, NM, 13-16 January 1986.

General Electric Company (1983) *SP-100 Conceptual Design Study, Final Report, Volume II - Reactor and Shield Design*, 83SDS4261, General Electric Company, Philadelphia, PA.

Johnson, E. (1985) "The Forming of Metal Components for Radioisotope Heat Sources," in *Space Nuclear Power Systems 1984*, M.S. El-Genk and M.D. Hoover, eds., Orbit Book Co., Malabar, FL, vol. 2, 341-347.

Raag, V. (1969) *Optimization of Thermoelectric Generators for Fixed Temperature and Fixed Heat Input Operation*, Memorandum #4, Jet Propulsion Laboratory, Pasadena, CA.

Seo, J.T. (1986) *SNPSAM - Space Nuclear Power System Analysis Model*, M.S. Thesis, Chemical and Nuclear Engineering Department, The University of New Mexico, Albuquerque, NM, May 1986.

Soo, S.L. (1968) *Direct Energy Conversion*, Prentice-Hall Inc., Englewood Cliffs, NJ.

Stapfer, G. (1985) *Personal Communication*, Jet Propulsion Laboratory, Pasadena, CA, March 1985.

APPENDIX A

Exact Solution for Transient Heat Conduction Problem with Isothermal Boundary Conditions

The governing equation and the boundary and initial conditions for the transient heat conduction problem in a TE generator with constant TE material properties are

$$\frac{1}{A}\frac{\partial}{\partial x}\left(kA\frac{\partial T}{\partial x}\right) + \rho J^2 + \tau J\frac{\partial T}{\partial x} = wc_p\frac{\partial T}{\partial t}\,, \qquad \text{(A-1)}$$

and

$$\begin{aligned} T(0,t) &= T_h\,, \\ T(L,t) &= T_c\,, \qquad \text{(A-2)} \\ T(x,0) &= T_0 = \text{constant}\,. \end{aligned}$$

The exact solution for the temperature distribution was given as (Cobble 1982)

$$\begin{aligned} T(x,t) &= \frac{T_h e^{\frac{-\mu}{L}x}\sinh\zeta(L-x) + T_c e^{\frac{\mu}{2}(L-x)}\sinh\zeta x}{\sinh\zeta L} \\ &+ \frac{2}{L}\sum_{n=1}^{\infty} e^{\frac{-\mu}{2}x}\sin\beta_n x\Bigg[e^{-\xi_n t}\left\{\frac{T_0\beta_n}{\gamma_n^2}\left(1-(-1)^n e^{\frac{\mu}{2}L}\right) - \right. \\ &\left.\frac{T_H\beta_n}{\gamma_n^2}\left(1-(-1)^n e^{\frac{\mu}{2}L}\frac{T_c}{T_h}\right)\right\} + \frac{\beta_n\nu}{\gamma_n^4}\left(1-(-1)^n e^{\frac{\mu}{2}L}\right) \\ &\left(1-e^{-\xi_n t}\right)\Bigg]\,, \qquad \text{(A-3)} \end{aligned}$$

where

$$\mu = \frac{\tau J}{k}\,, \quad \beta_n^2 = \left(\frac{n\pi}{L}\right)^2\,,$$

where

$$n = 1, 2, \ldots\ldots,$$

$$\zeta = \frac{\mu}{2}\,, \qquad \xi_n = a\,\gamma_n^2\,,$$

$$\nu = \frac{\rho J^2}{k}\,, \qquad a = \frac{k}{wc_p}\,,$$

$$\gamma_n^2 = \frac{\mu^2}{4} + \beta_n^2\,.$$

APPENDIX B

Exact Solution for Transient Heat Conduction Problem with Isoflux Hot Shoe and Isothermal Cold Shoe

The governing equation is the same as Equation (A-1), but the initial and boundary conditions are

$$\begin{aligned} q(0,t) &= q_0 t^{m/2}\,, \\ T(L,t) &= T_C = \text{constant}, \qquad \text{(B-1)} \\ T(x,0) &= T_i = \text{constant}\,. \end{aligned}$$

The exact solution was given (Carslaw and Jaeger 1973) as

$$T(x,t) = \frac{2^{m+1} q_0 \kappa^{1/2} t^{(m+1)/2} \Gamma\left(\frac{m}{2}+1\right)}{k} \sum_{n=0}^{\infty} (-1)^n \left[i^{m+1} erfc \left\{ \frac{(2n+1)L - x}{2(\kappa t)^{1/2}} \right\} - i^{m+1} erfc \left\{ \frac{(2n+1)L + x}{2(\kappa t)^{1/2}} \right\} \right], \quad \text{(B-2)}$$

where

$$\kappa = \frac{k}{wc_p},$$

$i^n erfc(x)$ = the complementary error function, and
Γ = gamma function.

Space Nuclear Power Systems 1986
Edited by M. S. El-Genk and M. D. Hoover

Chapter 18

Alkali Metal Thermoelectric Conversion (AMTEC) Technology Status Review

C. P. Bankston

Jet Propulsion Laboratory
California Institute of Technology
Pasadena, CA 91109

The Alkali Metal Thermoelectric Converter (AMTEC) or sodium heat engine has been the subject of experimental and systems investigations to assess its potential and feasibility for several space and terrestrial power applications. AMTEC converts thermal energy directly to electricity operating between a high temperature, liquid sodium reservoir at 900 K to 1400 K and a sodium vapor condenser at 400 K to 800 K. Efficiencies near 20% have been projected for actual devices operating at temperature ratios of 1.5 to 1.7. This paper reviews AMTEC operating principles, technical challenges, and reports on recent electrode research results. The mechanisms responsible for degradation of thin film molybdenum electrodes have now been elucidated and are discussed. Four alternate high power electrode compositions/morphologies have been identified and the results to date for these systems are summarized. Design, assembly and testing of self-contained AMTEC recirculating devices have continued. Emphasis on cell testing is now on electrode lifetime testing, and series, multi-cell operation. No new technical barriers have been identified. Finally, performance projections for AMTEC space power systems are reviewed and updated. The results continue to indicate that successful development of the AMTEC technology would provide a basis for the design and construction of an efficient, high power density space power system.

Introduction

The Alkali Metal Thermoelectric Converter (AMTEC) or sodium heat engine has been the subject of experimental and systems investigations to assess its potential and feasibility for several space and terrestrial power applications. AMTEC is a thermally regenerative electrochemical system operating between a high temperature reservoir at 900 K to 1400 K and a sodium condenser at 400 K to 800 K. Its operation is based on the sodium ion conductor beta″-alumina solid electrolyte (BASE), where thermal to electric conversion efficiencies of 20% to 40% have been predicted for practical systems. An efficiency of 19% has been demonstrated in a single cell laboratory recirculating device operating between 1073 K and 450 K (Hunt et al. 1981). Other AMTEC characteristics that make it attractive for space applications are compactness, light weight, no moving parts, modularity, and long lifetime potential.

The schematic of the AMTEC cycle is given in Figure 1. Sodium is contained in a sealed recirculating loop divided into high and low temperature regions. The high temperature (T_2) region comprises a heated liquid sodium reservoir that is contained by a BASE barrier and separated from the low temperature region by a sodium vapor space. The low temperature region is a sodium condenser. The high and low temperatures establish an effective pressure or chemical activity difference across the BASE, which has an ionic conductivity much larger than its electronic conductivity. Electrical leads make contact with the high temperature liquid sodium (negative electrode) and a (positive) porous electrode on the low pressure surface of the BASE. When the circuit is closed, sodium ions are conducted through the BASE due to the pressure difference, while electrons flow to the porous electrode surface through the load producing work. A return line and an electromagnetic pump circulate the sodium working fluid through the AMTEC. For more details on AMTEC operating principles and characteristics see Weber (1974), Hunt et al. (1978), and Cole (1983).

Systems studies (Bankston et al. 1983, Ewell 1983, Bankston et al. 1985, Ewell and Mondt 1985, Schmidt et al. 1987, and Chi et al. 1987) have shown

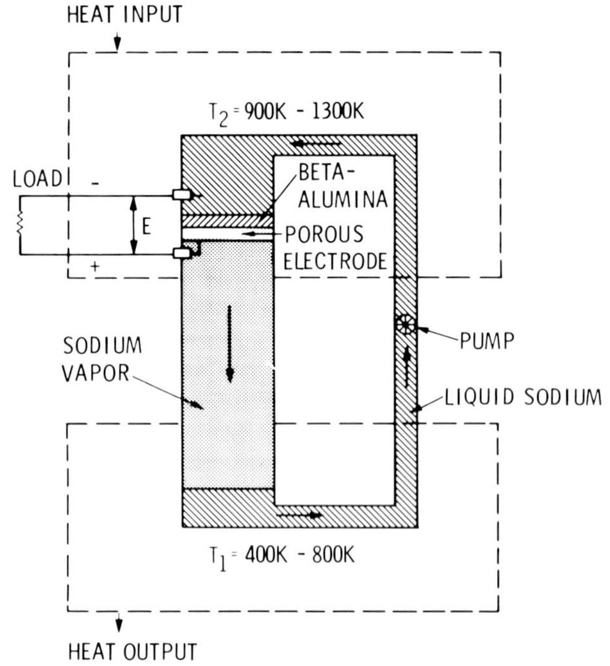

Figure 1 *AMTEC Cycle Schematic.*

that the AMTEC has significant potential benefits for space power applications. Laboratory work has continued on AMTEC technical issues and significant progress has been achieved. This paper briefly reviews AMTEC operating characteristics and technical challenges, reports on recent electrode research results, and summarizes the status of AMTEC device experiments at the Jet Propulsion Laboratory (JPL) and elsewhere. Also, updated projections for AMTEC space nuclear power system characteristics are discussed.

Operating Characteristics

Theoretical voltage versus current characteristics are given by the relationship (Weber 1974 and Cole 1983):

$$V = a - b\ln(I + \delta) - IR\ , \tag{1}$$

where a, b, and δ depend on temperatures T_1 and T_2, and R is the ohmic resistance of the BASE which is dependent on BASE thickness and T_2. The relationships for a, b, δ and R are given in the Nomenclature Section. Typical voltage-current curves for possible space power operating conditions are given in Figure 2. Efficiency is given by

$$\eta = \frac{IV - I^2R_L/N}{IV + (I/F)(L + C_p(T_2 - T_1) + (I^2R_L/2N) + Q_{\text{loss}}} \tag{2}$$

Q_{loss} is the sum of parasitic conductive and radiative heat losses given by

$$Q_{\text{loss}} = Q_{c,l} + Q_{c,s} + (\sigma/Z)(T_2^4 - T_1^4)\ . \tag{3}$$

Z is a radiation reduction factor, which depends on the emissivities of the surfaces and geometric view factors. $Q_{c,l}$ is the conductive electrical lead loss and $Q_{c,s}$ is the structural conductive loss, both depending on $(T_2 - T_1)$. Since AMTEC cells can be series connected in the high temperature zone, $Q_{c,l}$ can be minimized by series connection. Structural conductive losses will be design dependent. A family of effi-

NOMENCLATURE

English

- a: $2.441 \times 10^{-3}\, T_2 - 8.617 \times 10^{-5} T_2 \ln T_2 - 1.1 \times 10^4$, V
- b: $8.617 \times 10^{-5} T_2$, V
- *BASE*: β''-alumina solid electrolyte
- C_p: liquid sodium specific heat (J/g·mole·K)
- F: Faraday, 96,500 C/g·mole
- h: BASE thickness (cm)
- I: AMTEC cell current density (A/cm²)
- L: sodium heat of vaporization (J/g·mole)
- N: number of AMTEC cells in series
- P_f: Final power density (W/cm²)
- P_i Initial power density (W/cm²)
- $Q_{c,l}$: conductive electrical lead loss (W/cm²)
- $Q_{c,s}$: structural conductive loss (W/cm²)
- Q_{loss}: radiative plus conductive parasitic heat loss (W/cm²)
- R: $hT_2\,[4.03 \times 10^{-4} \exp(1420\ \text{K}/T_2) + 3.24 \times 10^{-7} \exp(4725\ \text{K}/T_2)]$, BASE bulk ionic resistivity ($\Omega \cdot$ cm²)
- R_L: AMTEC cell lead resistance ($\Omega \cdot$ cm²)
- T: temperature (K)
- V: AMTEC cell voltage (V)
- Z: radiation reduction factor

Greek

- δ: $2.00 \times 10^{12}\, T_1^{-1} \exp(-12{,}818/T_1)$ (A/cm²)
- η: efficiency
- σ: Stefan-Boltzmann constant, 5.67×10^{-12} W/cm² · K⁴

Subscripts

- 1: Low-temperature reservoir (condenser) condition
- 2: High-temperature reservoir condition

ciency versus current density curves are shown in Figure 3 for four parasitic heat loss levels.

Control of the radiative loss from the hot electrode surface to the condenser is important. For the conditions of Figure 3, Z must be greater than 10 to achieve curve C (assuming all other losses are negligible) and greater than 50 to achieve curve B. Calculations show that by fully using the reflectivity (~ 98%) of liquid sodium on the condenser surface, a Z factor near 50 may be possible assuming optimized geometric (area) view factors. Also, a radiation shield with remote condensing is a possible strategy for increasing the effective Z factor to 50 or greater. To date, laboratory recirculating cells have operated between curves C and D on Figure 3, where Hunt et al. (1981) reported an effective Z of 14.

In addition to minimizing parasitic thermal losses, AMTEC devices will require a high power, long life porous electrode on the vapor side of the BASE. To date, thin film, sputter deposited molybdenum electrodes have achieved 90% of the theoretical power densities predicted from the current-voltage characteristics in Figure 2. However, these power densities usually degrade by a factor of 2 to 5 in the first 1000 hr of operation at AMTEC temperatures. Thus, a major focus of research has been on understanding the degradation mechanisms of molybdenum electrodes and on identifying alternative electrode compositions which exhibit long life, high power operation. As will be discussed, the degradation mechanisms have recently been elucidated, and means to optimize molybdenum electrodes are now being studied.

Electrode Research Status

The AMTEC porous electrode must possess a combination of demanding characteristics including good electronic conductivity, a thermal expansion coefficient near that of BASE, non-reactivity with sodium, very low vapor pressure at AMTEC temperatures and capability of high rates of ionic or diffusive transport of sodium. Efforts have focused primarily on the refractory metals molybdenum and tungsten, and their alloys. Table 1 lists the present electrode systems under study and their performance characteristics. Sputter deposition is the method of choice for the application of molybdenum and tungsten, and their alloys, and titanium nitride. Chemical vapor deposition has also been used for molybdenum and tungsten. Some electrodes have been formed by applying a metal resinate ink followed by an oxidation and reduction cycle. While this technique does produce a porous, moderate to high power electrode, the poor durability of electrodes applied in this manner has been a problem. The cermet electrodes are usually composed of a mixture of BASE and molybdenum powders, and some success has been achieved with flame spray application (Hunt 1986a).

As shown in Table 1, porous, thin film molybdenum electrodes, 1 μm to 3μm thick, exhibit the highest initial power densities of all electrodes considered now or previously. These initial power densities, if sustained for long periods, would provide a basis for the development of high efficiency, high power density space power systems. However, the initial power densities exhibit a relatively sharp decrease in the first hours of operation with a relatively stable plateau (P_f in Table 1) being reached in less than 1000 hours and

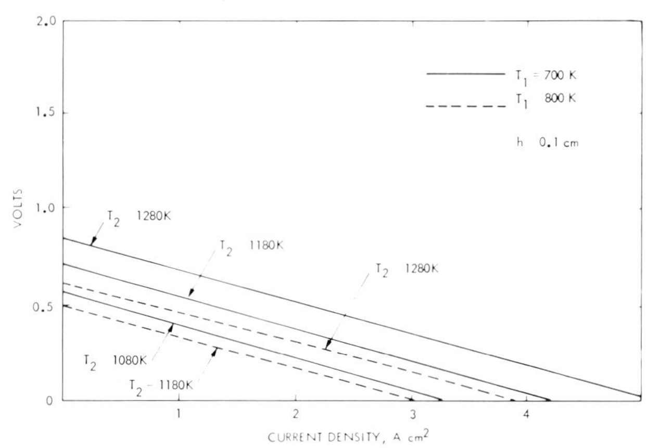

Figure 2 *Typical Voltage versus Current Density Curves for Different Temperature Conditions.*

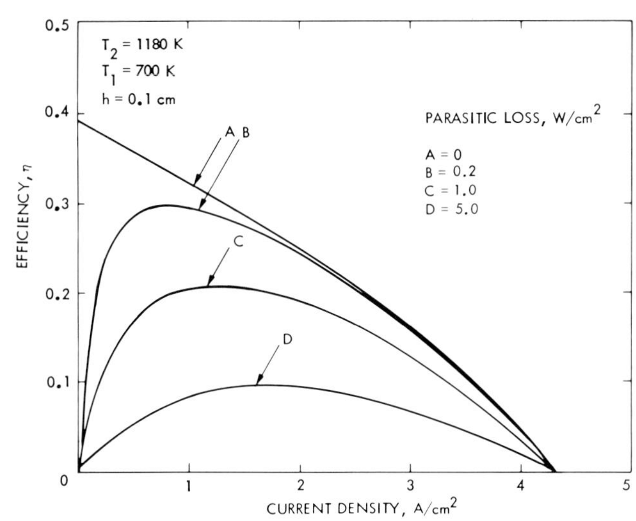

Figure 3 *Efficiency versus Current Density for Different Parasitic Heat Loss Levels.*

often less than 100 hours. Furthermore, the study of the performance of molybdenum electrodes has focused on two other important (previously reported) experimental observations.

First, measured current densities (near 1 A/cm^2) from relatively fresh molydenum electrodes are too large in relationship to the (small) measured pressure drop across the electrode if only sodium vapor flow through the pores of the typical electrode is considered. That is, a much larger pressure drop would be necessary to sustain sodium vapor flow in the pores at the measured levels (Bankston et al. 1985). In fact, the degraded performance of Mo electrodes is consistent with the performance calculated assuming only vapor transport through the electrode. Second, exposure of degraded molybdenum electrodes to oxygen usually results in regeneration of those electrodes to power densities near their original maximum power (depending on the conditions of oxidation (Williams et al. 1986). These observations led to the hypothesis that an unidentified sodium transport mechanism was responsible for enhanced transport in the electrode pores.

Subsequently, Williams et al. (1986) found that the intentional or unintentional exposure of porous molybdenum electrodes to oxygen, followed by drawing current through the electrode under AMTEC conditions, leads to the formation of sodium molybdate, Na_2MoO_4 in the pores. Sodium molybdate is a molten salt and excellent ionic conductor under AMTEC conditions, and the presence of this material in the pores leads to measured, enhanced transport of sodium to near the electrode surface. Thus, the sodium molybdate provides a favored transport path which reduces the effective electrode flow resistance by ~0.5 ohm at 1073 K and 1.0 A/cm^2.

Oxygenated species form in the electrode pores by several reactions, including

$$2\ Mo + 3O_2 \rightarrow 2\ MoO_3\ . \quad (4)$$

This occurs upon exposure to air or oxygen; and also by desorption of water from the BASE.

When current is drawn through the electrode, sodium molybdate (Na_2MoO_4) as well as other reduced compounds are formed:

$$8\ Na + 6\ MoO_3 \rightarrow 3\ Na_2MoO_4 + Na_2Mo_3O_6\ , \quad (5)$$

where $Na_2Mo_3O_6$ is an electronic conductor. The ionic conductivity of sodium molybdate at 1100 K is greater than 1.0 (ohm-cm)$^{-1}$.

Once sodium molybdate is formed in the pores, high power operation is possible for nearly 100 hr. Although sodium molybdate is chemically stable under typical AMTEC operating conditions ($\simeq$ 0.5 V), degradation (Williams et al. 1986) will occur due primarily to the evaporation and loss of Na_2MoO_4. At open circuit, Na_2MoO_4 is also lost by the interconversion reaction:

$$3\ Na_2MoO_4 + 3\ Mo \rightarrow 2\ Na + Na_2Mo_3O_6\ . \quad (6)$$

Thus, when Na_2MoO_4 is lost and power reduced, high power operation can be regenerated by oxygen treatment. However, this treatment will not be permanent and its effectiveness determined by the regeneration conditions. We now estimate that periodic regeneration of Mo electrodes may extend high power operation of 1 μm to 2 μm thick Mo electrodes to the 1000 hr range. Further lifetime testing and studies of regeneration conditions are needed to determine the ultimate potential of these electrodes. In particular, more lifetime testing under load conditions is needed since this is the most electrochemically stable situation for these electrodes. The lifetime of periodically regenerated electrodes will ultimately be limited by the irreversible loss of molydenum followed by an increase in electronic sheet resistance.

Table 1 *Electrode Compositions Under Study.*

	T(K)	P_i(W/cm^2)	P_f(W/cm^2)	hr
Mo, 1 μm to 3 μm	1175	0.8 - 1.0	~0.1 - 0.3	~200
Mo, 1 μm to 3 μm	1075	~0.7		
Mo, 1 μm to 3 μm	975	~0.4		
Mo, ≤ 1 μm	1200	0.5	0.4	90
Pt or Rh with W	1100	0.55	?	?
TiN	1173	0.45	0.23	458
TiN	1073	0.52	0.47	246
Cermet	1190	0.5	0.27	40

An alternate molybdenum electrode concept is now under study at JPL. It is composed of a very thin ($\leq 1.0\ \mu m$) sputter deposited layer with a heavy overlying current collection grid. Such an electrode is intended to function without transport enhancement. The very thin film is intended to minimize vapor flow pressure drop while the overlying grid compensates for the increase in sheet resistance. The grid must be very porous ($> 50\%$) to prevent voltage loss due to additional flow resistance. Typical sputter deposited Mo films are only 20% to 25% porous. Figure 4 presents initial results for two very thin grid electrodes tested on a 1.5-cm diameter BASE tube. These are the highest stable power densities yet achieved for molybdenum electrodes. Subsequent model calculations indicate that an optimized film/grid system may be capable of producing $> 0.6\ W/cm^2$ at 1150 K to 1200 K (or $> 1/2$ theoretical). This system is now undergoing detailed study and further lifetime testing.

A second alternate electrode system now being studied at JPL uses either a W/Pt or W/Rh electrode for which initial results are presented in Figure 5. This plot compares the performance of W/Rh deposited by application of resinate inks and W/Pt sputter deposited with W, Pt, and Rh alone. These data show a power density of approximately $0.6\ W/cm^2$ at 1100 K. This compares to a maximum theoretical power density of $0.8\ W/cm^2$ and compares favorably to the highest power density measured for Mo electrodes at this temperature. The mechanism of enhanced transport for these binary systems is not yet understood and lifetime data are needed. However, the measured power densities for W/Rh and W/Pt are higher than virtually all other electrode systems previously tested except molybdenum.

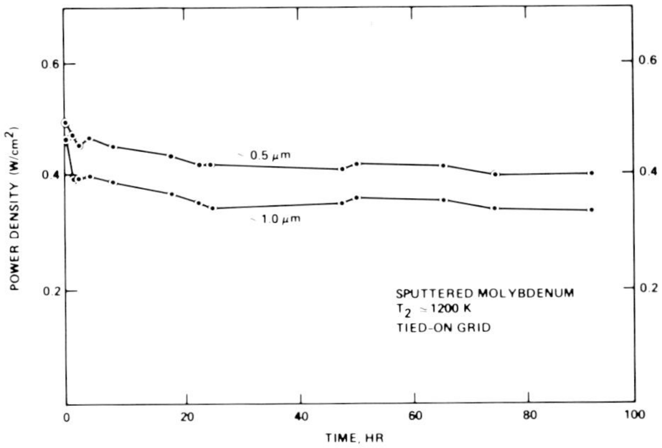

Figure 4 *Maximum Power Densities of Thin, Porous Molybdenum Electrodes with Tied-On Current Collector Grids.*

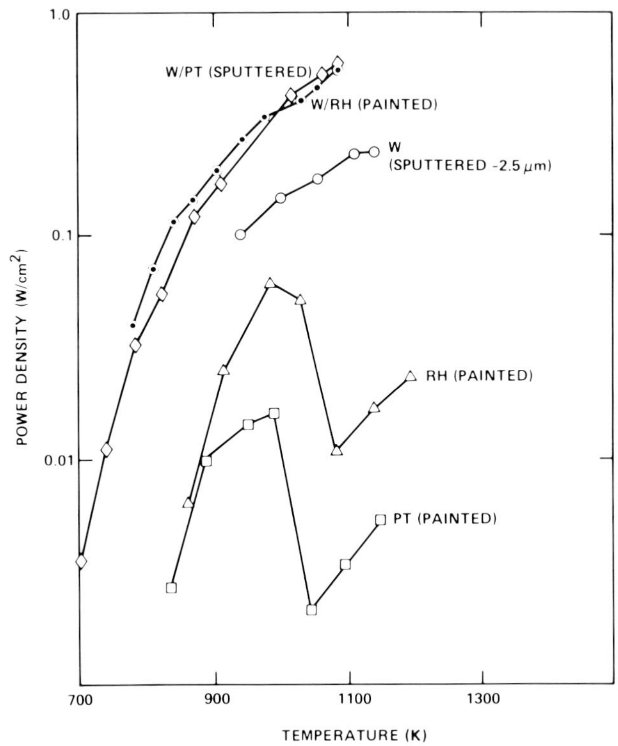

Figure 5 *Initial Power Densities versus Temperature for Tungsten and Noble Metal Electrodes.*

Two electrode compositions (Hunt 1986a) under study at Ford Motor Company's Dearborn Research and Engineering Laboratories may have potential in space power applications. Reactively sputtered, porous titanium nitride (TiN) electrodes, $1\ \mu m$ to $3\ \mu m$ thick, were deposited on BASE tubes and studied at temperatures up to approximately 1200 K. Titanium is chemically active and is often used as a constituent in metal to ceramic (BASE) brazes. Titanium based electrodes are expected to have extremely good adherence to the BASE substrate, although the electronic sheet resistance of these electrodes may exceed that of molybdenum films by a factor of 5 depending on the final film morphology. At 1173 K, a TiN electrode with initial power density of $0.52\ W/cm^2$, degraded by 10% in 246 hr with relatively flat power density at the end of test. At 1273 K, another TiN electrode with initial power density of $0.45\ W/cm^2$, degraded by a factor of 2 in 458 hr. To date, the mechanisms responsible for the performance of TiN electrodes are not yet known, however their initial power densities approach high power ($> 0.5\ W/cm^2$) operation and further studies are warranted. In particular, it was found that electrode performance is very sensitive to sputtering conditions.

Hunt has also investigated flame sprayed cermet

electrodes formulated with BASE and molybdenum powders (Hunt 1986a). This system forms a porous ionic and electronically conducting phase with a very high interfacial surface area between the BASE and the electronic charge carrier. The cermets were usually composed of 50% to 85% BASE powder and 50% to 15% molybdenum. A 55% BASE/45% Mo composition achieved a power density of approximately 0.5 W/cm^2 at 1170 K to 1190 K. This electrode degraded by a factor of 2 after 40 hr at high temperature, with power density being nearly constant for the last 20 hr. The Ford results indicate that cermets have potential for AMTEC operation; however, more research is needed to optimize cermet composition and morphology. Initial experiments at JPL have confirmed the sensitivity of cermet performance to the formulation and deposition processes.

Finally, AMTEC area power densities could be increased by increasing the active surface area of the ceramic by corrugation or grooves; a practice used in fuel cell technology. A technique for forming the BASE into more complex shapes has been demonstrated at Ford, however it has not been developed further.

Recirculating Cell Testing

The previously discussed electrode research experiments were typically conducted in demountable, electrode test cells. Such cells are designed specifically for electrode studies over relatively short periods with maximum instrumentation and flexibility. Run times are usually limited by the volume of sodium that can be loaded into a BASE tube (15 g), and sodium cannot be recirculated after condensing in the oversized vacuum chambers used for these experiments. Thus, parallel efforts have been carried out to design, fabricate and test self-contained recirculating devices in order to conduct long lifetime tests of electrodes and to demonstrate fabrication procedures, operating characteristics, and efficiencies in actual AMTEC devices.

Figure 6 shows a diagram of a JPL designed recirculating cell that will be utilized for long lifetime electrode testing. This design has several characteristics that are different from other recirculating cells designed and built at Ford. However the overall geometry is the same, and it incorporates the essential features typical of all recirculating cells. The containment vessel is a 3.81-cm (1.5-inch) I. D. copper tube, which also serves as the sodium condenser. A 1.5-cm diameter, 30-cm long BASE tube with 1.2-mm wall thickness contains the sodium reservoir that is heated by an electrical resistance heater inserted into a stainless steel heater well. The heater well extends the entire length of the BASE tube.

The top flange contains two insulated 50 A electrical feed-throughs capable of withstanding sodium exposure; thus, two electrodes can be tested simultaneously. The sodium feed line and a thermocouple well enter the containment vessel via a 1.59-cm (⅝-inch) collar in the top flange, which also supports the heater well. A gas vent connected to the sodium reservoir is also mounted in the top flange. The BASE tube is demountable and attached to the heater well by a modified, stainless steel miniconflat flange with a metal O-ring seal. The BASE-to-stainless steel transition is accomplished with a molybdenum transition sleeve sealed to the BASE tube with a 70% Ti/15%Cu/15%Ni braze, which is, in turn, sealed to a stainless steel sleeve with a nickel (73%)-chromium (14%) based braze. The copper containment tube/condenser is air cooled. A small electromagnetic pump will be utilized to recirculate the sodium. This recirculating cell is presently being put into operation at JPL.

A similar recirculating cell design at Ford (Hunt et

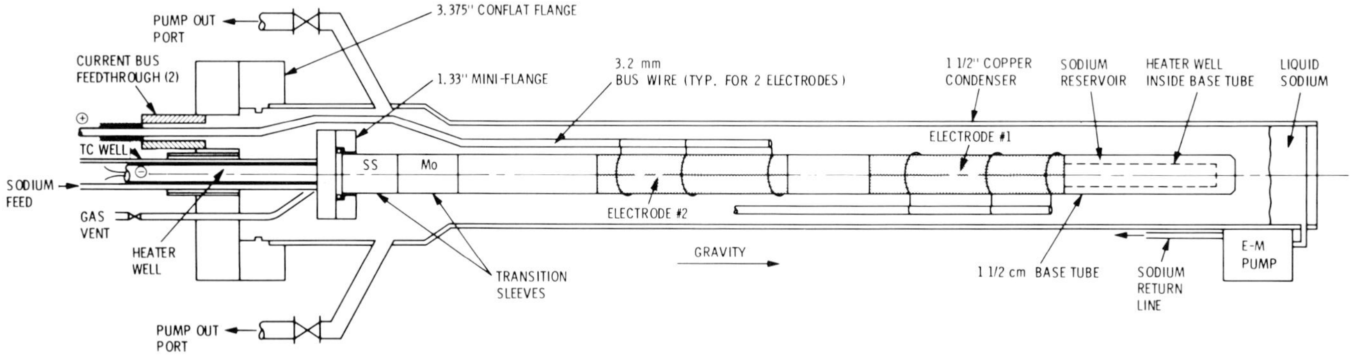

Figure 6 *AMTEC Recirculating Test Cell.*

al. 1981) has achieved a 19% thermal to electric conversion efficiency at 25 W. End-condensed cells have also been tested at Ford (Hunt 1982). In this design the internal heater well (see Figure 6) is removed and the sodium reservoir is heated radiatively by externally heating the containment vessel (stainless steel). A sodium vapor flow path is then provided to a remote location in a cooled zone where sodium condensation occurs for recirculation. A device of this design has achieved more than 10,000 hr at high temperature (above 1050 K) with a relatively low power cermet electrode (Hunt 1986b). (See Hunt 1982 for complete details of the Ford devices.) More recently, the Department of Energy's Sandia National Laboratories in Albuquerque has initiated assembly of a recirculating bench test AMTEC module (Martinez 1986). This is a new design which has the porous electrode deposited on the *interior* of the BASE tube. The BASE tube is immersed in the sodium reservoir and heated externally. This experiment will be used to study design concepts proposed for terrestrial solar thermal applications and it is being assembled at this writing.

A new multi-cell recirculating unit has also been recently placed into operation at Ford, Dearborn (Hunt 1986b). This cell is designed to accommodate six BASE tubes circumferentially mounted inside an annular containment vessel. The inner wall of the annulus is a heater well that heats the BASE tubes containing sodium by radiative heat transfer. The outer wall is cooled and is the sodium condenser. This device has demonstrated series operation of two cells and has produced a total power output of 112 W in multi-cell operation to date. Also, this device has been operated using combustion heating, thus becoming the first AMTEC device to be integrated with a practical heat source.

To date, no major technical issues have arisen in the fabrication, assembly and testing of AMTEC recirculating cells. However, control of radiative parasitic losses and a properly designed current bus network to minimize resistive losses are two important factors in cell design, operation and performance. Also, high temperature BASE-to-metal and metal-to-metal seal techniques may require further development depending upon the design or application considered. In summary, further progress has been achieved in the assembly and operation of self-contained laboratory AMTEC devices. Utilization of these devices will continue to emphasize electrode lifetime testing, design methods to minimize parasitic losses for optimum efficiency, and demonstration of series, multi-cell operation at increasing power levels. For space applications, designs for condensing, collecting and recirculating sodium in zero-g conditions must be identified and tested.

Space Power Systems

Previous studies (Bankston et al. 1983 and 1985 and Ewell 1983) demonstrated that projected AMTEC performance and operating characteristics make it a candidate for advanced space power systems from the one kilowatt range to megawatts. The results of the Bankston et al. studies (1983 and 1985) have recently been updated.

During 1985, JPL conducted an in house study (Awaya 1985) of power source options for unmanned planetary space exploration missions in the next 20 to 40 years. The power level used for comparison was 500 W with a 7-year lifetime, based on the requirements for 20 inner planet, outer planet, and small body exploration missions. In this study, AMTEC small power system performance projections of Bankston et al. (1983) were reviewed and considered with other factors, in comparison with thermoelectric, thermionic, and dynamic cycle conversion technologies. The criteria used for evaluation were relative costs, mass, dimensional compatibility with the spacecraft and mission, ability to meet a 98% reliability goal assuming successful development, modularity, effects on spacecraft systems, and the risk associated with developing the technology and achieving projected performance characteristics. At 500 W, the AMTEC converter would operate between temperatures of 1150 K and 600 K at 20% efficiency, and its mass would be 28 kg. It projected the highest ranking of all potential new radioisotope powered systems considered for the time frame of interest.

In the 100+ kW range, the study of Bankston et al. (1985) was extended to consider conductively coupled, liquid metal heat transfer from reactor to AMTEC conversion system. The results are given in Table 2, with a summary of the previously considered radiatively coupled projections (Bankston et al. 1985) and the reference SP-100 system, all at 100 kWe. The previous SP-100 AMTEC system was based on a UO_2 fueled reactor cooled by 120 lithium heat pipes that delivered their energy to AMTEC converters by radiation. In the revised projections for a conductively coupled system we have taken the baseline SP-100 liquid lithium reactor cooling system and replaced the thermoelectric converters with AMTEC cells in the same basic design concept as considered previously. The result is that the conductively coupled AMTEC projections increase system efficiencies by 2 to 5 percentage points under the same operating conditions and assumptions as the radiatively coupled projections. This is due to a reduction in parasitic losses due to a mismatch in AMTEC cell area with the hot shoe collector area required in a radiatively coupled system. That is, in the radiatively coupled design, a relatively small fraction of the required hot shoe area is covered with AMTEC cells, resulting in significant (increased) parasitic radiation losses from hot side to condenser. In a conductively coupled system, these parasitic losses are reduced and efficiencies are increased. For heat rejection, the AMTEC sodium condenser can be coupled with a heat pipe radiator; however, an integrated condenser/radiator is possible depending upon final design and performance requirements.

The operating regimes for the AMTEC system were also extended to consider the SP-100 baseline converter temperature (1260 K) at two radiator temperatures (600 K and 700 K, Table 2). The results show that further reductions in reactor thermal power and radiator area are possible. The projected radiator areas are less than one-half the area for the reference SP-100 system. Thus, a conductively coupled AMTEC space nuclear power system in the 100+ kWe range could reduce the reactor thermal power and lithium flow rates. Depending on operating temperatures, the option would exist to either reduce reactor temperature or eliminate the need for the deployable radiator envisioned for the SP-100 system. Also, the AMTEC system is a potentially low mass system as shown in Table 2.

Finally, note that the system projections in Table 2 will require an AMTEC development program of several years prior to entering the engineering design phase. To achieve the system efficiencies predicted, the combination of porous electrode and current collection network must approach the initial performance of state of the art molybdenum electrodes 1 μm to 3 μm thick, and the rate of degradation must be very small for long lifetime operation. The final electrode performance needed will depend on factors such as the ability to control parasitic thermal losses in a specific design, or the effectiveness of increasing the active surface area of the BASE as discussed earlier. A thorough study of potential design concepts is needed to resolve these, as well as other issues including zero-g effects, start up/transients, shelf life, and redundancy.

Concluding Remarks

This paper has described recent important progress in research and development on the Alkali Metal Thermoelectric Converter (AMTEC). Specifically, a detailed understanding of the mechanisms responsible for the initial high power performance and subsequent degradation of thin film molybdenum electrodes has been achieved. Based on these findings, work is proceeding on 1) very thin molybdenum, 2) binary platinum or rhodium with tungsten compositions, 3) titanium nitride, and 4) cermet electrode systems which have shown promise for long life, high power operation. Separate, self-contained recirculating AMTEC cells have now demonstrated 19% efficiency, 10,000 hr at high temperature, and multi-cell operation including series connection. A review and update of systems studies in the 0.5 kWe to 100 kWe range still show that AMTEC performance and oper-

Table 2 *AMTEC SP-100 Systems Comparisons.*

	Converter Temperature (K)	Radiator Temperature (K)	Efficiency (%)	Reactor Power (kWt)	Radiator Area (m^2)	Cell Area (m^2)	Mass (kg)
Reference Thermoelectric System	1260	825	4.9	2400	65	6	3000
Radiatively Coupled AMTEC[a]	1080–1180	700	12–16	950–710	60–45	35–15	<2500
Conductively Coupled AMTEC	1080–1180	700	14–21	815–540	50–30	35–15	<2500
"	1260	700	25	455	25	10	<2000
"	1260	800	18	630	20	20	<2000

[a] "Improved/advanced" technologies from Bankston et al. (1985).

ating characteristics make it a strong candidate for future space applications. Similar results have been obtained in a recent study of the AMTEC in a multi-megawatt system (Schmidt et al. 1987 and Chi et al. 1987). Development of a suitable electrode system should continue to be emphasized to achieve AMTEC performance projections.

Acknowledgments

The research described in this paper was carried out by the Jet Propulsion Laboratory, California Institute of Technology, and was sponsored by the National Aeronautics and Space Administration, U. S. Army Tank-Automotive Command, the U. S. Department of Energy's Sandia National Laboratories, Albuquerque, and the Air Force Office of Scientific Research. The author would like to thank Mr. R. E. Thompson of JPL for his assistance in updating the 100-kWe system projections.

References

Awaya H. (1985) "Evaluation of Space Power Supply Options," in-house study, Jet Propulsion Laboratory, Pasadena, CA.

Bankston, C. P., T. Cole, R. Jones, and R. Ewell (1983) "Experimental and Systems Studies of the Alkali Metal Thermoelectric Converter for Aerospace Power," *J. Energy*, 7(5):442.

Bankston, C. P., T. Cole, S. K. Khanna, and A. P. Thakoor (1985) "Alkali Metal Thermoelectric Conversion (AMTEC) for Space Nuclear Power Systems," in *Space Nuclear Power Systems 1984*, M. S. El-Genk and M. D. Hoover eds., Orbit Book Co., Malabar, FL, p. 393.

Chi, J. W. H, K. Kesavan, R. E. Morgan, and J. E. Schmidt (1987) "Comparison of Multi-Megawatt Space Nuclear Power Systems," in *Space Nuclear Power Systems 1986*, M. S. El-Genk and M. D. Hoover eds., Orbit Book Co., Malabar, Fl.

Cole, T. (1983) "Thermoelectric Energy Conversion with Solid Electrolytes," *Science*, 221(4614):915.

Ewell, R. (1983) "Energy Conversion for Megawatt Space Power Systems," in *Proceedings of the 18th Intersociety Conversion Engineering Conference*, AIChE, New York, p. 87.

Ewell, R. and J. Mondt (1985) "Static Conversion Systems," in *Space Nuclear Power Systems 1984*, M. S. El-Genk and M. D. Hoover eds., Orbit Book Co., Malabar, FL, p. 385.

Hunt, T. K., N. Weber, and T. Cole (1978) "Research on the Sodium Heat Engine," in *Proceedings of the 13th Intersociety Energy Conversion Engineering Conference*, SAE, Warrendale, PA, p. 2011.

Hunt, T. K., N. Weber, and T. Cole (1981) "High Efficiency Thermoelectric Conversion with Beta" -Alumina Solid Electrolytes, The Sodium Heat Engine," in *Fast Ionic Transport in Solids*, J. B. Bates and G. C. Farrington eds., North-Holland Publishing Co., Amsterdam, p. 263.

Hunt, T. K. (1982) *Research and Development Program on a Sodium Heat Engine, Final Report*, U. S. Department of Energy Contract No. DE-AC02-79ER10347, Ford Motor Company, Dearborn, MI, October 1982.

Hunt, T. K. (1986a) *Research on Materials Related to the Sodium Heat Engine: Final Report,* Report No. LBL-21232, Lawrence Berkeley Laboratory, Berkeley, CA, March 1986.

Hunt, T. K. (1986b) *Private Communication*, Ford Research and Engineering Staff, Ford Motor Co., Dearborn, MI.

Martinez, J. I. (1986) *Private Communication*, Sandia National Laboratories, Albuquerque, NM.

Schmidt, J. E., J. W. H. Chi, R. E. Morgan, J. D. Hanson, and T. K. Hunt (1987) "A Liquid Metal Cooled Reactor - Alkali Metal Thermoelectric Space Power System Concept for Multi-Megawatt Applications," in *Space Nuclear Power Systems 1986*, M. S. El-Genk and M. D. Hoover eds., Orbit Book Co., Malabar, FL.

Weber, N. (1974) "A Thermoelectric Device Based on Beta-Alumina," *Energy Conversion*, 14:1.

Williams, R. M., G. Nagasubramanian, S. K. Khanna, C. P. Bankston, A. P. Thakoor, and T. Cole (1986) "The Role of Oxygen in Porous Molybdenum Electrodes for the Alkali Metal Thermoelectric Converter," *J. Electrochem. Soc.*, 133(8):1587.

Space Nuclear Power Systems 1986
Edited by M. S. El-Genk and M. D. Hoover

Chapter 19

Monolithic Fuel Cell Based Power Source for Burst Power Generation

D. C. Fee, S. K. Bhattacharyya, D. E. Busch, L. W. Carlson, T. D. Claar, D. W. Dees, J. Dusek, T. E. Easler, W. A. Ellingson, B. K. Flandermeyer, R. J. Fousek, H. K. Geyer, J. J. Heiberger, K. D. Kuczen, S. Majumdar, C. C. McPheeters, F. C. Mrazek, J. J. Picciolo, and R. B. Poeppel

Argonne National Laboratory
Argonne, IL 60439-4837

A unique fuel cell coupled with a low power nuclear reactor presents an attractive approach for Strategic Defense Initiative (SDI) burst power requirements. The high power, long duration bursts, quoted in the open literature, (100 MWe, 200 s) appear achievable within a single shuttle launch limitation with appropriate development of the concept.

Introduction

A monolithic fuel cell/nuclear reactor system clearly possesses several advantages.

- The system has essentially no moving parts. A static power conversion system (thermoelectric) is used with the reactor (assumed to be the SP-100 reactor). The fuel cell is an all solid state device. Mechanical motion is required only during the power burst for pumping gases through the fuel cell. As a result, no large torques are added to the system. Furthermore, high system reliability is envisioned during extended standby operation over a period of years.
- The system is closed. No effluents are produced that might affect the sensitivity of on-board detectors and surveillance devices during system operation and testing. Complex thrust cancellation devices are not needed.
- The system is regenerative. Full power capability can be recharged following extended bursts under battle or test conditions. The length of the recharge time is determined by the power level of the nuclear reactor.
- The heat rejection requirements are minimal. The fuel cell system is highly efficient. Furthermore, the high operating temperature of the fuel cell (811 K (1000°F)) also reduces the weight and volume of the heat rejection device. This, of course, is the critical point in any closed system burst power operation.

The system performance advantages result from a significant breakthrough in fuel cell design. The monolithic design employs the same thin ceramic components used in other oxide fuel cells in a strong, lightweight honeycomb structure of small cells, and thus can achieve very high power per unit mass or volume. The light weight and low volume, as well as the efficiency and reliability of electrical systems, are advantageous in space applications.

Coupling a nuclear reactor with a monolithic fuel cell appears attractive for space based pulse power systems. It combines nuclear and electrochemical energy conversion to take advantage of their special features. In the preferred configuration, the monolithic fuel cell combines hydrogen and oxygen to produce the power pulse and the product water is stored onboard. Following a test power pulse, the fuel cell is run in reverse to regenerate the hydrogen and oxygen. The nuclear reactor provides the power input for the reverse operation. As a result of the reverse operation, the quantity of reactant which must be put into orbit is significantly reduced. There are other advantageous links between the nuclear reactor and the fuel cell subsystem. For example, reject heat from the reactor may be used to maintain the fuel cell in a "hot standby" mode in order to rapidly respond to demands for power. In addition, the reactor and fuel cell may use a common heat rejection system.

Since system level details are classified, the focus of this paper is on the fuel cell itself. A brief discussion of the system is presented.

Monolithic Fuel Cells

In the monolithic concept, fuel and oxidant are combined electrochemically in a ceramic cell at an operating temperature of 1100 K to 1300 K. Cell components are fabricated as one piece, much like a block of corrugated paper-board, as shown in Figure 1. Fuel and oxidant are conducted through alternating passages in the stack. These passages are formed from thin (0.025 mm to 0.100 mm) layers of the active cell components: electrolyte (yttria stabilized zirconia), cathode (strontium doped lanthanum manganite), anode (nickel-zirconia cermet), and the interconnection material (magnesium doped lanthanum chromite) that connects cells in electrical series (bipolar plate). The corrugations also form the gas seal at the edges of the structure. Advantage is taken of the ability to fabricate the solid electrolyte and other solid cell components into shapes that cannot be achieved in liquid electrolyte systems. In liquid electrolyte systems, much of the mass and volume goes into building the inert container for the liquid. Eliminating this unnecessary material gives the monolithic fuel cell a significant advantage in performance.

The potential for the high power density of the monolithic fuel cell results from the small cell size. Cells 1 mm to 2 mm in diameter or smaller are achievable when the inert container for electrolyte and the inert support for the thin active layers are eliminated (Fee et al. 1983). The small cell size increases the active surface area per unit volume of the cell. More importantly, the small cell size in the monolithic design reduces the voltage losses due to internal electrical resistance. This reduction is a significant consideration, because internal resistance is the principal dissipative loss for the ceramic materials and temperatures of interest. Decreasing the cell size decreases the current path length because current is carried in-plane by the electrodes in the monolithic design.

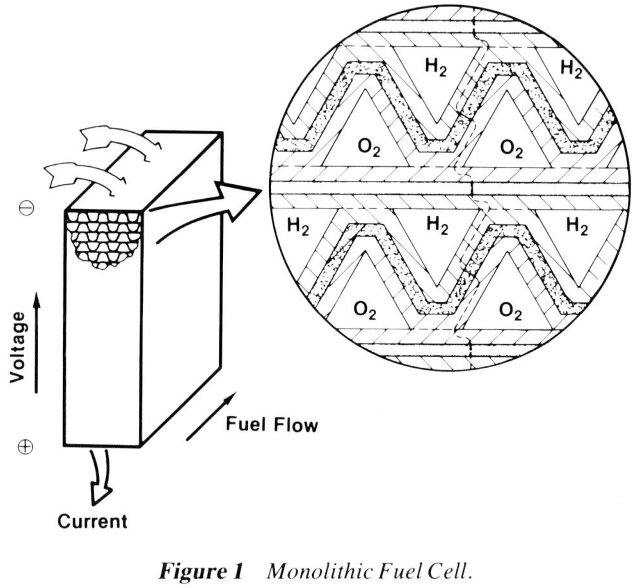

Figure 1 *Monolithic Fuel Cell.*

Fabrication

The principal challenge for the development of the monolithic fuel cell is fabrication of the intricate structures. Well developed ceramic fabrication techniques are being employed. Tape casting methodology is being optimized to fabricate thin, high density layers of electrolyte and interconnection material. Tape casting is used to form thin, high density layers in the ceramic capacitor industry. Tape casting is also being used to form the thin porous layers of anode and cathode. Alternate thin layer fabrication techniques, such as spraying, curtain coating, and hot roll milling, are also being considered.

The fabrication effort focuses on the critical three-layer composites: cathode-electrolyte-anode and cathode-interconnection-anode. These are the fundamental building blocks of the array. Considerable attention is given to achieving adequate bonding between the layers. In addition, methods have been developed to produce the desired corrugated structure. Thermosetting resins make the structure sufficiently rigid so that the corrugated tapes may be stacked in the green state.

All layers of the array are co-fired in air under the same conditions. The principal challenge for co-firing is to match the shrinkage rates of all four materials by tailoring the particle size and the binder content in each layer. Gas manifolding is added after the array is fired. The present effort employs cross-flow gas passages to reduce the complexity of manifolding.

Present Status

Cells

Cells of the monolithic design have been fabricated and operated. The cells were three layer composites of cathode/electrolyte/anode with an active area of 3 cm^2. The cell tests required a configuration different from that of a monolithic array. The cell geometry was flat to simplify gas manifolding. The unsupported three-layer composite was sealed to the end of a tube with a commercial zirconia/silica cement. For the cell test, air was supplied to one side of the cell and hydrogen to the other side. The electrical contacts applied to the cell for test purposes were a platinum paste and platinum screen. The cell and test apparatus were maintained in an isothermal zone.

Cell performance has been improved. The fabrication methodology has been upgraded so that cell performance more closely approaches the materials resistance, as shown in Figure 2 (for tests at 1273 K (1000°C) with hydrogen as fuel and air as oxidant). The principal increase in cell performance arises from reducing the resistance at electrode/electrolyte interfaces. In particular, tailoring the distribution of nickel at the fuel electrode/electrolyte interface is important to reducing interfacial resistance.

High current densities have been achieved. Cells of the monolithic design have been operated at 2.2 A/cm^2 with hydrogen as fuel and air as oxidant. This current density is about a factor of two higher than previously achieved with these conditions and materials (Nagata et al. 1983). The high current density results directly from the low internal resistance of the monolithic cells. Furthermore, the high current density clearly exceeds the monolith design goal of 0.6 A/cm^2. More importantly, the data indicate that there is no fundamental limitation to achieving high current densities with solid oxide fuel cell materials.

Arrays

Arrays of the monolithic design have also been fabricated and operated. An array has been operated for more than 650 hours, as shown in Figure 3. The array was two cells in electrical series; each with an active area of 9 cm^2. The operating conditions were 1273 K, 50 mA/cm^2, hydrogen as fuel, and air as oxidant. The performance demonstrates the feasibility of the monolithic fuel cell concept.

The array incorporated the cross-flow pattern of gas channels. The electrochemically active region was a three layer composite of cathode/electrolyte/anode, as shown in Figure 4 for an array of four cells in electrical series. A corrugated anode layer formed the fuel gas channels on one side of the three layer composite. Similarly, a corrugated cathode layer formed the oxidant gas channels on the other side of the three layer composite. The oxidant channels in the cathode are at right angles to the fuel channels in the anode, as shown by a cut at 45 degrees through an array in Figure 5. Consequently, the corrugations in the cathode do not show in the photograph of the anode face of the array in Figure 4. An interconnection layer separated the fuel and oxidant channels and linked the layers in electrical series.

The performance of the first monolithic arrays is similar to the performance of the first monolithic fuel cells. The array performance is expected to improve with time in much the same manner as the cell performance, as shown in Figure 2. In addition, a slight improvement is desired in matching the firing shrinkage to reduce the incidence of cracks in the electrolyte and interconnection layers, as shown in Figure 5. The cracks allow fuel and oxidant gas to mix. The mixing is the principal limitation on open circuit performance

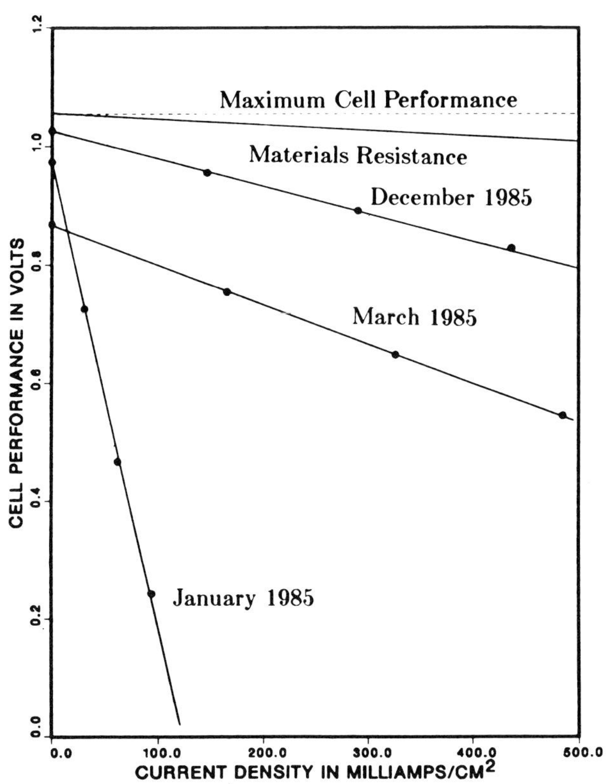

Figure 2 *Performance of Single Cells.*

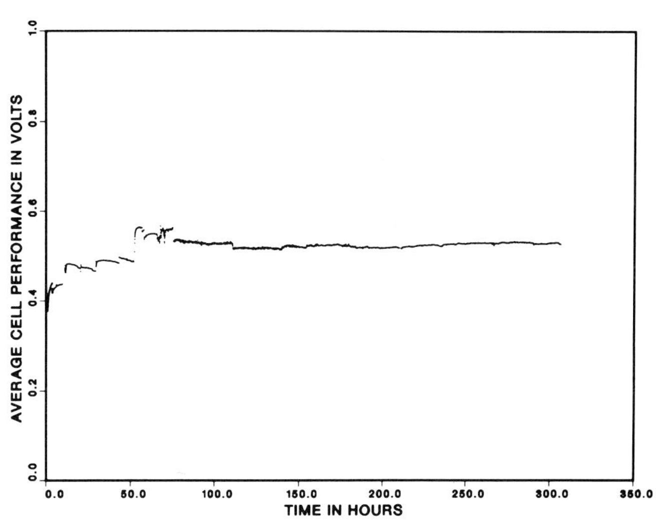

Figure 3 *Performance of Monolithic Fuel Cell Array.*

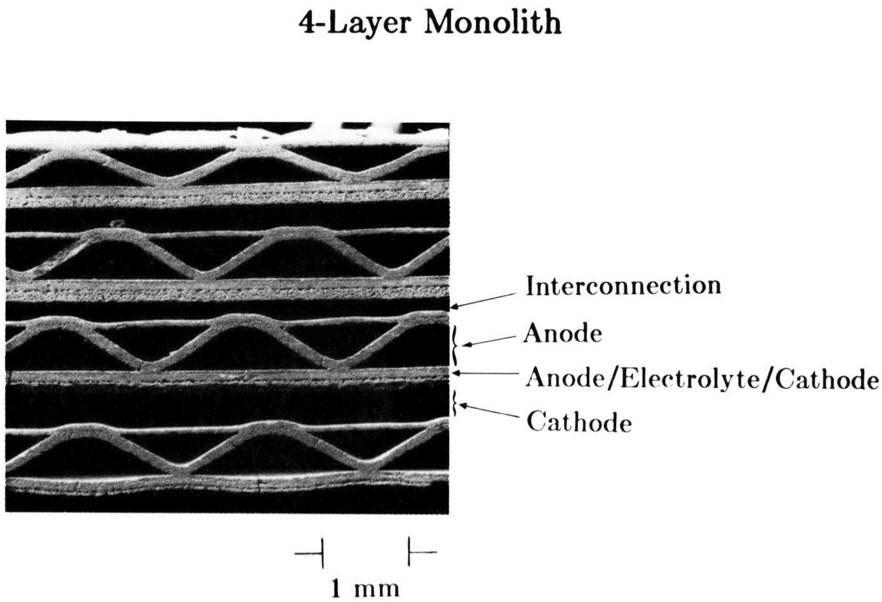

Figure 4 *Thermodynamics of Cell Operation.*

of the array. The internal resistance of the array is similar to the first cells tested.

Applications In Space

A regenerative monolithic fuel cell system looks attractive for space applications. The simplicity of an all solid state system and the minimal heat rejection requirements are key design features. The simplicity offers a high degree of system level flexibility. For example, the power and energy portions of the system can be sized independently. In addition, because the reactants are common with other components on the space platform, integration is possible with life support and propulsion systems, including open cycle operation (Hoberecht and Rieker 1985). Further, the system performance penalty associated with heat rejection is low. The fuel cell is highly efficient. The high operating temperature of the fuel cell (1273 K) also reduces the mass and volume of the heat rejection device.

The fuel cell electrochemically combines hydrogen and oxygen to produce electrical power. The reactants

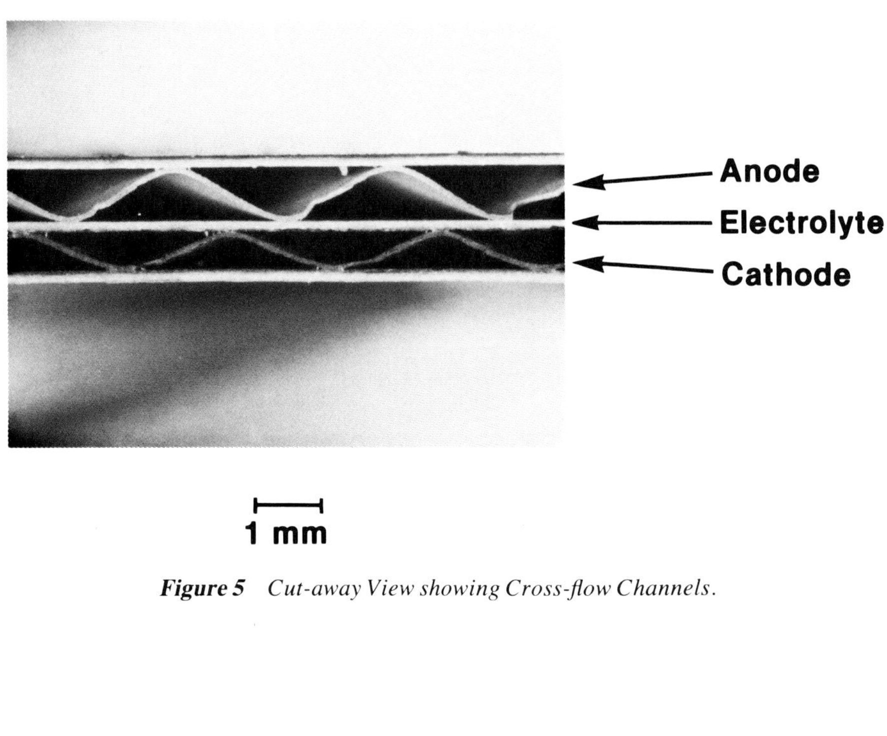

Figure 5 *Cut-away View showing Cross-flow Channels.*

are stored as gases in tanks at elevated temperatures and pressures, as shown in Figure 6. Emerging from the storage tank, oxygen stream passes through a blower and is preheated by mixing with an oxygen recycle stream before entering the fuel cell. Similarly, the hydrogen passes through a blower and preheater before entering the fuel cell. Within the fuel cell, DC power is produced as oxygen is consumed and hydrogen is converted to water. The heat generated in the process is rejected when the oxygen recycle stream passes through a radiator. After a reduction in temperature, the products are stored. In present concepts, a sliding seal is used to separate reactants and products within the same tank.

The system can be operated in the reverse mode to regenerate the power capability. The same monolithic fuel cell module can be run backwards for the regeneration; a separate electrolysis unit is not required. The length of time required for regeneration is determined by the size of the external power source. An obviously attractive choice is a link-up of the SP-100 thermoelectric system to a fuel cell device. This could be a nearer term burst capability than other alternatives. The key drawback of such a system is long recharge times. In the reverse operation, the fuel cell consumes water and DC electrical power (from the nuclear reactor) to produce separated streams of hydrogen and oxygen. Emerging from the storage tank, the water rich stream is preheated before entering the fuel cell. Power is provided to the cell from an external source to electrolyze the water and form hydrogen and oxygen. The heat generated in the electrolysis process is rejected when the gas streams pass through a radiator. Both the hydrogen and oxygen generated are stored in tanks.

The thermodynamics of regeneration are an important consideration in system design. Heat input must be provided for regeneration, as shown in Figure 7. For a fuel cell operating at 1273 K, 136 atmospheres pressure, and oxygen as oxidant, the open circuit voltage is 1.17 V for a fuel of 90 mol% hydrogen/10 mol% water. As the hydrogen is consumed and water is produced going down an anode channel, the local maximum theoretical voltage (Nernst voltage) decreases. At the outlet of the anode channel, the Nernst potential is 930 mV for a gas composition of 10 mol% hydrogen/90 mol% water. In the reverse mode, the reversible (electrolysis) cell voltage is slightly higher than the 1.17 V at the outlet of the steam/hydrogen channel because the steam/hydrogen electrode is a good conductor and approximates an isopotential surface. However, the heat of reaction is higher (the equivalent of 1.29 V), so heat must be supplied to remain thermoneutral. In theory, the additional heat

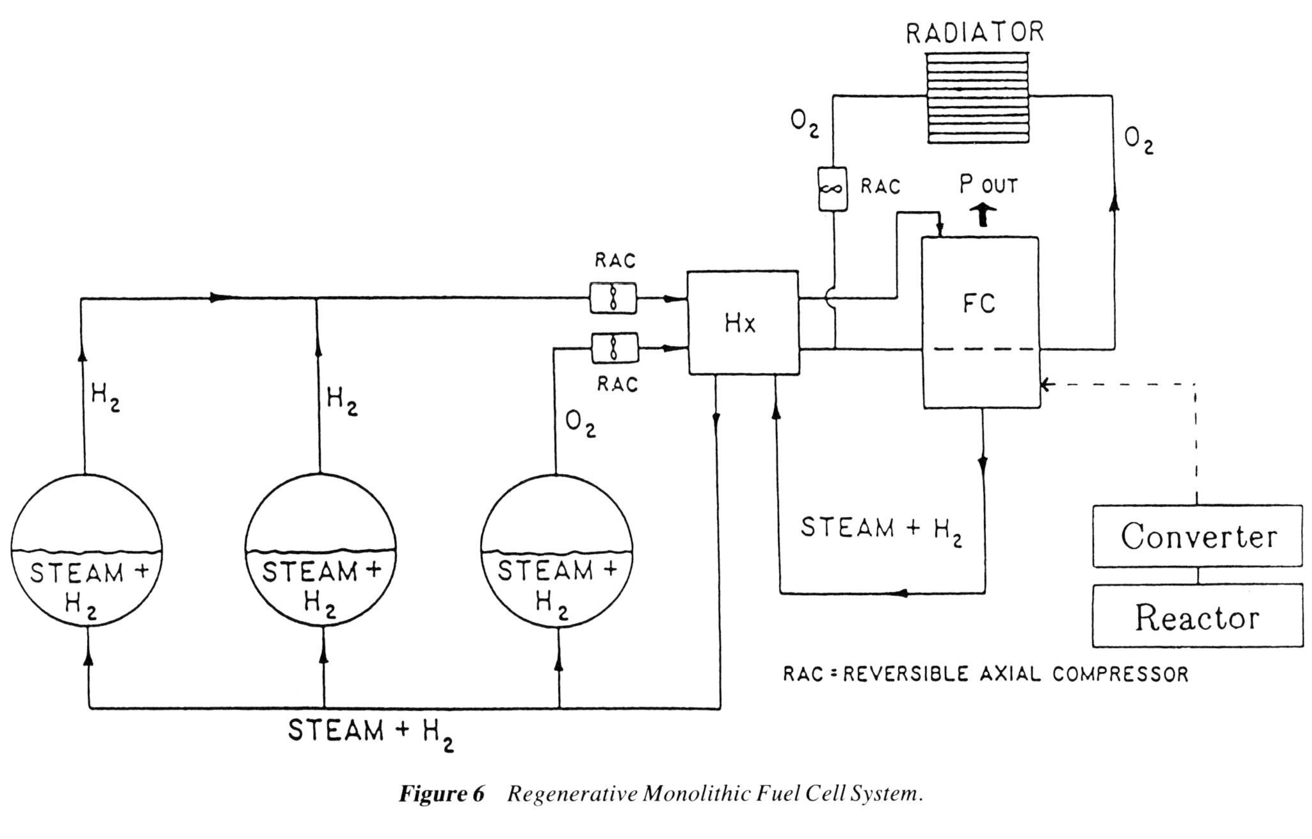

Figure 6 *Regenerative Monolithic Fuel Cell System.*

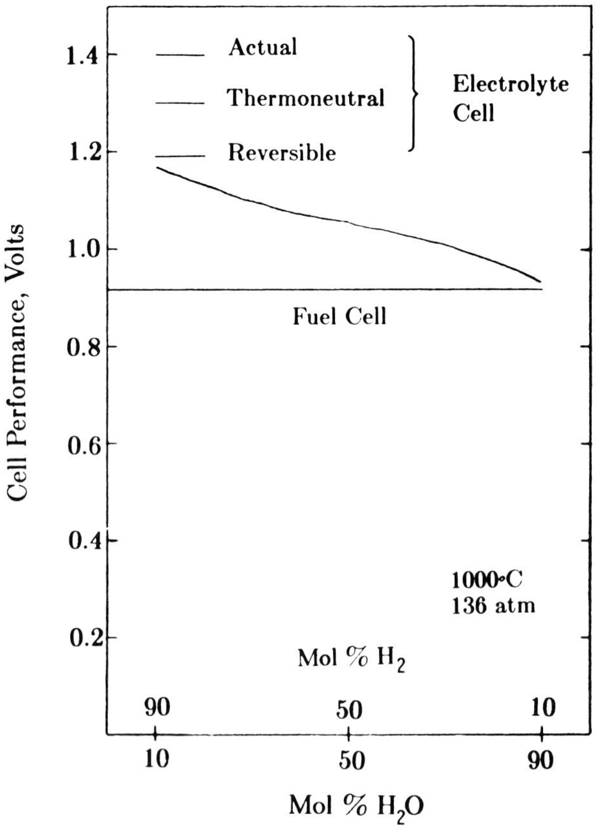

Figure 7 *Thermodynamics of Cell Operation.*

could be supplied from an external source and only 1.2 V supplied electrically. In practice, the heat is supplied in the form of dissipative losses (overpotentials) and 1.4 V is supplied to the electrolysis cell at a current density of 0.6 A/cm^2 (Doenitz et al. 1984 and Maskalick 1984).

The efficiency of the overall system is attractive. The projected performance of the system is to recover 72% of the heat of reaction in the fuel cell mode (0.92 V) and store 92% of the input electrical energy as chemical energy in the electrolysis mode. The overall round trip electrical efficiency is projected to be 66%.

Conclusion

The feasibility of the monolithic concept has been demonstrated. Small arrays of the monolithic design have been operated for hundreds of hours. The challenge is to improve fabrication methodology and demonstrate performance at a power level of 10s of kilowatts where other fuel cell technologies have already been tested. The development of the monolith is proceeding faster than an other fuel cell. An aggressive development program is recommended to ensure that the monolith system is ready for use in the 1990s.

Work reported in this paper addressed the tying in of an SP-100 class of thermoelectric reactor systems with a monolithic fuel cell burst power system. Recently, we have coupled a fuel cell device with higher dynamic power conversion systems. These power systems have been shown to be capable of producing very high power bursts for longer durations with significantly shorter recharge times.

Acknowledgments

This research was supported by the U.S. Department of Defense: Strategic Defense Initiative Organization and Defense Advanced Research Projects Agency. The submitted manuscript has been authored by a contractor of the U.S. Government under contract No. W-31-109-ENG-38.

References

Doenitz, W., E. Erdle, H. Hermeking, A. Koch, W. Kock, R. Roettenbacker, W. Schafer, R. Schamm, and R. Schmidberger (1984) *High Temperature Electrolysis of Water: HOT ELLY Phase II A,* Dornier Systems Final Report, BMFT-FB-T 84-032, Federal Republic of Germany.

Fee, D. C., R. K. Steunenberg, T. D. Claar, R. B. Poeppel, and J. P. Ackerman (1983) "Monolithic Fuel Cells," in 1983 *National Fuel Cell Seminar Abstracts,* Courtesy Associates, Orlando, FL, 13–16 November 1983, p. 74.

Hoberecht, M. A. and L. L. Rieker (1985) "Design of a Regenerative Fuel Cell System for Space Station," in *Proc. 20th Intersociety Energy Conversion Engineering Conference,* Society of Automotive Engineers, Miami Beach, FL, 18–23 August 1985, p. 2.202.

Maskalick, N. J. (1984) *The Electrolysis of Steam by Solid-Oxide Electrolyte Cells,* Report 83-9D12-WESOC-R1, Westinghouse Electric Company, Pittsburgh, PA, January 1984.

Nagata, S., Y. Ohno, Y. Kasuga, and H. Sato (1983) "Development of High Temperature Solid Electrolyte Fuel Cell at Electrotechnical Laboratory," in *1983 National Fuel Cell Seminar Abstracts,* Courtesy Associates, Orlando, FL, 13–16 November 1983, p. 165.

Part 7

CONTROL, POWER CONDITIONING, AND DISTRIBUTION

Space Nuclear Power Systems 1986
Edited by M. S. El-Genk and M. D. Hoover

Chapter 20

Power Conditioning for Space Nuclear Reactor Systems

Baruch Berman

Rockwell International
Satellite Systems Division
2600 Westminster Boulevard
PO. Box 3644
Seal Beach, CA 90740–7644

This paper addresses the power conditioning subsystem for both Stirling and Brayton conversion of space nuclear reactor systems. Included are the requirements summary, trade results related to subsystem implementation, subsystem description, voltage level versus weight, efficiency and operational integrity, components selection, and shielding considerations. The discussion is supported by pertinent circuit and block diagrams. Summary conclusions and recommendations derived from the above studies are included.

Introduction

The need for a higher level of power and energy, projected for future space vehicles, is creating substantial interest in space qualified, safe, and reliable nuclear reactor power systems. Application of these power systems is envisioned for both manned and unmanned missions, such as space-based laser (SBL), space-based radar, Space Station, and others. To direct, coordinate, and formalize many of the activities related to the nuclear reactor power source, the SP-100 Program was established under the auspices of the National Aeronautics and Space Administration (NASA), Department of Defense (DOD), and Department of Energy (DOE), with Jet Propulsion Laboratory (JPL) as the program manager. Rockwell International participated in the program contributing to novel technology concepts. Analytic studies and trades of the subsystems and systems design were performed. Optimized preferred options were selected and recommended for implementation.

A space qualified nuclear reactor power system (NPS) consists of the following major subsystems: nuclear reactor, shield, heat transport and rejection, power conversion, power conditioning and cabling, controls, and structure. The NPS and its subsystems possess unique characteristics requiring both a system approach and careful characterization and implementation of the subsystems. Nuclear reactor power sources are intrinsically base load type sources. Their output cannot be throttled or changed to comply with user load variations on a real time basis. Thus, the overall system must accommodate variations in user demand, while allowing for its own orderly start-up, on-line sequencing, steady state and transient performance, failure protection, and orderly shutdown.

Characterization, trade, and review of the NPS subsystems point to the power conditioning subsystem as the most suitable for matching user load demand to source output. The power conditioning subsystem must also regulate the output of the power conversion subsystem and deliver power of specified quality to the user. Prime power is supplied by the reactor to the conversion system. The raw unregulated electric power is conducted via special cabling to the power conditioning subsystem (PCS). The PCS processes and conditions the raw electric power for delivery, via dedicated programmed breakers, to the user at the specified voltage and quality.

The type and characteristics of the major PCS elements depend on the specific nature of the power conversion subsystem. This subsystem may be of the Brayton/alternator (ac), Stirling/alternator (ac), thermoelectric (dc), or thermionic (dc) type. However, all systems are subject to common underlying parametric drivers influencing selection, such as voltage levels, power quality, efficiency, power density, reliability, and survivability, among others.

Power Conditioning

Power Conditioner Derived Specifications

Specifications were established to scope the requirements imposed on the PCS. The pivotal parameter is power—25 kWe to the user per unit. This dictates that the PCS be designed to allow for regulation tolerances, ballast margins, PCS series inefficiencies, instrumentation power, battery charging, and housekeeping. The 400-Vdc level was selected based on sensitivity tradeoffs and program guidelines. Careful attention was given to isolation, EMI, and statics. MIL-type specifications were used to establish guidelines for design, and additional specifications will be added as the program progresses.

The derived electrical performance specifications are as follows:

- Net power output—25 kWe per unit, 100 kWe for subsystem;
- Efficiency—94 percent overall;
- Voltage—400 Vdc ± 5 percent, two wire;
- Load response—continuous load following of 0 to 100 percent;
- Load response rate—not to exceed TBD kWe/s;
- Ripple and transient limits
  - — 3.2 percent Vdc peak to peak narrow band (30 Hz to 70 Hz) falling 10 dB per decade to 1 percent Vdc peak to peak at 70 kHz; thereafter remaining constant to 400 MHz,
  - — Transient spacecraft system specifications per MIL-STD-704, and
  - — Prevention of electrical breakdown in spacecraft per N70-18656; and
- Environment
  - — MIL-E-6051, EMI requirement specifications,
  - — MIL-STD-1541, EMI characteristics requirement and equipment, and
  - — MIL-STD-461, 462 EMI characteristics and measurements.

The main functions of the PCS are to regulate the output of the alternators, ensure that power of a specified quality is delivered to the user, absorb excess power available from the system but not needed by the user, and transmit power from the engines to the power conditioner and user interface.

The selected PCS consists of five 25-kWe units (5/4 redundancy) connected in parallel. The elements constituting each unit are connected in series for fully redundant fault isolation. The PCS was designed with radiation tolerance in mind, and the most sensitive power components, the thyristors, have been tested to 10^{14} nvt and 4 x 10^{6} rad.

The PCS consists of the following major elements:

- The switching networks for power flow, turn on, turn off, and protection;
- The rectifier that converts the ac from the alternators to dc;
- The ballast equipment that controls and absorbs excess system power;
- The paralleling switches that connect the multiple units to the user bus;
- The source and load control equipment that optimizes overall PCS response and performance; and
- The cabling that distributes power throughout the system.

Power Conditioning Diagram

The overall PCS consists of five parallel units. The concept has a 5/4-redundancy and each unit can process 25 kWe. However, during normal operation, when all five are operating in parallel, each unit will process only 20 kWe net to the user, or 80 percent of its rating. Operation at 80 percent of the design rating for long periods further enhances the reliability of the PCS by reducing component stresses. All five units and the paralleling equipment are located in the PCS package at the end of the boom (nearest the payload). Each unit receives ac power from the alternators in the engine module. The ac power is carried along independent cables routed through the boom. Figure 1 shows the general system arrangement, interfaces, and elements in block diagram form.

Tradeoff for Method of Electrical System Interconnections

A trade was performed that considered series, parallel, and hybrid arrangements. The series arrangement requires additional active components and interdependency as does the hybrid. Parallel dc electrical connection of the five individual series string units was selected. The parallel connection allows maximum flexibility to connect or remove any of the five units. Any or all of the individual units may be operated at their set point and rating without imposing implementation requirements on the other units. In the parallel arrangements, if one of the units is shut down, the voltage of each unit will remain at the prior preset level and the current will be redistributed to provide the power demand. The five units may be started sequentially or in parallel. The user is redundantly protected from failures in the individual units due to the series arrangements of their elements.

In addition to the above trade, the concept of par-

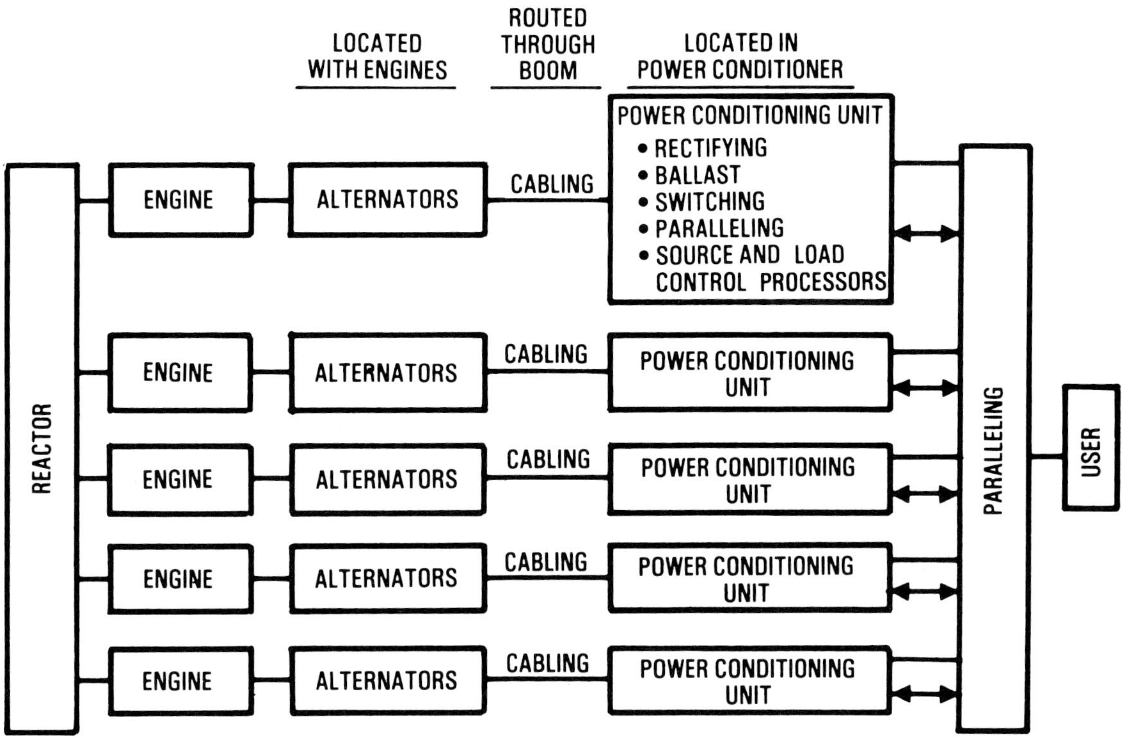

Figure 1 *Generalized Block Diagram of a Nuclear-Reactor Power System (NPS).*

alleling the free-piston Stirling engine alternators at the ac terminals was reviewed. A schematic of the concept is shown in Figure 2. Since power level, frequency and phase lock controls would be necessary the approach of paralleling at the ac terminals was not selected. Figure 3 shows the conceptual circuitry of the power semiconductors that implements the preferred selection for a Stirling system.

Stirling System Power Conditioner Operation

The raw ac power delivered by the series-connected linear alternator pair (in the engine module) is transmitted via a shielded, radiation-hardened twisted pair cable routed through the boom to the main power conditioning network. The PCS network uses thyristors (SCRs) to process and condition the raw power before delivery to the user. The operation of the conditioner unit is described below.

As the reactor temperature increases and reaches a preset level, the first Stirling engine module is pulsed (via the alternators) and begins oscillating at the nominal frequency The open-circuit voltage of the alternators builds up within a few cycles, and as it reaches the preset regulation level, the ballast control SCRs are triggered in a proportional manner to begin accepting the delivered power. As the reactor power increases, the ballast rejects more and more power. The ballast control SCRs are connected to the ac line and are naturally commutated. Each ballast control responds independently to its dedicated engine module. After all the engines are up to full power and stable equilibrium conditions have been verified via Sensors A and B (see Figure 4), the unit control processors, central control processor, and rectifier/decouplers are turned on. The rectifier/decoupler rectifies the ac from the alternator and also blocks and decouples faults that could occur and reach it from the user side. Sensor C verifies that nominal conditions are attained at the breaker terminals and primes the local and central processors to allow breaker initiation. No breaker is allowed to turn on unless the conditions for paralleling have been satisfied on at least four units. No unit is allowed to turn on unless its own conditions meet the prescribed central control algorithms. After all the breakers have been turned on satisfactorily Sensor D enables turn on of the main user breaker and power begins flowing to the user bus. Sensor E monitors alternator conditions, such as temperature and pressure.

Analysis of Stepped Ballasts

A single-stepped ac ballast introduces a certain amount of harmonics into the system and raises $(I_{rms}/I_{ave})^2$, the square of the form factor $(FF)^2$, by ap-

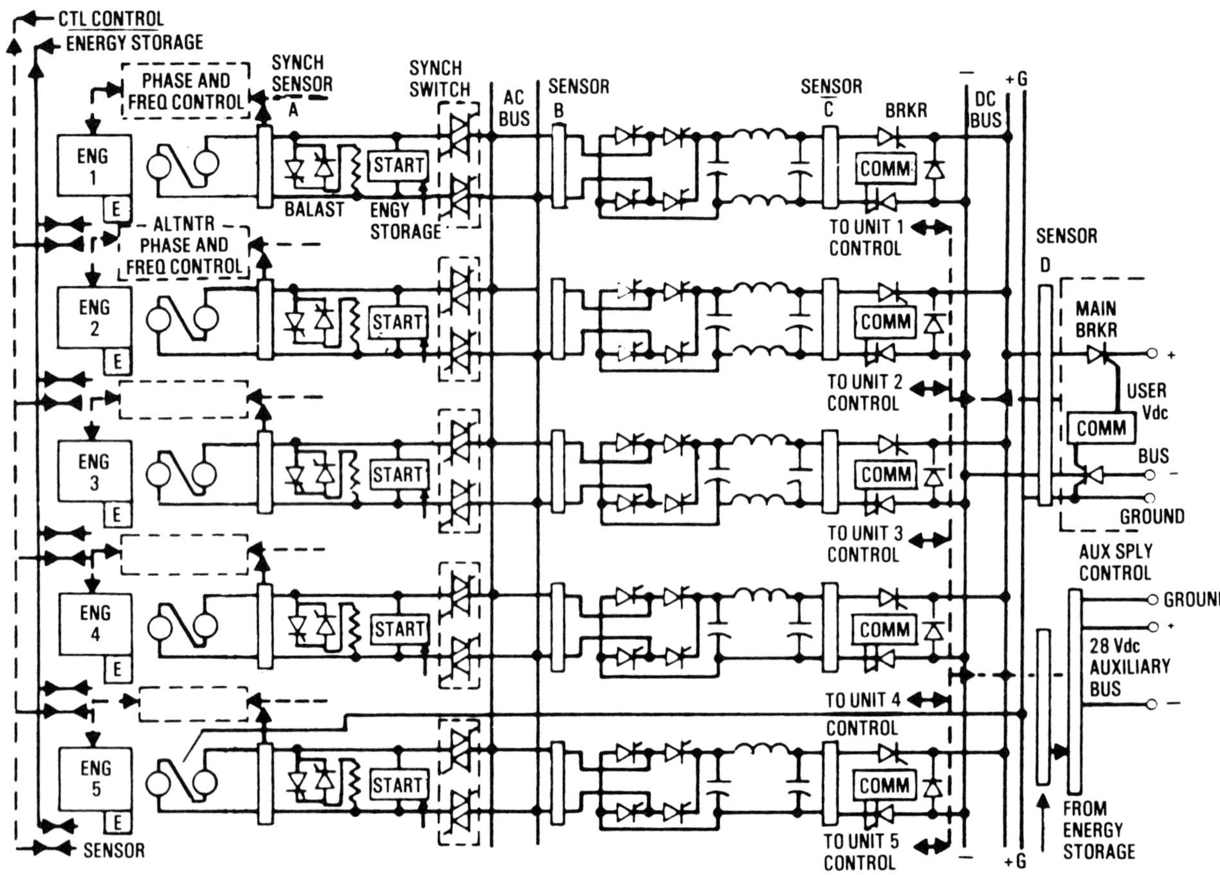

Figure 2 *AC Paralleling of Stirling Engine.*

proximately 13 percent. A multistepped system was also studied. It would reduce the harmonics but increase the complexity. At this point, the concept of a five-stepped ballast was selected. The single individual ac ballast is shown in Figure 5. The multistepped ac ballast is shown in Figure 6. A multistepped dc ballast concept is shown in Figure 7.

The selected ac five-stepped ballast will enhance stability and improve source system efficiency by maintaining the $(FF)^2$ at approximately unity with a power conditioner mass penalty of less than 1 percent. The $(FF)^2$ penalty drops as the number of steps increases, as shown below:

Ballast Steps	$(FF)^2$ Penalty (percent)
1	13
2	4
5	1

Thyristor Triggering and Low-Level Control

Several options for triggering the thyristors were examined. These included unijunctions, photocouplers, transistorized multivibrators, and pulsed actuation. Magnetic triggering was selected for its reliability, high tolerance to radiation, and continuous drive delivered to the thyristor gate. The magnetic circuitry is of the Remy type, which resets and responds within one-half cycle of the supply frequency. This type of trigger may be used for both continuous or step-type metered control via very low-level control components. The power gain of the trigger/SCR combinations is over 50 dB.

Semiconductors for low-level control may be of the discrete bipolar type of radiation-hard complementary metal oxide semiconductor (CMOS). The semiconductors possess an adequate radiation tolerance level (10^6 rad, Si, 10^9 rad/s, neutron fluence 10^{10} to 10^{14} nvt, or higher). These devices are further shielded, within their dedicated cavity for enhanced performance.

Power Conditioner Subsystem Layout

There are five power conditioner units (PCUs), each rated to process 25 kWe. The five units are

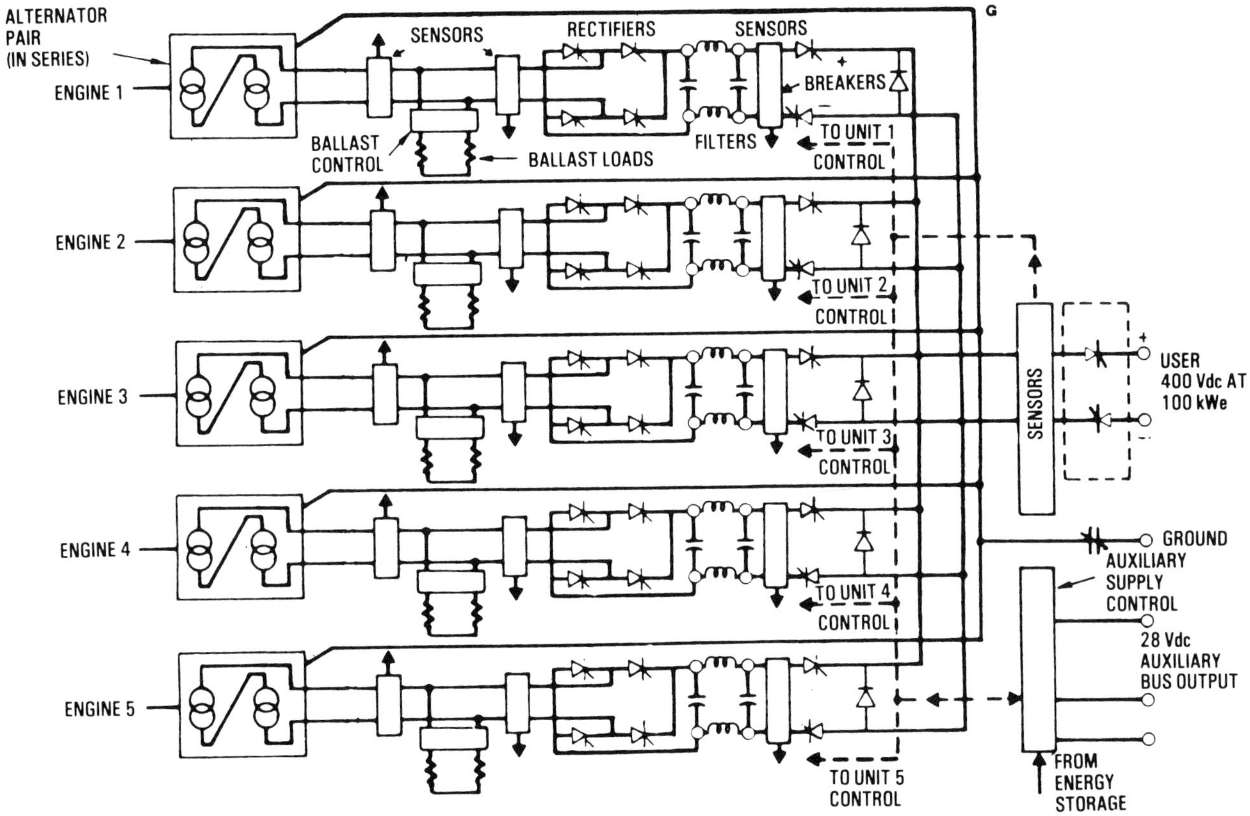

Figure 3 *Conceptual Circuitry of Selected Power Conditioner Subsystem for a Stirling System.*

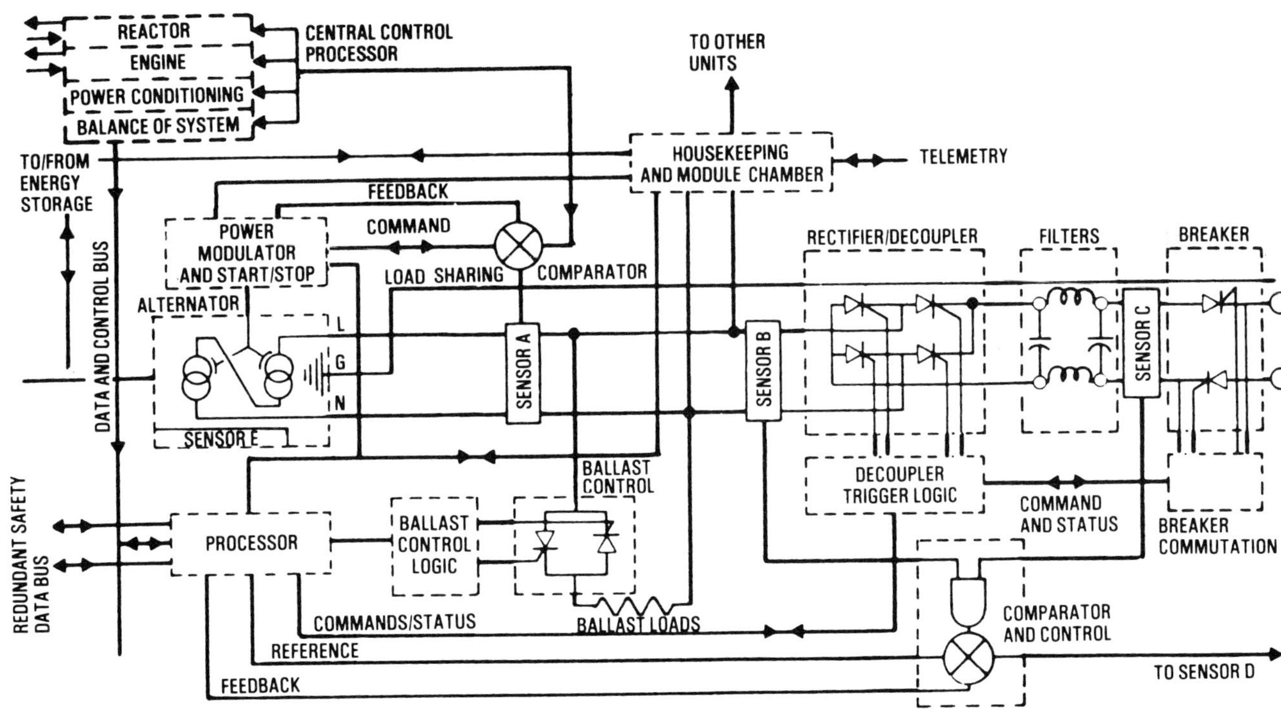

NOTE: SENSORS A, B, C, D, AND E SENSE VOLTAGE, CURRENT, POWER, TEMPERATURE, ETC. (AS REQUIRED)

Figure 4 *Simplified Schematic of a Single 25-kWe Power Conditioner Unit with Central and Individual Control Processors for a Stirling System.*

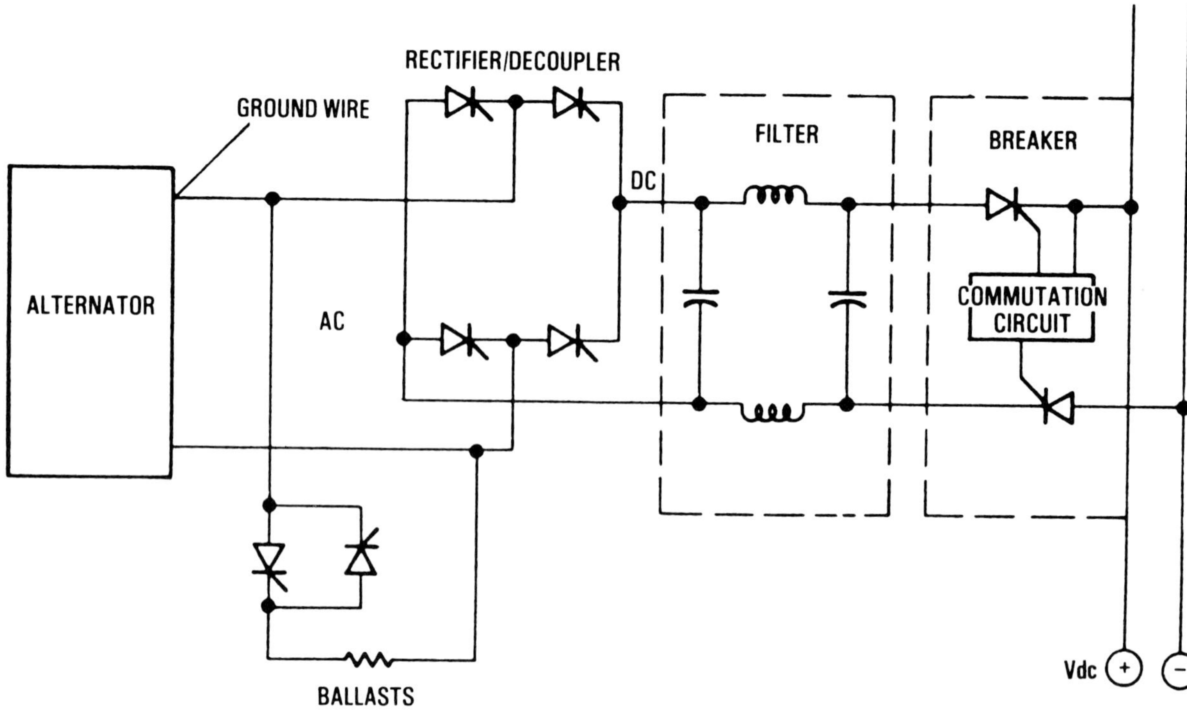

• EACH BALLAST CIRCUIT PROCESSES SINGLE ENGINE MODULE POWER
• CONTINUOUS CONTROL

Figure 5 *Power Conditioner Concept with a Single-Step AC Ballast Control per Engine.*

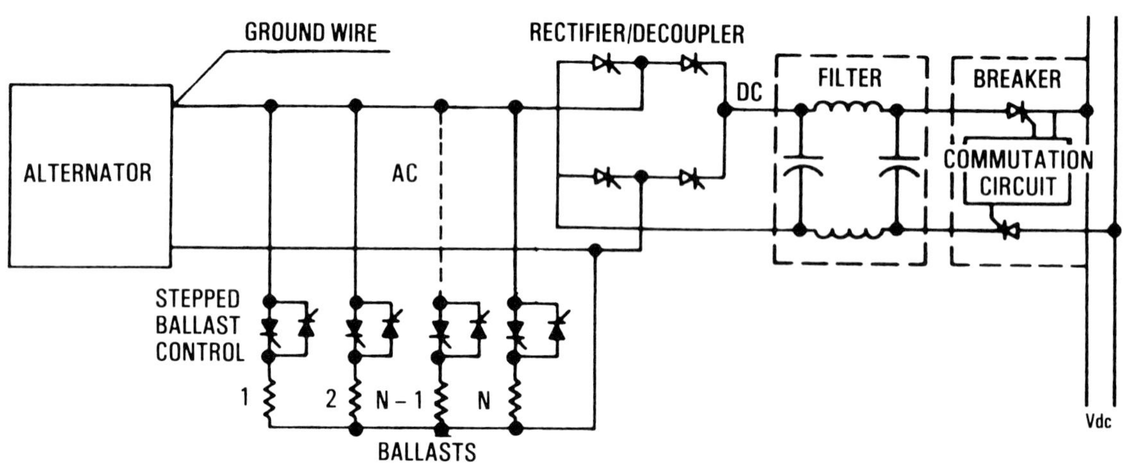

• EACH BALLAST CIRCUIT PROCESSES TOTAL POWER/N = 5 PERCENT
• CONTROL METHOD—CONTINUOUS CONTROL
• BENEFIT—NATURAL COMMUTATION WITH REDUCED HAMONICS
• PENALTY—N PARALLEL BALLAST LOOPS WITH INCREASED COMPLEXITY, COMPONENT NUMBER, AND SIZE

Figure 6 *Power Conditioner Concept with Multistepped AC Ballast (Selected).*

mounted to the electronic equipment radiator. Each unit dissipates approximately 1100 W (times four units operation); total PCU dissipation is approximately 4400 W. Sample PCS efficiency calculation is shown in Table 1. Efficiency was calculated using the nominal values of 25 kWe at 400 Vdc. Efficiency of power conditioners degrades as user voltage is reduced. Efficiency is defined as $P_o/(P_o + P_c)$, where P_o is output power (in kWe) and P_c is power conditioner and control losses. The temperature rise from

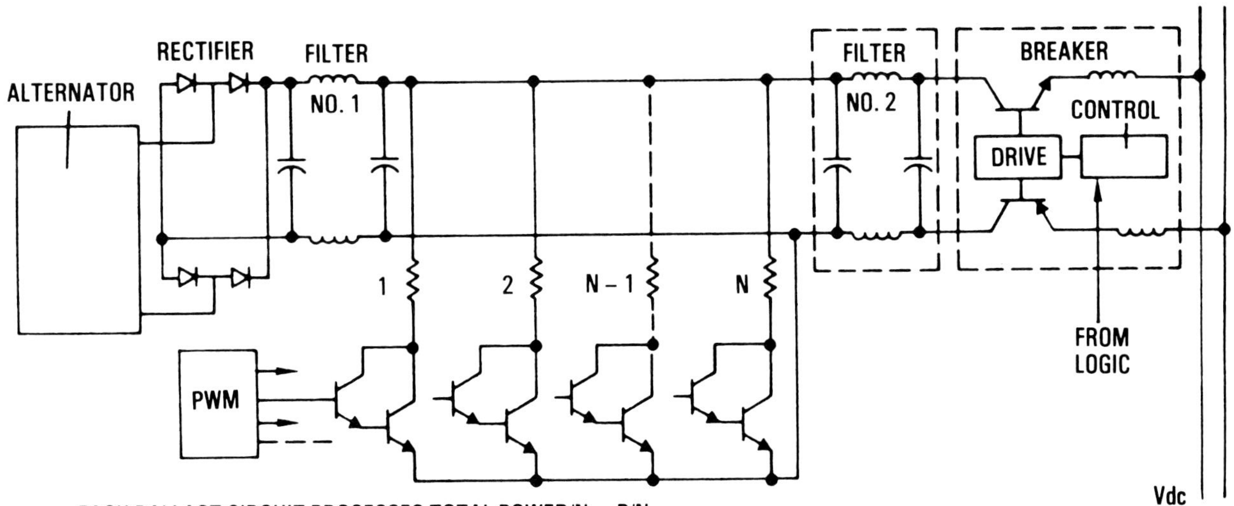

- EACH BALLAST CIRCUIT PROCESSES TOTAL POWER/N = P/N
- CONTROL METHOD—PULSE WIDTH MODULATION (PWM), CONTINUOUS
- BENEFIT—LOWER HARMONIC DISTORTION ON BOTH THE ALTERNATOR AND THE USER BUS
- PENALTY—TRANSISTOR DRIVE LEVEL CRITICAL, PWM GENERATORS REQUIRED, DOUBLE FILTER, INCREASED WEIGHT. NO DECOUPLING RECTIFIER, REDUCED RELIABILITY

Figure 7 *Power Conditioner Concept with Multistepped DC Ballast.*

the SCR base to the electronic radiator is approximately 20 K. The radiator is designed to maintain the SCR junction temperature well below the nominal allowable of 400 K (a radiator operating at 330 K will allow 350 K at the SCR base, leaving a margin of 50 K between the base and junction). The ballast resistors will be maintained at less than 600 K. For Nichrome resistors, the incipient glowing temperature is over 700 K.

Figure 8 shows the conceptual layout of the power conditioner assemblies and the ballast load, all sized at 400 Vdc. At 400 Vdc, the power conditioner dissipation is approximately one-half that at 200 Vdc, and the power conditioner radiator areas are smaller (Figure 9). Rockwell has selected 400 Vdc because it offers several advantages over 200 Vdc with no increase in technical risk.

PCS Conceptual Packaging

Various concepts with heat sinks and shieldings were evaluated, and one power conditioner packaging concept was selected. In the concept, the SCRs are mounted to the base that interfaces with the electron-

Table 1 *Power Conditioner Efficiency Sample Calculations (100 k We at 400 Vdc).*

Major Power Conditioner Component Losses (Nominal)	
P = 25 kWe at 400 *Vdc*, V_{rms} = 444 V_{rms}, I_{rms} = 56.3A_{rms}	
Ballast SCR-I_{rms} = $I_{pk}/2$ = 56.3 A_{rms} x 1.41/2 = 39.6 A_{rms}	
I_{ave} = $I_{rms}/1.57$ = 39.6 $A_{rms}/1.57 \cong 25\ A_{ave}$	
Rectifier/Decoupler SCR-I_{rms} = 39.6 A_{rms}	
I_{ave} = 25.0 A_{ave}	
Breaker SCR-I_{ave} = 25,000/400 = 62.5 A_{ave}	
Ballast SCR loses = 1.5 *V* x 25 A_{ave} = 37.5 W	2 x 37.5 = 75 W
Rectifier/decoupler losses = 1.5V x 25 A_{ave} = 37.5 W	4 x 37.5 = 150 W
Breaker SCR losses = 1.5V x 62.5 = 93.75	2 x 93.75 = 187.5 W
Filter loss = 4 V x 62.5 = 250.00	250 W
Cabling loss = 4 V x 62.5 = 250.00	250 W
Additional loss for ballast offset, sensors, processors, and commutation—2 percent	500 W
Loss in one train	1412 W
Total loss for four trains	5648 W
Efficiency of Four Trains	0.956
Mainbreaker 250 A x 3 V = 750W	
Overall power conditioner (100 kW) efficiency 750 + 5648 = 6398W	0.94

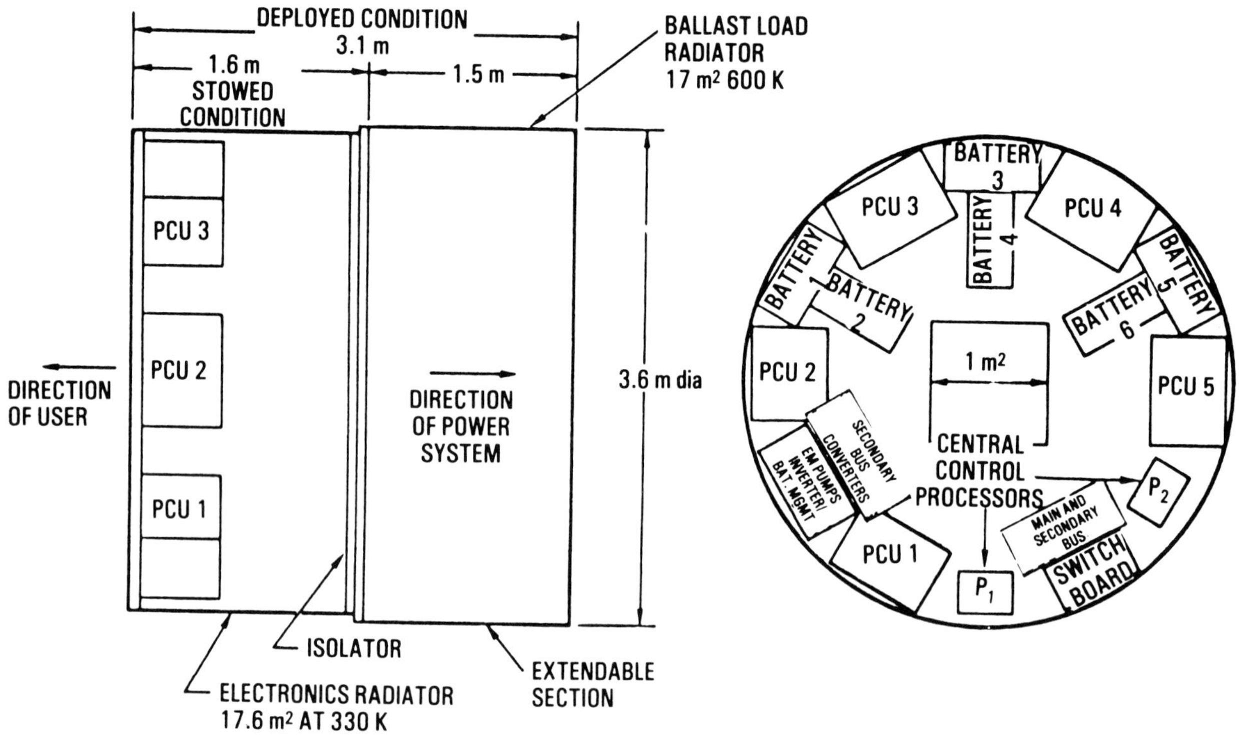

Figure 8 Packaging Concept for Power Conditioner Subsystem (Extended Configuration).

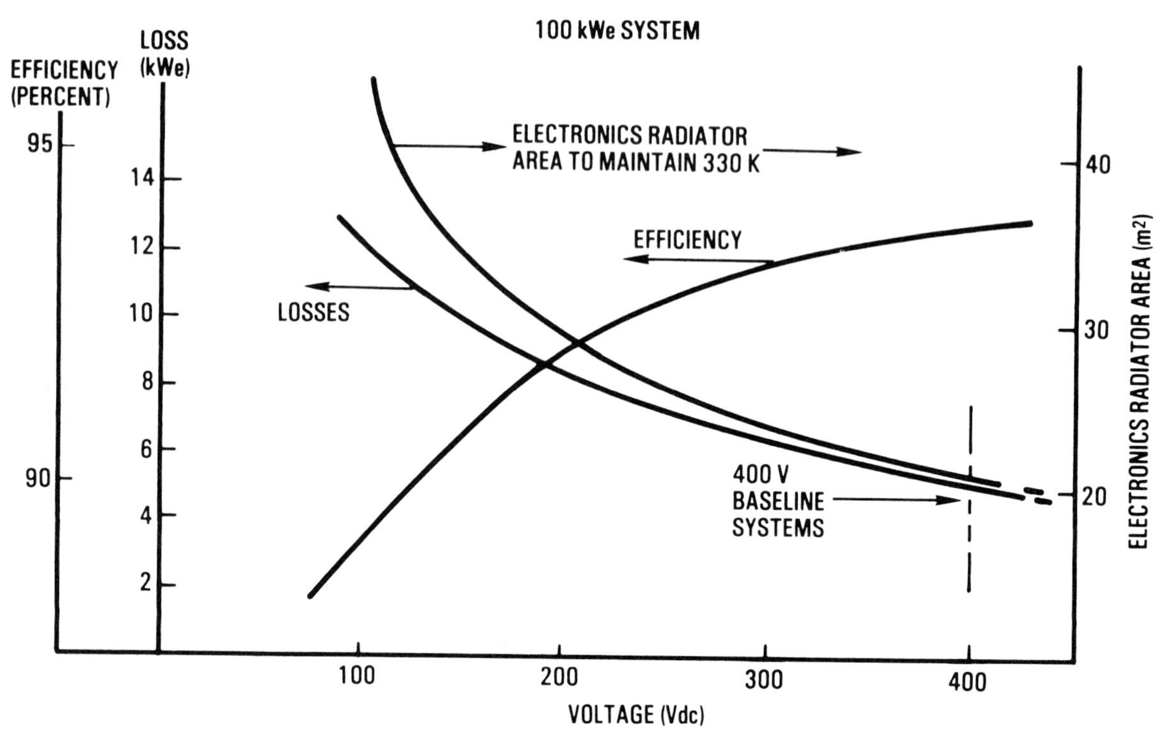

Figure 9 Power Conditioning System Efficiency increases at Higher Voltages.

ics radiator, and the electrical components are isolated. Thermal conduction is enhanced by mechanical pressure devices and a conductive coating such as beryllium oxide, aluminum oxide, modified mica, or powered diamond. Most of the low-power dissipation components, such as capacitors and controls, are mounted on the top part of the unit. The SCRs are also individually shielded to the dose requirements resulting from the worst-case Van Allen orbit. The box frames, made of aluminum, provide some additional radiation shielding.

The availability of ac at low frequency from the Stirling engines led to the selection of SCRs as the major power elements. Methods of heat sinking that also incorporate radiation shielding are being used. The conceptual arrangements for these methods are shown for both disc-and stud-type SCRs in Figure 10. These methods would also be directly applicable to power transistors for situations requiring dc power processing. The assemblies are electrically isolated but thermally conductive to maximize grounding flexibility.

Flexible Cable Selection

The flexible cable that conducts the ac power (down the boom) between the system and the PCS is shown in Figure 11. The cable is similar to the type used in nuclear power stations. Its materials are selected to withstand the expected environmental conditions, including radiation, and would meet and exceed IEEE Standard 383 requirements. Cable voltage rating is established at 1000 V, more than twice the operating voltage. The margin for radiation resistance, established by testing, will be five times the required tolerance level. The cable is not pressurized because tests have confirmed that detrimental discharge is liable to occur in the Paschen region if certain molecular gases, such as N_2, H_2, O_2, or CO_2 are present. The cable is flexible, and the bending radii for deployment with the boom have been set with margins. The cable is expected to withstand the environmental conditions of 575 K (at alternator), 9 x 10^9 rad (gammas), and 10^{15} nvt (neutrons). The copper stranding is Ni clad to withstand temperatures higher than 575 K without flaking. Inorganic material, such as Phosroc-C, serves as the basic insulation material and is rated at 1700 K.

Power Circuit Components

Design tradeoffs were performed to select reliable power components for the PCS. The SCRs have a comfortable rating margin when compared with their operating conditions. The selected rating is more than twice the requirement. The gamma radiation tolerance of the type of power SCRs used in the conditioner has been verified by ^{60}Co irradiation testing at Rockwell as part of an in-house program. A safety factor of five was employed on radiation requirements.

Mass Budget

The power conditioner masses were estimated for the 100-kWe baseline system from the component list shown in Table 2. Special shielding is not necessary for the reactor produced radiation, but tantalum was added to the aluminum for the nuclear threat X-rays and worst-case Van Allen conditions. Table 2 includes a comparison with the mass of Brayton PCS.

Cable Mass Sensitivity to Operating Voltage, Power, and Boom Length

Preliminary system studies were performed to determine the cable mass as a function of user voltage, power, and boom length. Figure 12 shows the mass factors for 1 and 4 percent cable losses.

Power Conditioning and Cabling Subsystem Capabilities

The power conditioning and cabling subsystem capabilities are as follows:

- Five parallel 25-kWe units
  - Each capable of operating independently processing or dissipating 25 kWe;
- Each of the five units is series connected
  - Fully redundant protection and
  - Dedicated breakers supply the user;
- 94 percent efficiency;
- Power conditioner and user voltage of 400 Vdc;
- Power conditioning equipment packaged at end of boom;
- Power semiconductors perform all major functions; and
- Radiation tolerance of the selected semiconductors is greater than 4 x 10^6 rad
  - No power conditioner shielding is needed for reactor-produced radiation,
  - Minor shielding (~ 10 kg) needed for nuclear X-ray threat, and
  - Additional shielding in selected locations for worst-case Van Allen radiation.

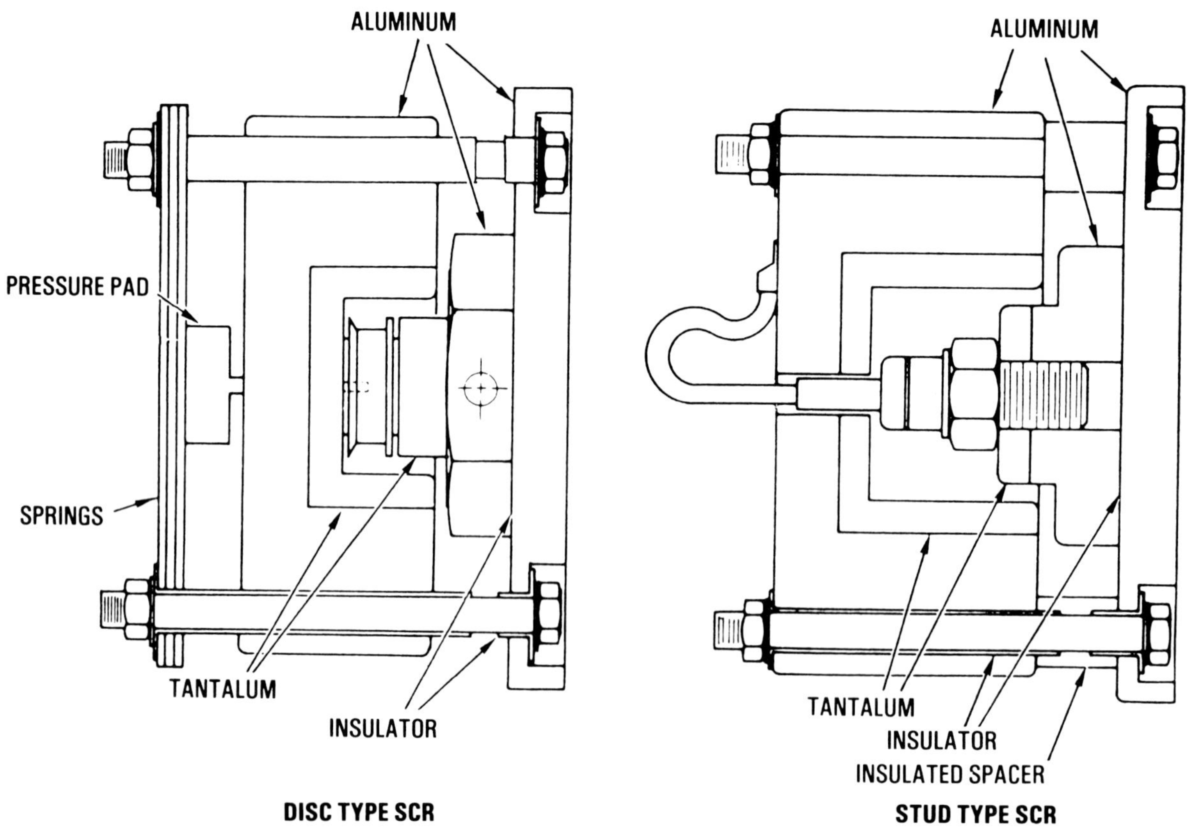

Figure 10 *Power SCRs Individually Shielded for Worst-Case Van Allen Radiation.*

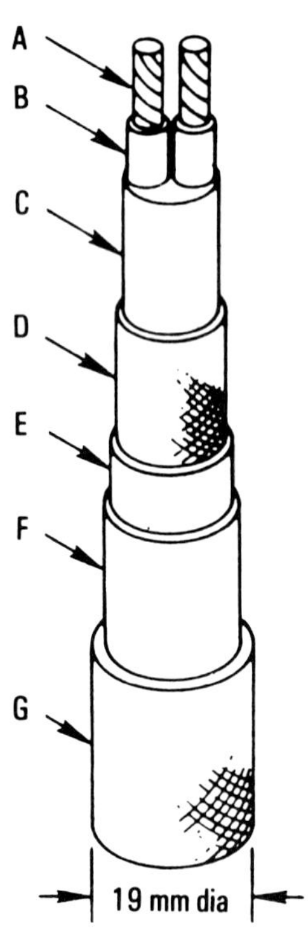

- FIVE CABLES USED FOR EACH ENGINE AT 400 Vdc USER LEVEL
- TWISTED PAIR OF NO. 6 AWG NICKEL CLAD COPPER CONDUCTORS STRANDED 133/0.0142 FOR FLEXIBILITY – (A)
- EACH CONDUCTOR INSULATED BY PHOSROC C BRAID RATED AT 1700 K – (B)
- HIGH TEMPERATURE MICA TAPE – (C)
- PAIR SHIELDING AND GROUNDING SHEATH OF ANNEALED NICKEL CLAD COPPER BRAID – (D)
- BRAIDED FILAMENT OF HIGH TEMPERATURE (1700 K) PHOSROC – (E)
- HIGH TEMPERATURE MICA TAPE – (F)
- OUTSIDE MECHANICAL SHIELD OF STAINLESS STEEL BRAID OR EQUIVALENT – (G)
- CLASS D MECHANICAL FLEXIBILITY WITH REELING CAPABILITY
- BENDING RADIUS IN DEPLOYMENT SET AT 12 TIMES CABLE DIAMETER (23 cm RADIUS)
- RADIATION RESISTANCE EXPECTED TO MEET OR EXCEED 9 x 10^9 rads GAMMAS
- CABLE WEIGHT 0.454 kg/m
- VOLTAGE RATING–1000 V FOR 200 Vdc; 2000 V FOR 400 Vdc

Figure 11 *Improved Power Cable Insulated and Shielded against Radiation, Temperature, and EMI.*

Table 2 Baseline 100-kWe Power Conditioning Elements Mass Breakdown Comparison.

Element	Stirling (400 Vdc) (kg)	Brayton (460 Vac) (kg)
Central control processor	10	10
Ballast controls and logic	22.5	20
Rectifier/decoupler thyristor controls	45	—
Filters	50	—
Breakers and commutators	40.5	—
Starters and compensating capacitors	10	—
Cyclo converters and filters	—	45
Field control amplifiers	—	15
Breakers (ac)	—	40
Start-up inverters (starters)	—	15
Unit processors, sensors, instrumentation comparaters, and controls	25.0	25
Housekeeping, module charger, and secondary supply	20	20
Boxes and structure	50	50
Power conditioning radiator	50	50
Total	323	290

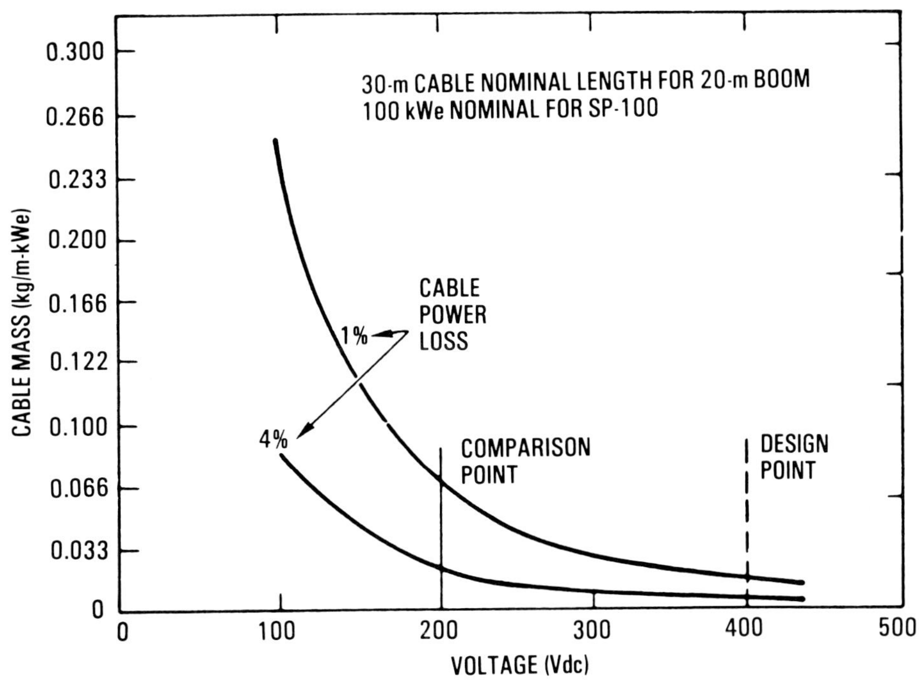

Figure 12 Power Cable Mass versus Voltage Level.

System Growth

Masses were calculated for the growth baseline systems at voltages that reflect the system size and reasonable cable losses. The power conditioning specific mass will decrease as voltage increases. The voltage level increases cause the specific mass to drop from 2.7 kg/kWe for 100-kWe, 400-V operation to 1 kg/kWe for 1000-kWe, 1000-Vdc operation (Table 3).

Power Conditioning for Brayton System

System Description

The requirements applicable to the PCS for the Brayton System are the same as those defined for the Stirling System. The basic consideration and trades are also similar. The salient difference is the use of a rotary alternator in the Brayton System versus the linear type alternator projected for the Stirling System. This leads to the production of a high frequency (approximately 2000 Hz) ac multiphase power in the Brayton System. The Stirling linear alternator output is single phase, at approximately 100 Hz. As stated previously the generalized block diagram is applicable to both system and is shown in Figure 1.

The overall basic power system is shown in block diagram form in Figure 13. The overall system consists of five 25-kWe alternators, each driven by a turbine. Each turbine output shaft also drives the compressors. A pony motor arrangement is coupled to the shafts to allow multiple startups of the system on command. The assemblies of the turbine, alternator, and compressor are referred to as the combined rotating units (CRUs). Prime power is supplied by the reactor to the turbines driving Rice alternators that deliver raw unregulated electric power through the cabling to the PCS. The PCS processes and conditions the raw power and delivers it to the user at 400 Vdc or three-phase, 400-Hz, 460-Vac line-to-line.

During normal operation, each CRU delivers approximately 20 percent of the user load (20 kWe). However, each generator and the associated ancilliary equipment is rated at the full 25-kWe capacity Thus, 5/4 redundancy is built into the system. Each CRU has its own dedicated power conditioning system (PCS) and full ballast loading capability.

Interconnection Trades

Various interconnection techniques were evaluated for the five PCSs. The possibilities included parallel strings, series strings, and hybrid combinations. Parallel electrical connection of the five individual series string units was selected. Parallel connection allows maximum flexibility to connect or remove any of the five units. Any or all of the individual units may be operated at their set point and rating without imposing implementation requirements on the other units. In the parallel arrangement, if one of the units is shut down, the voltage of each unit will remain at the prior preset level and the current will be redistributed to provide the power demand. The five units may be started sequentially or in parallel. The loads powered by the CRUs are continuously monitored. When an excess input is detected, the load control system acts to reduce the preset level of reactor output, thus, preventing damage or system runaway conditions. The load detector also senses insufficient power input to raise the power command level.

Electric Circuits and Implementation

The details of the electrical circuit are shown in Figure 14. The power processing system has voltage error preamplifiers, saturable reactors (SRs) for ballast control, SCR breaker assemblies, load sharing and control, magnetic power amplifiers, magnetic preamplifiers, ballast loads and radiators, switching, paralleling, protection, and the cycloconverter.

Magnetic components were selected wherever possible because of their inherent higher hardness to most radiation effects. Due to the high-frequency (2000-Hz) operation, there is no mass and size penalty. Discrete power semiconductors will be used dis-

Table 3 *Power Conditioner Mass Summary for Growth.*

System Power (kWe) / System Parameters	100			50	300		1000
System voltage (Vdc)	100	200	400	400	400	600	1000
Cabling mass (kg) 1 percent loss	700	175	35	18	105	50	100
PC system mass (kg) without radiators	615	395	273	188	709	549	1040
Cable + PC mass (kg)	1315	570	308	206	814	599	1140
Van Allen shielding	80	63	50	33	150	100	100
Electronic radiator	312	156	78	42	300	200	620
Resistor spool and ballast radiator (kg)	50	50	50	30	150	150	500
PC + cable + radiators (kg)	1677	776	436	278	1264	949	2260
PC specific mass (kg/kWe)	6.15	3.95	2.73	3.76	2.36	1.83	1.04
PC + cable + radiators specific mass (kg/kWe)	16.77	7.76	4.36	5.56	4.21	3.16	2.26
PC + cable + radiators + shielding specific weight (kg/kWe)	17.57	8.39	4.86	6.22	4.71	3.49	2.36

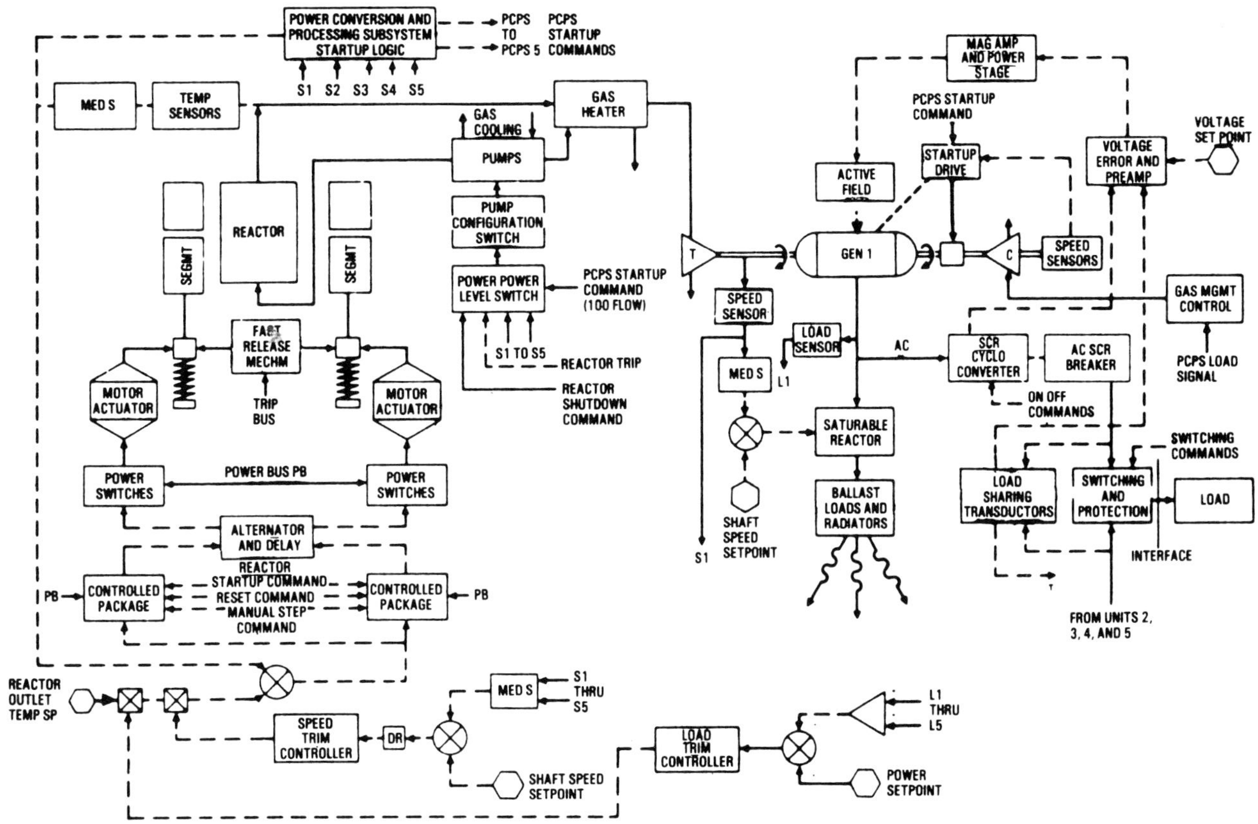

Figure 13 *Power Processing and Control Equipment Block Diagram for Brayton System.*

criminately to raise survivability levels above those available with CMOS and integrated chip technology The representative SCRs (thyristors) have been tested to 10^{14} nvt and 4 x 10^{6} rad by Rockwell.

Switching protection and load sharing are accomplished via the five fully rated (25-kWe) output SCR breakers, the current sharing transducers, and the switching network. The main breaker SCRs remain off until nominal condition is achieved, and both voltage and synchronization conditions are stabilized. When the paralleling comparators indicate acceptable conditions, the SCRs are commanded to turn on, and load current may begin to flow. The nominal voltage, frequency, and phase of all machines are now equal. The transductors compare alternator currents and adjust the field excitation to ensure equalization. Current transformers in the SR ballast loop act in a daisy-loop fashion to ensure equal overall torque and speed output.

Control of Ballast and Active Field of Alternator

Any deviation of alternator voltage from the preset reference level is detected by an error detector that commands the solid-state preamplifier to drive the magnetic preamplifier and the magnetic amplifier power stage to reduce this deviation to the predefined acceptable level. Voltage regulation is achieved by reducing the power to the active field in the alternator when the output exceeds the preset limits, or by increasing the excitation to the active field when the output voltage drops below those limits. Thus, the voltage regulator acts as a fast servomechanism to protect the user loads from damaging stressful conditions and compensates for speed variations, environmental effects, and transients caused by external load variations and input excursions. Figure 15 shows the relationships among the previously mentioned components.

The main smoothing effect on load variation is accomplished by the ballast load control. The ballast load control switching is done by a static SR magnetic system. The SR is essentially a linear controller that, like the magnetic amplifier, offers infinite resolution. The ballast control assists in voltage regulation but its main function is load control and, thus, frequency and speed regulation.

The power conditioning masses were estimated (Table 2) for the 100-kWe Baseline system. It was de-

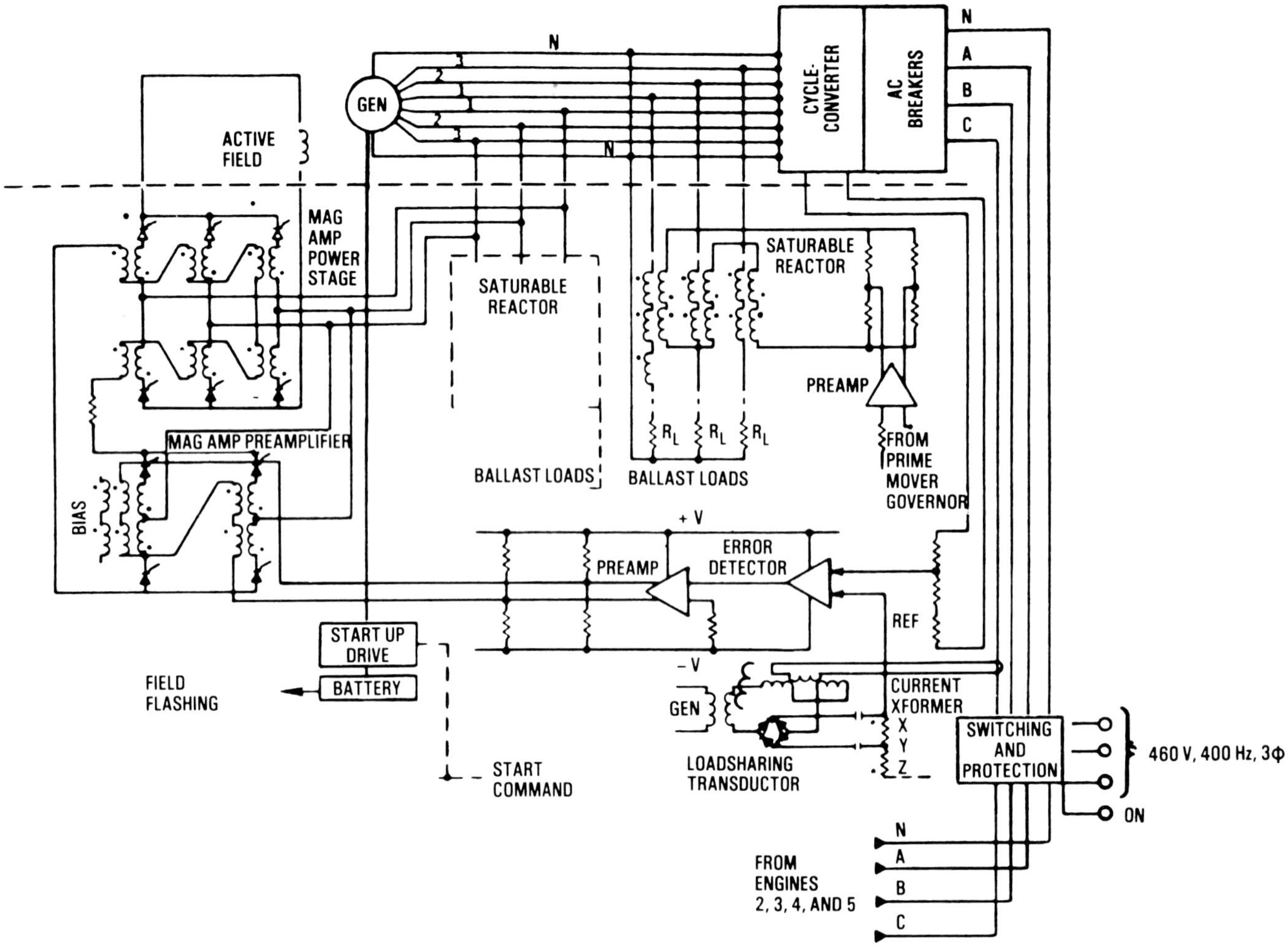

Figure 14 *Brayton System Electrical Circuit.*

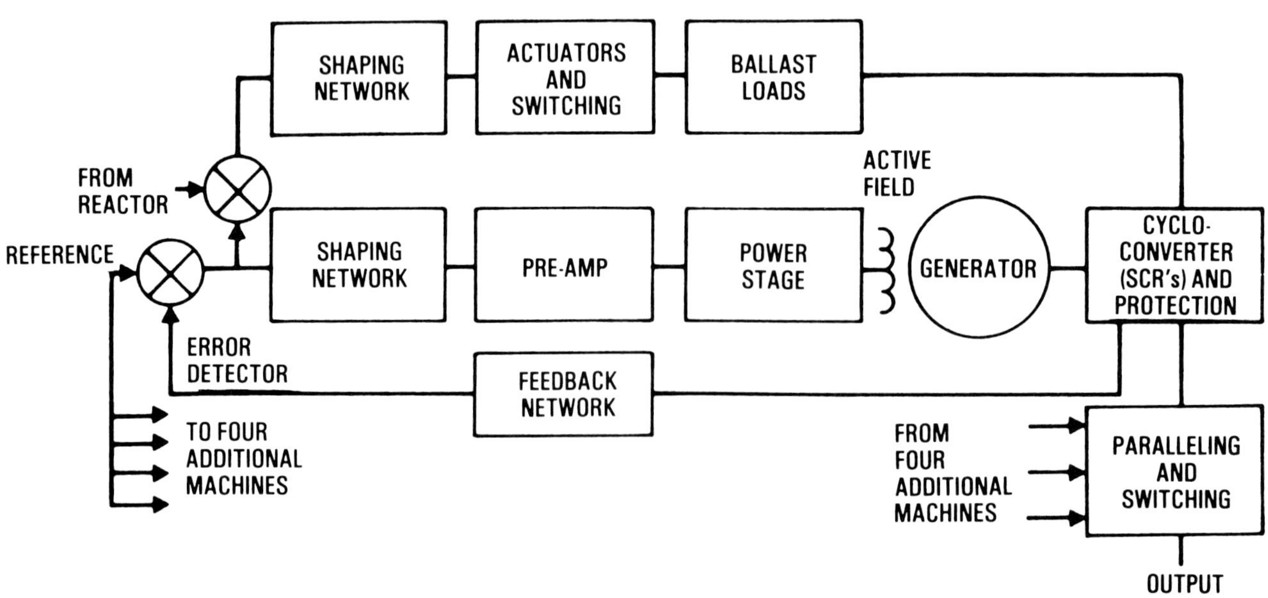

Figure 15 *Power Conditioner Controls Ballast and Active Field in Alternator of Brayton System.*

termined that no special shielding was necessary for the reactor produced radiation. Tantalum was added for the nuclear threat X-rays. Some shielding was also added for the worst-case Van Allen radiation. However, due to the use of magnetic material, the amount of shielding is minimal. Table 2 shows that, due to the high frequency operation of the Brayton System, it has an intrinsic weight and reliability advantage over the Stirling System while offering the additional flexibility of multiphase ac operation.

Conclusions

The optimum operating conditions and parameters for a space reactor power conditioning system are dependent on the required voltage, power level, and power quality specified by the user. Power, power quality and voltage levels dictate power conditioning design with regard to regulation tolerances, ballast margins, series inefficiencies, instrumentation power, battery charging, and housekeeping loads. General design factors such as isolation, electromagnetic interference, statics, and space environment must always be incorporated in the power conditioning design.

Acknowledgments

This summary presents conceptual design information developed for the Tri-Agency (DOD, NASA, and DOE) SP-100 Program under JPL Contract 956935. The permission of the SP-100 Program and Project Office to present this information is gratefully acknowledged.

Space Nuclear Power Systems 1986
Edited by M. S. El-Genk and M. D. Hoover

Chapter 21

Composite Aluminum Conductors for Pulsed Power Applications at Liquid Hydrogen Temperatures

J. C. Ho
Wichita State University
Wichita, KS 67208

C. E. Oberly
Aero Propulsion Laboratory
Wright-Patterson AFB, OH 45433

H. L. Gegel
Materials Laboratory
Wright-Patterson AFB, OH 45433

High-purity, stress-free aluminum has extremely low electrical resistance at liquid hydrogen temperatures and strong magnetic fields. A composite conductor constructed with this high-purity, fine-filament material can allow fast current penetration to the filaments embedded in a high-electrical, low-thermal-resistance matrix to minimize loss problems. This composite conductor can be effectively used in pulsed power applications where liquid hydrogen is available, and where liquid helium normally needed for superconducting devices is absent. Major manufacturing concerns require that the matrix material have workability compatible with that of aluminum and practically zero impurity diffusion to ensure the high purity of aluminum filaments in the final product. We report favorable results of a feasibility study on fabricating such a composite conductor. An aluminum-base alloy containing iron and cerium, recently developed as a high-temperature material with dispersion strengthening, was used as the matrix material. It satisfies most of the aforementioned requirements. Using a streamlined die, billets of this alloy with pure aluminum inserts were successfully extruded in steps. Processing parameters were selected based on results of computer-aided processing simulation.

Introduction

Nuclear reactor systems such as the ROVER Project utilized vast quantities of stored liquid hydrogen propellant to generate high specific impulse thrust in space. This same reactor concept could, in principle, deliver all or part of its hot gas effluent flow to a turbogenerator or a magnetohydrodynamic generator channel to generate enormous electrical power levels in space. The availability of vast amounts of stored liquid hydrogen permits the conceptual design of cryogenic power conversion and power conditioning equipment that could utilize the very low resistance performance of very pure metals. Potential applications of pure metal conductors in power devices that operated at cryogenic temperatures were considered many years ago by Arp (1969), Barthelemy (1966), and Oberly (1967).

In the late 1960s most efforts moved toward the multifilament superconductor due to difficulties in manufacturing and supporting low-strength, pure metal conductors at cryogenic temperatures. Superconductors achieved a development state that permitted successful engineering of magnets and electrical power equipment which were designed, built and tested to specification while retaining essentially zero electrical resistance in most applications. Depending on the application and the superconducting material, an operating temperature of 2 K to 10 K was mandatory to prevent superconductor quenching. Helium coolant and refrigeration cycles are required to maintain this very low temperature, which is below the freezing point of hydrogen. Although several successful cryogenic pure metal magnets have been built, superconducting magnets in the 1980s dominate every field and are considered to be less expensive, more dependable and less risky than pure metal magnets cooled by liquid hydrogen.

Unfortunately, the helium refrigeration cycle required for superconducting magnets in space power system applications has tremendous penalties associated with maintainability, reliability, electrical power consumption and heat rejection radiators. Utilization of available liquid hydrogen from space nuclear reactors such as ROVER can offer very attractive options for high field magnets, bus bars and other power conversion and conditioning apparatus in space. The requirement for a strong, low resistance, pure metal conductor is clear and the following discussion provides a means of meeting the requirement.

Low Loss, High Strength Conductors

Electrical resistivity and mechanical strength are the two main features to be provided in a lightweight conductor for application to space nuclear power systems. A residual resistivity ratio (RRR) can be defined as the ratio of the room temperature resistivity of a pure metal to the resistivity at some other temperature. At liquid nitrogen temperature (77 K) most pure metals have a RRR of 10 and perform quite well under these cryogenic conditions. With ohmic heating losses reduced by 10 times at 77 K, higher current densities and lighter coil windings result from pure metal applications. If extremely pure metals can be obtained (greater than 99.999% pure), the RRR at liquid hydrogen temperature (21 K) can be as high as 50,000!

If pure metal could be effected in a practical conductor, incredibly high-current density, lightweight power equipment could be developed. Unfortunately, these extremely pure metals have no strength (less than 7 MPa for some annealed metals). Strength can be added to pure metal composites with structural mate rials at some penalty in increased overall resistivity, but another problem occurs. Processing of composite metals where a structural material box may surround a pure metal or where many fine filaments are implanted in a structural matrix and extruded and drawn to long lengths requires heavy working and enormous strain and deformation of the pure metals in the composite. These large residual strains induced by work hardening cause a very large increase in electrical resistance. Annealing of the heavily worked composite materials can remove most of the residual strain, but this causes yet another problem for high conductivity. The long period of annealing at elevated temperature permits diffusion of impurities into the pure metals within the composite and ruins the high RRR. Many attempts have been made to alleviate these conditions including incorporation of diffusion barriers such as tantalum or niobium, but these very hard materials cannot be processed in a practical manner with the very soft, ductile pure materials in the composite. Catastrophic sausaging of pure metal filaments or fracture of the entire work piece usually results.

If a pure aluminum conductor could be successfully fabricated to perform the same job as superconductors, it might only need a RRR of 1000 to 2000. Moreover, an aluminum conductor would be only 40% of the weight of typical superconductors or other pure metals and could carry much higher current densities than the practical limit of 2×10^9 amperes/m^2 optimistically expected of superconductors. Also, aluminum has a saturating magnetoresistance that permits it to be used at very high magnetic fields without imposing large increases in resistivity. This work aims at developing a composite conductor usable at liquid-hydrogen temperature that is self supporting at high stress loads of 100s MPa (10s of ksi) while retaining excellent conductivity (RRR = 1000). At the outset of the work, we recognized that impurity diffusion barriers could not be successfully processed in the composite and that entirely new conceptual approaches to encasing the high purity aluminum would be required.

A new class of dispersion strengthened aluminums has been developed for high temperature (400 C to 500 C) high strength 280 MPa to 350 MPa (40 ksi to 50 ksi) applications. The best of these materials, Al-Fe-Ce alloy, was formulated by powder metallurgy techniques as reported by Griffith (1982) and Kirchoff (1982). We have successfully utilized this Al-Fe-Ce alloy as a matrix for filaments with reductions in size of greater than 10^6 during processing by extrusion and drawing to wire of less than 8×10^{-4}m (30 mils) diameter. Structural, thermal and electrical properties of these new conductor materials have been measured, as well as a RRR of greater than 1000 at 4 K. Measurement of RRR at higher temperature is difficult to accomplish on existing apparatus and has not been included here.

It appears that the Al-Fe-Ce material can provide a high strength matrix in which to embed pure aluminum filaments of very high electrical conductivity without severe degradation of conductivity due to metal processing requirements. In addition to the lightweight nature of the high strength Al-Fe-Ce matrix and the high conductivity pure aluminum filaments, the aluminum system will produce very short-lived activation due to close proximity to nuclear reactor fluence. The following discussion will detail our results in developing this new composite aluminum conductor systems for liquid hydrogen cooling situations based on the Al-Fe-Ce dispersion strengthened alloy matrix.

Conductor Selection and Description

Early studies (Arp 1969, Schauer 1975, Krevet 1976, Corruccini 1964, and Hartwig 1980 and 1984) provided a broad range of data in terms of purity, magnetic field, size, strain, and fatigue effect on low temperature resistivity of high purity aluminum. As a result of these studies, one would easily identify aluminum as a better cryoconductor than copper for applications where weight and magnetoresistance are among the main concerns. The only major disadvantage for aluminum is its extremely low strength. This fact is even more serious for such applications as pulsed power devices where, in order to allow full current penetration during a short pulse, the conductors must be in the form of fine filaments. One could conceivably braid the filaments with structural materials for mechanical support. However, a better approach would be to embed Al filaments in an appropriate matrix. Manufacturing technology similar to that for producing multifilament superconductors could also be employed if suitable matrix materials can be found. This report describes a new Al-Fe-Ce alloy that can be used effectively for this purpose, developed for high temperature structural properties by the ALCOA Technical Center, Alcoa Center, PA as reported by Griffith (1982).

High-purity, stress-free aluminum has extremely low electrical resistivity at cryogenic temperatures even under strong magnetic fields as demonstrated by Arp (1969). Saturating magnetoresistivity, as well as light weight, makes aluminum a better choice than copper for the high magnetic field applications under consideration. To allow fast current penetration in pulsed power devices, fine aluminum filaments are required. Their diameters are limited by the size effect of electrical resistivity. Since high-purity aluminum is very soft, the filaments need to be strengthened. This can be most conveniently done by embedding the filaments in a high strength matrix material. Meanwhile, the matrix should have the following properties:

- light weight,
- high strength,
- good thermal conductivity,
- reasonably high electrical resistivity to minimize eddy current loss and to enhance electromagnetic diffusion rate,
- workability compatible with that of high purity aluminum, and
- diffusionless impurities, if present.

The last requirement is most critical to enabling the high-purity aluminum filaments to persist in the final composite product. It is indeed this requirement that eliminates most known aluminum-based alloys from being considered here as a matrix for the high-purity filaments.

The breakthrough in this work came about when a new Al-Fe-Ce alloy was tested as the matrix, as reported by Ho (1985 and 1986). This lightweight material with favorable thermal and electrical properties was initially developed for high temperature applications. With 8.4 and 3.6 wt.% Fe and Ce, respectively, it derives its strength from densely dispersed fine intermetallic compounds of Al, Fe and Ce which have not been completely identified. Al-Fe-Ce ternary compounds are known to exist under equilibrium conditions. However, the alloy used in this effort was powdermetallurgically synthesized by rapid quenching. Consequently, its various phases do not necessarily match those under equilibrium conditions.

The new Al-Fe-Ce alloy (Griffith 1982), containing 8.4 wt.% Fe and 3.6 wt.% Ce, has a low density of 2.95 g/cm^3. Primarily developed for high temperature aerospace applications, it derives its strength from densely dispersed fine intermetallic particles yet to be identified. On the other hand, this material has been successfully used as the matrix in a feasibility study (Ho 1985) for fabricating multifilament aluminum conductors. Specifically, it was coprocessed with high-purity aluminum inserts by multistep extrusions with very uniform deformation which clearly indicated compatible workability. More importantly, significant diffusion of Fe or Ce from the matrix to the aluminum filaments was not detected through RRR measurements.

A complete documentation of physical and mechanical properties of this alloy undoubtedly will be needed for further development. Of particular interest is that such properties can be greatly affected in processing. Described below are results of some comparative experiments on the material before and after dynamic recrystallization.

Experimental Results and Discussion

Microstructures and Mechanical Properties

To successfully fabricate a multifilament aluminum conductor by extrusion, swaging and/or wire-drawing, the matrix material must have suitable microstructure and mechanical properties. This will ensure the integrity of the filaments in the product, and eliminate the possibility of crack formation. The starting Al-Fe-Ce material used in this work was prepared at ALCOA from powders through cold-compaction followed by vacuum hot-pressing (VHP). It was further recrystallized based on dynamic process modeling developed at the Air Force Materials Laboratory by Prasad (1984). Figure 1 shows the remarkably different microstructures before and after the dynamic recrystallization as demonstrated by Ho (1986). Elimination of prior powder particle boundaries in the VHP material yields a more homogeneous distribution of intermetallic particles in the recrystallized condition. As expected, this process results in a slight softening of the material. However, even with the associated decrease in room temperature yield strength (from above 410 MPa to 170 MPa) and ultimate tensile strength (from 480 MPa to 250 MPa), as well as an increase in elongation (from 8-10% to 15%), the alloy can still be considered a good structural matrix material for the multifilament aluminum conductors. It should also be noted that once the alloy is in the soft, recrystallized condition, it can be coprocessed with high-purity aluminum filaments at much reduced temperatures.

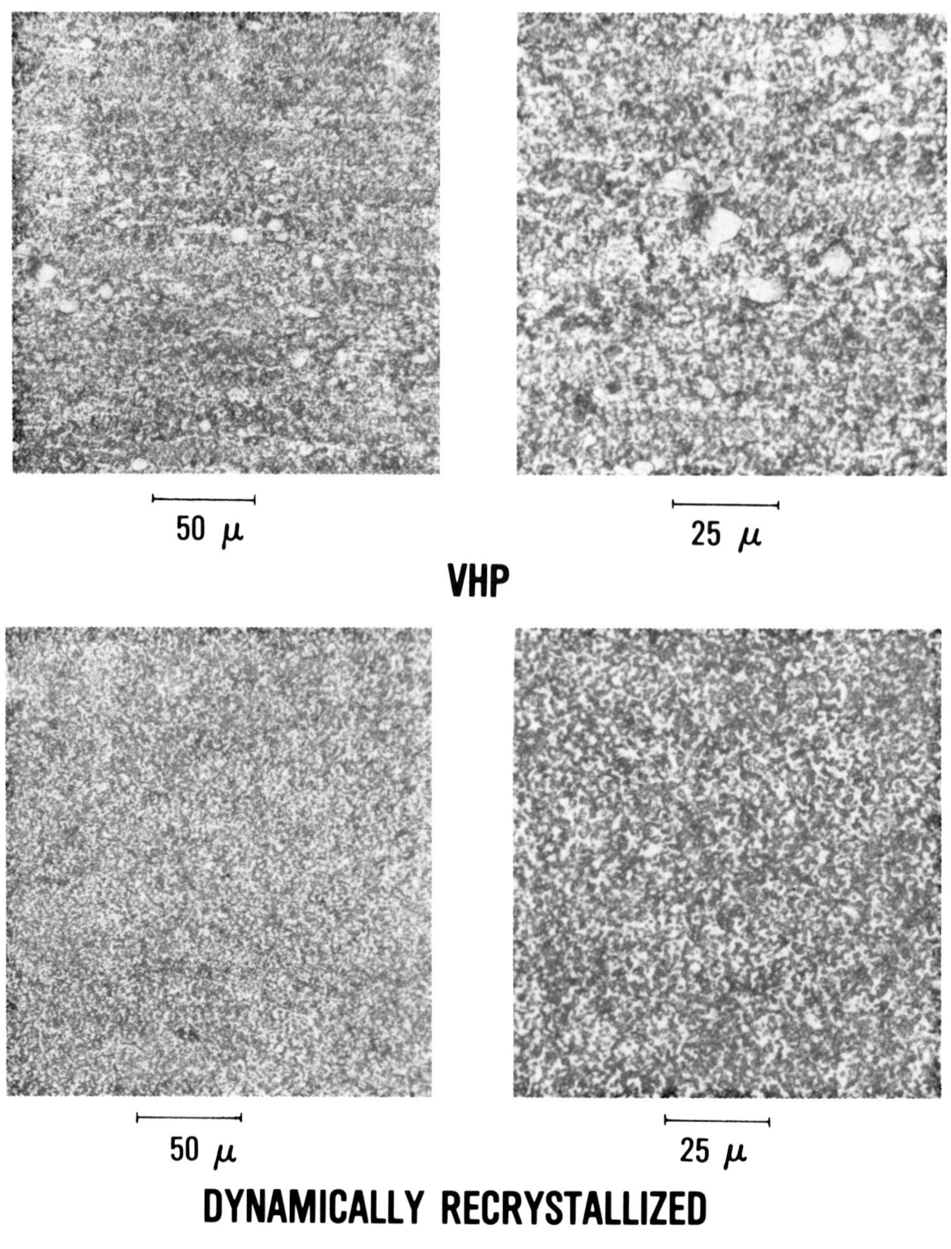

Figure 1 *Micrograph of Al-Fe-Ce Alloy before and after Dynamic Recrystallization.*

This should, in turn, minimize any possible impurity diffusion from the matrix to the filaments.

Electrical Resistivity

Using the standard four-probe method, electrical resistivity of both VHP and dynamically recrystallized samples was determined as given in Table 1 (1100 Al and 6061-T6 Al alloy data also included for comparison). The heavily alloyed Al-Fe-Ce in the dynamically recrystallized condition exhibits behavior similar to 1100-Al insofar as electrical conductivity is concerned. In practice, it is anticipated that, as the matrix material in a composite conductor, this alloy will have certain degrees of work-hardening when the conductor reaches its final wire form. The effect may not be removed even after the conductor is annealed (at relatively low temperatures) to relieve stress in the aluminum filaments. As a result, the alloy could have reasonably high electrical resistance at cryogenic temperatures needed for a short time constant for current penetration info filaments, as well as a low eddy current loss.

Low Temperature Heat Capacity

This study was for the purpose of identifying possible magnetic (Fe, Ce) clusters in the alloy. Typical metals and alloys have low temperature heat capacities, C, as the sum of a lattice term, C_l, and an electronic term, C_e, with different temperature, T, dependencies:

$$C = C_e + C_l = \gamma T + \beta T^3, \quad (1)$$

or,

$$C / T = \gamma + \beta T^2. \quad (2)$$

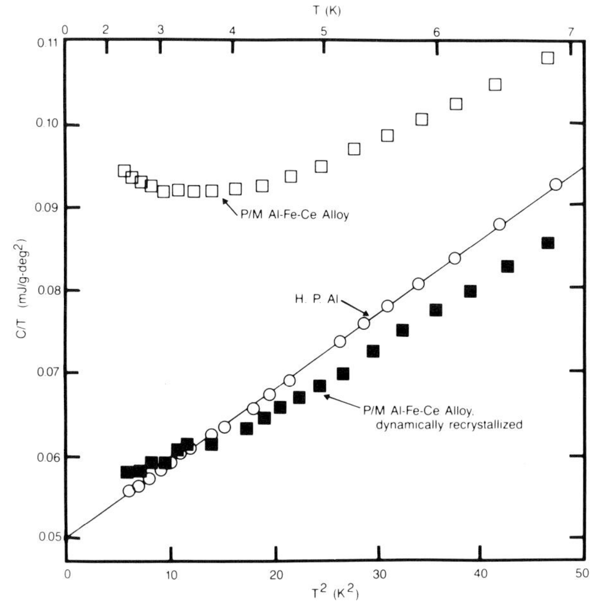

Figure 2 *Low Temperature Heat Capacity of High Purity Al and Al-Fe-Ce Alloy before and after Dynamic Recrystallization.*

This is true for our calorimetric data between 2 K and 7 K for high-purity Al, as indicated in Figure 2 by a linear fit of C/T versus T^2 where γ and β are constant parameters. However, for the VHP Al-Fe-Ce alloy, the observed nonlinearity in the same plot requires a third term to fit the data:

$$C = A + \gamma T + \beta T^3, \quad (3)$$

or,

$$C/T = A / T + \gamma + \beta T^2. \quad (4)$$

The temperature-independent term, A, responsible for the upturn in C/T as temperature decreases, has

Table 1 *Electrical Resistivities.*

	$\rho(\mu\Omega \cdot \text{cm})$		
	Room Temp.	Liq. Nitrogen	Liq. Helium
6061-T6 Al Alloy	3.9	1.9	1.4
VHP Al-Fe-Ce	5.5	1.1	0.66
Dynamically Recryst. Al-Fe-Ce	4.6	0.66	0.27
1100 Al	3.0	0.46	0.20

These lead to the following residual-resistivity-ratio (RRR) values:

	ρ(R.T.)/ρ (L. Nitrogen)	ρ(R.T.)/ρ (l. Helium)
6061-T6 Al Alloy	2.1	2.8
VHP Al-Fe-Ce	5.0	8.3
Dynamically Recryst. Al-Fe-Ce	7.0	17
1100 Al	6.5	15

been previously observed in other nonferromagnetic alloys containing one or more of the elements Fe, Co, and Ni. It is generally recognized as a magnetic clusters contribution by Schroder (1960). Basically, ferromagnetic clusters in a non-ferromagnetic matrix oscillate at low temperatures within directions given by the energy of crystal anisotropy. These extra degrees of freedom lead to a constant heat capacity contribution if the thermal energy is much greater than the energy associated with interaction between the magnetic moments of the clusters and the crystal field. In this regard, it is of interest to note that this effect was much reduced in the heat capacity data obtained from the dynamically recrystallized Al-Fe-Ce sample, again an indication of a more homogeneous microstructure.

Conductor Fabrication

The feasibility study involves basically extrusion and subsequent analysis of Al-Fe-Ce/pure-Al composites. The study was aimed at demonstrating (1) the compatible workability between the alloy and pure Al, and (2) the lack of impurity diffusion from the alloy matrix to the pure Al components. The experimental procedures and results can be summarized as follows: First, nineteen 1/4″-diameter rods of commercially-pure (99.8%) Al were inserted into drilledthrough holes of a 2″-diameter/4″-long Al-Fe-Ce alloy billet (Figure 3). The latter was cold-compacted from powders and VHP. The composite billet was hot-extruded with a streamlined die, as suggested by Gegel (1985), to an area reduction of 12 to 1. Processing parameters (temperature and strain rate) were selected based on dynamic materials modeling techniques developed by Gegel (1983) and Prasad (1984), such that dynamic recrystallization of the alloy matrix would occur. In comparison with the VHP condition, dynamic recrystallization resulted in a product reduced 60% in yield strength. The microstructure was much improved in terms of elimination of prior powder particle boundaries and a more homogeneous intermetallic particle distribution, which then allowed subsequent extrusions at lower temperatures. Indeed, this was done by restacking seven sections of the extrusion product into a commercially-pure Al ingot, followed by a second extrusion of 12 to 1 in area reduction. Figures 4 and 5, respectively, show the 133 Al inserts in the Al-Fe-Ce matrix before and after the second extrusion. The cross-section geometry remained practically the same, and the deformation of Al inserts was reasonably uniform. The micrographs in Figure 6 show well defined boundaries between the matrix and the inserts. The fact that no cracks developed was also encouraging.

All of the above observations suggest that the alloy and pure Al can be coprocessed successfully. In fact, part of the 133-filament composite has been further reduced (Male 1985) in multisteps to a final diameter of 0.79 mm (31 mil). This represents an overall area reduction of more than 100,000 for each Al filament. Detailed analyses of results are in progress.

A second billet was prepared in the same way as shown in Figure 1, except that the nineteen commercially-pure Al rods were replaced by seven high-

Figure 3 *Starting Materials of an Al-Fe-Ce/Al Billet.*

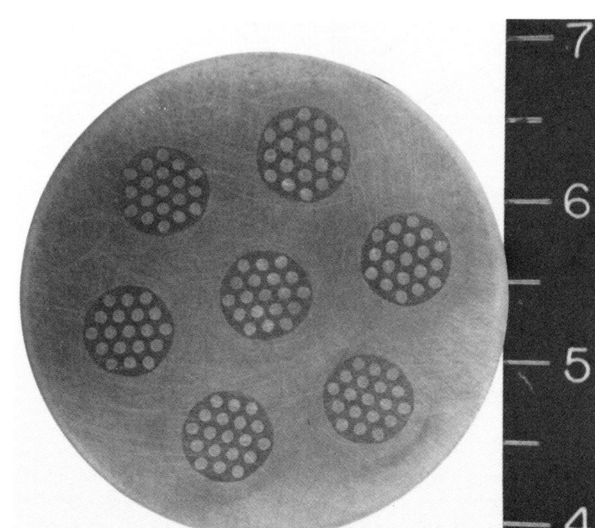

Figure 4 *Stacking of Extruded Al-Fe-Ce/Al Sections in an Al Matrix, ready for a Second Extrusion.*

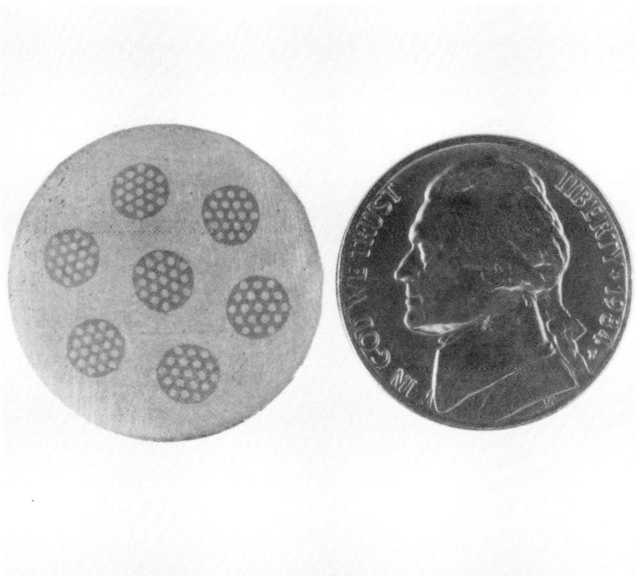

Figure 5 *Cross Section of Second Extrusion showing Little Change in Configuration of 133 Filaments.*

purity (99.998%) Al rods to evaluate impurity diffusion. Figure 7 shows the cross section of hot-extruded product with an area reduction of 16 to 1. No detectable Fe or Ce was found in the Al filaments by electron microprobe analysis, setting an upper limit of their concentrations at 100 ppm. Additional verification of impurity diffusion sensitivity was based on RRR determinations.

An electrical resistivity sample was prepared by turning down a 0.23-m long section of the extrusion product to an 8mm diameter, with the seven Al filaments comprising ¼ of the total cross section and the alloy matrix the remaining ¾. Using the standard four-probe method, electrical resistivity of the composite rod was obtained following a 2-hour annealing at 200 C:

$$\rho(297\ \text{K}) = 3.4 \times 10^{-2} \mu\Omega \cdot \text{cm},$$

$$\rho(78\ \text{K}) = 0.40 \times 10^{-2} \mu\Omega \cdot \text{cm},$$

$$\rho(4.2\ \text{K}) = 0.0088 \times 10^{-2} \mu\Omega \cdot \text{cm},$$

yielding an RRR value of 400. These results did not change after the sample was subjected to a longer (60-hour) annealing at 200 C.

The electrical resistivity of the matrix material itself was also measured (RRR = 17). By coupling these values with the relative cross-sectional areas of filaments and matrix (1:3), it was concluded that the RRR value for the Al filaments alone would be close to 900. This is comparable to that of stress-free Al with a 99.99 + % purity. At this purity level, the electrical resistivity at liquid hydrogen temperature should be only about twice as high as that at 4.2 K.

Conclusion

This work has demonstrated that the newly developed high-temperature Al-Fe-Ce alloy can be further processed to optimize its mechanical and physical properties for cryogenic electrical power conductor applications. Although resistance measurements have only been made at 4 K, liquid hydrogen resistance can be easily projected at the less restrictive temperature of 21 K for the composite aluminum conductor. The Al-Fe-Ce alloy has been demonstrated to be a suitable matrix material for a composite conductor with high-purity aluminum filaments which can be used in lightweight electrical power devices operating at liquid hydrogen temperatures. The alloy is lightweight and of high strength, but just as importantly, its major alloying elements (Fe and Ce) are practically diffusionless in Al as determined by resistivity measurements at 4 K, which are much more sensitive to impurity diffusion than liquid hydrogen measurements. The results support the expectations that impurity diffusion from the matrix to the high-purity filaments during processing was negligible, and that the high-purity Al filaments maintain their excellent electrical conductivity at low temperatures. It is also reasonable to conclude that internal stress in the filaments as induced by the extrusion deformation of the composite is not significant. Dynamic recrystallization processes were further developed at the Air Force Materials Laboratory, resulting in a more homogeneous micro-structure. This paper has described mainly the results of microstructural, electrical resistivity, and low temperature heat capacity measurements on samples before and after dynamic recrystallization.

It is interesting to note that this new aluminum conductor is easily processed since no annealing steps were required during the wire drawing stage. This lack of annealing during processing eliminates any potential impurity diffusion into the tiny pure aluminum filaments. The potential of the Al-Fe-Ce alloy to "lock up" troublesome Fe and Ce in dispersed inter-

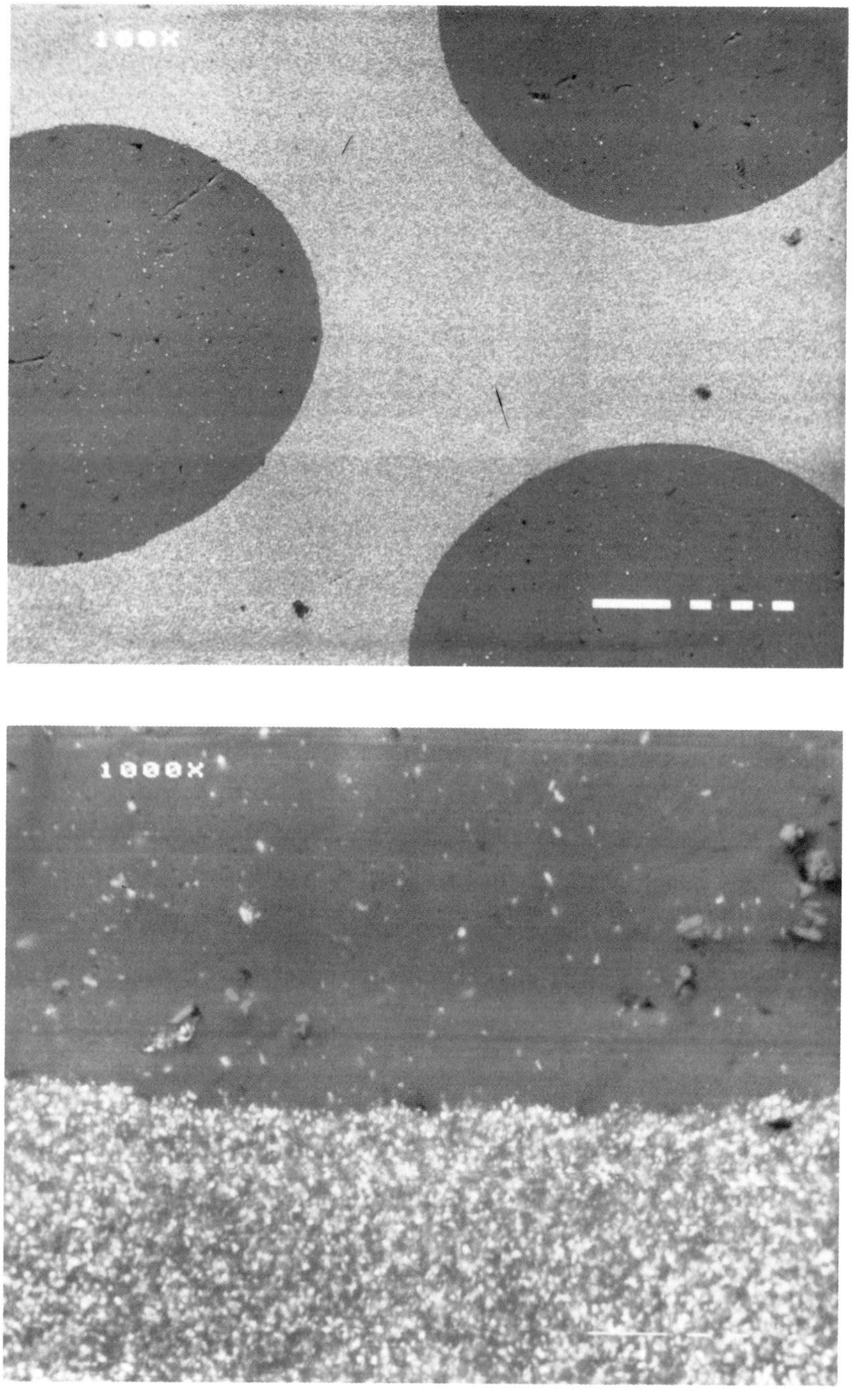

Figure 6 *Well-Defined Boundaries between Al Filaments and Al-Fe-Ce Alloy Matrix.*

metallic alloys is very beneficial to the composite in that no significant reduction in RRR could be achieved in post-processing annealing studies until a temperature of 500 C was maintained for 16 hours. This post-processing annealing is well in excess of the time and/or temperature which a prototype composite aluminum conductor would experience.

We believe this new conductor material performance is very impressive and will again become a serious competitor for large magnets and electrical power components for systems with large amounts of stored liquid hydrogen available. The aluminum conductor is also of interest for application to nuclear space power systems because of the fast decay of its

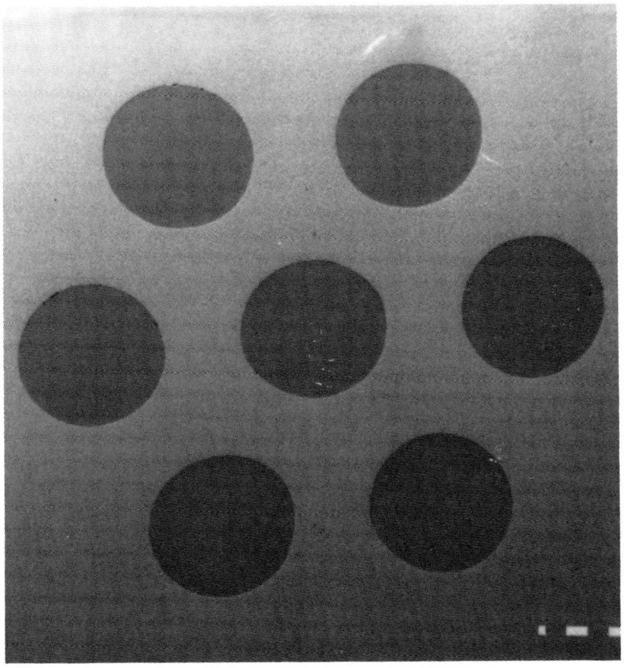

Figure 7 *Cross Section of 7 High-Purity Al Filaments in Al-Fe-Ce Matrix, after being Extruded to an Area Reduction of 16 to 1.*

activation products. One of the authors has already projected the application of this new composite aluminum conductor to lightweight generators (Schlicher 1985).

Acknowledgments

This work was supported by the Aero Propulsion Laboratory at Wright-Patterson Air Force Base. This in-house effort was performed almost entirely within the confines of the Aero Propulsion Laboratory and the Materials Laboratory at Wright-Patterson Air Force Base. We gratefully acknowledge the assistance of Dr. Alan Male and his group at the Westinghouse Research and Development Center in Pittsburgh, PA for the successful drawing of the composite extrusion to wire.

References

Arp, V. D., M. B. Kasen, and R. P. Reed (1969) *Magnet Energy Storage and Cryogenic Aluminum Magnets*, Air Force Aero Propulsion Laboratory Technical Report AFAPL-TR-68-87, Wright-Patterson AFB, OH, February 1969.

Barthelemy, R R. and C. E. Oberly (1966) "Optimum Magnets for MHD Generators," *International Symposium on MHD Electrical Power Generation*, held in Salzburg, Austria, 4–8 July 1966.

Corruccini, R. J. (1964) *The Electrical Properties of Aluminum for Cryogenic Electromagnets*, Technical Note No. 218, National Bureau of Standards, Boulder, CO.

Gegel, H. L. (1983) "Material Behavior Modeling - An Overview," in *Experimental Verification of Process Models*, C. C. Chen, ed., Am. Soc. Metals, Menlo Park, OH, pp. 3–32.

Gegel, H. L., J. C. Malas, and S. M. Doraivelu (1985) "Process Modeling of P/M Extrusion," in *Innovations in Materials Processing*, G. Bruggeman and V. Weiss, eds., Plenum Press, New York, pp. 137–159.

Griffith, W. M., R. E. Sanders, Jr., and G. J. Hildeman (1982) "Elevated Temperature Aluminum Alloys for Aerospace Applications," in *High-Strength Powder Metallurgy Aluminum Alloys*, M. J. Koczak and G. J. Hildeman, eds., the Metallurgical Society of AIME, pp. 209–224.

Hartwig, K. T. and P. G. Appius (1980) "Strength and Electrical Resistance of Cold Drawn Al and Cu," in *Proc. 8th Int. Cryogenic Eng. Conf.*, C. Rizzuto, ed., IPC Science and Tech. Press, pp. 720–725.

Hartwig, K. T., G. S. Yuan, and P. Lehmann (1984) "The Effects of Low Temperature Fatigue on the RRR and Strength of Pure Aluminum," *Appl. Supercond. Conf.*, held in San Diego, CA, 9–13 September 1984.

Ho, J. C., C. E. Oberly, H. L. Gegel, W. T. O'Hara, J. T. Morgan, Y. V. R. K. Prasad, and W. M. Griffith (1985) "Composite Aluminum Conductors for Pulsed Power Applications at Hydrogen Temperatures," *5th IEEE Pulsed Power Conf.*, held in Arlington, VA, 10–12 June 1985.

Ho, J. C., C. E. Oberly, H. L. Gegel, W. M. Griffith, J. T. Morgan, W. T. O'Hara and Y. V. R. K. Prasad (1986) "A New Aluminum-Base Alloy with Potential Cryogenic Applications." to be published in *Advances in Cryogenic Engineering (Materials) 32*.

Kirchoff, S. D., R. H. Young, W. M. Griffith and Y. W. Kim (1982) "Microstructure/Strength/Fatigue Crack Growth Relations in High Temperature P/M Aluminum Alloys," *High-Strength Powder Metallurgy Aluminum Alloys*, M. J. Koczak and G. J. Hildman, eds., the Metallurgical Society of AIME, pp. 237–248.

Krevet, B. and W. Schauer (1976) "Transverse Magnetoresistance and Its Temperature Dependence for High-Purity Polycrystalline Aluminum," *J. Appl. Phys.* 47:3656-3669.

Male, A. (1985) Personal Communication of effort volunteered at the metals processing research group at the Westinghouse Research and Development Center, Pittsburgh, PA.

Oberly, C. E. and R. R. Barthelemy (1967) "Low Temperature Magnets for MHD Power Generation," *12th Intl. Congress of Refrigeration*, held in Madrid, Spain.

Prasad, Y. V. R. K., H. L. Gegel, S. M. Doraivelu, J. T. Morgan, K. A. Lark, and D. R. Barker (1984) "Modeling of Dynamic Material Behavior in Hot Deformation: Forging of Ti-6242," *Met. Trans.*, 15A:1883–1892.

Schauer, W. (1975) "Conductor Materials for Cryomagnets," *Proc. 5th Intl. Conf. Magnet Technology*, N. Sacchetti, M. Spadoni, and S. Stipcich, eds., Laboratori Nazionali del CNEN, Frascati, Italy, pp. 629–643.

Schlicher, R. L. and C. E. Oberly (1985) "Cryogenic Aluminum Wound Rotor Concept," *5th IEEE Pulsed Power Conference*, held in Arlington, VA, 10–12 June 1985, pp.752–755.

Schroder, K. and C. H. Cheng (1960) "Correlation of Low-Temperature Caloric and Magnetic Effects in TiFe," *J. Appl. Phys.*, 31:2154–2155.

Space Nuclear Power Systems 1986
Edited by M. S. El-Genk and M. D. Hoover

Chapter 22

Cryogenic Aluminum-Wound Generator Rotor Concept for Nuclear Power Conversion

R. L. Schlicher

Air Force Weapons Laboratory
Kirtland Air Force Base, NM 87117

C. E. Oberly

Air Force Wright Aeronautical Laboratories
Wright-Patterson Air Force Base, OH 45433

This paper presents a design outline for a liquid hydrogen cooled generator rotor that could be used to fabricate a 20-megawatt cryogenic generator. The armature of an existing 20-megawatt superconducting generator could be utilized in this new cryogenic generator concept without electrical modification and with minimum modification to its housing. The acquisition and operating expense of liquid helium liquefiers, refrigeration requirements and the expense of fabricating a superconductorwound generator rotor make an aluminum-wound rotor a viable alternative. Ideally, the aluminum rotor could use the higher cryogenic temperatures of liquid hydrogen at 21 K as conductor coolant and not require the more difficult fabrication techniques of a superconducting generator rotor. A most likely conductor candidate is high purity aluminum which has 0.2% its room temperature resistance at liquid hydrogen temperatures. Recent research has indicated the feasibility of fabricating high-purity aluminum conductors in a composite conductor form.

Introduction

High purity aluminum conductors cooled by liquid hydrogen have long been recognized as significant to the development of lightweight, high power generators (Barthelemy and Oberly 1966 and Oberly and Barthelemy 1967). The aluminum rotor has two distinct advantages over the superconductor rotor for use in aerospace applications. Foremost is conductor stability. The aluminum coil does not have the inherent risk of a superconducting coil going "normal," that is, suddenly becoming non-superconducting. When this happens, the liquid helium coolant in the rotor assembly suddenly changes to helium gas and causes a lengthy delay in the recovery of superconductivity. However, cryogenic aluminum conductors have been shown to exhibit a critical current below which equilibrium can exist between ohmic losses and cooling rate for a given cryogenic coolant (Zhitomirsky et al. 1985). The second advantage is light weight. The mass of the alloying metals that constitute the superconducting wire used in a rotor coil is more than twice that of the equivalent aluminum wire.

To be competitive, the candidate aluminum wire for this alternative rotor design needs to be highly conductive in the temperature range around 21 K. The most likely aluminum conductor would be a composite containing filaments of high-purity aluminum, which at the cryogenic temperature of liquid hydrogen has a resistivity of less than 0.2% of the value of its room temperature resistivity. Hence, the best conductor candidate for this rotor design would be liquid hydrogen cooled aluminum. Several recent superconducting magnets (Hirabayashi et al. 1981 and Fast et al. 1985a and 1985b) have demonstrated a residual resistivity ratio of 1000 to 2000. These magnets were not operated at high magnetic fields and stresses; therefore, the magnetoresistance and work hardening potential of the rotor application is not present. Ideally, the aluminum conductor also would have high yield and creep strength, for example, the matrix material in the composite aluminum conductor discussed by Ho et al. (1985 and 1986).

Rotor Design

Since this paper presents a conceptual design, only the significant characteristics of this candidate rotor design will be described. The emphasis will be on the electrical design of the rotor and a general description of its physical characteristics. Because of the physical constraints imposed by using the 20-megawatt generator armature and housing (Gamble and Keim 1982), a highly conductive aluminum coil will be substituted for the superconducting niobium-tin coil, and liquid hydrogen flow instead of liquid helium coolant will be utilized to cool the rotor windings. Consequently, the overall dimensions of the aluminum rotor would be essentially the same as the superconducting rotor as illustrated in Figure 1. Several comprehensive references exist on synchronous machine design (McAbria 1979, Southall and Brockhurst 1979, and Jain 1968).

Design of the Aluminum Wound Coils

The physical constraints of the selected armature and housing require the overall physical dimensions of the aluminum rotor coils to be less than or equal to the overall physical dimensions of the superconducting rotor coils. Therefore, an aluminum coil will have to be designed to fit in the conductor slot of the superconducting rotor coil. A cross section of one pole of this slot geometry is illustrated in Figure 2. The implication of this assumption is that the current density of the aluminum conductor will equal or exceed that of the superconductor.

To simplify coil design and construction, a 20-turn coil of flat conductor is selected for this design. This provides maximum surface area for heat transfer of the ohmic heat to the liquid hydrogen coolant, as well as sufficient passageways for the hydrogen coolant. The maximum number of turns for a conductor with large cross section that would still allow sufficient room for coolant is about 20. The flat conductor itself would consist of multiple filaments of small diameter, pure aluminum bonded together to form the flat conductor geometry in Al-Ce-Fe matrix (Ho et al. 1985 and 1986).

In determining the dimensions of the conductor, we conservatively assume that 35% of the coil module cross section must be allowed for structural support material and coolant passage space (see Figure 2). The high-purity aluminum must be encased in a structural support material to maintain its conductive properties in order to prevent strains in excess of 0.1% on the pure aluminum filaments (Hartwig et al. 1985). Furthermore, a gap of 2.5 mm is assumed to separate the coil module from the slot walls. This gap provides room for spacers, coolant passages, and side insulation. Thus, we calculate the net conductor cross section per turn to be 3.77×10^{-3} m²/20 turns =

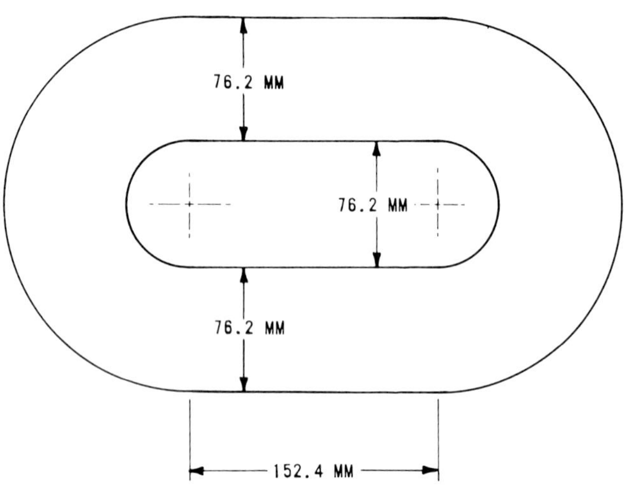

Figure 1 *Racetrack Module Dimensions.*

Table 1 *20-MW Superconductor Rotor Parameters.*

Number of Coils (Poles)		4
Number of Turns/Coil		980
Amount of Current/Coil		860 amps
Dimensions of Coil (Racetrack module):		
Cross Section	.0762 m × .0762 m	3″ × 3″
Active Length	.1524 m	6″
Inside Width	.0762 m	3″
Outside Width	.2286 m	9″
Inside Bend Radius	.0381 m	1.5″
Outside Bend Radius	.1143 m	4.5″

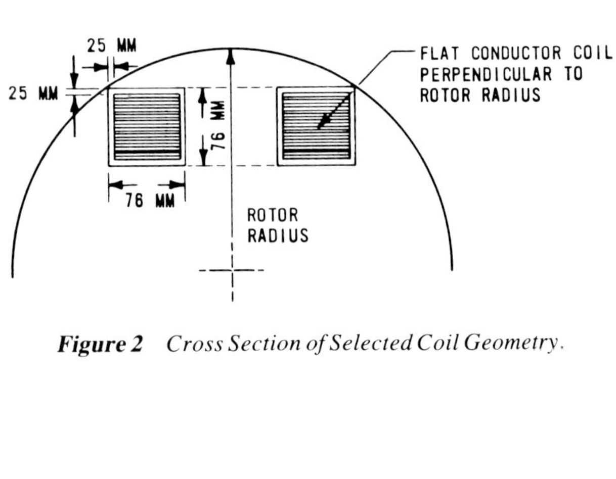

Figure 2 *Cross Section of Selected Coil Geometry.*

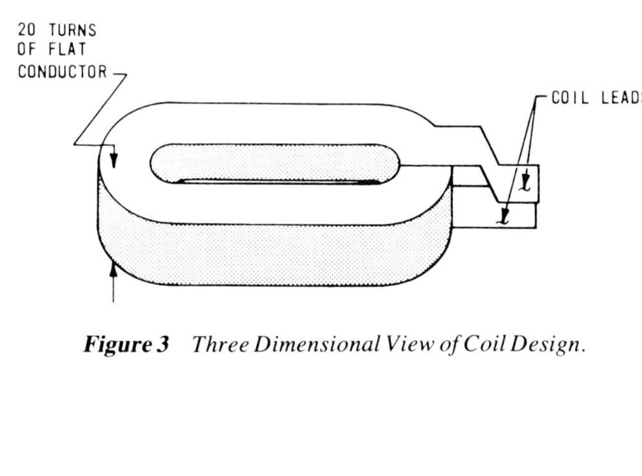

Figure 3 *Three Dimensional View of Coil Design.*

1.89×10^{-4} m²/turn. A .0712-m conductor width is selected with a thickness of 2.7 mm.

Composite Aluminum Conductors

Except for its high conductivity at cryogenic temperatures, the other properties of high-purity aluminum render it inadequate for use in the rotor coils of a high power alternator. In the annealed high-purity state, aluminum is mechanically too weak to withstand the internal mechanical stresses induced by the centrifugal forces of rotation and the back mmf of the stator under load (typically 7-MPa yield strength). Furthermore, high-purity aluminum work hardens, thereby lowering its conductivity. Embedding high-purity aluminum in a high strength matrix is a solution to these shortcomings.

Fabrication of a high-purity aluminum composite conductor requires that the matrix material have a workability that is compatible with aluminum, as well as practically zero impurity diffusion from the matrix into the high purity aluminum filaments. A feasibility study and experiment completed at the Air Force Wright Aeronautical Laboratories indicate that such a composite can now be fabricated (Ho 1985 and 1986).

Determining Ohmic Losses

For this conceptual design, we assumed that pure aluminum would be used as the filaments in an aluminum composite conductor with properties that have recently been determined (Ho et al. 1985 and 1986). If we assume that the resistivity ratio of this alloy at 21 K is equal to 500, then the estimated cryogenic electrical resistivity of the alloy is assumed to be 5.6×10^{-11} ohm-m.

Since our primary concern at this time is the steady state operation of the rotor coils, the following simple relation is used to calculate the net ohmic resistance of a single coil.

Conductor Electrical Resistance $= L/A$ where

- L = conductor length,
- ρ = electrical resistivity, and
- A = conductor cross section.

Referring to Figure 1, we calculate the net conductor length per turn to be .7836 m. Multiplying this by 20 turns and adding .3048 cm (12 inches) for the total length of the downleads provides us with the net conductor length for one coil, which is 15.98 m. Having calculated the conductor length per coil and knowing the resistivity and the conductor cross section, we can now estimate the steady state ohmic resistance of a single coil to be 4.73×10^{-6} ohms.

Since the basic magnetic geometry of the cryogenic aluminum wound rotor is the same as the superconductor rotor, the magnetomotive force (mmF) to generate the rotor magnetic field would be the same as in the superconductor rotor coils.

mmF/coil (superconductor rotor)
$$= 860 \text{ amps} \times 940 \text{ turns}$$
$$= 808,400 \text{ ampere-turns.}$$

Thus, the net current drawn per aluminum wound coil becomes

808,400 ampere-turns/20 turns per coil
$$= 40,420 \text{ amps per coil.}$$

Using the power relationship $P_c = I^2R$ watts for steady state DC operation, we have the net power dissipated per coil:

$$P_c = (40,420 \text{ amps})^2 \times 4.73 \times 10^{-6} \text{ ohms}$$
$$= 7.73 \times 10^3 \text{ watts per coil.}$$

Since there are four coils in the rotor, the net power dissipated in the rotor P_r becomes

$$P_r = 4 \times 7.73 \times 10^3 \text{ watts per coil}$$
$$= 3.09 \times 10^4 \text{ watts.}$$

Since the coils are to be connected in series and the downleads are relatively short compared to the coil conductor length, ohmic losses in the downleads were disregarded. Ohmic losses in the exciter circuit will be discussed later.

Determining the Conductor Heat Load

We selected the arrangement outlined in Figure 4 because of its simplicity. Strips of insulating material are alternately laid across the width of the flat conductor between each turn to form a coolant channel for the crossflow of liquid hydrogen. The width of the strips would depend upon the total number of strips and the total conductor area exposed to the liquid hydrogen. From our preliminary analysis and previous experience, at least one-half of the area on a given side of the flat conductor will be exposed directly to the hydrogen coolant cross-flow. Placing a coolant channel on the opposite side to an insulating strip will enhance the cooling process and minimize the possible occurrence of thermal hot spots under the insulating strips. Ultimately, the width of the insulating strips will not be determined by thermal analysis, but primarily by the results of mechanical analysis and fluid dynamics of the hydrogen flow.

In performing the preliminary analysis, we assumed that there was negligible insulation on the conductor matrix except for the insulator strips separating the coil turns. As previously stated, we assumed that one-half of the composite conductor outer surface is exposed to the cryogenic hydrogen coolant crossflow. With these assumptions, we calculate the thermal characteristics of the steady state behavior of rotor.

If we assume that 10% of the coil module cross section consisted of structural support material, then 25% of the cross section remains for the net coolant cross section separating the coil turns. The coolant cross-sectional area is 1.45×10^{-3} m^2. The cross section of each coolant channel separating the coil turns is 7.64×10^{-4} m^2, and the width of each coolant channel between conductor turns is 1 mm. The coolant orifice between conductor turns is

$$\begin{aligned} \text{Orifice} &= (\text{Length of turn} \times \text{channel width})/2 \\ &= 3.92 \times 10^{-4}\ \text{m}^2. \end{aligned}$$

The effective orifice of the coolant cross-sectional area depends on the fluid dynamic properties of the liquid hydrogen and its static pressure as it is forced between the coil turns. It is not the purpose of this paper to present a detailed thermal analysis of the rotor concept. Nevertheless, the orifice between each turn appears sufficient to permit effective coolant flow across the conductor surface.

The net heat transfer surface available in each rotor coil is 1.16 m^2. By dividing the ohmic power dissipated per coil by the heat transfer surface/coil, we arrive at a good approximation of the ohmic heat load per coil, which is 6670 W/m^2. A heat load of less than 10^4 W/m^2 of surface area is readily manageable. The temperature rise in the worst case in the aluminum under these heat flux conditions would be 3 mK. At this very low ΔT, the hydrogen flow requirements would be excessive, so a 3 K temperature rise in the hydrogen would allow a more desirable cooling flow less than 1 kg/s in the rotor.

Each coil has very low resistance, so to minimize the current required from the exciter, all four coils are wired in series. Since the lengths of the rotor downleads are relatively short compared to lengths of the windings in all four coils, the resistance of the downleads is disregarded. The resistance of the rotor, R, thus becomes the sum of the individual resistances of each coil. The exciter current, I, becomes the same value in each coil.

$$\begin{aligned} R_{net} &= 4\ \text{coils} \times 4.73 \times 10^{-6}\ \text{ohms/coil} \\ &= 1.89 \times 10^{-5}\ \text{ohms/rotor} \end{aligned}$$

Knowing the rotor resistance and the exciter current drawn by the rotor, we can now calculate the voltage, V_R, across the rotor downleads to be 0.76 volts.

Since the rotor is a low-voltage, high-current device, it seems suitable to utilize a homopolar generator mounted on the same generator drive shaft as the exciter current source. This would introduce an additional voltage drop (typically, on the order of 0.3 volts or less) due to the contact resistance of the homopolar generator current collector brushes. If we include brushes to transfer the exciter current from the homopolar generator to the rotor downleads, there is an additional voltage drop of 0.3 volts. Consequently, a third to almost one-half of the losses in the rotor exciter circuit could be attributable to the contact resistance of the current collector brushes of the homopolar generator and of the downleads. Consequently, a worse case exciter circuit utilizing a homopolar generator as the exciter current source would have the following ranges of characteristics:

- Output Voltage: 1.1 to 1.4 volts,
- Output Current: 40,300 to 40,500 amps, and
- Output Power Rating: 44,300 to 65,700 watts.

Depending on the efficiency of the exciter current source, the exciter power requirements could be as

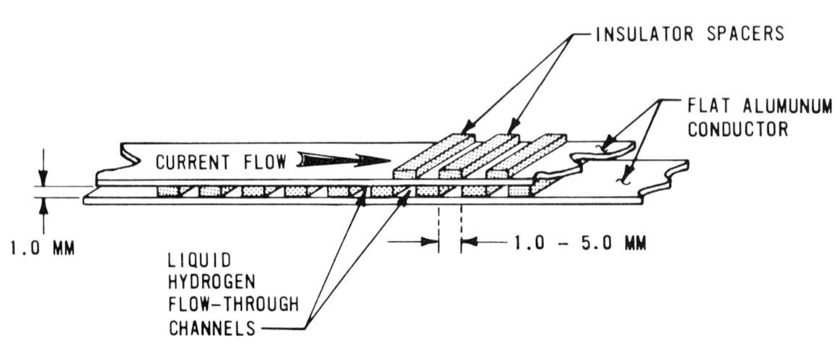

Figure 4 *Simple Conductor-Insulator Matrix.*

high as 100 kilowatts. For the rotor shaft mounted homopolar type exciter, this loss would figure directly into the overall efficiency of the generator. Figure 5 illustrates the complete cryogenic rotor concept.

Nuclear Power Conversion

Nuclear reactor systems such as the ROVER Project utilized vast quantities of stored liquid hydrogen propellant to generate high specific impulse space thrust. This same reactor concept could, in principle, deliver all or part of its hot gas effluent flow to a turbine which could then drive a lightweight alternator to generate enormous electrical power levels in space.

It would be very beneficial to a space nuclear power system if the light-weight alternator could utilize the tremendous reservoir of liquid hydrogen stored for the ROVER-type reactor before the hydrogen flow entered the reactor system. Since today's practical superconductors cannot exist at 21 K, it is propitious that a recently developed composite aluminum conductor can fully exploit the cooling capabilities available in the liquid hydrogen flow required to transfer heat from the reactor.

A major design consideration in utilizing this generator concept in a nuclear power system is the relative radiation hardness of the generator conductor coils. Because aluminum composite conductor is a low-Z material, heating due to the reactor fluence is negligible, provided that the generator is on the payload side of the reactor shield. Since hydrogen acts as a reactor moderator, placing the hydrogen reservoir between the reactor core and the generator would significantly reduce the nuclear heat deposition. Since the life expectancy of an operating, open-cycle, ROVER-type reactor is relatively short, aluminum alloy damage due to prolonged exposure to the reactor fluence is not a serious consideration.

Generator System Concept

This generator rotor design was predicated upon the assumption that the generator would be part of a hydrogen cooled turboalternator prime power system. The hydrogen fuel, in its liquid state, would be stored in cryostats. Upon system startup, the liquid hydrogen at 21 K would be pumped from the cryostats, through the generator rotor (perhaps including the stator) and then into the nuclear reactor. The hot gases generated in the nuclear reactor would power the turbine assembly turning the generator, as is illustrated in Figure 6.

Summary

Where a large reservoir of liquid hydrogen is already available, it appears feasible to utilize a hydrogen cooled aluminum generator rotor instead of a superconducting rotor. This would eliminate the risk of thermal instability associated with a superconducting rotor should one of its coils unexpectedly go "normal." Any number of mechanical and thermal stress combinations could cause this. Furthermore, weight and complexity would be reduced by eliminating the liquid helium reservoir and the associated liquid helium refrigeration and liquefication equipment required for superconducting rotors.

Acknowledgments

This work was supported by the Aero Propulsion Laboratory at Wright-Patterson Air Force Base as an in-house effort. The inspiration for this work was provided by the successful in-house development of a composite aluminum conductor for

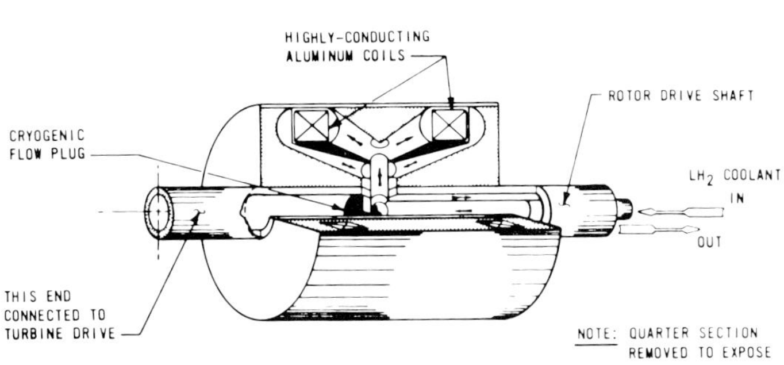

Figure 5 *Cryogenic Rotor Concept.*

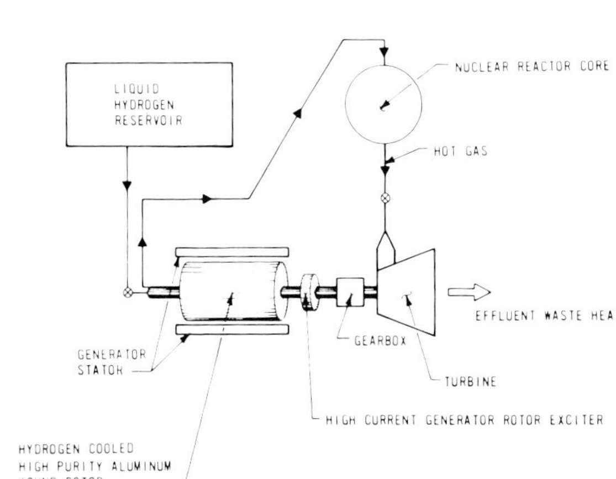

Figure 6 *Cryogenic Generator System.*

cryogenic applications by an in-house team of Aero Propulsion Laboratory and Materials Laboratory engineers at Wright-Patterson Air Force Base.

References

Barthelemy, R. R., and C. E. Oberly (1966) ''Optimum Magnets for MHD Generators,'' *Electricity from MHD*, Vol. III, Intl. Atomic Energy Agency, Vienna, pg. 153.

Fast, R. W. et al. (1985a) ''Fabrication and Testing of the 3m Diameter Superconducting Solenoid and Refrigeration System for the Fermilab Collider Detector,'' *IEEE Trans. Mag.*, 21:963.

Fast, R. W. (1985b) Personal Communication, CDF Magnet Test Achieved RRR = 1885 at 1–2% Strain Using Conductor Described in Reference by Hirabayashi 1981, and Fast 1985a.

Gamble, B. B. and T. A. Keim (1982) ''High-Power Density Superconducting Generator'', *AIAA J. of Energy*, 6:38.

Jain, G. C. (1968) *Design, Operation and Testing of Synchronous Machines*, ASIA Publishing House, New York, NY.

Hartwig, K. T., G. S. Yuan, and P. Lehmann (1985) ''The Effects of Low Temperature Fatigue on the RRR and Strength of Pure Aluminum'', *IEEE Trans. Mag.*, 21:161.

Hirabayashi, H. et al. (1981) ''Measurement of Propagation Velocities of the Normal Zone in a 1m x 1m Superconducting Solenoid Magnet,'' *Jap. J. of Appl. Phys.*, 20:2243.

Ho, J. C., C. E. Oberly, H. L. Gegel, W. T. O'Hara, J. T. Morgan, Y. V. R. K. Prasad, and W. M. Griffith (1985) ''Composite Aluminum Conductors for Pulsed Power Applications at Hydrogen Temperatures,'' *5th IEEE Pulsed Power Conference*, Alexandria VA.

Ho, J. C. , C. E. Oberly, H. L. Gegel, W. T. O'Hara, J. T. Morgan, Y. V. R. K. Prasad, and W. M. Griffith (1986) ''A New Aluminum-Base Alloy with Potential Cryogenic Applications'', *Adv. in Cryo. Eng. Materials*, 32:437.

McCabria, J. L. (1979) *Program for the Development of a Superconducting Generator*, Technical Report AFAPL-TR-79-2012, USAF Wright Aeronautical Laboratories, AFWAL/POOC, Wright-Patterson AFB, OH.

Oberly, C. E. and R. R. Barthelemy (1967) ''Low Temperature Magnets for MHD Generators,'' *12th International Congress on Refrigeration*, Madrid, Spain.

Schlicher, R. L. and C. E. Oberly (1986) ''Cryogenic Aluminum-Wound Rotor Concept,'' in *Proceedings of Fifth IEEE Pulsed Power Conference*, held in Arlington, VA, 10–12 June 1985, pp. 752–755.

Southall, H. L. and F. C. Brockhurst (1979) *Permanent Magnet and Superconducting Generators in Airborne, High-Power Systems*, Technical Report AFAPL-TR-79-2073, USAF Wright Aeronautical Laboratories, AFWAL/POOS, Wright-Patterson AFB, OH.

Zhitomirsky, I. S. , T. V. Bandos, and I. N. Nechiporenko (1985) ''Critical Currents in Hyperconducting Windings,'' *Cryogenics*, 25: 387.

Space Nuclear Power Systems 1986
Edited by M. S. El-Genk and M. D. Hoover

Chapter 23

Inorganic Dielectric Insulation for Superconductors and Conventional Conductors in Space Reactor Environments

T. K. Gupta

Alcoa Laboratories
Alcoa Center, PA 15069

W. N. Lawless

Ceram Physics, Inc.
Westerville, OH 43081

C. E. Oberly

Aero Propulsion Laboratory
Wright-Patterson AFB, OH 45433

G. J. Bich and G. R. Wagner

Westinghouse R&D Center
Pittsburgh, PA 15235

D. S. Holmes

Dept. of Nuclear Engineering
University of Wisconsin
Madison, WI 53716

The need for high current, stable superconductors had led to the development of Nb_3Sn as the most promising candidate for use in superconducting machines. However, its brittle nature of and the high reaction temperature (973 K) required to form it preclude the use of standard organic insulation systems. A recently developed class of high temperature dielectric materials, which are characterized by unusually large specific heats and thermal conductivities at cryogenic temperatures, offers increased enthalpy stabilization in a super-conducting winding, as well as the required dielectric strength. In recent years, the Air Force supported a series of programs to develop a film-type dielectric insulation system for superconductors using these materials incorporated in a glassy matrix. The development of this composite insulation system is described in this paper and its application is extended to ordinary conductors for space nuclear power systems.

Introduction

Reactors in space will be used to generate electrical power in an environment unfavorable to long-term reliability of conventional organic insulations. Ceramic insulations have been recognized as promising candidates for long-life dielectrics in regions of high fluence from nuclear reactors and in high temperature environments. These insulations have been used successfully on coil windings in reactor control mechanisms and high temperature aircraft generator windings. Only very recently has this same ceramic dielectric insulation approach proved of value in the cryogenic application of Nb_3Sn superconductors. These superconductors may be required in high power generators that can be turbodriven using fluids heated by reactors.

The need for high current, high field, low loss, stable superconductors has led to the development of multifilamentary Nb_3Sn as the most promising conductor for the use in superconducting machines. The Nb_3Sn multifilamentary wires are typically formed from thousands of Nb filaments of several microns diameter which are encased in a matrix of 13.5% tin-bronze. This matrix is surrounded by a refractory diffusion barrier with a thick layer of copper around the outside in wires of from 0.1 mm to 1 mm diamcter. Although considerable advances have been made in the development of multifilamentary Nb_3Sn wire, a suitable insulation system has not appeared to date. Major concerns in developing an insulation system are Nb_3Sn's high formation temperature (700 C to 800 C) and the fact that it is brittle after its formation and cannot be deformed. The insulation withstands this high formation temperature yet provides satisfactory electrical insulation at low use temperature (4 K to 8 K). Thus, the organics are ruled out from consideration. Furthermore, the insulation must absorb the thermal transients that are developed during fast charge and discharge; unless prevented, this may lead to an undesirable "quench."

The possible application of a new class of dielectric material as insulation coating for the Nb_3Sn wire offers a unique opportunity to achieve these goals. These materials are characterized by high specific heat and thermal conductivities at low temperature. Recent advances in their development as composites, as well as in combination with a glass, have proved that these enthalpystabilizing materials can be successfully applied very close to the potential heat source, for example, directly on the superconductor as an insulation coating (Bich 1984 and Gupta 1986).

With this objective in mind, the Air Force recently supported a series of programs for developing an enthalpy-stabilizing dielectric film-type coating for superconducting Nb_3Sn wire (Lawless 1982b and Gupta 1986). The broad picture is to incorporate the enthalpy stabilizing dielectric powders into appropriate glasses to coat Nb_3Sn wire, thereby imparting not only the electrical insulation, but also a significant enthalpy stabilization to the superconductor. This paper will present the state-of-the-art characteristic of this insulation system with special emphasis on properties and compatibility between the insulation and the wire. Since most superconductors have an exterior surface of copper, the same insulation approach can be extended to other conventional conductors in difficult environments involving high and/or cryogenic temperature and high nuclear reaction product fluence.

Critical Considerations for Developing Enthalpy Stabilized Dielectric Coating for Nb_3Sn Superconductors

Three critical considerations are required for designing the dielectric coating: 1) enthalpy stabilization of the coating, 2) high formation temperature (700 C to 800 C) of Nb_3Sn, and 3) compatibility between the coating and Nb_3Sn. These considerations are briefly described below.

Enthalpy Stabilization

Enthalpy stabilization is needed because thermal disturbances are present during the operation of the magnets. Two types of disturbances can be readily identified: point disturbances and distributed transient disturbances, and the latter presents the most severe instability problem. It has been estimated that the disturbance spectrum leading to quenches in medium-size magnets is in the range 10^3 to 10^4 J/m^3. Consequently, if the specific heat of the dielectric insulation is large enough to absorb this energy with minimal temperature rise, the magnet is "enthalpy stabilized," and the quench of the superconducting state is avoided.

Enthalpy stabilization refers to the ability of the coating to absorb thermal energy produced within the magnet winding. The coating acts as a buffer, and thermal energy is eventually conducted to the helium coolant. The time scale of the disturbance and the geometry of the winding determine which thermal properties of the coating material are most important. The spatial distribution of the disturbance does not affect which thermal properties are most important for the coating, but only the relative importance of the thermal properties of the coating and the wire. Three cases can be identified. The thermal relaxation time, τ, of the coating is needed to identify these cases and is given by

$$\tau = d^2/\alpha \qquad (1)$$

where d is the coating thickness, and α is the thermal diffusion coefficient. For disturbances shorter than the thermal relaxation time, the thermal energy absorbed by the coating is proportional to the thermal property group (Jacob 1957):

$$\eta = \sqrt{k\rho c} = k/\sqrt{\alpha} \qquad (2)$$

where k is the thermal conductivity, ρ is the density, and c is the gravimetric specific heat. The second case occurs when the disturbance time is longer than τ and the coating is not in direct contact with helium or any other thermal energy sink (for example, in a fully potted winding). The thermal energy absorbed by the coating is then proportional to the volumetric heat capacity, $C = \rho c$. The third case occurs when the disturbance time is longer than τ and the coating is in direct contact with a thermal energy sink such as a helium bath. The thermal energy absorbed by the coating can then be passed on to the heat sink, and the energy absorbed becomes proportional to the thermal conductivity, k, of the coating. To summarize, η, k, or c can be the most important thermal property of the coating, depending upon the temporal length of the disturbance and the winding geometry.

High Temperature Formation

The second constraint is imposed by Nb_3Sn's high formation temperature and its poor strain tolerance once formed. Nb_3Sn is formed by heat treating between 700 C and 800 C for many hours. Once formed, it is brittle, and straining the wire more than 0.1% degrades the critical current. This requires that the winding be carried out prior to formation reaction. Consequently, the insulation should be capable of being applied before the reaction, be able to withstand the time-temperature excursion during the reaction, and should exhibit no adverse effect on electrical or mechanical properties of the wire. The high reaction temperature rules out organics and prescribes inorganics.

Coating Compatibility

The third constraint is imposed by the compatibility requirement between the insulation and the superconductor. The insulation must be compatible with electrical, mechanical, thermal, and chemical properties of the Nb_3Sn's super-conductor. It must not degrade the superconducting current while exhibiting high electrical resistance, high breakdown voltage, and satisfactory enthalpy stabilization in zero and intense magnetic field. It is desirable that the coating's formation temperature and Nb_3Sn's reaction temperature be the same, and that no adverse chemical reaction occurs between the insulation and the superconductor. To achieve this, it was necessary to use a protective layer of nickel on the outer copper substrate of the Nb_3Sn wire, although other metals were also found satisfactory. Furthermore, the coating should not exhibit significant dielectric or mechanical property degradation when subjected to extensive durations of reactor fluence.

Experimental Results

A large number of materials and combinations thereof were studied, all of which cannot be described within the scope of this paper. We will, therefore, present a few examples of successful coating compositions and characteristics from a broad range of studies undertaken in the present program.

Two high specific heat ceramics reported here are SC1C, a chromite-spinel ceramic, and SC1A, a niobate-columbite ceramic. SC1C has an enormous specific heat maxima at 8 K of $2 \times 10^{-6}m^{-3}K^{-1}$, and SC1A has a specific heat maximum at 4.2 K of $0.3 \times 10^{-6}m^{-3}K^{-1}$. Of several glasses investigated, the one discussed here is labeled 3072, developed specifically for this study. In a previous study (Lawless 1984), two Corning glasses, 7570 and 7052, were investigated in making a composite with SC1C and SC1B, a ceramic analogous to SC1A. Adhesion of the Corning glasses to the wire proved unsatisfactory, so new glasses were developed which showed much improved adhesion. Pore-free bulk samples of cylindrical shape were prepared of various compositions by hot pressing at 800 C. The composite compositions reported in this paper for *bulk* samples are 50/50 weight ratios of SC1A/3072, SC1C/3072, and a ternary composite 32/18/50 SC1C/SC1A/3072. Simultaneous studies were conducted on composite compositions for *coating* and the one described here is 40/60 SC1C/3072. The reason for the higher glass content was to ensure better adhesion of the coating to the wire. This reduced the thermal characteristic only very slightly. The coating was applied by preparing a slurry of a mixture of glass, ceramic, organic binder, and solvent, and then drawing the wire through the slurry using the conventional wire enameling equipment at the Westinghouse R&D Center, Pittsburgh, PA. The coating developed, called the "green coating," is then heat treated at 700 C to 800 C to obtain what is known as "fired" or "vitrified" coating. At this stage, the glass has melted with the ceramic embedded in the glass and adhered strongly to the wire. The green coating is flexible enough to allow twisting, bending, and coiling. The test samples are prepared from the green coating in the form of a twisted pair, as specified in the IEEE standard for the organic enamel industry, and heat treated at various temperatures to obtain the vitrified test samples.

Bulk Samples

Specific heat and thermal conductivity data (zero field) were measured in the adiabatic calorimeter described elsewhere (Lawless 1976). Specific heat measurements in intense magnetic fields were performed at National Magnet Laboratory (MIT) by a drift technique using capacitance thermometry (Lawless 1982a).

Figure 1 shows specific heat data on pore-free SC1A/3072, SC1C/3072, and SC1A/SC1C/3072 composites. The sharp peak for SC1A and the broad-based peak for SC1C bulk ceramics at appropriate temperatures are maintained in the composites, indicating the absence of any adverse reaction between the ceramic and the glass. The broad maximum centered at approximately 8 K for SC1C/3072 composite may be due to correlation effects in the fine grain (approximately 1 micrometer) SCIC powder as observed previously for SC1C/7570 composite (Lawless 1984).

Figure 2 presents thermal conductivity data measured on the 50/50 ceramic/ glass composites. Also shown are the results on samples that had approximately 10% porosity. Porosity decreases the SC1C/3072 thermal conductivity by approximately 41% and the SC1A/3072 thermal conductivity by approximately 31%. Turning again to Figure 2, the thermal conductivities follow $K \cdot T^m$ where m = 1.5 and 1.7 for SC1A/3072 and SC1C/3072 composites, respectively. Initially, one would associate these exponents

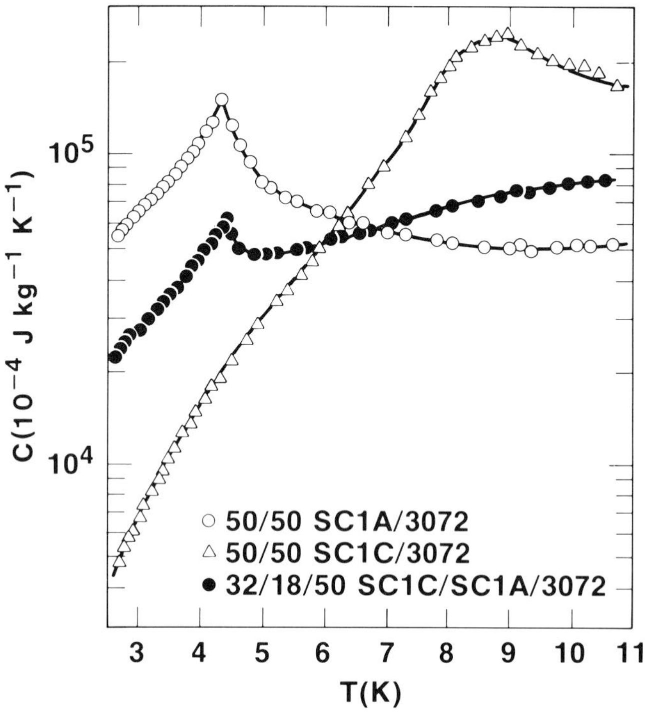

Figure 1 *Specific Heat of Pore-Free Ceramic Composites versus Temperature.*

to Kapitza-limited phenomena, but other studies have indicated that there is no Kapitza effects in SC1C/3072 composite. Note that the absence of Kapitza effect is one of the reasons for choosing SC1C/3072 composite as a leading candidate for wire coating.

Figures 3 and 4, respectively, provide specific heat data measured in intense magnetic fields on SC1A/3072 and SC1C/3072 composites. The two composites behave very differently in intense magnetic fields. The specific heat maximum is depressed with field for SC1A/3072 composite, whereas the broad-based maxima is still retained even at 10 T for SC1C/3072 composite. The reason for the peak-depression is not readily understood and would require fundamental studies.

Figure 5 shows the thermal conductivity data for these composites, along with the ternary composite. Experimental uncertainty for these measurements are shown by the error bars. The SC1C/3072 and SC1A/SC1C/3072 composites have apparently H-field independent thermal conductivities. However, the SC1A/3072 composite appears to have a slight H-field dependence unattributable to experimental uncertainties.

Coated Wire Samples

The coating studies can be broadly divided into two categories: those conducted on Cu wire and those on Nb_3Sn wire. The bulk of the coating evaluation was done on samples made from Cu wire, due to the fact that the outer layer of Nb_3Sn wire is Cu, and, an insulation system developed for Cu will apply equally well to Nb_3Sn. The other reason for testing Cu wire is the high cost and scarcity of Nb_3Sn wire. Both wires had Ni-plating on the outer surface. Tests were conducted on both ''green'' and ''vitrified'' twisted pairs. The thickness of the green coating was typically 5.5×10^{-5} to 7.5×10^{-5}m (2 to 3 mils) and that of the vitrified coating was typically 2.5×10^{-5} to 3.8×10^{-5}m (1 to 1.5 mils). As previously stated, vitrification was conducted at 700 C to 800 C in argon with a preheat treatment at 400 C in air to remove the organic binder.

Table 1 shows typical examples of mechanical and electrical characteristics of the ''green'' dielectric coatings applied on both Cu and Nb_3Sn wires. Since no standard test specifications exist for inorganic dielectric insulation, the test parameters were generated along the line of organic enamel coating modified by the brittle nature of the ceramic/glass coating. The test

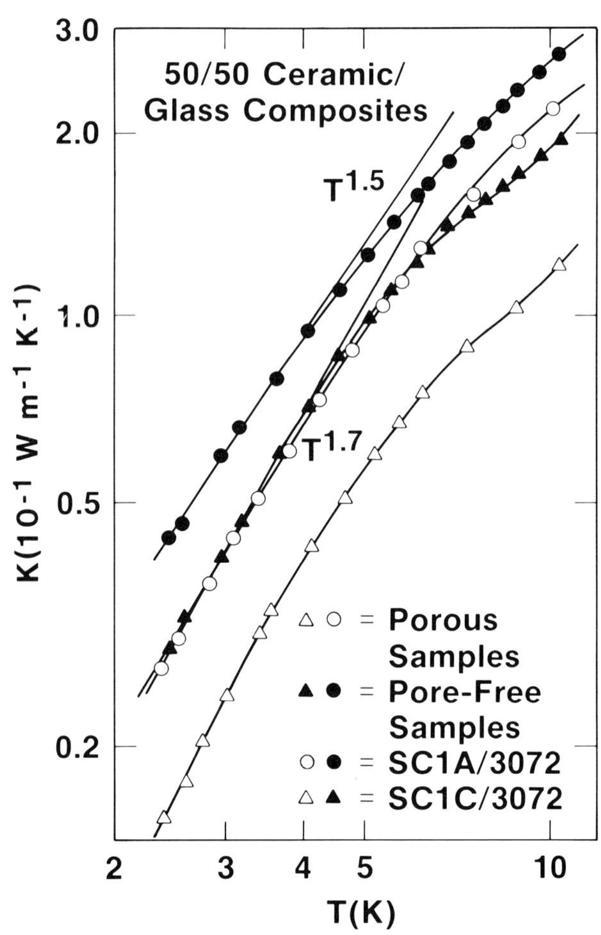

Figure 2 *Thermal Conductivity of Ceramic Composite versus Temperature.*

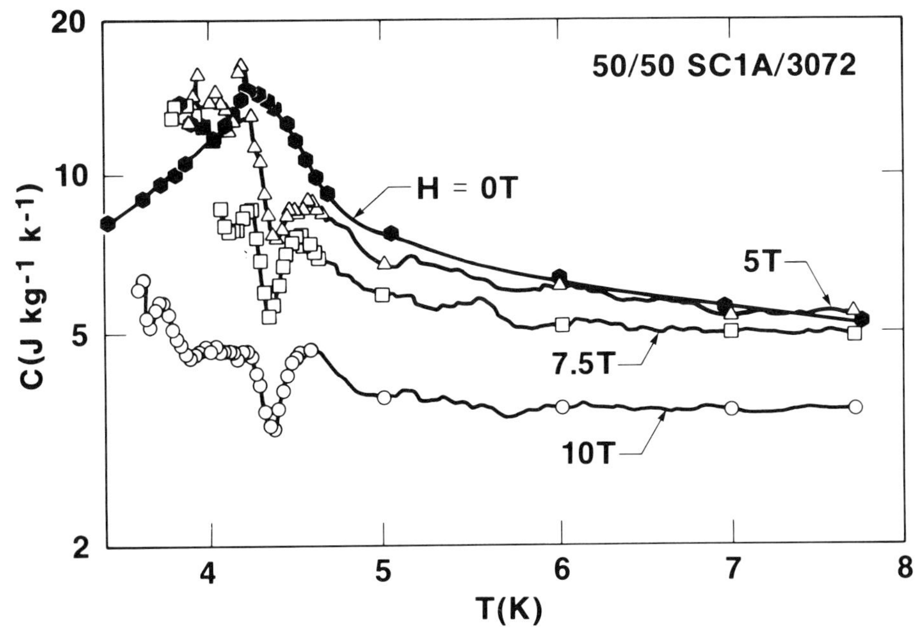

Figure 3 *Specific Heat Dependence on Magnetic Field versus Temperature for SCIA Ceramic Composite.*

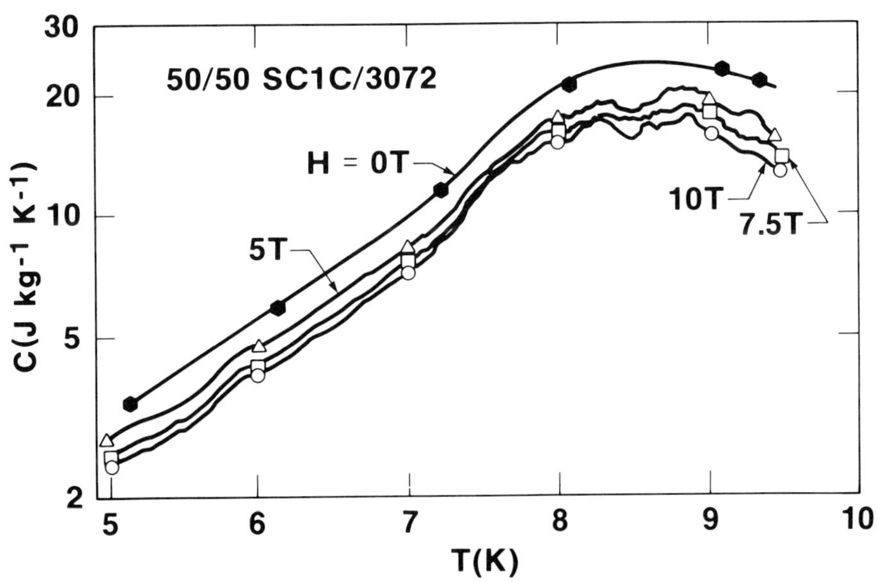

Figure 4 *Specific Heat Dependence on Magnetic Field versus Temperature for SCIC Ceramic Composite.*

data presented in Table 1 can be considered quite satisfactory. The coating exhibits high flexibility and high resistance to scrape the abrasion and high electrical strength. Figure 6 shows the breakdown strength of the vitrified coating as a function of vitrified temperature for Ni-plated Cu wire. These data represent one of the best breakdown voltages obtained in this program and are substantially in excess of the 500 volts targeted. These coatings also showed excellent adhesion to the wire after vitrification, and, in some cases, sand blasting was required to remove them.

Discussion

We have tried to demonstrate a novel method of enthalpy stabilization of superconducting magnet through the development of inorganic film-type coatings on Nb_3Sn wire, the coating having the characteristic of high thermal capacity combined with high electrical strength. Taking the case of enthalpy stabilization, it is instructive to evaluate the η-parameters for the composites from the thermal data of the bulk samples. The procedure followed was to compute η for $H=0$ and $H=7.5T$ using the appropriate

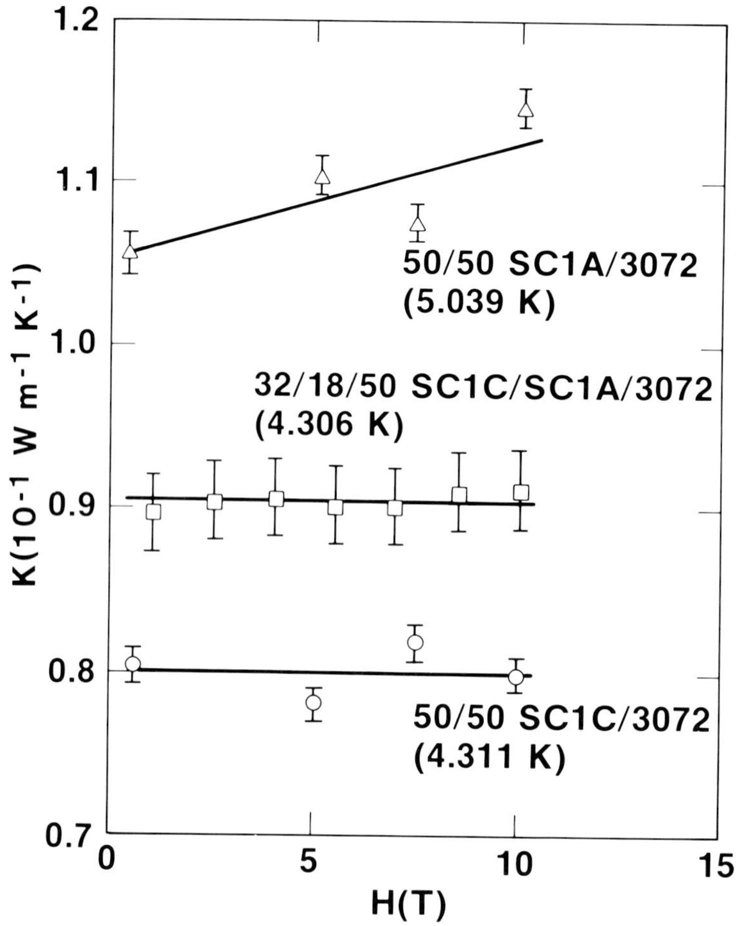

Figure 5 *Thermal Conductivity Dependence on Temperature versus Magnetic Field.*

Table 1 *"Green" Coating Characteristics.*

Coating Identification C03-310-00 Coating Composition 40/60 SC1C/3072		
	Ni-Plated Cu Wire	Ni-Plated Nb_3Sn Wire
Wire Size	#20 AWG	#20 AWG
Coated Surface	Fairly Smooth	Very Smooth
Build	5.0×10^{-5} m (2 mil)	$5.0–5.5 \times 10^{-5}$m (2.0–2.2 mil)
Flexibility:		
1. Quick Snap	passes	passes
2. Elongation + 5X Mandrel Wrap	20%	5%
3. e bend	passes	passes
Scrape Abrasion (No. of cycles @ 350 g lead)	33, 48, 56 52, 63, 17	12, 34, 24 36, 22, 18
Electric Breakdown (volts)	1000, 960, 820 790, 920, 990	1700, 1600, 1250 1200, 1100, 1500
Insulation Resistance megohms @ 50 volts (on twisted pair)	20,000	20,000

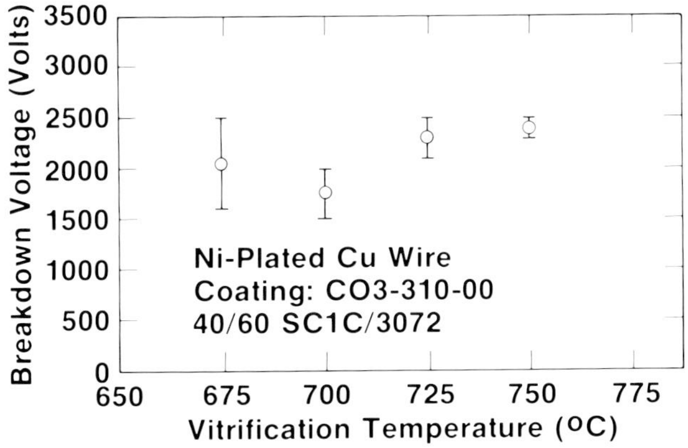

Figure 6 *Breakdown Voltage of Composite Ceramic versus Vitrification Temperature.*

specific heat and thermal conductivity data. Figure 7 presents the η-parameter data at 5 K to 8 K where the band of η-values (H = 0, 7.5T) is shown for each of the three composites discussed in this paper. Also shown for comparison are the η-data for glasses, the only other potential dielectric insulation for Nb_3Sn. The composites represent an improvement in the η-parameter = 20 times.

Turning to the dielectric coating, the inorganic filmtype coating consisting of a ceramic and a glass offers a very satisfactory solution to the conflicting requirements of a high reaction temperature of Nb_3Sn and a low use temperature (4 K to 8 K) of superconducting environment. An additional attractive feature is that the coating formation (vitrification) and Nb_3Sn reaction can be accomplished simultaneously. The vitrified coating shows a high degree of scrape and abrasions resistance combined with resistance to thermal shock upon immersion to liquid nitrogen or helium bath temperature. The vitrified coating maintains a high insulation resistance with values measured at $\rho \simeq 10^{-9} - 10^{-10}$ ohm-m and a high breakdown voltage with values in excess of 500 volts. Furthermore, the residual resistance ratio (RRR), measured for coated and uncoated wires that were heat treated similarly, showed no significant difference in values. The coating is very compatible with the Nb_3Sn wire.

Finally, the green coating is highly conformable and is extremely amenable to coil formation which can be vitrified to give final coil configuration. Figure 8, photograph of a green-coated unreacted wire wrapped on a 6.4×10^{-3}m (0.25 inch) mandrel, illustrates the flaw-free green coating. This then readily conforms to the "react-after-winding" (RAW) scheme now being pursued for making small coils of Nb_3Sn wire.

Space Power Applications

The nature of this glass-ceramic composite dielectric insulation system is much less conducive to dielectric property degradation in the hostile radiation environment of space in the vicinity of large space power nuclear reactors. Typical organic insulations decay quite ungracefully as the large organic molecules are shattered by various energetic fluences. In addition to the environmental stability displayed by these new composite dielectrics, they have been decisively proved under very low and very high temperature conditions, both of which may be required in space power situations. The only detriment to the wide scale application of these dielectric insulations is their inherent brittleness, unable to withstand strains of greater than 0.2%. If conductors can be wound prior to vitrification of the dielectric mate-

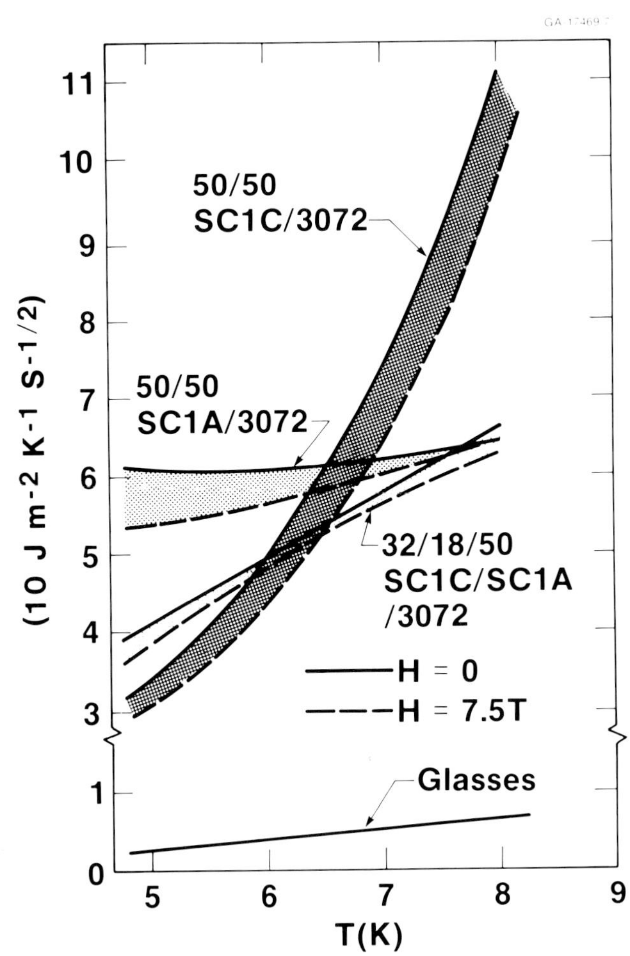

Figure 7 *Relative Thermal Property Performance versus Temperature for Ceramic Composites.*

rial, subsequent potting of the coil winding would remove this handling problem.

Figure 8 *Photograph of Green Coating on Coiled Wire.*

Acknowledgments

This effort has been jointly sponsored in its various phases by the Air Force Office of Scientific Research Physics Division at Bolling Air Force Base in Washington, DC, and the Aero Propulsion Laboratory at Wright-Patterson Air Force Base. The work has evolved from its basic to applied stages in several contracts with two different contractors and had benefited from significant assistance of in-house guidance on transient heat transfer analysis and dielectric properties measurement from the Aero Propulsion Laboratory at Wright-Patterson Air Force Base. T. K. Gupta conducted his efforts at the Westinghouse Research and Development Laboratory, Pittsburgh, PA. The effort relating to the dielectric enthalpy stabilization of Nb_3Sn superconductors appeared previously as an invited paper at the International Cryogenic Materials Conference held in Boston, MA, in August 1985 (Gupta 1986).

References

Bich, G. J. and T. K. Gupta (1984) *Electrical Wire Insulation and Electromagnetic Coils,* U.S. Patent No. 4,429,007, Westinghouse Corp., Pittsburgh, PA.

Gupta, T. K., W. N. Lawless, C. E. Oberly, G. J. Bich and G. R. Wagner (1986) "Inorganic Dielectric Insulation for Superconducting Nb_3Sn Wire," *Advances in Cryogenic Engineering (Materials)*, 32:225, A. F. Clark and R. P. Reed, eds., Plenum Press, New York.

Lawless, W. N. (1976) "Specific Heats of Paraelectrics, Ferroelectrics and Antiferroelectrics at Low Temperatures," *Phys. Rev.*, B14: 134.

Jakob, M. and G. Hopkins (1957) *Elements of Heat Transfer*, J. Wiley, New York, p. 65.

Lawless, W. N., C. F. Clark, and R. W. Arnez (1982a) "Method for Measuring Specific Heats in Intense Magnetic Fields at Low Temperature Using Capacitance Thermometing," *Rev. Sci. Instrum.*, 53: 1647.

Lawless, W. N., C. F. Clark, and R. W. Arenz (1982b) *Enthalpy-Improved Dielectric Insulation for Superconducting Wire, Final Report,* AFWAL-TR-82-2056, May 1982, Air Force Contract F33615-80-C-2022, Air Force Wright Aeronautical Laboratories, Dayton, OH.

Lawless, W. N., C. F. Clark, and T. K. Gupta (1984) "Ceramic Composites for Enthalpy Stabilization of Superconducting Wire," *Advances in Cryogenic Engineering (materials)*, 30:433 A. F. Clark and R. P. Reeds, eds., Plenum Press, New York.

Part 8

SYSTEMS INTEGRATION

Space Nuclear Power Systems 1986
Edited by M. S. El-Genk and M. D. Hoover

Chapter 24

Expert Systems for Space Power Supply—Design, Analysis, and Evaluation

Ralph S. Cooper

Apogee Research Corp.
10323 Rue Finisterre
San Diego, CA 92131

M. Kemer Thomson and Alan Hoshor

Q-Systems
10323 Rue Finisterre
San Diego, CA 92131

We evaluated the feasibility of applying expert systems to the conceptual design, analysis, and evaluation of space power supplies in particular, and complex systems in general. To do this, we analyzed the space power supply design process and its associated knowledge base, and characterized them in a form suitable for computer emulation of a human expert. The existing expert system tools and the results achieved with them were evaluated to assess their applicability to power system design. We applied some new concepts for combining program architectures (modular expert systems and algorithms) with information about the domain to create a "deep" system for handling the complex design problem. We developed, programmed and tested NOVICE, a code to solve a simplified version of a scoping study of a wide variety of power supply types for a broad range of missions, as a concrete feasibility demonstration.

Introduction

The selection and design of space power systems is complicated by the wide range of subsystems and performance characteristics, and by the variety of mission requirements and constraints. There is need for a validated data base of component state of the art and potential performance available, upon which to base selection, and a need for a methodology for using these data in the design and evaluation process.

The emerging techniques of "Artificial Intelligence" (AI), particularly expert systems, may form the basis for handling these complex decision problems. They are especially well suited to situations with uncertain and incomplete data, and with qualitative and conflicting requirements. The field of expert systems is growing rapidly, and its history, methods, and status are reviewed in many recent books, including "Expert Systems, Artificial Intelligence in Business," by Harmon and King (1985).

Selection, evaluation and conceptual design of space power supplies are conducted for many purposes, ranging from broad, long term planning, through development of particular subsystems and components. They all share certain basic functions including synthesis (design), analysis, criteria, and technology base. Therefore, a specific set of computer based tools can be applied to many of them, using similar data bases and methods specialized for each usage.

Current system analyses involve mathematical modeling of an entire system with given components, or more detailed computation of a specific subsystem or aspect such as the thermal or structural behavior. The selection of the components and configuration, as well as other considerations such as safety and survivability are made by the human experts from heuristic rules or separate analyses. The emergence of computer techniques for manipulating symbols rather than just numbers ("artificial intelligence, expert systems") allows the synthesis of designs and their evaluation, including the qualitative and uncertain issues, based on the knowledge of experts in the field.

A subsidiary, but valuable, adjunct would be an extensive, validated data base comprising the current and projected state of the art for component performance and cost. It can be useful in itself, containing information on many systems, derived from the literature, experiments and experts in varied fields. It may be "validated" through review by one or more agencies, and can include a range of opinions on uncertain items such as technology forecasts.

Approach

Expert systems have been applied in a variety of roles, depending on the sophistication and extent of interaction with the user. In space power systems, they are being developed for real-time autonomous control (Pistole 1982), to relieve the astronauts where possible and to assist them when their intervention is necessary. Our goal is to assist generalists, such as spacecraft designers, to identify candidate subsystems and technologies and to support specialist power supply designers. These two functions require different types of expert systems; however, there is some overlap, particularly in the data base and mission requirements areas.

The first, more general, application can use the more conventional expert system paradigm, in which expert opinion is codified to respond with a set of candidates and their performance in response to inquiries from the user. The second application is at least quantitatively different in that the expert will have a greater influence upon the cases investigated by the computer system. In this role the nomenclature "advisory system" is more appropriate. Moreover, the data base and parameters are much more detailed, as are the numerical analyses. In practice the scope of the system must be restricted, at least initially, to make the system developable in an acceptable time. For either application we plan to use "deep" systems, that use the fundamental laws used by the experts, as well as empirical rules derived from experience.

The Design and Selection Process

Even a conceptual design is a rather complex process, but we have been able to characterize it for a generic mission and power supply. The process (Figure 1) begins with the initial description of the mission requirements and decision criteria by the user, with interrogation by the expert to elicit issues of which the user may not be aware. The expert then considers the range of possible power supply types and major options based upon his knowledge of the field. This is a scoping study, that may be conducted very quickly (a few minutes to a few hours) or even unconsciously by the expert. This is the scale of problem that is most suitable to current conventional expert systems. The expert system output is a set of a few power supply candidates (with performance) for further consideration.

Once the major types have been identified, they must be described in greater detail for mission configuration and performance and cost computations. This is accomplished by breaking them down into main subsystems and defining each subsystem's parameters. The subsystem options and parameters are varied to optimize the total system performance against some set of the requirements and criteria, which usually is not simple and fixed. This raises the problem of establishing and weighting these various requirements and criteria, including such qualitative and complex issues as safety, technology risk, reliability, and cost. This process may take several iterations to produce acceptable results.

With the criteria at least tentatively fixed, the next step is to calculate the performance for the optional systems and subsystems. The technology is selected on the basis of an assumed IOC (initial operational capability) date, usually with some development and testing time included. The performance computations optimize the system mass or power output by varying

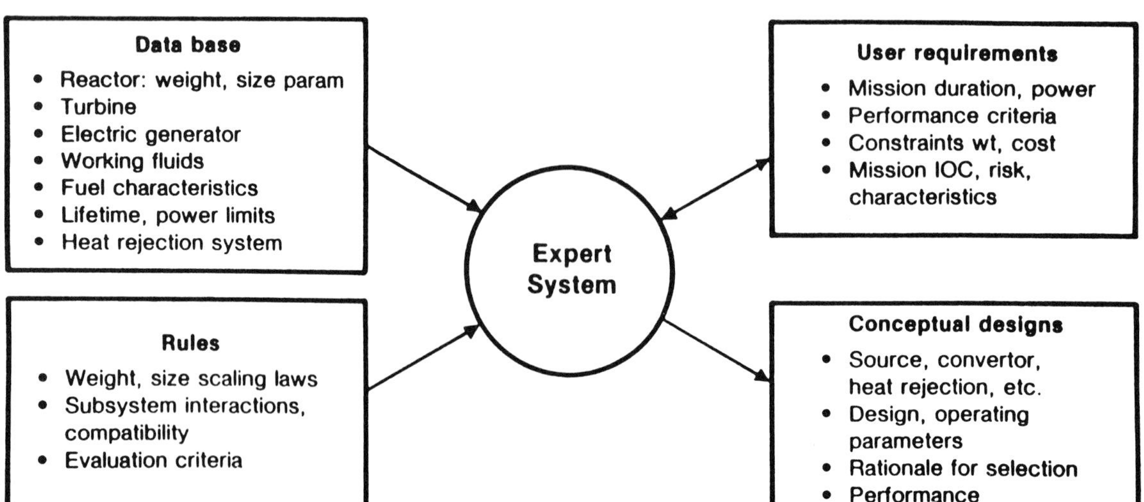

Figure 1 *Expert System/User Interaction to Produce Conceptual Design.*

the component parameters with such fixed constraints as dimensions, material temperature limits, lifetimes, and environments (source or natural radiation fluences).

Scoping Study Model

As one step in the feasibility evaluation, we defined a simple scoping model for construction of an expert system, named "NOVICE" to emphasize its limited capability. The missions were first characterized only by their power P and duration T, with the system mass as the primary criterion. Other considerations, such as the trajectory, manned status, and technology level, were incorporated later. Analytic models were generated for the mass in terms of P and T, with parameter values for 1990 technology and with assumed moderate risk. The power system mass M was given by

$$\begin{aligned} M &= M_b + C_p P^N + C_E PT \\ \text{where } M_b &= \text{base mass } (g) , \\ P &= \text{power } (W) , \qquad (1) \\ T &= \text{mission duration } (h), \text{ and} \\ C_p, C_E, &\text{ and } N \text{ are coefficients} \end{aligned}$$

Rules for the limits on P and T, were based on such factors as chemical corrosion and radiation degradation, and rules for the effects on the parameters by the trajectory, the manned status or the technology state. For example there were rules for the solar-based systems that altered the values of the power coefficients for medium altitude and geosynchronous orbits, to reflect the energy storage requirements and radiation effects. The initial set of parameters for nine representative power supplies is listed in Table 1.

The lower and upper bounds of acceptable values (subscripted l and u) were chosen for the power and duration of each power type, as representative ranges for practical, economic application. Although the basis for the limits, such as the area limitation on solar arrays, can be readily identified, there are certainly differences of opinion among experts as to the value for the limit. We determined the effects of trajectory and of the manned status by eliciting data from the user and employing rules for modifying the coefficients accordingly. These modifications also included by implication the safety, maintenance, radiation environment, and other factors.

The trajectory affected the solar systems mass through the need for energy storage due to occultation and the radiation degradation of photovoltaics, both of which are functions of orbit altitude. It affected the reactor systems through the safety criterion of the minimum safe altitude of 700 km. This is a simplistic rule that is used for illustrative purposes, and in more sophisticated codes would be replaced by more detailed considerations, including tethers, free flying systems, and so on. However it does point out the major effect that such rules and considerations can play in the selection of power supplies.

The manned status affected some systems favorably; they could be repaired, operate for longer periods and, consequently, have lower unit masses. The nuclear systems required heavier shielding and possibly greater structure for remote mounting.

Implementation of the NOVICE Code

Our review of the expert systems development tools commercially available in January 1985 indicated that they were either insufficiently flexible for our approach or prohibitively expensive for our level

Table 1 *Scoping Model Parameters for NOVICE Expert System Program.*

	Mass	Coefficients			Limits			
System	M_b (g)	C_p	N	C_E	P_l (W)	P_u (W)	T_l (h)	T_u (h)
1. Batteries	0	0	0	50	0	10^6	0	2×10^4
2. Solar Cells	0	50	1	0	0	10^6	0	10^5
3. Fuel Cells	0	20	1	1	0	10^6	0	2×10^4
4. RTG	5×10^3	500	1	0	10	10^4	10^3	10^5
5. Chemical Dynamic	10^4	1	1	2	0	10^{10}	0	2×10^4
6. Reactor Static	10^5	9×10^3	0.5	0	10^3	10^5	10^3	3×10^4
7. Reactor Dynamic (Closed cycle)	10^6	10^4	0.5	0	10^4	10^9	10^3	5×10^4
8. Reactor Dynamic (Open cycle)	2×10^5	2	1	1	10^4	10^{10}	10^{-3}	1
9. Solar Dynamic	5×10^5	40	1	0	10^3	10^7	10^2	2×10^4

of effort. We chose to generate our system in LISP on a 32 bit minicomputer, in order to gain experience in defining the knowledge base and inference engine (rule-invoking structure), and to assess the feasibility of our approach. We plan to again review the commercially available tools and select one of them when we are ready to implement a production level Space Power Expert System. The details of our NOVICE knowledge and rule compilers, rule syntax, and other features are presented in an appendix to our NASA Final Report (Cooper 1985). Operation and results of the NOVICE program are given in the RESULTS section.

Design Process for Power Systems/Subsystems

To investigate the next level of depth in the design process, a specific example was chosen: the nuclear reactor, dynamic conversion system, with the Growth Manned Space Station as the primary application. The power supply system was divided into eight major subsystems, with options defined for several of them (Table 2).

The design process proceeds to select a complete set of subsystem options and to calculate the performance. A set of rules, based on expert experience is used to reduce the very large number of possibilities to sets of compatible components. These rules may be in the form of: "If the Subsystem 1 is a, Then for Subsystem 3, a and b are acceptable and c is rejected." The rules may include preferences, multiple premises or conclusions, and even may be somewhat in conflict. The latter may be checked by the code. Examples of the subsystems, rules for generating configurations for a nuclear reactor system, and a sample session are given by Cooper (1985).

Results

NOVICE Program Operation

The program asks for input on mission power and duration, and on the trajectory and manned status. There may be additional questions for the latter two items, depending upon which power supply types are acceptable candidates for the input P and T. The code eventually provides a list of the acceptable types, sorted by mass. In addition, the code can respond with the rationale for its selections, such as omission of a particular type. It can redo the calculation for different values of the input parameters to obtain sensitivity and trade-off studies.

For a set of values of P and T, it generates plots identifying the first and second choices on maps of the form used in the literature (Figure 2). On these maps, the selected power system is shown by a one letter key (described at the top of the figure) for each combination of average power out P and mission duration T. This output form was useful in evaluating the accuracy of the knowledge base data, by comparison with the power systems in actual use for missions with these parameters, and with technology assessment plots in that form.

NOVICE Program Results

We ran a scoping analysis without inclusion of the qualitative factors, such as manned versus unmanned operation, and with initial rough estimates of the parameter values (Table 1). The program correctly showed the general regions (Figure 3) where each type is thought by experts to be the best candidate. Examination of these initial results showed areas differing from actual practice, particularly for fuel cells.

Table 2 *A Sample System and Representative Subsystem Options for a Nuclear Reactor with Dynamic Conversion.*

	Options[a]			
Subsystem	(a)	(b)	(c)	(d)
1. Source	ZrH/M	Fast/LM	Fast/gas	Particle bed
2. Shield	Shadow	2 Pi	4 Pi	
3. Heat Transfer	LM pumped	Heat pipe	Gas loop	
4. Converter	LM Rankine	Stirling	Brayton	Organic Rankine
5. Radiator	LM pumped	Heat Pipe	Liquid Drop	Belt
6. Distribution	Hi voltage	Low Voltage	Autonomous	
7. Structure	Close-in	Boom	Tether	Free flying
8. Auxiliary Systems	Safety	Start-up	En. Store	

[a]LM = Liquid Metal.

```
R: Reactor_dynamic_closed
C: Chemical_dynamic
F: Fuel_cell
P: solar_cell (Photovoltaic)
B: Battery
S: Solar_dynamic
I: rtg (Isotope)
T: reactor_static (Thermoelectric)
O: reactor_dynamic_Open

>>> manned       = t.
>>> trajectory   = leo.
>>> under_700km  = t.

=============== Best Power Supply ===============

          1.0e9 + C C C C C C C C C C C C C
          1.0e8 + C C C C C C C C C C C C C
          1.0e7 + C C C C C C C C S S S S S
          1.0e6 + B B B F F F F F S S S S S P P
(Watts)   1.0e5 + B B B F F F F F S S S S S P P
 Power    1.0e4 + B B B F F F F F P P P P P P P
          1.0e3 + B B B F F F F F F P P P P P P
          1.0e2 + B B B B F F F F F F P P P P P
          1.0e1 + B B B B B F F F F F F P P P P
                  +-^-+-^-+-^-+-^-+-^-+-^-+-^-+-^
                    |       |           |     |
                  1.0e-2   1.0        1000  1.0e5
                   Duration (Hours)

========== Second-best Power Supply ==========

          1.0e9 +
          1.0e8 +
          1.0e7 +                 C C C C C
          1.0e6 + F F F B P P P P P P P P P
(Watts)   1.0e5 + F F F B P P P P P P P P P
 Power    1.0e4 + F F F B P P P P F S S S S I I
          1.0e3 + F F F B B P P P P F F S S I I
          1.0e2 + F F F F B P P P P P F F F I I
          1.0e1 + F F F F F B B P P P P F F I I
                  +-^-+-^-+-^-+-^-+-^-+-^-+-^-+-^
                    |       |           |     |
                  1.0e-2   1.0        1000  1.0e5
                   Duration (Hours)
```

Figure 2 *Typical NOVICE Program Maps identifying First and Second Best Power Supplies for Low Earth Orbit, Manned Missions.*

These were not the preferred choice in the kilowatt range for 10 to 100 hours, in contrast to their use on many manned missions in that range.

Certain factors, such as the geometric and operational deployment problems of solar cells were not included in the analysis. However, consultation with a fuel cell expert (Hoberecht 1985) showed that our estimated efficiency of 45% was overly conservative. A superior model, based on data from Martin (1985) was incorporated in the code.

In addition improved models for the chemical and nuclear open cycle dynamic systems were obtained through Massie (1985). Several calculations were provided by Iden (1985) from a system analysis code for a nuclear driven turboalternator in the range of hundreds of seconds of operation at 100 to 200 MWe. Similar results were generated by Colgrove (1985) for LOX/LH2 combustion systems. These results were reduced to the form of Equation (1), an oversimplification consistent with our other models. The different boundaries for the regions (Figure 4) show better agreement with actual practice, particularly for the region of fuel cell used in the Apollo and Space Shuttle missions. Inclusion of the improved data is an example of the learning process with the expert system to improve the sophistication of the models and the accuracy of the data.

Next, two of the qualitative characteristics were incorporated: the manned/unmanned status and the trajectory, specifically the orbit altitude. The former affected the mass (still the sole criterion) through the shield mass and modified mass coefficients due to manned repair or replacement of parts. The knowledge was incorporated in the form of rules. For example, if the system was manned, the power coefficient for solar photovoltaic was reduced by a factor of 0.9; those for solar and chemical dynamic systems were reduced by a factor of 0.8 relative to the unmanned system value. This difference is intended to represent the effects of design and redundancy required for unmanned operation, which we assume would be larger for the dynamic converter. For manned spacecraft nuclear reactors, the base masses and power coefficients are increased about a factor of 10 to account for shielding. Again, these are just illustrative values for code development purposes.

The orbit altitude affected the reactor system's applicability (nuclear safe orbits of 700 km or more); it affected all solar system masses by energy storage

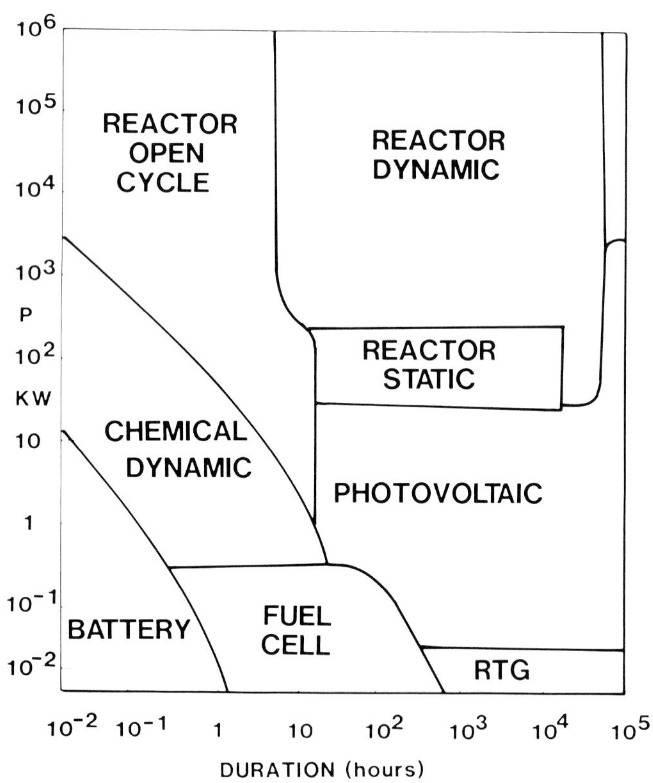

Figure 3 *Initial Scoping Study Results for Lowest Mass Power Supply.*

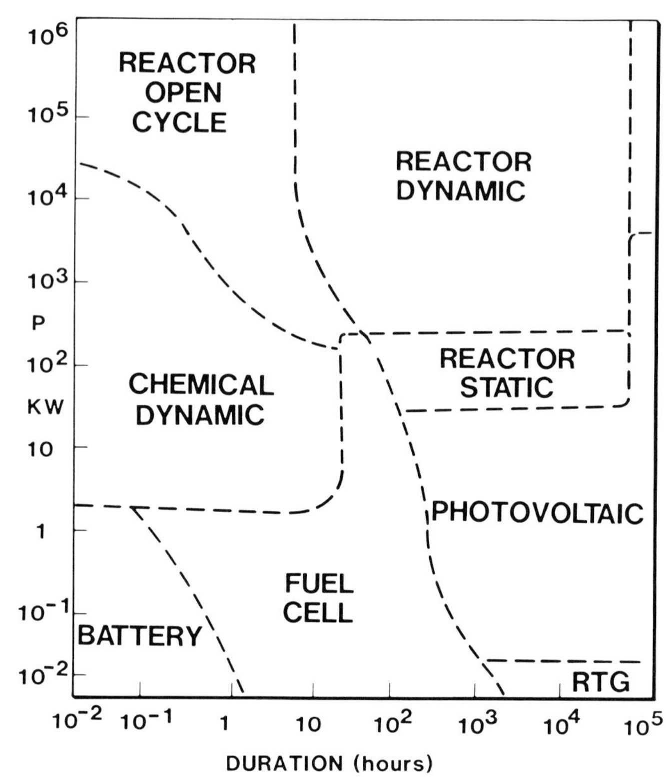

Figure 4 *Scoping Study Results with Improved Models for Fuel Cells and Dynamic Converters.*

requirements due to occultation, and solar cell system lifetime because of radiation damage at medium altitudes. For affected systems, the code requested the orbit altitude (low, medium, or geosynchronous) from the user. Reactor systems were rejected for low orbit (assuming this is below 700 km) although in practice there are options for their use. This is especially important for the Manned Space Station. The effects of orbit altitude on solar power supplies were estimated from the work of Rockey (1982). The results for a manned, low orbit power supply (Figure 5) are very different from the unmanned case.

Discussion

The feasibility decision is based upon two sets of information: the NOVICE code results and the study of the Expert System approach in conjunction with the nature of the space power design problem(s). We conclude that both bodies of evidence fully support the technical feasibility and, more significantly, the utility of this application, recognizing the different user requirements and the effort needed for implementation of the technique.

Conclusions from NOVICE

The NOVICE code, an "Expert" System with about 100 rules (the mass parameters, P and T ranges, and the orbit and manning consequences) makes reasonable estimates for system mass and applicability for all power supply types over the full range of power and duration, and begins to handle the special operating conditions such as trajectory. Yet it was developed with a small amount of effort. It easily is capable of much greater depth and accuracy in modeling and in breadth of environmental situations and constraints.

Examination of the Problem and Technique

As we have indicated, the processes vary with the objectives of the user: from simple identification of candidates to selection of design and operating parameters of components under hardware development. We conclude that all of these are amenable to expert systems techniques, especially if one views these techniques with Patrick Winston (1984) as "advanced design and analysis software". The associated applications have varying manpower requirements to reach expert status. For the simpler, more well defined problems, the development effort has been decreasing dramatically (Davis 1984 and Harmon and King 1985) as a result of tool development and better problem selection.

Several other conclusions were reached by consideration of the design application and review of the experiences gained in developing expert systems. First, it is important to select a manageably sized problem. In our case this would be a scoping study or conceptual design of one type of power supply. Second, it is necessary to work very closely with the users from the beginning. Their expertise is needed, but more importantly their acceptance and recognition of the system as a useful adjunct, rather than as a competitor, is required to support the system through its learning stages. Third, practical systems can reach prototype stages relatively quick, and enter service with five man-years of development. We believe that production expert systems with a significant domain range, for example nuclear reactor/dynamic power conversion, can be developed to a useful state in two years with about five man-years of effort. This effort is decreasing as more and better tools and hardware become available. Fourth, we concluded that the straight production rule method would not be adequate, but that the addition of a "frame" based structure (Minsky 1975) would provide an effective

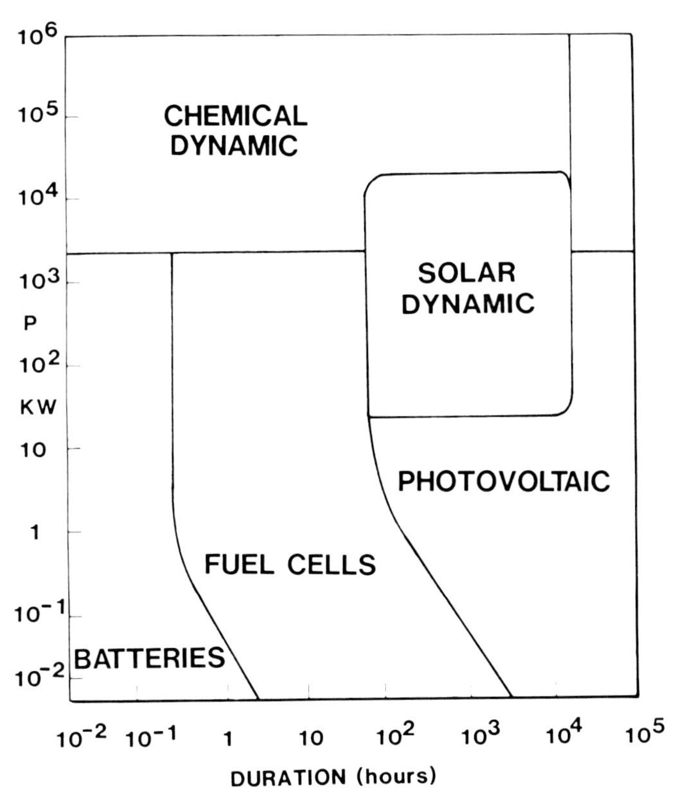

Figure 5 *Scoping Study Results for Low Earth Orbit, Manned Missions.*

descriptive model. Finally, we recognize the advantage of using algorithms, especially those already developed and tested, to conduct the bulk of the numerical analyses. Therefore, it will be necessary to interface and incorporate "number-crunching" routines implemented in FORTRAN or Pascal.

Conclusions

We found that expert and advisory systems can be practical and valuable tools, both for broad study applications and for designers and project managers working on specific power supplies and components. Somewhat different techniques are applicable in each case, but there is overlap in the data structures and information, and in the system requirements and criteria. Combining the symbolic processing with existing algorithms for the numerical analysis is the efficient method in both development and operating time. Both AI hardware and expert systems software are rapidly gaining power. Thus the approach will become more effective with time over the near future.

Acknowledgments

This work was supported by the NASA Lewis Research Center under Contract NAS-23900. The authors wish to thank Mr. Raymond Beach and Mr. Robert Stochl for their support and encouragement.

References

Colgrove, R. (1985) *Personal Communication*, Wright Aeronautical Laboratory, Dayton, OH, April 1985.

Cooper, R., M. K. Thomson, and A. Hoshor (1985) *An Expert System for Space Power Design*, Final Report CEFR-8507, Creative Enterprises, San Diego, CA, July 1985.

Davis, R. (1985) "Amplifying Expertise with Expert Systems," in *The AI Business, The Commercial Uses of Artificial Intelligence*, P. K. Wilson and K. A. Prendergast, eds., The MIT Press,Cambridge, MA., p. 26.

Harmon, P. and D. King (1985) *Expert Systems*, John Wiley & Sons, New York, p.9.

Iden, S. (1985) *Personal Communication*, Wright Aeronautical Laboratory, Dayton, OH, April 1985.

Martin, R. (1985) *Personal Communication*, United Technology Research Center, Hartford, CN, through Hoberecht, M. A. (1985) *Personal Communication*, NASA Lewis Research Center, Cleveland, OH, April 1985.

Massie, L. D. (1985) *Personal Communication*, Wright Aeronautical Laboratory, Dayton, OH, April 1985.

Minsky, M (1975) "A Framework for Representing Knowledge," in *The Psychology of Computer Vision*, P. H. Winston, ed., p. 211.

Pistoler C. and J. Bein (1984) "Investigation of Expert System Application to Spacecraft Power System Control," in *Proc. 19th Intersociety Energy Conversion Engineering Conf.*, held in August 1984, p. 379.

Rockey, D. E., R. M. Jones, and I. Schulman (1982) "Comparison of Evolving Photovoltaic and Nuclear Power Systems for Earth Orbital Applications", in *Proc. 17th Intersociety Energy Conversion Engineering Conf.*, held in August 1982, p.70.

Winston, P. H. (1984) "Perspective", in *The AI Business, The Commercial Uses of Artificial Intelligence*, P. K. Wilson and K. A. Prendergast, eds., The MIT Press,Cambridge, MA., p. 1.

Space Nuclear Power Systems 1986
Edited by M. S. El-Genk and M. D. Hoover

Chapter 25

User Interface and Operational Issues with Thermionic Space Power Systems

R. C. Dahlberg and C. R. Fisher

GA Technologies Inc.
P. O. Box 85608
San Diego, CA 92138

Thermionic space power systems have unique features which facilitate predeployment operations, provide operational flexibility and simplify the interface with the user. These were studied in some detail during the SP-100 program from 1983 to 1985. Three examples are reviewed in this paper: 1. System readiness verification in the prelaunch phase; 2. Startup, shutdown and dormancy in the operations phase; and 3. Part-load operation in the operations phase.

Introduction

The overall conceptual design of the thermionic power system in the SP-100 program integrates both user interface requirements and operational flexibility. Typical user interface and operational topics are

- Command and control requirements;
- Nuclear radiation limits at the payload;
- Voltage options available to the user;
- Control flexibility, including start-up, shutdown, load following, and part-load operation;
- Ground testing and servicing;
- STS integration; and
- Dormancy capability.

Each of these was studied for the pertinent life-cycle phases of a space nuclear power mission, including prelaunch testing, launch, ascent, and operations in orbit. The results of three of these studies were selected for presentation in this paper because of their importance in demonstrating the inherent operating flexibility of the thermionic power system. They are

1. System readiness verification in the prelaunch phase;
2. Start-up, shutdown, and dormancy in the operations phase, and
3. Part-load operation in the operations phase.

A prerequisite to understanding these topics is understanding the design features of a 100-kWe thermionic space power system. A brief review follows (Homeyer et al. 1985).

Thermionic Power System Description

The thermionic (TI) power system is shown in Figure 1. Its major subsystems are

- Reactor-converter subsystem,
- Shield subsystem,
- Heat rejection (coolant) subsystem, and
- Power conditioning and control subsystem.

The thermionic nuclear power system combines the nuclear heat source with the power converter into a thermionic fuel element (TFE). The close thermal contact between the power converter and the nuclear fuel quickly communicates changes in either heat supply or electrical demand throughout the system, allowing rapid and direct response to load changes. The reject heat from the energy conversion process is carried by flowing liquid metal from the reactor core to the radiator. The heat transfer loop is extremely simple, as shown in Figure 2. The efficiency of the energy conversion process is a relatively weak function of the temperature of the coolant. Therefore, for modest load changes, the temperature feedback from the heat rejection process is small and does not cause instabilities or control difficulties. The time response of the thermionic power system, results in a responsive and easily controlled power supply due to the aforementioned characteristics, as well as the small magnitude of the negative coefficient of reactivity.

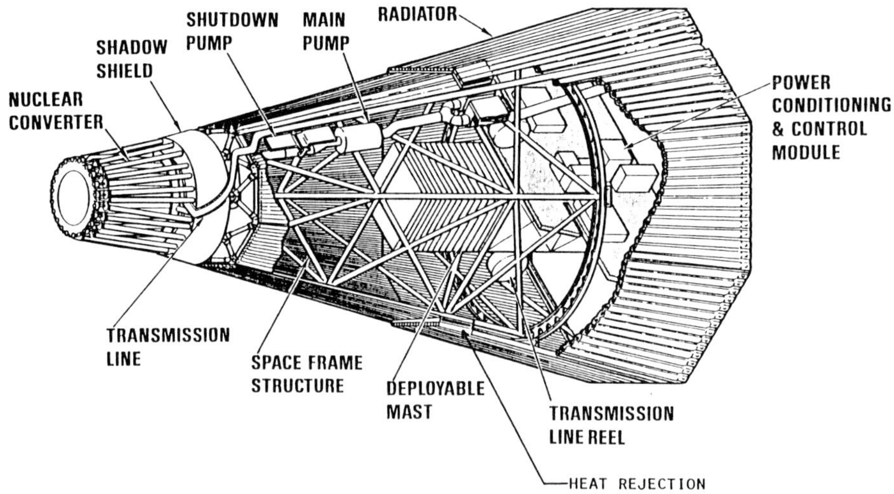

Figure 1 *100-kWe Baseline Power Plant (Stowed Configuration).*

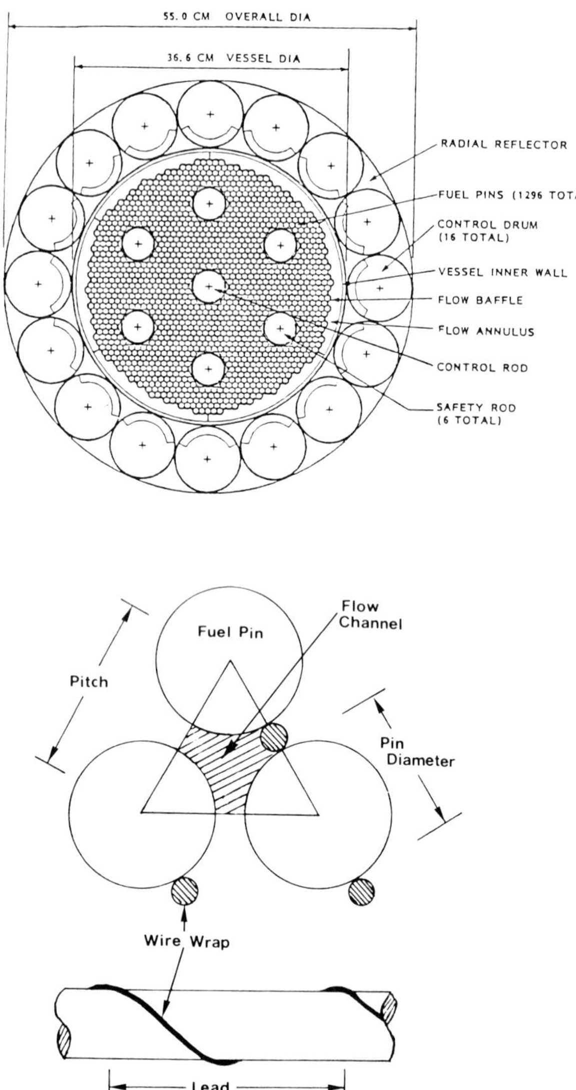

Figure 2 *Heat Transport in Thermionic System is Inherently Simple: Single Loop, Reject Heat Only, and NaK Technology.*

Thermionic Fuel Elements (TFE)

The reactor core is composed of TFEs, one of which is shown schematically in Figure 3. Its components are

- Six thermionic converters stacked like flashlight batteries in a structural sheath tube;
- The sheath tube, which is cooled by the NaK coolant at about 1000 K;
- Electrical leads, which carry the electrical power out of the attenuation; and
- A B_4C segment in each TFE to capture neutrons, reduce activation of the NaK coolant, and reduce the capture gamma source in the ZrH and LiH.

The configuration and thicknesses of materials in the shield subsystem are selected through trade studies, where the total power system mass is minimized consistent with radiation dose limits at the location of critical components.

Power Conditioning and Control Subsystem

The power conditioning and control subsystem is deployed near the payload, at the opposite end of the boom from the reactor, as shown in Figure 4. This reduces the radiation dose to the electronic components. The shunt regulator and radiator panels are at the forward end, followed by the transmission line reel and various electronics boxes located between the power conditioning radiator panels. Low-temperature heat pipes conduct the reject heat from the electronics to the radiator shown.

System Readiness Verification

Once the power system is manufactured, assembled and ready for launch, system readiness checks re-

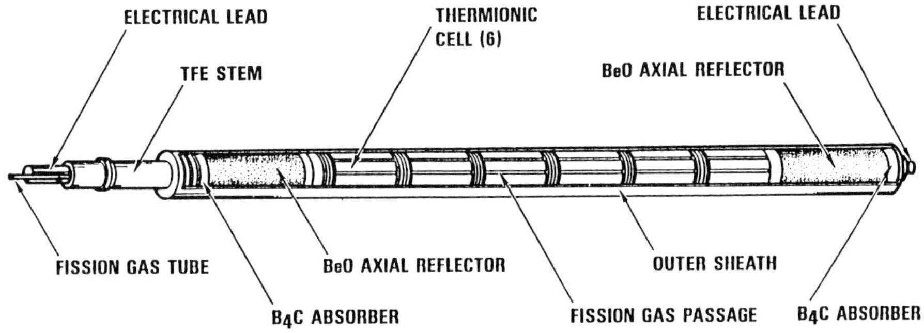

Figure 3 *Thermionic Fuel Element.*

duce the risk of discovering problems in orbit. Two examples of readiness verification for thermionic systems follow.

Power Conversion Subsystem

The combination of heat source, high-temperature heat transport system and the power conversion system in the thermionic fuel element facilitates readiness verification. The procedure is as follows:

1. A heater is attached to the inlet NaK coolant pipe to the reactor-converter and a cooler to the outlet NaK pipe in preparation for heating the reactor converter.
2. The reactor-converter subsystem is enclosed within a shroud and flushed with a high-purity inert gas, such as helium, to protect the refractory metal from oxidation during the heating process.
3. The NaK is heated to bring the reactor converter to a uniform temperature of approximately 800 K. This will result in a cesium pressure in the interelectrode gap of about 0.1 torr.

The work function of the emitter under these conditions will be about 1.6 eV. It will emit about 0.5 amperes when ignited. When a voltage is applied across the reactor electrical leads (available outside the reactor vessel), ignition will occur first across one converter, then another, and so on until all 12 converters in each TFE pair are passing approximately 0.5 amps. The successive ignition of the 12 converters in a TFE pair can be observed and the current-voltage curve for each converter measured. The performance capability of the power conversion function is verified by comparison with expected data. Detectable defects include cesium leaks, short and open circuits, and contaminated electrodes. All of these would have been checked after manufacture and prior to assembly into a reactor. However, the testing could be done at the launch site to assure the continuing performance capability of the power system.

Heat Rejection Subsystem

The heat rejection subsystem is extremely simple, as shown in Figure 2, and lends itself to diagnostic testing at the launch site. Since the NaK coolant is liq-

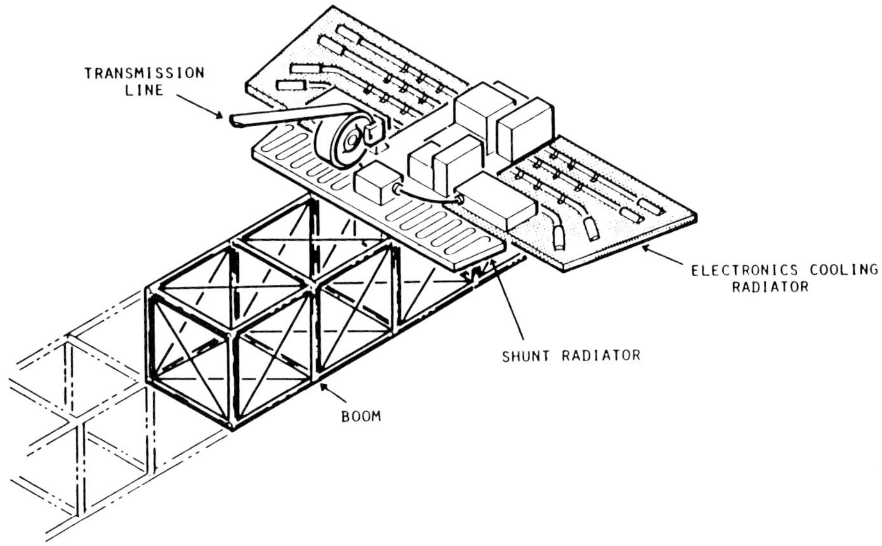

Figure 4 *Power Conditioning Control Module.*

uid above 253 K, coolant pumps can be run at any time to test pump performance. Pressure drop checks also can be made to assure coolant loop geometric integrity as it affects coolant flow.

Start-Up And Shutdown Operations

Ease of start-up and shutdown are key user interface issues. These are discussed in the following paragraphs along with dormancy options for TI power systems.

Initial Start-up

The thermionic power system is easy to start up because the coolant is liquid, and there is a simple single heat transport loop between reactor and radiator.

The coolant temperature through launch, ascent and separation from the shuttle is maintained at approximately 370 K. After separation from the orbiter, the NaK will gradually cool but remain liquid for about one week. During this interval, the power system and payload are taken to the operational orbit and deployed.

At the time of start-up, the TI power system will receive a wakeup command from the spacecraft. Six hundred W(e) of power will be drawn from the onboard primary battery to operate the coolant pump.

The subsequent rise to power is shown in Figure 5. During the first 2.3 minutes, criticality will be achieved as the shutdown rods are removed and the control drums are rotated to their critical position.

From 2.3 to 5.6 minutes, the reactor thermal power level increased to 60%, the electrical power to 2%, and the coolant temperature to near full power operating level. Fuel temperatures also approach full-power levels.

At about 6 minutes, the control mode changes to control on voltage-current (*V-I*) from control on neutron flux. At this point, the output voltage is approximately 10%, and reactor power and fuel temperatures are better controlled by measuring the output current and voltage. At about 8 minutes, full-power operation at design temperatures is reached.

If dormancy occurs after reactor operation, and assuming no special heat source, the decay heat from the reactor will assure a liquid coolant for time periods shown in Table 1.

A low reactor power level may be desirable during dormancy to generate electrical power to meet a small load demand. This is discussed in the next section.

Table 1 *Coolant Temperature after Shutdown.*

Reactor Operating Time	Time Period during which Coolant will remain Liquid: Sink Temperature 225 K	Sink Temperature 180 K
1 week	100 days	30 days
1 month	400 days	80 days
1 year	3000 days	600 days

Part-Load Operation

Capability for part-load operation is an essential flexibility for planning space missions. It is always possible to operate the reactor and power conversion system at 100% power and to reject any electrical power not needed by the payload in the power conditioning radiator. However, this results in continued

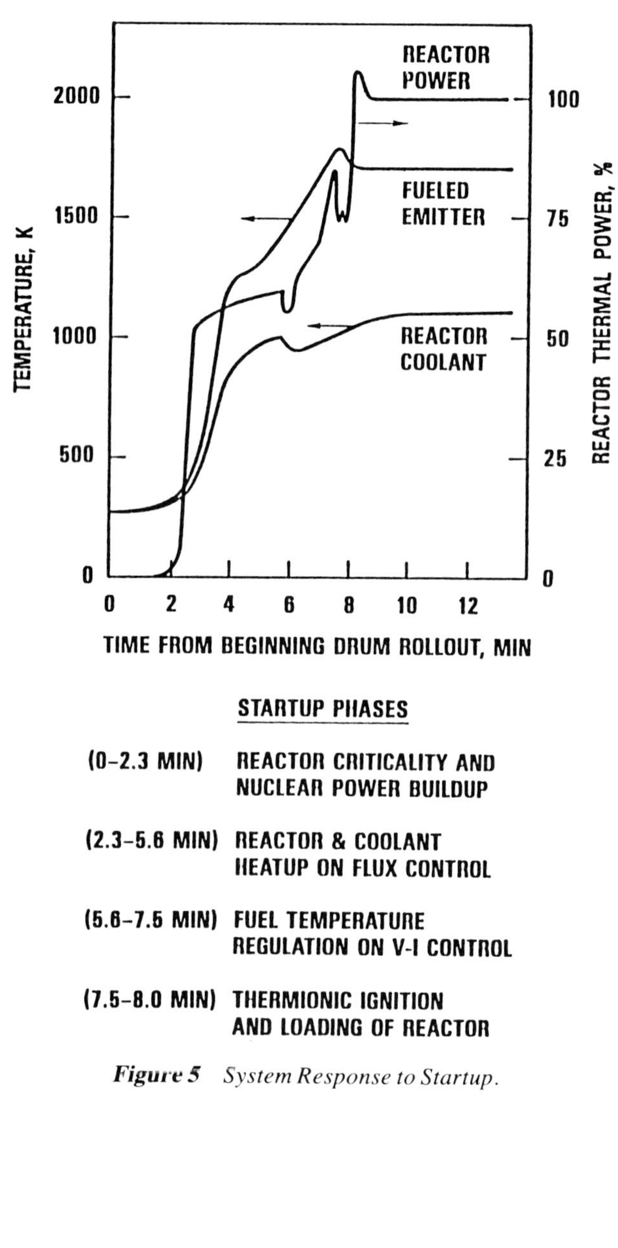

Figure 5 *System Response to Startup.*

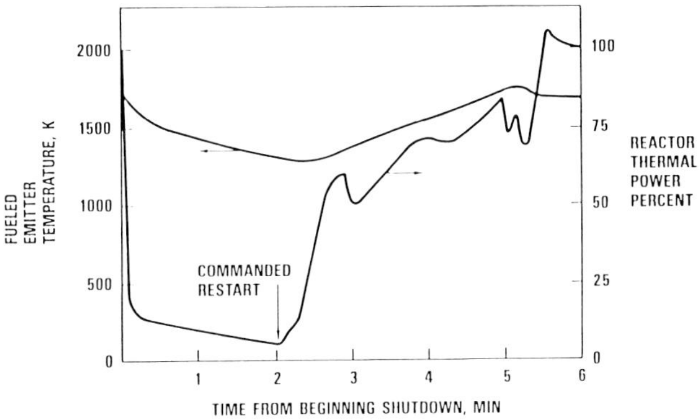

Figure 6 *Typical Commanded Shutdown and Restart Sequence.*

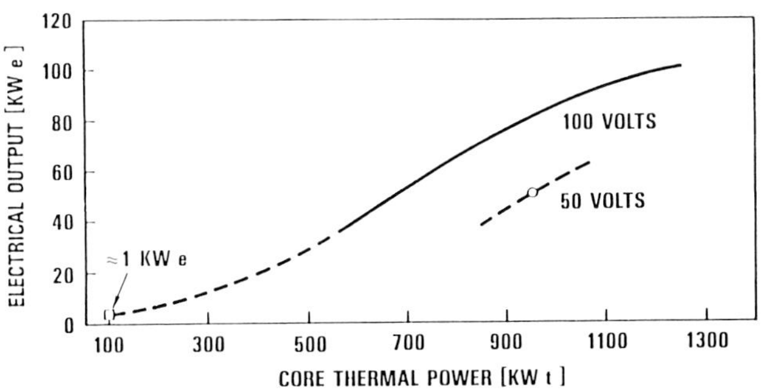

Figure 7 *Part-power Performance Characteristics.*

use of nuclear fuel with attendant continuing radiation to components; this may not be desirable for long periods. Moreover, if the reason for low power is to reduce the thermal signature, as a defensive measure, continued operation at high power would be unacceptable.

The part-load characteristics of TI power systems are illustrated in Figure 7. Operating the reactor at about 8% of design power results in one percent electrical power at 100 volts. Lower voltage operation is also possible.

Conclusions

The foregoing described some of the inherent and designed features of a nuclear thermionic space power system that contribute to its capability to perform a range of possible missions from a user viewpoint. These include

- Rapid response to load changes with easy controllability,
- Long-term operating capability at low loads while producing design voltage,
- Capability for operational readiness verification prior to launch, and
- Capability for long dormancy periods with little mass penalty.

Acknowledgments

The work described herein was done as a part of the SP-100 Program under JPL contract 956472.

References

Homeyer, W. G., M. H. Merrill, J. W. Holland, C. R. Fisher and D. T. Allen (1985) "Thermionic Reactors for Space Nuclear Power", in *Space Nuclear Power Systems 1984*, M. S. El-Genk and M. D. Hoover, eds., Orbit Book Co., Malabar, FL.

Part 9

REFRACTORY ALLOYS

Space Nuclear Power Systems 1986
Edited by M. S. El-Genk and M. D. Hoover

Chapter 26

Selection Of Structural Materials For Space Nuclear Power Generators

Alex F. Condliff and Robert J. Marsh

Teledyne Wah Change Albany
1600 N.E. Old Salem Road
Albany, OR 97321

During the 1960–1970 period, a high emphasis was placed by both govermental and industrial sponsors on the development of refractory alloys for aerospace applications. Hundreds of alloys of columbium (niobium), tantalum, tungsten, molybdenum, and rhenium were devised, prepared, and evaluated. Out of these development programs, only a very few were ever produced commercially. The alloys that evolved then, primarily columbium alloys, are still the alloys that are being used today and will remain so for near-future programs. Properties of these alloys are compared for space applications and one more recently developed alloy containing niobium, tungsten, and hafnium is described as is newly emerging powder metallurgy technology for processing this high performance alloy.

History

Beginning with its discovery by two separate scientists, at about the same time period, there has been a continuing controversy as to whether element 41 should be called niobium or columbium. In 1950, the International Union of Pure and Applied Chemistry adopted niobium. More recently several other technical societies of chemists and physicists have also adopted niobium. However, many United States metallurgists and metallurgical societies continue to use columbium. With the exception of when referring to alloy designations or referenced tables or figures, this paper will call element 41 niobium.

In the field of aircraft and space nuclear power, the same companies and military establishments that were active in refractory metal alloy development 20 years ago are essentially the same ones that are active today. Most of the early development work involved the following:

- CANEL (Connecticut Aircraft Nuclear Engineering Laboratory)-Pratt & Whitney Aircraft,
- Wright Patterson AFB (Wright Air Development Division),
- Boeing Airplane Company,
- General Electric Company,
- Westinghouse Corporation,
- Union Carbide Corporation,
- Wah Chang Corporation (now Teledyne Wah Chang Albany), and
- Fansteel Metals.

With the exception of Union Carbide, Haynes Stellite Division, currently part of Cabot Corporation, these same companies remain the primary thrust for the development of refractory metal alloy uses. Much of the work at CANEL concerned Nb-1%Zr and PWC-11. Wright Air Development Division developed the Ta base TCV alloy. Boeing and Teledyne Wah Chang Albany (TWCA) had a joint program wherein approximately 240 Nb alloys were devised and prepared by TWCA which Boeing then evaluated. Alloys C-103 and C-129Y evolved from this work. C-103 is the most widely used Nb alloy for aerospace applications today. General Electric developed AS-55, AS-30, F-48 and F-50, all Nb alloys. Union Carbide developed Cb752 while Westinghouse developed tantalum alloys T-111 and ASTAR 8llC. Table 1 shows some of the properties of the refractory base metals.

The bulk of this work was carried out during the mid-to-late 1960s while there was a high degree of interest, both governmental and privately, in the alloys for a variety of applications such as in jet engines and re-entry vehicles. Table 2 lists many of the Nb alloys devised or developed during this period.

However, as interest declined, so did the budgetary support. Apparently, it was felt that certain refractory alloys, such as C-103, would meet the requirements of the forseeable future. Thus, little development of re-

Table 1 *Comparative Properties of the Principle Refractory Metals*[a].

Property	Nb	Mo	Ta	W
Melting Point (°C)	2468	2610	2996	3410
(°F)	4474	4730	5425	6170
Density (g/cc)	8.57	10.22	16.6	19.3
(1b/cu.in.)	0.31	0.369	0.60	0.697
Thermal Neutron Cross Section (Barns)	1.1	2.4	21.3	19.2
Thermal Expansion (in/in/°F × 10^{-6})				
RT to 400°F	4.0	3.0	3.7	2.25
RT to 2000°	4.4	3.3	3.9	2.45
Thermal Conductivity (BTU/ft^2/hr/°F/ft)				
RT	26	80	33	80
2000°F	36	58	43	63
Elastic Modulus (PSI × 10^6)				
RT	15	46	27	58
2000°F	12	28	24	52

[a] From Frank 1968.

Table 2 *Densities of Niobium Alloys*[a].

1. High Strength		
Cb-1	10.293 g/cc	(0.372 lbs/in^3)
B-88	10.321	(0.373)
VAM-79	9.850	(0.356)
Cb-132M	10.653	(0.385)
AS-30	9.601	(0.347)
F-48	9.408	(0.340)
SU-31	9.546	(0.345)
2. Moderate Strength and Ductility		
SU-16	9.269	(0.335)
FS-85	10.598	(0.383)
D-43	9.020	(0.326)
Cb-752	9.020	(0.326)
SCb-291	9.601	(0.347)
C-129Y	9.491	(0.343)
B-66	8.439	(0.305)
AS-55	8.771	(0.317)
PWC-11	8.578	(0.310)
3. Low Strength and High Ductility		
Cb-753	8.384	(0.303)
C-103	8.854	(0.320)
B-33	8.467	(0.306)
D-14	8.578	(0.310)
D-36	7.914	(0.286)
Nb-1% Zr	8.578	(0.310)

[a] From Frank 1968.

fractory metal alloy technology for aerospace applications has been carried out for many years.

Traditional Metallurgical Processing

Ingots or Billets

Currently established commercial processing of tantalum and niobium alloys to produce useful mill products normally begins with electron beam melting of base metal ingots, followed by consumable electrode vacuum arc remelting to produce ingots of the desired alloys. In the case of tungsten and molybdenum alloys, base metal powders are mixed with the prescribed alloy powders, then compacted and extruded to billet at very high temperatures. Molybdenum alloys are also made by vacuum arc melting compacted powders.

Hot Working

The ingots or billets are then normally either extruded or forged after being surface conditioned and coated to reduce reaction with the atmosphere. Depending on the fabricability of the alloy, the extrusion and/or forging steps may be necessarily repeated, with interim conditioning and coating. Extrusion or forging temperatures and pressures are normally very high due, of course, to the high temperature strength of these alloys. Hot extruded slabs or sheet bar are normally further hot rolled to plate or sheet.

Tube hollows are hot extruded from drilled billets, either to final or intermediate size.

Cold Working

If the alloy exhibits sufficient cold ductility, it may be further reduced to sheet, strip, or tubing. Nb-1%Zr and PWC-11 are very ductile at room temperature and offer no severe problem areas in producing any size or shape desired.

Product Fabrication and Availability

The selection of refractory metal alloys for aircraft and aerospace applications currently is quite limited. A large percentage of the alloys that were devised and tested during the late 1960s and early 1970s, and exhibited excellent high temperature properties, unfortunately also exhibited limited lower temperature ductility and fabricability. Thus, their usefulness was either non-existent or very limited, when produced by traditional metallurgical processing. The diminishing interest in these alloys that followed this initial development phase is responsible for the paucity of useful test data on most of them, especially under irradiated conditions. Today, there are only a few refractory metal alloys available in most product forms. When considering Nb alloys, Table 3 lists those that are commercially available. These are all essentially lower strength, higher fabricability alloys.

Several Ta alloys, such as T-111 and TVC, exhibit excellent mechanical properties and lower temperature fabricability, however, the inherent high density and high neutron absorption of Ta alloys are disadvantages. TVC has the lowest density of the more fabricable Ta alloys. Figure 1 shows a strength comparison of a Ta, Nb, and Mo alloy.

SP-100 Program Applications

For the initial phase of the SP-100 program only Nb alloys are being considered for reactor and associated parts construction. Alloys will be required in a wide range of most common mill forms, plate, sheet, foil, tubing and possibly forgings. It is primarily due to this factor that the two Nb alloys that are currently favored are Nb-1%Zr and PWC-11 (Nb-1%Zr-0.1%C). Nb-1%Zr is readily available in all product forms from several commercial sources and has a significant data base available. Although PWC-11 has not been produced commercially in any significant quantity, the limited production carried out by Oak Ridge National Laboratories and several independent metal producers has proven this alloy to be very similar to Nb-1%Zr in fabricability. Although rather limited, there is also a data base available for this alloy. Nb-1%Zr has routinely been fabricated into tubing as small as 1/8″ OD × .010″ thickness and sheet as thin as .001″ foil for many years. PWC-11 has readily been fabricated into various tube sizes, as small as .230″ OD × .015″ thickness. This excellent fabricability is a definite advantage over many of the Ta, W, or Mo alloys. Another factor that provides an advantage to the use of Nb alloys is the density, being approximately one half that of Ta or W alloys.

The tensile and yield strengths and elongation of these alloys and other Nb base alloys are shown in Figures 2 and 3.

It is very evident that these alloys, although readily available, are at the lower end of the Nb alloy strength spectrum. Thus, follow-on space nuclear power generator programs will undoubtedly be seeking improved strength alloys.

Powder Metallurgy Technology

Although powdered refractory metals have long been used by consolidating for melting into ingots for further processing by traditional metallurgical operations, powder metallurgy (P/M), technology has now been developed to enhance the feasibility of employing some of the more difficult to fabricate alloys in order to take advantage of their better mechanical properties.

One area of development that TWCA has been pursuing for some time is in the production of ultra-pure spherical powders, particularly as applied to Nb alloys. TWCA has designed and constructed a prototype atomizing apparatus, utilizing electron beam melting, wherein sections of alloy ingot and/or selected scrap are drip melted by electron beam energy, in high vacuum, on to a disc spinning at high speed. As very small droplets of molten alloy are thrown from the spinning disc by centrifugal force, and shaped into minute spheres by surface tension forces, they solidify and cool very rapidly. In essence, each sphere is a micro-ingot exhibiting reduced oxygen content due to the high vacuum melting, and a very fine dispersion of phases such as oxides and carbides due to the rapid cool-

Table 3 *Commercially Available Niobium Alloys*[a].

Alloy Designation	Composition	Supplier
Nb-1%Zr	Nb-1%Zr	Fansteel,Cabot. TWCA[b]
C103	Nb-10%Hf-1%Ti	Fansteel,Cabot,TWCA
SCb291	Nb-10%Ta-10%W	Fansteel,Cabot,TWCA
C129	Nb-10%W-10%Hf-0.1%Y	TWCA
FS-85	Nb-27%Ta-10%W-0.8%Zr	Fansteel

[a]From Pionke and Davis 1979.
[b]Teledyne Wah Chang Albany.

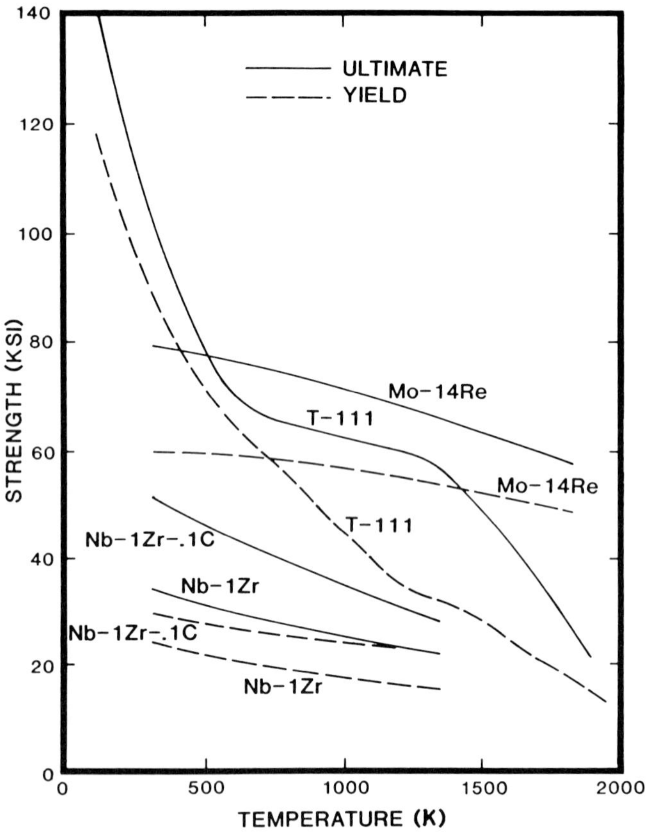

Figure 1 *Strength Comparisons of Ta, Nb, and Mo Alloy.*

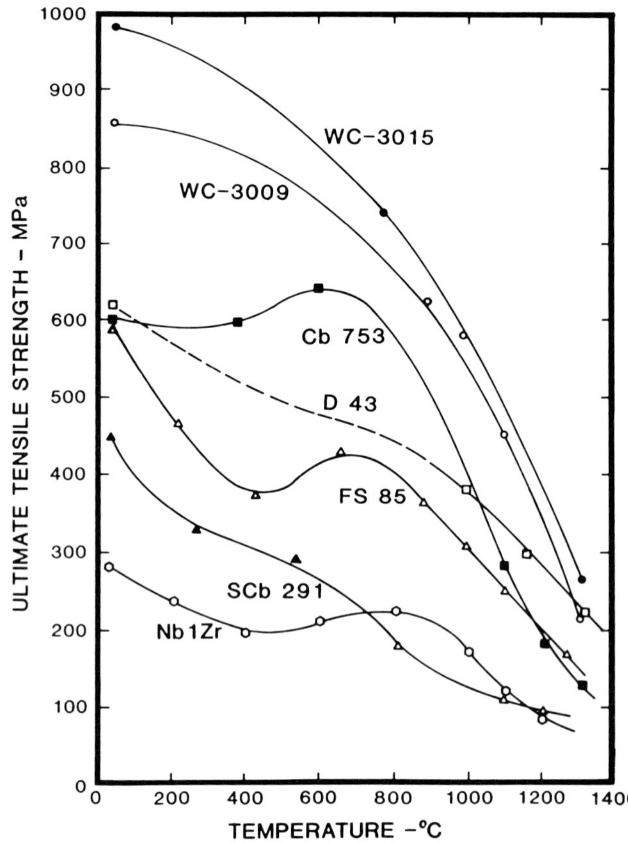

Figure 2 *Effect of Temperature on the Tensile Strength of Niobium Alloys.*

ing rate. Figures 4 and 5 respectively, are a sample of the spherical powder and cross sections of several individual spheres. Particle size distribution ranges from 150 to 850 micrometers.

Nb alloy WC-3009 has been produced at TWCA in small quantities by traditional metallurgical processing but because of its very high strengths at all temperatures to 1300°C (2400°F), it is quite difficult to fabricate into mill products (Figure 6). However, by eliminating the troublesome operations such as extrusion, forging, and most of the hot rolling operations, the feasibility of producing this alloy in all mill forms would be greatly enhanced. It is expected that this can be accomplished by turning ingot to spherical powder, then directly into sheet, bar, or tube hollow, by hot isostatic pressing. See Figures 7, 8 and 9.

Experimental work at TWCA has indicated the probability of success is favorable. Of course, ancillary parts such as valves, pumps, and flanges, as shown in Figure 10, can be made directly to net or near-net shape today. Unique structures such as heat pipes to condense and conduct liquid metal coolants are feasible through P/M technology, shown in Figure 11. It is felt the P/M approach, although having yet to be adequately demonstrated, offers another tool to improve the utilization and economics of using Nb alloys. This technology can, of course, be applied to many other refractory alloys and may well enhance the usefulness of the strong but hard-to-fabricate alloys.

Due to the relative lack of interest in Nb alloy development in the recent past, TWCA has only budgeted limited R&D funding in this direction. The resurgence of interest, specifically the SP-100 program, and less specifically a few other aerospace applications, has caused TWCA to re-evaluate R&D efforts in these alloys.

Conclusions

The selection of refractory metal alloys for structural applications in space nuclear power generators for the immediate future is still limited to the same alloys that were available 20 years ago. Alloys that can presently

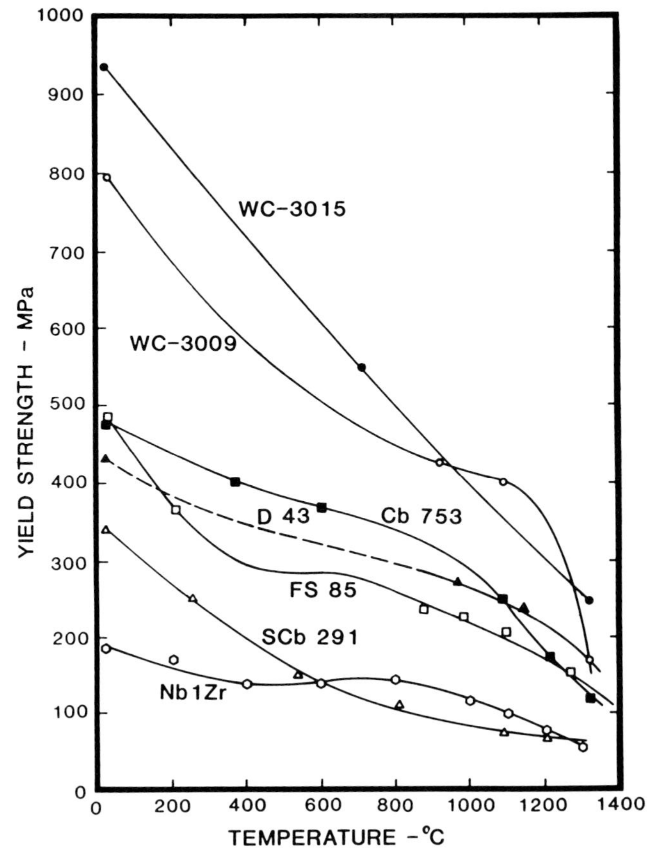

Figure 3 *Effect of Temperature on the Yield Strength of Niobium Alloys.*

Figure 4 *Electron Beam Atomized WC-3009 Nb Alloy Spherical Powder.*

be fabricated by contemporary methods and equipment into the many shapes and sizes required are unfortunately the lower strength alloys. This, of course, increases the overall weight of the system due to increased design sections. Even then the paucity of in-situ performance data on these available alloys leaves their application in long term space environments uncertain.

If significant strength improvements are required of structural alloys for the next generation of space nuclear power generators, it is most probable, in view of the many alloys that have been devised in the past, that these improvements will most likely be provided by innovations in process and production technology. Very pure alloys with very fine rapidly cooled microstructures such as electron beam atomized powders may provide one innovative approach to the production of final parts and mill products such as sheet and tube. Advanced technology in hot isostatic consolidation of these powders must be developed. The long period of low incentive on the part of refractory metals producers to develop new processing technology, and the highly competitive and difficult market place today, make many companies very wary of spending available R&D dollars in areas that are speculative. Since most of this type of work is ultimately funded by federal programs, many companies tend now to wait for the need and funding to become available before beginning innovative programs. Whether this fact is harmful or not is academic, it is a fact of corporate life. It is also probably the basic reason why, after the past 20 years of tremendous technological advancements in many fields, the refractory metal alloys and their properties that will be used in the immediate future, offer essentially no improvements over those of 20 years ago.

Producers must be able to expect technology development programs to lead ultimately to reasonable product sales and profit. The only alternative is direct federal funding.

Acknowledgments

The author would like to acknowledge Craig Wojcik, Research Metallurgist, Teledyne Wah Chang Albany, for preparing and processing the WC-3009 metal powders and for obtaining the test data presented.

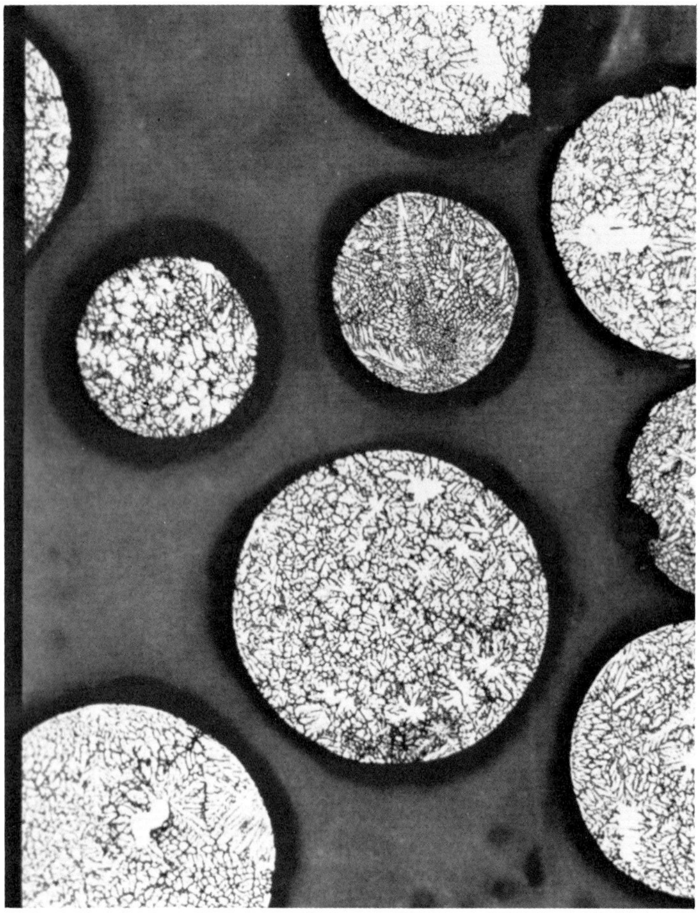

Figure 5 *Spheres of Electron Beam Atomized WC-3009 Nb Powder.*

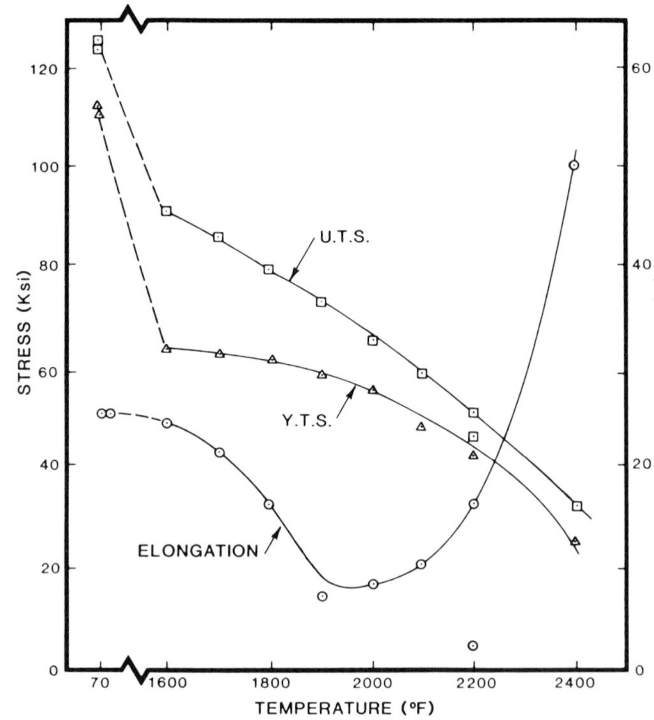

Figure 6 *The Effect of Temperature on the Strength and Ductility of WC-3009 Spherical Powder Consolidated by Hot Isostatic Pressing.*

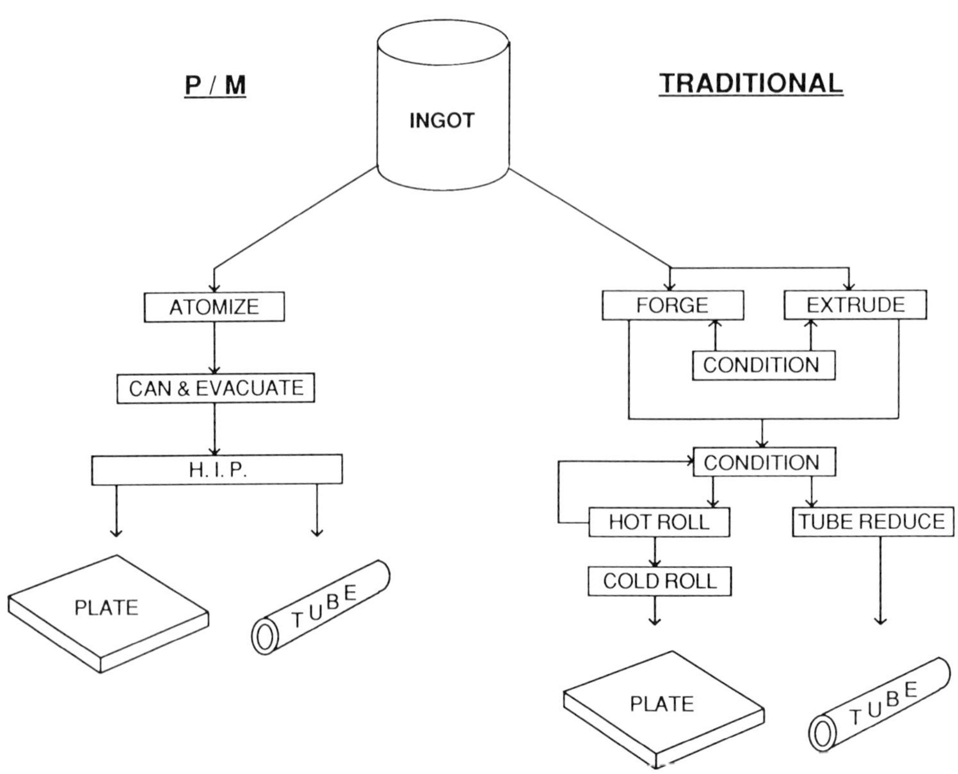

Figure 7 *Comparison of Process Steps using Traditional Ingot Metallurgy (I/M) and Powder Metallurgy (P/M).*

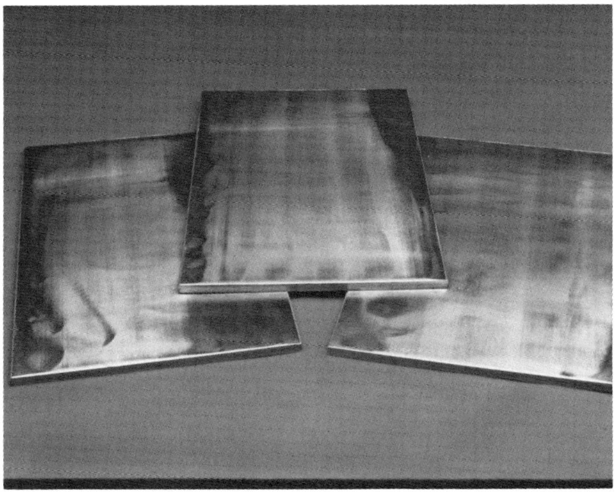

Figure 8 *Nb Sheet Bars made by Powder Metallurgy.*

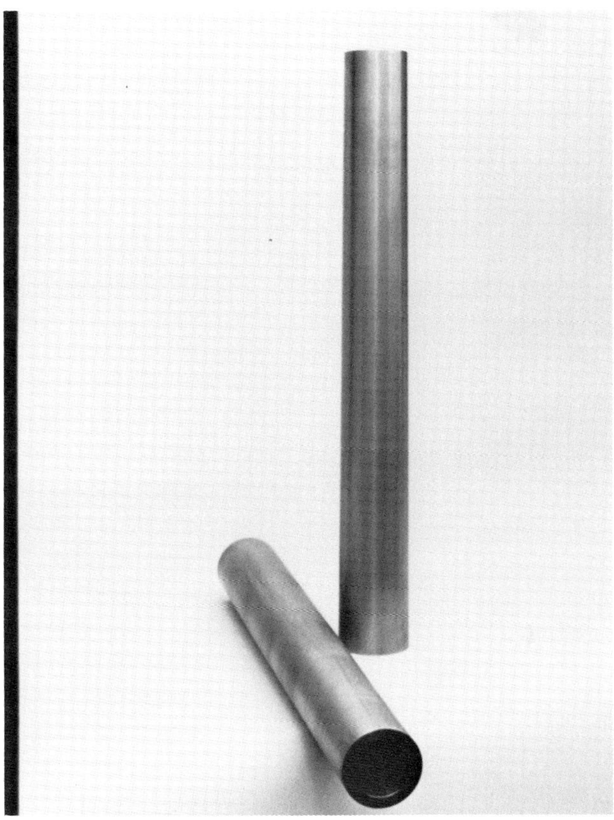

Figure 9 *Nb Alloy WC-3009 Tube Hollows prepared from Spherical Powder.*

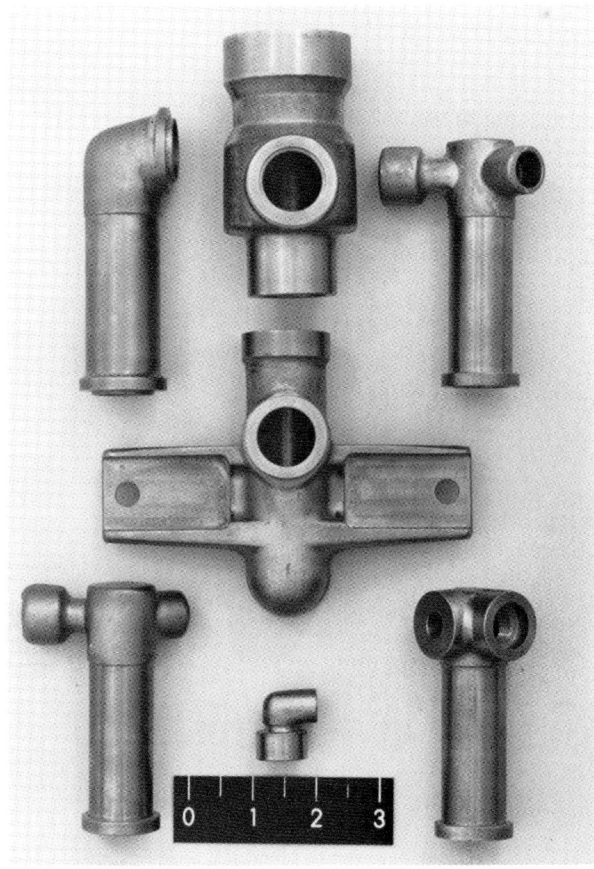

Figure 10 *Nb Alloy C-103 Parts made by Powder Metallurgy.*

Figure 11 *Nb Alloy C-103 Heat Pipe containing a Porous Nb Spherical Powder Lining and Three Return Arteries.*

References

Aerospace Structural Materials Handbook (1975) Volume 4, Belfour Stuler, Inc., Mechanical Properties Data Center, Traverse City, MI.

Frank, R.G. (1968) "Recent Advances in Columbium Alloys," in *AIME Proceedings on Metallurgy and Technology of Refractory Metal Alloys*, I. Machlin, R. T. Bagley, and C. D. Weisert, eds., held in Washington, DC, 25–26 April 1968.

Metals Handbook (1980) Volume 3, *Heat Resistant Materials*, Ninth Edition, American Society for Metals, Metals Park, OH.

Pionke, L.J. and J.W. Davis (1979) *Technical Assessment of Niobium Alloys Data Base for Fusion Reactor Applications*, COO-4247–2, McDonald Douglas Co., St. Louis, MO., August 1979.

Pratt and Whitney (1965) *Summary of Design and Analysis of The SNAP-SO/SPUR Reactor Coolant Pump*, PWAC-467, Pratt and Whitney Aircraft Co., Middletown, CT.

Space Nuclear Power Systems 1986
Edited by M. S. El-Genk and M. D. Hoover

Chapter 27

Observations On The As-Welded (GTAW) Bend Ductility Of Mo-13%Re Alloy

Robert L. Ammon and R. William Buckman, Jr.

Westinghouse Electric Corporation
Advanced Energy Systems Division
Pittsburgh, PA 15236

The ductile-to-brittle transition (DBTT) behavior of Mo-13%Re alloy sheet was evaluated in bending. Recrystallized material and as-GTA welded materials were tested. GTA welding raised the DBTT from 200 K to 450 K. This rise can be attributed primarily to the large grain size of the fusion/heat-affected zones. However, recrystallized 50%Mo-50%Re alloy exhibits minimal change in DBTT as a result of GTA welding. Although dilute rhenium additions to molybdenum enhance low-temperature ductility of both wrought and recrystallized material, the ''ductilizing'' effect appears to be minimized in the as-GTA welded condition.

Introduction

One of the refractory metal alloy compositions being considered for application in space nuclear power reactors is a binary alloy of molybdenum containing approximately 10 to 15 wt. % rhenium (Lundberg 1981a and 1981b and Merrigan and Lundberg 1983). Rhenium additions of up to 50 wt. % have been shown to improve the low-temperature ductility of recrystallized molybdenum (Geach and Hughes 1956 and Jaffee et al. 1961). Similar behavior in dilute molybdenum-rhenium alloy sheet and rod containing 11 to 14 wt. % rhenium has been shown by Klopp and Witzke (1972). Although rhenium additions to molybdenum have been demonstrated to be beneficial in lowering the DBTT temperature of material in both wrought/stress relieved and recrystallized conditions, little information is available on the behavior of welded material. The purpose of this study was to evaluate the effect of inert-gas-tungsten-arc welding (GTAW) on the DBTT behavior of a Mo-13 wt. % Re alloy.

Experimental Procedures

Material Preparation

The starting materials consisted of electron beam, float zone-purified molybdenum rod and high-purity CVD rhenium sheet. Four 100-gram ingots, approximately 4 cm dia × 0.8 cm thick, were prepared by nonconsumable electrode D.C. arc melting under an argon atmosphere pressure of 250 torr. Prior to melting, the chamber was evacuated to 1×10^{-5} torr, leak checked, and then backfilled to a pressure of 250 torr with high-purity argon. A titanium getter button then was melted and visually examined to assure an acceptably pure melting atmosphere. A sequence of sectioning, metallographic examination and remelting was performed to assure homogeneous Mo-13%Re alloy ingots.

Fabrication

Details for producing the 0.75-mm-thick sheet for welding evaluation are summarized in the process schedule in Figure 1. Rolling was done using a 200 mm × 200 mm one high Stanat mill. Billet heating at 755 K for rolling was done under a flowing argon atmosphere. The starting sheet bar was machined from the as-cast ingot and measured approximately 30 mm wide × 6.4 mm thick × 35 mm long. In-process and final annealing was accomplished in a tantalum resistance heated vacuum furnace operating at $\leq 1 \times 10^{-5}$ torr. Chemical cleaning prior to annealing was in a HNO_3/HF solution.

Inert Gas-Tungsten Arc Welding

Inert gas tungsten are welding (GTAW) was done under a high-purity helium atmosphere using the procedures developed by Stoner and Lessmann (1965).

The chamber was evacuated to $\leq 1 \times 10^{-5}$ torr, leak checked and then baked at 360 K overnight. After backfilling, the helium atmosphere contained less than 10 ppm total active impurities. Automatic autogeneous bead-on-plate welds were made using the arrangement shown in Figure 2 and the parameters listed in Figure 3. The weld bead was parallel to the rolling direction.

Bend Testing

Base metal and as-welded specimens were tested in bending following the procedures developed by the Materials Advisory Board for ductile refractory metal alloys (Material Advisory Board 1963). The arrangement is shown in Figure 4. The base metal specimen was perpendicular to the rolling direction. Similarly, the weld bead was parallel to the rolling direction and perpendicular to the axis of bending for the as-GTA welded specimens. The welded specimens were tested in the as-welded condition with the top of the weld bead, the tension side of the bend.

Results And Discussion

Sheet Fabrication

Sheet was produced successfully by doing the initial breakdown and intermediate rolling at 755 K. Ini-

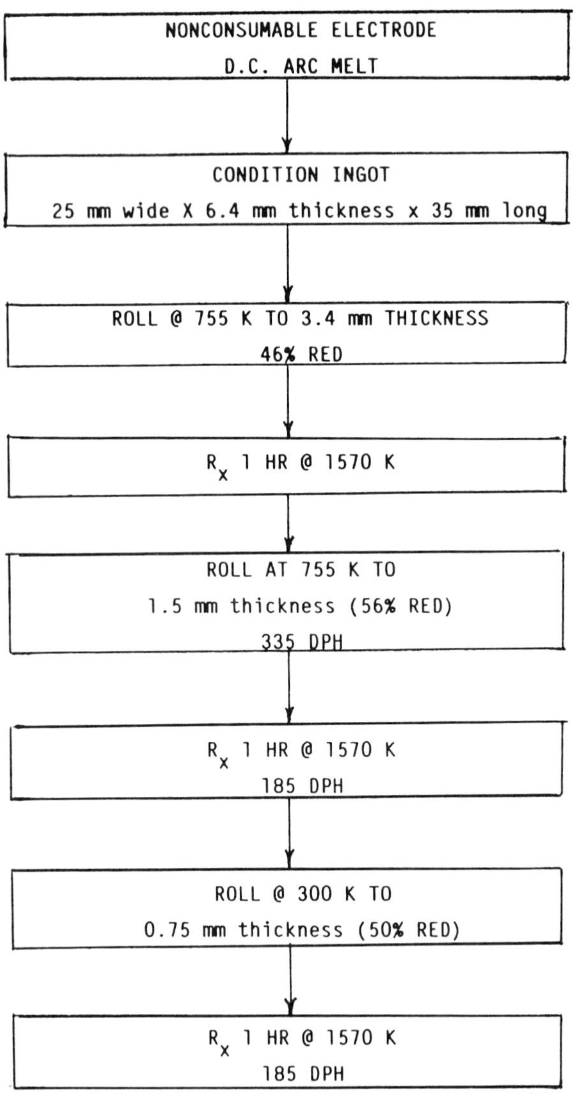

Figure 1 Processing Schedule for Mo-13%Re Alloy Sheet.

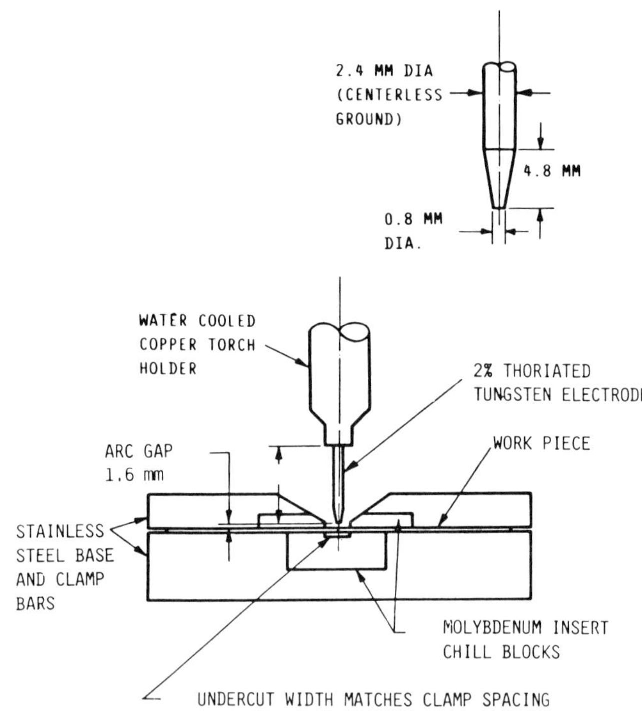

Figure 2 GTA Welding Arrangement.

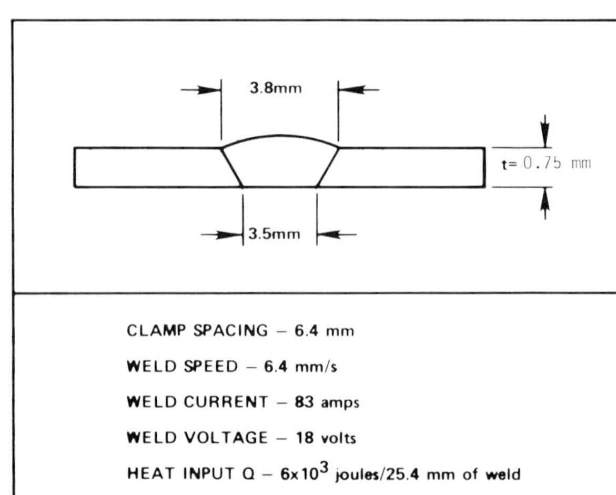

Figure 3 Weld Parameters for Mo-13%Re Alloy Sheet.

tial attempts to roll the as-cast sheet bar at room temperature were unsuccessful, with the sheet bar fracturing on the initial passes through the rolls. Similar behavior also was observed when the billets were heated to 370 K. Rolling at 755 K was sufficiently above the DBTT for the as-cast Mo-13%Re ingot material to allow fracture free initial breakdown. However, deformation processing at this temperature is still considered cold work, because significant recovery is unlikely at this fraction of the melting temperature (0.27 T_m). During rolling, catastrophic failure of the material occurred when the total reduction exceeded approximately 60%. Thus, reductions between in-process anneals were limited to approximately 50%. Rolling to final thickness was done at room temperature. The starting grain size for the as-melted material was large, with grain size on the order of the ingot thickness (6.4 mm). The combination of the large as-cast grain size and the high strain rate deformation during rolling raised the DBTT above 370 K. The recrystallized (1 hr @ 1570 K) grain size was ~ 33 μm at the 1.5 mm thickness and ~ 23 μm at the 0.75 mm thickness (see Figure 5).

The one-hour annealing temperature of 1570 K gave essentially a 100% recrystallized microstructure. Klopp and Witzke (1972) reported a one-hour annealing temperature of 1590 K for Mo-14%Re rod that had been swaged 91% at 1450 K to 1200 K with the as-swaged hardness 290 DPH and the recrystallized hardness 179 DPH. Considering the difference in processing history, this is comparable to 325 DPH (50% cold work) and 185 DPH (recrystallized) for the Mo-13%Re sheet product in this study.

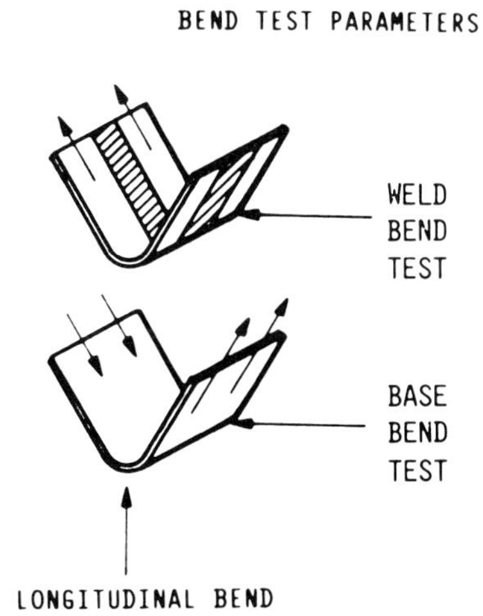

NOTE: ARROWS SHOW ROLLING DIRECTION

THICKNESS, t = 0.75 mm
WIDTH = 12 t
LENGTH = 24 t
TEST SPAN = 15 t
PUNCH SPEED = 0.4 mm
TEMPERATURE - variable
PUNCH RADIUS - 1 t

BEND DUCTILE TO BRITTLE TRANSITION TEMPERATURE = LOWEST TEMPERATURE FOR 90° + BEND WITHOUT CRACKING

Figure 4 *Bend Testing Details.*

Chemical Analysis

The carbon, oxygen, and nitrogen contents of the recrystallized base metal were determined. The results are tabulated in Table 1, along with chemistry of the as-GTAW weld metal. For comparison, interstitial chemistry data on Mo-Re alloy reported in the literature are included. The processing and subsequent welding during sample preparation did not appear to change the interstitial chemistry significantly. The material produced in this study generally had interstitial purity equivalent to that reported by Klopp and Witzke (1972) and Merrigan and Lundberg (1983).

GTA Welding

After preliminary welding trials to confirm the adequacy of the weld parameter selection, a full penetration automatic autogeneous bead-on-plate weld was made using the parameters listed in Figure 3. Visually, the weld bead had a smooth, bright surface, indicating the adequacy of the high-purity helium cover gas. A specimen was cross sectioned for metallographic examination and is shown in Figure 6. The metallography of the fusion zone and heat affected zone is what was expected for the given weld parameter/ configuration. The grain size in the fusion zone is large, as would be expected for a high-purity material. Expitaxial growth from the large-grained, heat-affected zone and the original fine grain microstructure near the restraining clamp is typical. A hardness traverse indicated that the hardness of the base metal (185 DPH), heat-affected zone (180 DPH), and the fusion zone (180 to 187 DPH) were equivalent. Considering the purity of the starting material and welding atmosphere, no excess interstitials were taken into or rejected from solid solution during the weld heating and cooling transient.

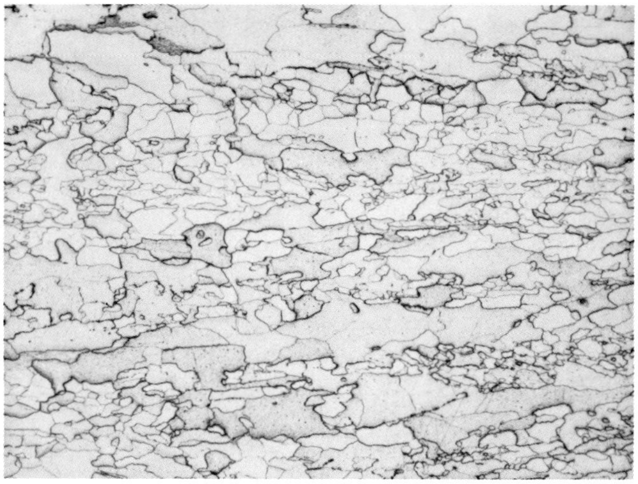

(a) 1.5-mm thick

(b) 0.75-mm thick

Figure 5 *Recrystallized (1 hr @ 1570 K) Mo-13%Re Sheet.*

Table 1 *Chemical Analysis of Mo-Re.*

Composition/Form	Wt %			Reference
	Carbon	Oxygen	Nitrogen	
Recrystallized base metal 0.030'' thick	0.0042	< 0.0007	N.D.[a]	Ammon and Buckman 1986
Weld metal - as GTAW welded	0.0052	< 0.0007	< 0.0005	Ammon and Buckman 1986
Mo-13%Re sheet/rod (electron beam melted)	0.0023	0.0027	0.0002	Klopp and Witzke 1972
Mo-13%Re tubing (E.B.and vac. arc melted)	0.0040	0.010	0.0019	Merrigan and Lundberg 1983
Mo-14.5%Re	< 0.0005	0.0025	0.0020	Merrigan and Lundbert 1983

[a]Not determined.

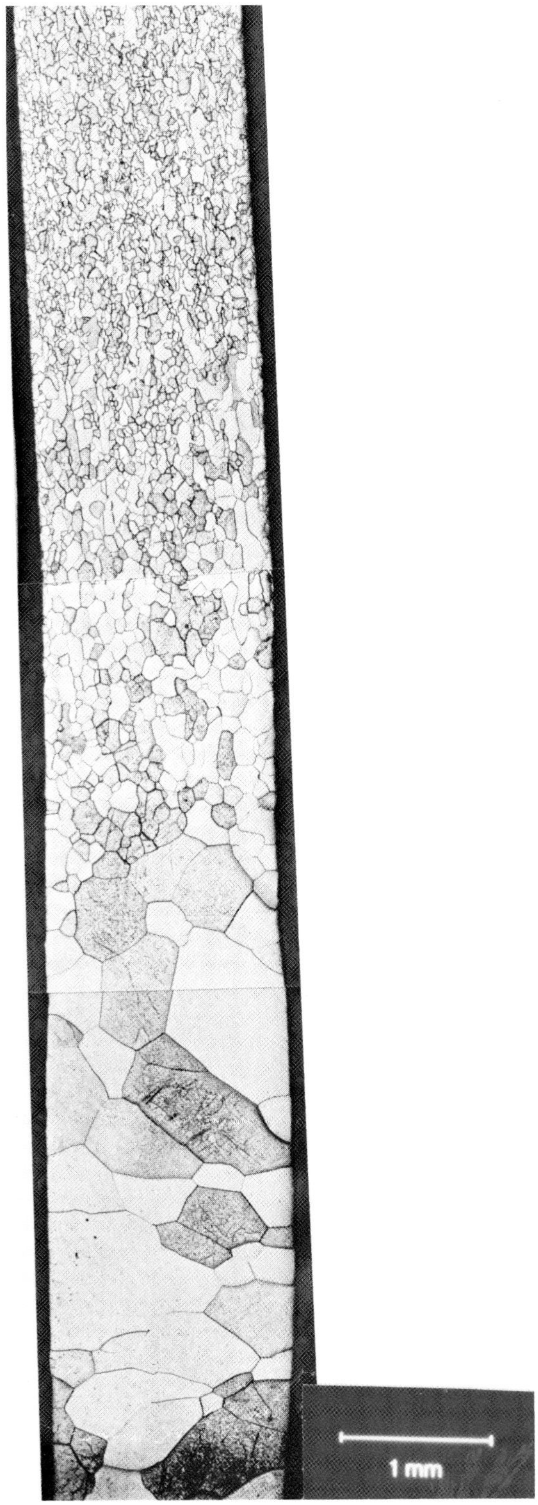

Figure 6 *Microstructure of Mo-13%Re GTA Weld. The upper region is the base metal (185 DPH hardness and ~ 0.020 mm grain size). The central region is the heat affected zone (180 DPH hardness and ~ 0.5 mm grain size). The lower region is the weld (180–187 DPH hardness and ~ 0.15 mm grain size).*

Bend Test

The DBTT behavior determined in bending for the recrystallized base metal and the as-GTA welded material is presented in Figure 7. At 250 K, the recrystallized base metal underwent a full 90° bend over a 1 t (0.75 mm) bend radius without any indication of failure; whereas at 200 K, failure occurred after bending through an angle of 50°, After GTA welding, heating to 450 K was necessary to obtain a full 90° bend. Klopp and Witzke (1972) report a DBTT of 150 K for Mo-13.8%Re for recrystallized material of similar purity and grain size, which is approximately 100 K lower than measured in this study. However, they were using a less severe bend radius (4 t) versus the 1 t used for this study. For a full 90° bend, the maximum outer fiber strain is ~ 10% over a 4 t bend radius and ~ 33% over a 1 t bend radius. Using the tensile reduction in area curve generated by Klopp and Witzke (1972) and assuming that the reduction in area was equivalent to the total tensile elongation at fracture (no necking), the temperature for 33% reduction in area reported for Mo-13.8%Re is 250 K, which corresponds to the bend DBTT determined in this investigation.

The effect of GTA welding on the DBTT of recrystallized Mo-13%Re is quite significant, causing an increase in DBTT behavior from well below room temperature to significantly above room temperature. Although significant grain size differences exist between the fusion, heat-affected zones, and base metal, the hardness level across all three zones essentially is identical, indicating their similar strength lev-

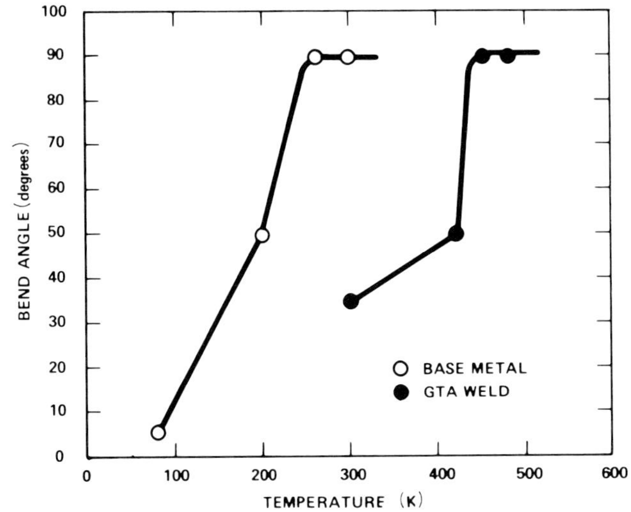

Figure 7 *Bend Ductile-Brittle Transition Behavior for Mo-13%Re Alloy Sheet.*

els. Wadsworth et al. (1983) tested GTA-welded wrought stress-relieved Mo-TZM and observed brittle fracture at room temperature for material tensile tested in the as-welded condition and ductile behavior for material tested in the as-welded plus annealed condition. Since the weld bead was transverse to the tensile axis and represented a small fraction of the total gage length, strain would be localized in the softer weld/heat-affected zone, giving rise to a high localized strain rate. For this investigation, the weld, heat-affected zone, and base were oriented such that all three zones experienced equivalent deformation rates during bend testing.

Thus, strain localization effects attributed to the room temperature brittle fracture of Mo-TZM by Wadsworth et al. (1983) are not considered to be a factor in the behavior of the Mo-Re alloy. SEM examination of the fractured surface of the recrystallized base metal tested at 77 K and the welded specimen tested at 298 K revealed a typical brittle cleavage fracture mode (see Figure 8). The difference in DBTT transition temperature between the recrystallized material and the as-GTA welded material is thought to be controlled predominately by the large grain size of the GTA weld zone. Referring to Figure 9, 4 t DBTT data for recrystallized and GTA welded Mo-50%Re alloy (Svedberg and Buckman 1979) are compared with the Mo-13%Re alloy behavior. Two features are apparent: (1) The DBTT for recrystallized Mo-50%Re is at 80 K; and (2) GTA welding had a minimal effect and raised the DBTT by only 100 K. Subsequent post-weld annealing for one hour at 1920 K resulted in about a 50 K decrease. This slight decrease in DBTT may be due to removal of residual stress induced during welding, since the edges of the weld specimen were restrained during the GTA welding operation. The Mo-50%Re alloy, although prepared by powder metallurgical techniques, had a purity equivalent to that for the arc-melted Mo-13%Re. No grain-size data were reported but the significantly higher annealing temperatures used on the Mo-50%Re should result in an equal or larger grain size. Thus, the apparently higher ductility (lower DBTT) and lesser change in DBTT due to GTA welding would indicate that the

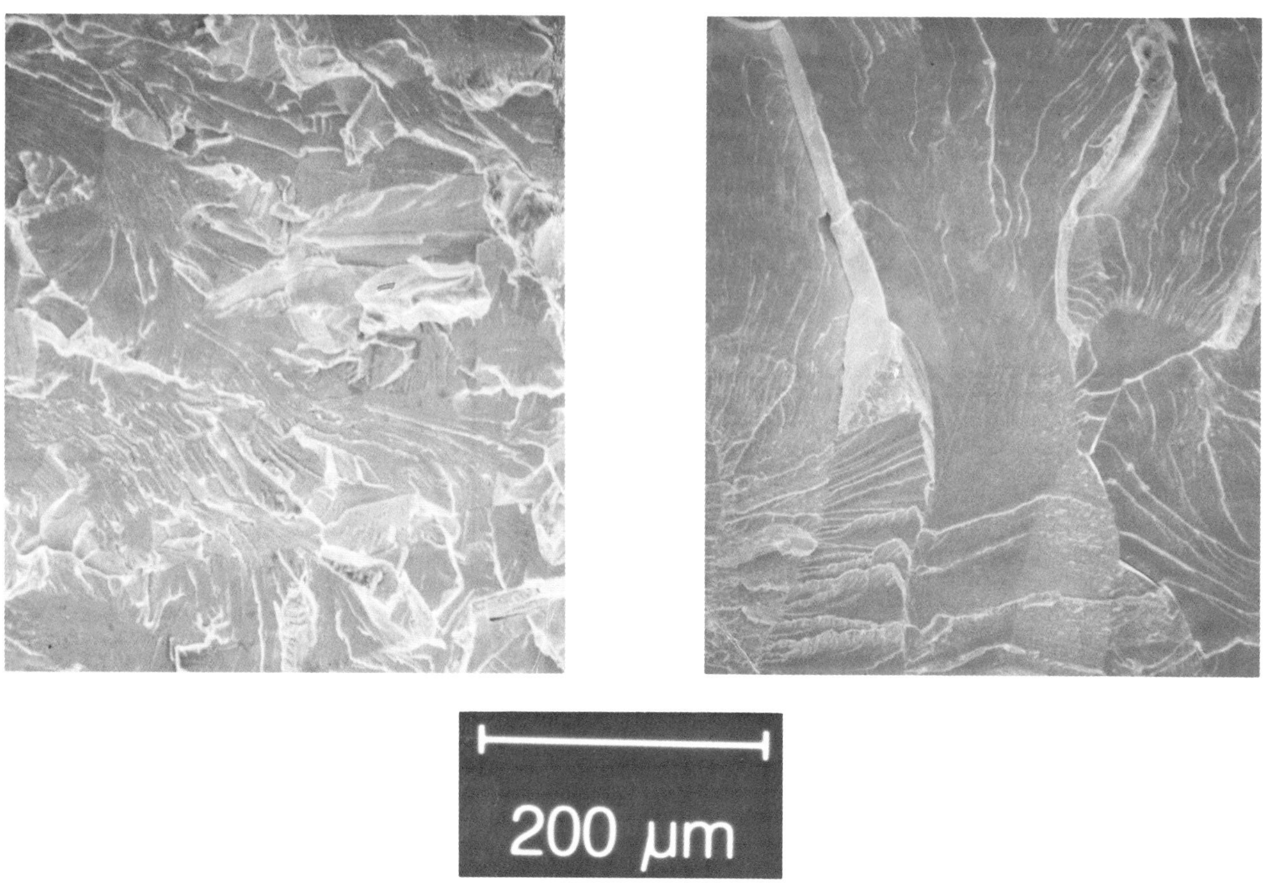

Figure 8 *Fracture Surface of Mo-13%Re Bend Specimens.*

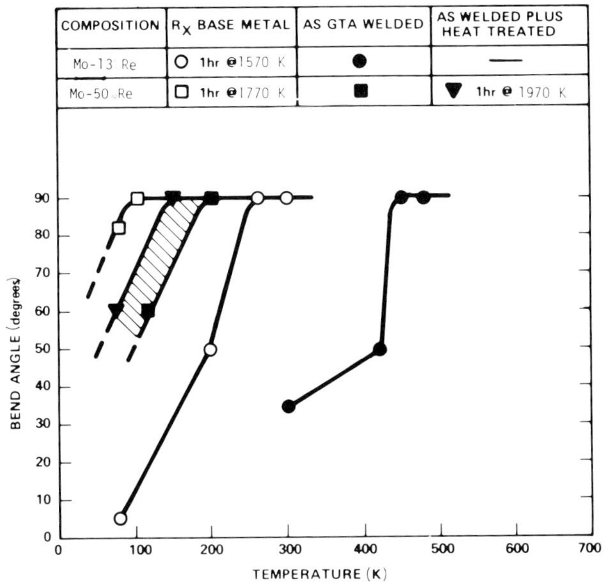

Figure 9 *Bend Ductility of Mo-Re Alloy Sheet.*

beneficial effects of rhenium on the low-temperature ductility of molybdenum are more pronounced at the higher rhenium content.

Conclusions

Although limited in scope, this study clearly demonstrated that: (1) gas tungsten arc welding significantly raises the DBTT for Mo-13%Re alloy; (2) increasing the rhenium content to 50 wt. % results in an as-GTA welded alloy that retains ductility in bending to well below room temperatures; and (3) before considering utilizing Mo-13%Re alloys in space nuclear power plant hardware, detailed investigation of its weldability should be accomplished.

Acknowledgments

The authors wish to acknowledge the efforts of H. J. Grubich and S. S. Laciak, for producing the sample material, G. O. Yatsko for conducting the bend tests, B. Chiappetta for manuscript preparation, and Westinghouse Advanced Energy Systems for supporting the work.

References

Geach, G. A. and J. E. Hughes (1956) "The Alloys of Rhenium with Molybdenum or with Tungsten and Having Good High Temperature Properties," in *Plansee Seminar*: De Re Metallic, 2nd edition, Pergamon Press, NJ, pp. 245–253.

Jaffee, R. I., D. J. Maykuth, and R. W. Douglass (1961) "Rhenium and the Refractory Platinum-Group Metals," in *Refractory Metals and Alloys 11*, 383–463, Interscience Publ., Inc., NY.

Klopp, W. D. and W. R. Witzke (1972) *Mechanical Properties of Electron Beam Melted Molybdenum and Dilute Molybdenum Rhenium Alloys*, NASA TM-2576, Lewis Research Center, National Aeronautics and Space Administration, Cleveland, OH.

Lundberg, L. B. (1981a) *A Critical Evaluation of Molybdenum and Its Alloys for Use in Space Reactor Core Heat Pipes*, LA-8685-MS, Los Alamos National Laboratory, Los Alamos, NM.

Lundberg, L. B. (1981b) *An Evaluation of Molybdenum and Its Alloys*, AIAA-81-1105, Los Alamos National Laboratory, Los Alamos, NM.

Material Advisory Board (1963) *Evaluation Test Methods for Refractory Metal Sheet Materials*, Reports 176M and 192M, Material Advisory Board, April 1963.

Merrigan, M. A. and L. B. Lundberg (1983) "An Initial Evaluation of Molybdenum Alloy for Reactor Heat Pipes." Paper No. Bo. 5, 18th Intersociety Energy Conversion Engineering Conference, pp. 1880–1886, American Institute of Chemical Engineers, NY.

Stoner, D. R. and G. G. Lessmann (1965) "Measurement and Control of Weld Chamber Atmospheres," *Weld. J. Res. Suppl.*, 44: 331-s-346-s.

Wadsworth, J., G. R. Morse, and P. M. Chewey (1983) "The Microstructure and Mechanical Properties of a Welded Molybdenum Alloy," *Mater. Sci. Eng.*, 59: 257–273.

Svedberg, R. C. and R. W. Buckman, Jr. (1979) *Personal Communication of Unpublished Data*, Westinghouse Electric Corporation, Advanced Energy Systems Division, Pittsburgh, PA.

Space Nuclear Power Systems 1986
Edited by M. S. El-Genk and M. D. Hoover

Chapter 28

Thermal Aging Effects in Refractory Metal Alloys

Joseph R. Stephens

National Aeronautics and Space Administration
NASA Lewis Research Center
Cleveland, OH 44135

The alloys of niobium and tantalum are attractive from a strength and compatibility viewpoint for high operating temperatures required in materials for fuel cladding, liquid metal transfer, and heat pipe applications in space power systems that will supply from 100 kWe to multimegawatts for advanced space systems. To meet the system requirements, operating temperatures ranging from 1100 K to 1600 K have been proposed. Expected lifetimes of these space power systems are from 7 to 10 yr. A program was conducted at NASA Lewis to determine the effects of long-term, high-temperature exposure on the microstructural stability of several commercial tantalum and niobium alloys. Variables studied in the investigation included alloy composition, pre-age annealing temperature, aging time, temperature, and environment (lithium or vacuum), welding, and hydrogen doping. Alloys were investigated by means of cryogenic bend tests and tensile tests. Results showed that the combination of tungsten and hafnium or zirconium found in commercial alloys such as T-111 and Cb-752 can lead to aging embrittlement and increased susceptibility to hydrogen embrittlement of ternary and more complex alloys. Modification of alloy composition helped to eliminate the embrittlement problem.

Introduction

Niobium and tantalum alloys are being considered for advanced space power systems applications such as nuclear fuel element cladding, liquid metal transfer, and heat pipes. Minimum requirements for these applications include high creep resistance in the temperature range 1100 to 1600 K, a ductile-brittle transition temperature (DBTT) well below room temperature, good formability and weldability, and corrosion resistance to liquid alkali metals such as lithium, sodium, and potassium. Results have shown that severe corrosion can occur in tantalum-10 tungsten (Ta-10%W) (weight percent) contaminated with oxygen concentrated at grain boundaries. This corrosion is caused by preferential attack of these high-oxygen content regions by liquid metals (Moss et al. 1970). Because of this potential corrosion problem, a reactive element such as hafnium is added to the Ta-base alloy to serve as a getter for oxygen.

The stable oxide that forms, hafnium dioxide (HfO_2), is not attacked by the alkali metal, thus permitting operation of Ta-base alloys in alkali metal environments even when oxygen contamination exists. Tests have shown that T-111 (Ta-8%W-2%Hf) can successfully operate for several thousand hours in an alkali metal environment with little or no corrosion (Moss et al. 1970). However, room temperature embrittlement has been noted in T-111 specimens aged for 1000 h or longer at 1275 K.

A program conducted at NASA Lewis by Watson and Stephens (1972) indicated that the brittle intergranular fractures observed in aged T-111 were caused by hydrogen embrittlement during post-aging handling operations. The results further suggested that the presence of HfO_2 at the grain boundaries of T-111 aged at 1315 K may be associated with its subsequent increased sensitivity to hydrogen embrittlement. A series of studies conducted at NASA Lewis by Stephens (1974, 1975a, 1975b, 1977a, and 1977b) investigated in greater detail the causes of embrittlement after long term aging and sought means of alleviating this problem through alloy modification.

Eight Ta-base alloys, three Nb-base alloys, and one Mo-base alloy were investigated. Test variables included pre-age annealing temperature, simulated butt welding, and aging conditions (time, temperature, and environment). Evaluation included bend testing and tensile testing to measure effects on ductility and strength.

Experimental

Materials

The chemical compositions of the alloys evaluated in this study are listed in Table 1. The alloys were obtained in the form of sheet ranging in thickness from 0.75 to 1.5 mm thick and in the form of tubing having a 19 mm o.d. and a 1.5 mm wall thickness. To evaluate the effect of welding, a gas-tungsten-arc bead-on-plate weld (to simulate a butt weld) was made along the longitudinal centerline of some of the Ta-base alloy specimens. A standard annealing treatment for the Ta-base alloys consisted of 1 h at 1925 K plus 1 h at 1590 K. Some specimens were also annealed at 2090 K and 2250 K instead of the 1925 K standard heat treatment. Resulting grain sizes of 26, 52, and 150 μm were achieved for annealing temperatures of 1925 K, 2090 K, and 2255 K, respectively. Niobium alloys were annealed at 1620 K for 1 h while the Mo-base alloy was annealed at 1700 K for 1 h.

Aging

The tantalum alloy T-111 was aged at 1315 K for 1000 h and 5000 h in vacuum and lithium and at 1200 K and 1425 K for only 1000 h in both vacuum and a lithium environment. All other Ta alloys were aged for 1000 h at 1315 K in vacuum and in lithium. Niobium and Mo alloys were aged only in vacuum for 1000 h at temperatures ranging from 975 K to 1300 K.

Hydrogen Doping

Hydrogen doping was accomplished by heating several specimens from each alloy in an evacuated furnace to 1315 K (Ta alloys) and 1100 K (Nb and Mo alloys), at which point hydrogen was introduced into the furnace to a pressure of 13 kN/m^2. The specimens were held at temperature for 10 minutes followed by cooling in a helium atmosphere.

Evaluation

Bend and tensile testing were used as the primary measurement to determine the effects of aging, alloy composition, and other variables on the mechanical properties of the three alloy systems. Bend testing was conducted at a crosshead rate of 2.54 cm per minute. A bend radius of 2t, where t is equal to the specimen thickness, and a total bend angle of 140° were used for all tests. After completing the bend test, all ductile specimens were flattened upon themselves with pliers at the test temperature. In effect, then, the tests were a 180°, 0t bend. Tensile tests were conducted at a cross-head speed of 0.1 cm per minute. Light and scanning electron microscopy along with chemical analyses were used to further characterize the alloys.

Results

At the time these programs were conducted, in the early 1970s, emphasis was primarily on tantalum-base alloys and niobium-base alloys. Interest in the molybdenum-base alloys as candidate space power system materials was minor because of the fabricability problems associated with forming heat pipes, fuel cladding configurations, and so on. Therefore, our work focused primarily on the Ta-base alloy T-111, and secondarily on niobium alloys because of their similarity to Ta-base alloys. It was desirable to determine if the aging embrittlement problem noted in

Table 1 *Composition of Ta, Nb, and Mo Alloys.*

Alloy	Ta	W	Mo	Hf	Nb	Zr	Ti	Re	C	H	N	O
	content, wt.%								content, ppm			
Ta-10%W	Bal.	10.35	---	----	----	----	---	---	15	1.0	40	21
Ta-8%W-0.5%Hf		7.8	----	0.51	0.04	0.02	----	---	50	1.0	9	30
Ta-8%W-1%Re-0.7%Hf-0.25% C (ASTAR 811 C)		7.5	----	0.65	---	----	----	1.0	260	3.0	32	25
Ta-8%W-1%Hf		8.0	----	0.9	0.04	0.03	----	---	50	1.6	5	24
Ta-8%W-2%Hf(T-111)		8.5	----	2.0	0.05	0.06	----	---	50	2.9	15	120
Ta-8%W-3%Hf		8.2	----	2.8	0.04	0.09	----	---	60	1.0	8	28
Ta-4%W-2%Hf		3.8	----	1.8	0.04	0.06	----	---	50	0.8	7	26
Ta-2%Hf		----	----	1.9	---	----	----	---	22	1.0	39	33
Nb-9.8%Hf - 0.5%Zr-1%Ti 0.4%Ta-0.4%W (C-103)	0.4	0.4	----	9.8	Bal.	0.52	0.95	---	<30	<5	40	130
Nb-10%W-2.5%Zr (Cb-752)	<0.05	9.8	----	<0.3	Bal.	2.7	----	---	38	3	100	126
Nb-1%Zr	----	-----	----	---	Bal.	0.95	----	---	100	3	20	73
Mo-0.5%Ti-0.1%Zr (Mo-TZM)	-----	-----	Bal.	---	----	0.49	0.1	---	175	1	5	20

some instances for the Ta alloys also occurred in Nb alloys under similar conditions. Molybdenum alloys were investigated to include a Group VI alloy with the Group V alloys, Ta and Nb. The results will be discussed in the order of emphasis at the time the research was conducted. Included among the Nb alloys was Nb-1%Zr, which is the primary alloy under consideration for current space power systems such as the SP-100 Program.

Tantalum-Base Alloys

Chemical Analysis

Interstitial impurities were determined for the alloys before and after aging. Results indicated that aging in lithium or vacuum had no apparent effect on the nitrogen content of the Ta alloys. In contrast, the oxygen data revealed that aging in lithium substantially reduced the oxygen content of the Ta alloys at 1315 K and 1425 K (for example from 65 to 33 ppm and from 33 to 8 ppm in other cases), and to a lesser extent at 1200 K (from 65 to 44 ppm). Aging in a vacuum of 1.3×10^{-7} N/m^2 did not reduce the oxygen content of the alloys except for T-111 which had been pre-aged annealed at 2090 K and 2255 K (to 52 and 34 ppm). The hydrogen analysis for doped specimens revealed a pickup of about 5 to 10 ppm hydrogen for all the alloys in the pre-age annealed condition and from 10 to 30 ppm hydrogen for aged specimens.

Bend Ductility

Figure 1 shows the effect of pre-age annealing temperature on the ductility of T-111 tube that was subsequently aged in lithium for 1000 h at 1315 K. Specimens were brittle at 77 K after all pre-age annealing treatments and, as the pre-age annealing temperature was increased to 2255 K, T-111 became brittle at room temperature. This is attributed in part to the large (150 μm) grain size of these specimens compared with the grain size of 26 and 52 μm for specimens annealed at 1925 K and 2090 K, respectively. Figure 2 illustrates the effects of aging temperature on the bend ductility at 77 K of T-111 tube. It is shown that the annealed specimens and those aged at 1200 K and 1425 K could undergo a full 180° 0t bend at 77 K without fracture or cracking while specimens aged at 1315 K fractured with essentially no ductility. This lack of ductility will be termed aging embrittlement and seems to be confined to specimens aged over a narrow temperature range near 1315 K. The remaining aging conditions investigated (time and environment) had very little effect on the bend ductility of T-111. Testing the tube specimens at room temperature resulted in ductile behavior, 180° bend. Extending the aging time to 5000 h at 1315 K resulted in brittle behavior at 77 K and ductile behavior at room temperature, results similar to those obtained after the standard 1000 h treatment.

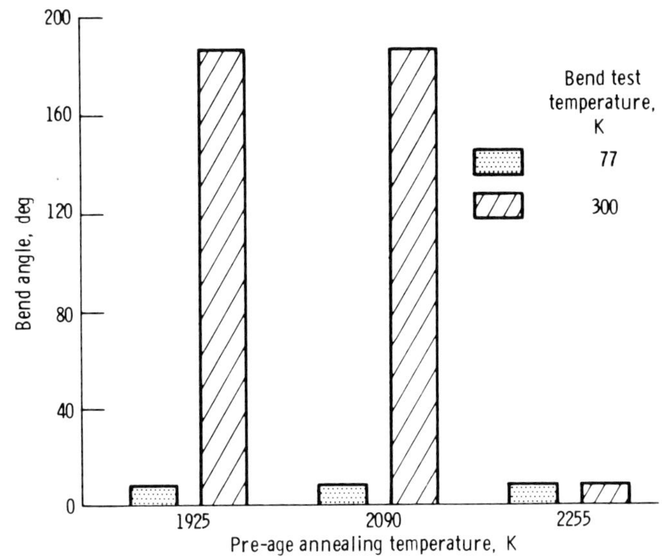

Figure 1 *Effect of Pre-age Annealing Temperature on Bend Ductility of T-111 Tube Aged in Lithium at 1315 K for 1000 hours.*

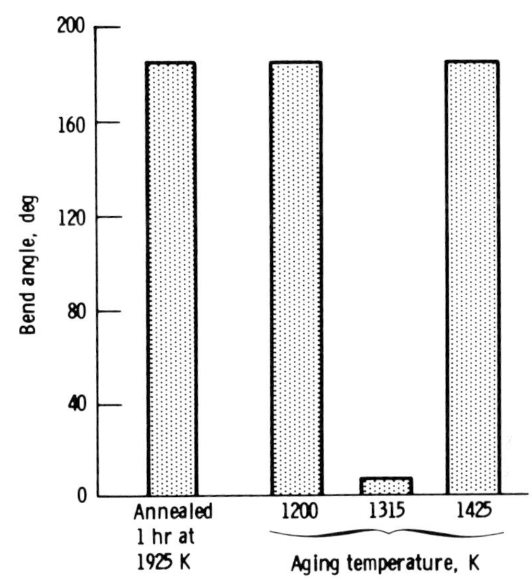

Figure 2 *Effect of Aging Temperature on 77 K Bend Ductility of T-111 Tube Annealed at 1925 K and Aged in Vacuum or Lithium for 1000 hours.*

Several of the sheet bend specimens were welded before the standard aging treatment and in the as welded condition were ductile at 77 K. Since the standard aging treatment produced embrittlement at 77 K, bend tests on welded and aged T-111 were conducted only at room temperature where all specimens which had been both welded and aged were brittle. A similar finding was reported previously for welded T-111 after aging for 1000 h to 10,000 hr over the temperature range 1255 K to 1425 K (Lessmann and Gold 1970). The large grain size developed in the weld area is assumed to contribute to the embrittlement. Bend test results are summarized in Figure 3 for T-111 sheet doped with hydrogen. It should be noted that sheet aged at 1200 K and then doped with 10 ppm hydrogen still exhibited some ductility at 77 K. In contrast, T-111 sheet aged at 1315 K and 1425 K and then doped with 10 ppm hydrogen was brittle at the less severe test temperature of 300 K. This corresponds to an increase of over 200 K in the ductile-brittle transition temperature (DBTT) of T-111 sheet and is in agreement with previous results reported by Watson and Stephens (1972). Similar results were found on tubing material also. However, this embrittlement is not solely related to the bulk hydrogen content of 10 ppm since varying degrees of ductility were observed depending on aging conditions.

The effects of alloy composition are most interesting and give some insight into the reasons for aging embrittlement and susceptibility to hydrogen embrittlement in Ta-base alloys. All compositions were ductile at 77 K in the pre-age annealed condition. Only two of the alloys, Ta-8%W-2%Hf (T-111) and Ta-8%W-3%Hf, were susceptible to aging embrittlement. These two alloys contain the greatest amount of W and Hf, suggesting that the combined levels of W and Hf have an important effect on the alloy's susceptibility to aging embrittlement. Bend data for hydrogen-doped Ta alloys are shown in Figure 4. In the annealed condition only T-111 indicated any evidence of loss in ductility at 77 K owing to the addition of approximately 8 ppm hydrogen. Aging and then doping with hydrogen resulted in embrittlement of all the alloys except the two binary alloys: Ta-10%W exhibited only slight surface cracks after a 180° bend at 77 K and Ta-2%Hf showed no loss of ductility at 77 K. Hydrogen embrittlement was most severe in the Ta-8%W-2%Hf (T-111) and Ta-8%W-3%Hf alloys, the specimens being brittle at 300 K after aging either in lithium or vacuum. In contrast, the ductility at 300 K of the remaining ternary alloys and ASTAR-811C showed a dependence on aging environment which in turn could be related to the hydrogen content of the specimens (12 ppm for vacuum aging and 25 ppm for lithium aging). The reduced oxygen content after lithium aging is believed to have permitted higher amounts of hydrogen to enter the specimens during subsequent hydrogen doping.

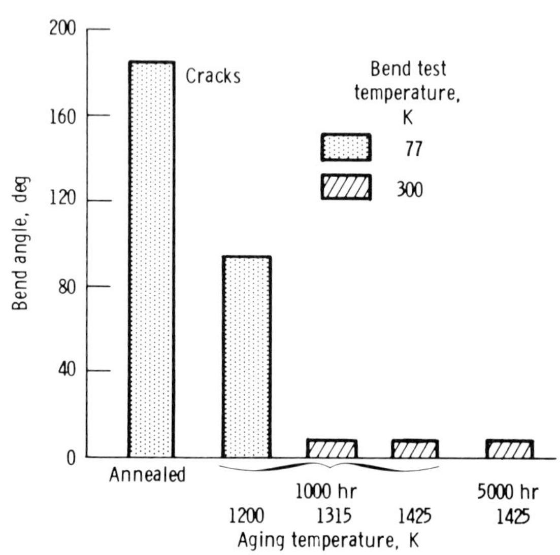

Figure 3 *Effect of Hydrogen Doping (10 ppm H) on Bend Ductility of T-111 Sheet.*

Tensile Properties

The tensile behavior of the Ta alloys was similar and was characterized by three distinct types of stress-strain curves depending on the test temperature, as shown in Figure 5. At lower test temperatures (from 300 K to 900 K) the alloys exhibited a yield point drop that is attributed to the locking of dislocations by interstitial impurities present in the alloys. At the highest test temperatures (1315 K to 1455 K) smooth stress-strain curves were observed, while at the intermediate test temperatures (primarily from 1025 K to 1300 K) serrated flow curves characterized the alloys. This behavior is attributed to dynamic strain aging, which is due to repeated breaking away of dislocations from impurity atmospheres and to subsequent locking of dislocations by impurity atmospheres. Oxygen has previously been identified by Sheffler et al. (1970) to be the responsible species.

Based on a comparison of the data for sheet and tubing, it is concluded that processing history does not have a significant effect on the tensile properties of T-111. The effects of aging time and environment on the tensile properties of T-111 sheet are shown in Figure 6. It may be seen that the various conditions had little effect on the yield strength and only a minor effect on ultimate tensile strength. Over the temperature range of about 875 K to 1425 K the ultimate tensile

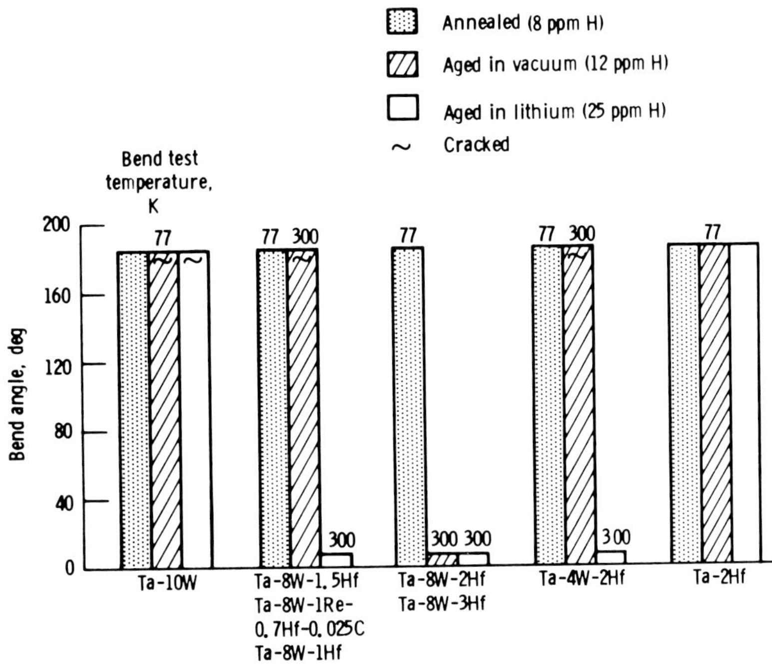

Figure 4 *Effect of Hydrogen Doping on Bend Ductility of Annealed 1000 hour at 1315 K and Aged Tantalum Base Alloys.*

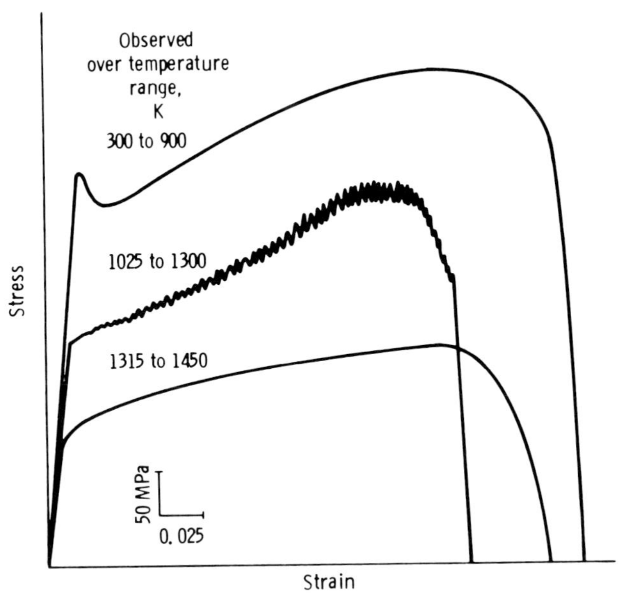

Figure 5 *Generalized Engineering Stress-Strain Curves for Tantalum Base Alloys illustrating Three Types of Flow Behavior Normally Observed.*

strength was lowered by minimizing dynamic strain aging.

The effect of aging on the tensile strength of each of the other alloys is shown in Figure 7. Note that both the ultimate and yield strength of Ta-10%W decreased substantially, with lithium having a greater effect than vacuum. Aging of Ta-2%Hf minimized the effects of dynamic strain age strengthening over the 875 K to 1315 K temperature range for the ultimate tensile strength and decreased the yield strength primarily over the 300 K to 775 K temperature range. In contrast, the remaining ternary alloys exhibited only minor variations in strength with aging conditions.

Scanning Electron Microscopy

Figure 8 shows scanning electron micrographs of fractured surfaces of the eight Ta alloys that had been aged, hydrogen-doped, and then bend tested at 300 K. It should be noted that Ta-10%W (Figure 8a) failed in a ductile manner (fracture achieved by repeated bending in opposite directions) and was free of precipitate particles at grain boundaries. With increasing Hf content in the Ta-8%W base alloys, brittle intergranular fracture dominates the microstructure with some evidence of ductile tearing, as shown in Figures 8b to 8d for Hf levels of 0.5, 0.7, and 1.0 percent, respectively. Alloys with larger Hf additions (2.0 and 3.0 percent) fractured in a completely brittle manner as shown in Figures 8e and 8f. Brittle intergranular failure also characterized the Ta-4%W-2%Hf alloy (Figure 8g). However, on complete removal of W from the alloy system, ductile failure characterized the Ta-2%Hf alloy (Figure 8h), and the grain boundaries were free of precipitate particles, as were those of Ta-10%W.

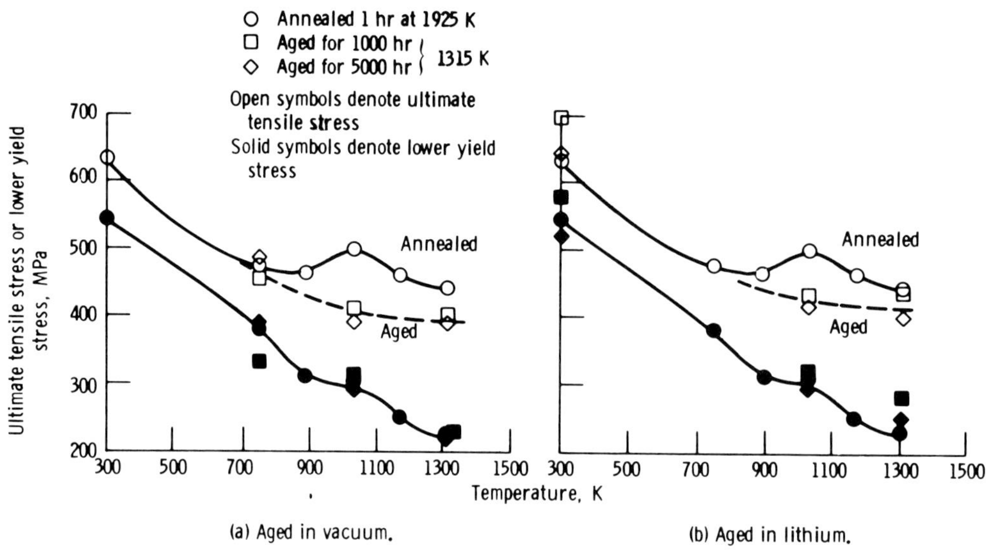

Figure 6 *Effect of Aging Time at 1315 K on Ultimate Tensile Stress and Lower Yield Stress of T-111 Sheet.*

Precipitate particles were observed in all the ternary alloys and in ASTAR-811C, which exhibited either predominantly or totally intergranular failure. The Ta-8%W-3%Hf alloy contained the greatest concentration of precipitate particles at the grain boundaries. As either the Hf content is reduced to 0.5 to 1.0 percent or the W is reduced to 4 percent, fewer particles were observed at the grain boundaries. It is this variation in amount of precipitates at the grain boundaries that is believed to account for the aging embrittlement exhibited only by the Ta-8%W-2%Hf and Ta-8%W-3%Hf alloys. The remaining ternary alloys and ASTAR-811C with fewer precipitates at the boundaries and the two binary alloys with an absence of precipitate particles did not exhibit aging embrittlement.

Upon doping with hydrogen, all the alloys with particles at the grain boundaries exhibited hydrogen embrittlement. The identity of the particles was achieved by performing characteristic x-ray analysis in the scanning electron microscope using an energy dispersive spectrometer. Figure 9a shows the fracture surface of a sheet T-111 specimen aged at 1315 K. Analysis of the grain boundary surface away from the particles is shown in Figure 9b while Figures 9c and 9d show analyses of particles A and B in Figure 9a located at a grain boundary intersection and on the grain boundary surface, respectively. The analysis demonstrates that the particles are Hf rich. In addition, by using a step scan technique, oxygen peaks were observed to be associated with Hf peaks at grain boundaries. Based on these results it is concluded that the particles are HfO_2. The results suggest that not only is composition of the Ta-base alloys important to the aging embrittlement problem, but aging temperature is of extreme importance as well. The results suggest that a critical temperature for Ta-base alloys is over the range 0.40 to 0.44 T_m on a homologous temperature basis.

Niobium-Base Alloys

As mentioned previously, primary emphasis was placed on Ta-base alloys because of their high priority at the time of these investigations. After analyzing the aging results obtained on the Ta alloys, a similar study was conducted on three Nb-base alloys. Alloys included were C-103 and Nb-1%Zr, both of interest for space power systems. In addition, Cb-752 was selected because of its composition. The combination of W and Zr in this alloy suggested that aging embrittlement would also occur in this alloy and thus would provide an opportunity to determine how widespread the aging embrittlement phenomenon is in refractory metal alloys. Nb-base alloys were evaluated by chemical analysis, bend tests, and metallography.

Chemical Analysis

Aging of the Nb-base alloys was conducted in vacuum only and chemical analyses before and after aging indicated no significant changes in interstitial content of the alloys. Hydrogen doping was achieved in a manner similar to that used for the Ta alloys except that a temperature of 1100 K was used for the Nb alloys which on a homologous temperature basis is

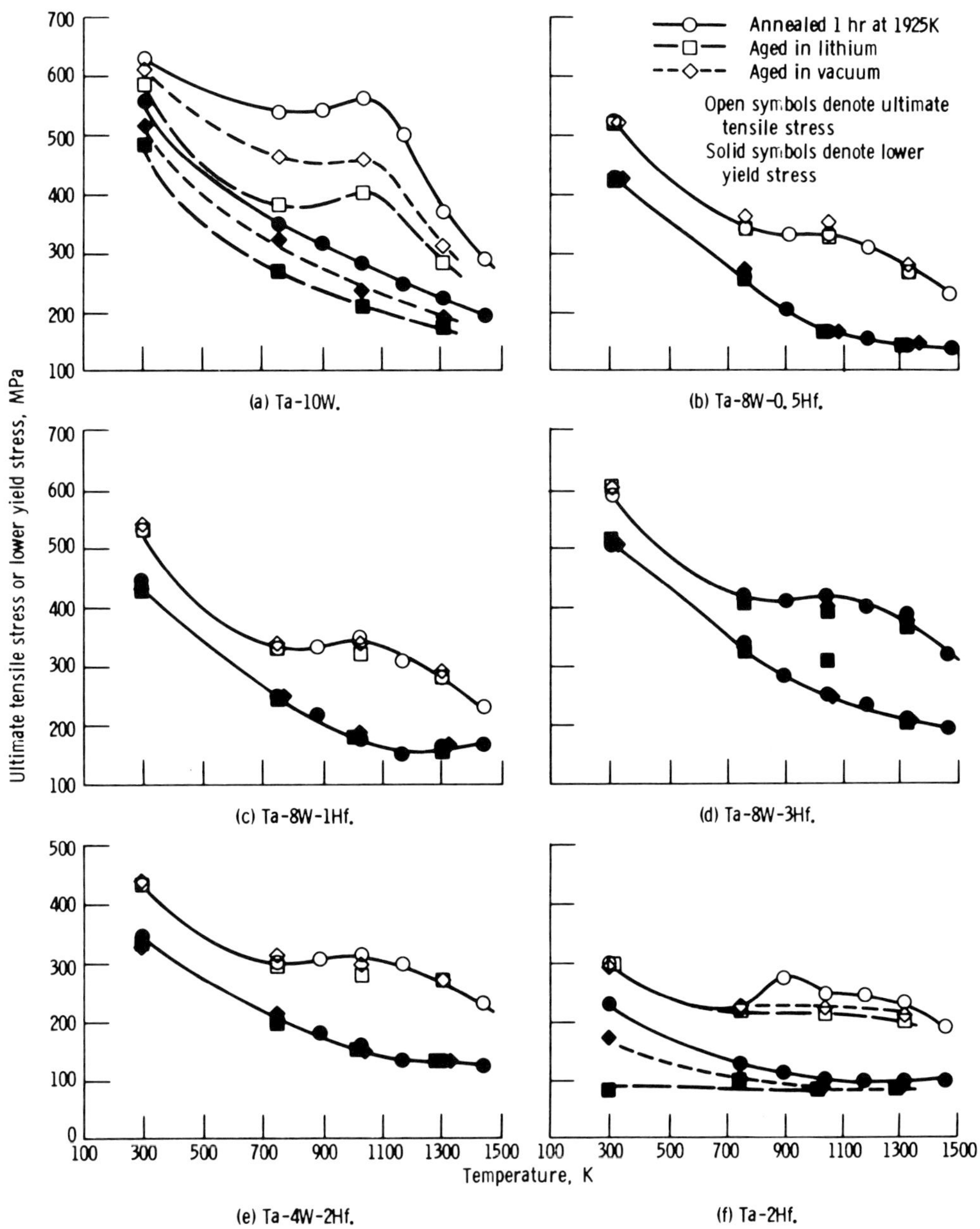

Figure 7 *Effect of Aging 1000 hours at 1315 K on Ultimate Tensile Stress and Lower Yield Stress of Tantalum Alloys.*

equivalent to that used for the Ta alloys. After doping, hydrogen contents of the alloys were about 60, 50, and 40 ppm for C-103, Nb-1%Zr, and Cb-752, respectively.

Bend Ductility

Bend tests were conducted on specimens in the annealed condition and after aging for 1000 h at temperatures ranging from 975 K to 1300 K. A comparison of the bend test results is shown in Figure 10. Note in Figures 10a and 10b that the bend ductilities for C-103 and Nb-1%Zr are not impaired as a result of aging over the critical temperature range 0.35 to 0.47 T_m. In contrast, for Cb-752, Figure 10c, aging at 1175 K (0.43 T_m) resulted in a ductile-brittle transition temperature of 125 K, an increase of at least 50 K. Thus aging embrittlement occurs in Cb-752 in a manner similar to the behavior observed in Ta-base alloys such as T-111.

A comparison of the effects of hydrogen doping on the ductility of the Nb alloys is shown in Figure 11. Hydrogen doping to an average content of 60 ppm in C-103 did not affect the ductility at 77 K (Figure 11a) for the annealed or aged conditions. The DBTT of the

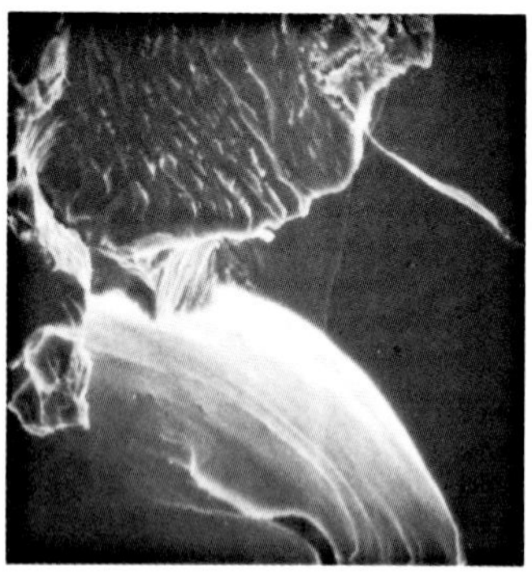

(a) Ta- 10W.

(b) Ta-8W-0.5 Hf.

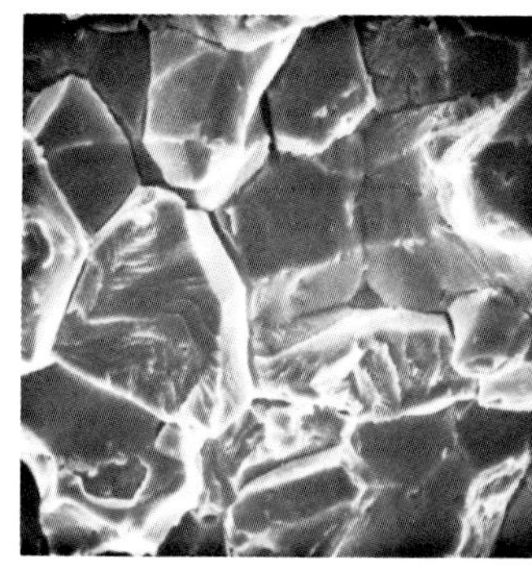

(c) Ta-8W - 1Re-0.7 Hf-0.25C.

(d) Ta - 8W - 1 Hf.

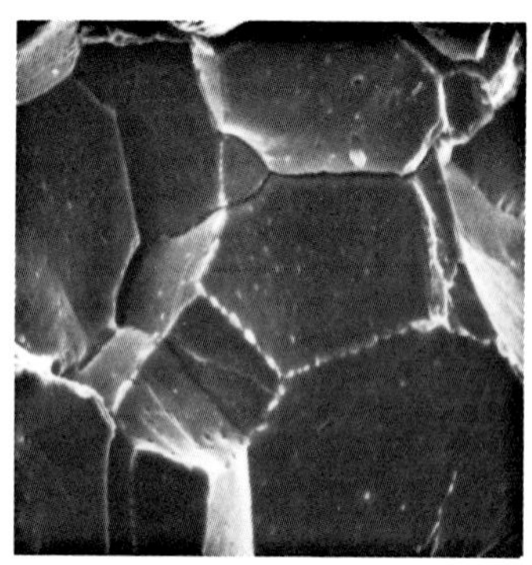

(e) Ta - 8W - 2 Hf.

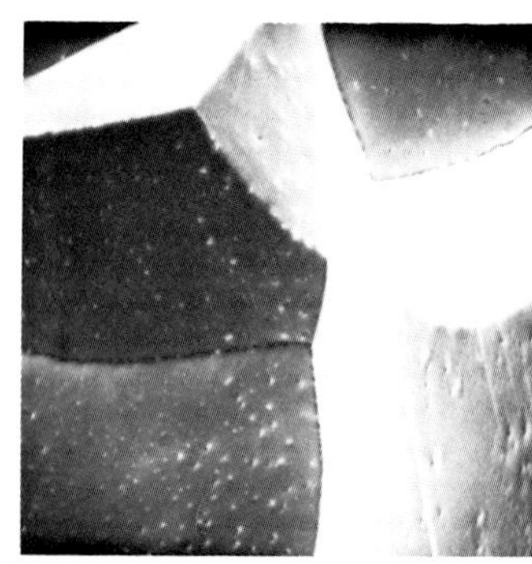

(f) Ta - 8W - 3 Hf.

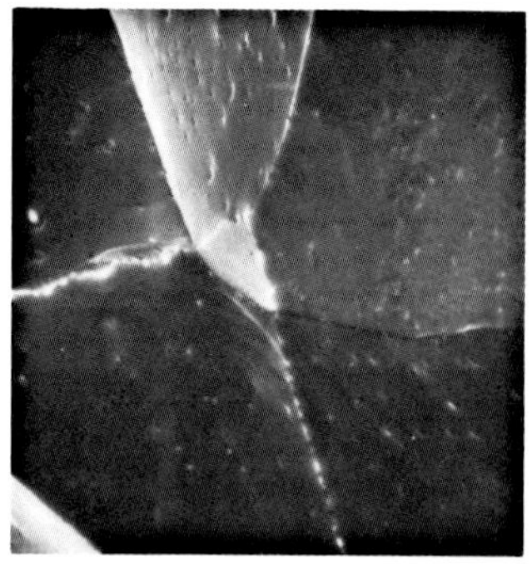

(g) Ta - 4W - 2 Hf.

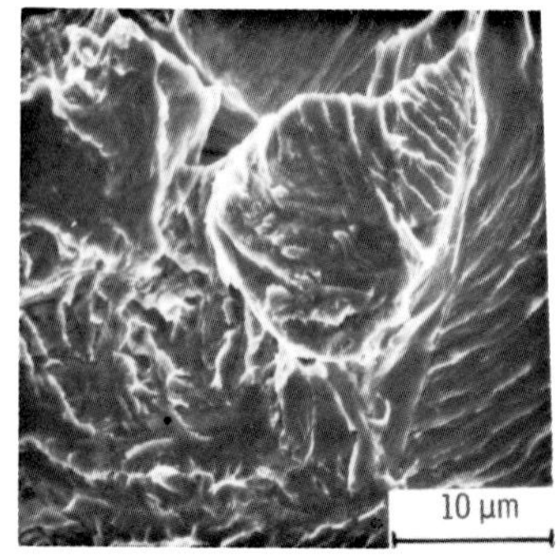

(h) Ta - 2 Hf.

Figure 8 *Scanning Electron Micrograph of Fractured Surfaces of Tantalum-Base Alloys Aged 1000 hours at 1315 K. Hydrogen Doped: Bend Test Temperature 300 K.*

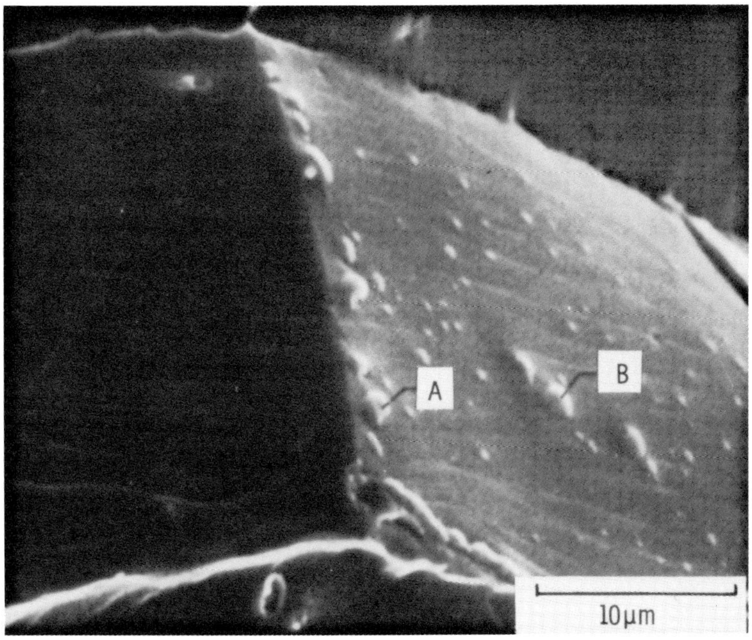

(a) Scanning electron micrograph.

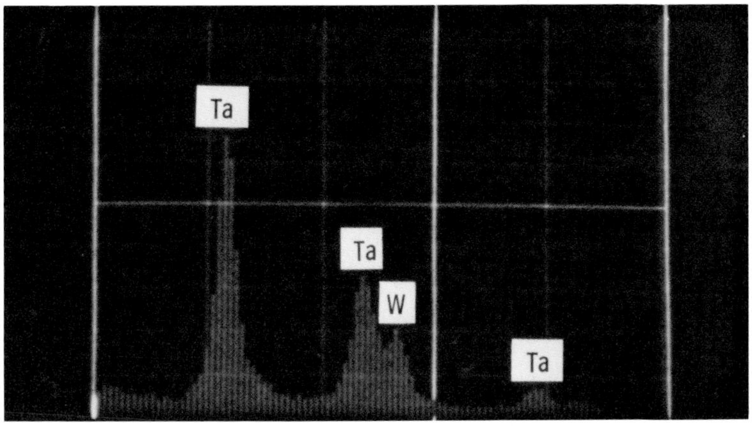

(b) Grain boundary surface.

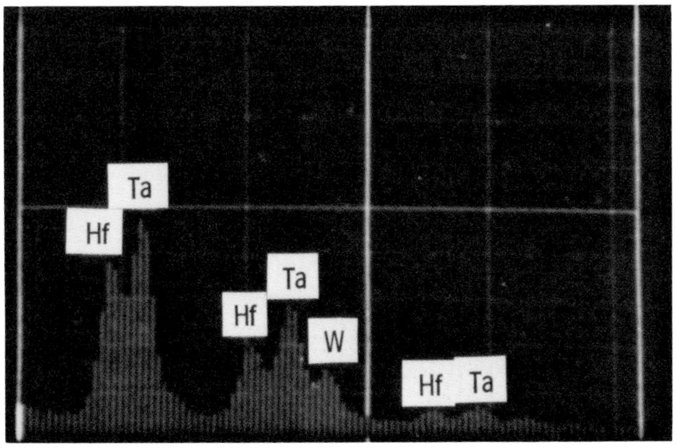

(c) Particle A.

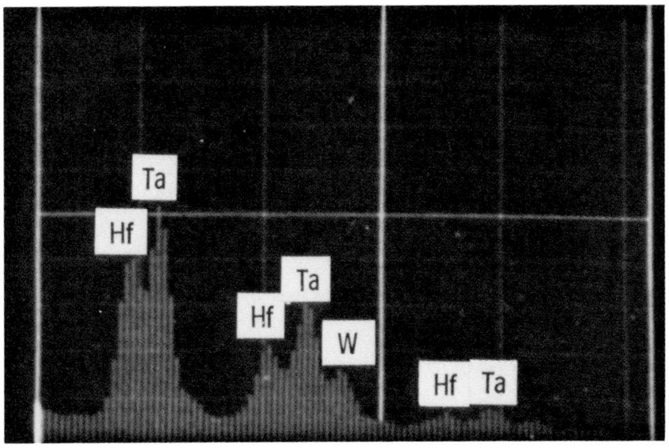

(d) Particle B.

Figure 9 *Scanning Electron Microscope Analysis of Precipitate Particles Observed in 1315 K Aged T-111 Sheet*

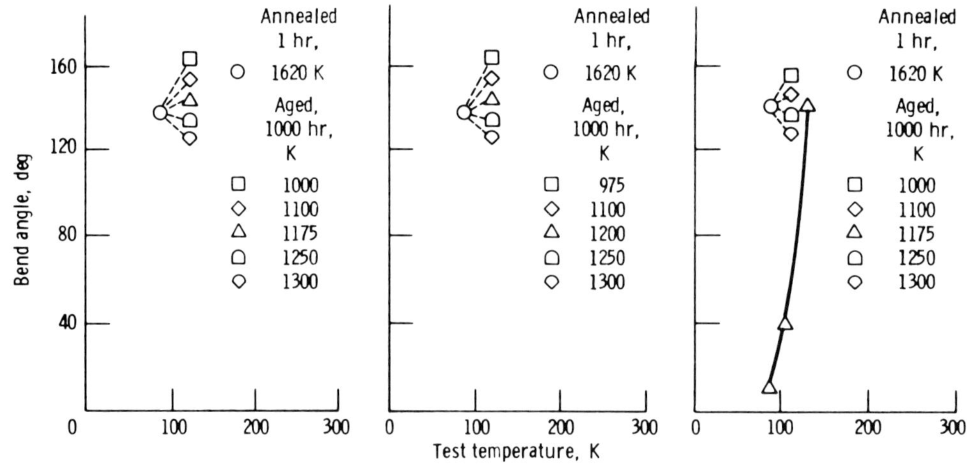

Figure 10 *Effect of Aging on the Ductile-Brittle Transition Temperature of Niobium Alloys.*

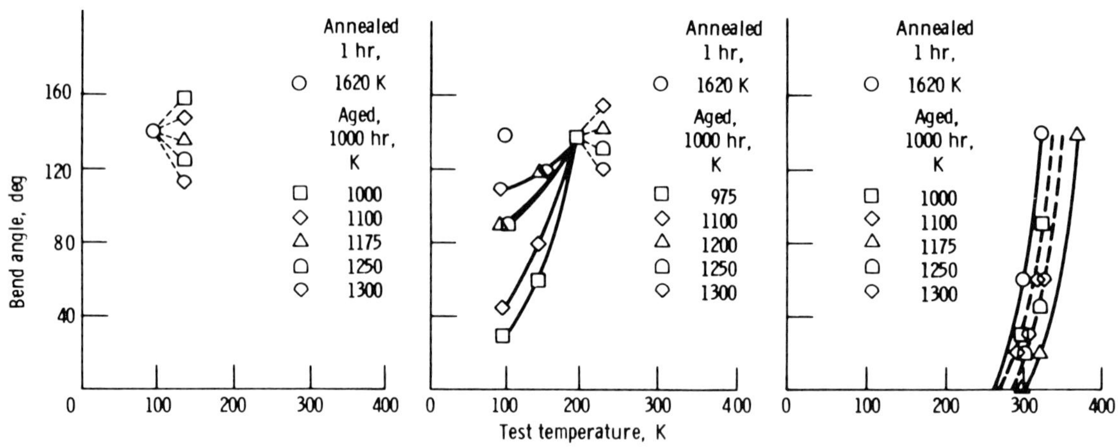

Figure 11 *Effect of Hydrogen Doping on the Ductile-Brittle Transition Temperature of Annealed and Aged Niobium Alloys.*

annealed Nb-1%Zr alloy was below 77 K after doping to about 50 ppm hydrogen. In contrast, the DBTT (that temperature where the limit of the bend test fixture of 140° was achieved) of all aged and hydrogen-doped Nb-1%Zr specimens was 175 K, as illustrated in Figure 11b. A general trend noted is that ductility at 77 K and 125 K increases with increasing aging temperature which may be attributed to a change in microstructure observed in these alloys. Aging at 975 K and 1100 K produced platelet-type particles primarily at grain boundaries and after aging at 1200 K these particles were primarily intergranular. In contrast, after aging at the higher temperatures of 1250 and 1300 K, the microstructures of Nb-1%Zr were free of precipitate particles. The presence of the precipitate particles combined with hydrogen may account for the low ductility for specimens aged at lower temperatures. Ductility increased for higher aging temperatures, which also resulted in the disappearance of the platelet particles. The effect of hydrogen doping on the DBTT of Cb-752 is illustrated in Figure 11c. Hydrogen doping to an average level of 40 ppm increased the DBTT of annealed Cb-752 to above room temperature, 325 K. The highest DBTT was for specimens aged at the intermediate temperature of 1175 K, the temperature where aging embrittlement occurred in Cb-752.

Scanning Electron Microscopy

Since aging embrittlement occurred in the Cb-752 alloy, results for this alloy only will be presented here. The fracture surface of a Cb-752 specimen aged at 1175 K and subsequently tested at 77 K is shown in Figure 12. Aging embrittlement that occurred at this temperature produced a brittle, intergranular fracture. Particles are observed on the grain boundary surfaces and at grain boundary intersections. Analysis of a grain-boundary surface area free of particles is shown in Figure 12b. Only Nb and W were detected in this area. In contrast, analysis of a grain-boundary region containing a precipitate particle exhibited peaks for Zr in addition to the Nb and W peaks shown in Figure 12c. As postulated, Zr-rich particles formed at grain boundaries in Cb-752 in a manner similar to the Hf-rich particles formed in the Ta alloys.

Molybdenum-Base Alloy, Mo-TZM

Since there is some interest in Mo-base alloys for space power systems, the commercial alloy Mo-TZM was chosen to round out the study of hydrogen embrittlement of refractory metal alloys. It is well known that hydrogen has a very low solubility in Group VI alloys such as Mo. In addition, the reactive metal (Zr and Ti) content in Mo-TZM is very low so that aging embrittlement and increased susceptibility to hydrogen was not predicted for this Mo-base alloy.

Chemical Analysis

As with the Nb alloys, Mo-TZM was aged only in vacuum. No change in interstitial content resulted from the 1000 h aging and no increase in hydrogen content was detected upon subsequent doping attempts. Hydrogen content after all the aging treatments was near 1 ppm.

Bend Ductility

Bend DBTT results are shown in Figure 13 for Mo-TZM in the aged condition. In the annealed condition the DBTT was 260 K. Aging at 975 K, 1100 K, and 1200 K resulted in a DBTT of 260 K while aging at 1250 K and 1300 K produced a slight increase in the

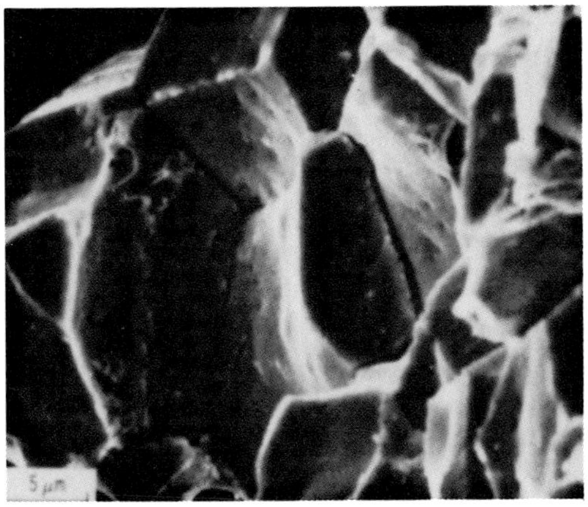

(a) Scanning electron micrograph.

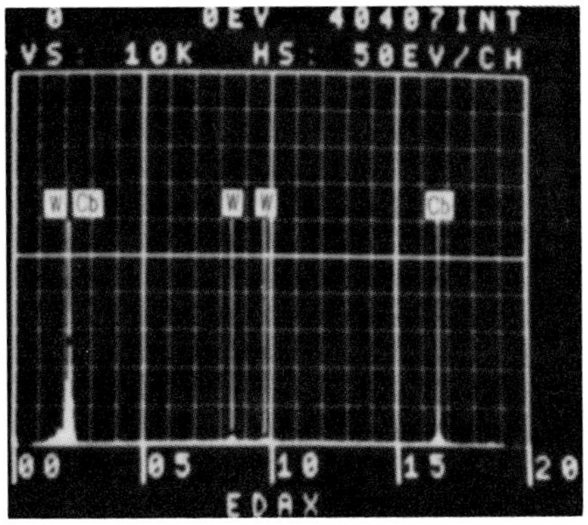

(b) Grain-boundary fracture surface analysis.

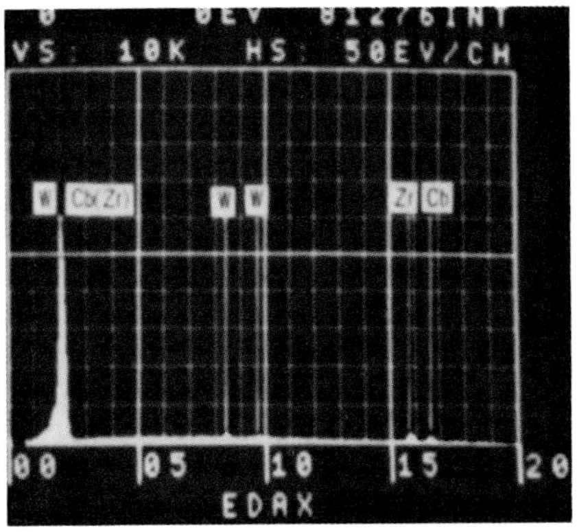

(c) Particle analysis.

Figure 12 *Scanning Electron Microscope Analysis of Fracture Surface of 1175 K Aged Cb-752. Test Temperature 77 K.*

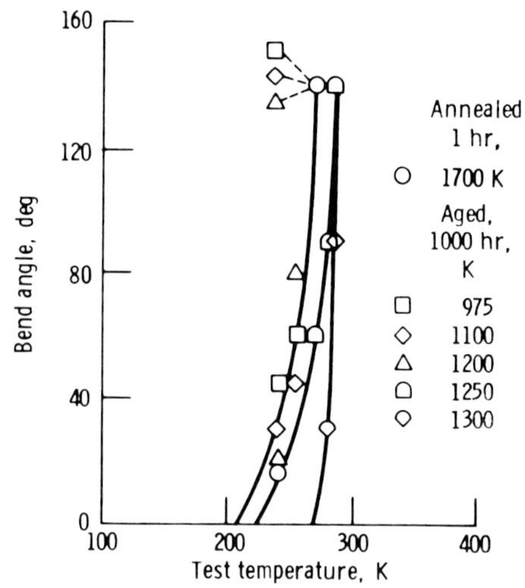

Figure 13 *Effect of Aging on the Ductile-Brittle Transition Temperature of Mo-TZM.*

DBTT to 275 K. This increase in DBTT is believed to be due to a change in carbide morphology rather than aging embrittlement as was observed in the Ta and Nb alloys. The DBTTs after hydrogen doping were similar to those in the aged condition which was expected since hydrogen content did not increase during doping.

Scanning Electron Microscopy

Scanning electron microscopy results for a Mo-TZM specimen tested at 260 K after aging at 1250 K are shown in Figure 14. The fracture appearance for this specimen (60° bend angle) reveals a mixed mode of failure involving areas of ductile tearing, transgranular cleavage, and grain-boundary fracture. Only Mo was detected in this specimen at the grain-boundary fracture area.

Discussion

The aging and hydrogen embrittlement of T-111 and other similar Ta base alloys is believed to be due to Hf segregation at grain boundaries. The absence of particle formation in Ta-2%Hf suggests that the presence of W in the ternary alloys affects the rate and degree of Hf segregation that is observed in these alloys. This may occur due to the lattice contraction that occurs upon adding W to Ta (Pearson 1974) causing the larger Hf atom to segregate to misfit or grain boundary areas. Competing with this equilibrium segregation process is diffusion which will tend to evenly disperse the solute at higher temperatures (McLean 1957). Hence, T-111 aged at 1590 K did not exhibit precipitate particles. Also, this material did not exhibit aging embrittlement and was not susceptible to hydrogen embrittlement.

The results for the Nb alloys indicate a similar type behavior as was observed for the Ta alloys. Segregation did not occur in the binary Nb-1%Zr alloy, which would be expected to behave similarly to the Ta-2%Hf alloy. In addition, Hf segregation did not occur in the C-103 alloy with 10 percent Hf. However, in Cb-752, which contains both W and Zr, grain-boundary Zr segregation occurred and produced about a 50 K increase in the DBTT of this alloy. Aging at a critical temperature of 0.41 to 0.43 T_m in these alloys produces the aging embrittlement. Aging at higher temperatures resulted in ductile behavior as did aging at lower temperatures. Homogenization by diffusion apparently occurs at the higher temperatures while longer times at the lower temperature will result in the aging embrittlement phenomenon occurring due to Hf or Zr segregation, since longer times are required for lower diffusion rates.

The results for Cb-752 are best compared with the Ta-4%W-2%Hf alloy. On an atom percent basis, Cb-752 contains 5.2%W and 2.9%Zr. Therefore, with aging temperatures based on a homologous temperature scale and compositions expressed on an atom basis, results indicate that the Ta-base and Nb-base alloy systems can be described by the same mechanisms. Although the alloying elements Ti and Zr in Mo have larger atomic radii than does Mo and would be expected to segregate at grain boundaries during long term aging, this phenomenon was not observed in Mo-TZM. Consequently, aging embrittlement did not occur in this alloy. The low concentrations of the reactive elements and the carbon content of Mo-TZM may account for this behavior. The reactive elements were present as carbides or in solution at these low levels.

Application of Results

The studies described herein have established that aging embrittlement is most pronounced as a result of long term aging near 0.41 to 0.43 T_m (but can occur over the temperature range 0.40 to 0.44 T_m) in Ta and Nb alloys with 4 to 8 atomic percent W and 2 to 3 atomic percent Hf or Zr, respectively. If the use of alloys with combinations of solutes in this concentration range is proposed for the critical temperature range, then extreme care should be taken during subsequent handling or maintenance after cool down to

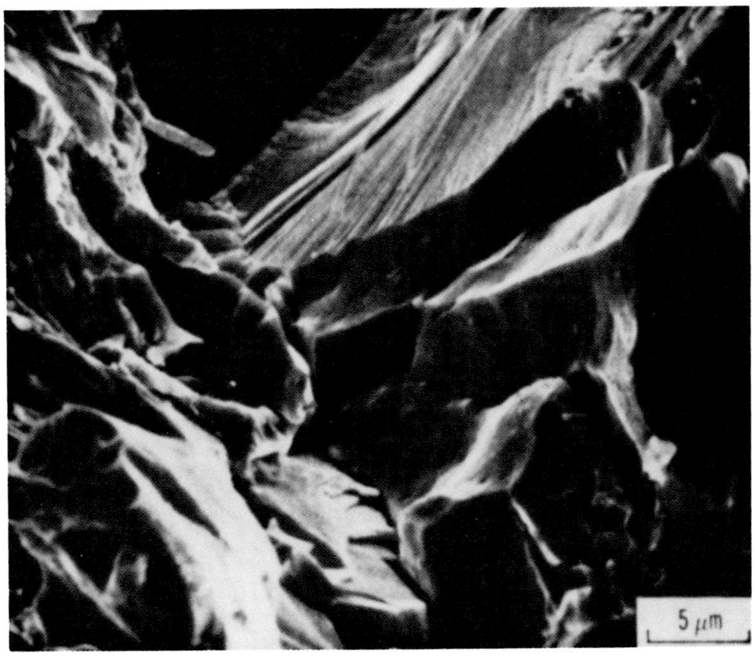

(a) Scanning electron micrograph.

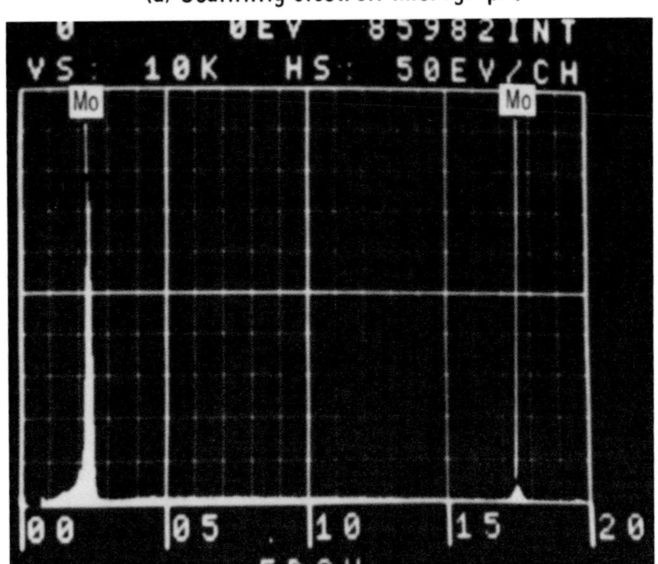

(b) Fracture surface analysis.

Figure 14 *Scanning Electron Microscope Analysis of Fracture Surface of 1250 K Aged Mo-TZM. Test Temperature 260 K.*

room temperature. Great care should also be exercised to prevent exposure to hydrogen.

Conclusions

Based on a study of the long term aging effects and susceptibility to hydrogen embrittlement of Ta, Nb, and Mo alloys the following conclusions are drawn

1. Aging embrittlement does not occur in the alloy Nb-1%Zr.
2. Aging embrittlement occurs in Ta and Nb alloys that contain a critical combination of W (4 to 8 atomic percent) and Hf or Zr (2 to 3 atomic percent) that have been aged for 1000 h to 5000 h in the homologous temperature range of 0.35 to 0.47 T_m.
3. Alloys exhibiting aging embrittlement are much more susceptible to hydrogen embrittlement as evidenced by a further increase in the ductile- brittle transition temperature of these alloys.
4. Scanning electron microscopy demonstrated that segregation of Hf and Zr to grain boundaries led to aging embrittlement and increased susceptibility to hydrogen embrittlement in these alloys.
5. The Mo-base alloy Mo-TZM did not experience aging or hydrogen embrittlement as a result of long term aging over a similar homologous temperature range.

Acknowledgments

This work was conducted at the NASA Lewis Research Center. The helpful discussions with Gordon K. Watson and Robert H. Titran are greatly appreciated.

References

Lesmann, G.G. and R.E. Gold (1970) *Determination of the Weldability and Elevated Temperature Stability of Refractory Metal Alloys. II - Long-Time Temperature Stability of Refractory Metal Alloys.* NASA CR-1608, Westinghouse Electric Co., Pittsburgh, PA.

McLean, D. (1957) *Grain Boundaries in Metals*, Clarendon Press, Oxford.

Moss, T.A., R.L. Davies, and G.J. Barna (1970) in *Recent Advances in Refractory Alloys for Space Power Systems*, NASA SP-245, NASA Lewis Research Center, Cleveland, OH, pp. 1–18.

Pearson, W.B. (1958) *A Handbook of Lattice Spacings and Structures of Metals and Alloys*, Pergamon Press, New York.

Sheffler, K.D., J.C. Sawyer, and E.A. Steigerwald (1970) in *Recent Advances in Refractory Alloys for Space Power Systems*, NASA SP-245, TRW, Inc., Cleveland, OH, pp. 75–125.

Stephens, J.R. in *Hydrogen in Metals*, ASM, Metals Park, OH, pp. 383–392.

Stephens, J.R. (1975a) *Effects of Alloy Composition in Alleviating Embrittlement Problems Associated with the Tantalum Alloy T-111*, NASA TN D-7838, NASA Lewis Research Center, Cleveland, OH.

Stephens, J.R. (1975b) *Effects of Long-Term Aging on Ductility of the Columbium Alloys C-103, Cb-1Zr, and Cb-752 and the Molybdenum Alloy Mo-TZM*, NASA TN D-8095, NASA Lewis Research Center, Cleveland, OH.

Stephens, J.R. (1977a) *"Effects of Long Term Aging on Ductility and Microstructure of Columbium and Molybdenum Alloys," Metallography*, 10: 1–25

Stephens, J.R. (1977b) *"Effects of Alloy Composition in Alleviating Embrittlement Problems Associated with the Tantalum Alloy T-111," J. Less-Common Metals*, 51: 93–111

Watson, G.K. and J.R. Stephens (1972) *Effects of Aging at 1040 °C (1900 °F) on the Ductility and Structure of a Tantalum Alloy, T-111*, NASA TN D-6988, NASA Lewis Research Center, Cleveland, OH.

Space Nuclear Power Systems 1986
Edited by M. S. El-Genk and M. D. Hoover

Chapter 29

Irradiation Swelling in Self-Ion Irradiated Niobium

Ram Bajaj and Stuart A. Shiels

Westinghouse Electric Corporation
Advanced Energy Systems Division
P. O. Box 10864
Pittsburgh, PA 15236

Barbara O. Hall

Westinghouse Electric Corporation
Research and Development Center
1310 Beulah Road
Pittsburgh, PA 15235

George R. Fenske

Argonne National Laboratory
9700 South Cass Avenue
Argonne, IL 60439

This paper presents initial results of an investigation of swelling mechanisms in a model body centered cubic (bcc) metal, niobium, irradiated at elevated temperatures (0.3 T_m to 0.6 T_m) where T_m = melting point in K. The objective of this work is to achieve an understanding of the elevated temperature swelling in bcc metals, which are the prime candidate alloys and composite matrix materials for space reactor applications. Niobium was irradiated with 5.3 MeV Nb^{++} ions, at temperatures ranging from 700°C to 1300°C, to a nominal dose of 50 dpa at a dose rate of 6×10^{-3} dpa/s. Swelling was observed over a temperature range of 700°C to 1200°C, with a peak swelling of 7 percent at 900°C. The microstructural data, obtained from transmission electron microscopy, were compared to the predictions of the theoretical model developed during this program. A reasonable agreement was obtained between the experimental measurements of swelling and theoretical predictions by adjusting both the niobium-oxygen binding energy and the incubation dose for swelling to realistic values.

Introduction

Advanced space power systems impose stringent service requirements on structural materials, including elevated temperature performance, high strength to density ratio, rigidity and low vapor pressure. Additional requirements of liquid metal coolant compatibility and neutron irradiation resistance are imposed for nuclear systems. Among the metallic materials, only refractory metal alloys and refractory metal-matrix composites based on niobium, tantalum, molybdenum, and tungsten are considered to approach meeting these requirements.

High accumulated neutron fluences, of the order of 10^{23} n/cm^2 ($E > 0.1$ MeV) at elevated temperatures to 1100°C to 1400°C, are expected in advanced multi-megawatt plants. Thus, the knowledge of neutron induced swelling behavior of these materials is vital to the design and operation of nuclear systems.

Data on the high fluence swelling behavior of refractory metals irradiated to elevated temperatures are not available because these materials were investigated primarily for their applications in fusion reactor first walls which are planned to operate at temperatures below 1000°C. Wiffen (1984) has provided an excellent state of the art review of the data on irradiation effects in refractory metals. The mechanisms of swelling in these materials, however, have not been fully understood. This paper presents the initial results of an investigation of elevated temperature swelling in a refractory metal with the purpose of achieving a better understanding of swelling mechanisms through theory-experiment coupling.

Objective and Approach

The primary purpose of this investigation is to further the understanding of high temperature cavity swelling microstructural response of refractory metals to neutron irradiation. High temperature is defined as being between 0.3 T_m and 0.6 T_m, where T_m = melting temperature in Kelvin. This goal is to be achieved by a close interaction of theory development and controlled experimental studies. The neutron induced swelling is simulated by single and dual ion irradiation with self-ions, and self-ions plus helium, respectively. Simulation techniques were chosen for this work because of their versatility, cost and time effectiveness and ease of control of irradiation variables and because they have been shown to be effective in simulating neutron induced swelling (Bleiberg and Bennett 1977, Bajaj et al. 1982, Wood et al. 1981, Johnston et al. 1976, and Bleiberg et al. 1976). Self-ions were chosen for irradiation to avoid chemical effects introduced by the injected interstitials. At the outset of the investigation, it was necessary to select a model material to fulfill the objectives effectively. The selection was based on three major considerations: (1) theoretical, (2) experimental, and (3) practical (those related to the engineering applications of these materials in space systems). Foremost among the theoretical considerations were the availability of parameters needed for theoretical modeling and the probability of formation of <100> type loops. The experimental factors considered were the availability of a self-ion beam with sufficient ion current in an energy range of 4 MeV to 8 MeV, the depth of penetration of the incoming self-ion, and susceptibility to contamination during irradiation. Application oriented considerations included the service temperature limits, fabrication and the effect of irradiation on ductile-to-brittle transition temperature (DBTT). Based on these factors, niobium was chosen as the material for study.

Theory

The aim of the theoretical portion of this investigation is to develop a model for cavity growth in body-centered cubic (bcc) metals and to calculate growth rates and swelling behavior for refractory metals. The basis for model development is provided by the chemical reaction rate formalism, which has been used extensively for treating nucleation and growth of cavities and dislocation loops in face-centered-cubic metals (Katz and Wiedersich 1971, Russell 1971, Burton 1971, Hayns 1975, Hall 1980, Wiedersich 1972, Brailsford and Bullough 1972, and Hall 1983). Since the model development has been discussed in detail elsewhere (Bajaj et al. 1985), only a brief description is presented here.

In the rate theory approach, fluctuations in the vacancy and interstitial (point defect) concentrations arising from random and inhomogeneous production processes and defect sinks are ignored, and spatially and temporally averaged vacancy and interstitial production and loss rates are assumed. The metal is, in effect, replaced by a continuum in which defects are continuously produced and annihilated at all points. Each point is characterized by averaged properties of the real material (dislocation density, grain boundary area, and solute concentration). Growth of cavities and dislocation loops are described by a set of coupled nonlinear differential equations from which the mean cavity and loop radii can be calculated as a function of time:

$$dr_c/dt = \Omega\ (D_v(C_v - C^{th}_{vc}) - D_iC_i)/r_c \text{ and} \qquad (1)$$

$$dr_1/dt = a^2\ (ZD_iC_i - D_v(C_v - C^{th}_{cl}))\ , \qquad (2)$$

where r_c is the mean cavity radius, r_1 is the mean loop radius, Ω is the atomic volume, D_v and D_i are the point defect diffusion coefficients, C_v and C_i are the point defect concentrations, C^{th}_{vc} is the thermal vacancy concentration at the surface of the cavity, C^{th}_{vl} is the thermal vacancy concentration at the surface of the loop, a is the lattice constant, and Z is a parameter referred to as the capture efficiency. The capture efficiency is given by the ratio of the actual point defect current to an ideal current, which would result if the sink were a perfect absorber and had no stress field to interact with that of the point defect. Theoretical capture efficiencies for both vacancies and interstitials have been evaluated for a variety of sink types (Bajaj et al. 1985), including loops and cavities. In numerical calculations, however, they are generally treated as parameters. Only the capture efficiency of interstitials at loops is retained (as above in Equation (2)) since the others are close to unity.

The swelling rate S is directly related to the time rate of change dr_c/dt of the mean cavity radius r_c by the equation:

$$S = 4\pi\ N_c r_c^2 dr_c/dt\ , \qquad (3)$$

where N_c is the cavity density.

As can been seen from the equations, the rates of change of the radii are proportional to the net flux of point defects to the cavity or loop and depend on the point defect concentrations, their diffusion coefficients, and the capture efficiencies. In the treatment of swelling caused by cavity growth, the vacancy and interstitial concentrations are generally assumed to be in a quasi-steady-state, because the relaxation times for changes in their concentrations are short compared with those required to change the microstructure.

The primary difference between the rate theory model for bcc and that for face-centered-cubic metals is that in the former case equations for both <100> and <111>–type loops is needed.

Since Little et al. (1980) have proposed a mechanism for the swelling resistance of ferritic steels that is based on the differing capture efficiencies of the two types of radiation-induced loops in body-centered-cubic metals, this aspect of microstructural development is being closely examined in the current model development (Hall unpublished results). In this paper, however, the emphasis is on theoretical and experimental swelling results.

Experimental

Experimental Facilities and Procedures

The investigation of high temperature swelling in refractory metals requires a facility capable of providing a high current beam of heavy ions and helium, which can simultaneously bombard a specimen heated to high temperature (up to 0.6 T_m, ~1370°C for niobium). Experimental facilities capable of meeting these stringent requirements, as well as with high vacuum capability, high isotopic purity of incoming ions, and diagnostics ability, are limited in number. A survey of existing accelerator facilities was conducted to evaluate the best facility for this work, and we concluded that the HVEM-Tandem facility at Argonne National Laboratory (ANL) satisfied the program requirements. ANL was therefore selected for the performance of irradiations. Briefly, the facility, located in the Materials Science and Technology Division at ANL, consists of a National Electrostatics 2-MV Tandem Type Universal Ion Accelerator, and a Texas Nuclear 300-kV Ion Accelerator. Together, the two accelerators provide ion beams of essentially all elements with energies from approximately 10 KeV to 8 MeV. The specimen heating is accomplished by an electron beam heater designed by ANL.

The heavy ions were provided by a ANL modified NEC/University of Wisconsin SNICS (Source of Negative Ions through Cesium Sputtering) source with a metallic niobium target. The source yielded an ion current of ~40 nA, which is too low to perform high fluence (~60 displacements per atom (dpa)) irradiations in a reasonable amount of time. Therefore, a novel approach was required to achieve higher ion output. After considerable effort, a procedure analogous to that used for production of titanium ions proved successful in achieving high ion currents of niobium. This procedure requires a hydride target material. A special hydriding method was developed at ANL to obtain NbH. The process involves heating the niobium sample to high temperature under vacuum, and backfilling the vacuum chamber with high purity hydrogen at room temperature to absorb the hydrogen and form a NbH compound. The output of the SNICS source increased from about 40 nA of Nb^- with a metallic source to about 5 to 10 μA of niobium ions Nb^{++} with a a hydrided target.

The energy deposition profile for niobium was calculated using Biersack's TRIM Monte Carlo program. The maximum energy deposition for 5.3 MeV Nb^{++} ions was computed to occur at a depth of ~900 nm. The calculated dpa values have an accuracy of ±10%. In calculations of dpa, a threshold displacement energy of 40 eV and an atomic density of 5.6×10^{22} atoms/cm^3 was assumed for niobium.

Niobium specimens used in this investigation were high-purity (Marz-grade) foil, 0.5 mm thick, obtained from the Materials Research Corporation. The foil was received in an annealed condition (1150°C to 200°C for 0.5 h). Chemical analysis conducted of interstitial elements showed C and O contents of 33 and 120 wppm, respectively.

The preparation of specimens for ion irradiation is a delicate and painstaking operation, requiring of a number of polishing steps and optical examination to achieve flat, scratch-free surfaces. This process is made more difficult because of the softness of niobium. Details of the mechanical polishing procedures are given by Bajaj et al. (1985). The polished specimens were individually annealed at 1200°C for 0.5 h in the vacuum chamber of the accelerator. This assured that the irradiated surface was in fully annealed condition, devoid of any deformation due to specimen preparation. The irradiations were performed with each specimen mounted in a tantalum holder in an all-metal chamber with a base pressure of 2.5×10^{-9} torr. The vacuum during the irradiation was 2.8×10^{-8} torr. The temperature was controlled and monitored by an infrared pyrometer (IRCON 300 HCS) coupled to the electron beam heater. A 5.3 MeV Nb^{++} ion beam was used to perform the irradiations, which resulted in the peak damage at ~0.9 μm below the specimen surface. All specimens except one (700°C) were irradiated to peak damage of ~60 dpa corresponding approximately to a fluence of $\sim 1.2 \times 10^{23}$ n/cm^2 ($E > 1.0$ MeV) based on stainless steel correlation. The temperature of irradiation varied from 700°C to 1300°C (0.35 T_m to 0.57 T_m).

The post-irradiation specimen preparation consisted of removing ~750 nm from the front surface by ion milling with 800 eV argon followed by back-thinning to performation. This allowed examination of material near the front surface by transmission electron microscopy. The back-thinning was accomplished by polishing in a solution containing NH_4F in methanol at $-30°C$.

Transmission electron microscopy (TEM) was used to analyze the microstructure of the irradiated specimens which consisted of voids and dislocation arrays. The voids were imaged in kinematic diffraction conditions in a slightly defocused condition to obtain the true size of the voids. The dislocations were imaged in two beam (dynamic diffraction) conditions. Stereopairs were used to determine the thickness of the foils. Using enlarged micrographs (500,000 X), the diameters of the voids, number densities, average swelling, and dislocation densities were obtained. The voids were assumed to be spherical in shape in calculating the swelling. Because of their relatively large sizes, especially at high irradiation temperatures, voids exposed at the foil surface during electropolishing quickly eroded to larger sizes and more irregular shapes, losing observable contrast. Therefore, a correction factor suggested by Spitznagel et al. (1980) which alleviates the shortcomings of ASTM-recommended correction factors, was used to account for these voids.

Experimental Results

The results from the TEM examination of niobium specimens irradiated from 700°C (0.35 T_m) to 1300°C (0.57 T_m) are presented in Table 1. This table lists the damage level, swelling, void number density, and void size for each of the specimens examined. Additionally, ranges in swelling, void number densities and void sizes are included. These ranges were obtained from the analysis of multiple areas of a specimen.

Figure 1 shows swelling as a function of irradiation temperature. Swelling was observed at temperatures from 700°C to 1200°C. No voids were observed in the specimen irradiated at 1300°C. As seen in Figure 1, the swelling increased from 700°C to 900°C, reached a peak of 7 percent at 900°C, and decreased slowly thereafter with increasing temperature to zero swelling at 1300°C.

The void number density, expressed as voids per cm^3, is plotted as a function of irradiation temperature in Figure 2. The maximum void density was observed at the lowest irradiation temperature of 700°C, where the density was 1×10^{16} cm^{-3}. The void density decreased gradually, to 1×10^{15} cm^{-3}, over a temperature range 700°C to 900°C. From 900°C to 1000°C, the void density decreased rapidly to 5×10^{13} cm^{-3} and then decreased gradually from 1000°C to 1200°C.

The decrease in void density was accompanied by an increase in the average void size as shown in Figure 3. At 700°C, the average diameter of voids was 10 nm. The void diameter increased to ~45 nm at 900°C and then increased to 110 nm at 1000°C. Thereafter, the diameter increased slowly with increasing temperature. At 1200°C, a large scatter in void diameter was obtained from area to area in the specimen. This scatter is partially attributed to the enlargement of surface voids due to the attack of the electropolishing agent on the large voids shown in Figure 3.

Figure 4 shows a typical photomicrograph of voids in niobium irradiated at 800°C. Dislocations are also shown in the inset in this figure. The dislocations were imaged in a two beam condition, mostly with a diffraction vector $\vec{g} = \langle 110 \rangle$.

Micrographs from specimens irradiated at 700°C, 800°C, and 900°C were carefully examined under a stereoscope for void ordering (void lattice formation). No ordering was observed in any of the specimens. Void ordering was also absent in specimens irradiated at higher temperatures.

Table 1. *Summary of Swelling Data for Self-Ion Irradiated Niobium.*

Irradiation Temperature (°C)	Dose (dpa)	Swelling Range (%)	Swelling Average (%)	Number Density Range (cm^{-3})	Number Density Average (cm^{-3})	Diameter Range (nm)	Diameter Average (nm)
700	32 ± 4	0.6–0.8	0.7	$1–1.2 \times 10^{16}$	1.1×10^{16}	10.5–10.8	10.7
800	54 ± 9	2.3–2.9	2.6	$4.1–6.9 \times 10^{15}$	5.9×10^{15}	19.3–22.5	20.7
900	50 ± 9	2.8–7.9	7.0	$1–1.8 \times 10^{15}$	1.4×10^{15}	44.1–49.9	47.0
1000	57 ± 10	3.6–4.9	4.3	$4.1–6.3 \times 10^{13}$	5.52×10^{13}	107.5–118.3	115.9
1100	53 ± 9	2.4–3.2	2.7	$2–3.4 \times 10^{13}$	2.8×10^{13}	118.2–135.4	125.5
1200	52 ± 9	1.24–2.74	2.15	$1.9–3.1 \times 10^{13}$	2.75×10^{13}	97.7–140.8	116.5
1300	54 ± 9	—	—	—	—	—	—

Specimen irradiated at incidence angles of 10° and 45° for equal times.

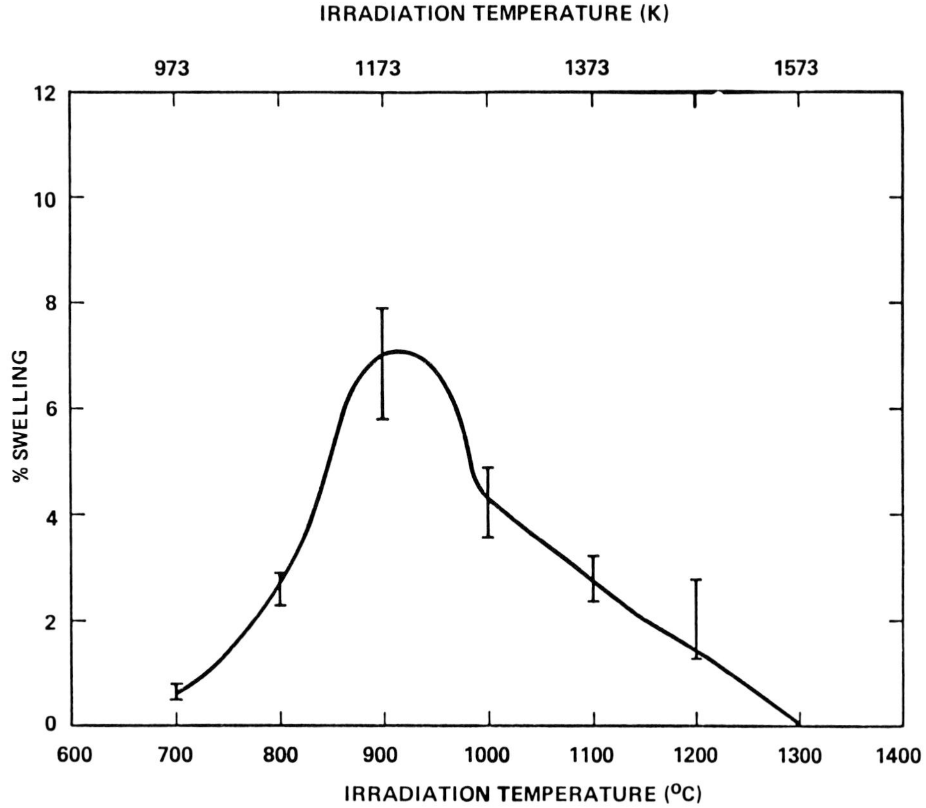

Figure 1 *Swelling as a function of Irradiation Temperature for Self-Ion Irradiated Niobium at Irradiation Exposures near 50 dpa.*

Bajaj et al. (1985) made a comparison of the present data with those reported in literature.

Discussion

Data and Model Calculations

In this section, the experimental results are compared with model calculations in order to test the validity of the theoretical assumptions, to evaluate values of critical model parameters, and to determine areas where experiments that might lead to refinement of the model can be performed.

The results of two sets of calculations are presented. The first set, a determination of the critical cavity size required for growth as a function of temperature, provides an explanation for the high-temperature cutoff observed for swelling. In the second set, calculations of swelling as a function of temperature show that swelling depends strongly on interactions of impurities with point defects and development of the dislocation structure before the onset of swelling.

Critical Size for Cavity Growth

The differential equation that describes the rate of change of the cavity radius with time contains two terms: the rate at which the cavity radius increases due to vacancy absorption, and the rate at which the radius decreases due to interstitial absorption per unit time increment. In order for a cavity to grow, the cavity radius must be greater than the ''critical radius'' r_c^{crit}, defined here as the radius at which the vacancy and interstitial absorption terms balance, and is given by

$$r_c^{crit} = \frac{kT}{2\,Y\Omega} \ln \frac{(D_vC_v - D_iC_i)^{-1}}{D_vC_{vc}^{th}} \tag{4}$$

where k is Boltzman's constant, T is temperature, and Y surface energy. The value of the critical radius depends on the sink structure, the bias parameters, dose rate, and temperature. Using the average void densities given in Table 1, and the experimental temperatures and dose rate, the critical radius was calculated as a function of temperature. The results are shown in Figure 5 (curve labeled 0).

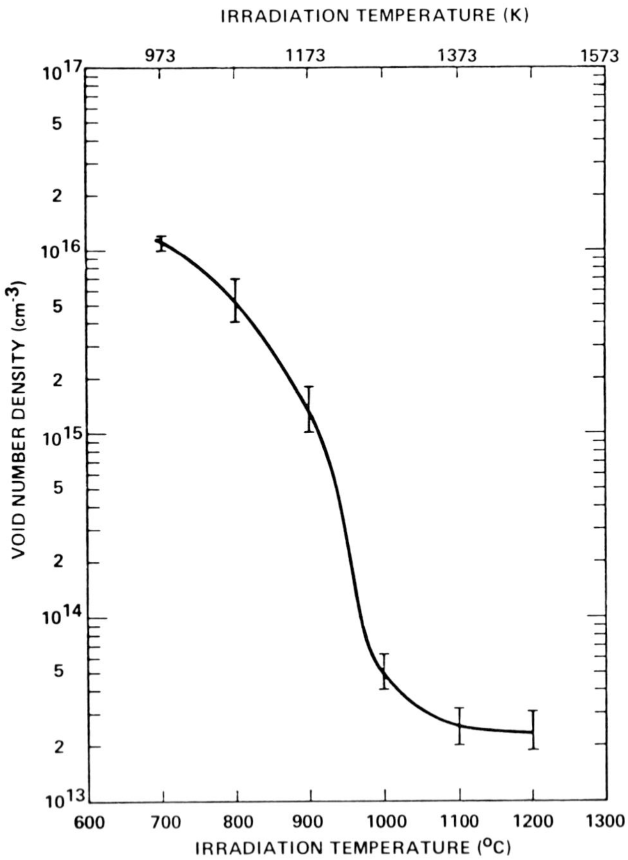

Figure 2 *Void Number Density as a function of Irradiation Temperature for Self-Ion Irradiated Niobium at ~50 dpa.*

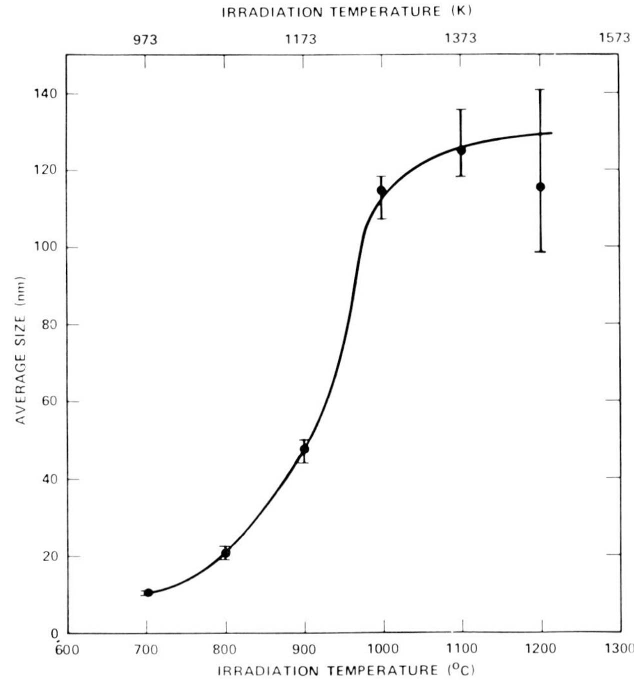

Figure 3 *Average Void Size as a function of Irradiation Temperature for Self-Ion Irradiated Niobium at ~50 dpa.*

Evaluation of the critical radius is important for analysis of the swelling data. First, it establishes the end of the nucleation process for a given set of experimental conditions. Once a cavity has attained this size, we assume it will continue growing at a rate defined by Equation (4). In a practical sense, this means that the value of r_v^{crit} is used as the initial cavity radius when the swelling equation is solved. Second, an estimate of the incubation period for swelling at each temperature can be obtained if the measured swelling values are fit to those calculated using the model equations. The difference between the actual irradiation times and the times at which the model calculations reach the measured values provide these data. Finally, the rapid increase in the critical radius at high temperatures explains the observed absence of cavities and the lack of swelling at 1300°C. No cavity embryo can reach the critical radius at this temperature in the absence of a large quantity of gas acting to stabilize the cavity by reducing thermal emission of vacancies. Therefore, the cavity nucleation rate drops to zero.

Calculated Swelling Curves

Theoretical swelling curves were obtained from the model equations for three different sets of assumption. Each will be discussed to indicate how the analysis developed.

For the first calculation, the experimental dislocation density was taken as the value for the network density with an interstitial bias equal to 1.05. The average cavity density was used for N_c, and loop densities were assumed to be zero. The total swelling time was taken equal to the irradiation time; thus, nucleation is assumed to occur at $t = 0$. The results of this calculation are given in Figure 6 (curve labeled 0). The experimental results are indicated in Figure 6 by the dashed curve, with the measured points and error bars shown. It is apparent that the calculation does not reproduce the data, overpredicting the low temperature results by an order of magnitude, and the high temperature results by roughly a factor of three.

The first modification of the initial assumptions was the inclusion of an incubation time for swelling which is commonly seen in face-centered-cubic metals. At the onset of irradiation, a sequence of events occurs to prepare for the swelling phenomenon. This sequence includes establishment of steady state point defect concentrations, loop nucleation, loop growth,

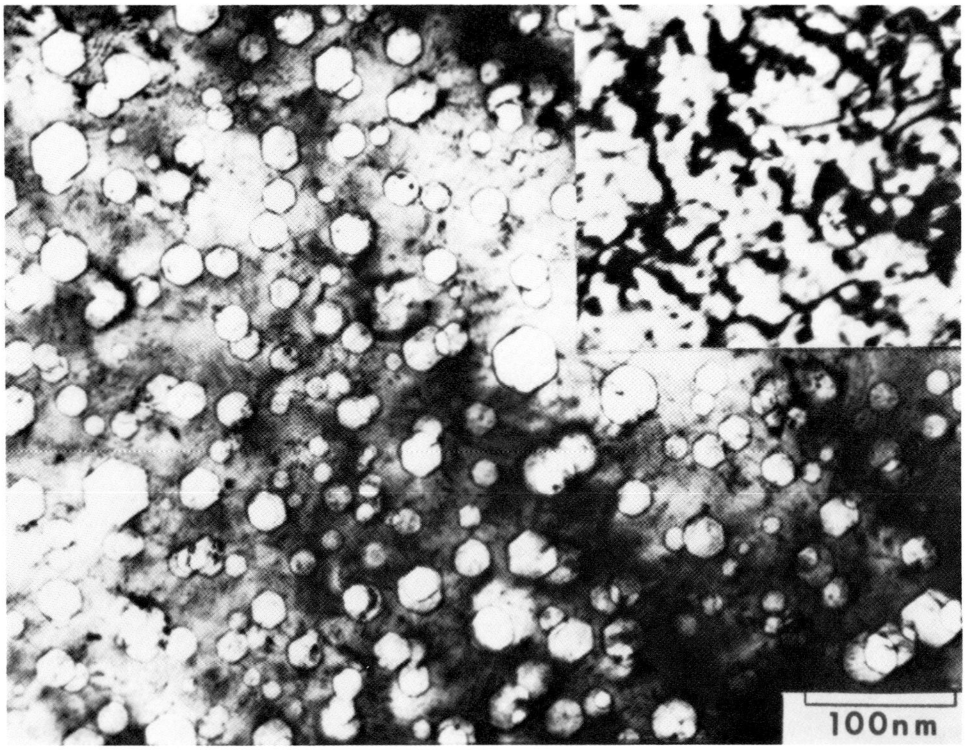

Figure 4 *Voids and Dislocations (inset) in Niobium Irradiated to 50 ± dpa at 800°C. For voids $\vec{Z}$ ~ [111], for Dislocations $\vec{g}$ = [110], $\vec{Z}$ ~ [111].*

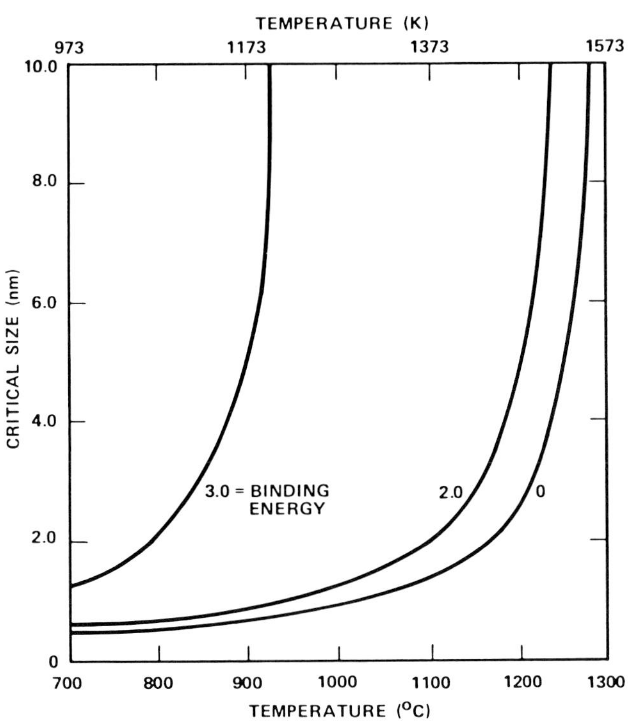

Figure 5 *Critical Radius as a function of Temperature for Three Values of the Interstitial-Oxygen Binding Energy, E_B. Curves are labeled by E_B in eV.*

and cavity nucleation. Loop growth and dislocation network formation may be completed within the initial transient period, or may extend into the cavity growth regime. Cavity growth is significant only after completion of this sequence. The initial transient period depends on temperature, dose rate, and impurity concentrations. We expect that the same phenomena also occur in bcc metals and, therefore in the present work, we assume that the initial transient period is of the order of 1–10 dpa.

To examine the effect of an incubation period on the calculated swelling, four incubation times, corresponding to doses of 1, 5, 10 and 15 dpa, were assumed, and the calculations redone. The results are shown in Figure 6, where the curves are labeled by the incubation dose. The curves show that inclusion of an incubation period does reduce the calculated swelling, but the reduction is not large enough to produce agreement between the calculations and the data. For example, with an incubation dose of 10 dpa, the calculated swelling is reduced by roughly 25 to 30 percent over the entire temperature range. At this point, two major types of effects remain to be considered within the context of the model as it has been developed to date: those due to the two types of loops and

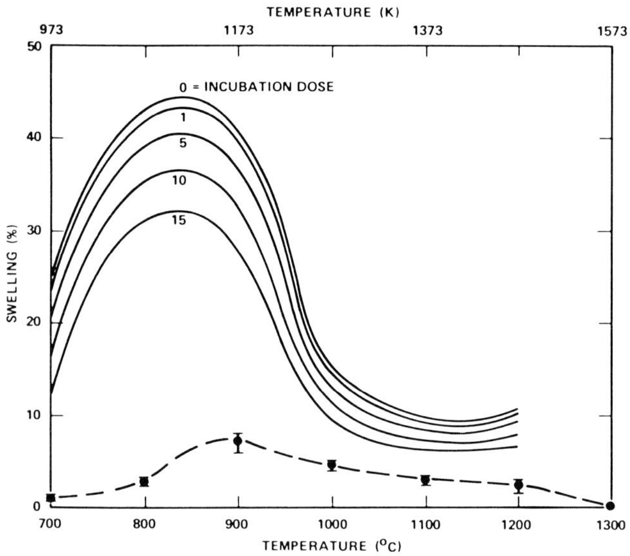

Figure 6 *Calculated Swelling as a function of Temperature. Curves are labeled by the Assumed Value of the Incubation Dose in dpa. Experimental Data are given by the Dashed Curve.*

their differing biases for interstitials, and those due to impurities. At the doses examined, it is apparent from the micrographs that the dislocation structure has evolved into the stage where the network dominates the microstructure, and data on loop types, densities, and sizes cannot be obtained. Calculation of the impact of loop properties on swelling is therefore difficult and will be deferred until low dose data are available. The third set of swelling calculations focused on impurity effects. Trapping of point defects by impurities can have a significant effect on the cavity growth rate and, hence, the swelling rate. Measured oxygen concentrations in the material used in this work are roughly 0.07 at percent. Work by Loomis and Gerber (1981 and 1982) has shown that concentrations of this order are sufficient to affect swelling, leading to a reduction in swelling relative to that in low-oxygen material. If we assume that the mechanism involved is trapping of niobium interstitials by oxygen, the diffusion coefficient D_i appearing in the model equations must be replaced by an effective diffusion coefficient given by Hall and Potter (1978) as:

$$D^+_{i,v} = D_{i,v}/(1 + 4\,\pi\,C_T R_T D_{i,v}(\tau_T - \tau_H)) \quad (5)$$

where C_T is trap concentration, τ_H is the time spent at a host atom site during the diffusion process, τ_T is the time spent at a trap site, and R_T is the capture radius of the trap. A square-well interaction potential is assumed, with a capture radius equal to a single jump distance. The time τ_T and τ_H are given by:

$$\tau_T^{-1} - \nu_t = \nu_o \exp\left(-(E^m_{i,v} + E_B)/K_T\right) \text{ and} \quad (6)$$

$$\tau_H^{-1} = \nu_H = \nu_o \exp\left(E^m_{i,v}\, T/K\right) . \quad (7)$$

Use of Equation (5) introduces the interstitial-oxygen binding energy E_B as a parameter in the model, since its value has not been measured.

Preliminary calculations showed that binding energies of the order of 2 eV are needed to produce calculated swelling curves that correspond to the experimental curves. A series of calculations was made with the binding energy ranging between 1.75 and 2.1 eV. Two incubation doses were used: 0 and 10 dpa. Each experimental point was then used, along with its error bar, to determine the range of E_B values that would produce a fit. Results for all temperatures were obtained for zero incubation dose, and for an incubation dose of 10 dpa. There is some scatter in the results, with an average value of binding energy of 1.995 eV for 0 dpa and 1.943 eV for 10 dpa. These average values were used in the model equations and swelling calculated as a function of temperature, as shown in Figure 7. Triangles indicate an incubation dose of 0 dpa, and circles indicate 10 dpa.

The two combinations of incubation dose and binding energy give approximately the same fit to the data. Significant discrepancies occur at two temperatures, 800°C and 1000°C. At 800°C, the swelling is overpredicted by roughly a factor of two; at 1000°C,

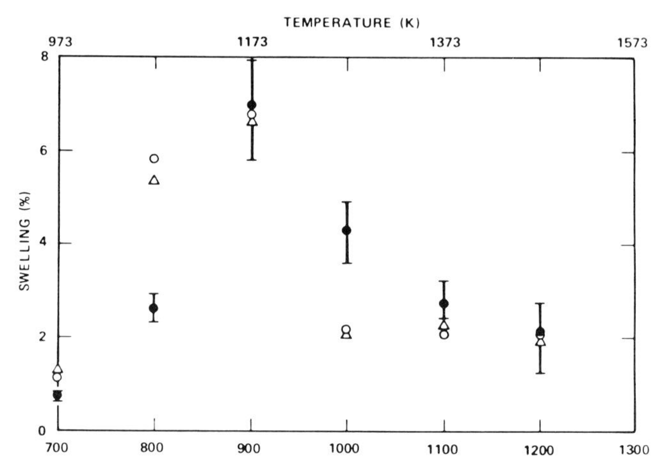

Figure 7 *Calculated Swelling as a function of Temperature for Two Combinations of Incubation Dose and Interstitial-Oxygen Binding Energy: Δ (0 dpa, 1.995 eV); 0 (10 dpa, 1.943 eV). Experimental Points are shown with Error Bars.*

swelling is underpredicted by the same factor. These discrepancies are not believed to be caused by the presence of grain boundaries.

Whether the values obtained for E_B are reasonable can be tested to a limited extent by observing at the effect of the binding energy on the critical radius. Trapping of interstitials by oxygen reduces the diffusivity of interstitials and changes r_v^{crit} (see Equation (4)). Calculations of r_v^{crit} as a function of temperature were done for a range of assumed binding energies, and r_v^{crit} increases as binding energy increases for a fixed temperature. Values of r_v^{crit} are plotted in Figure 5 as a function of temperature for binding energies of 2 eV and 3 eV. As can be seen from the figure, there is a large shift of the high temperature portion of the curve toward lower temperatures as the binding energy increases from 2 eV to 3 eV. The shift when E_B changes from 0 eV to 2 eV is relatively small in comparison. For example, any increase in assumed binding energy that would provide a better fit to the 800°C data point, would result in unacceptably large critical radii at high temperatures. Because cavities are observed at 1200°C, r_v^{crit} must remain relatively small until this temperature is exceeded. For binding energies near 2 eV, r_v^{crit} is 5 nm, a somewhat high value. Therefore, a reduced binding energy is more reasonable than an increased one.

Summary and Conclusions

In this study, a theoretical model based on the chemical reaction rate formalism was developed specifically for body centered cubic metals. Calculations were performed for a selected metal; that is, niobium. A special source preparation method was developed to produce a niobium hydride target material. The use of this target in the ion source resulted in an output of niobium ions of sufficient energy and flux to permit, for the first time, ion bombardment studies of niobium with self-ions. The experimental results showed:

1. A peak swelling of 7 percent was obtained at 900°C in niobium (0.07 at percent oxygen) irradiation with 5.3 MeV Nb^{++} ions at a nominal dose of 50 dpa and a dose rate of 6×10^{-3} dpa/s, and
2. Swelling occurred over a temperature range of 700°C to 1200°C. (Swelling was not observed at 1300°C.)

A comparison of model calculations with the experimental results showed a reasonable agreement between the calculated and experimental swelling values when trapping of interstitials by oxygen was included. An interstitial-oxygen binding energy of approximately 2 eV provided the best fit to the data.

Acknowledgments

This work is the result of efforts of a great many people at ANL and Westinghouse. Noteworthy amongst them are Jack Wallace, Pete Baldo, Bob Erek, Ed Johanson, Jim Emerson, Roy Kentra, Ben Loomis, Beth Beiversdorf, and Bernei Kestel of ANL, and Janet Lauer, Wesley Hughes, Jim Haugh, Bob Reese, Sam Laciak, Bob Muir, and Jim Greggi of Westinghouse. Management support provided by R. W. Buckman, M. L. Bleiberg, and S. Wood is appreciated. This work was sponsored by the U.S. Air Force Office of Scientific Research under Contract No. F49620-83-C-0120.

References

Bajaj, R., B. 0. Hall, and G. R. Fenske (1985) *An Investigation of the Irradiation Swelling Mechanisms in Refractory Metals at High Temperatures*, WAESD-TR-85-005, Westinghouse Electric Corporation, Pittsburgh, PA.

Bajaj, R., S. Diamond, and M. L. Bleiberg (1982) "Optimization of Precipitation Hardened Fe-Cr-Ni Alloys Using Nickel Ion Bombardment Simulation Studies," in *Effects of Radiation on Materials, Eleventh Conference*, ASTM-STP-782, pp. 856–884, H. R. Brager and J. S. Perrin, eds., American Society for Testing and Materials.

Bleiberg, M. L., R. Bajaj, I. M. Baron, et al. (1976) "Swelling and Gamma-Gamma-Prime Particle Stability of Ion-Bombarded Iron-Chromium Nickel Al loys," in *Irradiation Effects on the Microstructure and Properties of Metal*, ASTM-STP-611, pp. 315–336, American Society for Testing and Materials.

Bleiberg, M. L. and J. W. Bennett, eds. (1977) *Proceedings of International Conference on the Radiation Effects in Breeder Reactor Structural Materials*, Metallurgical Society of AIME, held in Scottsdale, AZ.

Brailsford, A. D. and R. Bullough (1972) "The Rate of Swelling Due to Void Growth in Irradiated Meta ls," *J. Nucl. Mater.*, 44: 121–135.

Burton, J. J. (1971) "Effect of Mobile Interstitials on the Nucleation of Voids," *Scripta Met.*, 5: 449–458.

Hall, B. O. (1986) "Conditions for Interstitial Loop Growth/Shrinkage in Body-Centered Cubic Meta ls," submitted for publication in the Journal of Nuclear Metals.

Hall, B. O. (1983) "Helium Partitioning Between Voids and Dislocations," *J. Nucl. Mater.*, 117: 89–95.

Hall, B. O. (1980) "Point Defect Clustering During Irradiation," *J. Nucl. Mater.*, 91: 63–72.

Hall, B. O. and D. I. Potter (1978) "Microstructural Development During Low Dose Irradiation," in *Effects of Radiation on Structural Materials*, ASTM-STP-683, pp. 32–45, J. A. Sprague and D. Kramer, eds.

Hayns, M. R. (1975) "The Nucleation and Early Growth of Interstitial Dislocation Loops in Irradiated Materials," *J. Nucl. Mater.*, 56: 267–274.

Johnston, W. G. , T. Lauritzen, J. H. Rosolowski, et al. (1976) "The Effect of Metallurgical Variables on Void Swelling," in *Radiation Damage in Metals*, pp. 227–266, N. L. Patterson and S. D. Harkness, eds., ASM, Metals Park, OH.

Katz, J. L. and H. Wiedersich (1971) "Nucleation of Voids in Materials Supersaturated with Vacancies and Interstitials," *J. Chem. Phys.*, 55: 1414–1425.

Little, E. A., R. Bullough, and M. H. Wood (1980) "On the Swelling Resistance of Ferritic Steels." in *Proceedings Royal Soc. London A*, 372: 565–579.

Loomis, B. A. and S. B. Gerber (1981) "Similar Dependence on Dilute O Concentration of Void Formation in Ion-Irradiated Nb and Some Properties of Unirradiated Nb," *J. Nucl. Mater.*, 97: 113–125.

Loomis, B. A. and S. B. Gerber (1982) "Swelling of $^{58}Ni^{+}$ and $^{3}He^{+}$ Ion-Irradiated Nb and Nb Alloys," *J. Nucl. Mater.*, 103 & 104: 1193–1197.

Russell, K. C. (1971) "Nucleation of Voids in Irradiated Metals," *Acta Met.*, 19: 753–758.

Spitznagel, J. A., S. Wood, and W. J. Choyke (1980) "Irradiation Response of Materials," in *Damage Analysis and Fundamental Studies, Quarterly Progress Report, April-June 1980*, pp. 103–114, DOE/0046, U. S. Department of Energy, Washington, D.C.

Wiedersich, H. (1972) "On the Theory of Void Formation During Irradiation," *Rad. Effects*, 12: 111–125.

Wiffen, F. W. (1984) "Effects of Irradiation on Properties of Refractory Alloys with Emphasis on Space Power Reactor Applications," in *Proceedings of the Symposium on Refractory Alloy Technology for Space Nuclear Power Applications Conference*, R. H. Cooper, Jr. and E. E. Hoffman, eds., held in Oak Ridge, TN, August 1983.

Wood, S., J. A. Spitznagel, and W. J. Choyke (1981) "Microstructural Development in Dual-Ion Bombarded 316 Stainless Steel," in *Effects of Radiation on Materials: Tenth Conference*, ASTM-STP-725, pp. 455–469, David Kramer, H. R. Brager, and J. S. Perrin, eds., American Society for Testing and Materials.

Part 10

RADIATION EFFECTS

Space Nuclear Power Systems 1986
Edited by M. S. El-Genk and M. D. Hoover

Chapter 30

Radiation Damage Analysis for State-of-the-Art Microelectronics

Maximo S. Lazo, David M. Woodall, and Patrick J. McDaniel

The University of New Mexico
Chemical & Nuclear Engineering Department
Albuquerque, NM 87131

Space nuclear power systems will demand a high degree of reliability and operability, requiring advances in the performance of electronic subsystems. The single most important factor limiting the lifetime and operability of such components is the radiation environment they will encounter. In this paper we address the improvement of computational capabilities to determine the physical damage due to radiation in microelectronic materials. Using the NJOY nuclear data processing code, we predict the neutron radiation damage response function of semiconductor materials. This includes the prediction of displacement and ionization damage due to the interaction of high-energy neutrons or gamma radiation. We have made calculations of neutron response functions for pure silicon and for silicon dioxide, and the results are presented here. The methodology developed included using the Lindhard electronic screening theory to compute the distribution of damage between ionization and displacement. The variation of those results with a primary knock-on atom (PKA) is of significance for a molecule such as SiO_2. The methodology of doing such molecular damage calculations is presented here. This work was performed on a VAX 11/780 computer using the VMS operating system.

Introduction

Long lifetimes, in excess of seven years, are anticipated for space nuclear power systems. This fact, coupled with the requirement for a high degree of reliability and operability, demands significant advances in the performance of electronic systems used on such spacecraft. Specifically, the electronic systems will be exposed to long-term irradiation by neutrons and gamna rays from the reactor. In addition, these systems will experience a flux of cosmic rays and energetic particles from the solar wind. Since these radiations are harmful to most electronic systems, they must be eliminated by shielding or by physical separation from the reactor, or the effects of such radiation must be mitigated by radiation-tolerant design procedures. Note that shielding may add significant mass to the space platform and physical separation may result in a configuration that is difficult to control.

Along with the desire for a rad-hard electronic system is the requirement that all of the functionality of state-of-the-art very large scale integrated (VLSI) circuit technology be available. This places a real challenge before the designers of electronic components, particularly in the microelectronics area. As integrated circuits become smaller, with smaller feature sizes, the interaction of radiation with individual features becomes significantly more complicated. To continue the miniaturization of microelectronics, better correlations are needed for the damage and failure mechanisms caused by radiation.

The assumptions of classical radiation damage analysis and the correlation of integrated circuit radiation damage effects with displacement damage in silicon may not be adequate. Traditional neutron radiation damage analysis (ASTM 1985) assumes the damage in silicon irradiated by a flux of neutrons can be represented by infinite medium silicon damage due to a 1.0-MeV equivalent neutron beam. The damage produced by this beam can be then correlated with effects observed in reactor irradiation of components (Gover 1985), such as modifications in minority carrier lifetime, majority carrier concentration or carrier mobilities. The conversion to a 1.0-MeV equivalent basis allows for the intercomparison of damage due to different neutron sources. We question the validity of this approach. While the relationship probably holds for large discrete components based on conventional bipolar technology, it is of questionable validity for state-of-the-art microelectronic components with micrometer-sized features. Furthermore the approach does not address the effect of damage in molecular systems, such as SiO_2.

For the purpose of radiation effect analysis, a microelectronic system is a mixture of metallic and nonmetallic monoatomic and polyatomic solids such as Si, Ge, Au, Al, SiO_2, GaAs and GaP. The physical dimension of the features of a typical microelectronic device are approximately 1 μm to 3 μm in width and only 10 nm to 100 nm in depth.

Neutron radiation damage in solids results from nuclear collisions and reactions that produce energetic recoil atoms. The recoiling atoms generate electronic excitation, or ionization, in the host material and elastic and inelastic collisions that may result in displacing additional host atoms from their lattices sites. Thus, the two features of such damage are residual ionization and atomic displacement damage. The structure of displacement cascades in solids can be studied via a Monte Carlo simulation (Robinson 1974 and 1985 and Mueller 1980) or via a transport calculation.

Nuclear collisions and reactions are characterized by neutron cross sections, or interaction probabilities. These parameters are measured experimentally in the range of neutron energies from a few eV to several MeV for most elements in the periodic table and for some compound materials. Due to limitations in time and techniques, there are regions for which the cross sections are not available (MacFarlane 1984 and Greenwood 1985). Therefore, theoretical analysis of nuclear reactions is an important tool, both for evaluating the present measured values and for the interpolation and extrapolation of new values. The reactions between the neutron and heavy nuclei can be modeled successfully using a phenomenological model based on experimental information about the nuclear structure and an average potential, the ''local optical potential'' (Young 1977 and 1985).

The cross section for the interaction between a neutron and a nuclei in a solid structure is usually of the order of 10^{-8} Å^2 and, thus, much smaller than the area of the elementary unit lattice cell, which is about 5 Å^2. The mean free path of neutrons is large, typically many centimeters, and PKAs are created uniformly in the solid. The development of a cascade started by a PKA is not modifed by the presence of other cascades until it reaches very high neutron fluences. The probability of multiple neutron interactions in silicon integrated circuits has recently been measured and modeled (Srour et al. 1985).

In this paper the neutron radiation damage is calculated done via the NJOY computer code (MacFarlane 1977). All nuclear reaction data are obtained from ENDF/B-V, distributed by Brookhaven National Laboratory (Kinsey 1979), and from the theoretical models included in the NJOY code. The NJOY code uses the Lindhard model to calculate the partition of recoil energy between nuclear and electronic effects for the recoiling atoms.

Theory

The probability of displacing atoms from their lattice sites depends on the incident neutron energy (E) and cross section ($\sigma(E)$), the primary recoil atom energy (E_R), and the partition of recoil energy between electronic excitation and atomic displacement ($P(E_R)$). The damage function is given by

$$D(E) = \sum_k \sigma_k(E) \int dE_R f_k(E, E_R) P(E_R) , \qquad (1)$$

where the sum is over all reactions, and $f_k(E_R)$ is the probability that a neutron with energy E in a reaction k produces a recoil with energy E_R:

E = incident neutron energy (laboratory system),
E_R = recoil nucleus energy (laboratory system), and
$\sigma_k(E)$ = nuclear cross section for reaction k at energy E.

Partition of the Recoil Energy: LSS Theory

The partition function or ''damage eficiency'' is given by the relation derived by Robinson, based on the electronic screening theory of Lindhard (1963):

$$P(E_R) = \frac{E_R}{1 + F_L(3.4008\epsilon^{1/6} + 0.40244\epsilon^{3/4})} , \qquad (2)$$

if $E \geqslant 25$eV, and $P(E_R) = 0$ otherwise, where

$$\epsilon = \frac{E_R}{E_L}$$

$$E_L = 30.724\, Z_R Z_L (Z_R^{2/3} + Z_L^{2/3})^{1/2}(A_R + A_L)/A_L , \qquad (3)$$

and

$$F_L = \frac{0.0793\, Z_R^{2/3} Z_L^{1/2}(A_R + A_L)^{3/2}}{(Z_R^{2/3} + Z_L^{2/3})^{3/4} A_R^{3/2} A_L^{1/2}} . \qquad (4)$$

Z_i and A_i refer to the charge and atomic number of the lattice nuclei (L) and the recoil nucleus (R).

Note that the original paper by Lindhard (1963) treated the derivation of the partition function for the case $Z_R = Z_L$ thoroughly, but presented only an outline for the case $Z_R \neq Z_L$. It is important to point out that in a series of papers by Parkin and Coulter (1977, 1979, and 1980), the corresponding integral equations for the radiation effects derived by Lindhard have been numerically solved for monoatomic, diatomic and polyatomic systems. These equations are treated considering the universal scattering cross section and electronic stopping power formula of Lindhard. In this way, the damage, partition functions or "damage efficiency" functions are generated. These functions are fitted then to analytic expressions of the Robinson type, which can be used in the NJOY program replacing the Lindhard's damage efficiency proposed by Robinson.

The relationship between neutron energy and recoil energy can be described by one general equation derived from conservation of energy and momentum in the center of mass system:

$$E_R(E, E_x, \theta) = U_3E_x + U_1U_4E - 2(EE_xU_1U_3U_4)^{1/2}\cos\theta , \quad (5)$$

where:

$E_x = U_2E + Q$ = total energy,

Q = energy-mass conversion in the nuclear reaction,

θ = angle between the incident neutron direction and the recoil atom direction,

$U_i = \dfrac{m_i}{m_1 + m_2}$,

m = mass, and i = 1, 2, 3, or 4 for the neutron, target atom, emitted particle, and recoil atom, respectively.

The evaluation of the damage cross section for binary materials requires knowledge of the damage efficiency for each pair of elements:

$$D_{i,j}(E) = \sum_k \sigma_k^i(E) \int dE_R f_k^i(E,E_R) P_{i,j}(E_R) . \quad (6)$$

An average over all possible pairs weighted by atomic fraction (fr) gives an approximation to the damage cross section for the compound:

$$< D > = \sum_{i,j} fr_i fr_j D_{i,j}(E) . \quad (7)$$

As an example, for SiO_2, it is necessary to determine the number of oxygen atoms displaced by silicon PKAs, which is proportional to D(Si,O), as well as by oxygen PKAs, which is proportional to D(O,O). The results are then combined according to the atomic fraction of each species of PIA. In this case

$$fr_{Si} = 1/3, \text{ and} \quad (8)$$
$$fr_O = 2/3 .$$

This methodology is applied to the results generated within the NJOY calculation to obtain the corrected values for the molecule.

Calculational Results

The results of our NJOY calculations are presented here for both pure silicon and silicon dioxide. The multigroup cross-sectional set chosen for these results is the 37, group DNA set which has been extensively used in the radiation damage literature. We also have performed this analysis for a much finer (300 group) cross-sectional set, but for the sake of comparison with published results, only the 37 group results are presented.

Figure 1 displays the displacement damage in silicon as a function of neutron energy on a linear plot. The elastic scattering and total cross sections are displayed in two curves. Note that the components other than elastic scattering are dominant above 2.0 MeV. The same data are plotted over a much larger energy range in a log-log plot in Figure 2. At high energy, both inelastic scattering and particle-producing reactions play a significant role in the displacement damage cross section.

The displacement damage cross sections for the components of SiO_2 are shown in Figure 3. Note that the oxygen PKA produces lower damage than the silicon PKA. Only values for the oxygen-on-oxygen (O—O) and silicon-on-silicon (Si—Si) damage are shown in that figure. As is apparent from Figure 3, the high energy limit is only about 75 MeV-mb, for the O—O damage and about 200 MeV-mb for the Si—Si damage. The net effect is that there is substantially less displacement damage in SiO_2 than in pure silicon. The difference between the Si—Si and the silicon-on-oxygen (Si—O) damage, as well as the difference between the O—O and the oxygen-on-silicon (O—Si) damage, is shown in Figure 4.

The ionization damage cross sections for SiO_2 and its components are shown in Figure 5. A comparison of the displacement damage cross section and the ionization damage cross section for SiO_2 is plotted in Figure 6. On this linear scale, the displacement damage is shown to be an order of magnitude smaller at high energy than the ionization damage. Figure 6 also includes the Si—Si and O—O values, from which one can determine the ratio of Si damage to SiO_2 damage at a given energy.

The summary comparison of the 1-MeV equivalent

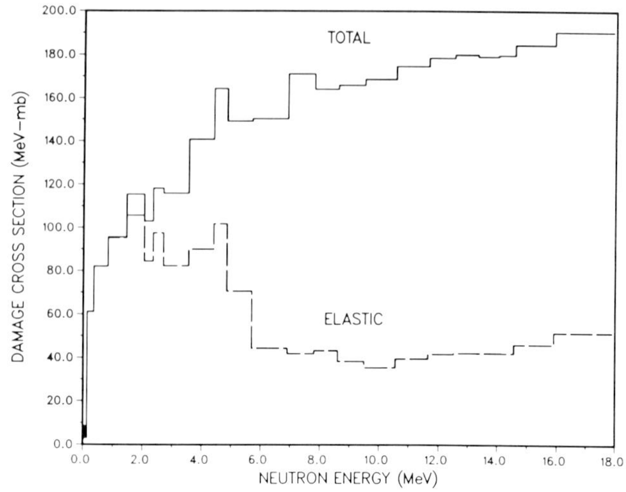

Figure 1 *Displacement Damage Cross Sections for Silicon as a function of Neutron Energy, using 37 Neutron Groups (the DNA Group Structure).*

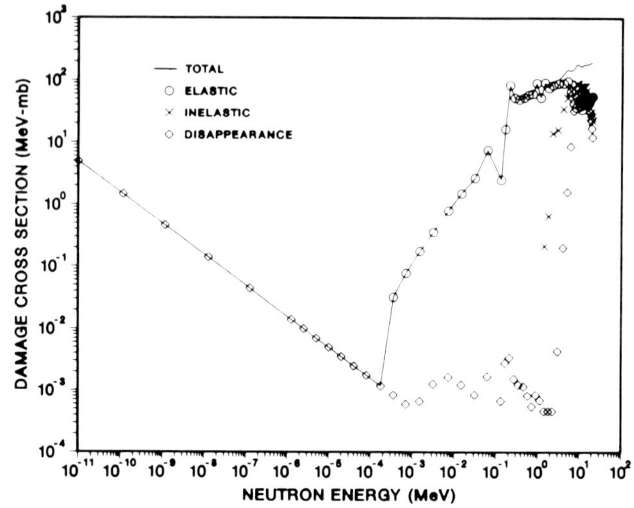

Figure 2 *Radiation Damage Cross Sections for Silicon, as a function of Energy, with the Components of the Damage Identified.*

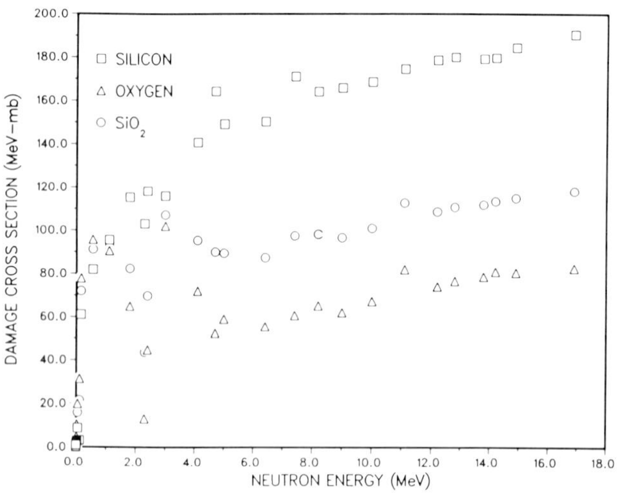

Figure 3 *Displacement Damage Cross Sections for Silicon Dioxide, displaying the Si—Si and the O—O Components.*

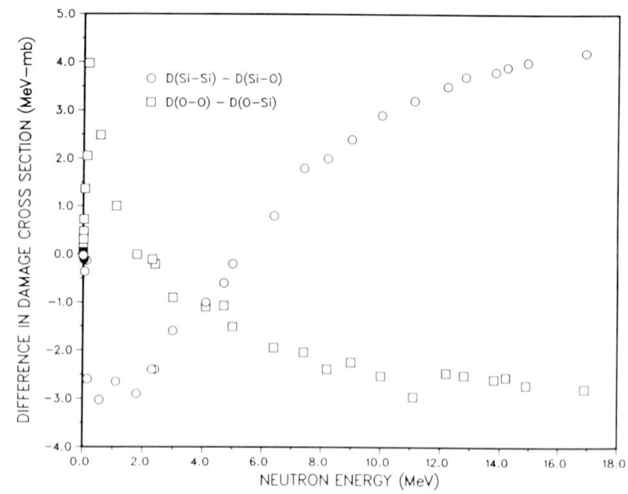

Figure 4 *Difference in Displacement Damage Cross Sections for Atomic Components of SiO_2.*

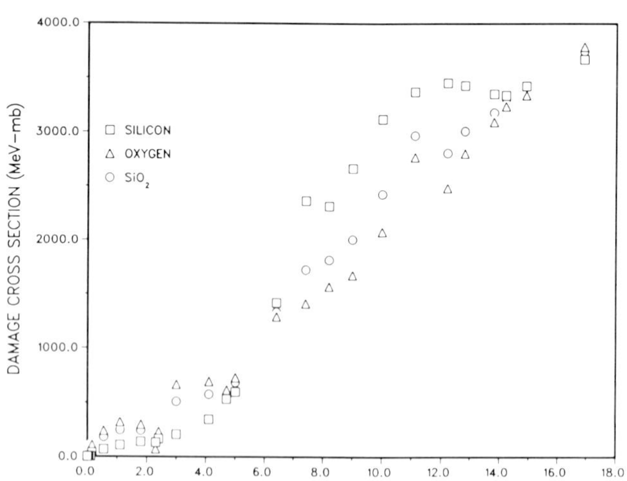

Figure 5 *Ionization Damage Cross Sections for Silicon Dioxide, displaying the Si—Si and the O—O Components.*

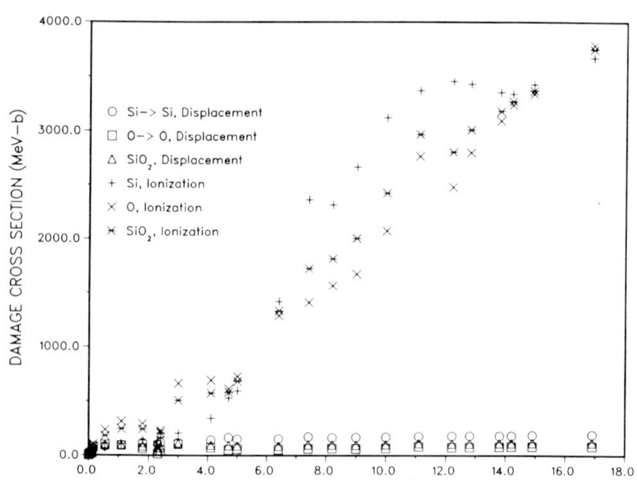

Figure 6 *Displacement and Ionization Damage Cross Sections for SiO_2 and its Components.*

Table 1 *Comparison of 1-MeV Silicon Displacement Damage Functions.*

E Range (MeV)	NJOY	ASTM[a]	Rogers[b]	Messenger[c]
14.2–13.8	2.00	2.26	2.25	2.24
11.1–10.0	1.88	2.07	2.12	2.18
4.97–4.72	1.83	2.03	1.79	1.94
1.83–1.11	1.07	0.926	0.913	1.25
1.11–0.550	0.914	0.866	0.873	0.838

[a]From ASTM 1985.
[b]From Rogers 1975.
[c]From Messsenger 1968.

displacement damage response function for silicon is shown in Table 1, in which the results obtained using NJOY are compared with results from other authors. A caution is in order with respect to using these results. The silicon cross section has a resonance in the vicinity of 1.0 MeV. Thus, the values obtained from this type of analysis are sensitive to the group structure and group averaging scheme used. Nevertheless, it is noteworthy that the 14-MeV to 1-MeV ratio obtained from NJOY (300 group) agrees with the best experimental values obtained for such damage.

Conclusions

The NJOY system provides a consistent and convenient method to compute damage cross sections for both single species materials, such as Si, and compound materials, such as SiO_2. The computational results presented in this report are the most complete available for these materials. They indicate that currently used values may be in error by approximately to 10-20% for Si and substantially more for SiO_2. Note that the current results are the first effort to use the most up-to-date (ENDF/B-V) cross-sectional information consistently to compute the ionization and displacement damage for Si and SiO_2. These results should be used in place of previous calculations, which were generated with various cross-sectional sets and various model "fix-ups" for the molecular effects. Results obtained using the NJOY system also assure consistency between damage and heating cross sections for these materials.

The key feature of the application outlined in this report is the availability of the most up-to-date cross sections through the ENDF/B-V library. This methodology can be applied to other molecular materials of interest for radiation damage in microelectronic materials. Further work is suggested for materials such as GaAs; however, additional effort may be required because the cross sections in the ENDF/B-V library might be inadequate. Cross sections generated by this method may be used in transport calculations or Monte Carlo calculations to determine the actual damage effects in the various components of a typical microelectronic device.

Acknowledgments

This research has been sponsored by Sandia National Laboratories under Contract No. SNL-21-6559 to the University of New Mexico. Conversations with R. M. MacFarlane and P. Young (LANL) and L. R. Greenwood (ANL) are gratefully acknowledged.

References

ASTM (1985) *Characterizing Neutron Energy Fluence Spectrum In Terms Of An Equivalent Monoenergetic Neutron Fluence For Radiation-Hardness Testing Of Electronics*, ASTM Standard E-722-85, February 1985.

Coulter, C. A. and D. M. Parkin (1980) "Damage Energy Functions for Polyatomic Materials," *J. Nucl. Mat.*, 88: 249–260.

Dell, G. F. et al. (1980) "Damage Parameters for Non-metals in a High Energy Neutron Environment," *Symposium on Neutron Cross Sections From 10 to 50 MeV*, BNL-NCS-51245, Vol. II, Brookhaven National Laboratory, Brookhaven, NY, July 1980.

Gover, J. E. (1983) *Basic Radiation Effects in Electronics Technology*, Course Notes For INTEC SP712, Sandia National Laboratories, Albuquerque, NM, October 1983.

Greenwood, L. R. (1985) *Private Comminication*, Argonne National Laboratory, Argonne, IL.

Greenwood, L. R. (1985) *Specter: Neutron Damage Calculations for Materials Irradiations*, ANL/FPP/TM-197, Argonne National Laboratory, Argonne, IL, January 1985.

Kinsey, R., ed. (1979) *ENDF/B Sumary Documentation*, BNL-NCS-17541(ENDF-201), 3rd ed., Brookhaven National Laboratory, Brookhaven, NY.

Lindhard, J., V. Nielsen, M. Scharff, and P. V. Thomsen (1963) "Integral Equations Governing Radiation Effects," *Kgl. Danske Videnskab. Selskab, Mat.-fys. Medd., 33* (10):2–42.

MacFarlane, R. E. (1979) *NJOY A Code System For Producing Pointwise and Multigroup Neutron and Photon Cross Sections from ENDF/B-IV and -V Evaluated Nuclear Data*, RSIC, PSR-118, Los Alamos National Laboratory, Los Alamos, NM, March 1979.

MacFarlane, R. E. and D. Graham Foster, Jr. (1984) "Advanced Nuclear Data for Radiation Damage Calculations," *J.Nucl. Mat.*, 122: 1047.

MacFarlane, R. E. (1985) *Private Communication*, Los Alamos National Laboratory, Los Alamos, NM.

Messenger, G. C. (1968) ''Radiation Effects on Microcircuits,'' *IEEE Trans. Nuc. Sci.*, NS-13(6): 141.

Mueller, G. P. (1980) ''Simulations of Cascades Damage in Silicon,'' *IEEE Trans. Nuc. Sci.*, NS-27(6): 1474.

Parkin, D. M. and C. A. Coulter (1977) ''Damage Energy Functions for Compounds and Alloys'', *Trans. Am. Nuc. Soc.*, 27.

Parkin, D. M. and C. A. Coulter (1979) ''Displacement Functions for Diatomic Materials,'' *J. Nuc. Mat.*, 85 & 86:611–615.

Robinson, M. T. and I. M. Torrens (1974) ''Computer Simulation of Atomic-Displacement Cascades in Solids in the Binary-Collision Approximation,'' *Phys. Rev.*, B9:5008.

Robinson, M. T. (1976) ''The Theory of Radiation Induced Defect Production,'' in *Radiation Damage in Metals*, N. L. Peterson and S. D. Harkness, eds., American Society for Metals publication, pp. 1–27.

Robinson, M. T. (1985) *Private Communication*, Oak Ridge National Laboratory, Oak Ridge, TN.

Rogers, V. C., L. Harris, D. K. Steinmam, and D. E. Bryan (1975) ''Silicon Ionization and Displacement Kerma for Neutron from Thermal to 20 MeV,'' *IEEE Trans. Nucl. Sci.*, NS-22(6):2326.

Srour, J. R. et. al. (1985) ''Effects of Single Neutron Interactions In Silicon Integrated Circuits,'' *IEEE Trans. Nucl. Sci.*, NS-32(6): 4195.

Young, P. G. and E. D. Arthur (1977) *GNASH: A Pre-equilibrium Statistical Model Code for Calculation of Cross Sections and Emission Spectra*, LA-6947, Los Alamos National Laboratory, Los Alamos, NM.

Young, P. G. (1985) *Private Communication*, Los Alamos National Laboratory, Los Alamos, NM.

Space Nuclear Power Systems 1986
Edited by M. S. El-Genk and M. D. Hoover

Chapter 31

Radiation And High Temperature Effects On MOSFETS

T. J. Lin, C.H. Wu, And A.S. Kumar
University Of Missouri-Rolla
Rolla, MO 65401

The effects of gamma-ray irradiation and high temperature environments on MOSFETs (Metal-Oxide-Semiconductor Field Effect Transistors) were investigated to a maximum gamma dose of 10 Mrad(Si) and a maximum temperature of 200°C. The drain-source current and gate voltage characteristics were studied before and after irradiation and after annealing. A kinetic model of recovery by thermal annealing that takes into account the random potential barriers between silicon-oxide interface is presented. The model is different in concept from the previous thermally activated tunneling model of Winokur et al. (1984). Our kinetic model explains satisfactorily the annealing behavior of radiation damage in MOSFETs as observed in our experiments as well as those of a previous study by Danchenko et al. (1980).

Introduction

Investigation and modeling of radiation-induced damage and high temperature effects in electronics are important for the design of hardened devices for space nuclear power systems. The tolerance limits for the high neutron and gamma radiation from a reactor core in space are of the order of 10^{16} n/cm^2 and 10 Mrad(Si), respectively (Gover and Srour 1984). The electronics further may be required to operate reliably and predictably up to 300°C for 1000 hrs. (Shaucair et al. 1984).

From a radiation standpoint, a balance must be reached between the mass of the shielding required to reduce radiation levels and the practical hardening levels of the electronic systems. Due to the systems' requirements in space, use of radiation-hardened electronics is preferred over heavily shielded electronics. However, radiation-hardened integrated circuits are in short supply and the replacement of radiation-soft circuits by hardened circuitry usually is impractical.

Therefore, it is essential to model the radiation damage and its annealing behavior at elevated temperatures. Since MOSFETs, used for the current study, are majority carrier devices, they do not depend strongly on the minority carrier life time. Hence, MOSFETs are relatively unaffected by neutron irradiation in comparison to the total dose effects of gamma irradiation. The model presented here for the annealing of radiation damage in MOSFETs is different from the previous thermally activated tunneling models, and is not restricted by the assumption of a smooth interface between the semiconductor and the oxide. The high temperature experiments in the current study were performed to determine safe annealing temperatures for the devices studied.

Experimental

The electrical characteristics of devices ($I_d - V_g$, $I_d - V_{ds}$, and so on, where I_d is the current flowing from the source to the drain, V_g is the gate voltage, and V_{ds} is the voltage drop across the source and the drain) were studied using the setup in Figure 1. The HP4140B pA meter is addressable and is interfaced with an IBM PC. High temperature behaviors of the devices were measured by placing the chips in a thermal chamber. The temperature was monitored by a chromel-alumel thermocouple. Temperature control was within ± 0.5 C. Co-60 gamma-ray irradiation of the test chips was performed in the Research Reactor Facility of the University of Missouri. The test chips (TA629) were made by Sandia National Laboratories. MOSFET transistors within this chip were tested individually.

Results And Discussions

Electrical Characteristics of MOSFETs at Elevated Temperatures

$I_d - V_g$ characteristics for each MOSFET were investigated at various elevated temperatures (25, 50, 100, 150 and 175°C). Effects of temperature on some

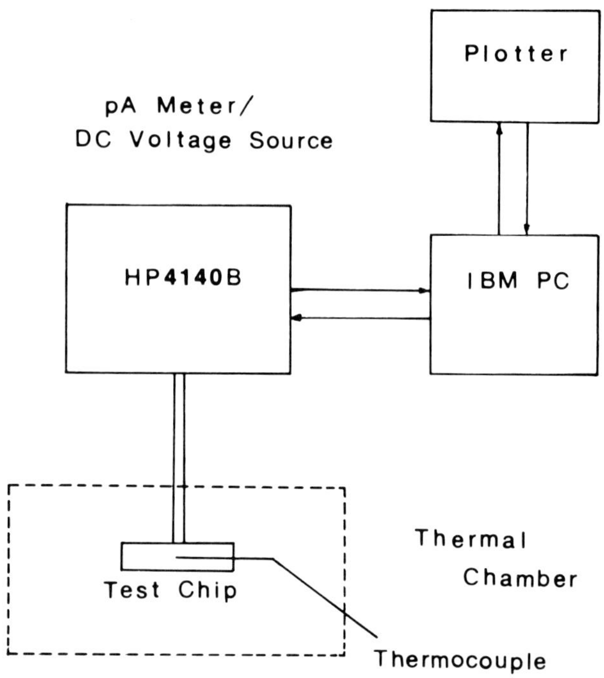

Figure 1 *Schematic Diagram of the Experimental Setup for Measuring the Electrical Characteristics of Devices.*

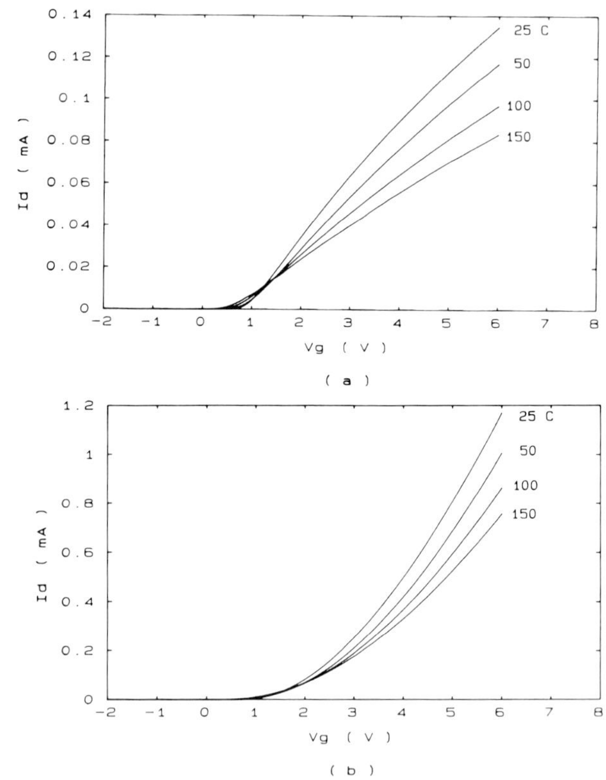

Figure 2 *$I_d - V_g$ Curves for an n-Channel MOSFET (W/L = 23/3) at Various Temperatures: (a) V_{ds} = 0.2 V and (b) V_{ds} = 6.0 V.*

of the parameters, such as threshold voltage, (V_{th}) and drain-body leakage current, are presented here.

When drain-source voltage (V_{ds}) is very small, the drain current is given by (Shourcair et al. 1984):

$$I_d(\mathrm{T}) \simeq \mu(T)(W/L)C_{ox}(V_g - V_{th}(T))V_{ds}, \quad (1)$$

where μ(T) is the carrier mobility at temperature T, W is the channel width, L is the channel length, and C_{ox} is the capacitance of the oxide layer. Figure 2 presents the $I_d - V_g$ curves for an n-channel MOSFET (W/L = 23/3) for V_{ds} = 0.2 and 6.0 volts, respectively. The change in the threshold voltage with temperature is shown in Figure 3. As can be seen, the threshold voltage decreases linearly as the temperature increases.

Results of a p-channel MOSFET (W/L = 16/4) are shown in Figures 4 and 5. Similar to an n-channel MOSFET, the magnitude of the threshold voltage decreases linearly with the temperature for the p-channel MOSFET as shown in Figure 5. The variation due to temperation, as shown in the slopes of the $I_d - V_g$ curves in Figures 2 and 4, is caused by the temperature dependence of carrier mobilities, as indicated by Equation (1).

The leakage current flowing across the reverse biased drain-body junction is the saturation current level at which the MOSFETs are off ($|V_g| < |V_{th}|$). Drain-body junction leakage currents were measured at various temperatures for an n-channel MOSFET. Figure 6 shows the effect of temperature on the leakage current. The slope is smaller at low temperatures (below 60°C) than at temperatures above 60°C. This can be explained by a well-known p-n junction theory (Shoucair et al. 1984); that the total reverse p-n junction current can be approximately given by the sum of the diffusion current component in the neutral region and the generation-recombination current component in the depletion region. The diffusion component is proportional to the square of the concentration of the intrinsic carrier, $n_i^2(T)$, while the generation component is proportional to $n_i(T)$ only. For Si (n_i is very small), the generation current tends to dominate in the low temperature range (below 60°C in our devices); whereas, at sufficiently high temperatures (above 60°C) the diffusion component dominates.

Figures 2 and 4 show the zero-temperature-coefficient (ZTC) bias points for n-channel and p-channel MOSFETs, respectively. Each device behaves the same at various temperatures when the gate

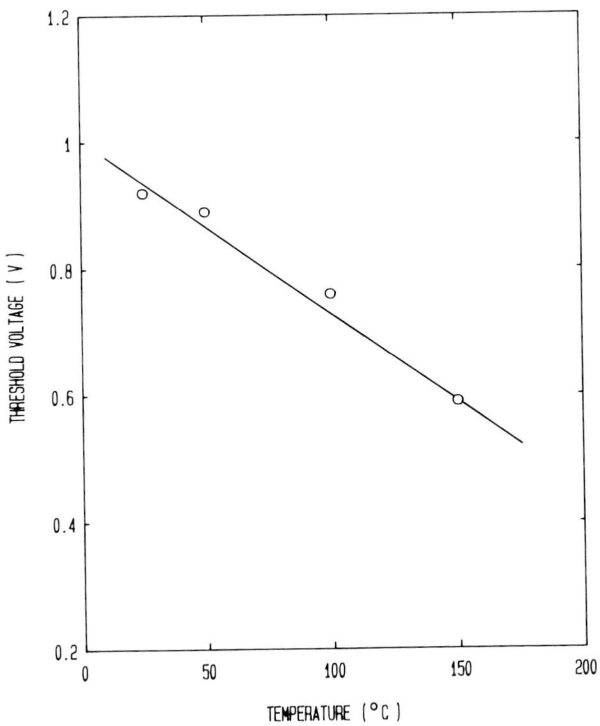

Figure 3 *Threshold Voltage versus Temperature Relationship for an n-Channel MOSFET (W/L = 23/3).*

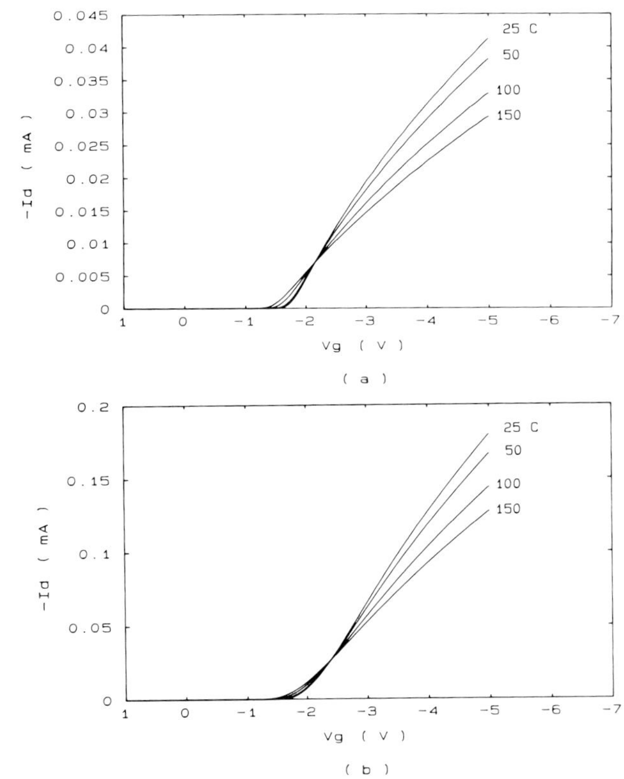

Figure 4 *$I_d - V_g$ Curves for a p-Channel MOSFET (W/L = 16/4) at Various Temperatures: (a) $V_{ds} = -0.2$ V and (b) $V_{ds} = 1.0$ V.*

is biased at these points. (In other words, temperature has no effect on the $I_d - V_g$ characteristics under the ZTC conditions.) These points are at $V_g = 1.4$ V and $I_d = 0.014$ mA, and at $V_g = -2.15$ V, and $I_d = 0.0067$ mA for the n-channel and p-channel devices, respectively.

When the temperature was raised to 175°C, both p-channel and n-channel devices lost their normal characteristics as shown in Figures 7a and 7b. The $I_d - V_g$ characteristics at low values of V_d indicate the oxide breakdown.

Gamma-Ray Radiation Effects on the Characteristics of MOSFETs

Two sets of the chips were irradiated by the Co-60 gamma-ray source to a total dose of 9.42×10^6 rads(Si) and to 7.33×10^5 rads(Si). The $I_d - V_g$ characteristics of each MOSFET in the chips was studied before and after the chips were irradiated. There is an obvious shift in the turn-on characteristics when the devices are subjected to the radiation, as can be seen in Figure 8.

Annealing makes the devices ''rebound'' to their initial settings or even bounce back beyond their original settings (Winokur et al. 1984). One of the heavily-irradiated chips (9.42×10^6 rads(Si)) was placed in a circulation oven at 125°C for 40 hours. $I_d - V_g$ data were taken after this annealing. Figures 8a and 8b reveal this evident ''rebound'' effect. For the n-channel MOSFETs (Figure 8a), there is a shift in threshold voltage (+1.0 V, from 0.9 V to 1.9 V) for the irradiated devices; the threshold voltage shifts back to 0.75 V when the chip is annealed at 125°C for 40 hours. Figure 8b shows the result for the p-channel MOSFETs. Radiation has a greater effect on the turn-on characteristics for the p-channel MOSFETs than for the n-channel MOSFETs. The threshold voltage decreases from −2.6 V to −5.9 V when irradiated, but it bounces back to −2.35 V when annealed. Radiation has a greater effect on the threshold voltage shift for the p-channel devices, while annealing has a greater effect on the recovery of the threshold voltage for the n-channel devices. Leakage current had obvious increases when the devices were irradiated by gamma-rays.

One of the heavily-irradiated chips stayed at the room temperature (25°C) for 28 days. The room temperature annealing effects on the $I_d - V_g$ behavior

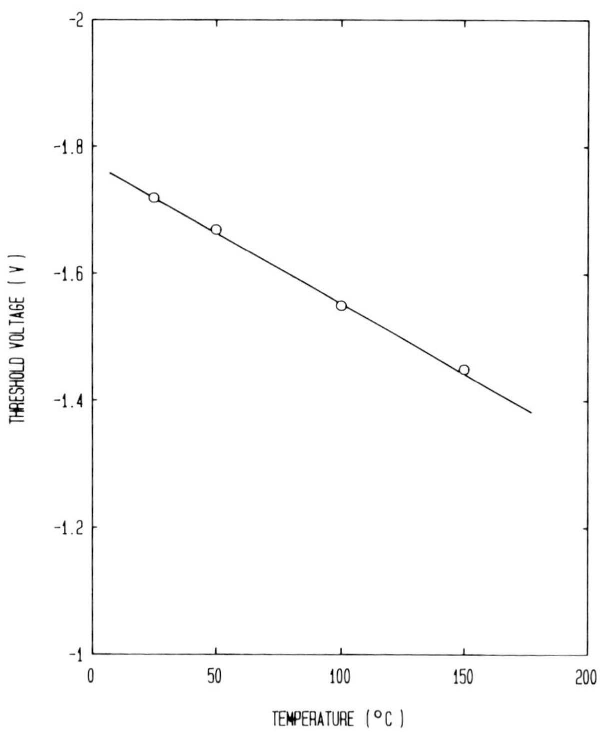

Figure 5 *Threshold Voltage versus Temperature Relationship for a p-Channel MOSFET (W/L = 16/4).*

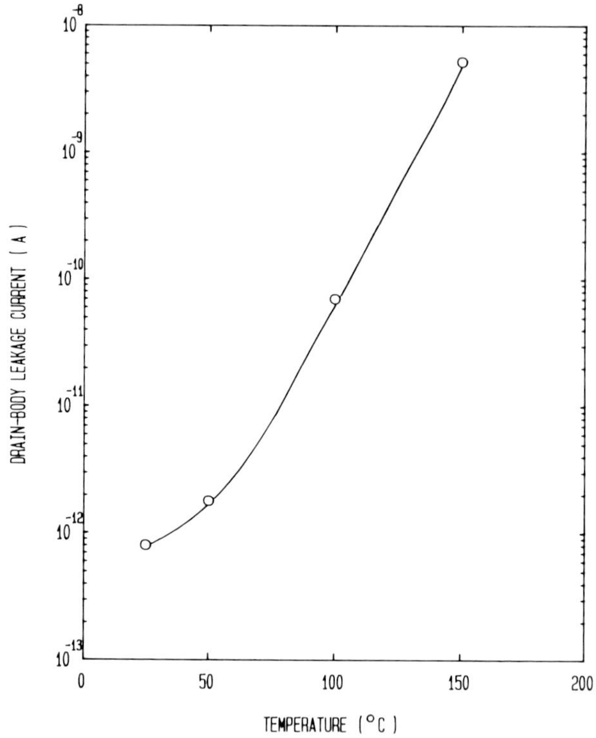

Figure 6 *Drain-body Leakage Current as a function of Temperature for an n-Channel MOSFET (W/L = 23/3).*

were observed. Figure 8c shows that there is little recovery for n-channel MOSFETs when they are subjected to long-term room temperature annealing. However, slight recovery is found for the p-channel MOSFETs (Figure 8d). Figures 8e and 8f present the results of a lightly-irradiated chip (7.33×10^5 rads(Si)). The $I_d - V_g$ characteristics vary proportionally with the gamma-ray dosage. Smaller dosage results in small changes, while heavier dosage has greater effects on the electronic behavior.

Kinetic Model of Recovery By Thermal Annealing

The increase in threshold voltage of p-channel MOSFETs, or the decrease in the case of n-channels, is caused by the buildup of positive charges at the Si-SiO_2 interface. High energy gamma-rays generate hot electron-hole pairs uniformly throughout the oxide layers. However, most are swept out of the oxide by diffusion and recombination processes. Only a fraction of these holes drift toward the Si-SiO_2 interface and are trapped at impurity centers or eventually trapped at the interfacial states. It is the buildup of these trapped holes that causes the change in threshold voltage and decrease of $I_d - V_g$ slope. Thus, the removal of these trapped holes will result in recovery of the MOSFET's original settings.

Using a p-channel MOSFET as an example, thermal annealing provides a mechanism for electrons in silicon side to diffuse into the interfacial region and to recombine with the trapped holes. The Si-SiO_2 interface is not atomically smooth, and the roughness of 0.4 to 0.8 nm has been observed by the use of high-resolution transmission electron microscopy (Su et al. 1985). The topographical variation of the interface implies that the structure of Si atoms near Si-SiO_2 interface is amorphous; hence, there is a wide range of potential barriers associated with electron diffusion from Si side to the oxide.

The thermally assisted tunneling process (Schwank et al. 1984) modeled earlier assumes that the interface is uniform. In this case, the unannealed fraction of trapped holes decreases exponentially in time, as will be the case of classical diffusion. This is in contradiction to our results and those of Danchenko et al. (1980). Our model does not use the thermally assisted tunneling concept and permits the Si-SiO_2 interface to be rough. Therefore, our model is distinctly different from previously classical diffusion models. In the

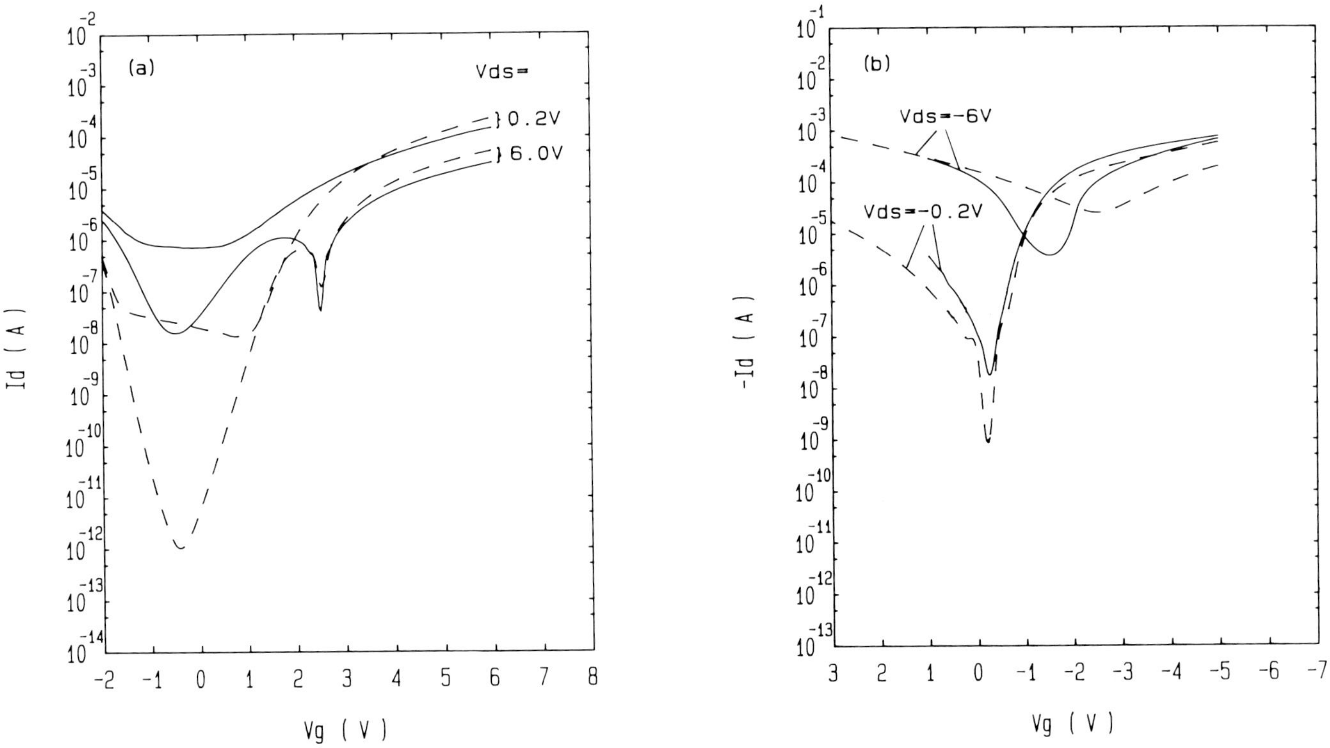

Figure 7 *$I_d - V_g$ Characteristics for (a) an n-Channel MOSFET (W/L = 23/3) and (b) a p-Channel MOSFET (W/L = 23/3) at 175°C (solid) and at 25°C after 175°C Heating (dashed).*

case of uniform interface, an electron at the bottom of silicon conduction band, E_c, will need an activation energy E to move into the oxide region. However, electrons may be at an energy lower than E_c because interface states exist between E_c and the top of the valence band, E_v. Thus, activation energy E and interfacial energies form a set of energies, or an energy ladder. At a given temperature, an electron will gain or lose energy randomly and, thus, move randomly along the energy ladder. This random walk is characterized by a waiting time distribution function, $P(t)$; the probability at time t for an electron to move into the nearest energy levels. Thus the first passage time, the probability that an electron starting at an energy level in the energy ladder will reach the top of the ladder for the first time, determines the rate of recombination with trapped holes in the oxide.

If the interface is uniform, the probability $P(t)$ at time t for an electron to move along the energy ladder will be associated with a constant $\lambda(T)$ for a given temperature T such that:

$$P(t) = \lambda e^{-\lambda t} \qquad (2)$$

$P(t)$ given by Equation (2) is a result of the fact that there is only one type of potential barrier and one set of energy states or one type of energy ladder. However, since the Si structure is amorphous near the interface, there will be many different types of energy ladders associated with the random character of the interface. This results in having a long tail of the waiting-time distribution function for an electron to move along the energy ladder. This is an analog of a diffusion-controlled chemical reaction (Bordewijh 1975). The theory of such diffusion-controlled reactions for a continuous-time random walk, has been given by Helman and Funabashi (1977), Hamill and Funabashi (1977) and Shlesinger and Montroll (1984). Such a long-tail behavior is characterized by the probability function $P(t)$ having an inverse-power tail and is given by:

$$P(t) = \frac{\alpha A t^{-1-\alpha}}{\Gamma(1-\alpha)} \quad (0 < \alpha < 1) \, , \qquad (3)$$

where A is a constant. $P(t)$ is such that, while being properly normalized, the average time of diffusion, $\langle t \rangle$, becomes infinite. Equation (3) was first used in the theory of charge transport in amorphous materials (Scher and Montroll 1975).

The survival probability of trapped holes, $\phi(t)$, at the thermal annealing time t is related to the first passage time $F(E,t)$ by:

$$\phi(t) = \phi(0)\ \exp(-c\int_0^t I(t)dt), \tag{4}$$

where I(t) is related to F(E,t) by:

$$I(t) = \Sigma F(E,t) \text{ for } E \neq 0. \tag{5}$$

$F(E,t)$ is the probability that an electron, starting originally at energy E (negative) of the energy states, will move to the top of the potential barrier and into the oxide side for the first time. The structure constant c, is related to energy states. Here, we assume that most of the recombination time is consumed by the electron to overcome that potential barrier at Si-SiO_2 interface.

The first passage time $F(E,t)$ has been calculated by Montroll (1964) and is related to $P(t)$. The survival probability, ϕ(t) can be written as (Montroll and Weiss 1965):

$$\phi(t) = \phi(0)\ \exp\ (-(bt)^{\alpha}) \tag{6}$$

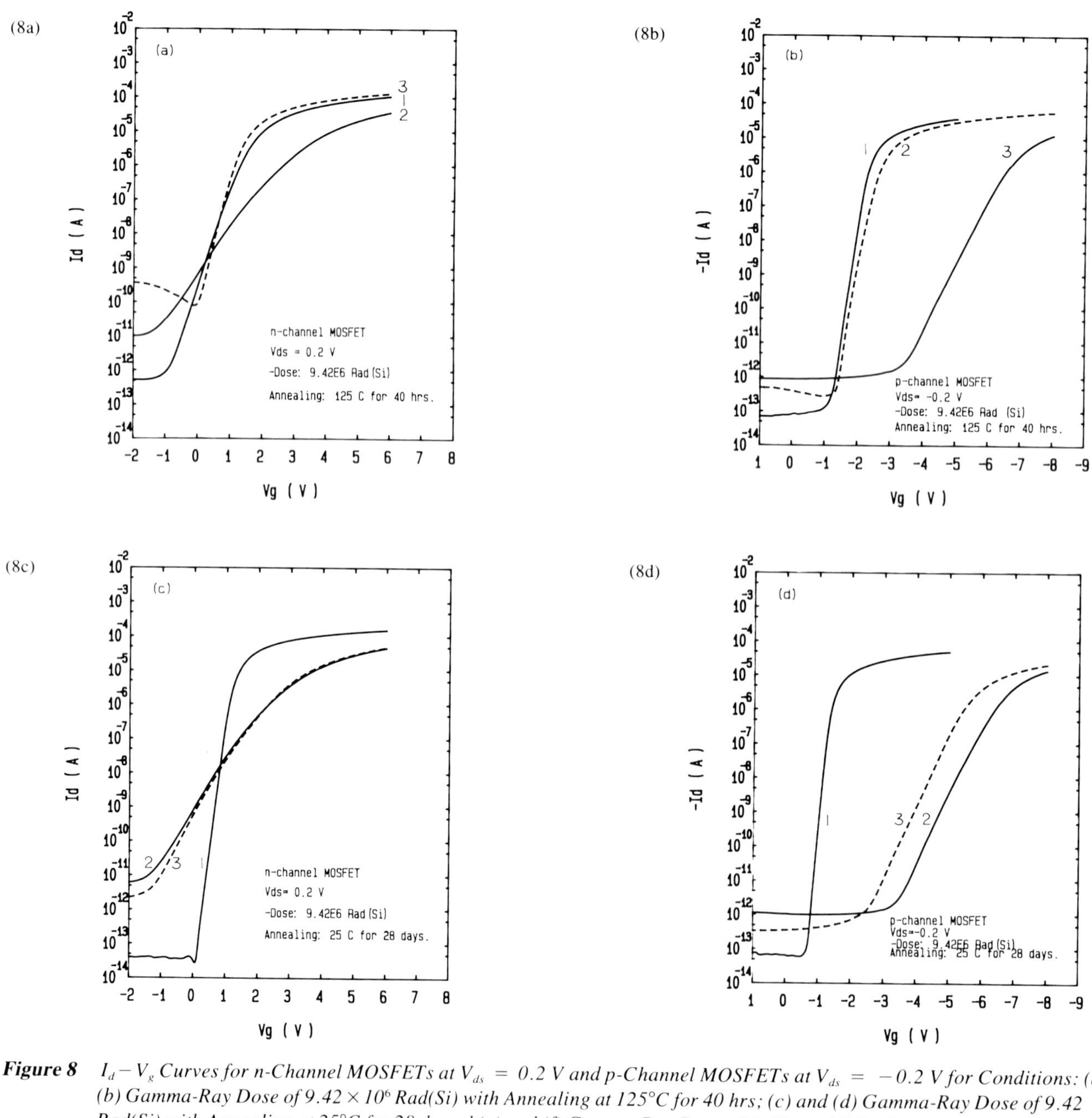

Figure 8 *$I_d - V_g$ Curves for n-Channel MOSFETs at V_{ds} = 0.2 V and p-Channel MOSFETs at V_{ds} = −0.2 V for Conditions: (a) and (b) Gamma-Ray Dose of 9.42×10^6 Rad(Si) with Annealing at 125°C for 40 hrs; (c) and (d) Gamma-Ray Dose of 9.42×10^6 Rad(Si) with Annealing at 25°C for 28 d; and (e) and (f) Gamma-Ray Dose of 7.33×10^5 Rads(Si) with Annealing at 125°C for 48 hrs. (1) Preirradiation, (2) Irradiated, and (3) Annealed for Irradiation.*

(8e)

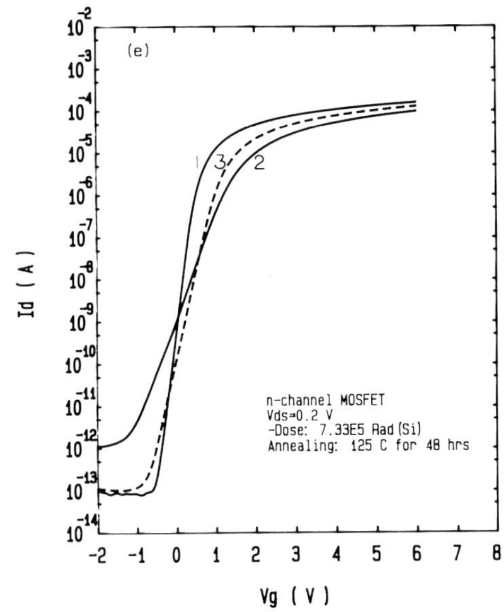

(8f)

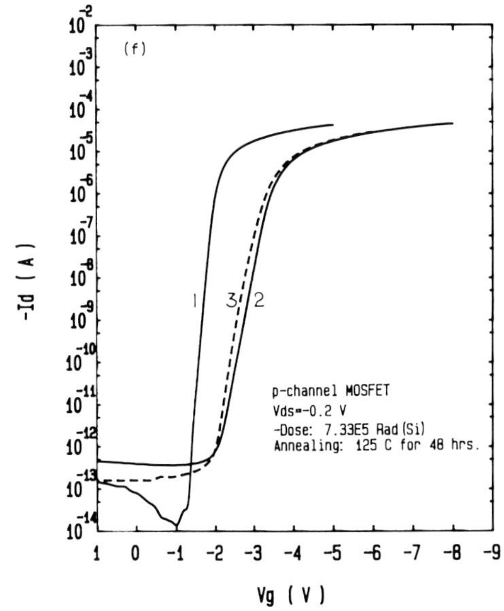

Figure 8 *$I_d - V_g$ Curves for n-Channel MOSFETs at V_{ds} = 0.2 V and p-Channel MOSFETs at V_{ds} = −0.2 V for Conditions: (a) and (b) Gamma-Ray Dose of 9.42 × 10^6 Rad(Si) with Annealing at 125°C for 40 hrs; (c) and (d) Gamma-Ray Dose of 9.42 × 10^6 Rad(Si) with Annealing at 25°C for 28 d; and (e) and (f) Gamma-Ray Dose of 7. 33 × 10^5 Rads(Si) with Annealing at 125°C for 48 hrs. (1) Preirradiation, (2) Irradiated, and (3) Annealed for Irradiation.*

if $P(t)$ of Equation (3) is used. The constant b is related to the structure constant c, coefficient A from Equation (3) and the initial energy of the electron. Note that if $P(t)$ of Equation (2) is used, the survival probability reduces to classical diffusion (α = 1 in Equation (6)). In the α = 1 case, the exponential decay indicates a faster recovery than our experimental data. Equation (6) is the William-Watts function (1970), originated from the study of polarization decay in disordered and polymeric materials.

ϕ(t) of Equation (6) is directly related to the unannealed fraction of trapped holes, $N(t)$. We fit Equation (6) with our experimental results and those of Danchenko et al. (1980). An excellent agreement was found with α = 0.21 and b = 1/(62 hrs) (Figure 9) for our own experiments and α = 0.45 and b = 1/(240 hrs) for Danchenko's experiments (Figure 10). The continuous lines in Figures 9 and 10 represent our model. The experimental data are presented by scatter graph. Generally, constant b depends on temperature (through coefficient A of Equation (3)). At room temperature, b is very small, which reflects a very slow recover. The discrepancy between our b values and that of Danchenko is due to the differences in MOSFET geometry.

Conclusion

The above kinetics model agrees very well with the thermal annealing of the p-channel MOSFET. However, we expect that in the n-channel case the situation will be quite different (Danchenko et al. 1980) because there are very few electrons available to recombine with trapped holes. The diffusion time of electrons on the silicon side, as well as that at oxide layer after overcoming the interface barrier, will not be negligible. Hence, the model presented here is suitable for p-channel MOSFETs only.

Acknowledgments

The authors would like to thank Dr. J.E. Gover for the valuable discussions and Sandia National laboratories for the supply of test chips. We are also grateful to Ms. Vicki Spate of University of Missouri Research Reactor for her assistance in performing the gamma-ray irradiations.

References

Bordewijk, P. (1975) "Defect-Diffusion Models of Dielectric Relaxation," *Chem. Phys. Lett.*, 32:592–596.

Danchenko, V., E.G. Stassinopoulos, P.H. Fang, and S. Brashears (1980) "Activation Energies of Thermal Annealing of Radiation-Induced Damage in n- and p-Channels of CMOS Intetrated Circuits," *IEEE Trans. Nucl. Sci.*, NS-27:1658–1664.

Gover, J. E. and J. R. Srour (1984) *Basic Radiation Effects in Nuclear Power Electronics Technology*, Sandia National Laboratories, Albuquerque, NM.

Hamill, W.H. and K. Funabashi (1977) "Kinetics of Electron Trapping Reactions in Amorphous Solids; A non-Gaussion Diffusion Model," *Phys. Rev.*, B16:5523–5527.

Helman, W.P. and K. Funabashi (1977) "Continuous-Time-Random-Walk in Three Dimensions as a Model for Electron Scav-

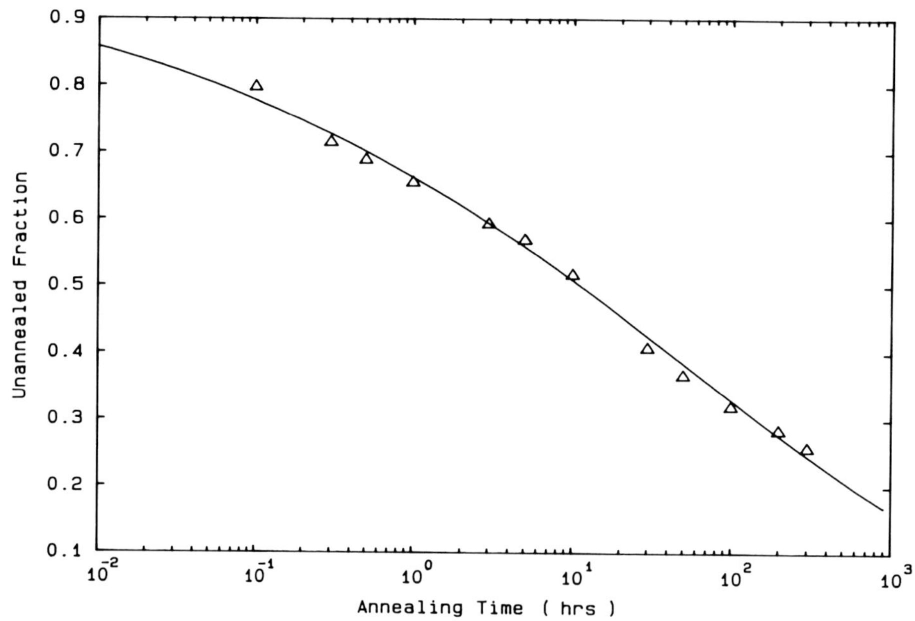

Figure 9 *Unannealed Fraction Changes during the Prolonged Thermal Annealing. p-channel MOSFET (W/L = 23/3). Triangles are the Experimental Data and Solid Curve is Our Model for* $\alpha = 0.21$, $b = 1/(62\ \text{hrs})$.

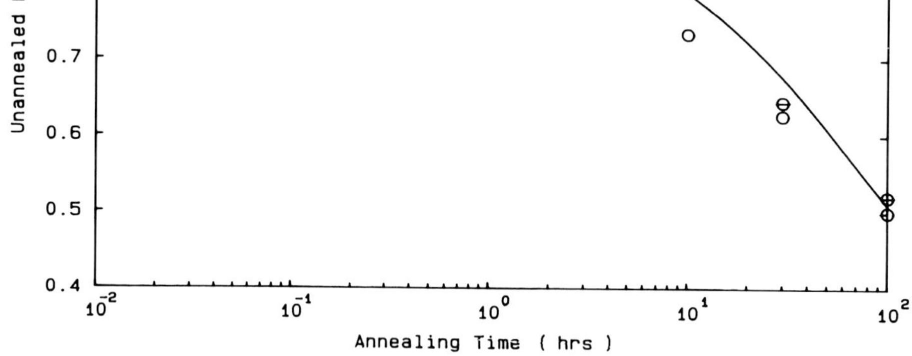

Figure 10 *Unannealed Fraction as a function of Thermal Annealing Time (hours). Data Points are from Danchenko (1980). Solid Curve represents Equation (6). p-Channel MOSFET.* $\alpha = 0.45$, $b = 1/(240\ \text{hrs})$.

enging and Ion Recombination in Liquids,'' *J. Chem. Phys.*, 66:5790–5797.

Montroll, E.W. (1964) ''Random Walks on Lattices,'' *Proc. Symp. Appl. Math.*, Am. Math. Soc., 16:193–198.

Montroll, E.W. and G. Weiss (1965) ''Random Walks on Lattices, II,'' *J. Math. Phys.*, 6:167–181.

Neamen, D.A. (1984) ''Modeling of MOS Radiation and Post Irradiation Effects,'' *IEEE Trans. Nucl. Sci.*, NS-31:1439–1443.

Scher, H. and E.W. Montroll (1975) ''Anomalous Transit-Time Dispersion in Amorphous Solids,'' *Phys. Rev.*, B12:2455–2477.

Schwank, J.R., P.S. Winokur, P.J. McWhorter, F.W. Sexton, P.V. Dressendrofer, and D.C. Turpin (1984) ''Physical Mechanisms Contributing to Device 'Rebound' ,'' *IEEE Trans. Nucl. Sci.*, NS-31:1434–1438.

Shlesinger, M. and E.W. Montroll (1984) ''On the Williams-Watts Function of Dielectric Relaxation,'' *Proc. National Acad. Sci.*, 81:1280–1283.

Shoucair, F., W. Hwang, and P. Pain (1984) ''Electrical Characteristics of Large Scale Intergation (LSI) MOSFETs at Very High Temperatures, Part I: Theory,'' *Microelectronics Reliability*, 24:465–485.

Su, H.Q., C.C. Wei, and T. P. Ma (1985) ''Mobility Degradation in Very Thin Oxide p-Channel MOSFETs,'' *IEEE Trans. Electron. Dev.*, ED-32:559–561.

Williams, G. and D.C. Watts (1970) ''Non-Symmetrical Dielectric Relaxation Behavior Arising from a Simple Empirical Decay Function,'' *Trans. Faraday Soc.*, 66:80–85.

Wilson, C.L. and J.L. Blue (1984) ''Two-Dimensional Modeling of n-Channel MOSFETs Including Radiation-Induced Interface and Oxide Charge,'' *IEEE Trans. Nucl. Sci.*, NS-31:1448–1452.

Winokur, P.S., J.R. Schwank, P.J. McWhorter, P.V. Dressendrofer, and D.C. Turpin (1984) ''Correlating the Radiation Response of MOS Capacitors and Transistors,'' *IEEE Trans. Nucl. Sci.*, NS-31:1453–1460.

Space Nuclear Power Systems 1986
Edited by M. S. El-Genk and M. D. Hoover

Chapter 32

Gigarad-Tolerant Power Switches and Memory Elements

W. T. Joyner, G. F. Becknell, and J. M. A. Donelson

Department of Physics
Hampden-Sydney College
Hampden-Sydney, VA 23943

A new class of silicon devices, reported by Lampert and Mark (1970), Henderson, Asbrock, and Kapoor (1978), Kapoor and Henderson (1980), Tuntasood (1982), and Joyner (1984), operating without p-n junctions has been studied for radiation hardness. The electric field in these devices is uniform over most of the device, thus causing high-voltage breakdown to increase with electrode spacing. Deep acceptor levels are created in the silicon by diffusing gold atoms into the lattice, or by electron radiation. These acceptor levels, near the center of the band gap, trap out electrons so that the low resistivity of the n-type doped wafer can be raised to intrinsic resistivity. Switching between a high-resistivity state at low currents and a low-resistivity state at high currents occurs at a definite threshold voltage, exhibiting characteristics similar to silicon-controlled rectifiers. Compensating 0.1 ohm-meter (10 ohm-cm) silicon wafers requires electron fluxes of 10^{23} electrons per square meter (10^{19} electrons per square centimeter). Experiments demonstrating gamma tolerance to 1 Gigarad(Si) are described. Calculations for maximum tolerable neutron fluence are shown, and a phenomenological explanation for these results is presented.

Introduction

Double Injection in Silicon

Double-injection diodes have been prepared by making diffused planar contacts, one p+ and one n+, onto n-type silicon, described by Whitson (1985). These electrodes (p+ and n+) allow easy injection of holes and electrons, respectively. Deep traps are then produced throughout the base silicon wafer by diffusing in an impurity (Au in this case) or by irradiating the wafer with 2-Mev electrons. The resulting device is referred to as a double-injection deep impurity or $(DI)^2$ device. As the concentration of deep traps is increased, the resistivity increases due to trapping of majority carriers. With care, near-intrinsic resistivity has been attained with starting material as low as 0.1 ohm-meter (10 ohm-cm).

The behavior of the resulting diode can be described by considering three regions. At low injection levels, it appears to be ohmic with a relatively high resistance. At intermediate injection levels, it exhibits a space charge-limited current due to the increased current density. In both of these regions, conduction is by majority carrier only. The holes injected at the p+ electrode are trapped by the deep levels in a very short distance, so that electron injection is responsible for the current flow. (It is fundamental to the properties of these devices that the cross section for hole capture by a deep trap is considerably greater than the cross section for electron capture as reported by Holonyak et al. (1962).) Finally, as the injection level continues to increase, the holes begin to diffuse into the silicon. They can now recombine with the injected electrons. The device then switches into a low-resistance, high-conduction states. Gated devices have also been studied, including both MOS gates by Tuntasood (1982) and injection gates by Kapoor and Henderson (1980). In general switching from the off state to the on state is easy with either type of gate. Switching from the on state to the off state is also possible provided current density is not too large. Switching times are in the nanosecond region for small low-power devices, and below 1 microsecond for large thousand-volt, 10-ampere devices as reported by Joyner (1984).

Radiation Effects

Since radiation damage is essentially caused by increasing numbers of donor-acceptor sites created by the radiation, the deep impurity traps should inhibit the onset of radiation effects just as they do for temperature effects. In Figure 1, the resistivity is plotted as a function of total electron fluence for several wafers that were compensated by electron irradiation, described by Whitson (1985), rather than Au diffusion. The increasing radiation dose required to compensate the majority carriers for silicon of increasing carrier concentration (lower initial resistivity) is clearly evident. The measurements extend down to 0.1 ohm-meter (10 ohm-cm) material. Extrapolation to 0.01 ohm-meter (1 ohm-cm) is shown by the curve nearest the horizontal axis. An electron dose of about $10^{23}/m^2$ ($10^{19}/cm^2$) will be required to compensate the 0.01 ohm-meter (1 ohm-cm) silicon wafer. The neutron dose required to produce the same effect on silicon can be approximated by using published data on radiation effects on silicon transistors, using the excellent review by Vavilov (1977). This yields a conversion factor of about 100; that is, 100 electrons at 2 Mev produce the same damage as one fission spectrum neutron. An alternative approach is to use NASA (1982) data on solar cell damage; a conversion factor of about 700 results from this approach. In view of the uncertainty of these data (Brucker et al. 1982 and Nichols et al. 1982), a conversion factor of 1000 is used for the numbers shown in Figure 1. This indicates that double-injection, deep-impurity devices prepared on silicon wafers with 0.01 ohm-meter (1 ohm-cm) resistivity might tolerate radiation doses up to 10^{20} fission spectrum neutrons/m^2 ($10^{16}/cm^2$), or ^{60}Co gamma ray doses above 1 Gigarad (Si).

Temperature Effects

The concentration of carriers as a function of temperature for intrinsic silicon is shown in Figure 2. Since these devices were initially doped to a carrier concentration of $10^{20}/m^3$ ($10^{14}/cm^3$), intrinsic carrier concentration begins to exceed the doping density at about 470 K (Figure 3). This would indicate device failure at higher temperatures. Choosing lower initial resistivity material should allow operation at higher temperatures. Results at the University of Cincinnati (Henderson 1985) indicate the Au wafer diffusion technique cannot revert silicon with initial resistivity lower than 1 ohm-cm to silicon with intrinsic resistivity. Using 1 ohm-cm as the lowest resistivity suitable, extrapolation indicates that operation up to 720 K should be possible with this material.

Methods

Two-lead devices prepared at the Westinghouse Research Laboratories, reported by Whitson (1985), were used in this study. The device geometry is shown in Figure 4. These diodes are planar. (Sandwich devices with electrodes on the front and back of wafers of varying thickness have also been studied.) Figure 5 shows the I-V characteristics of these devices. The

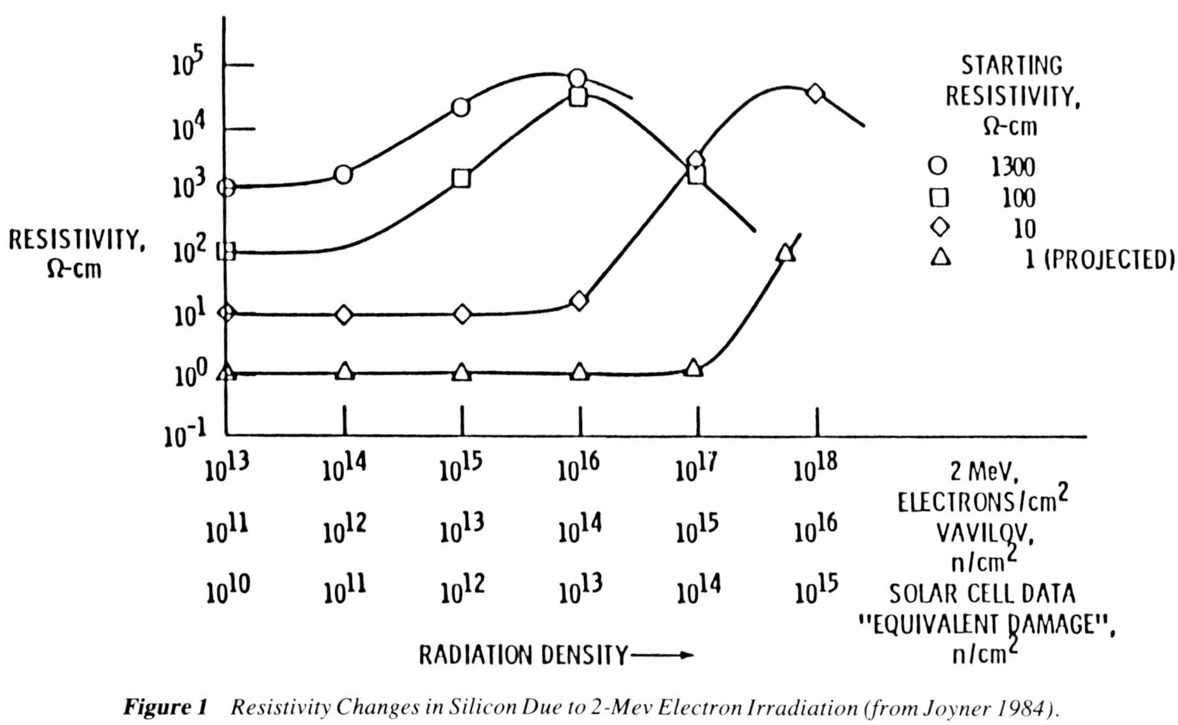

Figure 1 *Resistivity Changes in Silicon Due to 2-Mev Electron Irradiation (from Joyner 1984).*

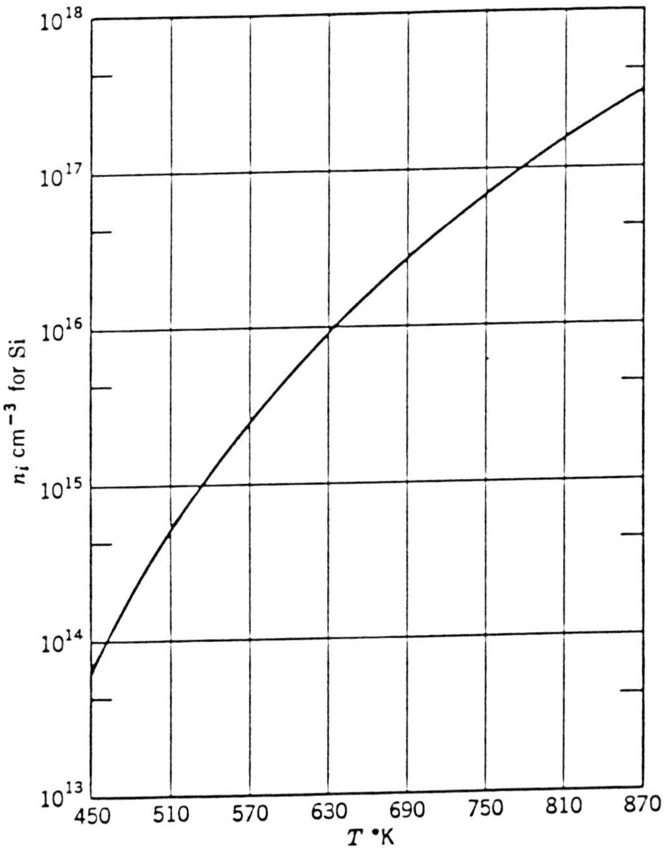

Figure 2 Carrier Concentration as a function of Temperature for Intrinsic Silicon (from Adler et al. 1964).

impedance in the off state is seen to be a few thousand ohms. The impedance in the on state is a few ohms. Sixty-four were prepared on one 2-inch wafer. Device properties are remarkably uniform. All 64 are within 20% of the parameters shown in Figure 5. The initial resistivity of the silicon wafer was 0.65 ohm-meter (65 ohm-cm). Au diffusion was continued until all majority carriers were trapped out, producing near-intrinsic resistivity 4000 ohm-meter (400,000 ohm-cm) at room temperature. If the diffusion is continued past this point, the material becomes p-type and the resistivity begins to fall again. Since the resistivity is changing so rapidly in this region, it is difficult to control the trap density exactly. Electron irradiation offers a more controllable means of inserting the deep traps. Unfortunately, the resulting device properties are not as desirable as those produced by Au diffusion.

With increased injection, the traps fill, and the device switches to the high conduction or on state. As injection is increased above the threshold level, the time required to fill the traps decreases, and the delay before switching decreases as shown in Figure 6.

The threshold voltage and the holding voltage were also determined using a pulsed-power supply, using the circuit of Figure 7. The resulting oscilloscope time is included in that figure.

The effects of gamma irradiation were studied by exposing 12 pairs of devices from the common wafer to the National Bureau of Standards ^{60}Co gamma cell. For each device, measurements on the DC current-voltage characteristic, the pulsed threshold and holding voltages, and the time delay before firing were taken before and after irradiation. All devices were found to be stable both before and after irradiation for periods of several months. The effect of temperature was also measured up to 700 K. Current measurements in the on state were carried up to 100 amperes with pulse widths up to a few tenths of seconds.

Results

Figures 8 through 10 show the effects of gamma irradiation up to a maximum of 800 Megarad(Si). In Figure 8, the holding voltage is seen to vary slightly at levels below 1 Megarad(Si), with little change for higher radiation levels. The delay time before firing is affected in roughly the same fashion, as shown in Figure 9. In Figure 10, the current-voltage characteristic is plotted before and after an irradiation of 800 Megarad(Si); the threshold and the holding voltages are relatively unaffected, changing by about 20%. A generalization of the data indicates that the device properties are not affected by gamma irradiation up to 800 Megarad(Si), to plus or minus 20%.

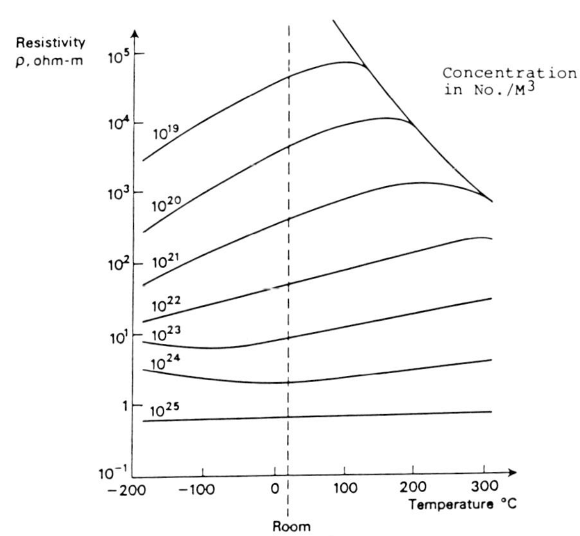

Figure 3 Carrier Concentration as a function of Temperature for Several Levels of Doping (from Seymour 1981).

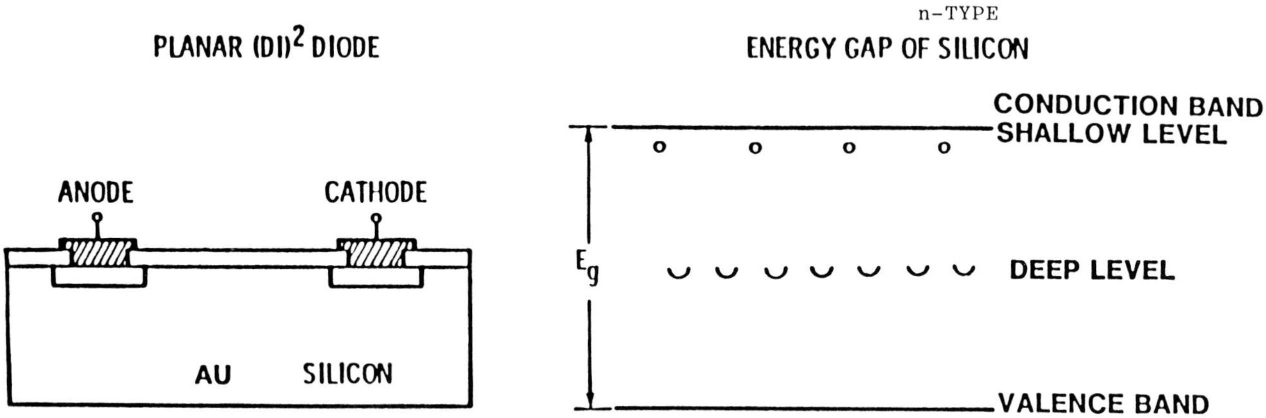

Figure 4 *Typical Deep-Impurity, Double-Injection Device (from Sundberg 1986).*

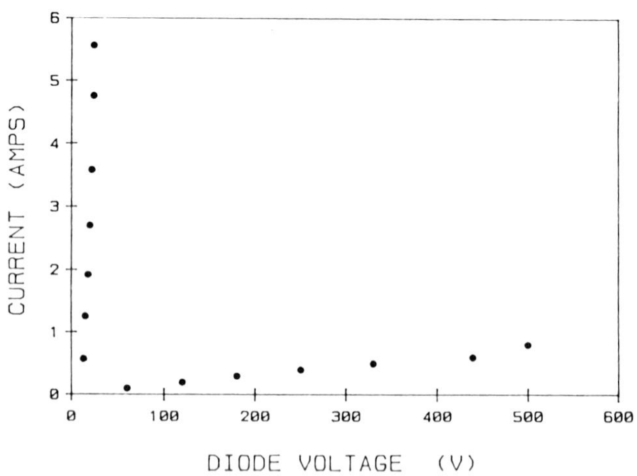

Figure 5 *Current-Voltage Characteristic for a (DI)² Gold-Activated Silicon Device with a Radius of 1450 Micrometers (57 mils) (from Joyner 1984).*

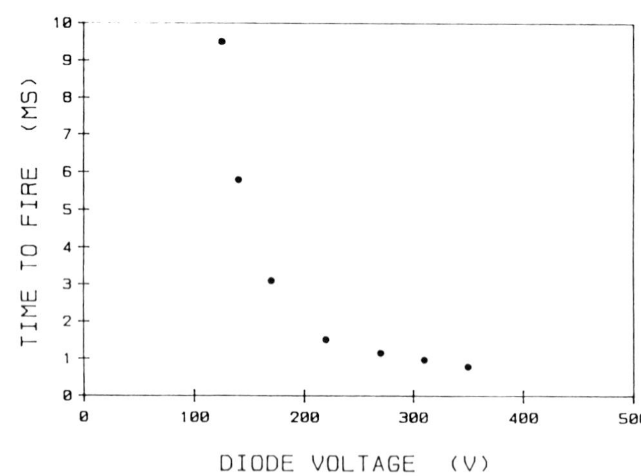

Figure 6 *Delay between Application of Power and Device Turn-On as a function of Applied Voltage (The Delay is Due to the Time Required to Fill the Deep Traps.) (from Joyner 1984).*

Current measurements in the on state were extended up to the 100-ampere level to study changes in the dynamic impedance. Figure 11 shows that the dynamic impedance decreases below 1 ohm for such large currents. For a practical device, the holding voltage must be lowered to a few volts. The dynamic impedance of this particular device is already inadequate for many applications, even though the radius of the device is only 1450 micrometers (57 mils) and it is a planar, or one-sided, device. In Figure 12, the leakage current for one device is plotted as a function of temperature. There is little change up to 620 K, and the device should be usable at lower temperatures. Thus, the power handling capability of a particular sized device is increased considerably beyond that of a thyristor or silicon-controlled rectifier because of the much larger temperature rise that can be tolerated.

These particular devices will tolerate DC power up to 30 watts (limited by the particular power supply used). Using the area occupied by the planar device, the power density is found to be more than 4 million W/m^2 (400 W/cm^2), a relatively large value for power semiconductor switches of any type. This is obviously not the maximum limit on power dissipation.

Discussion

The gold-activated silicon devices do indeed exhibit the predicted radiation tolerance. At the maximum dose of 800 Megarad(Si), little change is evident for any of the measured parameters; and, even more importantly, the slope of all parameters is horizontal. There is no indication of failure due to the gamma irradiation up to that level. Operation well into the Gigarad(Si) region seems probable.

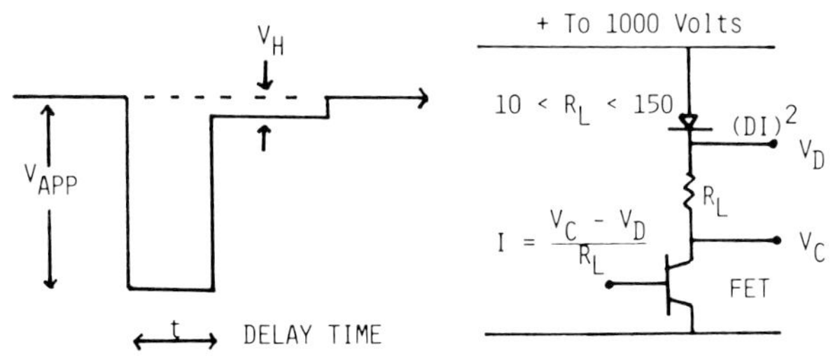

Figure 7 *Applied Voltage and Holding Voltage as Displayed on the Oscilloscope, Using the Circuit Shown (from Joyner 1984).*

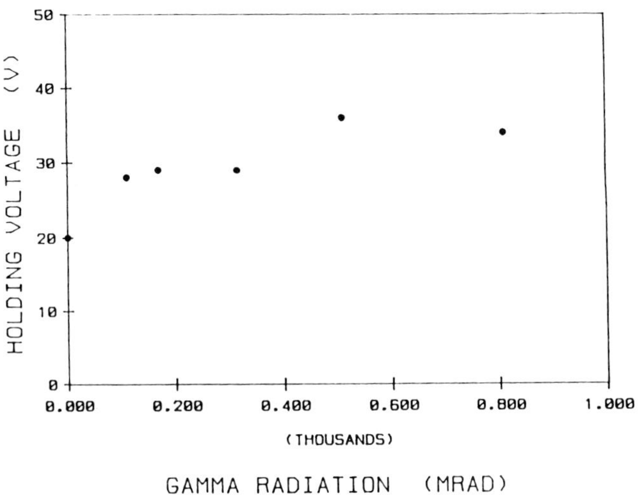

Figure 8 *The Effect of ^{60}Co Gamma Irradiation on the Holding Voltage of a (DI)^2 Silicon Device (from Joyner 1984).*

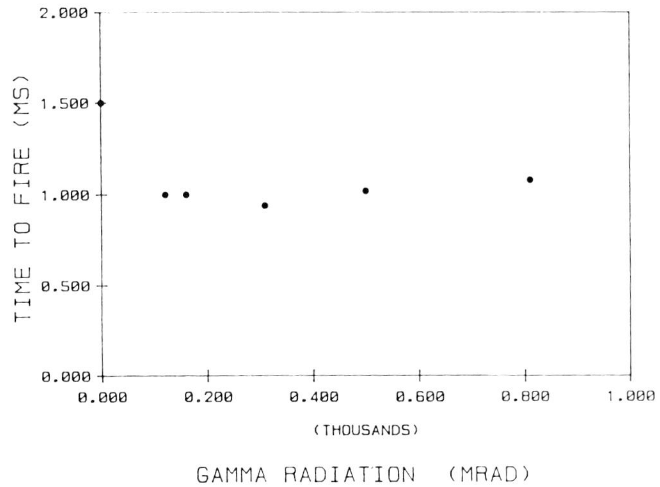

Figure 9 *The Effect of ^{60}Co Gamma Irradiation on the Time Delay between Application of Voltage and Turn-On (from Joyner 1984).*

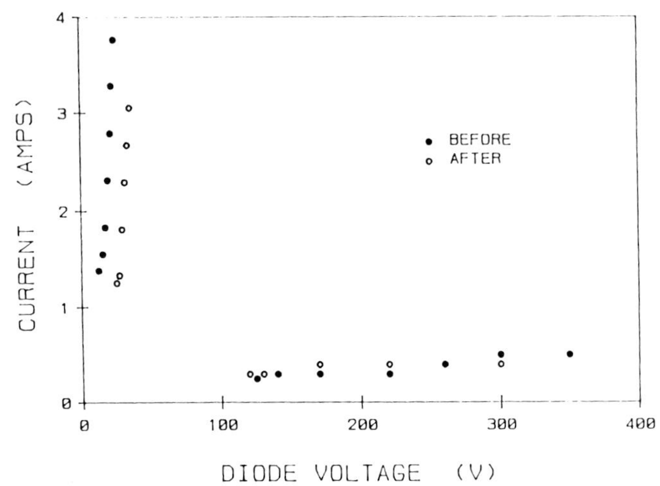

Figure 10 *Current-Voltage Characteristics of a (DI)^2 Silicon Device before and after ^{60}Co Gamma Irradiation at 800 Megarad (Si) (from Joyner 1984).*

The effect of temperature on leakage current, holding and threshold voltages, and delay time shows that the gold-activated devices are capable of operating in environments up to 200 K above usual power semiconductor devices.

A simple phenomenological model might view both the radiation tolerance and high temperature operation as resulting from the very large number of traps deliberately inserted in the silicon. Damaging the device by creating additional traps with radiation requires several Gigarads(Si) of gamma irradiation to disturb the existing trap density significantly, and high temperature operation is allowed by the suppression of majority carriers by the same high trap density.

The use of a wafer with an initial resistivity of 0.01 ohm-meter (1 ohm-cm) should increase both the radiation tolerance and the maximum operating temperature (see Figure 1), provided enough gold can be introduced into the silicon to compensate all of the majority carriers and bring the silicon wafer back to a near-intrinsic resistivity.

Experiments to extend gamma irradiation to device breakdown and to measure the effects of neutron irradiation up to a total fluence of 10^{20} neutrons per square meter (10^{16} neutrons per square centimeter) are planned for calendar year 1986. Figure 13 gives an overview of the (DI)^2 device parameters as presently understood.

Conclusions

Silicon devices utilizing deep impurity levels are radiation-tolerant beyond 800 Megarad(Si). Leakage

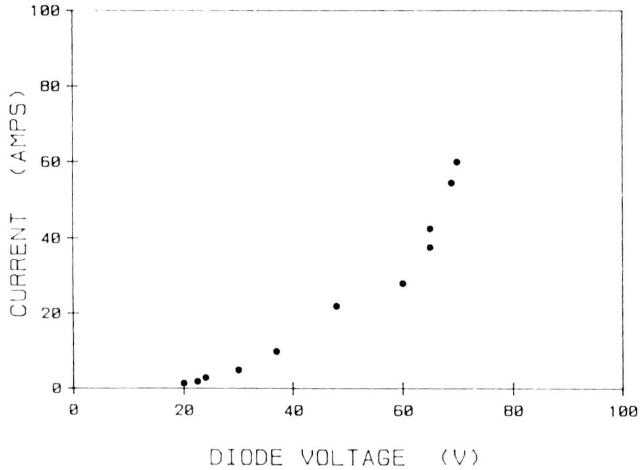

Figure 11 Current-Voltage Characteristics for a $(DI)^2$ Silicon Device Extended to the 100-Ampere Region (Note the increasing slope (decreasing dynamic resistance) at high currents.) (from Joyner 1984).

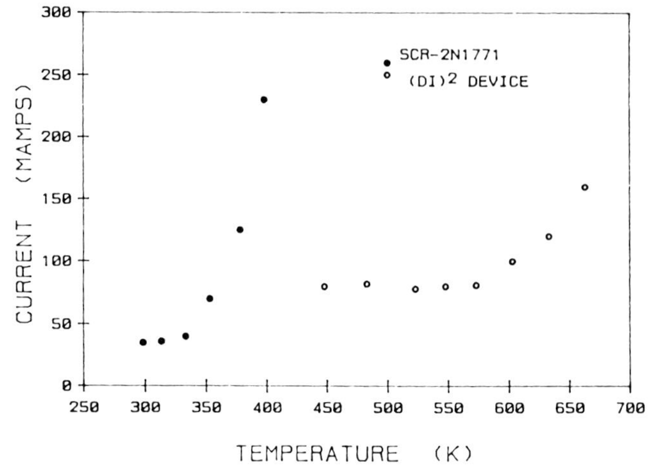

Figure 12 Leakage Current as a function of Temperature for a $(DI)^2$ Silicon Device. (For Comparison, Normalized Leakage Current for a Thyristor is also Shown.) (From Joyner 1984).

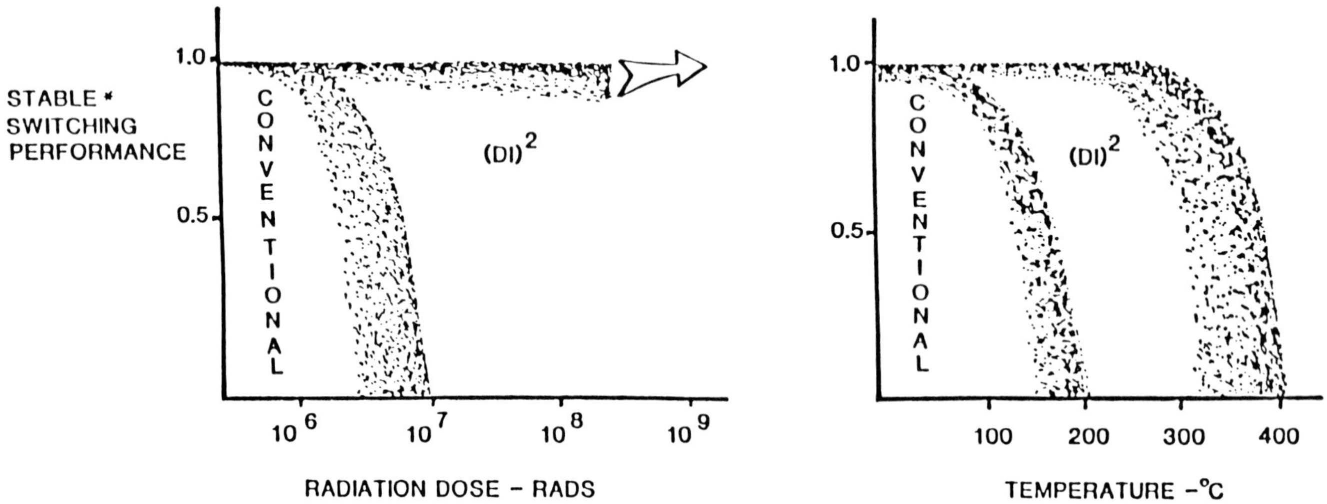

Figure 13 Overall Outline of Temperature and Radiation Capabilities of Present $(DI)^2$ Silicon Devices (from Sundberg 1986).

current is actually less after irradiation than before. The impedance in the on state increases slightly at 800 Megarad(Si), but the ratio of off-state impedance to on-state impedance is still greater than 100. Operation into the Gigarad(Si) region is probable. Devices with 1400 micrometer (50 mil) spacing have been operated in the 1000 volt 100 ampere region in pulse mode, and at temperatures up to 670 K.

Low-power devices switching at a few volts and a few hundred microamperes have been made by Henderson's group at Cincinnati, reported by Summe (1983 and 1985). Maintaining the same spacing of a few hundred micrometers and reducing the lateral dimensions should result in devices carrying a few microamperes and switching at a few volts, giving power dissipation levels in the microwatt region with Gigarad(Si) radiation tolerance.

Device properties seem best suited to high-power switching, extremely low-power memory units, and voltage-controlled delay units.

Acknowledgments

This research was performed at Hampden-Sydney College under NASA Grant 3-605. The support of Gale Sundberg and Ira Myers is gratefully acknowledged.

References

Adler et al. (1964) *Introduction to Semiconductor Physics*, John Wiley & Sons, New York. p. 43.

Brucker, G. J. et al. (1982) "The Damage Equivalence of Electrons, Protons, and Gamma Rays in MOS Devices," in *IEEE Transactions on Nuclear Science*, Vol. NS 29, No. 6, December 1982.

Henderson, H. T. (1984) *Personal Communication*, University of Cincinnati, Cincinnati, OH.

Henderson, H. T. et al. (1978) *Tech. Digest* IEDM, 680.

Holonyak, N., Jr. et al. (1962) "Double Injection with Negative Resistance in Semi-insulators," *Phys. Rev. Lett.*, *8*:426-428.

Joyner, W. T. 1984 "$(DI)^2$ Switching Studies at NASA-Lewis: Radiation Hardened Implications for Space Power," NASA Seminar on Novel Electronic Devices Using Deep Impurities in Semiconductors, NASA-Lewis Research Center, October 1984.

Kapoor, A. K. and H. T. Henderson (1980) *IEEE Trans. Elec. Dev.*, ED-28:1268.

Lampert, M. A. and P. Mark (1970) *Current Injection in Solids*, Ch. 14, Academic Press, New York.

NASA (1982) *Solar Cell Radiation Handbook*, Jet Propulsion Laboratory, Pasadena, CA. Prepared under contract with NASA, 1982.

Nichols, D.K., W. E. Price, and M. K. Gauthier (1962) "A Comparison of Radiation Damage in Transistors from Cobalt-60 Gamma Rays and 2.2 MeV Electrons," *IEEE Trans. on Nuclear Science*, NS-29, 6 December 1962.

Seymour J (1981) *Electronic Devices and Components*, John Wiley & Sons, New York.

Stassinopoulos, E. G. et al. (1983) "The Damage Equivalence of Electrons, Protons, Alphas and Gamma Rays in Rad-Hard MOS Devices, *IEEE Trans. on Nuclear Science*, NS-30, 6.

Summe, R. A. (1981) "A Voltage Controlled Oscillator Implemented with Double-Injection Diodes," M.S. Thesis, University of Cincinnati, Cincinnati, OH.

Sundberg, G. R. (1986) *Deep Impurity, Double Injection $(DI)^2$ Electronic Devices*, Power Technology Division, NASA Lewis Research Center, Cleveland, OH.

Tuntasood, P. (1982) *Development and Modeling of a New Family of Gated Double-Injection MOS Silicon Devices*, Ph.D. Dissertation, University of Cincinnati, Cincinnati, OH.

Vavilov, V. S. (1977) *Radiation Effects in Semiconductors and Semiconductor Devices*, 2nd ed., Plenum Press, New York.

Whitson, D. W. (1985) *Deep-Level, Double-Injection Voltage Switch Development Final Report*, NAS 3–23882, Westinghouse Research and Development Center, Pittsburgh, PA, 28 March 1985.

Part 11

TESTING

Space Nuclear Power Systems 1986
Edited by M. S. El-Genk and M. D. Hoover

Chapter 33

Recent BeO-Reflector-Controlled Reactor Experiments in ZPPR

Harold F. McFarlane, Stephen B. Brumbach, Stuart G. Carpenter, Peter J. Collins, and Richard D. McKnight

Argonne National Laboratory
P.O. Box 2528
Idaho Falls, ID 83403-2528

Integral reactor physics measurements were performed on a BeO-reflected fast reactor assembly in the ZPPR facility during January and February of 1985. The measurements emphasized power distributions and reflector control worths in two different critical states. The measurements have been analyzed using three-dimensional deterministic and Monte Carlo methods and the ENDF/B-V.2 nuclear data library. Together the measurements and analyses form a modern, reliable, benchmark data set for testing calculational methods that will be used in predicting some of the design parameters for future space reactors.

Introduction

Benchmark reactor physics experiments were conducted on a BeO-reflected, uranium-oxide-fuelled fast reactor assembly in the Zero Power Plutonium Reactor Facility (ZPPR) during January and February of 1985. These experiments, known as the ZPPR-14 Program, provide integral physics data on a reactor system with physics characteristics typical of some reactors recently proposed for space power applications. The measurements have been analyzed using the most powerful techniques presently available to fast reactor physicists. The recency of these experiments is significant in that measurement capabilities have evolved substantially in the intervening years since any previous relevant experiments were done. Only a few earlier experiments addressed the combination of a beryllium reflector and a fast reactor core (Stewart et al. 1958, Butler et al. 1966, and Callen et al. 1965).

The ZPPR facility (Lawroski et al. 1972), located at the Idaho National Engineering Laboratory (INEL) in eastern Idaho, is operated by the Applied Physics Division of Argonne National Laboratory (ANL). ZPPR has been used principally in the Department of Energy's (DOE) liquid metal fast breeder reactor (LMFBR) development program. After a loading period of several weeks, fast-neutron-spectrum reactors in ZPPR have their physics characteristics determined through a series of measurements that may include criticality, power distributions, control system worths, reactivity coefficients, reaction rate ratios, shielding effectiveness, neutron and gamma fluences, spectra, and breeding, among others. Generally, the reactors are built by loading "plates" of discrete materials into a ZPPR subassembly called a "drawer," and then arranging the drawers in a large stainless steel matrix. The facility is quite flexible, accommodating other loading arrangements, reactors of all sizes, and substantial variation in composition.

Part of the ZPPR matrix is shown in Figure 1 with one of the ZPPR-14 configurations loaded. Since ZPPR is of the split-table design, the very center of the reactor core is visible in the photograph. The ZPPR-14 assembly was based on a conceptual design for a small terrestrial liquid metal fast reactor. The design borrowed several features, including the reflector control concept, from the Systems for Nuclear Auxiliary Power (SNAP) space reactor program of the 1960s and 1970s. In the ZPPR-14 Program, the BeO reflector was divided into four segments that were moved axially and radially. In Figure 1, the reflector segments have been moved radially out away from the core in a simulated secondary scram condition. The photograph was taken during one step in a series of measurements to determine the reflector worth characteristics of the system.

The ZPPR-14 Program was quite brief. The intent was to obtain benchmark data for just the basic core parameters—critical loading, power distribution and

Figure 1 *The ZPPR-14B Assembly with BeO Reflectors in Secondary Scram Configuration.*

control worth. There was not time to measure other key parameters such as reactivity coefficients.

The SP-100 Program provides the immediate focus for space nuclear power. Comparisons between some reactor physics properties of a "typical" SP-100 design and ZPPR-14 are presented in Table 1 and Figure 2. The SP-100 data have been calculated for a 100-kWe reactor which is substantially smaller than ZPPR-14. However, with the recent increase in size of SP-100 from 100 kWe to 300 kWe, the SP-100 parameters should move closer to the ZPPR-14 values.

Considering first the neutron balance comparison in Table 1, the key item is the comparison of the radial leakage terms, which substantially govern the reflector control properties of each system. For consistent designs, the SP-100 will have less radial leakage in the 300-kWe size, making the comparison with ZPPR-14 reasonably favorable. The fission terms agree because they are both ^{235}U systems that have been normalized by the fission source. Capture in ZPPR-14 is substantially greater because the ^{238}U fraction is more than an order of magnitude greater. There is a large difference in axial leakage because

Table 1 *Comparison of ZPPR-14 and Typical SP-100 Neutron Balance.*

SP-100		ZPPR-14
	Production	
1.0000	Fission Source[a]	1.0000
0.0007	(n,2n)	0.0012
	Losses	
0.1052	Capture	0.2329
0.3991	Fission	0.3896
0.3946	Radial Leakage	0.3468
0.1019	Axial Leakage	0.0320

[a]Normalization.

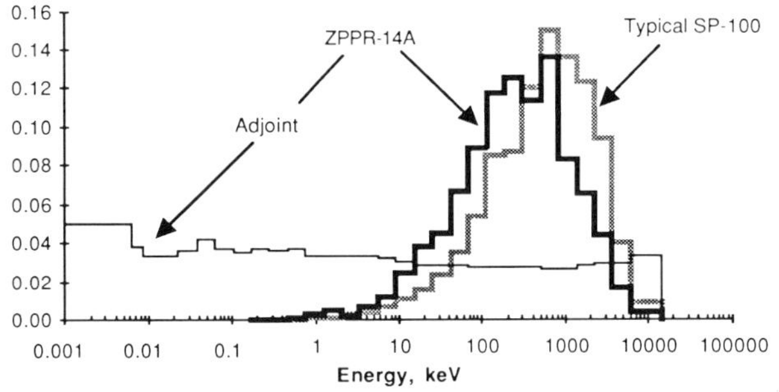

Figure 2 *Comparison of ZPPR-14 and SP-100 Neutron Spectra.*

the length to diameter ratio of ZPPR-14 is about 2.5 times that of SP-100. Axially varying parameters in ZPPR-14 are thus not very representative of proposed SP-100 designs.

The neutron spectra comparison in Figure 2 shows the SP-100 spectrum to be notably harder than that in the ZPPR-14 spectrum, as would be expected because of the factor of three difference in average enrichment. The spectral difference is most apparent in differences between the two cores with respect to ^{238}U fission and capture rates, neither of which is very important in the fully enriched SP-100 core. A larger SP-100 core probably would have a substantially higher ^{238}U fraction, hence more inelastic scattering and a softer spectrum, closer to that of ZPPR-14. Since the SP-100 was modeled with nitride fuel, its spectrum does not exhibit the large oxygen resonance that is apparent in the ZPPR-14 spectrum. The flat adjoint spectrum shown for ZPPR-14 is typical for a ^{235}U-fueled fast reactor.

Analyses of the ZPPR-14 measurements utilized advanced methods most likely to be of interest to designers. All calculations were done using cross sections derived from the Evaluated Nuclear Data Files (ENDF/B) Version V, Revision 2. Because of the effort and expense involved, some measurements have not yet been analyzed, and some have been calculated only with the simpler analysis models.

Program Description

Reactor Description

The ZPPR-14 assemblies constructed for the benchmark experiments consisted of two critical and numerous subcritical configurations representative of a low-power density, 40-MWt, sodium-cooled fast reactor. A cross-sectional view of ZPPR-14 is shown in Figure 3. The reactor core was constructed from unit cells (subassemblies) of the type shown in Figure 4. The 93%-enriched ^{235}U plates are combined in proper proportion with the depleted U_3O_8 plates to give the correct average enrichment for the critical core loading. With the other materials in the cell, the average composition is close to that of a UO_2 subassembly in a sodium-cooled fast reactor. The heterogeneity effect of using plates rather than pins is on the order of one percent of the critical eigenvalue. When inserted in the matrix, the typical dimension of one of the cells is about 55 mm. With 12 such cells to span the core, the diameter was about 660 mm. The core had an overall height of almost 1.8 m.

The downcomer region, representing both the so-

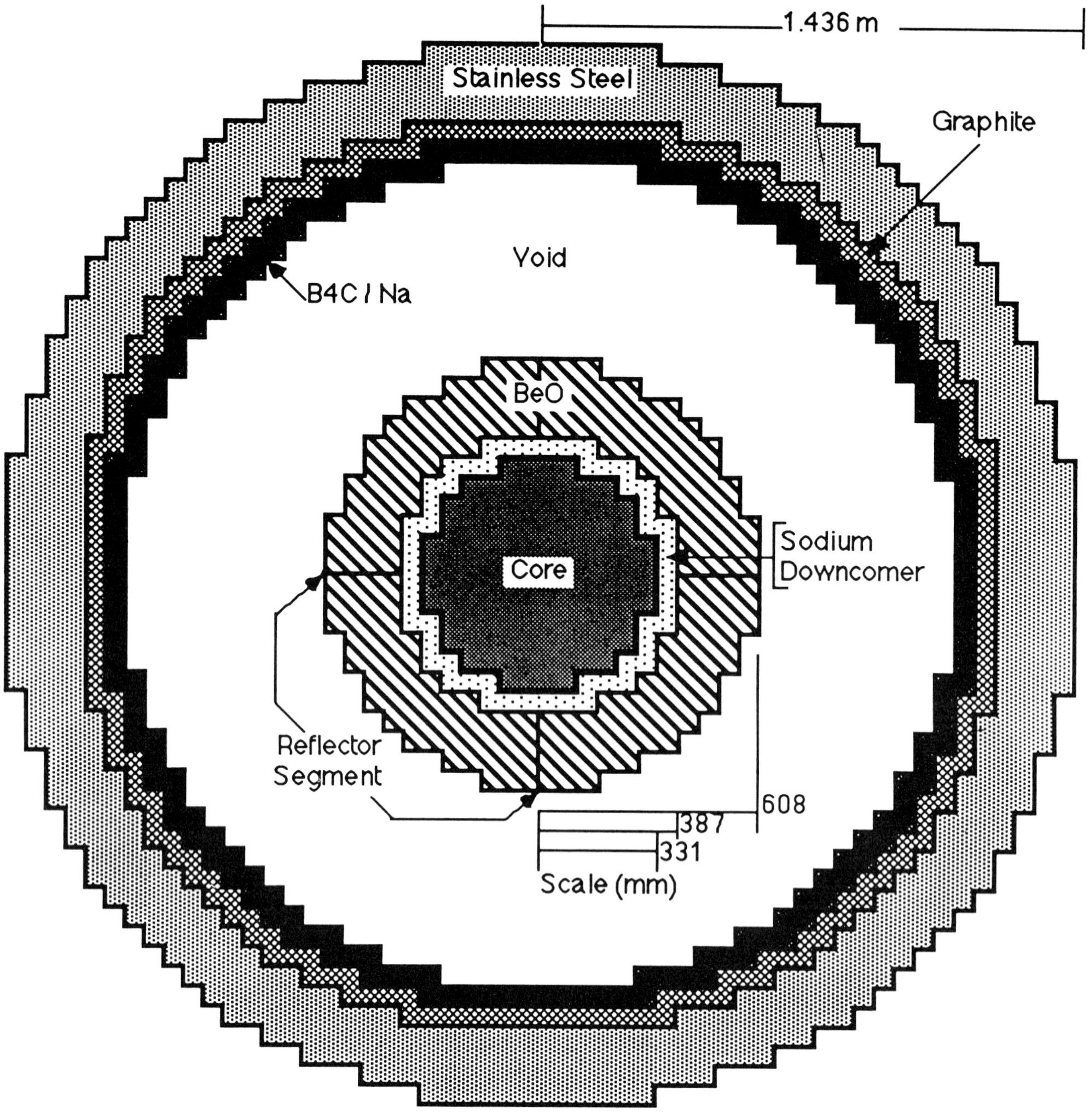

Figure 3 *Interface Diagram of ZPPR-14.*

dium downcomer and the core barrel, was only one cell thick. It was composed of sodium, stainless steel and void.

The reflector region had an average thickness of 205 mm, composed principally of BeO in stainless steel drawers and matrix. However, in the first critical configuration (ZPPR-14A), which simulated an end-of-cycle condition with the reflector inserted over the full core height, there was insufficient inventory to build the reflector entirely from BeO. Therefore, BeO was used in the two rings of cells closest to the core, Be metal was mostly used in the third ring, and graphite was mostly used in the fourth ring. In the ZPPR-14B configurations, representing a mid-cycle state with the reflector half inserted, BeO was used to construct the entire reflector.

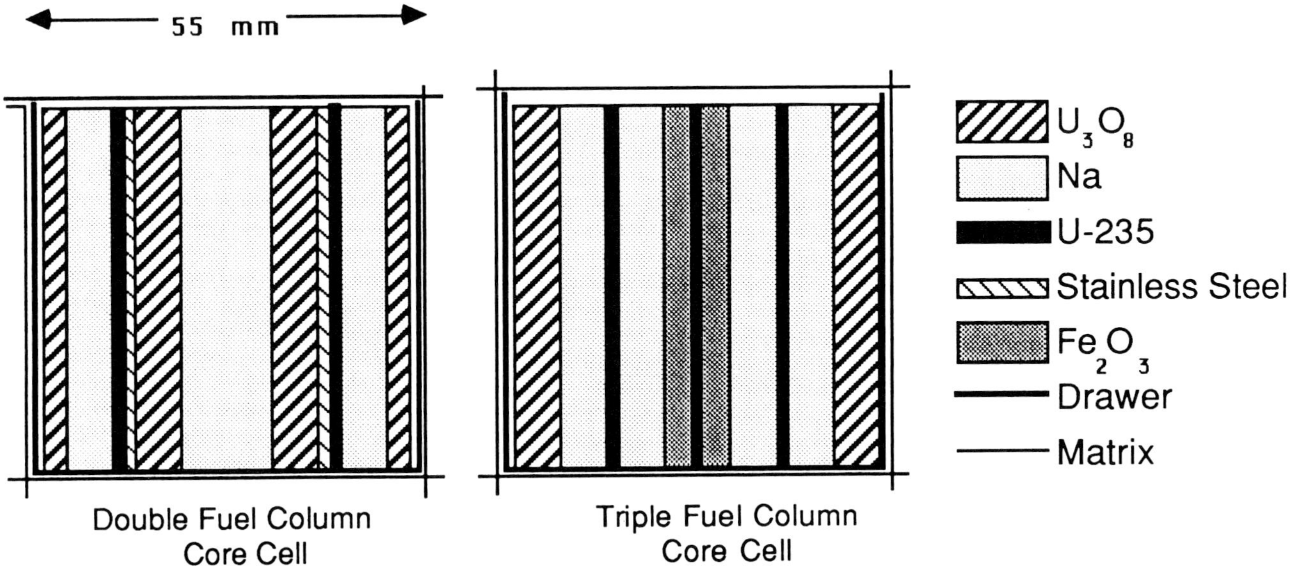

Figure 4 *Two Basic Core Cells of ZPPR-14.*

Outside the reflector was a large void region consisting of empty ZPPR matrix tubes. This region extended about 560 mm beyond the reflector.

Terminating the void was a massive, three-region, radial shield. The inner cell of the shield consisted of B_4C, sodium and stainless steel. The next layer of the shield was graphite, followed by more than 150 mm of stainless steel.

There was also extensive axial shielding of the ZPPR-14 assemblies. Each end of the reflector and void regions was shielded by 150 mm of depleted uranium, followed by 310 mm of graphite and another 150 mm of stainless steel. At either end of the core, there was a sodium and stainless steel plenum region extending for 432 mm. These plena were capped with 150 mm of solid stainless steel, which extended out to the inner edge of the radial shield region. The downcomer extended over the full axial height between the two stainless steel planes.

There were two reference configurations for the ZPPR-14 Program. In the ZPPR-14A reference loading, the reflector surrounded the downcomer over the full core height. The effective enrichment was 30% with a ^{235}U loading of about 500 kg. There were also 40 kg of plutonium (27% ^{240}Pu) in the core to provide a distributed neutron source for the subcritical configurations. In the second critical reference configuration, the BeO reflector covered only half the core height. Void remained in place of the part of the reflector that was removed. A schematic of the ZPPR-14B core configuration is shown in Figure 5. In this configuration the effective enrichment was increased to about 34% to maintain system criticality. Core-average enrichment in ZPPR-14 was adjusted by varying the ratio of the two core cell types shown in Figure 4.

Measurements

Although the measurement program was brief, a fairly extensive data base was acquired. The critical loading conditions for the two reference configurations were determined. Measurements of reflector control worth were emphasized, since that represented the area of principal design concern. Detailed spatial maps of many key reaction rates were also obtained, and ^{235}U fuel worth was measured as a function of radius.

Reflector Control Worths

For the control worth measurements, the reflector was assumed to be built in four independently movable segments. The segments were assumed to be able to move axially and to be able to pop radially away from the core barrel, roughly 250 mm outboard from their reference positions. The latter is the designated secondary scram mode. Figure 6 shows the types of altered reflector configurations that were allowed.

In the configuration on the left side of Figure 6, neutron streaming gaps of about 110 mm have been created at four symmetrical locations along the full height of the core. This was done in part to model structural gaps that would be present with operational

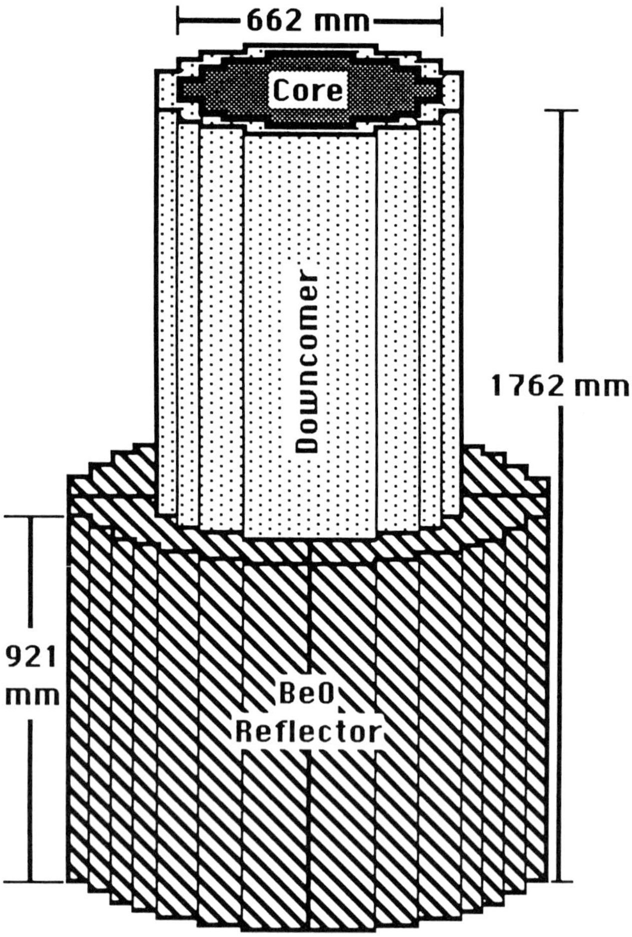

Figure 5 *Schematic of the ZPPR-14B Reference Configuration.*

reflector segments. (The 424 ZPPR drawers that composed the reflector had to be moved individually for changes in reflector configuration.) However, it was also felt that this case might prove useful as a diagnostic example for analysis of the secondary scram configurations.

In the configuration in the center of Figure 6, one of the reflector segments has been completely removed. This simulates full axial withdrawal of the reflector segment, the primary scram mode. Measurements were made with one, three and all four segments removed in this manner. In ZPPR-14A, the worth of putting the reflectors in the ZPPR-14B reference state was also measured.

In the configuration on the right side of Figure 6, one of the segments has been moved into the secondary scram position. Measurements of this type were only made relative to the ZPPR-14B reference configuration. One, three and all four segments were moved in this manner.

The reflector worth measurements were made using the modified source multiplication technique (Carpenter 1976) that has been substantially refined over the years at ZPPR. Sixty-four ^{235}U fission chambers were distributed throughout the assembly to obtain the data needed for the worth measurements. The fission chambers are built into 150-mm-long stainless steel cans that normally contain sodium, so that they fit into a normal unit cell with a minimum of perturbation. Although the data from the chambers were taken in order to measure the reflector worths, they also provide a coarse power distribution map for a number of configurations.

In measuring the change in system reactivity, it is necessary to know approximately how the relative power distribution changes between the perturbed and the reference states. This is a principal reason for using such a large number of in-core chambers in ZPPR. Figure 7 shows the relative change in ^{235}U fission distribution when one of the reflector segments was removed from ZPPR-14. There is a 5% change between the contours in the figure, starting at 1.15 and going down to 0.30. The values for the contour lines are the ratio of two ratios. The numerator is the ratio of the fission rate at a point to the total fission integral in the perturbed system. The denominator is the same type ratio, but for the unperturbed reference system. It is evident from Figure 7 that there is large relative power redistribution associated with the reflector movement. For example, the changes observed in the ZPPR-14 measurements are considerably larger than are observed for B_4C control rod measurements in LMFBR assemblies. Such large flux redistributions not only make analysis of the measurements difficult, but they make it essential to have spatially distributed instrumentation in order to achieve reasonable accuracy in the measurements themselves.

Reaction Rates

Reaction-rate distributions in both reference critical configurations of ZPPR-14 were measured for ^{235}U fission, ^{239}Pu fission, ^{238}U capture and fission, and gamma-ray heating. The neutron reaction rates were measured using thin foils of about 250 mg mass distributed throughout the reactor by slipping them in between plates of normal material (Brumbach et al. 1982 and 1985). The gamma heating rates were measured in stainless steel with thermoluminescent dosimeters (TLDs).

Uranium-235 fission rates were measured throughout the ZPPR-14 assembly. Data are available for 185

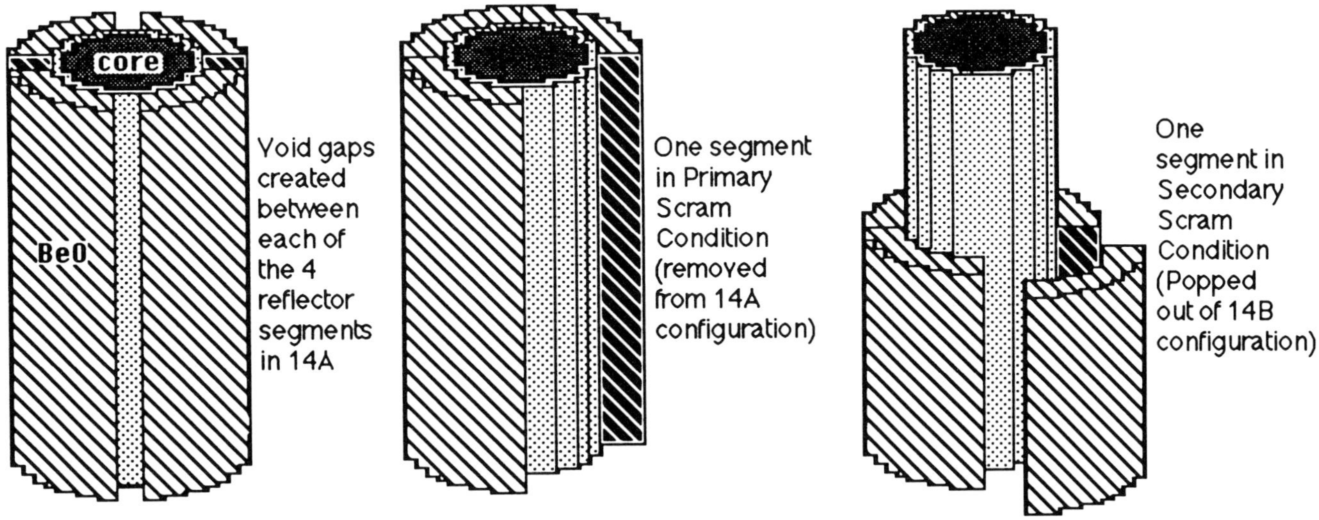

Figure 6 *Representative Reflector Perturbations used in the Worth Measurements.*

positions in ZPPR-14A and 174 positions in ZPPR-14B. There were detailed radial, axial, and azimuthal core maps. Measurements were also made in the downcomer, reflector, void zone, radial shield, and axial plena. The data outside the core zone provide a partial test for ex-core shielding calculations.

Measurements of ^{238}U and ^{239}Pu reaction rates were restricted to the core region. ^{238}U measurements were made at 72 locations in ZPPR-14A and 42 locations in ZPPR-14B. Plutonium-239 measurements were made at 24 locations in ZPPR-14A and 29 in ZPPR-14B.

Gamma-ray dose rates were measured with ^{7}LiF and CaF_2 TLDs. The TLDs are 1 mm square by 6 mm long. They fit snugly in 6.4 mm by 12.7 mm stainless steel cylinders, that are placed inside a special sodium plate in the core cells. Outside the core, in downcomer and plenum regions that contain sodium plates, the TLDs were contained in 12.7 mm by 3 mm stainless steel cylinders which were placed in the small air cooling gap between cells. Radial measurements extended out beyond the radial shield. The gamma heating map of the ZPPR-14A assembly included 182 TLDs, while 196 TLDs were used in the ZPPR-14B assembly.

Analysis Methods

The ZPPR-14 analysis has been done by diffusion theory, transport theory and Monte Carlo calculations. Cross section data for both the deterministic and the Monte Carlo calculations were derived from the latest nuclear data library release, ENDF/B-V.2. In all cases, only three-dimensional geometrical modeling was used. Thirty energy groups were used for all the deterministic calculations.

Multigroup Cross Section Preparation

The ETOE-2/MC^2-2/SDX (Toppel et al. 1978) code package was used to process the ENDF/B basic nuclear data into a 30-group cross section set. The ETOE-2 code takes the basic nuclear data for all the relevant isotopes from the ENDF/B files and prepares files that can be used directly by MC^2-2 and SDX. MC^2-2 is a dimensionless slowing-down code that is used to calculate an ultra-fine group spectrum. SDX is a one-dimensional code that is used to collapse (in space and energy) the fine group isotopic data to the homogenized multigroup cross sections.

MC^2-2 was used to calculate a 2082-group spectrum for the homogeneous composition of the ZPPR-14 double-fuel-column cell with a buckling search to critical. Thermal group cross sections, not normally available in this fast-spectrum calculational path, were added to the MC^2-2 files at the ultra-fine group level. Cross sections for such materials as plutonium that were in the reactor assembly, but not physically present in the reference core cell, were prepared by adding infinitely dilute material to the cell model. The isotopic cross section data were collapsed to 226 groups using the MC^2-2 calculated spectrum.

Cell calculations at the 226 fine-group level were made in the SDX code for the fuel cells using a buckling search to critical, and for the depleted uranium part of the axial shield using zero buckling. Group collapse to form a 30-group library was done using the asymptotic cell spectrum calculation as one of two ap-

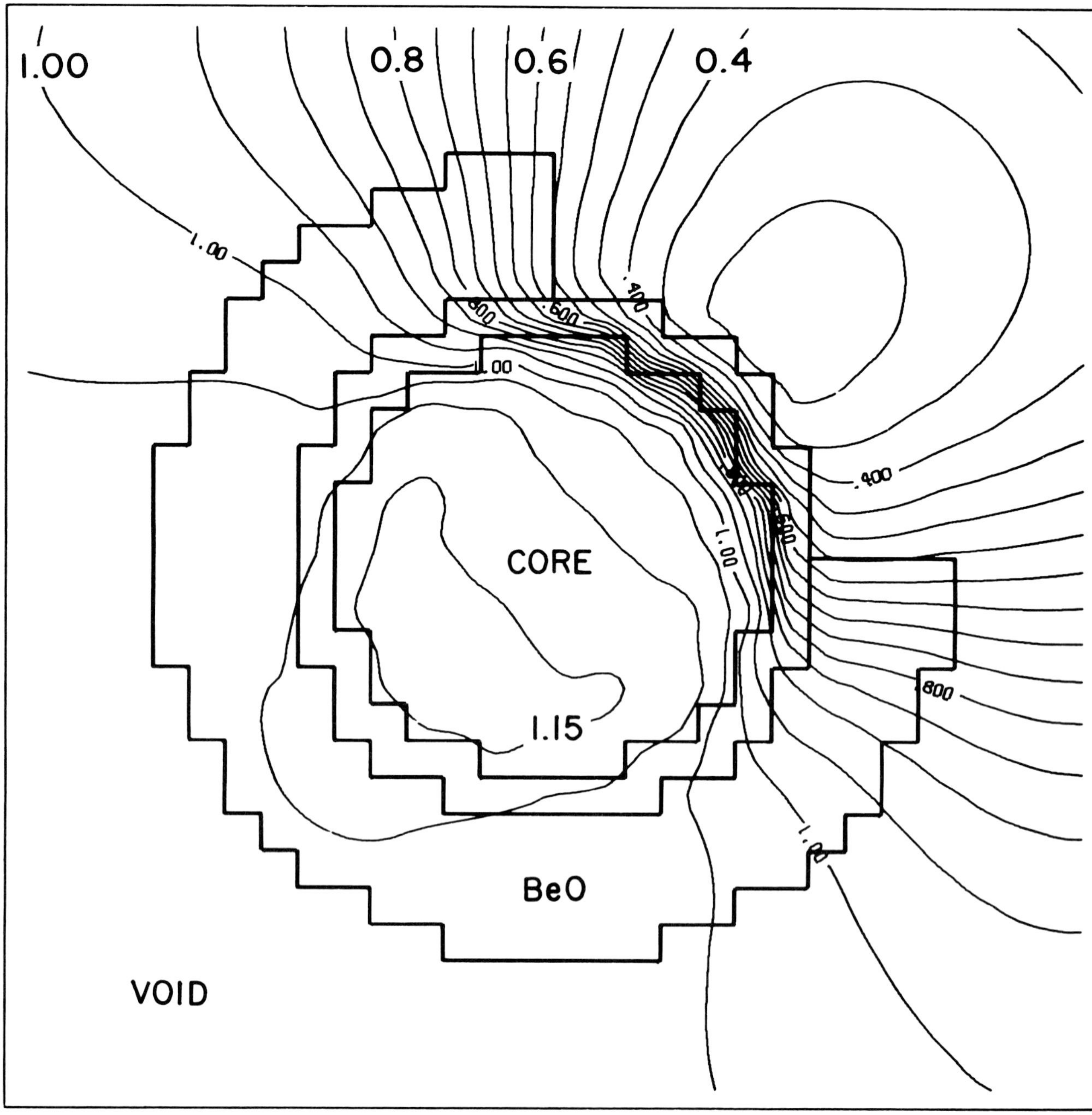

Figure 7 *Change in Power Distribution after Removal of Reflector Segment.*

proaches. The other approach used a one-dimensional reactor model to calculate the 226-group spectrum in different regions, with the group collapse being done within defined regions. Data from the latter approach were used for reference calculations in the ZPPR-14 Program. One model represented the core, downcomer, and reflector A second model represented the core, downcomer, and void (typical of the top half of ZPPR-14B). A third model represented the core and the sodium-filled region above the core. In each model, the data for the outer region of the core (~50 mm) were collapsed separately from those in the interior. This was done simply in recognition of the very rapid spectrum transition between the core and the beryllium reflector. The 30-group structure also represented a concession to the lower-energy neutrons returning from the reflector. The last group of the standard 28-group ZPPR structure was subdivided into three groups that included a thermal group.

An RZ model of ZPPR-14 was used to test the sensitivity of the calculations to the different group collapse schemes. Nine calculations were run with varying combinations of asymptotic and spatially collapsed cross section data. The difference in calculated k_{eff} for the system between using all asymptotic cross sections and all spatially collapsed cross sections was 4.9%. The effect was made up principally from the radial reflector cross sections (3.7%) and the downcomer cross sections (1.2%). The effect of switching core data sets was less than 0.1%. The high sensitivity to cross sections in the regions where the spectrum is softer gives rise to some concern about the adequacy of the thermal group data that were added to the the MC^2-2 files, since they were not generated specifically for this reactor.

Deterministic Calculations

All deterministic calculations for ZPPR-14 were made with the DIF3D code (Derstine 1982 and Lawrence 1983) using Cartesian geometry and 30 energy groups. In this geometry, the code has options for finite-difference diffusion theory calculations, nodal diffusion theory calculations, and nodal transport theory calculations. Each of these options was used in part of the ZPPR-14 analysis. A mesh spacing of 55 mm, corresponding to unit cell dimensions, was used in the XY plane. The axial mesh spacing was about 51 mm over most of the core height.

The finite-difference diffusion calculations were used in a truncated model that included only part of the empty matrix beyond the radial reflector, but retained the full axial dimension of the system. The X and Y boundary conditions were chosen so as to reduce the effect of the truncation on the core fission distribution with the reflectors removed. (The effect was trivial with the reflectors in place.) The truncated model was used for a number of reasons. First, calculations with quick turn-around time were required for use in reducing data from the measurements (basically for interpolating fission distributions between measured points). The truncated model had only one- quarter of the XY-mesh storage requirement of the full radial model. Since many of the perturbed reflector configurations were not symmetric in the XY plane, the compromise was necessary in order to get the necessary calculations done. Second, since this model was only used for diffusion calculations, including that part of the assembly beyond the reflector, it was not felt to be sufficiently meaningful to justify the extra expense.

The nodal calculations have only been used for measurements that preserved symmetry in the XY plane. The models used were

- One-quarter XY, half Z for ZPPR-14A, and
- One-quarter XY, full Z for ZPPR-14B.

The full radial extent of the assembly, out through the last stainless-steel shield zone, was included in each of these models.

The nodal transport calculational method used here is an approximate method in a developmental version of the DIF3D code, and is available only for Cartesian geometry (Lawrence 1984). It has been tested in LMFBR systems built in ZPPR and shown to compare favorably with low-order S_n solutions.

Monte Carlo Calculations

The Monte Carlo solutions have been computed with the VIM code (Blomquist 1980). VIM is capable of explicit three-dimensional modeling of all spatial details of the assembly. It is a continuous-energy code, so that there are no multigroup approximations. Also, special BeO nuclear data, including thermal upscatter, were used in the VIM calculations.

VIM was used to calculate the two reference configurations and three of the symmetric perturbed reflector configurations. The VIM models were slightly idealized in that some details of the experiment were omitted for simplicity (detectors, operational shim rods at the axial edges of the core, minor temperature differences). Small corrections (0.0005-0.0035 Δk) for the effects of these approximations were made using deterministic results. Two-hundred thousand neutron histories were run for the reference cases and 100,000 for the perturbed cases. With this many histories, the eigenvalue statistics are about 0.002 for one standard deviation.

Results

Critical Configurations

The results of eigenvalue analysis for the two reference critical configurations are presented in Table 2. The measured k_{eff} values listed for the two assemblies are for states with the operational shim rods withdrawn to the maximum reactivity positions. The experimental uncertainties are dominated by limits on the ability to describe the composition of the system.

The Monte Carlo results for ZPPR-14 of 0.9963 and 0.9992 for the ratio of the calculated to measured (C/E) values of k_{eff} are consistent within the statistics of the calculations. Values near unity are also consistent with other Monte Carlo calculations for ^{235}U-fueled fast reactors when using the ENDF/B-V nuclear data library.

Table 2 *Analysis of ZPPR-14 Critical Configurations.*

	ZPPR-14A	ZPPR-14B
Experimental k_{eff}	1.0001 ± 0.0005	1.0004 ± 0.0005
	C/E	
3D Nodal Diffusion[a]	0.9921	0.9796
3D Nodal Transport[a]	0.9964	0.9874
Monte Carlo[b]	0.9963 ± 0.0019	0.9992 ± 0.0020

[a]DIF3D, 30 Groups.
[b]VIM.

The deterministic calculations indicate a notable discrepancy between the ZPPR-14A and ZPPR-14B results, with the transport C/Es differing by 0.9% and the diffusion C/Es differing by 1.3% for the two cases. The evidence suggests that the problem lies in the multigroup cross section processing, most likely for the reflector and downcomer regions.

The 0.8% difference between the diffusion and transport C/Es in the ZPPR-14B case is typical for high-leakage fast reactor systems. It is interesting to note that in the fully reflected case, the transport correction to the diffusion calculation is only about 0.4%. The relative adequacy of diffusion theory calculations in the 14A assembly also shows up in the reaction rate analysis.

Table 3 gives the ^{235}U fuel worth as a function of radius in ZPPR-14A. Although the measurements were done with quarter-core symmetry to facilitate analysis, no calculations of these data are yet available. The measurements were done by removing the ^{235}U from four symmetric double-column cells of the type shown in Figure 4.

Reaction Rates

The ZPPR-14 Program contained too much reaction rate data to be presented in any detail here, so the results have been reduced to summaries of the power distributions and the reaction rate ratios.

Figure 8 shows the radial reaction rate distributions in the fully reflected ZPPR-14A core. The measured data are indicated by markers, while the lines are the results from the nodal transport calculations. The radial distributions are flatter than might have been expected, as some beryllium-reflected designs have shown considerably more power peaking near the core edge. There is some upturn in the ^{239}Pu and ^{235}U fission rates near the boundary, but apparently the downcomer region provides a substantial buffer. The calculated rates match the measured ^{235}U fission distribution well over the interior of the core, but are off by 6% for the two outermost positions. The calculation of ^{239}Pu fission does not track the upturn nearly as well. The reason for the disparity is in the difference in which the 30-group cross sections were prepared. The plutonium cross sections were collapsed from 226 groups to 30 groups in the asymptotic cell spectrum, whereas the uranium cross sections were collapsed in the one-dimensional reactor calculation with two core regions. Also, there were no thermal data for plutonium available at the time the cross sections were being prepared, and the default thermal-group values that were used are unrealistically low.

The in-core distribution of gamma heating rates in stainless steel are also shown on the same plot in order

Table 3 *^{235}U Fuel Worth in ZPPR-14A.*

Average Radius (mm)	Worth[a] ($/kg)
39	0.164 ± 0.003
141	0.145 ± 0.002
250	0.100 ± 0.002
305	0.099 ± 0.002

[a]18.68 kg ^{235}U, 1.10 kg ^{238}U, 1.61 kg stainless steel removed from core. Normalization is to ^{238}U mass.

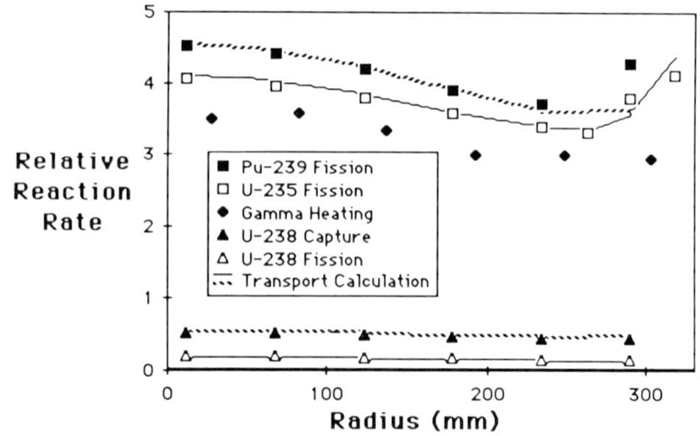

Figure 8 *Radial Neutron and Gamma Reaction Rates in ZPPR-14A.*

to compare to the distribution shapes. All the data have been normalized to an average core power of one watt. In Figures 8 through 10, the units for the neutron reaction rates are 10^{-17} reactions per atom per second; the units for the gamma heating measurements are 10^{-5} Gy/s. No calculations are available for any of the gamma heating measurements.

The well-behaved power distribution in the core may be contrasted with the radial fission rate plot for the whole assembly, shown in Figure 9. Two radial traverses are shown for ZPPR-14B, one at an axial position such that the traverse went through the reflector, the other starting from a spot higher on the core, so that it went through additional void zone rather than the reflector. Again, the solid lines indicate the results of a transport calculation. The calculation is clearly unable to track the fission rate through the reflector; beyond the reflector the C/Es are on the order of 2.0. Similarly, the calculations overpredict the reaction rates in the unreflected traverse by about 60% at large radius. Part of the problem is the difficulty in computing the self-shielding for the foils used in the measurements beyond the core region.

Figure 10 shows axial power distribution near the central axis in the half-reflected ZPPR-14B core. The shift in power to the reflected half is substantial. Unlike the situation in the fully reflected case, the calculations were unable to track the measurements inside the core. The C/E ratios are about 1.08 in the reflected half and about 0.93 in the unreflected half. The results shown are from the transport calculations, but they offer no substantial improvement over the diffusion theory results. The measured power contribution from gamma heating is shown in the same plot. The principal difference in shape occurs in the plenum region, as expected.

Axial power peaking is summarized in Table 4 for both reference configurations. The first entry under ZPPR-14B corresponds to the data plotted in Figure 10. In ZPPR-14A, there was little radial variation in the axial peak/average power ratio. The nominal value of 1.3 was well predicted by the transport calculation. Considerably more radial variation can be observed in the ZPPR-14B axial peak/average factors, but the predictions provided by the transport calculations are reasonably consistent. The overpre-

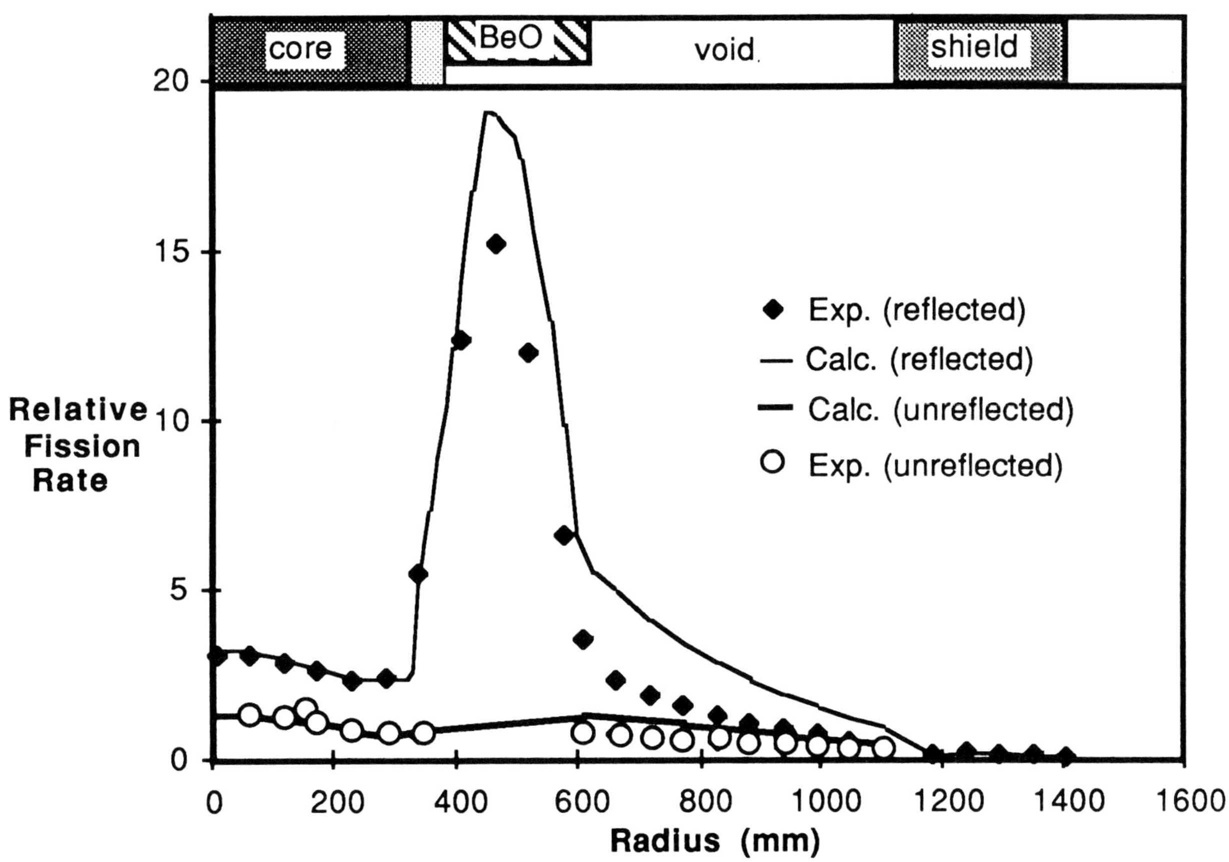

Figure 9 *Radial Fission Distribution in ZPPR-14B.*

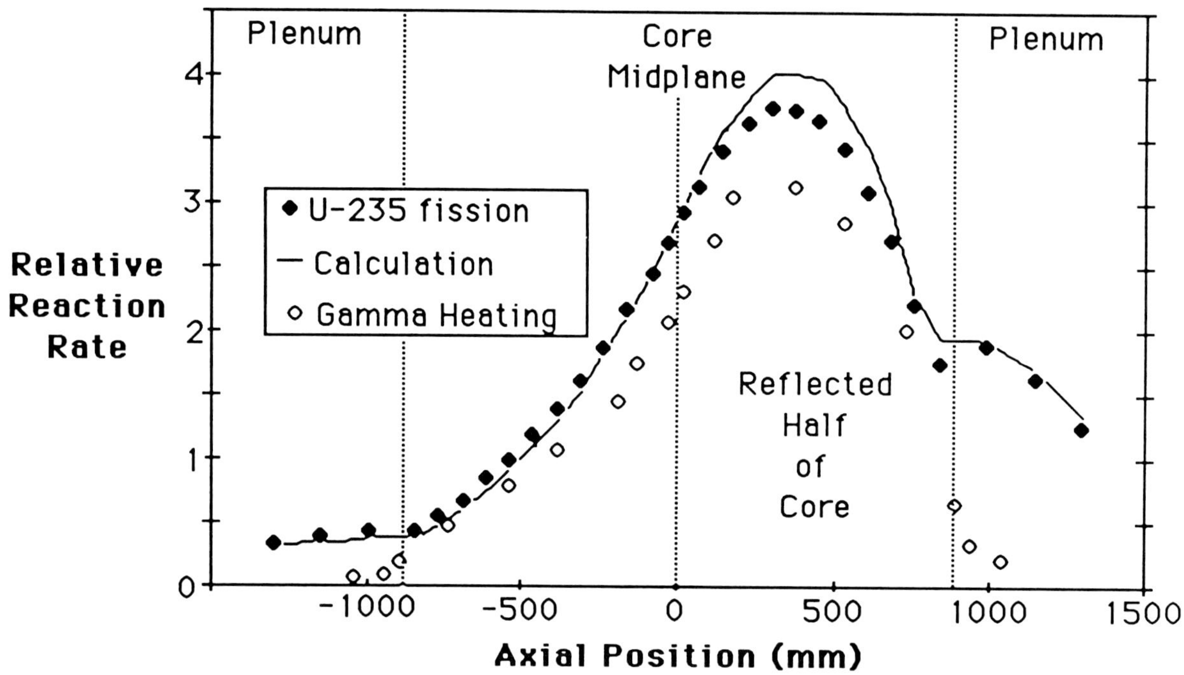

Figure 10 *Axial Fission and Gamma Heating Distribution in ZPPR-14B.*

diction of 3% to 4% in the 14B case can be correlated with the overprediction of reflector worth by deterministic methods.

The reaction rate ratio results are summarized in Table 5. As might be anticipated from Figure 7, they form a fairly consistent set. Also, since all the data were taken in the core, there are no major discrepancies between the measurements and the transport calculation results. The increase in the fission ratios from ZPPR-14A to ZPPR-14B is evidence of the hardening of the spectrum with the increase in average enrichment. There is a concomitant decrease in the capture-to-fission ratio. The C/E ratios are consistent with others that have been noted for calculations of fast reactor systems with ENDF/B-V.2 nuclear data.

Reflector Control Worths

There were five perturbations to the reference reflector configuration in each of the two reference assemblies. Briefly, the perturbations were chosen to simulate the following conditions:

- Full primary scram mode,
- Full secondary scram mode,
- Single segment scram in either mode,
- Single segment failure to scram in either mode,
- 14B configuration relative to 14A reference, and
- Gaps between reflector segments.

The worths of these perturbations ranged from 4 dollars to 30 dollars. They are summarized in Table 6, along with the results of the calculations. Only three

Table 4 *Axial Peak/Average Power in ZPPR-14.*

Radius of Traverse (mm)	ZPPR-14A Peak/Average[a]	ZPPR-14A C/E[b]	ZPPR-14B Peak/Average[a]	ZPPR-14B C/E[b]
13	1.313	0.996	1.634	1.034
69	1.312	0.997	1.627	1.044
124	1.308	1.000	1.656	1.035
179	1.316	0.993	1.682	1.034
234	1.303	1.002	1.751	1.026
290	1.305	1.000	1.880	1.017

[a]From measured $^{235}U(n,f)$ values along axial traverse within ± 839 mm of core midplane.
[b]Ratio of calculation to experiment. (XYZ nodal transport calculation in 30 groups).

cases have been analyzed with the calculational methods being tested. The complete column of C/Es on the far right-hand side of Table 6 is from the calculations done with the model that was truncated in the XY plane. Since the boundary condition in this model was chosen to preserve core reaction rates and not eigenvalue, these results were not expected to be meaningful. Nevertheless, they form a surprisingly consistent set with one exception. The exception is the case where the narrow gaps were formed, and is clearly not amenable to diffusion theory modeling.

Although both the nodal diffusion and transport calculations overpredict the reflector worths, the transport solutions clearly form a more consistent set. This is expected since there is a sizable transport correction for the reference ZPPR-14B eigenvalue calculation. However, the diffusion theory results may look favorable enough to be deemed adequate for some conceptual modeling calculations.

The Monte Carlo calculations slightly underpredict the reflector worths, but form by far the most consistent set of analyses. The degree of underprediction is almost insignificant when considering the 3% statistics on the calculations.

Conclusions

The ZPPR-14 Program has provided high-quality benchmark data for some of the key reactor physics parameters of a BeO-reflected fast reactor. Analyses of these data can provide useful guidance in the selection of calculational tools to be used in the design of space reactors of this type. The data are not an acceptable substitute for integral physics measurements for a specific design such as SP-100, but they can identify some of the problems that need to be avoided.

The Monte Carlo calculations with ENDF/B-V.2 nuclear data performed well in all aspects of the analysis, even relative to previous test results obtained for more typical fast reactor systems. However, the exclusive use of such calculations in a design program is not practical, since so much effort is required to set up the models and essentially no spatial detail is obtained from the whole-core calculations. The calculations already completed need to be edited further to check

Table 5 *Summary of Core Reaction Rate Ratios in ZPPR-14.*

Ratio	ZPPR-14A Measured[a]	ZPPR-14A C/E[b]	ZPPR-14B Measured[a]	ZPPR-14B C/E[b]
$^{239}Pu(n,f)/^{235}U(n,f)$	1.102 ± 0.010	0.991 ± 0.008	1.127 ± 0.010	0.997 ± 0.010
$^{238}U(n,\gamma)/^{235}U(n,f)$	0.1241 ± 0.0008	1.033 ± 0.010	0.1218 ± 0.0012	1.015 ± 0.009
$^{238}U(n,f)/^{235}U(n,f)$	0.0412 ± 0.0007	0.992 ± 0.020	0.0455 ± 0.0008	1.018 ± 0.026

[a]Mean value and one standard deviation of 10-20 measurements.
[b]Mean value and one standard deviation of ratios of calculations to experiments (XYZ nodal transport calculation in 30 groups).

Table 6 *Summary of ZPPR-14 Reflector Worth Results.*

Reflector Condition	Worth ($)	C/E Nodal Diff.	C/E Nodal Trans.	C/E Monte Carlo	C/E F.D. Diff.[a]
	ZPPR-14A				
1 Segment Removed	6.29 ± 0.12				1.22
3 Segments Removed	20.56 ± 0.39				1.19
4 Segments Removed	29.45 ± 0.62	1.08	1.05	0.969 ± 0.017	1.10
4 Segments Half Withdrawn	9.12 ± 0.17				1.22
4 Void Gaps	6.06 ± 0.11				1.72
	ZPPR-14B[b]				
1 Segment (SS[c])	3.93 ± 0.07				1.40
3 Segments (SS)	11.35 ± 0.32				1.29
4 Segments (SS)	14.42 ± 0.42	1.33	1.17	0.999 ± 0.029	1.22
1 Segment Removed	4.44 ± 0.09				1.33
4 Segments Removed	17.22 ± 0.51	1.20	1.07	0.923 ± 0.030	1.14

[a]Finite difference diffusion calculation in truncated model.
[b]Relative to reference configuration with all 4 segments half withdrawn. (Figure 5)
[c]Secondary Scram configuration, refer to Figure 6.

such parameters as the ratio of powers in the reflected and unreflected halves of ZPPR-14B.

Extreme care must be taken in the prepartion of multigroup cross sections to be used in deterministic design analysis for a reactor of this type. Collapse to the final group structure must be done in a calculation that models each specific region in the system as a whole. Despite the precautions taken in this work, many questions remain unanswered relative to errors in the cross sections. The effect of more careful selection of the thermal-group cross sections is unknown. Because of the high sensitivity of the calculations to group collapse in the reflector, it has been suggested that Monte Carlo calculations be used to generate absorption cross sections for the reflector and possibly for the downcomer.

The source of concern about multigroup cross sections is that the neutron spectrum changes so substantially over such a short distance between the core and the BeO reflector. This problem will be worse in most space reactor designs, since the reflector will be closer to a more highly enriched core.

Acknowledgments

This work was done at Argonne National Laboratory, supported by the U.S. Department of Energy, Nuclear Energy Programs, under contract No. W-31-109-Eng-38. The authors also wish to acknowledge their colleagues at the ZPPR facility for their very substantial role in the success of the ZPPR-14 Program. T. Misawa, a student from Japan, helped with the Monte Carlo calculations during his assignment to Argonne.

References

Blomquist, R. N., R. M. Lell, and E. M. Gelbard (1980) "VIM—A Continuous Energy Monte Carlo Code at ANL," in *Proc of Seminar/Workshop on A Review of the Theory and Application of Monte Carlo Methods*, held 21–23 April 1980, ORNL-RSIC-44, August 1980.

Brumbach, S. B. and D. W. Maddison (1982) *Reaction Rate Calibration Techniques at ZPPR for ^{239}Pu Fission, ^{235}U Fission, ^{238}U Fission and ^{238}U Capture, ANL-82–38*, Argonne National Laboratory, Argonne, IL.

Brumbach, S. B. and J. M. Gasidlo (1985) *In-Cell Reaction Rate Distributions and Cell-Average Reaction Rates in Fast Critical Assemblies*, ANL-85–44, Argonne National Laboratory, Argonne, IL.

Butler, D. K., R. C. Derner, and W. G. Knapp (1966) "Measurements and Analysis of Al-, Al_2O_3- and BeO-reflected Fast Critical Experiments," in *Proc. of International Conference on Fast Critical Experiments*, held at Argonne National Laboratory, Argonne, IL, 13–16 October 1966.

Callen, R. C., P. S. Check, R. E. Kearney, and W. F. Walsh (1965) *SNAP-50 Critical Experiments and Analysis*, PWAC-487, Pratt and Whitney Aircraft Corporation, Middletown, CT, September 1965.

Carpenter, S. G. (1976) "Measurement of Control Rod Worths Using ZPPR," in *Proc. of Specialists Meeting on Control Rod Measurement Techniques: Reactivity Worth and Power Distributions*, NEACRP-U-75, C.E.N., held in Cadarache, France, 21–22 April 1976.

Derstine, K. L. (1982) *DIF3D: A Code to Solve One-, Two-, and Three-Dimensional Finite-Difference Diffusion Theory Problems*, ANL-82–64, Argonne National Laboratory, Argonne, IL.

Lawroski, H., R. G. Palmer, F. W. Thalgott, R. N. Curran, and R. G. Matlock (1972) *Final Safety Analysis Report on the Zero Power Plutonium Reactor (ZPPR) Facility*, Argonne National Laboratory, Argonne, IL, June 1972.

Lawrence, R. D. (1983) *The DIF3D Nodal Neutronics Option for Two- and Three-Dimensional Diffusion Theory Calculations in Hexagonal Geometry*, ANL-83–1, Argonne National Laboratory, Argonne, IL.

Lawrence, R. D. (1984) "Three-Dimensional Nodal Diffusion and Transport Methods for the Analysis of Fast-Reactor Critical Experiments," in *Trans. of Topical Meeting on Reactor Physics and Shielding*, held in Chicago, IL, 17–19 September 1984.

Stewart, H. B. and M. L. Storm (1958) "The Critical Assembly Program," in *Naval Reactor Physics Handbook*, Vol III, U.S. Atomic Energy Commission, Washington, D.C.

Toppel, B. J., H. Henryson II, and C. G. Stenberg (1978) "ETOE-II/MC^2-2/SDX Multigroup Cross Section Processing," in *Proc. of RSIC Seminar Workshop on Multigroup Cross Sections*, held at Oak Ridge National Laboratory, Oak Ridge, TN, 14 March 1978.

Part 12

NUCLEAR ELECTRIC AND NUCLEAR PROPULSION

Space Nuclear Power Systems 1986
Edited by M. S. El-Genk and M. D. Hoover

Chapter 34

A Nuclear Electric Mars Payload Transfer Vehicle

William F. Lyon III

Department of Nuclear Engineering
Texas A & M University
College Station, TX 77843

Wayne Black III

Department of Aerospace Engineering
Texas A & M University
College Station, TX 77843

Shyam Bhaskaran

Department of Aerospace Engineering
The University of Texas at Austin
Austin, TX 78712

This paper summarizes the conceptual design of a mercury ion propulsion system for an unmanned Mars Payload Transfer Vehicle (MPTV) and compares its performance to a chemical propulsion system. This vehicle is a component in the mission scenario developed by the Texas A&M University and the University of Texas at Austin Manned Mars Mission Study conducted for the Universities Space Research Association at the National Aeronautics and Space Administration (NASA) Johnson Space Center in Houston, Texas. This scenario employs two vehicles, an MPTV that will carry the bulk of the equipment required for Mars proximity operations and a Rapid Manned Transfer Vehicle (RMTV) that will rendezvous with the MPTV in Mars orbit. Based on earlier studies (Spacecraft Design Group 1985), the conceptual design of the MPTV began with the requirement that it possess the capability to transfer a minimum of 378 tonnes into Mars orbit. The MPTV is powered by a 40-MWt liquid-metal cooled, fast nuclear reactor. A two-loop, alkali-metal, Rankine cycle power conversion system produces approximately 7.5 MWe to power the thruster and onboard systems. The ion engine utilizes mercury as its propellant and provides a thrust of 245 newtons. After launching from low Earth orbit (LEO), the MPTV begins a geocentric spiral out from Earth. Upon reaching escape velocity, it proceeds on a heliocentric transfer to Mars orbit insertion. With an initial mass of 750 tonnes in LEO, the ion propulsion system has the capacity to deliver a payload of 383 tonnes into Mars orbit yielding a mass fraction, based on the ratio of payload to initial mass, of 0.51. To deliver an equivalent payload of the same mass, a chemical propulsion system utilizing a Hohmann transfer (with 400-s specific impulse engines) requires a total mass in LEO of 1479 tonnes and yields a mass fraction of only 0.20.

Introduction

The unmanned nuclear electric Mars payload transfer vehicle (MPTV) is a component in the mission scenario developed by the Texas A&M University and the University of Texas at Austin Manned Mars Mission Study conducted for the Universities Space Research Association at the National Aeronautics and Space Administration (NASA) Johnson Space Center in Houston, Texas. This Scenario employs two vehicles, and MPTV that will carry the bulk of the equipment required for Mars proximity operations and a Rapid Manned Transfer Vehicle (RMTV) that will rendezvous with the MPTV in Mars orbit. Based on earlier studies (Spacecraft Design Group 1985), the conceptual design of the MPTV began with the requirement that it possess the capability to transfer a minimum of 378 tonnes into Mars orbit. The following sections summarize the conceptual design of a mercury ion propulsion system for the MPTV and compare its performance to a chemical propulsion system.

Ion Propulsion System

Nuclear Reactor

A compact, liquid-metal cooled, fast reactor is chosen as the power supply to minimize mass and volume requirements and to allow high temperature operation. This reactor is based upon a design developed at Texas A&M University and has an output of 40 MWt, which will provide electric power for the ion engines and onboard systems (Becker et al. 1985). Heat will be transferred by a pumped liquid lithium-7 coolant loop to the Rankine power conversion system.

Lithium-7 is utilized due to its many advantages over other coolants. As a liquid metal, lithium possesses high thermal and electrical conductivities that permit efficient heat transfer and electromagnetic pumping. In addition, lithium-7 has a moderately low melting point and vapor pressure, as well as the highest volumetric specific heat of any liquid metal. Thus, lithium requires only about one-half to one-third the usual coolant volume (McLain and Martens 1964).

The fuel of a nuclear reactor is of primary consideration when designing a nuclear power system. In selecting a fuel system, the fissile atom density, fuel operating temperature, fuel swelling, fission gas release, fuel-cladding compatibility and the consequences of cladding failure must all be considered (Cox et al. 1984). Uranium mononitride (UN) is chosen as the fuel since it exhibits many advantages over uranium oxide (UO_2) and uranium carbide (UC). For example, it possesses a fissile atom density some 30% higher than that of UO_2 (Cox et al. 1984). It also demonstrates a significantly lower fission gas release than either UO_2 or UC at high temperatures (Cox et al. 1984). UN does, however, have problems associated with fuel manufacture, stoichiometry, the characterization of swelling at high temperatures and burnups and, in general, lacks experimental and testing data in the required operating regime. A tungsten-rhenium (W/Re) alloy which demonstrates chemical compatibility up to 1700 K with UN fuels (Cox et al. 1984) is chosen as the cladding material. This alloy possesses a high thermal conductivity, low neutron cross section, good mechanical strength and less susceptibility to irradiation embrittlement than stainless steels or nickel based alloys.

At temperatures ranging from 1000 K to 2000 K, reactions between UN and lithium are not thermodynamically feasible, while both UC and UO_2 will react either indirectly or directly. UC fuels allow excess carbon, readily soluble in lithium, to react with cladding and structural materials. The reaction of UO_2 with lithium is a significant problem in that the formation of U_3O_8 and free uranium, in combination with cladding and structural materials, could result in the production of a low melting point eutectic (Cox et al. 1984).

The fuel assembly is arranged in a square lattice comprising 2745 fuel elements. This assembly will be positioned in the reactor vessel by the upper and lower support plates. Structural support for the assembly is provided by four honeycomb grid spacers placed evenly between the support plates. The grid spacers will allow the thermal expansion of the fuel elements in both radial and axial directions. In addition, they will provide an excellent turbulent flow field through the core and enhance interchannel mixing of the coolant. The entire fuel assembly is contained within the pressure vessel, which is the primary structural boundary for the reactor.

The reactor vessel is composed of a niobium-tungsten (Nb/W) alloy which possesses an ultimate tensile strength of 348 kPa at 1475 K (McLain and Martens 1964). The lower plenum of the reactor vessel is equipped with mixing vanes to uniformly distribute the coolant, reducing the thermal stresses in structural components and improving the heat transfer characteristics of the reactor. An upper plenum also will be used to ensure uniform coolant exit temperatures. Figure 1 is a schematic diagram of the reactor vessel and internal components.

The reactor will be controlled with a reflector/absorber control drum system. A total of eight drums are positioned symmetrically about the circumference of the core within the graphite reflector. They will be placed in a dry well that permits cooling by the liquid lithium coolant. Each will comprise one-half beryllium oxide (BeO, reflector) and one-half boron carbide (B_4C, absorber). By rotating the drums in tandem, which varies the ratio of reflector and absorber influencing the core reactivity, the reactor power level can be controlled. In addition to the control

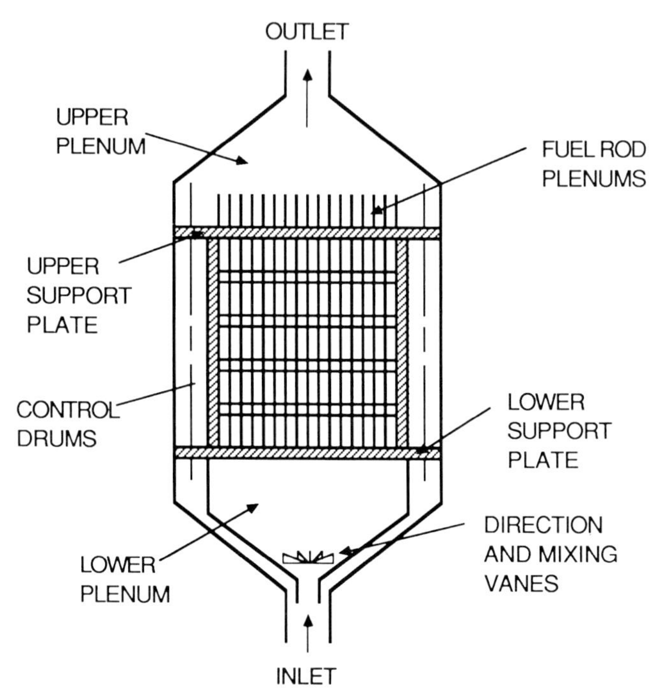

Figure 1 *Cut-Away View of Reactor Vessel.*

drums, five safety/shutdown rods are placed symmetrically within the lattice. During reactor operation these rods will be positioned outside the core with their utilization limited to use during start-up and shutdown. The design of the safety/shutdown rods is such that the core can be shut down regardless of the control drum positions. Figure 2 is a sectional view of the reactor vessel.

The reactor analysis included core critical calculations as well as thermal hydraulic analyses. A summary of the core parameters is provided in Table 2.

Power Conversion System

In this study, the components and performance of the power conversion system are not explicitly defined. Due to changes in the original design (Becker et al. 1985), at this point in the analysis estimates of an overall efficiency (19%) and mass (see Table 1) are used to characterize its performance. This system is modeled after a potassium, Rankine power conversion system developed by NASA's Lewis Research Center (Heller et al. 1969 and English 1982). A system of this type has the potential for very high performance when compared to other power conversion systems and thus is chosen for this design. For example, at a given peak cycle temperature, the radiator size required is only one-third that of the Brayton concept (English 1984). This is due primarily to the fact that systems of this type can operate at extremely high temperatures (up to 1500 K).

Figure 3 is a schematic diagram of the system. The primary coolant, lithium-7, is electromagnetically pumped through the nuclear reactor where it is heated but does not boil. The lithium then transfers its energy by way of a one pass, shell-and-tube, counterflow, potassium boiler. The lithium passes through the tubes and transfers heat to the potassium in the outer shell. The potassium is then vaporized to a superheated state and passes through a multistage potassium turbo-alternator. Due to operational limitations, the potassium vapor exiting the multistage turbine will not maintain 100 percent quality (approximately 80 percent or lower). This may pose a problem in that significant erosion of the turbine blades may occur. One solution to this problem is to place moisture separators in front of the turbo-alternator as well as behind each turbine stage. After exiting the turbo-alternator, the potassium is condensed to a subcooled liquid in the condenser assembly.

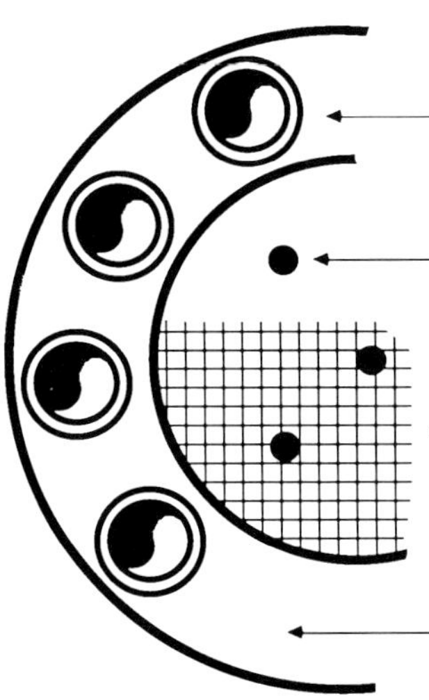

Figure 2 *Cross-Sectional View of Reactor Vessel Detailing Control Drum and Safety Rod Placement.*

Table 1 *System Mass Summary of Mars Payload Transfer Vehicle.*

System	Mass(kg)
Reactor/Power Conversion[a]	80,000
Ion Thruster Assembly	20,000
Structure	75,000
Propellant[b]	192,000
Payload	383,000
Total	750,000

[a]With non-manrated shielding system.
[b]Includes 10% additional fuel for orbital maneuvers.

Table 2 *Summary of Reactor Operational and Structural Parameters.*

Parameter	Value
Power	40 MWt
Fuel	Uranium mononitride
Coolant	Lithium-7
Fuel/Coolant Ratio	0.60
Active Core:	
Radius	0.31 m
Height	0.62 m
Critical Mass	865.60 kg
Fuel Element:	
Cladding	Tungsten/Rhenium
Thickness	0.51 mm
Rod Outer Radius	3.82 mm
Rod Length	0.96 m
Reactor Assembly:	
Pitch to Diameter Ratio	1.44
Total Number of Rods	2745
Reactor Vessel:	
Material	Nobium/Rhenium
Thickness	10.00 mm
Reactivity Control: Control Drums	
Reflector	Beryllium oxide
Absorber	Boron carbide
Safety Rods	Boron carbide

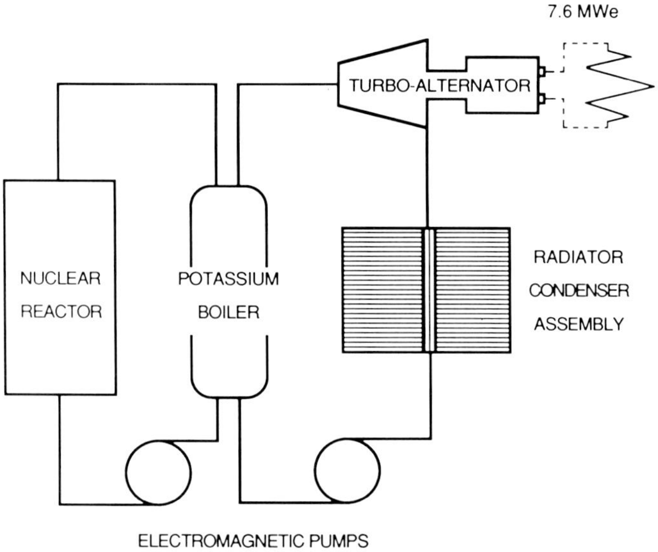

Figure 3 *Thermodynamic Cycle Schematic.*

The condenser is composed of molybdenum/lithium (Mo/Li) heat pipes inserted into a central channel. As the potassium passes through the central channel it loses heat and the lithium in the end of the heat pipe vaporizes. The vapor diffuses throughout the heat pipe and condenses upon its cold surfaces. The condensate is then transported back to the hot end of the pipe by capillary action along the stainless steel mesh lining the Mo tube. After passing through the condenser, the potassium is electromagnetically pumped into the boiler to continue the cycle.

Ion Thrusters

An ion propulsion system creates thrust only through the electrostatic forces resulting from the interaction between charged particles and an electric field. For the thruster for the MPTV, mercury is utilized as the propellant in the following process. The liquid mercury is vaporized by an electric heater and transferred to a diffusor screen to ensure uniform vapor quality and distribution. The vapor is then ionized by a contact potential ion source. The ionization process results in the emission of singly ionized mercury atoms (Corliss 1960).

Once ionized, the mercury enters the accelerator and is acted upon by an electric field. This design employs an acceleration-deceleration system to minimize thruster mass while still achieving the required current density, voltage and exhaust velocity. In this system, the ion thruster is operated at a very high voltage with a correspondingly large current density to minimize the size and mass of the accelerator grid. However, this configuration does not produce the desired exhaust velocity. Therefore, before the ions are emitted into free space they are subjected to a decelerating field that slows them to the desired exhaust velocity (Stuhlinger 1964). To reduce sputtering, the interaction of ions with the engine structure, the beam is concentered by focusing fields placed before and after the decelerator.

Neutralization is required to prevent a flow reversal of the accelerated ions resulting from the space charge field of the beam (Stuhlinger 1964). This is accomplished by bulk conduits attached to the accelerator exhaust port. The conduits are split into neutralizer probes which, under the influence of the space charge field, emit electrons into the beam. Figure 4 is an illustration of an integrated thruster module. The thruster assembly will consist of clusters of these modules. The performance characteristics of this system are presented in Table 3.

Trajectory Analysis

This section presents a preliminary trajectory analysis from LEO to Mars for a vehicle employing the aforementioned ion propulsion system. The purpose of this analysis is not to designate a nominal trajectory but to provide preliminary data, by means of a simple calculation, on the performance capabilities of the engine. Thus, the resultant trajectory is not optimized.

To simplify the computational analysis, the trajectory is divided into two phases. Phase One is the geocentric spiral out of LEO until the attainment of escape velocity. Phase Two is a "targeted" heliocentric transfer from Earth to Mars. The targeting is accomplished by varying the time and direction of thrust so

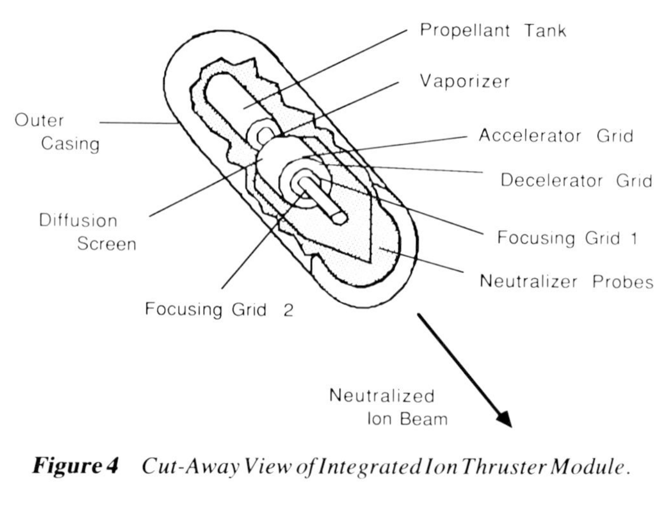

Figure 4 *Cut-Away View of Integrated Ion Thruster Module.*

Table 3 *Summary of Ion Thruster Assembly Performance Parameters.*

Parameter	Value
Exhaust Power	6.00 MW
Specific Impulse	5000 s
Mass Flow Rate	5.0×10^{-3} kg/s
Exhaust Velocity	49 km/s
Thrust	245 N

that at arrival the radial velocity is zero and the tangential velocity is equal to that of Mars.

Several general assumptions, applicable to both phases, are made in integrating the equations of motion used in the determination of the resultant trajectories. One assumption is that the vehicle is treated as a point mass with the thrust and gravitational attraction of the central body as the only external forces. A second assumption is that all motion takes place in the same plane. Third, when thrust is applied, it is instantaneous and at a constant 245 newtons. Also, launch window considerations are not taken into account.

Applying the above assumptions, the equations of motion are integrated using a Runge-Kutta fourth order integration scheme. Given the initial conditions, the resultant trajectory is found by iterating forward through time.

The state equations of the transfer trajectory in polar coordinates are as follows:

$$\overset{*}{u} = v^2/r - \mu/r^2 + a \sin\alpha$$

$$\overset{*}{v} = -uv/r + a \cos\alpha$$

$$\overset{*}{r} = u$$

$$\overset{*}{\theta} = v/r$$

$$a = T/[m_o - m(t - t_o)]$$

where the variables are defined as:

- r: Sun-spacecraft distance,
- θ: angle between the Sun-spacecraft line and the Sun-Earth line at launch,
- u: velocity component along the Sun-spacecraft line,
- v: velocity component perpendicular to the Sun-spacecraft line,
- μ: center body gravitational constant,
- T: thrusting force magnitude,
- m_o: initial spacecraft mass,
- m: mass flow rate, and
- α: thrust direction angle.

The initial conditions for the two phases are as follows:

- Phase One, the geocentric spiral out of earth parking orbit,
  —Starts in a circular parking orbit of 500 km, and
  —Has a thrust vector perpendicular to the radius vector.
- Phase Two, the heliocentric transfer from Earth to Mars,
  —Assumes circular orbits for both Earth and Mars,
  —Has an initial spacecraft velocity equal to the termination velocity of Phase One in addition to the Earth's orbital velocity, and
  —Has an initial spacecraft radius from the Sun equal to that of Earth's.

The heliocentric transfer has three distinct subdivisions. The initial portion of the trajectory has the thrust directed along the flight path. When the apoapsis of the transfer ellipse is equal to the heliocentric radius of Mars, the second portion of the trajectory begins as thrust is terminated and the vehicle is allowed to coast. The thrust is again turned on for the final portion of the transfer to target for Mars orbit insertion by matching the vehicle's heliocentric velocity to that of Mars. For the final portion of the transfer, the direction of thrust and the ''thrust on'' point are determined by iteration until the desired final conditions are obtained. The results are listed below.

- Phase One, the geocentric spiral out of Earth parking orbit, has
  —Initial mass = 750,000 kg,
  —Time to escape = 252.52 days,
  —Distance at escape = 970,610 km,
  —Radial velocity at escape = 0.57298 km/s,
  —Tangential velocity at escape = 0.70223 km/s,
  —Theta at escape = 6448.76134 rad, and
  —Propellant consumption at escape = 109,088.64 kg.
- Phase Two, the heliocentric transfer, has
  —Initial mass = 640,911.36 kg,
  —Thrust-off point = 325.61 after launch (73.09 days after escape),
  —Radius at thrust-off = 0.163212×10^9 km,
  —Theta at thrust-off = 1.25601838 rad,
  —Thrust-on point = 472.60 days after launch (220.08 days after escape),
  —Radius at thrust-on = 0.22027569×10^9 km,
  —Theta at thrust-on = 2.9652045 rad, and
  —Alpha at thrust-on = -21.73°.
- At Mars arrival, the vehicle has
  —Heliocentric radius = 2,279,400 km (+/− 10,000 km),
  —Radial velocity = 0.02077 km/s,
  —Tangential velocity = 24.01878 km/s,
  —Theta = 3.65902 rad, and
  —Final mass = 575,359.68 kg.

Comparison To Chemical System

To demonstrate the advantages of the ion propulsion system over a chemical propulsion system for this type of mission, propellant consumption comparisons were made with a two-stage, chemical system having a specific impulse of 400 s.

To place a total payload of 383,000 kg in Mars orbit, the following results were obtained.

- The ion engine has
  —Total time of flight of 551.25 days,
  —Total propellant consumption of 174,640 kg,
  —Mass fraction (defined as payload mass/initial total mass) of 0.51, and
  —Total initial mass in LEO of 750,000 kg.
- The chemical engine, using a Hohmann transfer trajectory, has
  —Total time of flight of 260 days,
  —Total propellant consumption of 986,152 kg,
  —Mass fraction of 0.26, and
  —Total initial mass in LEO of 1,478,724 kg.

Conclusion

The results of the trajectory analysis demonstrates that NEP systems may offer performance advantages when compared to conventional, chemical propulsion systems for Mars payload transfer missions. The NEP system in this study delivered the prescribed payload requiring 82% less fuel and 50% less mass in LEO. However, these conclusions should be tempered by the fact that a comprehensive analysis was not performed and many simplifying assumptions were utilized in the study. Additionally, the systems used in the analyses are quite speculative and, in some instances, beyond current technology. Areas that should be subjects of further study include iteration upon the conceptual reactor design; specifically concerned with increasing power, fuel pin integrity at high burnup and temperatures for extended periods of time, reactor shielding, and reactivity control. Pertaining to power conversion, analysis of a potassium Rankine cycle of the power and size utilized in this study should be performed, specifically considering two-phase flow and heat transfer of liquid metals in a microgravity environment. Additional study should be concentrated on refractory metal technology development relevant to high-temperature nuclear reactor and power conversion performance. For the ion thruster system, it is recommended that design and testing of prototypes at the power and voltage levels required be undertaken, specifically addressing anode/cathode erosion and overall thruster assembly endurance.

As mentioned previously, optimization of the trajectory was not included in this study. Further analysis should include the utilization of a two-point boundary value technique to determine a trajectory that would minimize propellant consumption. Additional refinements can incorporate the application of a planetary ephemeris to eliminate the assumptions of circular and coplanar planetary orbits.

Acknowledgments

This research was performed at NASA Johnson Space Center in the Advanced Programs Office and was a part of the Universities Advanced Space Design Program funded by the Universities Space Research Association. The following people made significant contributions to this study: John Alred, Barney Roberts, and Brigid Kranz of NASA Johnson Space Center and Gus Babb of Eagle Engineering, Houston, Texas.

References

Becker, D., D. Brozak, D. Eubanks, and W. Lyon (1985) *Mercury Ion Propulsion System for a Martian Transfer Vehicle, Final Report*, Nuclear Engineering 410 Design Project, Texas A & M University, College Station, TX.

Corliss, W.R. (1960) *Propulsion Systems for Space Flight*, McGraw-Hill, New York, NY.

Cox, C.M., D.S. Dutt and R.A. Karnesky (1984) "Fuel Systems for Compact Space Reactors", in *Space Nuclear Power Systems 1984*, M.S. El-Genk and M.D. Hoover, eds., Orbit Book Co., Malabar, FL.

English, R.E. (1984) "Dynamic Power Systems for Power Generation", in *Space Power*, NASA Conference Publication 2352, Lewis Research Center, Cleveland, OH.

English, R.E. (1982) *Power Generation from Nuclear Reactors in Aerospace Applications*, NASA-TM-83342, Lewis Research Center, Cleveland, OH.

Heller, J.A., T.A. Moss, and G.J. Barna (1969) *Study of a 300 Kilowatt Rankine Cycle Advanced Nuclear Electric Space Power System*, NASA TM X-1919, Lewis Research Center, Cleveland, OH.

McLain, S. and J.H. Martens (1964) *Reactor Handbook*, 2nd ed., Interscience Publishers.

Spacecraft Design Group (1985) *A Manned Mars Mission : Preliminary Design Report 2 and Final Report*, The University of Texas at Austin, Austin, TX.

Stuhlinger, E. (1964) *Ion Propulsion for Space Flight*, McGraw-Hill, New York, NY.

Space Nuclear Power Systems 1986
Edited by M. S. El-Genk and M. D. Hoover

Chapter 35

The Pegasus Drive— A Multimegawatt Nuclear Electric Propulsion System

Edmund P. Coomes, Judith M. Cuta, and Brent J. Webb

Pacific Northwest Laboratory
Richland, WA 99352

David Q. King

Jet Propulsion Laboratory
Pasadena, CA 91109

A propulsion system (The PEGASUS Drive) consisting of a magnetoplasmadynamic (MPD) thruster driven by a multimegawatt nuclear power system is proposed as the propulsion system for a manned Mars mission. The propulsion system described is based on a mission profile containing a 510-day burn time (for a mission time of approximately 1000 days). Electric propulsion systems have significant advantages over chemical systems, because of high specific impulse, lower propellant requirements, and lower system mass. The thermal power for The PEGASUS Drive is supplied by a boiling liquid-metal fast reactor. The system consists of the reactor, reactor shielding, power conditioning, heat rejection, and MPD thruster subsystems. It is capable of providing a maximum of 8.5 megawatts of electrical power of which 6 megawatts is needed for the thruster system, 1.5 megawatts is available for spacecraft system operations and inflight mission applications, leaving the balance for power system operation.

Introduction

With the Space Transportation System (STS), the advent of space station Columbus, and the development of expertise at working in space that this will entail, the gateway is open for missions to Mars. The missions are possible with state-of-the-art hydrogen/oxygen propulsion, but would be greatly enhanced by the higher specific impulse of electric propulsion. This paper presents a concept that uses a multimegawatt nuclear power plant to drive an electric propulsion system. The concept has been named PEGASUS (*P*ow*E*r *G*ener*A*ting *S*ystem for *U*se in *S*pace) Drive, and is intended as a "work horse" for general space transportation needs, both long- and short-haul missions.

The advantages of electric propulsion are well known in the aerospace community. But this high specific impulse, propellant-saving, potentially cost-saving, and mission-enabling technology has not received serious consideration for spacecraft propulsion primarily because of the lack of a suitable, light-weight electric power source. The recent efforts of the SP-100 program indicate that a power system capable of producing upwards of 1 megawatt of electric power should be available in the next decade. Of greater interest are efforts in other areas which indicate that a power system with a constant power capability an order of magnitude greater could be available near the turn of the century. With the advances expected in megawatt-class space power systems, the high specific impulse electric propulsion systems, such as a MPD thruster powered by a nuclear electric power plant, must be reconsidered as potential propulsion systems. A conceptual drawing of a manned Mars spacecraft powered by the PEGASUS Drive is shown in Figure 1. The electrical power for the propulsion system is provided by a nuclear multimegawatt fast reactor space power system. This power system is capable of meeting both the propulsion system and spacecraft power requirements. The size and mass limitations of the STS are a prime consideration in the design, and the collapsed system will be capable of lower earth orbit insertion by two shuttle missions. Development of this power system could be completed by the mid 1990s and the system available near the turn of the century. Since this is an advanced system concept some development efforts are still needed in the fuels, heat rejection, and turboalternator areas.

Mission Analysis

Any number of mission profiles may be devised for a manned Mars expedition, depending on the assumptions made concerning desired objectives and available technology. This paper uses a previously developed set of mission parameters (Atkins et al. 1983) as a basis for

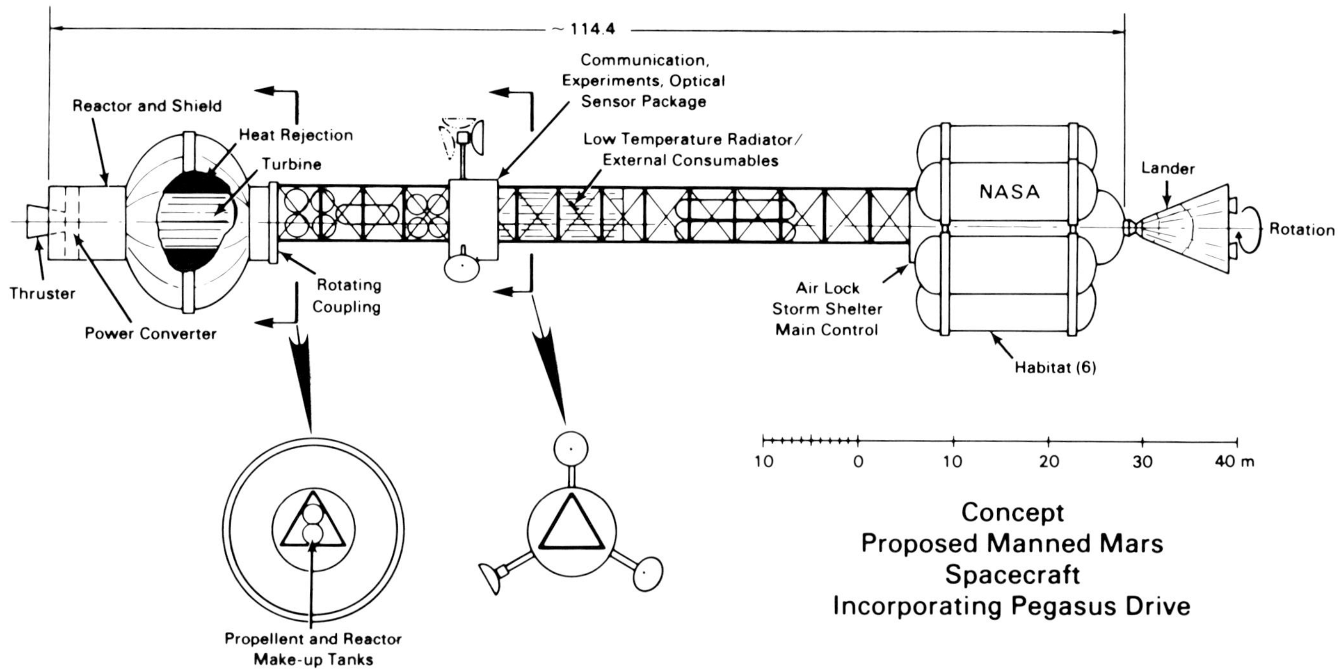

Figure 1 *Concept of Proposed Manned Mars Spacecraft with Pegasus Drive.*

design. However, the propulsion system concept is adaptable to a wide range of mission profiles. The parameters defining this Mars mission consist of a 344 metric ton initial mass vehicle in Low Earth Orbit (LEO) that autonomously raises itself through the radiation belt to Geosynchronous Earth Orbit (GEO), whereupon a crew of three boards. The trip to Mars takes 601 days, exploration 100 days, and the return trip 268 days. Propellant consumed is 111 metric tons with a 5% reserve. The return mass to GEO is 103 tonnes, 76 tonnes having been left on Mars. The allowance for power and propulsion system was 34 tonnes, including 11 tonnes for tankage. Figure 2 shows how vehicle mass M (excluding only propellant) can be traded against power, P, overall efficiency, η and burn time, t, to provide a desired specific impulse. Improvements to the propulsion system would be reflected in higher values of η, and have the same effect as reductions in vehicle mass M in reducing the required burn time. Actual trip time for a mission is a function of many variables, but it is clear from Figure 2 that the more power developed by the propulsion system (all other things being approximately equal), the shorter the trip time. The timesaving can be accomplished either by the shorter burn times required to produce the same velocities as projected for the chemical fueled system, or by longer burn times that produce higher velocities in the coast phase. Such tradeoffs in power, burn time, and coast time would have to

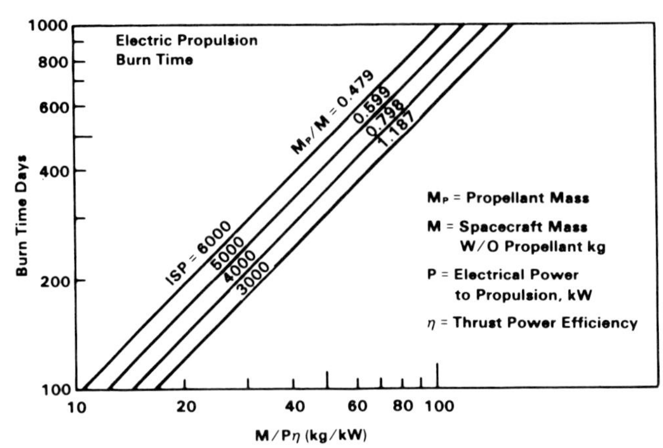

Figure 2 *Electric Propulsion Burn Time.*

be studied extensively to determine the optimum combination, but it is clear that electric propulsion systems are greatly mission-enhancing in this regard.

Electric propulsion also shows advantage for this mission in the modest propellant requirements as compared to chemical systems. While the scope of this paper excludes a detailed comparison, the propellant required is characterized by the specific impulse of the propulsion system. Since electric engines have a specific impulse that is 10 to 20 times that of hydrogen-oxygen chemical systems, the propellant required for electric systems can be from ⅓ to ⅒ that of chemical propellant.

Propulsion System

Electric propulsion systems have not been seriously considered for use with large spacecraft due to the lack of a suitable electric power source to drive them. However, recent efforts to develop megawatt-class space power sources show such systems to be technologically feasible, and a multimegawatt lightweight nuclear electric propulsion system would enable missions of almost any conceivable duration and scope. The manned Mars mission would be well within the capabilities of such a system. The electric propulsion could be provided by a magnetoplasmadynamic (MPD) thruster coupled with a nuclear power plant that provides the required power. Such a system is described in the following sections.

MPD Thruster System

The MPD system is composed of thrusters, propellant tanks, and thermal control subsystems. A schematic of the major components of the MPD thruster system is shown in Figure 3. Three phase ac at 10 to 20 kV and 1500 Hz is fed through a transformer and rectifier to the MPD thruster. Power control and regulation is confined to adjustments of the ac output from the alternator. A more detailed description of this component is given later in the section covering the power conversion subsystem. The high tension buss runs from near the turboalternator past the reactor shield to the transformer and thruster. This 15-meter long cable is insulated and electrically shielded to reduce electromagnetic immitance and has a mass of 69 kg.

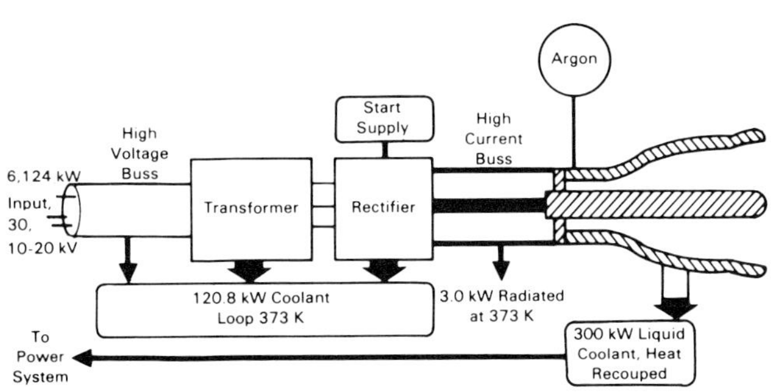

Figure 3 *MPD Thruster System Components.*

The transformer mass has been estimated from design equations and weighs only 1911 kg (McLyman 1978). (Efficiency of 99% is a reasonable design goal for this delta-delta transformer.) The rectifier is designed to use 3000 amperes diodes derated to 1500 amperes. Eight diodes are connected in parallel to handle 12,000 amperes. Six units are required for full wave rectification of three phase AC. The 48 diodes and cooling assembly is estimated to mass 200 kg, and will dissipate 59.5 kW when rectifying 25,000 ADC. The transformer, rectifier, and high tension buss require 120.8 kW of heat to be removed by a cooling plant associated with the cryogenic superconducting alternator.

The current buss to the MPD thruster is designed to be self-radiating at 373 K, and has a mass of 817 kg, which includes 0.5 cm of insulation between the coaxial conductors on the outside. An alternative that may reduce overall mass at the cost of complexity is to reduce the cross section of aluminum which increases the dissipation and provides active cooling. Also, 3 meters were allowed to electrically connect the MPD thruster, which operates at about 1600 K, and the rectifier assembly which is designed for 373 K, thus providing thermal isolation.

The MPD thruster assembly consists of seven engines which are used one at a time, for a 510-day burn. Based upon ongoing work at the Jet Propulsion Laboratory to evaluate thruster lifetime with subscale devices, and the present understanding of cathode physics, thruster lifetime (which is limited by the cathode) is estimated to be 2000 hours. But life test of a multimegawatt MPD thruster can only follow a vigorous development program, which is not presently planned in NASA's research program.

Figure 4 shows a schematic of the MPD thruster which accelerates propellant by means of a magnetic body force. Propellant is injected through the rear insulator and is ionized by a high current, diffuse discharge. The large multi-kiloampere current creates an azimuthal magnetic field which interacts with the current to produce a magnetic body force directly on the ionized gas. Exhaust speeds of 15,000 m/s to 80,000 m/s have been measured on a thrust stand using a variety of propellants (Burton et al. 1983). Thruster conversion efficiency of electrical power to directed kinetic energy has been measured at 35% on a thrust stand and analysis

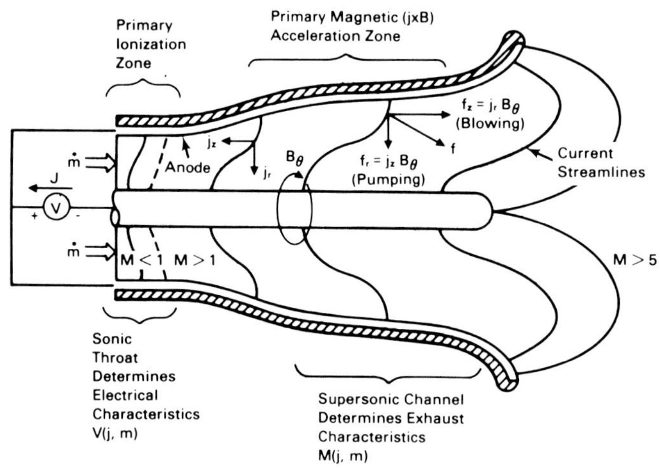

Figure 4 *MPD Thruster.*

suggests that 50% to 60% is ultimately possible (Burton et al. 1983 and King et al. 1981).

All seven engines are connected in parallel to the high current buss, but each has a separate propellant valve and a contactor on the cathode current feed. Though performance development may change the electrode shape somewhat, the overall dimensions of a multimegawatt thruster will remain about the same (King and Vondra 1984). Each engine is assumed to have a center-body cathode 3 cm in diameter and 10 cm long, with an anode about 12 cm inside diameter, and of a comparable length. The most massive part of the MPD engine is the anode heat removal system, which must remove approximately 300 kW. An allowance of 700 kg is made for the assembly of seven engines, contactors, and thermal control.

The MPD thruster is started by application of 500 V and a low propellant flow which results in a glow discharge. As the power in this glow discharge is increased to 2 to 5 kW, the cathode is heated to incandescent temperatures. High temperatures result in prodigious thermionic cathode emission and the arc jumps to a low impedance arc mode. The alternator output can then be ramped up along with the propellant flow rate to full power. The MPD thruster exhaust velocity and thrust may be throttled down to half of nominal by adjusting the mass flow and power.

A summary of the mass breakdown and power consumption is shown in Table 1. A contingency of 10% of the system mass is included, and 20% on top of that added for structure. The net propulsion system mass excluding power and the cooling plant is 5144 kg. The electrical system is 98% efficient while the MPD thruster is assumed to have a 50% conversion efficiency. An interesting point on the waste heat of the MPD thruster is that since the engine operates a high 1600 K temperature, the 300 kW of heat rejected can be recouped to the advantage of the power conversion system.

The MPD thruster has been studied extensively with argon propellant, and the range of specific impulse available with this propellant is well suited to this mission. Cryogenic storage of argon has been considered in detail elsewhere and presents no unusual constraints or problems. A 16,301 kg storage tank has been designed with a mass of 721 kg (Eberhardt et al. 1981). This tank is shuttle compatible and operates in an environment of 293 K while allowing only 0.14 W of heat energy to leak into the fluid. Additional heat is needed to drive off gas at 0.0025 g/s at 2 atm for the MPD thruster. Seven of these tanks are required to store 111,000 kg of propellant for the mission for a total tankage mass of 5047 kg.

Potassium, which is the working fluid in the dynamic conversion equipment, has nearly the same atomic mass as, and one more proton than, argon. Since it's ionization potential is about one-fourth of argon, potassium should provide a higher thrust efficiency than argon. Potassium offers a simpler storage problem in that it can be stored in a solid form and then liquified at 335 K. Since the density of solid potassium is comparable to liquid argon, and mercury tanks for ion propulsion have a tank fraction comparable to 4.4% for liquid argon, this study will assume that the tank mass for potassium will be the same as for argon.

There is a concern that the small fraction of the exhaust that flows back toward the spacecraft may plate out on many surfaces after 111 metric tons of potassium are exhausted. Though potassium may be a better propellant from a performance and storage standpoint, con-

Table 1 *Manned Mission to Mars Propulsion System Summary.*

Item	Mass	Power Dissipated
MPD Thruster Assembly	700 kg	6000.0 kW
– 7 engines required for 510 day burn		
– 2000 hr life per engine assumed		
– 240 VDC, 25,000 ampere input		
– 5000 s specific impulse		
– 50% thrust efficiency		
– 2.5 g/s propellant flow rate		
Engine Startup and Control Subsystem	50 kg	
High Current Buss	817 kg	3.0 kW
– 25,000 amp insulated coaxial conductor		
– 0.36 m dia by 3 m length		
– self-radiating at 373 K		
Rectifier	200 kg	59.5 kW
– 48 diodes, 3,000 amp rating each		
– heat sink requires active cooling at 373 K		
Transformer	1911 kg	60.0 kW
– 6123 kW, 1500 H, 3 phase, 10 to 20 kV input		
– 99% efficiency		
– 0.5 m^3 volume		
High Voltage Buss	69 kg	1.3 kW
– 10 to 20 kV, 3 phase, 500 A rating		
– insulated, shielded		
– self-radiating at 373 K		
– 10 cm dia. by 50 m long		
Thermal Control	100 kg	
– liquid coolant to MPD anode, recoups 300 kW at up to 1600 K		
– coolant to diodes and transformer and high voltage buss	50 kg	
Subsystem Total	3897 kg	6123.8 kW
Contingency Mass 10%	390 kg	
Structrual Mass 20%	857 kg	
Propulsion System Total Mass	5144 kg	
Specific Mass	0.840 kg/kW	
Power Processing Electrical Efficiency	98.0%	

tamination of the spacecraft surfaces may require argon propellant.

PEGASUS

PEGASUS, the proposed power source for the MPD thruster system, is an 8.5-MWe boiling liquid-metal, space-based nuclear power system. The system employs a direct Rankine power cycle and is designed to meet the power requirements for a 6-MWe electric propulsion system with an additional 1.0 to 1.5-MWe available for mission-specific tasks and experiments. The balance is used for the operation of the power system itself.

The power system is comprised of five major subsystems or components as shown in Figure 5. These are a cermet fueled, boiling liquid metal fast reactor; a four-pi contoured man-rated reactor shield; two radial flow Ljungström turbines with four superconducting alternators for power conversion; a power conditioning subsystem; and a heat rejection/thermal control subsystem. A schematic diagram of the system is shown in Figure 6. Table 2 provides a mass breakdown by component and shows the anticipated division for each shuttle load required to place the PEGASUS Drive in orbit. A description of these subsystems is given in the following sections.

Reactor Subsystem

The reactor selected for the PEGASUS system is a fast reactor using a boiling alkali metal coolant and cer-

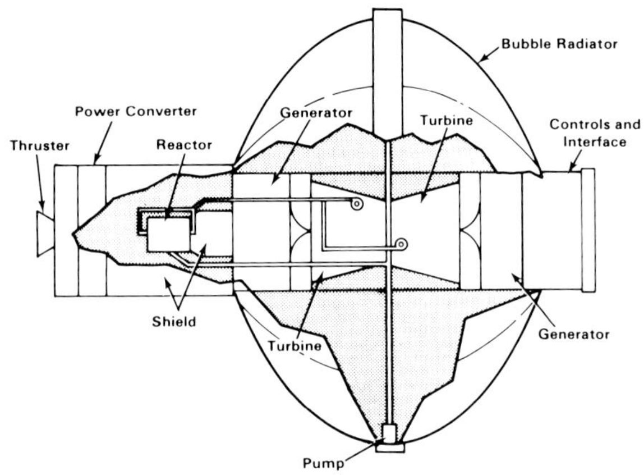

Figure 5 *Schematic of Pegasus Drive System.*

met fuel. The cermet fuel is composed of a refractory metal matrix, such as tungsten, with 93% enriched uranium nitride as the fuel material (Barner et al. 1987). A cermet fuel was selected because of its superior thermal characteristics and its capability to withstand the high stresses imposed by enhanced fission gas release and fuel swelling. A strong, relatively ductile metal matrix with a high thermal conductivity is well suited to the needs of a reactor where rapid power swings are necessary to meet varying power demands. Also the enclosed coolant channel configuration eliminates cross flow instabilities that occur during boiling in open lattice cores.

The core has an average power density of 103 kW/L and produces 26 MW of thermal energy. The reactor shown in Figure 7 is a split core 63 cm high and 73 cm in diameter. Operational reactor control is accomplished by movable reflector segments with backup and safety control accomplished by moving one half of the core relative to the other to ensure a safe shutdown margin under launch or accident situations. Each core half is comprised of five levels of fuel elements containing 151 fuel elements. Each fuel element, as shown in Figure 8, is a hexagonal design 6.0 cm high, 3.5 cm across, and contains 169 coolant channels. Potassium has been selected as the alkali metal coolant because of its compatibility as both a working fluid for the turbine and as coolant for the core. To minimize coolant activation the potassium is depleted with respect to potassium-41.

A preliminary thermal analysis of the cermet fuel, assuming full power operation and stable nucleate boiling, indicates that a peak centerline fuel temperature of 1650 K can be expected for a coolant channel bulk temperature of 1400 K. It should be noted, however, that detailed thermal analysis of this system (or any system involving two-phase flow) will require additional experimental and theoretical work. Table 3 gives a mass breakdown of the reactor system.

Reactor Shielding Subsystem

The PEGASUS system incorporates a four-pi contoured shield to reduce radiation from the reactor to safe levels during full-power operation. The shield is a composite material of depleted LiH, W, and natural LiH (Barattino 1985). To maintain the physical integrity of the shield, it is cooled by the reactor inlet coolant. The mass of the shield requires that the major portion of it be

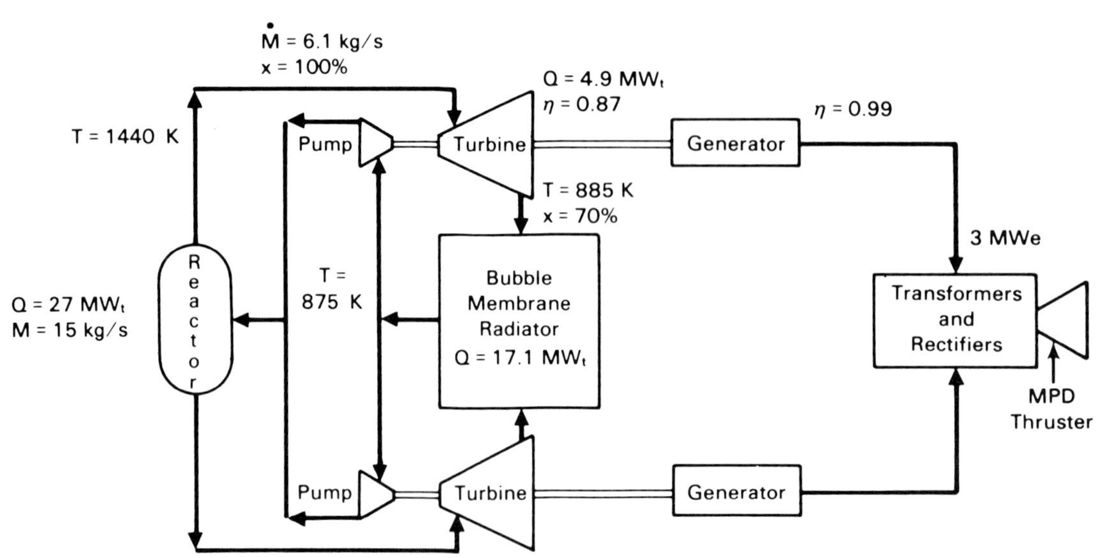

Figure 6 *Pegasus Drive System Component Locations.*

Table 2 *Power System Mass Summary.*

System		Mass(kg)
Reactor	28.5%	5,332
Shadow Shield	26.9%	5,032
Ljungström Turbines	13.1%	2,444
Alternators	10.1%	1,890
Membrane Radiator	7.9%	1,479
Auxiliary Cooling	4.9%	907
Support Structure	4.6%	854
Potassium Inventory	4.0%	744
Main System Mass		18,682
Propulsion System Mass		5,144
1st Shuttle Load		23,826
2nd Load Four-Pi Shield		26,942

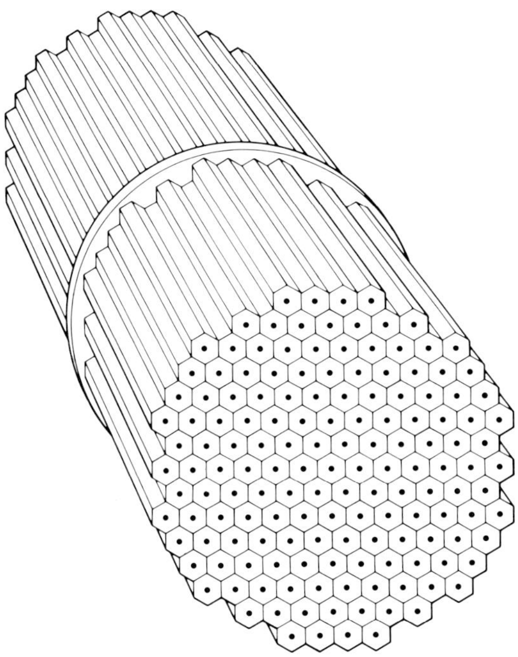

Figure 7 *Split Core Design.*

delivered to orbit on a separate shuttle load, to be installed as the system is deployed.

Power Conversion Subsystem

Power conversion is accomplished by a pair of turboalternators. The turbine designs incorporate a multistage radial out-flow design developed by Berger Ljungström (Ljungström 1949) in 1910. This turbine design, appropriately named the Ljungström Turbine, is an ideal design for space applications. This is a very compact high power density turbine. Sixty-five-Mw units of this design have demonstrated the ability to go from cool rest to full power ready for loading in 5 minutes (Ljungström 1949). The counter rotating shafts of this design provide a means to balance rotational inertia effects in a single machine. The radial out-flow design gives the turbine good moist vapor handling characteristics which are needed for a saturated potassium vapor turbine design. The turbines are designed to operate with an inlet pressure of 1.5 MPa and an exit quality of 70%. Optimum turbine efficiency is obtained with blade speeds of 300 m/s to 600 m/s, and system shaft speeds will be determined through trade-off studies involving stress analysis of the turbine components at desired operating temperatures and size optimization of the turboalternator system.

Superconducting alternators are chosen to develop the electrical power because of their high power-to-mass ratio. Each superconducting alternator is expected to operate at 1500 Hz and 10 kV to 20 kVac and be capable of providing a continuous power output of 2.5 MWe. Before actual system design is selected a trade off study is needed. This study would examine the turboalternator system, the transmission lines and the power conditioning system with respect to operating frequency and voltage output to determine the optimum system configuration to provide a least mass system design. The overall power conversion subsystem is expected to have a specific mass on the order of 0.43 kg/kW and an efficiency of 85%.

Power Conditioning Subsystem

The requirements for the power conditioning system have been specified in the description of the MPD thruster system. It is recognized that these requirements lead to a high speed alternator design. The final requirements will be derived from the study previously mentioned.

Heat Rejection and Thermal Control Subsystem

Heat rejection for the PEGASUS is accomplished by means of both a high and a low temperature heat rejection subsystem. The high temperature subsystem handles waste heat rejection from the turbines. The low temperature subsystem takes care of waste heat from the alternator and other components requiring cooling.

The high temperature heat rejection system consists of a bubble membrane radiator, associated pumps, piping, and structure. The radiator is sized to reject 17 MW of waste heat from the system during full-power operation. This system consists of a 15-m diameter thin film ellipsoidal envelope, coolant stowage tanks, structural piping, thermal insulating material, and associated pumps. Turbine exhaust vapor enters the radiator and is condensed on the inner surface of the bubble membrane. System rotation causes the condensed fluid to travel along the inner surface of the membrane to the

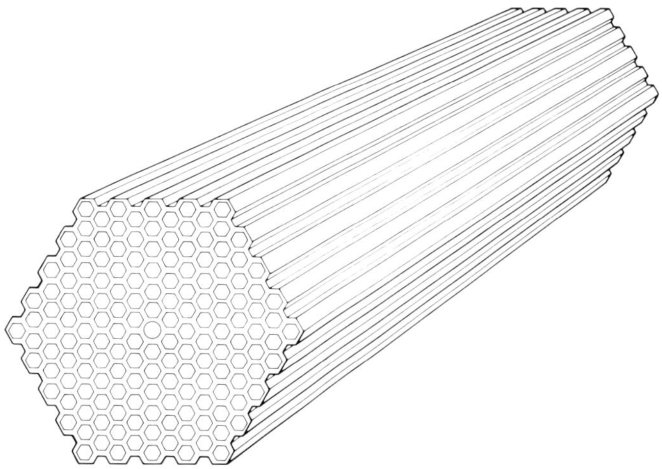

Figure 8 *Fuel Element.*

Table 3 *Reactor System Mass Summary.*

Component		Mass(kg)
UN Fuel	10.0%	533.9
Tungsten	49.8%	2669.6
Vessel	16.9%	904.8
Reflector	13.9%	743.3
Structural	8.9%	480.1
Coolant	0.6%	33.0
TOTAL		5364.7

Table 4 *Membrane Radiator Mass Summary.*

Component		Mass (kg)
Sheathing	36.5%	714.3
Coolant	24.5%	479.8
Pumps	23.2%	107.2
Structure	10.3%	455.3
Piping	5.5%	202.6
TOTAL		1959.2

gravity trough at its equator. Once collected, low pressure pumps are used to return the coolant fluid to the turbine driven high-pressure reactor feed pumps. A mass breakdown of the membrane radiator system is given in Table 4.

Because of the nature of the radiator design, meteoroid survivability is an area of primary importance to the overall system design. Preliminary calculations were performed to determine the probability of meteoroid interactions with the membrane radiator and to assess the membrane thickness necessary to assure system

Table 5 *Manned Mission to Mars Overall Mass Summary.*

Spacecraft, habitat, lander, etc.	143,310 kg
Power plant summary	45,624 kg
Propulsion system	5,144 kg
Propellant, tankage	116,047 kg
Total mass in GEO	**310,125 kg**

reliability during the course of the mission. These calculations based upon NASA's Near Earth to Lunar Surface Meteoroid Environmental model (Smith et al. 1983) indicate that the bubble membrane radiator will receive approximately 120 meteoroid impacts during a 5-year period. Of these impacts, the majority of the incident meteoroids will possess masses of less than 0.0001 grams, while those of greater than 0.01 grams will comprise less than 0.02 per cent of the accumulative total.

If the membrane radiator's operational reliability can be defined as its ability to resist meteoroid penetration, then system reliability can be calculated by subtracting the percentage of all meteoroids designed against from 1. Or conversely, by specifying the design reliability, the largest meteoroid that must be designed against can be determined.

The thickness of the membrane radiator can then be estimated based upon its cross-sectional area, the operational lifetime, the design reliability, the estimated meteoroid threat, and the material selected. For PEGASUS, the design reliability of the radiator system has been specified as 0.996; and the corresponding design meteoroid mass limit has been determined to be 0.0025 grams. The corresponding membrane thickness to meet these criteria has been calculated to be approximately 0.50 mm.

The low-temperature heat rejection system consists of an auxiliary cooling system designed to reject waste heat produced within the alternator and other equipment operating at much lower temperatures. The auxiliary cooling system has its own working fluid, pumps, and radiator. This system utilizes helium as its working fluid and is composed of a Stirling cycle cryogenic cooler, an auxiliary chiller, a low-temperature radiator, and associated pumps and piping. The cryogenic cooler removes heat from the liquid helium and transfers this heat to the auxiliary coolant in the chiller. A closed loop system is used to pump this coolant through the auxiliary cooling radiator located around the perimeter of the main radiator. Before returning to the chiller this coolant is used to cool pumps and other components.

Conclusion

The trajectory used for the manned Mars mission described here has not been optimized for the MPD thruster system and serves only as an example of how electric propulsion would greatly enhance the mission over chemical systems. However the PEGASUS Drive as described here clearly demonstrates the ability of a nuclear electric propulsion system as mission-enabling for the proposed manned Mars mission. With the advances expected in megawatt-class space power systems, the high specific impulse propulsion systems, such as the MPD thruster powered by a nuclear electric power plant, must be reconsidered as potential propulsion systems for this and other long duration missions.

Acknowledgments

This work was a joint effort performed at the Pacific Northwest Laboratory supported by the U.S. Department of Energy under Contract Number DE-ACO6-76RL01830 and at the Jet Propulsion Laboratory supported by the National Aeronautics and Space Administration under Contract Number NASA7–918.

References

Atkins, K. L. et al. (1983) *Expedition to Mars*, Jet Propulsion Laboratory, Pasadena, CA, 10 February 1983.

Barner, J. 0., E. P. Coomes, R. E. Williford, and L. A. Neimark (1987) "Cermet Fuels for Space Power Systems", in *Space Nuclear Power Systems 1986*, M. S. El-Genk and M. D. Hoover, eds., Orbit Book Co., Malabar, FL.

Barattino, W. J. (1985) *Thermal and Stress Analysis of Radiation Shield for Space Reactors*, Ph.D. Dissertation, Department of Chemical and Nuclear Engineering, University of New Mexico, Albuquerque, NM, May 1985.

Burton, R. L., K. E. Clark, and R. G. Jahn (1983) "Measured Performance of a Multimegawatt MPD Thruster," *J. Spacecraft and Rockets*, 20(3):299–304.

Eberhardt, R. N., D. A. Fester, and J. C. Aydelott (1981) "Shuttle Compatible Cryogenic Liquid Storage and Supply Systems," AIAA Paper No. 81–1509, *17th Joint Propulsion Conference*, Colorado Springs, CO, July 1981.

King, D. Q. and R. J. Vondra (1984) "Development Status and Projected Power Requirements of MPD Thrusters," presented at the Space Nuclear Propulsion Conference, Los Alamos, NM, December 1984.

King, D. Q., R. G. Jahn, and K. E. Clark (1981) *Magnetoplasmadynamic Channel Flow for Design of Coaxial MPD Thrusters*, Ph.D. Thesis, Dept. MAE Report No. 1552, Princeton University, Princeton, NJ, December 1981.

Ljungström B. (1949) "The Development of the Ljungström Steam Turbine and Air Preheater" in *Proc. International Mechanical Engineering Conference*, April 1949, Vol. 160, p.262.

McLyman, C. W. T. (1978) *Transformer and lnductor Desiqn Handbook*. Marcel Dekker Inc., New York.

Smith, R. E. and G. S. West (1983) "Space and Planetary Environment Criteria Guidelines for Use in Space Vehicle Development", NASA Technical Memorandum 82478, Jet Propulsion Laboratory, Pasadena, CA, January 1983.

Space Nuclear Power Systems 1986
Edited by M. S. El-Genk and M. D. Hoover

Chapter 36

Particle Bed Reactor Propulsion Vehicle Performance and Characteristics as an Orbital Transfer Rocket

F.L. Horn, J.R. Powell, and O.W. Lazareth

Brookhaven National Laboratory
Department of Nuclear Energy
Upton, NY 11973

The particle bed reactor designed for 100-MW to 300-MW power output using hydrogen as a coolant is capable of specific impulses up to 1000 seconds as a nuclear rocket. A single space shuttle compatible vehicle can perform extensive missions from LEO (Low Earth Orbit) to 3 times GEO (Geosynchronous Earth Orbit) and return with multi-ton payloads. The use of hydrogen to directly cool particulate reactor fuel results in a compact, lightweight rocket vehicle whose duration of usefulness is dependent only upon hydrogen resupply availability. The LEO to GEO mission has a payload capability of 15.4 tonnes with 3.4 meters of shuttle bay. To increase the volume limitation of the shuttle bay, ammonia is used in the initial boost phase from LEO to give greater payload volume with a small decrease in payload mass, 8.7 meters and 12.7 tonnes. The reactor fuel elements consist of HTGR-type fuel particles 700 micrometers in diameter held between coaxial porous metal or ceramic frits forming annular elements through which coolant gas flows radially. Nineteen of these elements are arranged in a reactor core matrix with a canned ^{7}LiH moderator between them, forming a thermal reactor. The nominal core size is 40-cm diameter by 60-cm length, with a 5-cm liquid H_2 reflector. The incoming hydrogen gas flows through holes in the moderator cooling the moderator before entering the thin inlet plenum around each fuel element. The gas flows radially inward through the frits and packed particle bed exiting along the central channel of each element. The neutronic studies indicate a fuel loading of about 4.5 kg of ^{235}U when the reflector hydrogen temperature is maintained at 112 K.

Introduction

The gas-cooled particle bed reactor can be used for direct heating of the propulsion gas for an orbital transfer rocket. When hydrogen is used to cool the nuclear reactor, the specific impulse approaches 1000 seconds, which is more than twice that of chemical rockets. Reactor power levels of 100 MW to 300 MW are possible from a small particle bed reactor, which provide thrusts high enough to perform true Hohmann orbit transfers.

Reactor Design

The reactor design that produces the high gas temperatures of up to 3500 K is of the particle bed fuel technology design. A typical fuel is made up of particles 700 μm in diameter each with a 300-μm kernel of zirconium carbide that contains several percent of uranium carbide. The uranium carbide is in solid solution in the zirconium carbide and has a melting point of about 3600 K. The 300-μm kernel is coated with alternating layers of porous carbon to contain the fission products, a pyrocarbon barrier, silicon carbide, and zirconium carbide for hydrogen resistance. The superior heat transfer of the small particles, which allows the high power densities of 10 kW/cm^3 is readily evident from their large surface area (86 cm^2 per cm^3), and the heat transfer coefficient of the high velocity gas that flows over them. The film temperature difference is about 100 K. The fuel also has been tested at Brookhaven National Laboratory in beds at high power densities. The fuel also has been tested as HTGR fuel at Oak Ridge National Laboratory, General Atomic Company, and Los Alamos Scientific Labora tory (Hollabough et al. 1977, Kaae et al. 1977, and Morgan and Malinaukas 1977).

The particle bed is held between concentric porous cylindrical frits. The coolant gas flow is radially inward to the central channel, exiting axially to the rocket nozzle (Figures 1 and 2). The fuel is held in 19 fuel elements arranged in a triangular pattern within the zirconium hydride or lithium-7 hydride moderator (Figure 3). The reflector around the core has been varied from a stainless-steel vessel, to beryllium metal, to zirconium hydride to cold hydrogen gas, depending upon the reactor mass desired and reactivity control method. The core size is optimized at a 40-cm diameter and 60-cm length with the overall reactor size at a 56-cm diameter and a 100-cm

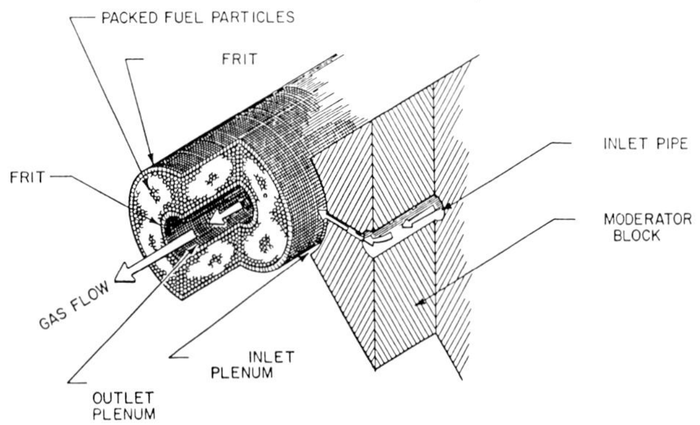

Figure 1 *Baseline Fuel Element and Moderator Block.*

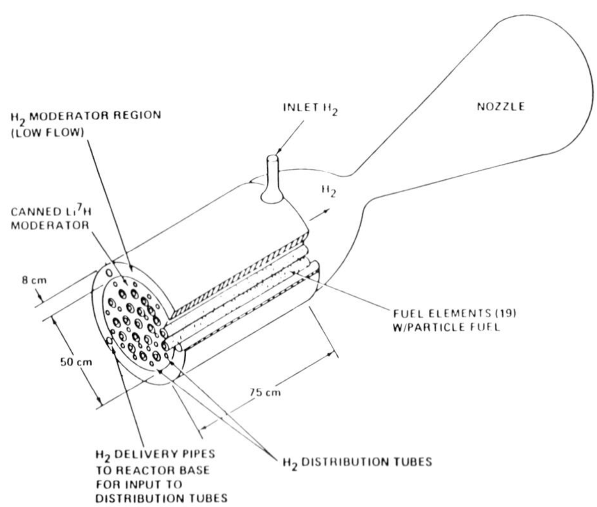

Figure 2 *Reactor with Rocket Nozzle.*

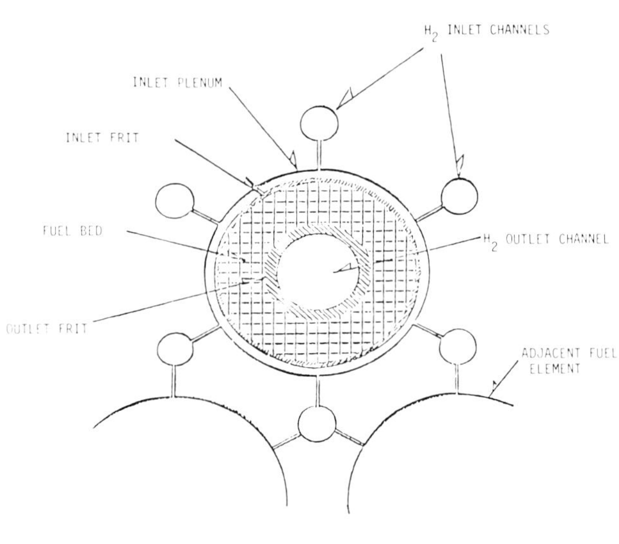

Figure 3 *Baseline Fuel Element for Particle Bed OTV.*

length with a minimum mass of 300 g (Figure 4). The design parameters for the base case reactor are

Number of Fuel Elements: 19,
Fuel Particle Size: 700 μm,
Fuel Volume Fraction of Core: 25%,
Range of ^{235}U (93.5% Enrichment) Loadings: 4 to 10 kg,
Poison Fraction of Control Rod: 33%,
Number of Control Rods: 12,
Fuel: UC in ZrC,
Poison: B_4C,
Inlet Frit: inconel, 50% porosity,
Outlet Frit: rhenium, 50% porosity,
Moderator: $ZrH_{1.7}$ Canned in 10 mil SS, and
Reflector: stainless steel.

The Mark-2 reactor using reflector reactivity control is shown in Figure 5. The cold hydrogen gas at 900 psi is fed to the outside region of the reactor where radiation heating warms the hydrogen from 30 K to about 100 K. The hydrogen then passes upward through the fuel elements to be heated to 2750 K before reaching the rocket nozzle. The reactivity of the reactor is controlled by the temperature of the hydrogen in the reflector. As shown in Figure 6, the critical multiplication constant for a 4.0-kg uranium-235 loading requires that the average hydrogen temperature be about 50 K. A 4.5-kg uranium-235 loading is critical when the average hydrogen temperature is 110 K, while a 5.0-kg uranium-235 loading cannot be controlled with a 5-cm H_2 reflector. The characteristics of the Mark-2 reflectivity control reactor are

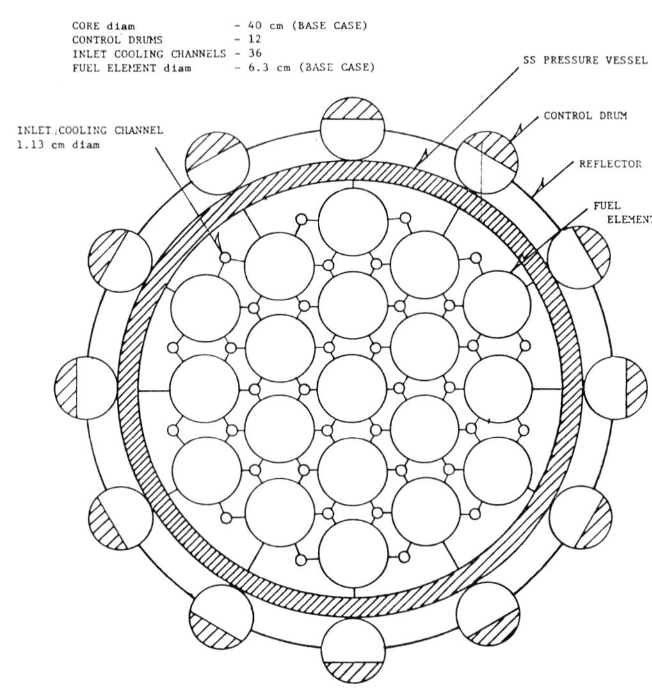

Figure 4 *Particle Bed OTV (19 Fuel Element).*

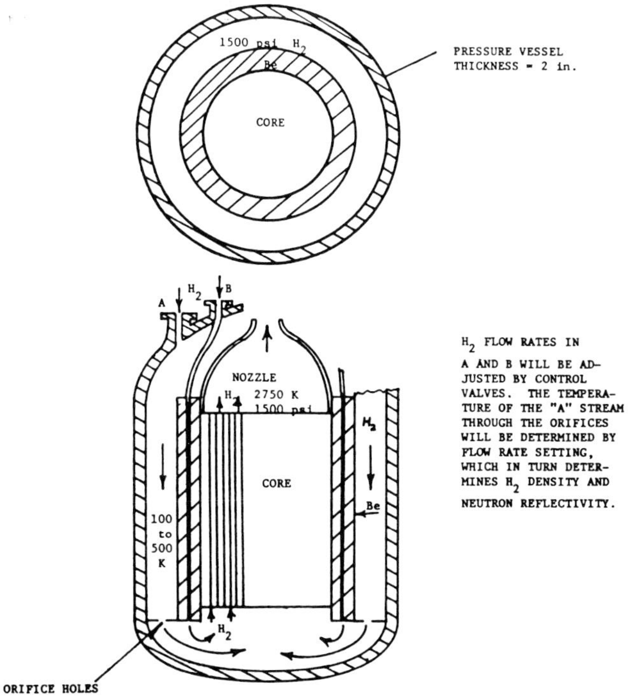

Figure 5 *Mark-2 OTV with H_2 Control/Reflector.*

Power: 200 MW,
Gas Outlet Temperature: 2750 K,
Coolant Pressure: 60 atm H_2, 80 atm NH_3,
Power Density: 10 kW/cm^3,
Fuel Bed Volume: 25% of reactor core,
Fuel Bed Material: 700 μm, ZrC-coated, UC_2 spheres,
Number of Fuel Elements: 19,
Diameter of Core: 40 cm,
Length of Core: 60 cm,
Moderator: ^{7}LiH (canned),
Reflector: cold hydrogen,
Thickness of Reflector: 5 cm,
Reactivity Control: reflector hydrogen density temperature control,
Aluminum Pressure Vessel: 3 cm thick,
Overall Size: 100 cm × 56 cm diameter, and
Overall Mass: 300 kg.

Mission Study

For the study, the orbital transfer vehicle (OTV) begins operation at an altitude of 150 nautical miles above the earth in LEO and reaches GEO at 19,365 nMi within 5.28 hours (Perkins 1983). The volume and lift capability of the space shuttle are used as limitations on the various missions. The bay cross section available is 4 m in diameter with a length of 17.7 m (Figure 7). The shut-

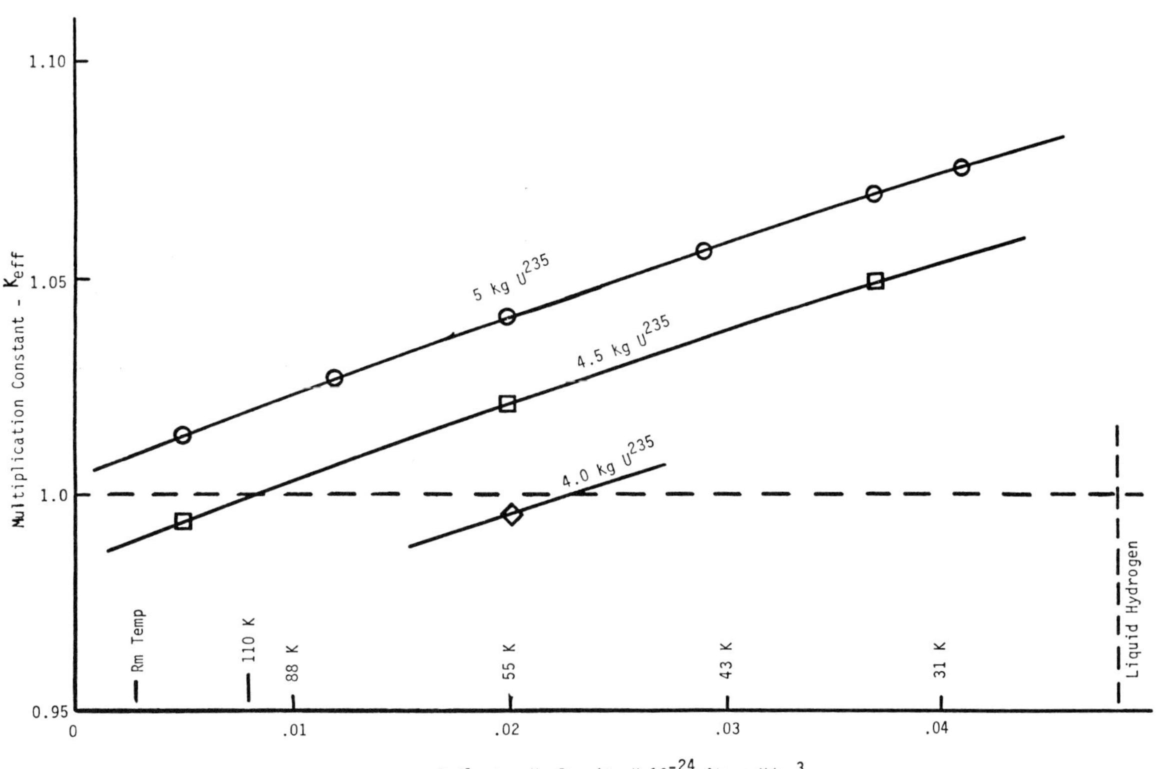

Figure 6 *Dependence of the OTV Reactor Multiplication Factor on the Reflector H_2 Density for an OTV Reactor with a 40-cm^2 Diameter by 60-cm Long Core and a ^{7}LiH Moderator with a 5-cm H_2 Reflector at 60 atm.*

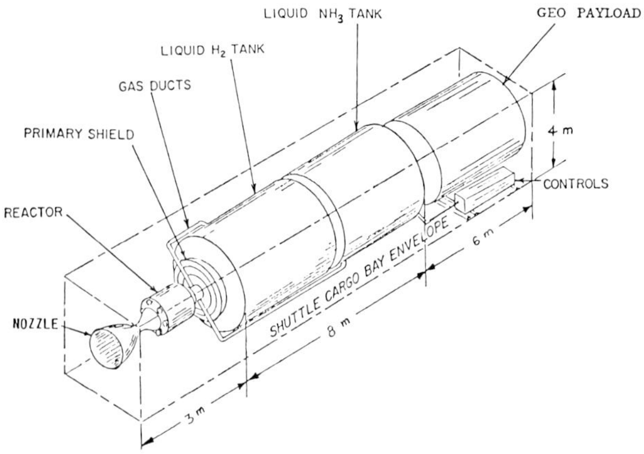

Figure 7 *Orbital Transfer Vehicle with Reactor shown to Approximate Scale for 200 MWt.*

tle lift capability of 29,500 kg is used to carry the OTV and its payload to LEO in a single mission.

Five basic missions were calculated:

- LEO to GEO;
- LEO to Polar orbit;
- LEO to GEO and drop package, go to three times GEO altitude, back to GEO with payload;
- LEO to 3x GEO, back to GEO with payload, and
- LEO to GEO, drop package, reverse direction with payload.

The studies for the five missions assumed the necessary fuel volumes as well as the equipment that fit within the shuttle bay volume. One of the limitations is that the bay is volume limited, rather than mass limited, when using hydrogen as a propellant. To allow more volume payload, the more dense coolant ammonia was substituted for hydrogen in the initial boost phase from LEO orbit, and hydrogen was used for all subsequent propellant thrusts. An additional option to the missions was the ability to jettison the empty ammonia tank. The option of using ammonia as a propellant causes a small increase in reactor pressure drop over the hydrogen propellant from 0.5 atm to 1.4 atm.

The volume available for the payload is reported as the bay length in meters because the shuttle bay cross section is constant at a 4-m diameter (allowing side clearance). The payload mass is the difference between the shuttle lift capability (65,000 lbs.) and the mass of the propellants, reactor, rocket chamber-nozzle, tanks, and structures required reported in metric tons (2,200 lbs).

Results

The number one mission, LEO-to-GEO, using all hydrogen propellant has the highest payload of any mission—15.4 tonnes (34,000 lbs) and an available bay length of 3.37 meters (11.0 ft). Figure 8 compares to the Centaur G′ rocket that has a payload of 6.0 tonnes and payload length of 9.15 meters. The other missions of increasing complexity are also shown with decreasing payloads and lengths. The payload length becomes negative for the fourth and fifth missions, which are not possible with hydrogen.

The use of ammonia as the initial coolant-propellant in the reactor reduces the payload by 17% for the first mission as shown in Figure 9; however, the payload length increases to 8.68 m. All five missions now become possible and are compared in Table 1. The payload length is a minimum of 4.5 m on the most complex missions, and this may be sufficiently large. Payload masses and sizes will require that the propellant quantities of NH_3 and H_2 be proportioned to fit the particular case. The jettisoning of the empty ammonia tank is an important saving of vehicle mass in the multiple thrust missions.

In addition to the five basic missions, two missions were added that returned the OTV to the so-called high LEO. Hi LEO is a 250 nMi high orbit with a long orbit decay time (hundreds of years) and is considered a nuclear safe orbit with sufficient fission-product decay time for the very low fission-product inventory. This orbit will allow the possibility of refilling propellant tanks or the discharge of reactor fuel prior to returning the OTV to Earth.

The first of the two is the single LEO-to-GEO mission with return to Hi LEO. Using all hydrogen propellant, the payload is high (7.5 tonnes) but with no room in the shuttle bay for payload. This configuration could be used for satellite retrieval from orbit. When ammonia is used prior to the hydrogen, room becomes available at the cost of 1 to 2 tonnes of payload mass (Figure 10). A mid-LEO jettison of the ammonia tank allows a 5.5 tonne payload with 2.2 meters of shuttle bay room. An after-LEO jettison of the ammonia tank gives a 4 tonne payload, but a large 6.5-m shuttle bay length.

The second mission of the two is LEO-to-GEO with the deployment of a 2 to 3 tonne payload, continue on to three times GEO altitude with the rest of the payload, whose mass varies with reactor gas temperature and mode of tank jettisoning, either mid or after-LEO ''burn'' (Figure 11). A mid-LEO tank jettison provides very little room for payload (0.2m to 0.4 m), while after-LEO provides 4.5 m to 5.0 m. After-LEO reduces the balance of payload to about one tonnes with two tonnes deployed using reactor temperatures of 3250 K and 3500 K. The return from 3 GEO-to-GEO allows the pickup of the original deployed payload including exchange of on-

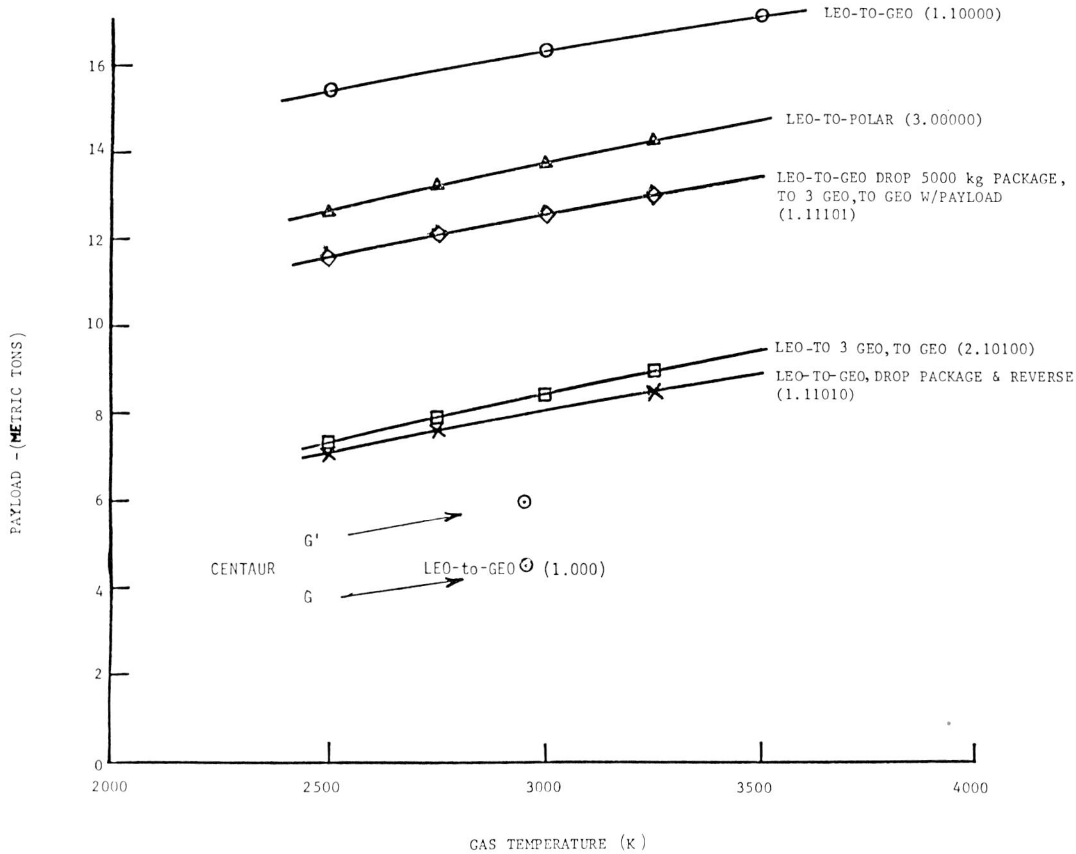

Figure 8 *Comparison of Five Missions with $H_2 + 0_2$ Rocket: Reactor Power-100 MW, Propellant-Hydrogen.*

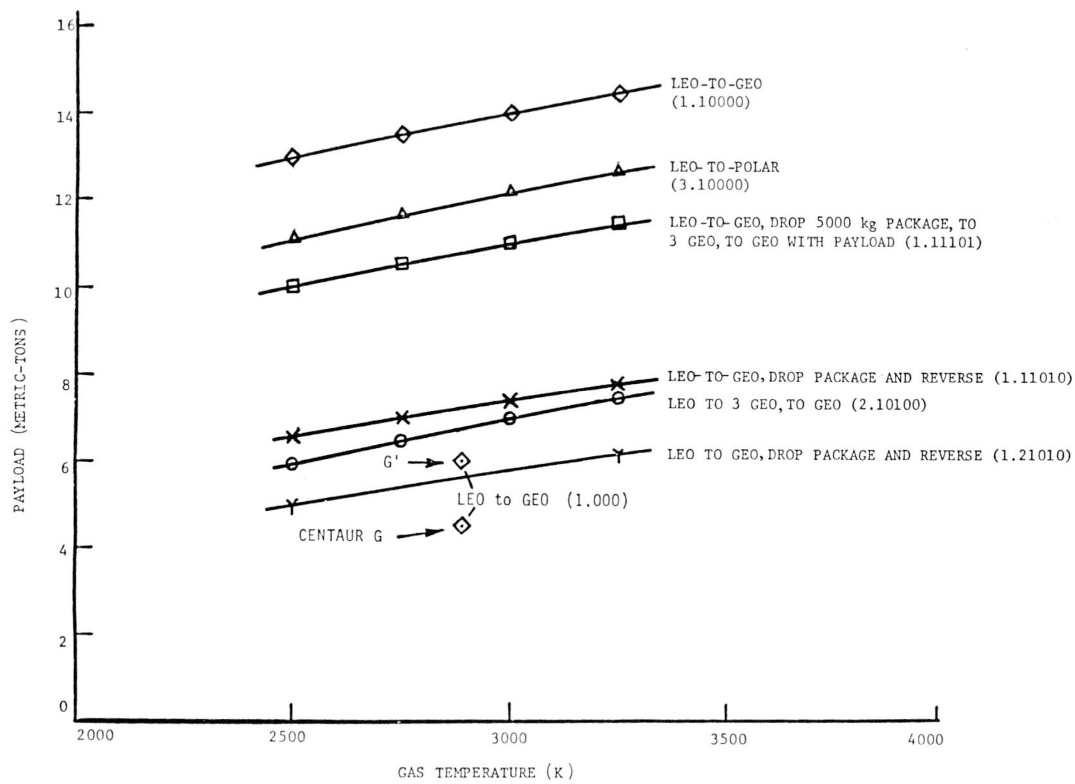

Figure 9 *Comparison of Five Missions with $H_2 + 0_2$ Rocket for Reactor Power of 100 MW and NH_3 and H_2 Propellant.*

Table 1 *Mission Performance for a 100-MW Reactor with 2750-K Gas Outlet Temperature.*

BASIC MISSION	H_2 PROPELLANT Payload Mass (kg)	H_2 PROPELLANT Payload Length(m)	$NH_3 + H_2$ PROPELLANT[a] Payload Mass (kg)	$NH_3 + H_2$ PROPELLANT[a] Payload Length (m)	H_2/O_2 CENTAUR G′ Payload Mass (kg)	H_2/O_2 CENTAUR G′ Payload Length (m)
LEO-TO-GEO (equatorial)	15,422	3.37	12,748	8.68	5,993	9.2
LEO-TO-POLAR	13,268	1.36	10,927	6.98	3,000	9.0
LEO-TO-GEO, TO-3 GEO, TO-GEO (equatorial)	10,143	1.56	8,288	4.52	–	–
LEO-TO 3 GEO-TO-GEO (inclined)	12,976	0.97	9,468	6.77	–	–
LEO-TO-GEO AND REVERSE (equatorial)	3,828	7.45	2,126[b]	4.54	–	–

[a] Jettison of NH_3 tank at mid-LEO.
[b] Jettison after LEO.

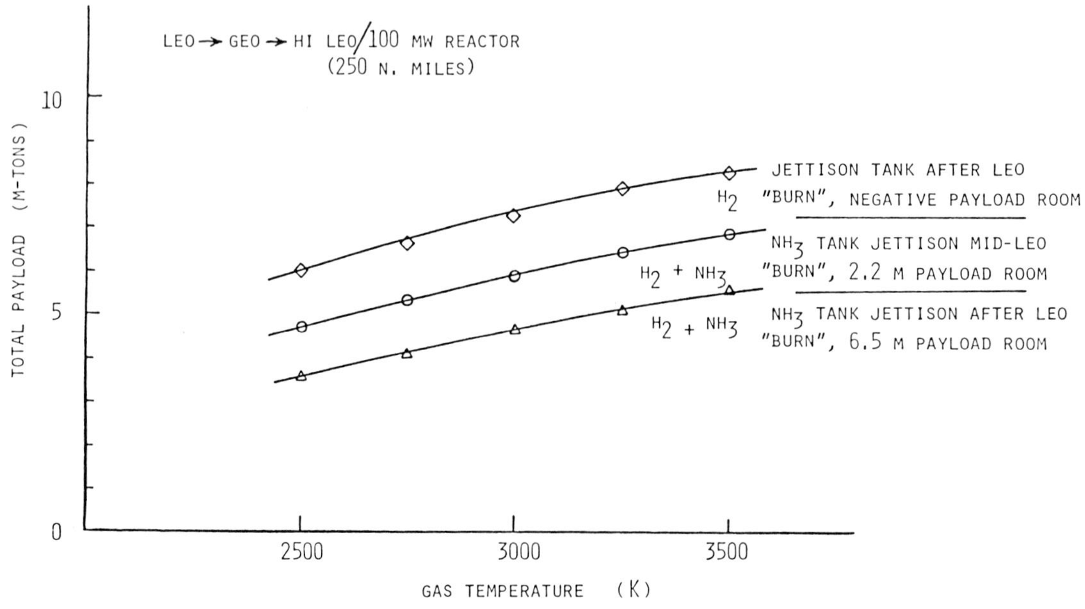

Figure 10 *LEO-to-GEO to Hi LEO.*

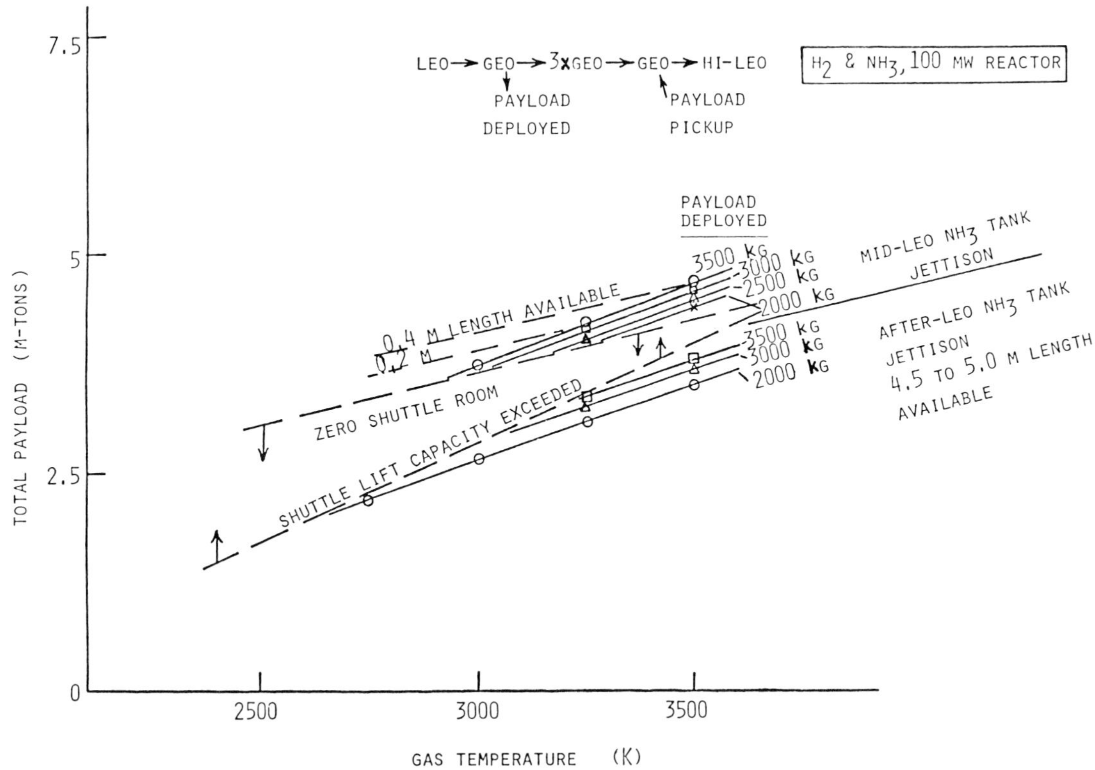

Figure 11 *LEO-to-3 GEO and Return.*

board payload and return to Hi LEO. An area in Figure 11 indicates conditions that exceed the shuttle capability, and that becomes smaller as the reactor temperature rises. The return to Hi LEO allows the discharge of payload for shuttle pickup, and possible reuse of the OTV.

Summary and Conclusions

This analysis indicates that a nuclear-powered OTV greatly increases payloads for GEO and extra-GEO missions by two to three compared to a chemical hydrogen-oxygen rocket. The nuclear OTV enables new extra-GEO missions to be performed, including parking in ultra-high orbits and returning to GEO with a spare satellite, as well as the transfer to retrograde GEO orbits upon command. When shuttle-compatible conditions are imposed, reasonable payload volumes become available with the use of ammonia propellant for the initial thrust and hydrogen propellant for the balance of the thrusts, at only a 15% reduction in payload.

The reactor used is small and lightweight, of the HTGR particle bed technology with a critical loading of about 4 kg of uranium-235. The total reactor mass of 300 kg can produce 200 megawatts of power at a gas temperature of 2750 K (4490°F). The short operating time, about an hour, produces negligible radiation damage, fuel burnup, and fission product buildup. The reactor can be disposed of, after use, in a non-returning orbit.

Acknowledgments

This research was carried out under the auspices of the U.S. Air Force, Rocket Propulsion Laboratory. We wish to acknowledge with appreciation the extensive computer calculations of R. Benenati of N.Y. Polytechnic Institute, and the help and drawings of M. Solon of Grumman Aerospace Corporation. We also appreciate the help of Susan Signorelli and Olga Perez in assembling this paper.

References

Hollabough, C.M., L.A. Wahman, R.D. Reisurg, R.W. White, and P. Wagner (1977) "Chemical Vapor Deposition of ZrC Mode by Reactions of $ZrCl_4$ with CH_4 and with C_3H_4" *Nuclear Technology*, 35 (2):527–535.

Kaae, J.L., S.A. Sterling, and L. Yang (1977) "Improvements in the Performance of Nuclear Fuel Particle Offered by Silicon Alloyed Carbon Coatings," *Nuclear Technology*, 35 (2):536–547.

Morgan, M.T. and A.P. Malinaukas (1977) "Cesium Release and Transport in Biso-Coated Fuel Particle," *Nuclear Technology*, 35 (2):457–464.

Perkins, D.R. (1983) *Preliminary Analysis and Comparison of Recoverable Space Based Orbit Transfer Vehicle for LEO to GEO Missions*, AFRPL-TR-83-029, Air Force Rocket Propulsion Laboratory, Edwards AFB, CA.

Space Nuclear Power Systems 1986
Edited by M. S. El-Genk and M. D. Hoover

Chapter 37

Space Based Nuclear-Microwave Electric Propulsion

Gerald D. Nordley

Applied Research in Energy Storage Organization
Air Force Rocket Propulsion Laboratory
Edwards AFB, CA 93523

William C. Brown

Consultant
6 Perry Lane
Weston, MA 02193

Electric propulsion can move more mass through space than chemical propulsion due to the higher exhaust velocities generated by electric propulsion devices. This performance is achieved at the expense of very heavy power sources or very long trip times, which, in turn, create technical and economic penalties of varying severity. These penalties include higher operations costs, delayed availability of the payload, and increased exposure to Van Allen Belt radiation. It is proposed to reduce these penalties by separating the power source from the propulsion device and using microwave energy beaming technology as an extension cord. This technology was recently explored and partially developed/tested for Solar Power Satellite concept studies. The advantages of this approach include reducing the specific power of the electric propulsion system power supply to 2kg/kW or less, reducing the vulnerability to radiation in comparison with solar cell electric propulsion systems, sharing the cost of the power generation hardware among many missions without returning the hardware each mission, eliminating or reducing the shadowing power losses in low orbit, and eliminating the safety concerns inherent with a maneuvering nuclear power source. These benefits come at the expense of building one or more power plants in a shuttle accessible orbit, including some rather large transmitter arrays or antennae. This paper summarizes the state of the art of the technology needed for space-based beamed microwave power transmission, takes a preliminary look at some of the cost/performance trade-offs involved with the use of beamed microwave/electric propulsion for some typical orbit transfer missions, and offers some suggestions for addition work.

Introduction

This paper addresses the concept of using a space-based nuclear power facility to generate and beam power at microwave frequencies to a spacecraft equipped with a suitable receiver and an electric propulsion system for orbit transfer. There are several movitations for taking this approach. Electric propulsion systems can lower the cost of space transportation by using their inherently high specific impulse to reduce the mass of fuel required for orbital maneuvers (Goddard 1925). However, this benefit is limited by the mass of the systems needed to provide power for electric propulsion. Thus, lighter power sources are desired. The thin film rectenna (The term "rectenna" is a contraction of "rectifying antenna."), which can convert beamed microwave energy into DC power at 1 to 2 kg/kWe (Brown 1985), offers substantial advantage over solar electric and nuclear electric propulsion systems weighing 10 to 30 kg/kW of delivered electrical power. Thin film inflatable concentrators offer similar power gains. Concepts using ground-based transmitters have been studied (Brown and Glasser 1983 and Brown 1969) and appear promising. However, ground-based transmitters have siting and visibility limitations that result in relatively longer trip times for orbit raising missions. Since visibility improves with altitude, one consideration is to base the transmitter in space

In such a scenario, the benefits of achieving shorter mission times with high specific impulse propulsion must be balanced against the cost of developing, producing and deploying the system of transmitters. The advent of the Strategic Defense Initiative (SDI) has spurred the development of space-based power reactor technology. These efforts may significantly reduce the development cost of propulsion systems. Inflatable mirror technology, being developed for solar rockets (Sercel 1985a), will enable the construction of large microwave concentrators and focusers at very low masses. Electric propulsion research is producing devices that can maintain approximately half a pound of thrust for hundreds of hours using tens of kilowatts. These developments, taken together, increase the attractiveness of microwave-supported propulsion with space based transmitters.

This paper's purpose is to generate interest in this concept and the related technologies by providing an overview of the system components, a first order estimate of economic benefits, and recommending potentially profitable future research. No particular technology is advocated above others because the field is not sufficiently mature to make this choice with any degree of confidence. This analysis will assume space based nuclear powered transmitters and rectennas, and focus on systems issues as a matter of personal interest in the technologies involved.

System Description

Overview

Figure 1 shows the basic components of a space-based beamed microwave electric propulsion system. An earth-to-orbit transportation system places the power stations into a low orbit. Electric propulsion can raise the stations to a more suitable orbit. The power stations comprise the nuclear power plant, the transmitter(s), and an aperture consisting of a focusing antenna or an electrically steerable array. The microwaves are beamed to the using spacecraft, which may include a concentrator, a rectenna for generating DC power on the spacecraft, and an electric propulsion system. The sizing of such a system is deceptively complex, and a rigorous optimization has not been attempted. For example, if the spacecraft rectenna receives 300 W/m^2 maximum, the transmitter must provide a beam of roughly this intensity over the full surface of the rectenna to fully utilize the potential mass savings. This requirement allows a wide range of transmitter powers and antenna configurations which, in turn, can trade off in mass with the power plant power level and radiator size. This, in turn, can trade off with the economics of the spacecraft mission timelines. The reasonable, optimal size of the system will be the subject of further research.

Note that orbit transfer missions that show economic benefit from use of electric propulsion are

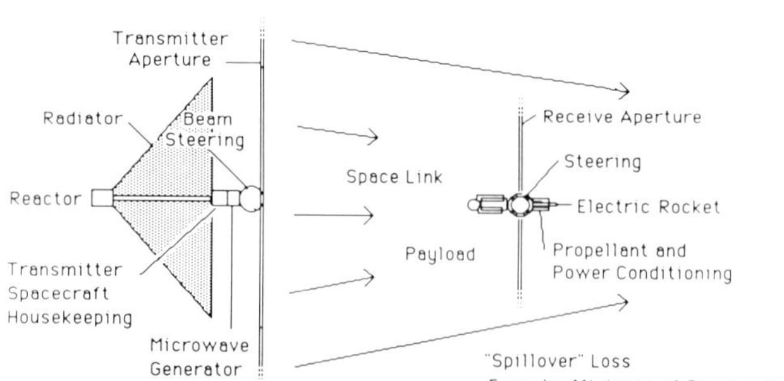

Figure 1 *Beamed Microwave Propulsion Systems Components.*

generally those to higher energy orbits, such as geosynchronous orbit, and that practical space shuttle payload capacity is about 25,000 kg. For such missions, and electric propulsion system in the 300-kWe class appears attractive (Figure 2). This is fortuitously the same order of magnitude being considered for the initial SP100 nuclear power systems. Thus, a 300-kWe rectenna at 300 W/m^2 would need to be 1000 m^2, or roughly 18 m in radius. This is not an unreasonable size for a "gossamer" spacecraft structure weighing less than half a kilogram per square meter. This is a fairly small target at synchronous orbit altitudes and would require a very large transmitter aperture. It might be feasible to use on the spacecraft an inflatable collector of substantially larger dimensions to concentrate the beam onto the rectenna. (This concentrator need not be attached to the spacecraft; consideration of that option would unnecessarily complicate the present first order analysis.) It should be possible to connect the rectenna elements to provide a simple electric propulsion device, such as an arcjet, with appropriate voltage and current levels with a minimum of power conditioning mass. Electric propulsion system specific powers in the range of 3 to 4 kg/kWe appear achievable in the near term (Vondra and Cassady 1984), and 2 kg/kWe is possible with further development.

Electric Propulsion Selection

The selection of an electric propulsion device involves a trade-off between thrust and payload based on economics or mission requirements. Essentially, greater fuel efficiency requires higher exhaust velocities. The power requirements increase along with the square of the exhaust velocity. Therefore, thrust must be decreased to maintain the power balance. Lower thrust means longer trip times and increased operational costs, which must be balanced against the earth-to-orbit savings from the reduced propellent load. Figure 2 illustrates a cost curve for such a trade-off, generated from a cost model (Nordley and Vondra 1985) for parametric comparisons of electric propulsion systems. On the one hand, expense increases due to the cost of lifting the orbit transfer propulsion fuel up to low earth orbit; on the other hand, expense increases due to trip time costs. The location of the minima (the selection range) depends on a complex relationship of various cost and physical parameters, the most important being the total velocity change the system is required to produce and the assumed cost of mission time. As shown, the thruster

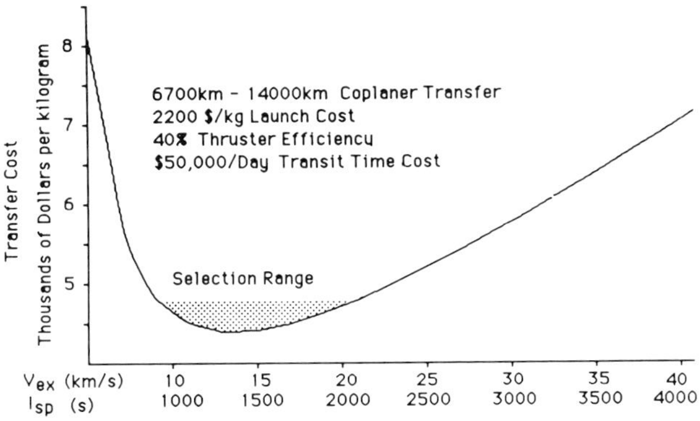

Figure 2 *Orbit Transfer Cost per kilogram of Payload Delivered to Mission Orbit versus Exhaust Velocity (V) and Specific Impulse (I_{sp}).*

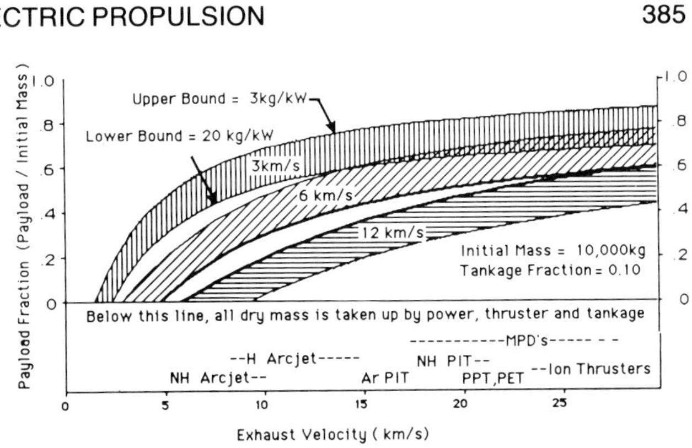

Figure 3 *Payload Fraction as a function of Exhaust Velocity for Different Mission Velocity Change (Delta V) Requirements.*

should have an exhaust velocity in the range of 8 to 20 km/s (the shaded area in Figure 2). Also, because of the low thrust and long trip times in this power regime, a noncryogenic propellant would be desired. Note that the cost minima are broad and that other considerations, such as reliability and producibility, may dictate the choice of thruster.

In Figure 3, the three broad curves each represent a different total velocity change requirement. These curves are generated by expressing the rocket equation, velocity change = exhaust velocity × natural log (initial mass/final mass), in terms of its component masses, solving for payload mass and dividing by initial mass for each exhaust velocity. The vertical axis represents payload fraction, which is that fraction of the initial mass (wet mass) of the orbit transfer vehicle which is payload. The top of each curve represents a propulsion power associated mass of 5 kg/kWe. The bottom of each curve is for 20 kg/kWe. Intermediate values lie in between, in the shaded region. Note that the curves rise quickly at first and then level off to the right. This is because as I_{sp} increases, the propellant mass becomes a smaller and smaller portion of the whole and proportional reductions in this mass have less effect. This rule of diminishing returns for increasing Isp explains why operational costs associated with longer mission times outweigh the savings in lower propellant mass. Usually, a thruster should operate near the broad "knee" of these curves; in this case, a thruster should have approximately 1000 s I_{sp} for the 3 km/s mission, 1500 s for the 6 km/s mission and 2500 s for the 12 km/s mission. The approximate exhaust velocity/specific impulse ranges of several types of electric thruster are indicated (Vondra et al. 1984). These are the ammonia propellant arcjet, the hydrogen arcjet, the Pulse Inductive Thruster (PIT) with argon and ammonia, the Pulse Plasma Thruster (PPT, currently available in only very low power), the Pulse Electrothermal Thruster (PET, experimental), Magnetoplasmadynamic thrusters (MPDs), and ION thrusters.

Regardless of the mission and thruster selected, the orbit transfer will take weeks at these power levels. While future microwave transmitters might operate at much higher power levels, allowing other trajectory options, the initial application trajectory will probably be the standard low thrust spiral. The transit time will be minimized if a transmitter is continuously in view.

Power Reception

The mass of the power reception system depends on the maximum power level. If reflectors are used to concentrate power onto the receiver, then the mass of the reflectors also must be considered. Figure 4 (Friese et al. 1982) shows the relationship of concentrator mass to concentrator diameter for several technologies. Figure 5 shows how receiver (r_r) aperture depends on distance (R) and wavelength (L) at a given transmitter aperture (r_x) and space link efficiency (τ). This relationship is approximated by (Sercel 1985b):

$$E \approx 1 - \exp(-\tau^2) \text{ with } \tau = \pi r_x r_r / LR \quad (1)$$

where E is space link efficiency. Figure 6 shows how efficiency varies with distance and wavelength for a fixed pair of apertures. The shorter wavelength can form a tighter beam and thus is more efficient at distance. We have assumed a transmitter (effective) radius of 1 km. A single reflector this size would weigh approximately 20 tonnes, using deployable mesh technology (Friese et al. 1982). As the proposed col-

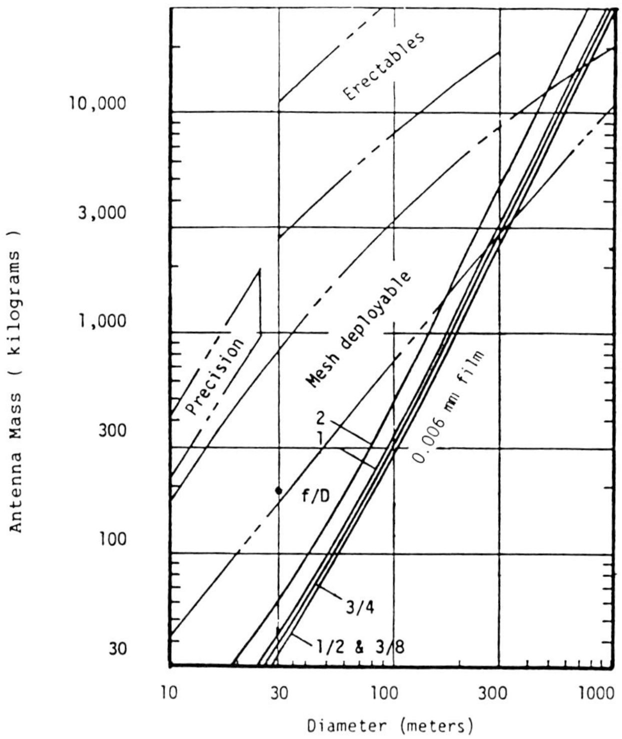

Figure 4 *Mass of Antennas of Different Construction versus Antenna Diameter. (From Friese et al. 1982).*

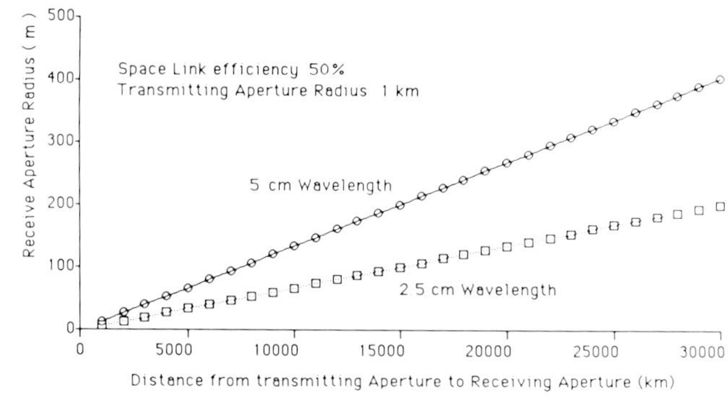

Figure 5 *Receiver Aperture Radius as a function of Distance at a Fixed Efficiency and Transmit Aperture for Two Wavelengths.*

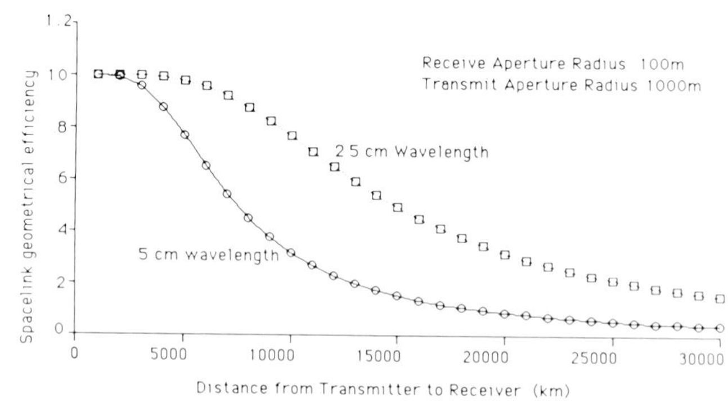

Figure 6 *Space Link Efficiency as a function of Distance for Constant Apertures at Two Different Wavelengths.*

lector radius approaches 100 m, its specific power at 300 kWe reaches ≈1 kg/kWe. From this point, the mass of the receive system will be dominated by the collector and the rule of diminishing returns. At 300 m radius, specific power rises to 10 kg/kWe, and the economic advantage of low mass power is almost lost; however, other advantages, such as greater radiation hardness, remain versus solar cells. A solar thermal propulsion system with collectors of this size would operate at more than 100 *mega*watts at much higher thrust (though probably lower Isp). A detailed trade-off between beamed microwave and solar thermal propulsion is beyond the scope of this paper. Moreover, long distance transmission space link trade-offs will probably not favor increases in receiver aperture. Note that link efficiency depends only on geometry and not power; therefore, the specific power (kg/kWe) of the collectors decreases as the power level increases. As shown in Figure 3, the payload fraction increases as well. Higher power systems will be more economical than lower power systems. This suggests not limiting the mission to a single shuttle payload, but rather collecting payloads in orbit and moving several at once.

Assuming the power is concentrated, it must be received and converted into electricity for use with an existing electric propulsion device, or coupled directly to the propellant in a microwave heating scheme. The main concern of this paper is the electrical device, which is more near term and much less demanding of attitude control. However, direct coupling of microwave energy is a promising future technique (Asmussen and Whitehair 1984). Microwave energy can be converted to electricity in one step by the thin film rectenna. Table 1 presents a brief history of actual experiments in rectenna technology (Brown 1969 and 1984 and Dickinson 1975).

The following material has been previously published (Brown 1985) and is summarized for the benefit of the Space Nuclear Power community. It describes the general features of the thin film rectenna for space use.

Physical Description of the Thin-Film Rectenna

The physical format of the rectenna consists of a foreplane and a reflecting plane that is positioned approximately one quarter wavelength behind the foreplane. The reflecting plane is a completely passive

Table 1 *Brief Summary of Recent Microwave Power Transmission History*[a].

	Intensity (kW/m^2)	RCV Efficiency (%)	Specific Power (kW/kg)
1965 Helicopter	≈ 1[b]	≈ 70	.31
1968 Diode Array	≈ .22	≈ 75	1.0
1975 Goldstone	≈ 1.7[b]	83	N/A
1983 Thin Film	≈ 1.2[b]	85 +	≈4

[a]From Brown 1969 and 1984, and Dickinson 1975.
[b]Air Cooled.

element. For use in space it can consist of a thin metallic film on a thin plastic film such as Kapton.

The rectenna foreplane is characterized electrically as shown in Figure 8. It consists of a collection of rectenna elements combined with a DC collection bus that performs some microwave circuit functions. The DC power collection busses can be combined in series or parallel, pending certain restrictions. The outputs of a minimum of five adjacent rows of elements should be connected in parallel to form an elemental assembly of rectenna elements. Then, these outputs can be combined in either series or parallel to provide the desired voltage-current relationships.

The rectenna element consists of a half wave dipole, one rectifying diode, and filter circuits as shown in Figure 7 and in more detail in Figure 8. The electrical operation of the rectenna element is complex but it has been mathematically modelled in great detail with computer simulation. Results of the simulation agree remarkably well with experimentally observed operation as reported by Brown (1977). The physical construction of a thin-film rectenna element is shown in Figure 9.

Features of the Thin-Film Rectenna for Space Use

Mass to Power Output Ratio (Specific Mass)

This is expected to be no more than 2 kilograms per kilowatt of DC power output and probably considerably less with additional development specifically for space use.

Efficiency

The rectenna is highly efficient. From 80% to 85% of the power in the portion of the microwave beam that impinges upon the rectenna is converted into well-filtered DC power.

Disposal of Heat resulting from Inefficiency

Only a small amount of heat is generated per unit area due to the high efficiency of the rectenna. This feature plus the ability of the film and the rectifying diodes to operate at about 100 K over earth orbital environment thermal equilibrium allows direct, passive radiation of this heat into space.

Power Handling Capability per Unit Area

Heat disposal considerations could allow operation up to 400 watts of DC power/square meter output.

Physical Flexibility

The rectenna can be easily rolled up and then unrolled for deployment. Studies indicate that one space shuttle flight could carry aloft enough rectenna for two 10,000-kWe interorbital vehicles.

Manufacturability

Starting with a material consisting of Type F Kapton bonded to copper film on both faces, the manufacturing process uses simple masks and standard etching procedures. The only subsequent procedure is to bond the diode to the circuit.

Life and Reliability

Although a life test of the rectenna has not been performed in a vacuum environment, projections made on the basis of theoretical calculations and stagnant air tests indicate lifetimes of tens of years in space.

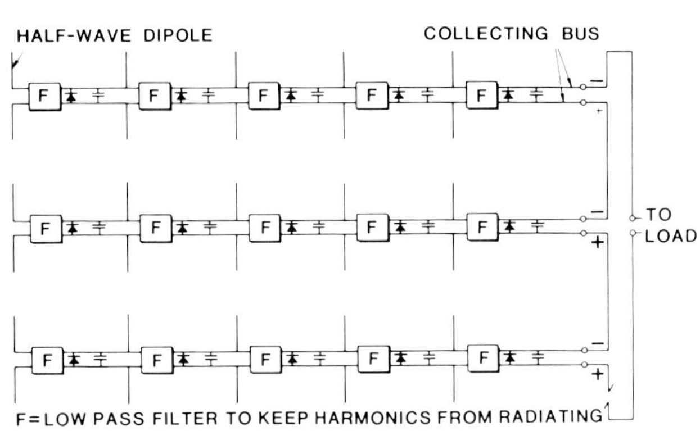

Figure 7 *Circuit Schematic of Rectenna Foreplane.*

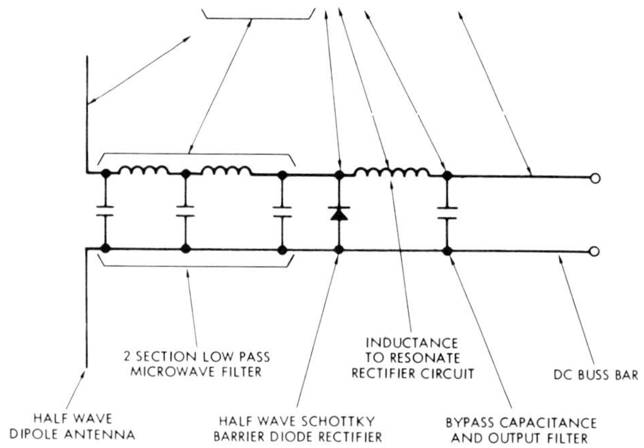

Figure 8 *Simplified Electrical Schematic for a Rectenna Element.*

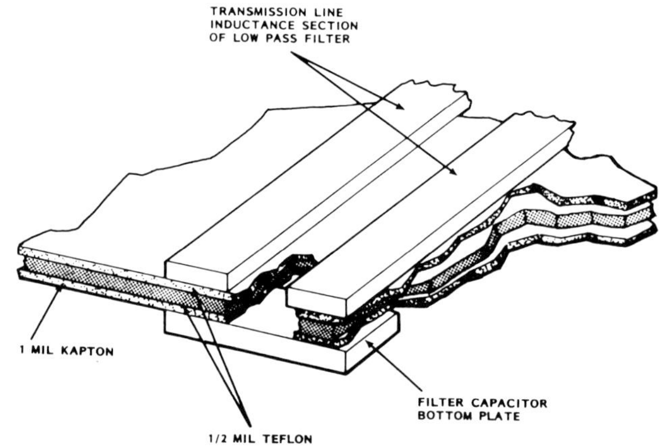

Figure 9 *Rectenna Element Construction showing Etched Metal Portions and Thin Plastic Film.*

Kapton, the thin-film used in the rectenna, is well known for its resistance to ultraviolet radiation and its successful use in space. The copper- (or aluminum) etched circuits will not deteriorate. The rectifying diodes are made from GaAs (gallium arsenide) and are so small in area and volume that they can be readily shielded without significant increase in rectenna mass. Both Kapton and GaAs diodes will operate reliably at the temperatures needed. The diode in the rectenna element will fail during the rectification cycle if operating conditions cause the reverse voltage to exceed the reverse voltage breakdown rating for the diode. Thus, open circuits should be avoided. The protective procedure that has been used successfully is a simple crowbar that places a short circuit on the output if the DC voltage output level is exceeded.

Sensitivity to Incoming Microwave Power

One of the important features of the rectenna is its non-directive properties. The angle of incidence to the arriving microwave beam may be varied as much as 30 degrees from normal incidence with no more than 5% reduction in efficiency. The rectenna is relatively insensitive to moderate changes in the wavelength or frequency of the microwave beam. This allows relatively loose mechanical tolerances in the construction of the rectenna. It also allows some change in the transmission frequency which may be used as a means of controlling the direction of the microwave beam.

Sensitivity to Polarization

To date, work has been done with linearly polarized transmitters and rectennas which must be kept in approximate alignment. While circularly polarized elements are possible in principal, it is difficult to predict whether the simplicity and low cost aspects of the linearly polarized format could be duplicated. In transmission geometries where polarization may be a problem (such as large inclination changes), mechanical steering of either the transmitter feed or the rectenna could be done. Alternatively, a cross polarized rectenna could be constructed, by superimposing foreplanes rotated at 90 degrees. Since power reception is limited by the thermal limits of individual diodes, this doubling of the number of diodes would permit absorbtion of equal amounts of power in both planes of polarization, and may not increase net weight.

Output Voltage Level

The typical voltage output of a single rectenna element at one watt is 15 to 25 volts. They can be combined in a series-parallel arrangement and in the numbers necessary to provide almost any desired combination of current, voltage, and DC power.

Input-Output Relationships

The rectenna tends to operate with constant DC power output with constant microwave power input. Unlike solar arrays, the load impedance can be varied over a wide range without much change in efficiency. The rectenna requires a minimum of power conditioning between the rectenna and its load. For example, an electric motor can be operated directly from the rectenna outputs. The resulting power output level can be directly controlled by varying the power level of the incident microwave beam.

Deployment

It would be fairly straightforward engineering to deploy an array capable of receiving 200 to 300 kWe at 2 kWe/kg, if constructed with inflatable technology. A "windowshade" type system also shows promise, and some preliminary work has been done in this area (Brown 1985). Goals for growth in this technology include higher power densities, by using shorter wavelengths for smaller arrays, and more efficient radiative cooling or active gas cooling, by circulating gas between the foreplane and backplane of the array. Currently, the rectenna specific power is about the same as that of the electric thrusters; essentially the point of diminishing returns for further mass reductions.

Power Transmission Technology

The current amount of space traffic destined for high orbit is probably not enough to amortize the cost of developing a space-based power transmission station in a time of commercial interest. Thus, the first such stations will have to "piggyback" on higher priority projects, Fortunately, three programs show promise of generating near term power transmission technology: the SP100 program, the Space-Based Radar, and the Space Station. A study effort on microwave-supported aircraft also may generate applicable transmission technology data.

Powell et al. (1981) have described space-based nuclear power for microwave power generation at a level (20 to 1000 MWe) higher than might be required for transfer of typical payloads with a realistically sized electric propulsion system. The particle bed reactor and power generation equipment can scale down to the powers of interest fairly readily. Since the size of the transmitter effective aperture is controlled by geometrical considerations, it is possible to trade off the size of the transmit aperture for a higher power reactor. The link efficiency may be low, but this is a secondary cost consideration.

Due to the high cost of putting mass into earth orbit, a main consideration in costing a microwave-supported propulsion system is to minimize the overall mass (and, thus, cost) in orbit as opposed to maximizing transmission efficiency. For instance, it is pointless for a 10-tonne power plant to feed a 50-tonne focusing array if the same power density at destination could be achieved by a 20-tonne power plant and a 10-tonne array. By making a number of simplifying assumptions, it is possible to generate the curve in Figure 11. This indicates a space link transmission efficiency for propulsion as low as 10% at the destination altitude at minimum cost. Some previous microwave-supported propulsion studies required that efficiency be 90% for the entire transfer.

With these considerations in mind, the model depicted in Tables 2 and 3 was constructed with TK! Solver@ software run on an Apple Macintosh personal computer. The model incorporates the assumption that the development and production costs of space-qualified hardware systems scale as their mass. Cost model experience reported by Space Division (1981) has shown that this is generally true. However, to simplify the equations and preserve generality, this model was not rigorously incorporated, and the costs developed are only approximate. Table 2 incorporates the equations used in the model and Table 3 lists the variables and their definitions. The mission is a medium orbit-raising mission with no inclination change. A standard low thrust equivalent velocity change equation for spiral orbits was used. The equation for mass of the transmitting aperture was fitted to the mid-range (Figure 4) for mesh deployables. The equation for space link efficiency (LE) was taken from Sercel (1985b) and provides a good approximation for efficiencies up to 90%. Table 3 also includes input values and the output values for one point on the curve. This curve was driven by the one input variable, RX, with an L at the start of its line. Thus, the model was calculated for each of twenty-two different transmitter radii and the output of the "COST" variable graphed in Figure 10. The efficiency variable was used as the x axis. The cost increases with efficiency toward the right of the graph due to the increase in transmitting aperture weight.

This model was iterated four times for successive minimum cost points of EP power (PWR), transmitter radius (RX) and receiver radius (RR)— enough to indicate that the proposed geometry was reasonable for the system. A rigorous optimization study should use a more precise cost model and incorporate sensitivity analysis on the other values. The results of the iteration with this model are listed in Table 4. Figure 11 shows the first iteration on power.

A space based radar is a microwave power transmitter that locates and tracks a target. Unlike propulsion-oriented hardware, it has pulse forming electronics not needed for simple power transmission and it probably requires a lower average power density on target (a more diffuse beam). To modify a space-based radar into a propulsion power transmitter the average power should be increased and the pulse formation eliminated and substituted with a large (20 to 30 tonne) steerable focusing aperture. Slew rates would be low due to orbital geometry.

The SP100 project promises a power system in the 300-kWe class for various applications. At the low end of efficiency and power, this is not sufficient for full power operations at synchronous distances. How-

Table 2 *Equations for Microwave Electric Space Transportation (In TK! Solver@ Printout Format).*

```
S Rule
* DELV = 20000000*SQRT( 1/R1 + 1/R2 - 2*COS(PI*DELI/2)/SQRT(R1*R2) )
* Q = EXP(DELV/VEX)
* MPROP = ( (Q*Q-1)*(A*PWR + MGCS)  + (Q-1)*MPAYL )/( 1 + (Q*Q-1)*TF)
* MRA =  SIG*PI*RR*RR
* MTUG = MPROP*TF + A*PWR + MGCS + MRA
* F = 2*ETA*(PWR*1000)/VEX
* MSS = MPAYL + MPROP*(1+TF)
* TT = MPROP*VEX/(86400*F)
* LE = 1 - 1/EXP( PI*RX*RX * PI*RR*RR / (WL*R2)^2 )
* MPP = 1.2*( MMPP + AP*PWR/(LE*ETAC) )
* MXA = 62.4*(2*RX)^.7936
* KPOW = (KD+2*KP)*MPP + (3*KPI)*MXA + L*3*(MPP+MXA)
* RCOST = ( L*MSS + ( KOPS + KP*MPAYL*INT )*TT ) / MPAYL
* FCOST = ( KPOW/N + ( KD + KP*NTUG)*MTUG / N ) / MPAYL
* COST = FCOST+RCOST
* MO = MTUG + MPROP + MPAYL
```

Table 3 *Parameters for Microwave Electric Space Transportation Cost Model (TK! Solver@ Printout Format).*

St	Input	Name	Output	Unit	Comment MWCOSTmesh
		DELV	2651.9758	M/S	DELTA V PER MISSION ONE WAY
		Q	1.3036885		MASS RATIO
L		MPROP	6176.281	KG	MASS OF PROPELLENT/PAYLOAD
L		MO	27356.637	KG	MASS OF LOADED TUG
L		MPAYL	18514.905	KG	MASS OF PAYLOAD
		MRA	1256.6371	KG	MASS OF RECEIVEING APERTURE
L		MTUG	2665.4511	KG	DRY MASS OF TUG
L		TT	29.785306	DAYS	THRUST TIME
L		KPOW	3.51668E9	$	FIXED COST OF POWER PER MISS
L		FCOST	1075.6558	$	FIXED COST PER MISSION PER K
L		RCOST	4412.7133	$	RECURRING COST PER MISSION P
L		COST	5488.3691	$	TOTAL COST PER KILOGRAM
L		F	24	NEWTONS	THRUST
L		MXA	25994.531	KG	MASS OF TRANSMITTING APERTUR
L		MPP	8384.5188	KG	MASS OF POWER PLANT SYSTEM
L		LE	.15260189		LINK EFFICIENCY
	300	PWR		KW	POWER DELIVERED TO THRUSTER
	10000	VEX		M/S	EXHAUST VELOCITY
	6674000	R1		M	STARTING RADIUS
	15440800	R2		M	DETINATION RADIUS
	0	DELI		RAD	INCLINATION CHANGE
	200	N			NUMBER OF LAUNCHES
L	1000	RX		M	RADIUS OF TRANSMITTING APERT
	100	RR		M	RADIUS OF RECEIVING APERTURE
	.05	WL		M	WAVELENGTH OF MICROWAVE BEAM
	2400	MMPP		KG	MINIMUM POWERPLANT MASS
	2.1	AP		KG/KW	SPECIFIC MASS OF MARGINAL PO
	100000	KD		$/KG	COST OF DEVELOPMENT + FIRST
	25000	KP		$/KG	COST OF PRODUCTION / KG
	25000	KPI		$/KG	COST OF CONCENTRATOR MATERIA
	100000	KOPS		$/DAY	SYSTEM OPERATIONS COST/DAY
	.00027	INT			DAILY INTEREST RATE
	.04	SIG		KG/M^2	AREAL DENSITY OF CONCENTRATO
	3000	L		$/KG	LAUNCH COST PER UNIT MASS
	3	NTUG			NUMBER OF TUGS
	3	A		KG/KW	PROPULSION SYSTEM SPECIFIC P
	.9	ETAC			MICROWAVE POWER CONVERSION E
	.4	ETA			ROCKET EFFICIENCY
	.05	TF			TANKAGE FRACTION
	200	MGCS		KG	MASS OF G&C + STRUCTURE
	25000	MSS		KG/UNIT	MASS LIFTED PER SHUTTLE
	3.1415927	PI			PI

Table 4 *Example of Iterative System Sizing with a Cost Model.*

Iteration	Power (kWe)	Transmitter Radius (m)	Receiver Radius (m)	Cost/kg ($)
1	300	1000[a]	100	5430
2	300	1000	140[a]	5400
3	350[a]	1000	140	5370
4	350	800[a]	140	5350

[a]Independent variable during iteration.

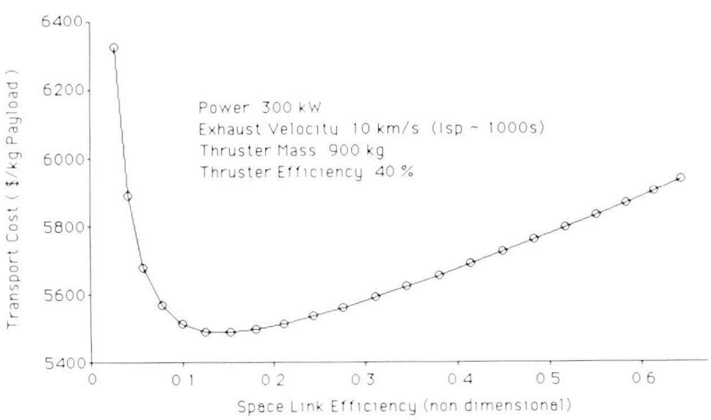

Figure 10 *Cost as a function of Geometrical Efficiency at a Destination Altitude for Variable Transmit Aperture.*

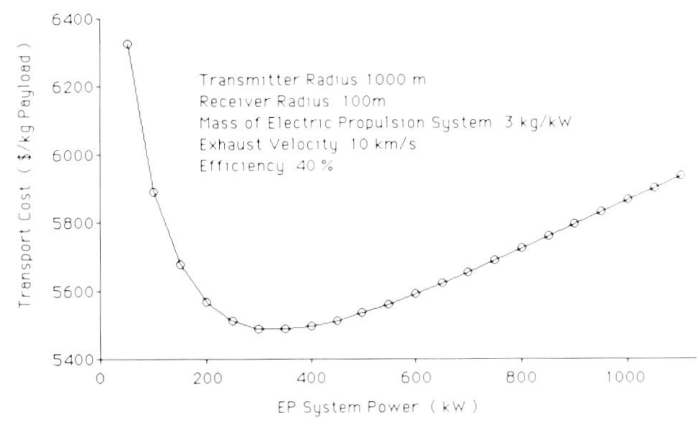

Figure 11 *Transportation Cost as a function of EP System Power with a Fixed Antenna Aperture.*

ever, SP 100 studies have gone up to a megawatt and beyond. Thus, operation at lower powers at long distances (by reducing the number of thrusters, for instance) can be tolerated because most of the orbit raising work is done in the first few thousand kilometers. A modified Space-Based Radar with an SP 100 power source could provide a near-term microwave power transmission capability, much the way the Goldstone 26-m antenna was used in 1974 (Dickinson 1975). Its potential depends on how much is invested in the transmitting aperture.

The third program, the Space Station, could provide a logistical base for maintaining the space-based transmitters, which might otherwise require periodic, expensive shuttle visits. The station may also create a demand for beamed microwave power by playing host to a variety of platforms. It may prove less complicated to beam power to these platforms rather than provide each with a separate sun-oriented power system.

The impact of these developments is that a microwave-supported propulsion system may not have to bear the entire cost of development of the power transmission facility. If the additional cost for propulsion purposes can be lower, then the mass cost savings of fewer missions will be needed to amortize the transmission costs and to make the system more attractive to potential users.

Power Transmission Orbits

Convincing analysis of ground based microwave transmitters has been done elsewhere (Brown and Glasser 1983). Space-based transmitters, if affordable, may offer several advantages, chief among these being visibility.

Steering a one-kilometer radius transmitting array by other than electronic means presents interesting engineering challenges. To approximate the order of magnitude of angular rates required, a simulated mission was run on a personal computer. The mission updated position, velocity, and acceleration at 5-minute intervals with a Keplerian trajectory assumed between updates. The transmitters are assumed to be in a two-hour orbit with a 28.5 degree inclination. (Only one transmitter was used in the model, the concern being rates at close approaches.) The mission is to move a 15-tonne payload from an initial low earth orbit (6700 km radius) to an orbit with a radius of 8500 km, and from an inclination of 28.5 degrees to 50 degrees. Out-of-plane thrusting was restricted to 45 degrees on either side of the nodes. Figure 12 shows the lines of sight from transmitter to receiver projected on

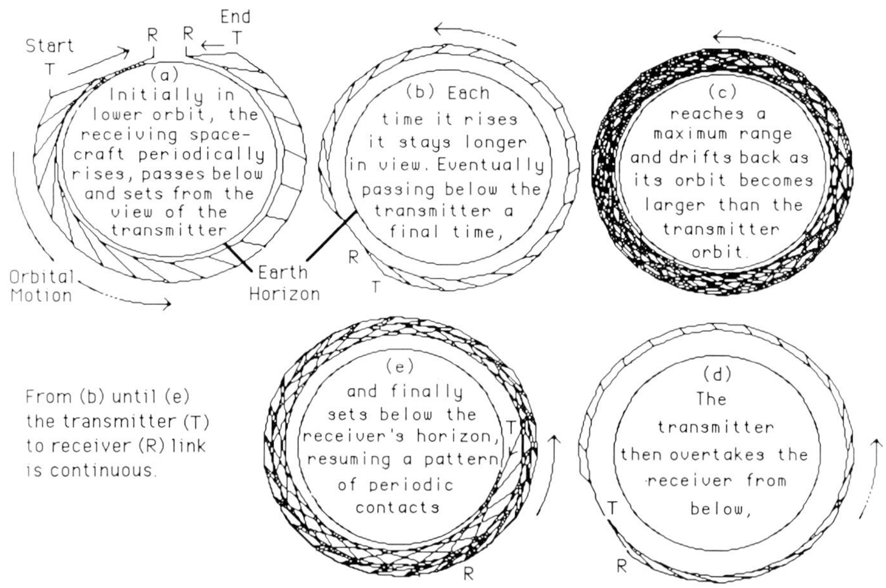

Figure 12 *Position Plots as 5 min Intervals during Part of a Simulated Beamed Microwave Orbit Transfer Mission with One Transmitter.*

the transmitter orbit plane. This vector rotates slowly in inertial space, the worst case being about 3.2 degrees per minute. Moreover, the slewing rate is very steady, with acceleration reaching only. 1 degree per minute per minute within 10 minutes of close approaches. Table 5 is a printout of the time, distance and angular rates during a close approach.

The low space-to-space angular rates are not remarkable if one considers the geometry in the moving, three dimensional, frame of reference (Figure 13). At similar altitude, the transmitter and spacecraft have very nearly the same orbital velocity. If the nodal relationship is chosen so that the positions are far apart in declination when close in right ascension, then they never come very close even if the orbits cross.

Physical means of directing a transmitter array may devolve from any of several pointing technologies being considered for the Strategic Defense Initiative. The accelerations and masses involved seem lower than for concepts that require large high-quality laser mirrors or steerable electromagnetic guns. A number of fairly self-explanatory ideas are listed in Table 6 to give an idea of the range of potential solutions to this problem. Note that these devices need to be around two kilometers (20 football fields) in diameter and should maintain their figure to a few centimeters.

Some of these concepts have field of view restrictions that are subject to cost and link efficiency trade-offs. For example, is it better to have one full-size ''dragon scale'' array covering ± 60 degrees or to have four small ones covering 4π steradians and compensate for lost efficiency by increasing power output? Optimizing the average transmission distance is beyond the scope of this paper; however, 10,000 km is a good point for ''back of the envelope'' sizing estimates. With more transmitters, this could be lower. Figure 14 shows a 3 transmitter system.

The example used for Table 5 combines continuous thrusting and ''pumping'' strategies. The required altitude is obtained early on and the inclination is 'pumped' at the nodes thereafter. Pumping can also be used to gain altitude. Figure 15 shows an example of pumping at perigee to increase apogee and then at apogee to circularize the orbit. This is a time-consuming strategy for low thrust engines. However, it has several mass saving advantages, principally lower average distance (assuming carefully worked out phasing, and a transmitter near the destination orbit) and quasi impulsive mission delta v requirements. The decrease in total delta v can have a dramatic impact on payload fraction in marginal circumstances, as shown in Figure 3.

System Summary

As previously discussed, a fully optimized mission scenario must consider a daunting number of variables and trade-offs. There are some ''common sense'' considerations for planning such a system. A large mission model will be dominated by recurring costs; therefore, the marginal cost of one additional transmitter may be small compared to operational cost (time) savings. It depends on the size of the model. The orbits should be selected to avoid very close approaches, but to keep the average distance low during the most energy demanding parts of the transfer. At first guess, it appears the system might work well with a 1-km radius class transmitter and a receiver with a 100-m radius class concentrator and a 20-m (1256 m^2, 376 kWe) rectenna. With an advanced ammonia arcjet array and further work on rectenna mass, the propulsion associated dry mass could be as low as 3 tonnes; about the same dry mass as a Centaur upper stage, but with twice the fuel efficiency.

Research Recommended

The Raytheon/NASA rectenna experiments, the Jet Propulsion Laboratory (JPL)/Air Force Rocket Propulsion Laboratory (AFRPL) arcjet performance demonstration, the SP 100 space nuclear power studies, the commercial availability of rugged kilowatt class magnetrons, and the work of AFRPL and others on lightweight collectors and concentrators constitute a set of technologies that clearly demonstrate the feasibility of near-term space to space microwave power beaming. Still needed is an intensive look at

Table 5 *Printout of Time Range and Angular Rates for a Close Approach Between a Transmitter and Receiver.*

1265	MIN	1632.5629152311	KM	2.9366261960075	DEG/MIN
1270	MIN	1586.3017127352	KM	2.9405160751529	DEG/MIN
1275	MIN	1540.8263372608	KM	2.9377788879535	DEG/MIN
1280	MIN	1495.9739196457	KM	2.926699977503	DEG/MIN
1285	MIN	1451.0695764019	KM	2.907088764242	DEG/MIN
1290	MIN	1405.121152918	KM	2.8799025931711	DEG/MIN
1295	MIN	1356.7943414586	KM	2.8480201897889	DEG/MIN
1300	MIN	1304.8385165772	KM	2.8148191702104	DEG/MIN
1305	MIN	1248.2932497944	KM	2.7832891584719	DEG/MIN
1310	MIN	1186.6729940472	KM	2.755343489619	DEG/MIN
1315	MIN	1120.0461795675	KM	2.7312307522679	DEG/MIN
1320	MIN	1048.3280119668	KM	2.709778618462	DEG/MIN
1325	MIN	974.101178209	KM	2.6874983043044	DEG/MIN
1330	MIN	897.23418613775	KM	2.6575008701301	DEG/MIN
1335	MIN	819.6533371817	KM	2.6096617550498	DEG/MIN
1340	MIN	742.6303746408	KM	2.5302588108033	DEG/MIN
1345	MIN	667.1533496252	KM	2.4015471525302	DEG/MIN
1350	MIN	593.96816152965	KM	2.2011004901824	DEG/MIN
1355	MIN	523.7665989368	KM	1.8955602807193	DEG/MIN
1360	MIN	457.85259189022	KM	1.4542399331189	DEG/MIN
1365	MIN	399.31996037965	KM	.8836218132186	DEG/MIN
1370	MIN	354.54091930648	KM	.67584201810528	DEG/MIN
1375	MIN	333.65098911794	KM	1.3965714645031	DEG/MIN
1380	MIN	346.40610970851	KM	1.9618163502631	DEG/MIN
1385	MIN	394.3165408612	KM	2.0137302958792	DEG/MIN
1390	MIN	470.3713928174	KM	1.9082992497406	DEG/MIN
1395	MIN	565.3864857301	KM	1.9500786491377	DEG/MIN
1400	MIN	672.08403264445	KM	2.0829830276398	DEG/MIN
1405	MIN	785.49125015345	KM	2.2206661434255	DEG/MIN
1410	MIN	902.3964698437	KM	2.3389019025782	DEG/MIN
1415	MIN	1020.7154480161	KM	2.4392951778511	DEG/MIN
1420	MIN	1139.6956901324	KM	2.5263482738515	DEG/MIN
1425	MIN	1259.5096049299	KM	2.6021837112805	DEG/MIN
1430	MIN	1381.0207190241	KM	2.6671249874424	DEG/MIN
1435	MIN	1505.5063996354	KM	2.7204293068982	DEG/MIN
1440	MIN	1634.3521317722	KM	2.7612665396572	DEG/MIN
1445	MIN	1768.7619187509	KM	2.7895166420257	DEG/MIN
1450	MIN	1909.6392084432	KM	2.8061634925858	DEG/MIN
1455	MIN	2057.2560985952	KM	2.8128995152354	DEG/MIN
1460	MIN	2211.3166944416	KM	2.8125775388626	DEG/MIN
1465	MIN	2371.0623643153	KM	2.8082973343038	DEG/MIN
1470	MIN	2535.4432713923	KM	2.802795278321	DEG/MIN
1475	MIN	2703.3043840869	KM	2.7981400324827	DEG/MIN
1480	MIN	2873.4345363934	KM	2.7958503222134	DEG/MIN
1485	MIN	3045.0300871064	KM	2.7963447379364	DEG/MIN
1490	MIN	3217.5957222425	KM	2.7993234773137	DEG/MIN
1495	MIN	3391.7487357146	KM	2.8022748526668	DEG/MIN
1500	MIN	3567.1530985264	KM	2.8052309515355	DEG/MIN
1505	MIN	3743.4178228272	KM	2.8078780021033	DEG/MIN
1510	MIN	3920.0304017245	KM	2.8100285782282	DEG/MIN

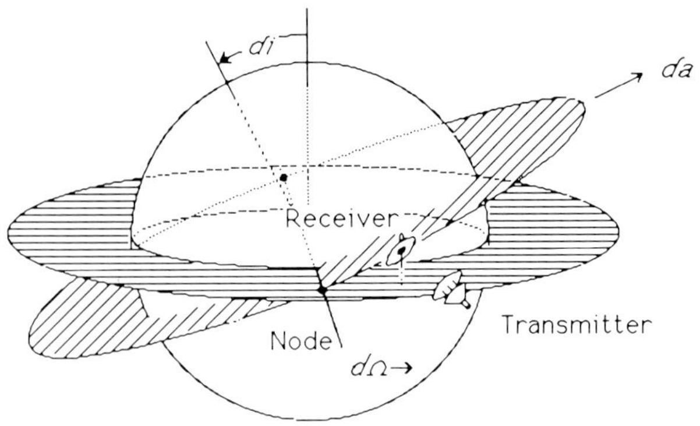

Figure 13 Orbital Geometry of Transmitter and Receiver in Close Encounter.

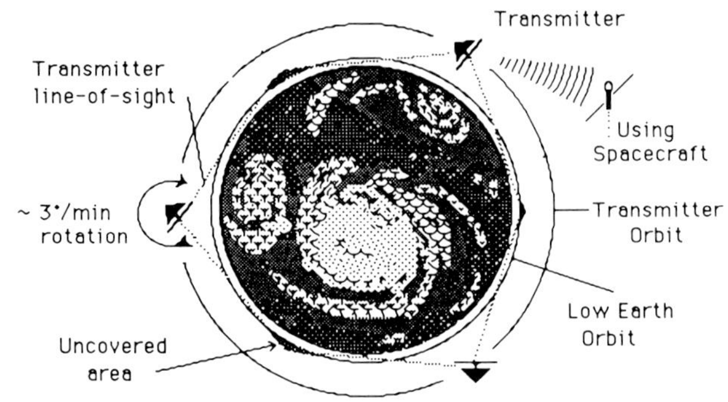

Figure 14 Microwave Orbital Power Transmitter Constellation.

system designs, costs and benefits versus various mission models and advance development projects. This should be done on a subsystem level, to assure prospective users that the risk is very small. Implementation of such a system would be a significant change to existing orbit transfer architecture, and, therefore, would require significant 'pump priming' from the research and development community before user demand can be depended on to drive its development. At first look, there may be significant benefits over chemical or even solar electric orbit transfer options.

Specifically, the space nuclear power community should consider:

1. A one man-year mission benefits study covering a wide range of potential beamed microwave system architectures and low, medium and high mission models. This should include transmitter power mass versus link efficiency trade studies.
2. A one-kilowatt microwave rectenna powered arcjet ground demonstration as a follow-on to work being done at NASA /Lewis Research Center. This might be done for less than $1 million given the present state of technology.

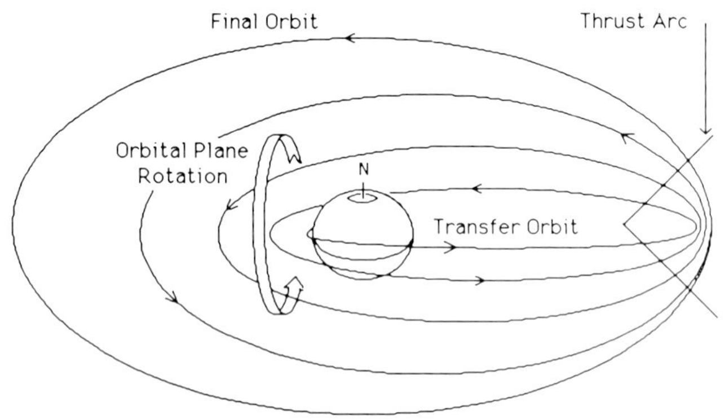

Figure 15 Orbit Circularization with Electric Propulsion (Every Tenth Orbit Shown).

3. A serious look at a 300 kWe continuous microwave power transmitter as a secondary payload on SP 100. In addition to providing a redundant demonstration of SP100 success, this would provide a start at beamed power technology with a wide range of applications, and contribute significantly to the perception of SP 100 as useful technology beyond purely SDI applications. It would serve as a load for times when the primary propulsion load is not

Table 6 *Concepts for Transmitter Aperture Construction.*

Name	Brief description
• Multiple small dishes	Many small, easily steered dishes are combined like radio telescope arrays.
• "Dragon scale" flat mirror array	Electrostatically controlled flat mirror array like that used to concentrate sunlight for the Solar One power plant.
• Inflated sphere	Huge, inflated, counterweighted sphere, with light mesh on one hemisphere and a linear transmitter array in the center. 'Optics' resemble the Aricebo radio telescope.
• Mesh dish	Made as light and strong as possible, counter-weighted, active tensioning for figure control

needed, making it less important for an electric propulsion primary load to operate an exceptionally long time.

In the far-term, research into the following would be productive:

- Higher frequency rectennas to reduce aperture sizes,
- Advanced lightweight electrical beam steering technology, and
- Alternative large-structure deployment technologies.

Acknowledgments

This paper was an unfunded project undertaken by the authors largely on their own time. We would like to thank Gayle Wiesner for typing and the Air Force Rocket Propulsion Laboratory and Raytheon Co. for logistic and graphics support.

References

Asmussen, S. and S. Whitehair (1984) "Experiments with a Microwave Electrothermal Thruster Concept." AIAA paper No. IEPC 84-74, in *Proceedings JSASS/AIAA/DGLR 17th International Electric Propulsion Conference*, held in Tokyo, Japan, 28-31 May 1984.

Brown, W. C. (1969) "Experiments Involving a Microwave Beam to Power and Position a Helicopter," *IEEE Transactions on Aerospace and Electronic Systems*, Vol. AES-5, No. 5, September 1969.

Brown, W, C. (1973) "Adapting Microwave Techniques to Help Solve Future Energy Problems," *IEEE Transactions on Microwave Theory and Techniques*, Vol. MTT-21, December 1973.

Brown, W. C. (1977) *Electronic and Mechanical Improvement of the Receiving Terminal of a Free-Space Microwave Power Transmission System*, NASA CR 135194, Raytheon Corp., Waltham, MA, August 1977.

Brown, W. C. (1981) "Status of the Microwave Power Transmission Components for the Solar Power Satellite," *IEEE Transactions on Microwave Theory and Techniques*, Vol. MTT-29, No. 12, December 1981.

Brown, W. C. (1984a) "Performance Characteristics of the Thin-Film, Etched Circuit Rectenna," *1984 IEEE MTTE International Microwave Symposium Digest*.

Brown, W. C. (1984b) *The History of Power Transmission by Radio Waves*, PJ64S2, Raytheon Co. New Products Center, Watham, MA.

Brown, W. C. (1985) "Earth to Space DC to DC Power Transmission System Utilizing a Microwave Beam as Source of Energy for Electric Propelled Interorbital Vehicles," Preprint of Paper presented at the AIAA International Electric Propulsion Conference held in Alexandria, VA 30 September to 2 October 1985.

Brown, W. C. and P. E. Glasser (1983) "An Electric Propulsion Transportation System from Low Earth Orbit to Geostationary Power," *Space Solar Power Review*, 4: 119-129.

Dickinson, R. M, (1975) *Evaluation of a Microwave High Power Reception—Conversion Array for Wireless Power Transmission*, NASA TM 33-741, September 1975.

Friese, G. J., G. D. Bilyeu, and M. Thomas (1982) *Initial 80's Development of Inflated Antennas*, LTR 82-GF-107, L'Garde, Inc., for NASA Contract NAS1—16663, Langley Research Center, VA, December 1982.

Goddard, R. H. (1925) *Electrostatic Repulsion of Matter in Connection with the High Altitude Rocket Principal*, Clark College, MA, August 1925.

Nordley, G. D. and R. J. Vondra (1985) "Electric Propulsion for New Earth Orbital Transfer". AIAA 85 1151, *AIAA/SAE/ASME/ASEE Joint Propulsion Conference*, held in Monterey, CA, July 1985.

Powell, J., T. E. Botts, and A. Hertzberg (1981) *Applications of Power Beaming from Space Based Nuclear Power Stations*, BNL 29502, Brookhaven National Laboratory, Upton, NY.

Sercel, J. C. (1985a) "Microwave Electric Propulsion for Orbit Transfer Applications," in *JANNAF Joint Propulsion Conference Transactions*, held in San Diego, CA, December 1985.

Sercel, J. C., (1985b) "Solar Thermal Propulsion for Planetary Spacecraft," in *JANNAF Joint Propulsion Conference Transactions*, held in San Diego, CA, December 1985.

Space Division (1981) *Unmanned Spacecraft Cost Model*, SD-TR-81-45, USAF Space Division, El Segundo, CA.

Vondra, R. J., J. Cassady, and G. Nordley (1984) *Presentation to Air Force Space Technology Center Vice Commander*, Edwards AFB, CA, June 1984.

Part 13

MULTIMEGAWATT SYSTEMS

Space Nuclear Power Systems 1986
Edited by M. S. El-Genk and M. D. Hoover

Chapter 38

A Pellet Bed Reactor for Multi-Modal Space Power

David Buden, Kenneth Williams, Peter Mast, and James Mims

Science Applications International Corporation
2109 Air Park Road, S.E.
Albuquerque, NM 87106

A review of forthcoming space power needs for both civil and military missions indicates that power requirements will be in the tens of megawatts. The electrical power requirements are envisioned to be twofold: long-duration lower power levels will be needed for station keeping, communications, and/or surveillance; short-duration higher power levels will be required for pulsed power devices. These power characteristics led the authors to propose a multi-modal space power reactor using a pellet bed design. Characteristics desired for such a multimegawatt reactor power source are standby, alert, and pulsed power modes; high-thermal output heat source (approximately 1000 MWt peak power); long lifetime station keeping power (10 to 30 years); high temperature output (1500 K to 1800 K); rapid-burst power transition; high reliability (> 95 percent); and stringent safety standards compliance. The proposed pellet bed reactor is designed to satisfy these characteristics.

Background

The operational characteristics of the power subsystem include: (1) station keeping power levels of 0.1 MWe to 1 MWe (nominal value, 0.5 MWe); and (2) burst/engagement mode of 100 MWe to 1000 MWe (nominal value, 200 MWe). Furthermore, the transition time from the station keeping mode to the burst/engagement mode should be less than 5 seconds. Other parameters of importance in power source selection relate to environmental characteristics, such as contamination, thermal radiation, and platform stability issues. The use of open- or closed-cycle power subsystem results in significant differences in contamination and platform stability issues. The effluents from open-cycle power subsystems can deteriorate sensors, lenses, coatings, and other surfaces. In addition, the effluent creates thrust as it is expelled, leading to platform stability problems. Closed-cycle power subsystems eliminate effluent problems, but introduce thermal radiation issues from high temperature radiation surfaces.

Additional power subsystem requirements in selecting a heat source include the following: (1) a high probability (reliability > 95 percent) that the power plant will operate when called upon; (2) a design accounting for repairability or modernization either by manned or robotic systems; (3) operation in natural, self-induced and hostile environments; and (4) lifetimes of 10 to 30 years for the station keeping mode (nominal value, 20 years) and 100 s to 2000 s for the burst mode (nominal value, 1000 s). Repairability includes the capability for self-diagnostic of the status of the heat source. The lifetimes assume that the space platform is a capital asset, much like a B-52, and will be upgraded periodically to avoid obsolescence. A desirable total mass goal is to launch the power subsystem in the equivalent of one or two shuttles, or have a mass of 25 to 50 metric tons. The power subsystem should be low mass and packaged in a configuration compatible with the launch vehicle in order to minimize transportation costs.

Safety considerations govern any reactor heat source design. In particular, the heat source must address issues of unplanned criticality, fuel dispersal, and final reactor disposal. Table 1 summarizes these power subsystems goals.

Potential Candidates for Multimegawatt Power Subsystems

Candidates under consideration to meet the goals in Table 1 are

- **Open-Cycle Chemical**—Versions of these subsystems are being developed to 40 MWe, based on the F-1 engine design. The main limitations are large quantities of propellants for long burst-power modes, effluent platform contamination, effluent stability effects, and the need for a separate power subsystem for station keeping.

- **Open-Cycle Nuclear**—Nuclear rocket technology could be used as a heat source. Twenty reactors were operated to 2450 K, 4200 MWt, and 1-hour duration (not all by one reactor) using hydrogen as the coolant. These designs need modification for dual-mode operation. The platform contamination, stability, and propellant makeup are similar to open-cycle chemical power subsystems issues.
- **Low Power with Storage**—A steady-state solar or nuclear power source for station keeping plus batteries, fuel cells, or flywheels for burst power could be used to satisfy the power needs. The major limitations are the very heavy mass and large volume for long-duration, burst-power operation.
- **Closed-Cycle Nuclear**—Numerous reactor options exist using various fuel configurations (pin, hexagonal, particles, pellets, liquid, and gaseous) that can be configured for single- or multi-mode. Solid fueled reactors probably could operate at exit temperatures of 2700 K to 3000 K, while gaseous cores could operate up to 6000 K. To meet the power needs in Table 1, the configurations are primarily conceptual. The major limitations are the state of technology, material uncertainties, and large heat rejection elements.
- **Others**—Closed-cycle chemical and your favorite concepts.

Pellet Bed Reactor Concept Description

The idea of a pellet bed for space reactors originated with David Buden in 1983. A schematic is shown in Figure 1. The reactor uses nuclear fuel in the form of pellets. These fuel pellets, about 0.5 to 2 cm in diameter, are confined in a cylindrical liner whose end plates contain holes to allow the coolant to circulate through the core. The spherical pellets are enriched with uranium-235 in the form of an oxide, carbide, or nitride encapsulated to minimize erosion and corrosion, and to avoid materials interaction problems. Two or more pellet sizes might be used to control the void fraction in the core. Surrounding the core is a reflector layer that incorporates control elements in the form of rotatable drums with neutron absorber segments, sliding reflector segments, or rotatable reflector segments. The reflector is cooled by the reactor inlet coolant, or radiation to space. The coolant flow paths determine the arrangement of the thermal barrier between the core and reflector, and how the containment vessel is cooled.

Typical materials for the fuel might be UC_2-coated with 25μm to 50μm of pyro-graphite for helium-cooled cores, or UN clad with W for Li-cooled cores; a refractory metal for the cylindrical core containment or core periphery components; Be or BeO for the reflector; and B_4C for the control reactivity neutron absorber. The pressure vessel is cooled or insulated from the high-temperature core region, allowing the use of refractory metal alloys with a ceramic liner.

Depending on the particular lifetime requirements

Table 1 *Power Subsystems Goals for a Multi-Modal Space Power Reactor.*

	STATION-KEEPING MODE	BURST/ENGAGEMENT MODE
Power Level (MWe)	0.01 to 1 (Nominal Value 0.5)	100 to 1000 (Nominal 200)
Duration	10 years growth to 30 years	100 to 2000 seconds
Startup	Several hours	< 5 seconds
Contamination	Avoid degrading platform elements beyond specifications	
Thermal Radiation	Reduce to acceptable levels for platform	
Platform Stability	Reduce to level to meet platform specifications	
Reliability	95-percent probability of success for stated mission lifetimes. Avoid single failures that result in significant loss of design power	
Safety	Reactor shall meet DOE safety criteria including: • Remaining subcritical if immersed in water or if exposed to other fluids, • Having an effective negative temperature coefficient, • Avoiding any credible accident that leads to a critical or super-critical configuration, • Having two independent systems to maintain subcriticality, • Having an independent shutdown heat removal path for decay heat, and • Presenting no significant environmental hazards from unirradiated fuel.	
Survivability	Hardened to operate in natural environments; Survive specified hostile actions.	
Repairability	• Defective elements can be replaced prior to launch, • Design for robotic or manned replacement of selected parts in space, and • Provide for self diagnostic of condition of power subsystem during operation.	
Mass	25 to 100 metric tons (nominal value, ~50 tonnes)	

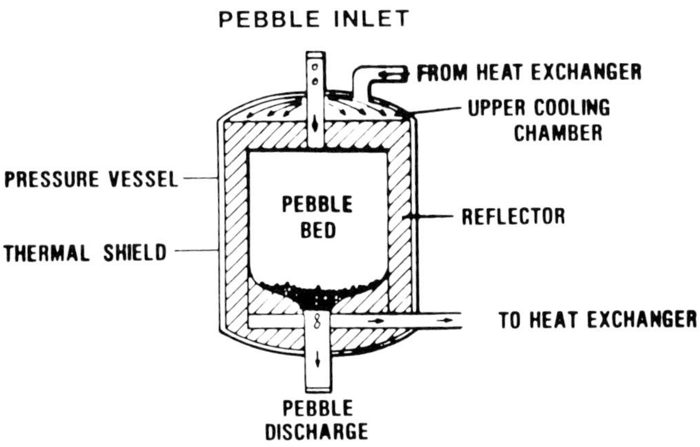

Figure 1 *Diagram of Pebble Bed Reactor.*

and the method selected to meet certain safety standards, the pellets either can remain relatively fixed in the core or be slowly removed. If removed, new pellets can be added to provide a long-lifetime system with low built-in excess reactor reactivity, or the used pellets can be reinserted in the core to provide relatively uniform burnup. In the weightlessness of space, proper design of the fluid hydraulic loop forces the flowing coolant to move the pellets forward.

Figure 2 shows schematics for various configurations of pellet reactor concepts. Figure 2a shows a non-circulating fuel configuration in which the pellets are located inside a pressure containment vessel that is cooled by the inlet fluid. A baffle arrangement distributes the flow before entering the core through the end plate, and the end plate hole arrangement controls the flow. The coolant is then heated by the fuel pellets, and the fluid exits through the top end plate. To reduce the pressure vessel size and mass, the reflector is located (as shown) outside the pressure vessel. Insulating multi-foils wrapped around the vessel can be used to minimize heating the reflector by conduction. The reflector is cooled by conduction of heat to the outer surface and radiation of the heat to space. The control elements are located within the reflector.

Figure 2b is a schematic of a once-through, then-out pellet fuel configuration. Extra storage volumes are located at the top and bottom of the reactor. The fuel is slowly moved through the reactor by hydraulic forces. New fuel is located at the bottom of the reactor; and used fuel, at the top. Longer lifetimes are expected from such a configuration relative to the non-circulating fuel case. In addition, the lack of a large amount of excess reactivity having to be built into the reactor is a safety advantage.

The design in Figure 2 incorporates a continuous circulating fuel arrangement to provide a uniform and maximum fuel burnup. It can also include a reserve of fuel pellets outside the core region for longer lifetimes. A means is needed to transfer the pellets from the relatively lower pressure outlet stream to the higher pressure inlet stream without significant bypassing of fluid. A method similar to a water wheel ar-

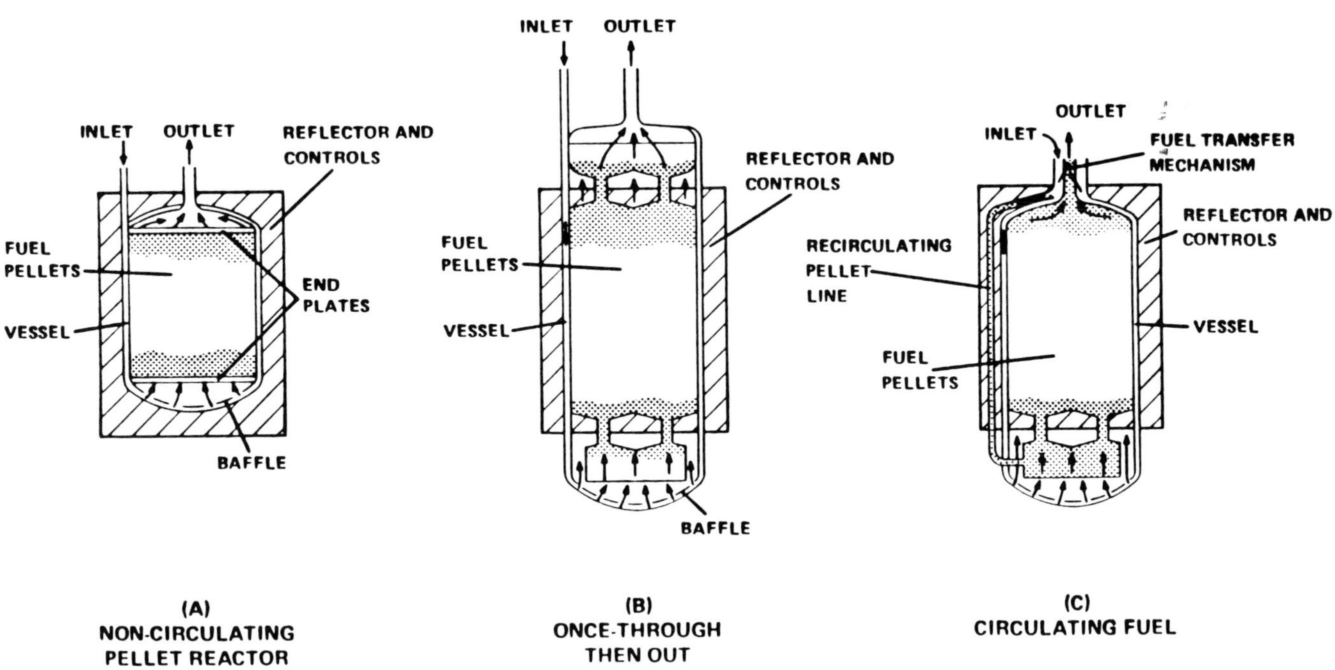

Figure 2 *Pellet Bed Reactor Schematics.*

rangement or a root valve can be used. Pellets exiting the core would be caught in the turning member, thereby rotating from low- to high-pressure sides. The higher pressure inlet flow is split to carry the pellets from the top of the reactor to the bottom and to cool the pressure vessel. For very high powers and long lifetimes, this concept has many advantages. Approximately 15 kg of uranium-235 will be consumed for a reactor that runs 1 year at 40 MWt (10 MWe). At 5-percent burnup, this implies a need for 290 kg of fuel. The ability to add new fuel solves criticality, safety, and the overall mass limitation problems.

Related Experience

Over the past 20 years, pebble bed fuels have been developed to a high degree of sophistication and predictability (Ziermann 1984, Nabrelek 1984, and Wittchow et al. 1984). Each fuel element is approximately 6 cm in diameter and consists of a coated fissile particle embedded in a graphite matrix contained within a graphite outer shell. The coated-particle concept is shown in Figure 3. These particles use a porous, inner pyrolytic carbon buffer layer designed to accommodate kernal swelling to absorb fission recoils, and provide void volume for the accumulation of gaseous fission products. The outer layer serves as a fission gas barrier and pressure vessel, and generally consists of dense isotropic pyrolytic carbon sometimes used in combination with chemically vapor-deposited silicon carbide. The silicon-carbide layer acts as a barrier to the release of metallic fission products and provides mechanical redundancy to the coatings. Thus, the basis for the use of coated-particle fuels in advanced nuclear systems is well established.

Pebble bed reactors have been successfully operated in terrestrial power systems. The 15-MW AVR experimental reactor demonstrated the feasibility of spherical fuel elements and high operating temperatures (1225 K) (Ziermann 1984). It has been in operation since 1967, with an availability of 78 percent. About 75,000 fuel pebbles are in the reactor, and some reached burnup of 185,000 MWD/tonne of heavy metal. The number of pebble bed fuel elements used in AVR (given in Table 4) is more than 265,000 (Nabrelek 1984). The successful experience with AVR led to the construction of a 300-MWe THTR demonstration plant. This reactor has now successfully reached nuclear criticality (Wittchow 1984).

Some modifications of current coated-particle technology are desirable for space power or other high-performance nuclear systems. Two key factors are materials compatibility and fuel loading. Coatings of ZrC, tungsten, or some other refractory metal may be more compatible.

The other key factor in the use of coated-particle fuels in advanced nuclear systems is fuel loading. Fuel loadings are maximized by both minimizing coating thickness and the interstitial void space in the bed of fuel particles. Both of these classes of thermochemical phenomena are sufficiently well understood to be treated quantitatively in design. They are particularly important for high-temperature fuel application. Sufficient analysis has now been performed to conclude that the burst-power mode will establish the core size to meet the thermalhydraulic needs; while the continuous mode will establish some flexibility in coating thickness and density without enlarging the core.

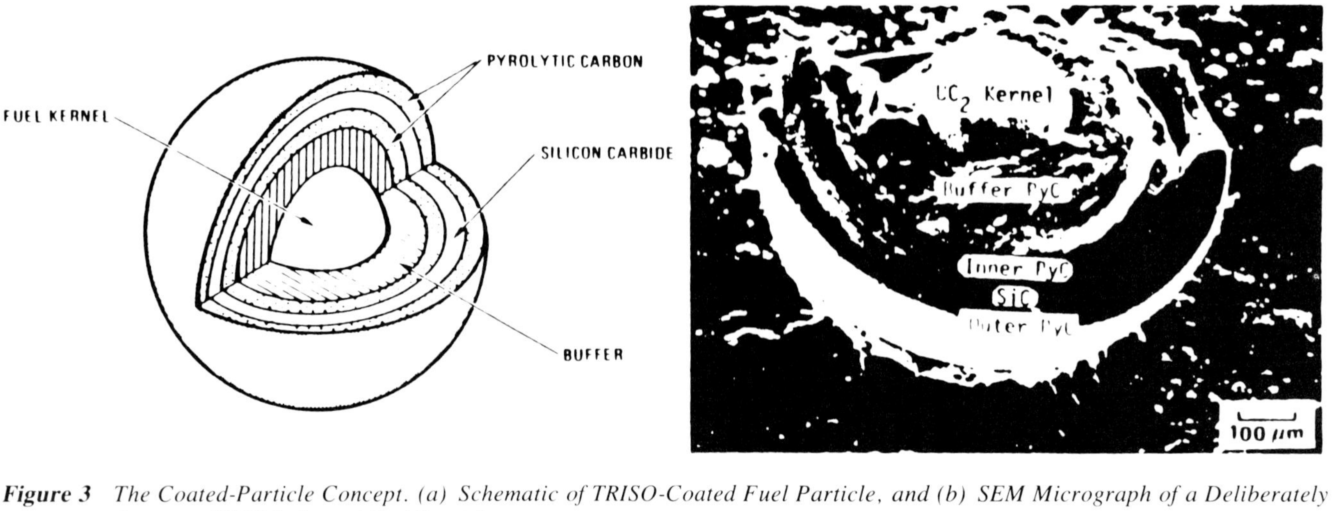

Figure 3 *The Coated-Particle Concept. (a) Schematic of TRISO-Coated Fuel Particle, and (b) SEM Micrograph of a Deliberately Fractured TRISO-Coated Fuel Particle.*

The reflector can be modeled after the SP-100 multi-hundred kilowatt program designs and previous space reactor experience. Various materials will be evaluated, but an extensive data base already exists. The longest operation times for beryllium reflectors were demonstrated in U-ZrH-fueled reactors. These reflectors were cooled by NaK. For example, the SNAP-8 Demonstration Reactor (S8DR) reflector reached a neutron dose of 1×10^{20} nvt and a gamma dose of 1×10^{11} rad (10^9 grays) (Angelo and Buden 1985). During steady-state operation, this reflector operated at about 600 K. The beryllium was anodized to provide oxidation protection and emittance enhancement. In a 1-year long vacuum test, the reflector drive operated satisfactorily, and no signs of self-welding or sticking were displayed by any component. Bearings made of a solid carbon-graphite ball were operated successfully for 7000 h in the S8DR test. Four individual bearing sets completed 12,000 h of vacuum testing at 895 K and 0.001 Pa or lower pressure. In all, more than 100,000 h of tests were accumulated on the bearings (Angelo and Buden 1985).

Design Considerations for Closed-Cycle Nuclear System

Multi-Modal Power

The reactor can be sized to provide 2500 MW or more thermal power, depending on the selected design point. Because the duration of the burst-power range (nominal, 1000 s at 200 MWe) is short compared to the lower ranges (nominal, 10 years at 500 kWe), the long-duration, lower range will establish the fuel loading requirements. Assuming 20-percent converter efficiency, the fuel burnup for 200 MWe burst power is 12 g. For station keeping, assuming 20-percent efficiency and 10-year operation, the fuel burnup is 9 kg. Detailed studies are needed to evaluate the amount of fuel to build into the reactor in order to account for such factors as criticality, burnup, refueling, complexity, and disposal.

Reactor Outlet Temperatures

The reactor outlet temperature is a compromise among the heat rejection elements, the power conversion equipment, and the reactor design. The desire is to push for closed-loop systems for burst power in order to eliminate platform stability, spacecraft contamination, and resupply problems. Also, as battle strategies evolve and change, the burst-power mode durations may become significantly longer. The power system should not be the limiting spacecraft operational factor. The heat rejection system size is minimized by higher temperatures (proportional to the inverse fourth power of temperature). Power conversion equipment efficiency is approximately proportional to the temperature ratio of the hot-to-cold temperatures. However, materials tend to limit the operational temperature level. Thermal stress at higher operating temperatures may lead to thicker lines and heavier components; therefore, higher temperature systems are not necessarily lighter.

Heat rejection temperatures for the burst-power mode need to be as high as feasible, around 1400 K to 1500 K. Based on projections from Figure 4, the heat rejection element would be more than 30,000 kg. Advanced short-life heat rejection designs could substantially reduce this mass. Materials and coolants (plus physical laws) will define the conversion system temperatures and cycle efficiency. Tables 2 and 3 present the Carnot and converter efficiencies, respectively. Turbomachinery technology may be able to operate over 2000 K for the short-burst times by use of ceramics and advanced design techniques. MHD can possibly operate even higher due to the heat transport arrangements. The result of analysis points to the fact that burst-power temperatures must be pushed to the 2500 K range to achieve a reasonable mass, multi-megawatt power subsystem.

High-Thermal Power Heat Source

The reactor with the highest power level desired and the longest lifetime without refueling is sized using neutronic and thermalhydraulic models. Criticality and core pressure drops will determine the final size for a given set of power needs. Either liquid metal or gas can be used as the coolant. Liquid metals have the advantages of better heat transfer, and minimum pressurization requirements, and, thus, lead to smaller and lower mass reactors. In addition, liquid metals need smaller heat transport lines, heat exchangers, and pumps external to the core. Their major disadvantage is that liquid metals such as lithium are more reactive with the fuel. Gas-cooled cores using helium reduce possible interaction problems between the fuel and coolant, but the pressurized primary loop is considerably heavier and larger. However, when replacing a heat exchanger or compressor requiring the opening of lines, it would be considerably easier to refill the system in a gas-cooled configuration.

The size of the reactor fuel pellets is selected to minimize the primary loop mass. Small spherical fuel

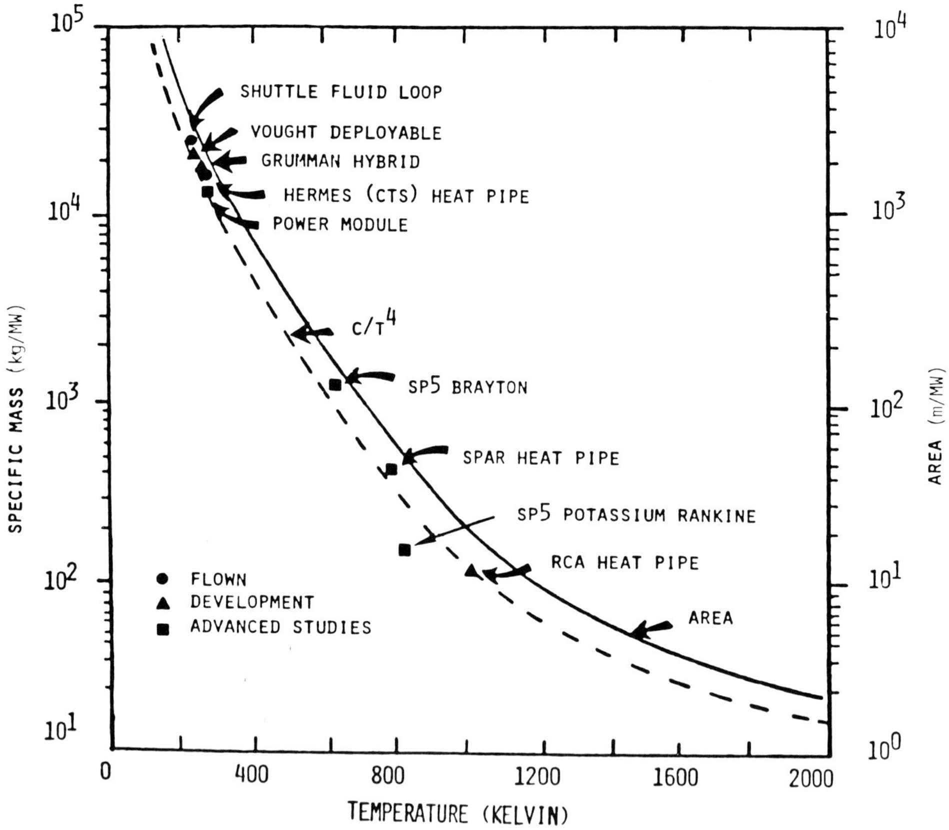

Figure 4 *Radiator Specific Mass and Surface Area as a function of the Radiator's Surface Temperatures.*

particles have low fuel loadings (leading to relatively large-size reactors). On the other hand, they provide a larger specific fuel heat transfer area and lower pressure drops in the core. Large pebble fuels, such as the 6-cm size in terrestrial pebble bed reactors, have smaller specific heat transfer area. It is estimated that 0.5 cm to 2 cm size fuel element may be best for this application. A method to regulate the void fraction is the use of two or more sizes of pellets.

Long Lifetimes

Long lifetimes are achieved by having the reactor both refuelable and maintainable. Also, the pellet reactor core eliminates extensive internal structural elements. Failures of structural elements in other designs could limit the lifetime of the reactor. The arrangement of the pellets in the core acts as the structural member. The reflector, another major life-limiting factor in the reactor, can be maintainable by designing it outside the core pressure vessel. Actuators and controls will be remotely located and designed to be maintained as replaceable units.

Rapid Burst Startup Times

Startup times are probably not critical for long-duration power. However, for burst power, startup

Table 2 *Efficiency versus System Operations Temperatures corresponding to a Carnot Efficiency of 25 Percent.*[a]

Sink Temperature T_L (K)	Source Temperature T_H (K)	Carnot Efficiency (%)
700	933	25
900	1200	25
1200	1600	25
1500	2000	25
1600	2133	25

[a] Carnot Efficiency $= 1 - T_L/T_H$.

Table 3 *Approximate Representative Values of Thermal-to-Electric Converter Efficiencies.*

	Temperature Ratio	Carnot Efficiency (%)	Device Efficiency (%)	Overall Efficiency (%)
Brayton	3:1	65	40	25
Rankine	1.8:1	45	45	20
—Superheat	2.3:1	55	45	25
Stirling	1.8:1	45	65	30
Thermionics	1.7:1	40	40	15
AMTEC	1.7:1	40	50	20
MHD				
—1500 K	1.5:1	35	70	25
—3000 K	1.2:1	15	70	10
—6000 K	1.3:1	25	60	15

times of seconds may be necessary. Significant factors on the length of time to achieve burst power levels are the number of decades of power to be traversed and the thermal stress on materials. To overcome these factors, the multi-modal reactor approach is recommended. If the reactor is already operating at a steady power level near a megawatt and at the operating temperature required for burst power, the flow rate and power level can be rapidly increased to meet burst-power needs. Power levels need to be increased by only a few decades using the multi-modal configuration.

High Reliability

Reliability, maintainability, and availability will be important factors in advanced space reactor designs. Having the fuel in the strongest possible geometry, minimizing the number of structural elements in the design, arranging the control drums and actuators in an easily maintainable manner, arranging the fuel in a configuration for replacement, and keeping the design as simple as possible will greatly improve the reliability, maintainability, and availability.

Compliance With Stringent Safety Standards

Safety is a major consideration in any space reactor, and the proposed design has some potentially unique aspects. It could be launched with all or part of the fuel outside the core in a geometric or separation arrangement that could not be made critical by water or fluid immersion or by compaction due to a launch accident. Disposal is facilitated by having only the fuel to consider. If reprocessing of the fuel in space or returning it to Earth is desired, the quantity of material is small enough to make these options possible. Thus, being able to remove the fuel can simplify disposal and open up new approaches to resolution of the safety problem associated with radioactive materials reentering the biosphere.

Repairability Prior to Launch

With the heat source designed for refuelability and maintainability, any repairs found necessary prior to launch can be readily accomplished. These are most likely to be reactor actuators or control element failures. As such, they are located external to the reactor.

Power Plant Configurations

In the most straightforward arrangement of a power plant, shown in Figure 5, the heat source (the pellet bed reactor) coolant is transported to the electric conversion elements (turboalternators) with the exhaust going through a heat rejection heat exchanger and completing the loop through a compressor or pump. The limitations of such an arrangement include the difficulties of having power conversion elements that can operate over the extreme ranges of station keeping to burst power and the differences in requirements among heat rejection elements for the different power modes.

Figure 6 shows an alternative arrangement. Here, separate loops are shown for the station-keeping and burst-power modes to allow optimum sizing for each power conversion loop.

Summary

Table 5 is a summary assessment of a multi-modal pellet bed reactor. A multi-modal pellet bed reactor is

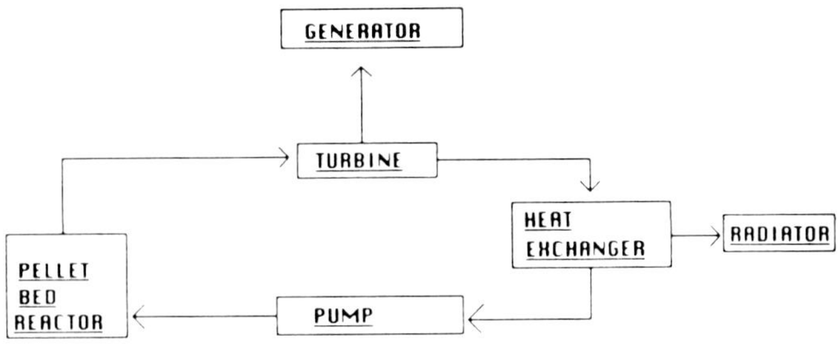

Figure 5 *Direct-Cycle Loop Power Plant Arrangement.*

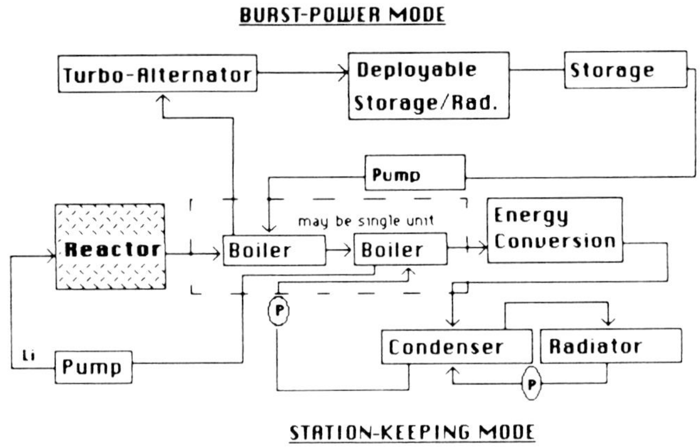

Figure 6 *Alternate Power Plant Arrangement.*

Table 4 *Spherical Fuel Elements manufactured by NUKEM/HOBEG for AVR Reloads.*

AVR RELOAD		FUEL PARTICLES			FUEL ELEMENT			
Manufacturers Designation	Operators Designation	Coated-Particle Type		Uranium Enrichment	Fuel Element Graphite	Firing Temperature	Number of Fuel Elements	Insertion Date
0[a]	UCC	$(Th,U)C_2$	HTI BISO	93%	Graphite	1450°C	30,155	Jul '66
1	T	$(Th,U)C_2$	HTI BISO	93%	Graphite	1450°C	7,510	Oct '68
3	GK	$(Th,U)C_2$	HTI BISO	93%	A3–3[b]	1800°C	17,770	Apr '69
4	GK	$(Th,U)C_2$	HTI BISO	93%	A3–3	1800°C	6,210	Jul '70
5–1	GK	$(Th,U)C_2$	HTI BISO	93%	A3–3	1800°C	25,970	Nov '70
5–2	GO 1	$(Th,U)O_2$	HTI BISO	92%	A3–3	1800°C	20,825	Dec 71
6–1	GO 1	$(Th,U)O_2$	HTI BISO	92%	A3–3	1900/ 1800°C	11,000	Oct '73
6–2	GLE 1	UO_2	LTI BISO	15%	A3–3	1900°C	2,446	Dec '73
		UO_2	LTI BISO	0.7%				
7	GO 1	$(Th,U)O_2$	HTI BISO	93%	A3–3	1900°C	7,840	Jan '73
8–1	GFB 1	UO_2	LTI BISO	93%	A3–3	1900°C	1,440	May '74
		ThO_2	LTI BISO	—				
8–2	GFB 2	UO_2	LTI TRISO	93%	A3–3	1900°C	1,610	May '74
		ThO_2	LTI BISO	—				
9	THTR 1	$(Th,U)O_2$	HTI BISO	93%	A3–3	1950°C	5,145	Sep '74
10	THTR 2	$(Th,U)O_2$	HTI BISO	93%	A3–3	1950°C	10,000	Dec '74
11	THTR 2	$(Th,U)O_2$	HTI BISO	93%	A3–3	1950°C	5,000	Dec '74
12	GO 1	$(Th,U)O_2$	HTI BISO	93%	A3–3	1950°	11,325	Mar '76
13–1	GFB 3	UC_2	LTI TRISO	90%	A3–27[b]	1950°C	6,077	Dec '77
		ThO_2	LTI BISO	—				
13–2	GFB 4	UC_2	LTI TRISO	90%	A3–27	1950/ 1800°C	5,861	Jul '80
		ThO_2[c]	LTI BISO	—				
13–3	GFB 5	UCO	LTI TRISO	92%	A3–27	1950°C	5,354	Dec '77
		ThO_2	LTI TRISO	—				
14	GO 1	$(Th,U)O_2$	HTI BISO	93%	A3–27	1950°C	9,930	Nov '76
15	GO 2	$(Th,U)O_2$	LTI TRISO[d]	93%	A3–27	1950°C	6,087	Feb '81
18	GO 3	$(Th,U)O_2$	HTI BISO	93%	A3–27	1950°C	11,547	Jul '81
19	GLE 3	UO_2	LTI TRISO	10%	A3–27	1950°C	24,615	Jul '82
20	GO 2	$(Th,U)O_2$	LTI TRISO	93%	A3–27	1950°C	11,854	Planned 1985
21	GLE 4	UO_2	LTI TRISO	16.7%	A3–27	1950°C	ca. 20,000	Planned 1984

[a] First core manufactured by Union Carbide.
[b] A3–3 and A3–27 are NUKEM manufactured matrix graphites.
[c] ThO_2 kemels with fission product gettering additives.
[d] Heavy-metal contamination $< 10^{-4}$.

Table 5 *Assessment of a Multi-Modal Pellet Bed Reactor.*

OPERATIONAL CHARACTERISTICS

Station-Keeping < 0.5 MWe
Burst Mode > 200 MWe
Rapid transition to burst power due to reactor being near operating temperature

MEETS ENVIRONMEMTAL CHARACTERISTICS

Contaminant Free: Closed-cycle system
Thermal effects controllable
Platform Stability: Closed-cycle system eliminated thrust

MEETS ABILITIES

Repairability: Simplified reactor design and could remove fuel
Reliability: Simplified reactor design, but higher operating temperatures
Long Lifetime: Refuelability

MEETS SIZE/MASS/COST

High operating temperatures produces high efficiency (lower mass)
Thermal management minimized due to high heat rejection temperature
Simplified core and fuel element design

MEETS SAFETY

Could fuel and refuel in orbit

a viable and attractive concept for multimegawatt space power systems because of the following features:

- User goals can be satisfied;
- Station-keeping, alert, and burst-power modes are obtainable with a single heat source;
- Rapid transition to burst-power mode due to the reactor being near steady-state operating temperatures is feasible; and
- Reduced size, mass, and costs are achievable by high operating temperatures, simplified core design, and refuelability.

Acknowledgments

This work was sponsored by the Science Applications International Corporation (SAIC).

References

Angelo, J. and D. Buden (1985) *Space Nuclear Power*, Orbit Book Co., Malabar, FL.

Nabrelek, H. et al. (1984) "Fuel For Pebble-Bed HTRs," *Nucl. Engr. and Design*, 78: 155–166.

Wittchow, G., E. Baust, and J. Schoning (1984) "Status of the Construction of the THTR-300 MW and the Design of a 500-MW Follow-On Plant," *Nucl. Engr. and Design*, 78: 109-117.

Ziermann, E. (1984) "The AVR Nuclear Power Facility—Progress Report," *Nucl. Engr. and Design*, 78: 98–108.

Space Nuclear Power Systems 1986
Edited by M. S. El-Genk and M. D. Hoover

Chapter 39

A Liquid Metal Cooled Reactor-Alkali Metal Thermoelectric Space Power System Concept for Multimegawatt Applications

J. E. Schmidt, J. W. H. Chi, R. E. Morgan, and J. P. Hanson

Westinghouse Electric Corporation
Advanced Energy Systems Division
P.O. Box 158
Madison, PA 15663

T. K. Hunt

Ford Motor Company
Ford Scientific Research Laboratory
Dearborn, MI 48121

This paper presents a study of a lithium cooled reactor Alkali Metal Thermoelectric (AMTEC) power system for multimegawatt space power applications. AMTEC is a thermally regenerative electrochemical device, a static energy converter that permits high power conversion efficiencies at moderate operating temperatures. Scoping designs and parametric analyses were carried out to establish the waste heat rejection temperature that yields the minimum system mass. The advantages of this power source concept are presented. They include a system that has no moving parts and total power system mass that is substantially less than those of systems using other static energy converters. The key technical issues are identified and the technology development requirements are discussed.

Introduction

Strategic Defense Initiative (SDI) space power requirements include multimegawatts of long-term, steady state electric power. The generic requirements for such power sources are low system mass, long life (7 to 10 years), and high system reliability (greater than 0.95). A system that has no moving parts could be inherently more reliable and should have better potential for meeting the optimum lifetime and reliability requirements. However, most static energy conversion systems have low power conversion efficiencies, making the mass of the total power source systems greater than those using dynamic converters. One exception is the alkali metal thermoelectric converter (AMTEC). This is illustrated in Figure 1, which compares the net efficiencies of various power conversion systems and their operating temperature ranges. It is apparent that AMTEC can yield relatively high power conversion efficiencies at moderate temperatures when compared to thermoelectric, thermionic, and magnetohydrodynamic power conversion systems. In addition, it can be operated at sufficiently high temperatures such that waste heat rejection can be carried out at reasonably high temperatures to minimize radiator mass consistent with minimizing the total system mass.

AMTEC is a thermally regenerative electrochemical system that provides direct thermal-to-electrical power conversion through an ion conductive ceramic material: beta alumina solid electrolyte. The basic principles of AMTEC have been described elsewhere (Hunt and Weber 1981 and Cole 1983). The feasibility of AMTEC was demonstrated in 1968. The development of AMTEC has been carried out continuously since that time at the Ford Scientific Research Laboratory, supported by DOE funding. More recently, AMTEC technology development has been initiated at the Jet Propulsion Laboratory and the Sandia National Laboratories. Power source systems using AMTEC were first considered for terrestrial applications (Subramanian and Hunt 1982), but are now being developed for space nuclear power systems as well (Bankston et al. 1983 and 1985). A 10-MWe system using a liquid metal cooled reactor integrated with AMTEC converters was compared with other types of space nuclear power systems (Chi et al. 1986). The comparison showed that for 10-MWe systems, a power source using AMTEC has the lowest total system mass. This paper describes this power source system concept, discusses the key features of such a

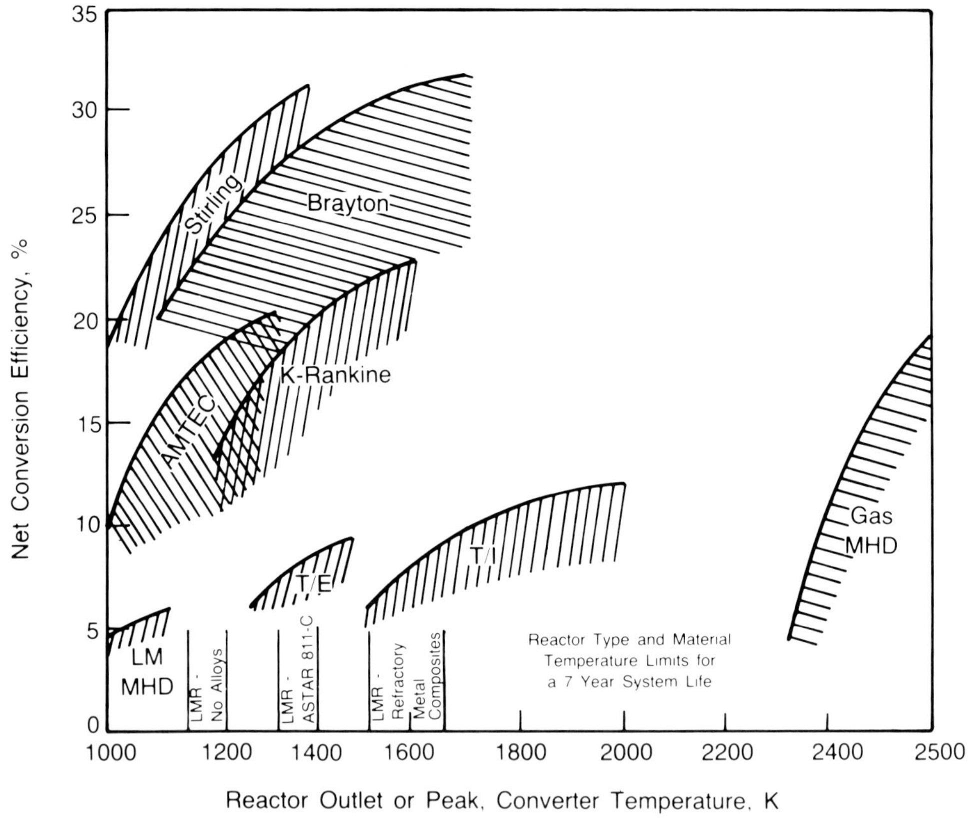

Figure 1 *Comparison of Potential Power Conversion System Efficiencies.*

system, and identifies the key development requirements for space power applications.

Power Source System Concept

Figure 2 is a simplified schematic of the power source system concept. The system consists of a fast spectrum, liquid metal cooled reactor integrated with alkali metal thermoelectric converters (AMTEC), where the converters are integral parts of the waste heat rejection system. The reactor can be cooled by either lithium or sodium. An intermediate heat exchanger (IHX) transfers the reactor power to the AMTEC working fluid (sodium). The IHX is necessary to preclude activation of the AMTEC working fluid. The key features of the reactor for the liquid metal cooled reactor-AMTEC (LMR-AMTEC) power conversion system are shown in Table 1.

The long life and high power requirements of a multimegawatt lithium cooled reactor result in high fuel burnup, significant fission gas generation and fuel swelling, all of which affect the fuel pin design. Specifically, a fuel clad material with the highest creep strength is needed to preclude the need for vented fuel and to minimize the mass of the reactor. Unvented fuel is possible because of the relatively low reactor operating temperatures made possible by AMTEC power conversion. UN fuel is selected because it has superior thermal conductivity, high uranium density, good fission gas retention and relatively low fuel swelling. These features can yield fuel pin designs with the lowest mass.

Two separate control systems are required to meet the OSNP-1 (U.S. DOE 1982) requirements. A secondary shutdown system is needed to maintain the reactor subcritical under all postulated credible accident conditions such as those resulting in water immersion and core compaction. The two control systems incorporated consist of radial control drums and in-core control rods. Permanent, irreversible shutdown of the reactor following the end of reactor life or mission life can be accomplished by placing neutron poisons in the core. Following final reactor shutdown and sufficient decay of the afterheat, lithium-6 is introduced into the core, where it is allowed to solidify.

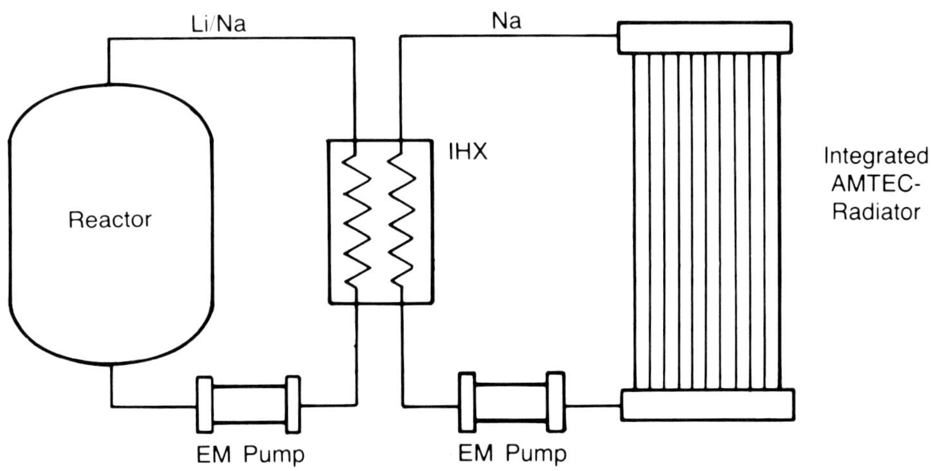

Figure 2 *Schematic of LMR-AMTEC Space Power System.*

AMTEC Converter-Radiator Concept

The AMTEC converter system is an integral part of the radiator, as illustrated in Figure 3. This configuration was selected following trade-off studies in which the alternative approaches investigated were

1. A converter-heat exchanger system in which the waste heat is transported by a pumped loop to the heat pipe radiator, and
2. A converter-heat exchanger in which the sodium vapor from the converter is transported to the radiator where it is then condensed and returned to the converter-heat exchanger.

Both cases turned out to be more massive than the reference concept selected. In the integrated AMTEC-radiator system, the sodium vapor from the converter condenses directly on the radiator tube, providing a constant radiator fin temperature. The condensate is collected on the inner surfaces of the radiator tubes by screened wick and returned to the IHX by electromagnetic pumps.

AMTEC System Performance

A scoping design on a 10-MWe system was made and mass models of the subsystems and components

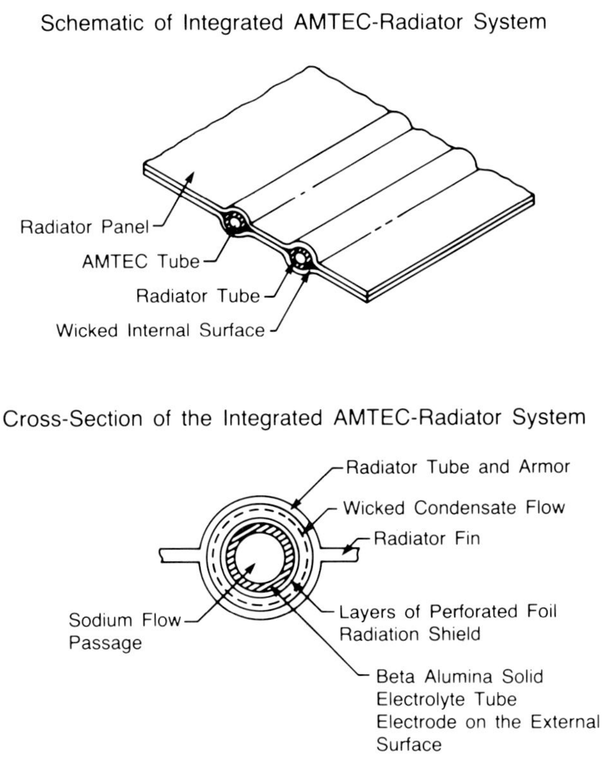

Figure 3 *Illustration of Integrated AMTEC-Radiator Concept.*

Table 1 *Key Features of the Reactor Subsystem.*

Feature	Description
Nuclear fuel	UN in refractory metal alloy cladding
Fuel pin type	Unvented
Fuel cladding	ASTAR 811-C
Independent, emergency decay heat removal system	Passive system consisting of radial conduction and inert gas controlled heat pipe-augmented decay heat transport
Reactor shutdown systems	Control drums and control rods
Post-mission permanent shutdown	Control rod insertion and replacing the normal coolant with lithium-6

were developed. Using these models, parametric analyses were carried out on the system operating temperatures to determine the operating conditions for minimum system mass. For this analysis, the AMTEC power conversion system was assumed to have the performance shown in Figure 4 (Subramanian and Hunt 1982). The converter was assumed to have an internal resistance of 0.18 ohms per cm^2, radiative loss factor (Z) of 20, a power density of 0.6 W/cm^2 and a 1% degradation per year (7-year total design life). The results of the parametric analysis are summarized in Figure 5. The analysis shows that the total system mass decreases with increasing converter hot side temperature. However, the condenser temperature for minimum total system mass is insensitive to the hot side temperature, remaining relatively constant at ~ 800 K over a fairly wide range of hot side temperatures.

A reasonably achievable hot side temperature was selected. The converter parameters and the key system state points are

- Converter cell power density = 0.6 W/cm^2,
- Converter cell specific mass = 0.26 W/g,
- Converter degradation = 1% per year,
- Radiator area/converter cell area = 1.92,
- Sodium at inlet to AMTEC tubes = 1300 K,
- Sodium condensing temperature = 800 K,
- Reactor outlet temperature = 1320 K,
- Reactor inlet temperature = 1270 K, and
- System efficiency = 14%.

Comparison with Other Systems

The optimum system mass was compared with those of other 10-MWe space power system concepts. The results are shown in Figure 6. The mass of the 10-MWe systems were estimated based on a common set of system requirements that include 7-year design life and > 0.95 system reliability. The comparison shows that the LMR-AMTEC system has the potential for lower mass than any of the other systems when state-of-the-art radiators are assumed. The figure also compares the total mass of the systems when advanced radiators are assumed. The latter shows that with the assumptions of advanced radiators, the LMR-AMTEC system mass exceeds only that of the NERVA Derivative Reactor-Brayton system. These results, coupled with the potential for high system reliability, renders the LMR-AMTEC concept highly attractive for SDI applications, where low system mass, long system life, and high system reliability are required.

While most of the parameters assumed for AMTEC performance are highly conservative, the rate of degradation assumed has not yet been demonstrated. Power densities as high as 1 W/cm^2 have been demonstrated in the laboratory; however they degrade relatively quickly and stabilize at 0.2 W/cm^2 in early studies (Bankston 1985).

However, the basic cause(s) for electrode degradation has become better understood and significant progress has been made in the development of im-

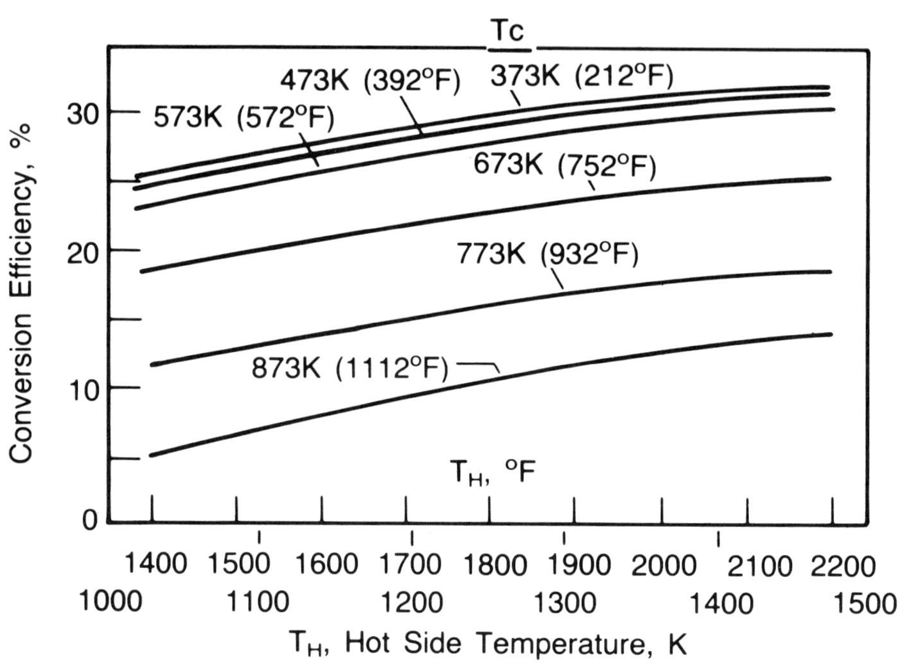

Figure 4 *Effect of Operating Temperatures on AMTEC Converter Efficiency.*

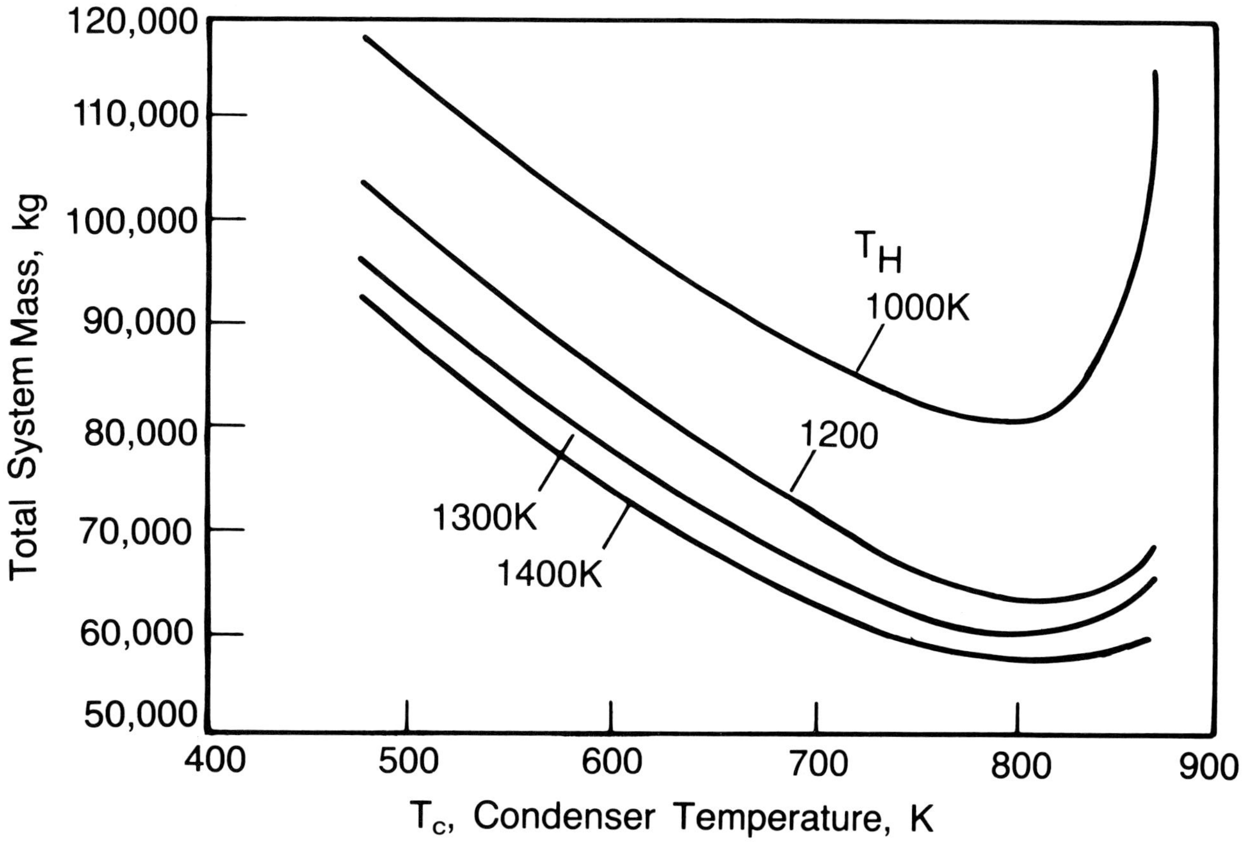

Figure 5 *Effect of Heat Rejection Temperature at Constant Hot Shoe Temperatures on the Total System Mass.*

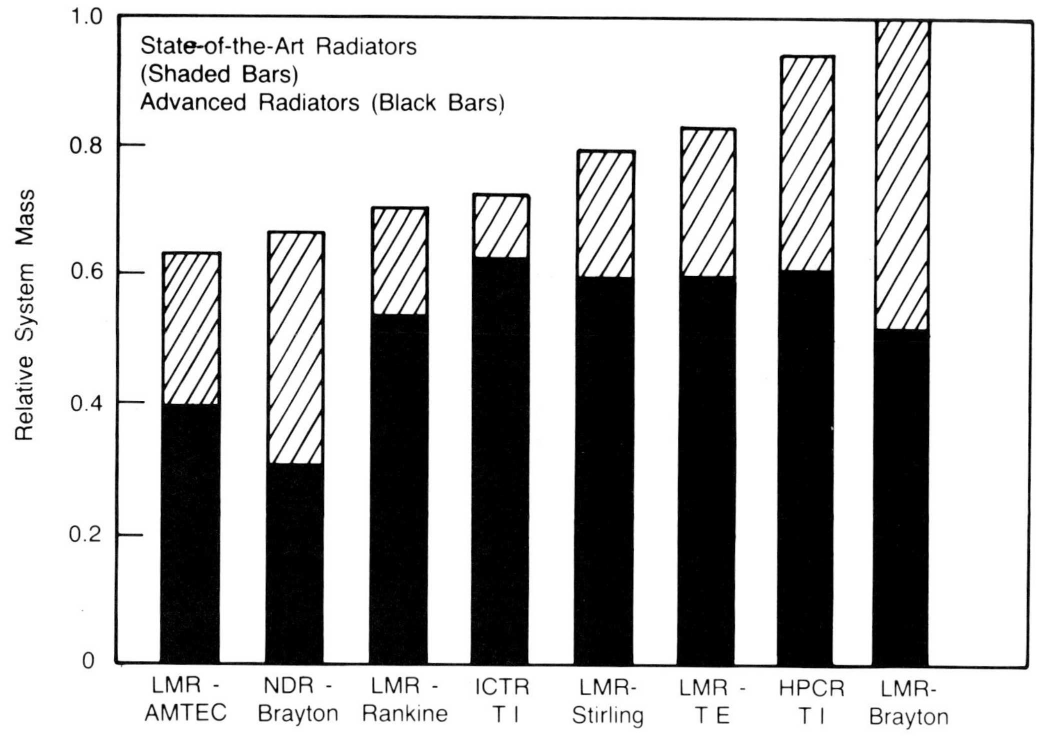

LMR	–	Liquid Metal Cooled Reactor	HPCR	–	Heat Pipe Cooled Reactor
NDR	–	NERVA Derivative, Gas Cooled Reactor	T I	–	Thermionics
ICTR	–	In-Core Thermionic Reactor	T E	–	Thermoelectric

Figure 6 *Comparison of the Mass of 10-MWe Space Power Systems.*

proved electrodes such that a stable power density of 0.43 W/cm^2 has been achieved using a titanium nitride electrode (Hunt 1985). Therefore, the 0.6 W/cm^2 assumed is a reasonable extrapolation.

It should be pointed out that current collection and power conditioning requirements were not taken into account in this study, because they depend largely on the user/mission requirements. It is recognized that these requirements can have significant impacts on the total system mass and on the relative masses of the power source systems compared.

Development Requirements

In addition to the electrode development, a number of other key technology developments can be identified. These are listed in Table 2. All of these are critical developments that impact either the basic feasibility of the power source concept or the system mass. A detailed design of a coupled AMTEC converter-radiator system has not yet been made. This is necessary to determine the feasibility of the concept and to identify other engineering developments required. Converter-radiator design issues which need to be addressed include design of the power leads and the interconnections between cells to minimize heat losses and resistive losses. A means of providing series electrical connection between cells must be examined. The feasibility of collecting the condensate in microgravity and returning it to the heat exchanger must be established. Finally, a thermal radiation shield system must be defined and the method for system startup (with the sodium initially frozen) must be worked out.

Reactor Development Issues

The concept proposed to ensure reactor breakup during reentry will require considerable development, to establish the basic feasibility of the approach, and to follow this by an experimental demonstration. The proposed passive mode of decay heat removal through the use of inert gas controlled heat pipes requires design, analysis and experimental verification. Design for survivability must consider

Table 2 *Key Technology Development Requirements For a Multimegawatt Liquid Metal Cooled Reactor—AMTEC System.*

Technology Development	Elements of a Development Program
Converter Related:	
High Power Density, Long Life Electrode	• Electrode research • Long-term performance demonstration • System design
AMTEC-Radiator Integration	• System conceptual design • Laboratory experiments • Analytical verification • Simulation testing
Startup in Space	• System design • Transient thermal analysis • Simulation experiments
AMTEC Irradiation Behavior	• System conceptual design • Irradiation experiments • Analytic determination
Reactor Related:	
Capability for Reentry Breakup and Fuel Dispersal	• Metallurgical fuse development • Laboratory testing • System conceptual design • Reentry analysis • Simulation experiments
Independent Decay Heat Removal Systems	• System conceptual design • Analytical verification • Simulation testing
Capability for Permanent Reactor Shutdown	• System conceptual design • Neutronics analysis • Experimental verification

the effects of natural and hostile threats to system operation and devise approaches to counter such threats.

The feasibility of using lithium-6 for permanent reactor shutdown requires detailed evaluation. It may be possible to incorporate the capability for permanent shutdown into the secondary control system, which has the primary function of ensuring a subcritical core under all credible accident events.

Recommended LMR-AMTEC System Technology Development Programs

The development elements of Table 2 can be grouped into several technology development programs consisting of

- Concept definition and system conceptual design,
- AMTEC-radiator subsystem development, and
- Reactor preliminary conceptual design.

The development element common to many of the development requirements is a preliminary system conceptual design. The reactor design effort will address the many issues related to the space nuclear safety requirements. The AMTEC-radiator subsystem development program includes the AMTEC-radiator integration, operation in zero gravity, and startup in space, as well as AMTEC irradiation behavior.

In view of the attractiveness of the AMTEC nuclear power source system, it is recommended that the critical technology development programs be initiated in parallel with a system conceptual design effort. The system conceptual design will provide a focal point for the resolution of the technology issues and provide the vehicle for evaluating design/engineering solutions.

Acknowledgments

This work was sponsored by the Westinghouse Electric Corporation.

References

Bankston, C. P. et al. (1983) ''Experimental and Systems Studies of the Alkali Metal Thermoelectric Converter for Aerospace Power,'' *J. Energy*, 7 (5): 442.

Bankston, C. P. et al. (1985) ''Alkali Metal Thermoelectric Conversion (AMTEC) for Space Nuclear Power Systems,'' in *Space Nuclear Power Systems 1984*, M,S. El-Genk and M. D. Hoover, eds., Orbit Book Company, Malabar, FL.

Chi, J. W. H., K. Kesavan, R. E. Morgan, and J. E. Schmidt (1986) ''Comparison of Multi-Megawatt Space Nuclear Power Systems,'' in *Space Nuclear Power Systems 1986*, M. S. El-Genk and M. D. Hoover, eds., Orbit Book Co., Malabar, FL.

Cole, T. (1983) ''Thermoelectric Energy Conversion with Solid Electrolytes,'' *Science*, 221 (4614): 915–221.

Hunt, T. K. and N. Weber (1981) ''High Efficiency Thermoelectric Conversion with Beta Alumina Electrolytes, the Sodium Heat Engine,'' *Solid State Ionics*, 5:263–266.

Hunt, T. K. (1985) *Personal Communications*, Ford Scientific Research Laboratory, Dearborn, MI, September 1985.

Subramanian, K. and T. K. Hunt (1982) ''Solar Residential Total Energy System Using the Sodium Heat Engine - A Concept Study,'' *17th IECEC*, pp. 1474–1480.

U.S. DOE (1982) *Nuclear Safety Criteria and Specifications for Space Nuclear Reactors*, OSNP-1, U.S. Department of Energy, Office of Space Nuclear Projects, Washington, D.C.

Space Nuclear Power Systems 1986
Edited by M. S. El-Genk and M. D. Hoover

Chapter 40

Comparison of Multimegawatt Space Nuclear Power Systems

J. W. H. Chi, K. Kesavan, R. E. Morgan, and J. E. Schmidt

Westinghouse Electric Corporation
Advanced Energy Systems Division
P.O. Box 158
Madison, PA 15663

This paper evaluates and compares a number of nuclear-electric space power systems that produce 10 MWe of steady state power for 10 years. Eight candidate systems were selected and their total masses were estimated and compared for two cases: one based on the use of state-of-the-art radiators and the other based on the use of advanced radiators. The results showed that a liquid metal cooled reactor—AMTEC system and a NERVA derivative reactor—Brayton system are the most promising concepts.

Introduction

Strategic Defense Initiative (SDI) space power requirements include multimegawatts of long-term steady state electric power. While the specific requirements and constraints for such power sources are mission-dependent and have not been fully established, they are expected to include low system mass, long life (7 to 10 years), high system reliability (greater than 0.95), and compatibility with existing and/or projected launch capabilities. All of these factors affect the life cycle cost of the power source and, hence, the cost of the mission. Westinghouse has evaluated and compared a number of different nuclear power source system concepts for multimegawatt space power applications. The purpose was to identify the most attractive concept(s). Scoping designs, parametric analyses and optimizations (mass minimization) were carried out for 10-MWe steady state power source systems. The optimized systems formed the basis for the comparison of the sizes and masses of the systems. This paper describes the various power source concepts considered and discusses the results of the comparative evaluations.

Power Source System Concepts

In the absence of specific mission definitions, the following set of power source requirements and constraints was assumed:

- 10 MWe of steady state power output,
- 7 year full power lifetime,
- Nuclear shielding criteria of 5×10^5 rads gamma and 10^{13} nvt neutrons in silicon on a 15-m-diameter dose plane 30 m from the reactor core midplane,
- System reliability > 0.95, and
- Initial Operating Capability (IOC) by the year 2000.

A large number of different types of reactors and power conversion systems can be considered for power sources in this application. Given the assumed deployment time frame and other requirements and constraints, the nuclear reactor technologies that can be considered are

- Liquid metal cooled, refractory metal fast reactors (LMR),
- Gas cooled solid core graphite reactor based on the NERVA technology—the NERVA derivative reactor (NDR),
- In-core thermionic reactor (ICTR), and
- Heat pipe cooled reactor (HPCR).

All of these reactors have been designed and developed to various degrees such that they can be evaluated and compared.

The power conversion systems that can be considered are shown in Figure 1, where the ranges of potential net conversion efficiencies are compared as functions of the peak conversion temperature or the reactor outlet temperature. The power conversion systems include both dynamic and static systems. The dynamic systems have the potential for higher power conversion efficiencies. Liquid MHD has the lowest conversion efficiencies; moreover, its operating temperatures are so low that waste heat rejection would incur substantial penalties on system mass (due to the large radiator area required). Therefore, this energy conversion system was not considered further. Alkali metal thermoelectric (AMTEC) power conversion is highly attractive because it has the potential for relatively high conversion efficiencies while operating at moderate temperatures; yet, the operating temperatures are sufficiently high that waste-heat rejection can occur at reasonably high temperatures to achieve reasonable radiator mass.

Shown in Figure 1 are the approximate temperature ranges within which various reactor technologies are suitable for a 7-to-10 year reactor life. While the NERVA reactor technology had been demonstrated at a 2500 K reactor outlet temperature, the demonstrated fuel lifetime was only 10 hours, limited by hydrogen corrosion. A 7-to-10 year reactor life in the steady state power source system considered here would not be limited by corrosion because the reactor coolant would be an inert gas. However, the reactor may be limited by graphite swelling. Data on irradiation swelling of reactor-grade graphite materials suggest that for a 7-to-10 year reactor life, the reactor outlet temperatures should be limited to less than 1400 K. This leaves no reactor technology for suitable integration with the gas MHD System for the long life steady state power source of interest in this study.

Following the preliminary screening of potential power source concepts, eight candidate reactor-power conversion systems were selected for further consideration. These are summarized in Table 1. Four of the power conversion systems are dynamic systems; the other four are static systems. The principal materials for the candidate system concepts were selected and are summarized in Table 2. Based on the materials selected and the reactor design lifetime, the maximum reactor outlet temperatures were specified, as given in Table 3. These, in turn, affected the peak power conversion cycle temperatures. Scoping designs of the subsystems and components of the candidate power source concepts were then made and mathematical models on the sizes and masses of the components/subsystems were developed. The models were then used in the parametric analyses aimed at establishing the system operating temperatures that yield the minimum total system mass. The major subsystems are the

- Reactor assembly,
- Nuclear radiation shadow shield,

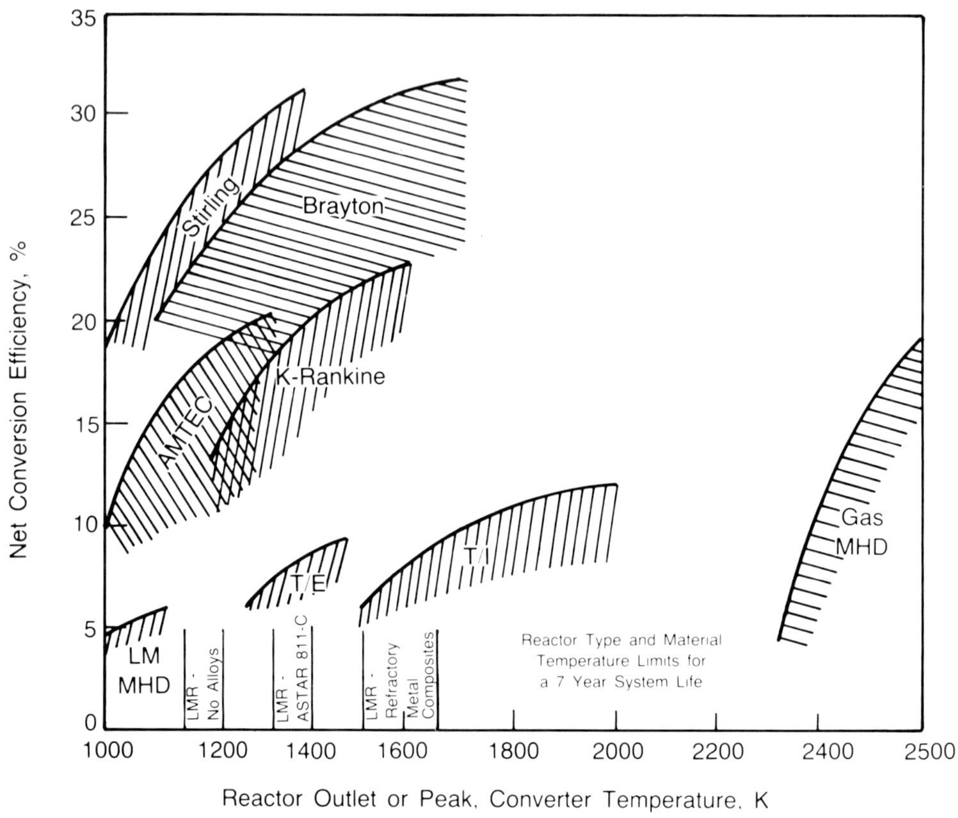

Figure 1 *Comparison of Potential Power Conversion System Efficiencies.*

Table 1 *Candidate 10-MWe Space Nuclear Power Systems.*

Nuclear Subsystem		Energy Conversion		
Reactor	Technology Base	PCS	Type	Technology Base
Gas-cooled Reactor (NDR)	NERVA Developments	Brayton	Dynamic	Extension of existing technology
Liquid Metal cooled Reactor (LMR)	Snap-50, SP-100	Stirling	Dynamic	Development initiated under SP-100 program
LMR	Snap-50, SP-100	Brayton	Dynamic	Extension of existing technology
LMR	Snap-50, SP-100	Potassium Rankine	Dynamic	Extension of Snap-50 Technology
LMR	Snap-50, SP-100	AMTEC	Static	Under development (DOE, NASA)
LMR	Snap-50, SP-100	TE	Static	Developments initiated under SP-100 program
In-core Thermionic (ICTR)	TFE, SP-100	Thermionic	Static	Development to be continued
Heat pipe cooled (HPCR)	New technology	Out-of-core thermionics	Static	New developments needed

Table 2 *Key Materials for the Candidate System Concepts.*

		Reactor				Power Conversion System	
Reactor	PCS	Fuel	Clad	Coolant	Structure	Working Fluid	Structural Material
NDR	Brayton	UC	PyC-ZrC	Helium	Refractory Metal	Helium	Refractory Metal
LMR	Stirling	UN	ASTAR 811-C	Lithium	ASTAR 811-C	Helium	Refractory Metal
LMR	Brayton	UN	ASTAR 811-C	Lithium	ASTAR 811-C	Helium	Refractory Metal Composites
LMR	K-Rankine	UN	ASTAR 811-C	Lithium	ASTAR 811-C	Potassium	ASTAR8II-C
LMR	AMTEC	UN	ASTAR 811-C	Sodium	T-111	Sodium	T-111
LMR	TE	UN	ASTAR 811-C	Lithium	ASTAR 8II-C		Refractory Metal Composites
ICTR	TI	UO_2	Tungsten	NaK	Nb-1%Zr	Cesium	Refractory Metal
HPCR	TI	UO_2	ASTAR 811-C	Li HP	ASTAR 811-C	Cesium	Refractory Metal Composites

Table 3 *Optimum System Operating Temperatures.*

		Temperatures (K)		Reactor
Reactor	PCS	Reactor Outlet	Peak Cycle	Power (MWt)
NDR	Brayton	1500	1500	45
LMR	Stirling	1370	1300	36
LMR	Brayton	1370	1300	47
LMR	Rankine	1370	1300	56
LMR	AMTEC	1300	1200	72
LMR	TE	1470	1420	100
ICTR	TI	1970	1970	85
HPCR	TI	1650	1640	85

- Power conversion and/or heat transport system,
- Radiator system, and
- Structures and miscellaneous.

Power conditioning equipment is not included as it is highly mission-dependent. The last item includes control system components.

For each system, two types of radiators were considered: state-of-the-art radiators and advanced radiators such as the Curie Point Radiator (CPR) currently under study at Westinghouse. The masses of the radiators were estimated based on assumed relationships between the radiator specific mass and the heat rejection temperature (radiator inlet temperature), shown in Figure 2. The curve for the assumed state-of-the-art radiators was based on data compiled by Buden and Lee (1985). The curve for the advanced radiators is optimistic, based on a number of advanced concepts, the feasibility of which have not been demonstrated.

It was recognized that the use of these curves would not give the correct absolute masses of the radiator subsystems because they are based on data for small power source systems. The mass of the radiator is expected to increase with increased waste-heat rejection. Nevertheless, for the purpose of the comparative evaluation, the curves can be used to determine relative masses of the radiator and the total systems. In general, the system operating temperatures for minimum total system mass did not correspond to the maximum conversion efficiency that can be obtained within the temperature limits of the materials of construction selected for each system. This is due to the T^4 effect on the radiator area, and, hence the radiator mass and to the fact that the radiator represents a large fraction of the total system mass in all the systems. The system operating temperatures for minimum total system mass when using state-of-the-art radiators were different from those using an advanced radiator.

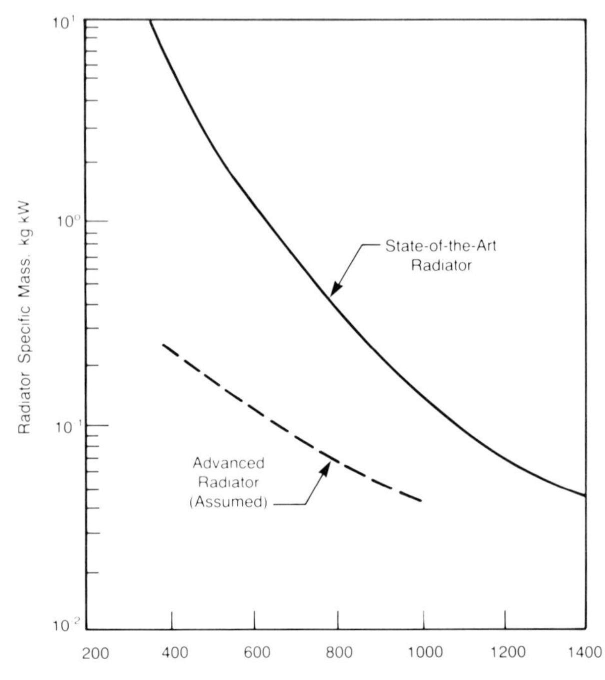

Figure 2 *Assumed Effects of Heat Rejection Temperature on the Specific Masses of Radiators.*

Comparisons of the Masses of the Power System Concepts

The results of the study are summarized in Figures 3 and 4. Figure 3 compares the total mass of the candidate space power systems that use state-of-the-art radiators. Figure 4 compares the masses of the major subsystems and the total system mass, rank ordered based on increasing subsystem mass. From Figure 3, it is noted that LMR-AMTEC and NDR-Brayton have the lowest total system mass. Within the uncertainties of the system mass estimates, there is no significance in the small difference between the masses of these two systems. LMR-Rankine, ICTR-Thermionic (ICTR-TI) and LMR-Stirling can be considered the group of concepts that has the second lowest mass. LMR-Thermoelectric (LMR-TE), HPCR-TI and LMR-Brayton have the highest total system mass.

Although LMR-AMTEC has a lower power conversion efficiency than the NDR-Brayton system, waste-heat rejection occurs at a constant temperature (condensing sodium working fluid). This contrasts with the significant temperature drop in the waste-heat exchanger of the NDR-Brayton system. Consequently, advantage of the higher power conversion efficiency of the Brayton cycle is partially overcome by the relatively low effective radiator temperature.

Stirling engines have capability for relatively high power conversion efficiencies. However, the optimum mass per unit power output is limited to about 50 kWe per engine (by the unique characteristics of the engine). Consequently, a large number of engines are needed to supply 10 MWe of output and to provide sufficient redundancy to achieve the system reliability goal. Although potassium Rankine power conversion can be relatively efficient, the optimum system mass occurs at a power conversion efficiency far less than the maximum attainable value within the cycle

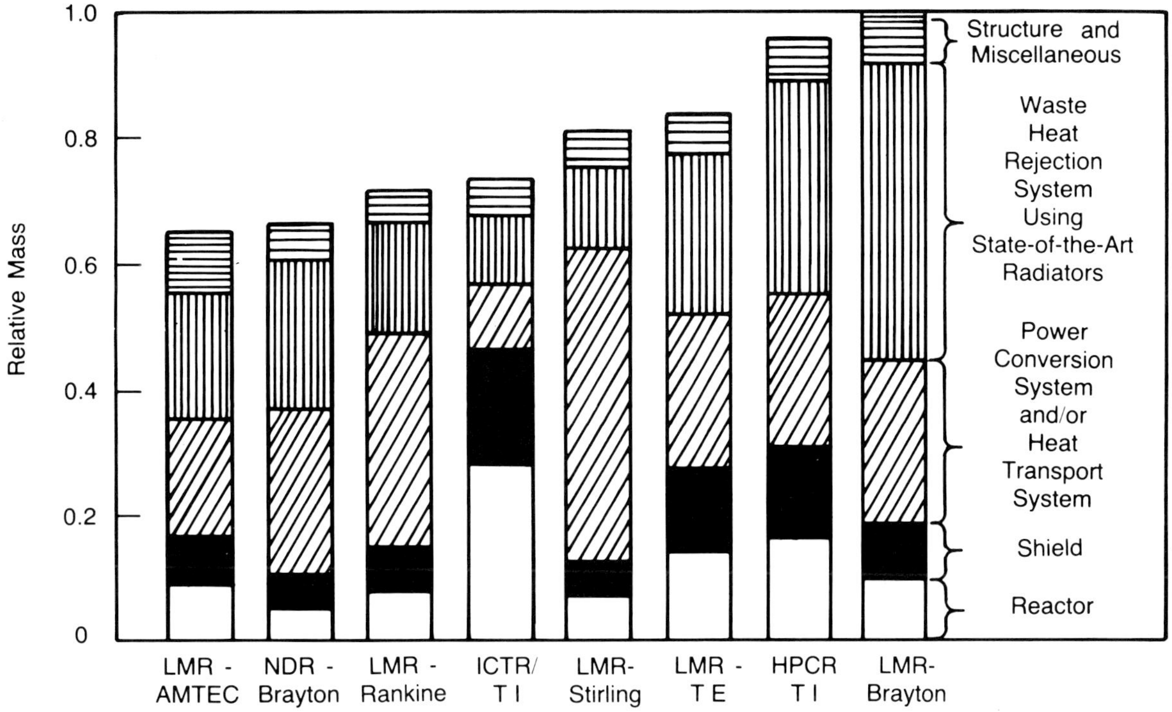

Figure 3 *Comparison of the Masses of 10-MWe Nuclear Space Power System Concepts.*

temperatures constrained by the materials temperature limitations.

A number of factors contributed to the relatively low optimized mass of the NDR-Brayton system. They include a relatively low reactor mass (Figure 4a) and no recuperator. The relatively low reactor mass can be attributed to the graphite core structure and the absence of metallic cladding materials. However, the solid core graphite reactor is larger than the liquid metal cooled reactors. Therefore, the shielding can be heavier (see Figure 4b).

The mass of the in-core thermionics reactor is far greater than those of the other reactor types. This is inherent to the ICTR concept, because the converters are integral to the reactor, resulting in a larger and heavier reactor assembly. Furthermore, high density refractory metals of construction are associated with the converter. As a consequence of the larger reactor, the required nuclear shielding is also larger and more massive (see Figure 4b).

The systems with static power conversion have much lighter power conversion systems. Among them the ICTR-TI system has the lightest power conversion subsystem (Figure 4c) followed by the AMTEC, out-of-core thermionics, and thermoelectric, respectively. The LMR-Stirling system has the heaviest power conversion subsystem for the reasons already discussed.

The ICTR-TI system also has the lightest radiator (see Figure 4d) because the thermionic converter operates at relatively high temperatures and the optimum system mass is obtained at a relatively high heat rejection temperature.

The thermionic converters in the HPCR-T/I system operate at temperatures lower than those of the ICTR-TI, limited by the maximum heat-pipe temperature capability, assumed to be 1640 K for an optimistic heat pipe lifetime of 7 years. Consequently, optimum system mass occurs at a waste-heat rejection temperature lower than that of the ICTR-TI system, resulting in a much heavier radiator.

The LMR-Brayton system has the heaviest radiator subsystem because the Brayton cycle peak temperature is limited by the LMR reactor outlet temperature and by the temperature drop across the liquid metal-to-gas heat exchanger. Within the maximum reactor temperature constraint, the indirect Brayton cycle must operate at temperatures substantially lower than those of the direct cycle of the NDR-Brayton system. Therefore, the optimum system mass occurs with a waste heat rejection temperature lower than that of the NDR-Brayton system, resulting in a heavier radiator.

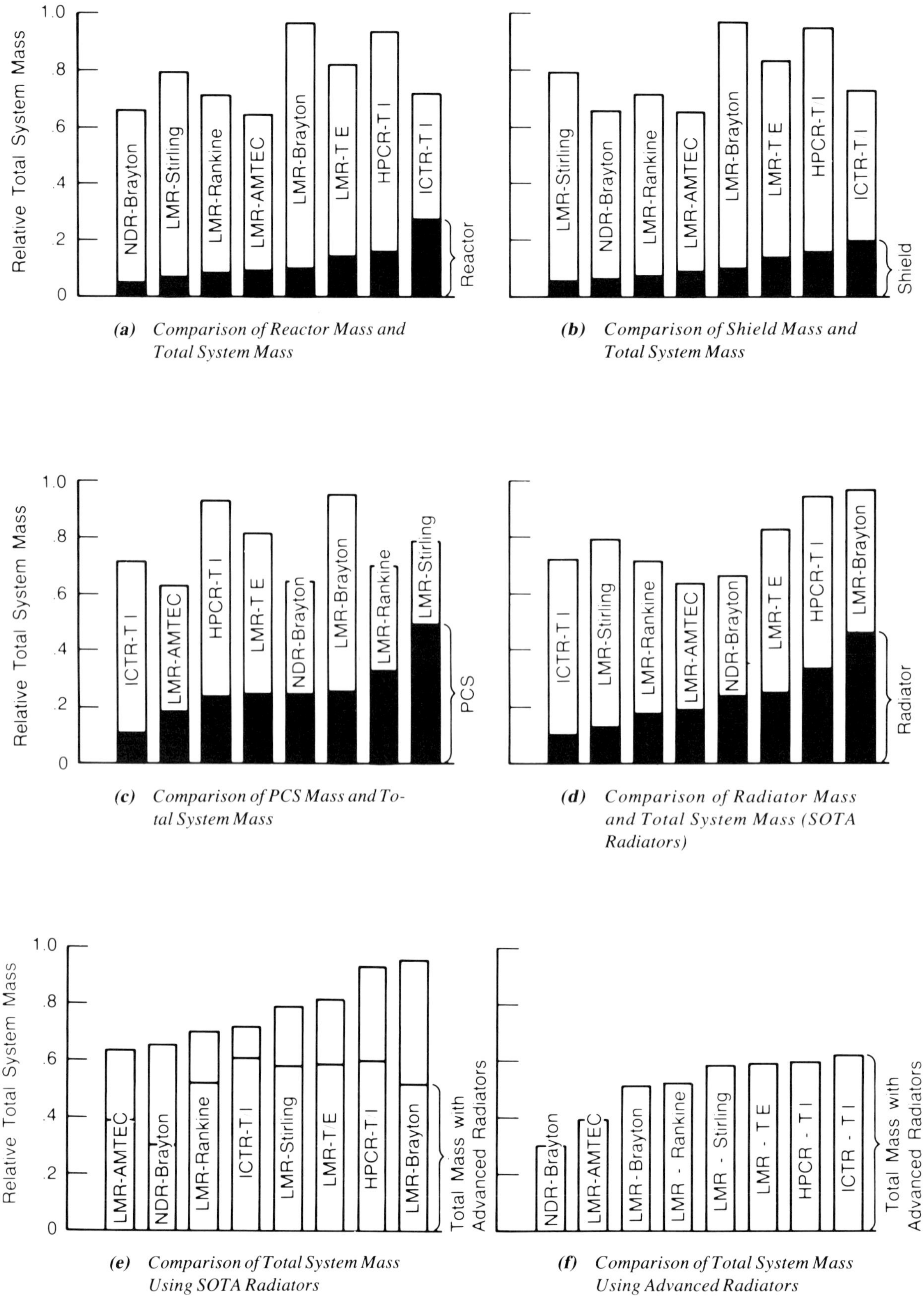

Figure 4 *Rank Order of Power System Concepts according to the Relative Mass of Major Subsystems.*

It is noted from Figure 4d that the radiator mass can represent 13% to 42% of the total system mass. Thus, if a lightweight radiator system can be developed, substantial mass reductions can be realized in some of the systems. This is illustrated in Figures 4e and 4f, which show the rank order of the systems when an advanced radiator is assumed. These results show that, as expected, those systems with the radiator contributing to a proportionately large fraction of the total system mass would benefit the most from an advanced radiator. Thus the systems using Brayton power conversion benefit the most; the NDR-Brayton system becomes the lightest power source concept, closely followed by the LMR-AMTEC and LMR-Brayton systems. The other five systems have approximately the same total system mass.

The results of the mass comparisons show that, while the mass of the radiator can represent a relatively large fraction of the total system mass, it is not the overriding factor in some of the system concepts. As a consequence, the masses of other subsystems are equally, if not more, important to the determination of the minimum achievable total power source system mass. It should be pointed out that the masses of the power conditioning subsystems were not included in this study, because these are highly mission specific. However, the masses of this subsystem can have a substantial impact on the total system mass and also on the relative masses of the systems compared.

Comparison of Other Factors

Each of the power source system concepts has unique characteristics and key technical issues that require resolution. The systems using static power converters do not have spacecraft stabilization problems and have potential for high system reliability. The dynamic systems have key issues involving shock and vibration and the need for active means to provide spacecraft stabilization. The concepts using liquid metal cooled reactors (including the ICTR and the HPCR) have a common technical issue involving the capability for reentry breakup and fuel dispersal.

It is assumed that for large reactors such as those encountered here, the design for intact reentry and intact impact would incur a large mass penalty that has not been taken into account in the scoping reactor designs. With the exception of the ICTR and the HPCR, the relatively large LMRs can encounter difficulties with startup in space with the coolant initially in the solid state.

While the dynamic power conversion systems tend to have lower reliability than the static systems, the same system reliability can be achieved if the associated mass penalties are acceptable. Brayton turbo-generators have been successfully tested and demonstrated with up to 43,000 hours of continuous operation. In contrast, potassium Rankine power conversion system development is in a relative state of infancy.

All of the static energy converters require significant further developments. While thermionic diodes are relatively well developed and have been demonstrated more than 100,000 hours (Yang and Fitzpatrick 1982), the long-term demonstrations apply principally to the diodes. When integrated with the fuel, the converter life can be severely limited due to fuel swelling and the closing of the inter-electrode gap. Another key issue is the need for an acceptable high temperature electrical insulation material. The individual thermionic converters must be electrically insulated; otherwise there would be no voltage rise among converters placed in series. The technology for these high-temperature insulators does not exist at present.

Both thermoelectric and AMTEC converters have degradation and performance issues that require further development. For the thermoelectrics, principal development efforts should be aimed at improving the converter efficiency. Significant progress has been made in AMTEC development. While the power density of early electrodes had a maximum stable value of 0.2 W/cm^2 (Bankston 1985), recent results using titanium nitride electrodes have shown stable values up to 0.43 W/cm^2 (Hunt 1985).

Although the mass of the power source is of primary importance, other factors important to the deployment of a power source should be considered as well. Table 4 summarizes the results of comparing the candidate power system concepts against six major criteria. Growth potential refers to the flexibility of the power source technology in terms of scalability to both higher and lower power levels, and to other modes of power production such as burst power and rocket propulsion. The different power source concepts were ranked using a set of weighting factors. The weighting factors, the calculated relative ratings (based on assigned relative ratings) and finally the rank of the concepts so determined are given in Table 5. The results show that the NDR-Brayton system ranks first, while ICTR-TI and LMR-ATMEC share, for all practical purposes, the second spot.

Table 4 *Summary of System Evaluations (Favorable Features Underlined).*

Reactor	PCS	Relative System Mass	System Reliability	Radiator Size	Development Cost	Development Risk	Growth Potential
NDR	Brayton	Low	High	Moderate	Low	Low	High
LMR	Stirling	High	Low	Small	High	High	Moderate
LMR	Brayton	Moderate	Moderate	Large	Moderate	Low	High
LMR	K-Rankine	Moderate	Low	Small	High	Moderate	High
LMR	AMTEC	Low	Moderate	Moderate	Moderate	Moderate	High
LMR	TE	High	High	Large	Moderate	High	Low
ICTR	TI	Moderate	High	Small	Moderate	High	Low
HPCR	TI	High	High	Large	High	High	Low

Conclusions and Recommendations

Based on the results of this study, it is concluded that the NERVA derivative reactor, gas-cooled, solid-core graphite reactor with closed-Brayton power conversion is the most attractive concept for multimegawatt steady state space power applications. For a 10-MWe space power system using heat-pipe radiators, it has a total system mass comparable to the lowest among all the systems concepts considered (LMR-AMTEC, which has virtually the same total system mass within the uncertainties of the mass estimates). However, if an advanced radiator can be developed, the NDR-Brayton system can benefit the most—the system mass can be reduced substantially more than the other system concepts to yield the system with the lowest total mass.

Between 1955 and 1972, $1.4 billion dollars (in as-expended dollars) had been invested in the Rover-NERVA program. The technology developed is a valuable national resource that must not be lost. The NERVA reactor was developed for nuclear propulsion. Substantial differences exist in the requirements of a reactor for electric power production, and these can have large impacts on the reactor design. It is therefore recommended that a preliminary conceptual design of a NDR-Brayton MultiMegawatt power system be made to provide concept definition and a focus for technology developments. Such a program should be initiated soon if the option for early deployment of this technology is to be kept open.

While the ICTR-TI and the LMR-AMTEC are rated as close seconds, ICTR-TI technology (in the form of the thermionic fuel element) development is being resumed. This should establish whether an ICTR with 7-to-10 year life can be successfully developed. While high-temperature, liquid metal cooled reactors also are being developed under the SP-100 program, there are significant differences between a reactor designed for a 100-to-300 kWe power source and one designed for multimegawatts of electric power; therefore, a conceptual design and development of a liquid-

Table 5 *Relative Ratings and Rank of Candidate Multimegawatt Space Power System Concepts.*

		Weighted Relative Ratings							
Criteria	Weighting Factor	NDR-Brayton	LMR-Stirling	LMR-Brayton	LMR-Rankine	LMR-AMTEC	LMR-T/E	ICTR-T/I	HPCR-T/I
System Mass	.30	.048	.033	.036	.042	.051	.030	.040	.020
System Reliability	.20	.027	.014	.022	.012	.028	.030	.032	.035
Development Risk	.20	.036	.020	.032	.018	.022	.025	.025	.022
Radiator Size	.15	.017	.021	.016	.020	.018	.016	.027	.015
Development Cost	.10	.016	.012	.014	.011	.012	.013	.012	.010
Growth Potential	.05	.008	.006	.007	.007	.007	.006	.004	.005
Composite Rating		.152	.106	.127	.110	.138	.120	.140	.107
Rank		1	5	3	5	2	4	2	5

metal cooled reactor for multimegawatt applications should be initiated. Although research on AMTEC has been carried out continuously over the past decade or more, extremely low levels of efforts have been expended to date (relative to TE and TI). In view of the significant potential of this power conversion system, more accelerated development efforts are needed. In addition, a preliminary conceptual design of a LMR-AMTEC space power system is recommended to provide concept definition and a focus for the development efforts. It is also recommended that these power source concepts be further evaluated and compared with others in a multi-mode application and in an SDI application framework to determine whether they remain attractive when synergistic effects are taken into account.

Acknowledgment

This work was sponsored by the Westinghouse Electric Corporation.

Reference

Bankston, C. P. (1985) "Alkali Metal Thermoelectric Conversion (AMTEC) for Nuclear Power Systems," in *Space Nuclear Power Systems 1984*, M. S. El-Genk and M. D. Hoover, Eds., Orbit Book Co., Malabar, FL.

Buden, D. and J. H. Lee, Jr. (1985) "Space Reactor System Selection and Design," in *Space Nuclear Power Systems 1984*, M. S. El-Genk and M. D. Hoover, Eds., Orbit Book Co., Malabar, FL.

Hunt, T. K. (1985) *Personal Communication*, Ford Scientific Research Laboratory, Dearborn, MI, September 1985.

Yang, L. and G. Fitzpatrick (1982) "Thermionic Conversion for Space Power Application," in *Proceedings of the AFOSR Special Conference on Prime Power for High Energy Space Systems*, 2, P. J. Turchi, ed., R&D Associates.

Space Nuclear Power Systems 1986
Edited by M. S. El-Genk and M. D. Hoover

Chapter 41

Burnup versus Criticality Limitations for a Potential Space Reactor Design in the 1 to 100 MWt Range

Susan S. Voss
Los Alamos National Laboratory
Los Alamos, NM 87545

Patrick J. McDaniel
Sandia National Laboratories
Albuquerque, NM 87185

David M. Woodall
University of New Mexico
Chemical and Nuclear Engineering Department
Albuquerque, NM 87131

In the design of large multimegawatt nuclear reactors for space a number of variables need to be optimized to achieve both the desired lifetime and size goals. The main purpose of this study was to identify the main lifetime limitors to the reactor core at enrichment levels of 45%, to 97%, for both beryllium and graphite reflectors and for power levels of 1, 10, and 100 MWt. A particulate fuel design was used for the study because it could easily span the intended power range and operate at the required high temperature levels. At higher power levels loss of reactivity control was shown to be the major lifetime limitor, not the fuel burnup. Possible ways of increasing lifetimes include incore breeding, burnable poisons, and internal control elements.

Introduction

As the U.S. continues to explore space, the need for higher power sources will become more apparent. To meet rising power needs within launch limitations, nuclear power is the obvious, if not the only, long-term solution. Therefore, it is important to identify the major lifetime limitations early into the program and to begin providing means of extending such limits with engineering solutions. The two major reasons a reactor core will be lifetime limited are: 1) it exceeds the maximum fuel burnup limit established for a particular fuel design and 2) it displays loss of reactivity control from the external control elements. Burnup limitations are determined by the initial design of the core, taking into account trade-offs among fuel container strength and thickness, fission product release, and the amount of allowable fuel distortion. If the burnup exceeds the maximum value for which it is designed, there is the potential for fuel distortion and possible blockage of the coolant flow path. The loss of reactor control is dependent upon the maximum amount of reactivity control available at the beginning of life and the change within the reactor fissile material over its operating lifetime. Lifetime limitations change with power level, enrichment and reactor design.

The purpose of this paper is to determine what the primary limitations are over the reactor lifetime for a single reactor design. The power level was varied from 1 to 10 and then 100 MWt to ascertain how the limit changes with power level and burnup. The ^{235}U enrichment was varied from 45% to 97% to examine how the lifetime varies for enrichment at each power level. Thus, an optimum enrichment can be obtained with respect to lifetime. The results allow the system designer to determine if reactivity control, burnup or enrichment play a major role in their design and in the overall system optimization.

The reactor chosen for this study is a bed of particulate fuel which has a UC_2 kernel coated with a thin layer of ZrC. The fuel is optimized with respect to the maximum fuel volume loading per particle. A Brayton cycle is assumed for the energy conversion to simplify the reactor design and minimize the coolant effects.

Studies with both beryllium and graphite reflectors examined the reactivity control difference over time. The core was sized at the beginning of life by ensuring full shutdown with the reflector poison, B_4C, fully inserted. All criticality computations were done with FEMP1D, a one-dimensional neutron transport code (Lazo et al. 1985). ORIGEN (Bennett 1979) was used to obtain the fission product and fuel depletion inventory used in obtaining the core isotope density data as a function of time. The output from ORIGEN was input into FEMP1D, where *k*-effective (k_{eff}), the multiplication constant, was determined as a function of time.

Fuel Design

Methodology

The unique space environment, coupled with the unavailablity of hands-on maintenance, makes space-based reactor design an exceptional challenge in comparison with ground-based reactor design. Major design parameters include power, physical size, and lifetime. Data are not generally available for the operating conditions encountered in space, this forces the designer to look beyond the usual materials or concepts.

Although particulate fuels may have the general characteristics needed for space applications, they have not been tested in as severe of an environment. Particulate fuel is advantageous over the traditional fuel pin because it retains fission products, does not swell or change in shape and can operate at very high temperatures. The fuel design for High Temperature Gas Reactors (HTGRs) is tailored for very high burnup (20% to 75% fissions per initial metal atoms (fima)), operating temperatures of approximately 1223 K, an epi-thermal neutron environment and low power densities. HTGR fuel is designed with a high safety factor to prevent failure of the fuel, where the fuel-to-coating ratio is small to ensure the containment of fission products inside the coating. Failure of the particulate fuel occurs when the coating is cracked or otherwise damaged and fission products are released into the coolant system. HTGRs are known for their exceptionally low exposure rates to personnel.

The typical operating environment for space reactors may include relatively low fuel burnup, high temperatures, high power densities and a fast neutron spectrum, which are all dictated by the size and mass of the reactor system. Traditional HTGR fuel is not suitable for space reactors and design changes are required.

Present HTGRs use TRISO fuel, which consists of four coatings around the fuel kernel to retain fission products and maintain its physical integrity. The main structural member and barrier against the release of fission product migration is SiC. However, SiC does not appear to be a primary candidate for space reactors for long lifetimes because of chemical attack of the coating at high temperatures and damage by fast neutron irradiation above 900 K to 1000 K (Price 1977).

The main alternative being studies under the HTGR program is replacement of the SiC coating with ZrC. Compared with SiC (Price 1977), ZrC has a higher melting temperature 3540 K versus 2700 K for SiC and a reduced tendency to form low melting point eutectics with certain metallic fission products. It also appears to maintain mechanical stability under irradiation (Reynolds et al. 1976). Note that although a great deal of work has been completed on ZrC coatings, its physical characteristics as a function of stoichiometry are not well understood. This is true where the amount of free carbon within the ZrC will affect how well the coating retains fission products (Fukuda et al. 1981).

Two points in which ZrC appears favorable are crushing strength and high-temperature fabrication. In its initial fabricated condition the whole particle strength of the ZrC-TRISO particle is less then that of SiC-TRISO. However, by annealing at 2073 K, the ZrC-TRISO particles become stronger than SiC particles (Ogawa and Ikawa 1981). The crushing strength is important in determining the failure fraction during particle transfer, fuel compact and subsequent irradiation. ZrC can also increase the temperature to which a graphite matrix containing the TRISO beads can be heated during the fabrication process. This, in turn, produces a fuel body matrix with higher thermal conductivity (Ogawa and Ikawa 1981), which improves heat transfer in the core and can yield a higher coolant temperature for a given coated particle operating temperature.

In conclusion, a particle based upon HTGR-TRISO fuel appears feasible if the SiC coating is replaced by ZrC and if the fuel can be maximized with respect to the coating to minimize the reactor core size. In the design incorporated for this study, ZrC was chosen as the primary barrier material and the additional layers of PyC were eliminated where possible to maximize the fuel to coating ratio.

The model and fuel design used in optimizing the fuel sphere design with respect to maximum fuel loading was taken from GA Technologies (1983). UC_2 fuel was chosen because it has high thermal conductivity and a large data base available at high temperatures and burnups. Sources of information include the Rover (Taub 1975) and the HTGR (Gulden and Nickel 1977 and Shamasunder 1982) programs.

For this study a large fuel kernel is used to ensure that the maximum amount of fuel per particle is obtained for each enrichment. The value of 800.0 μm was chosen as the maximum fuel diameter since this is a little beyond what is currently fabricable. The fuel design incorporates a thin layer of PyC around the fuel kernel to protect it from corrosive gases present during deposition of the ZrC coating. A 5.0 μm coating of PyC is used, which is the minimum amount needed to protect the kernel (GA Technologies 1983). It is not used to maintain the integrity of the ZrC pressure vessel, but is only used as a means of fabricating the particles. The thickness of the ZrC coating is dependent upon the internal pressure generated by the fission products. Therefore, it is different for each set of operating conditions, lifetime, and fuel enrichment considered.

The fuel kernel is designed to have internal void space to accommodate fission products and fuel swelling. This is a different approach from the SiC-TRISO particles, in which the fission gases are accommodated within a porous coating of PyC, commonly referred to as the buffer layer. The design for this study is changed to maximize the fuel-to-coating ratio, which is crucial in minimizing the reactor size. The fuel is illustrated in Figure 1. The methodology used in designing the fuel particles is taken from GA Technologies (1983).

The following assumptions were used:

1. All of the fission gas atoms are released to the kernel porosity and not retained within the fuel particle,
2. The porosity is interconnected, and
3. The resulting gas pressure is exerted on the ZrC coating.

All of assumptions over-estimate the potential gas pressure on the ZrC coating and are therefore conservative. Beginning with the basic pressure-volume relationship:

$$P = \frac{nRT}{V} , \tag{1}$$

where: T = temperature of the fuel (K),
n = the number of moles of gas, which is the number of atoms created per fuel atom consumed by fissions (n_1) times the moles of atoms consumed (n_2),
$= n_1 \times n_2$.

For uranium the number of gas atoms created per fission (n_1) is 0.26. the number of moles consumed by fission (n_2) is given by

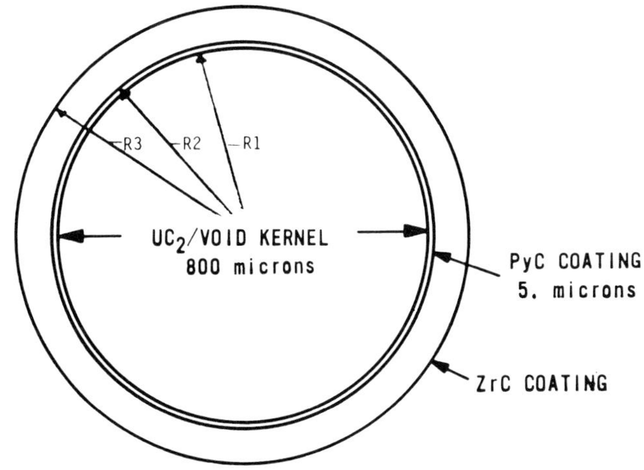

Figure 1 *Particulate Fuel Design.*

$$n_2 = \frac{4}{3}\pi r^3 \rho \kappa B_u / M , \tag{2}$$

where: r = kernel radius (cm),
ρ = density of the fuel,
B_u = burnup in percent ^{235}U, and
M = molecular weight of the fuel.

The volume available for the gases, V, is:

$$V = \frac{4}{3}\pi r^3 f_p , \tag{3}$$

where: f_p = the fraction porosity in the fuel kernel.
The density of the fuel is

$$\rho = \rho_t (1 - f_p) , \tag{4}$$

where ρ_t is the theoretical density of the fuel.

The final expression for internal pressure, P, and tangential stress, σ_t, as taken from GA Technologies (1983) is defined as

$$P = \frac{82\, n_1\, \rho_t\, (1 - f_p) B_u T}{M f_p} \text{ (atm)} \tag{5}$$

and

$$\sigma_t = \frac{P(2r_2^3 - r_3^3)}{2(r_3^3 - r_2^3)} \text{ (atm)} \tag{6}$$

where r_1 = the radius of the fuel kernel (μm),
r_2 = the outer radius of the PyC coating, and
r_3 = the outer radius of the ZrC coating.

The above correlation does not take into account the number density of the uranium versus the burnup of the ^{235}U. Therefore, the pressure must be modified by multiplying it by the enrichment of the ^{235}U to take into account the amount of fissionable material within the fuel kernel. Note that the conversion of ^{238}U to ^{239}Pu and the corresponding effect is not incorporated into the initial pressure model. This will not affect the higher enriched cases but will affect the lower enriched ones.

The pressure is therefore approximately give by

$$P = Px \text{ enrichment } (235) \text{ (atm)} \qquad (7)$$

This model assumes that all internal pressure is transmitted to the ZrC coating

Results

For design purpose the maximum allowable stress for the end-of-life is assumed to be 100 ksi, which is a conservative estimate based upon test results. Using this as a guideline, estimates of the burnup as a function of power were used. For 1 Mwt, 10 MWt and 100 MWt, the estimated burnup is 5%, 10%, and 25%, respectively.

Using the above correlations a program was written to calculate the stress as a function of burnup, temperature, enrichment and void fraction per fuel kernel. The results were then plotted to determine the maximum fuel fraction versus ZrC coating thickness. For each power and each enrichment the tangential stress was plotted versus the thickness of the ZrC coating and against the percent volume fuel per particle. Using 100 ksi as the maximum allowable stress, the corresponding fuel volume loading and ZrC coating thickness are determined. Both the fuel volume loading and the coating thickness are a direct function of the void fraction within the fuel. Therefore, it is important to optimize between the amount of void space and the coating thickness, as shown in Figures 2 and 3.

Using the information obtained from the graphs for each power level and enrichment, the corresponding number densities were calculated for each case. The basic assumptions made with respect to the core design are

1. The maximum fuel-to-coolant packing ratio for a bed of spheres is 60%,
2. The reactor length-to-diameter ratio was set at 1:1, and

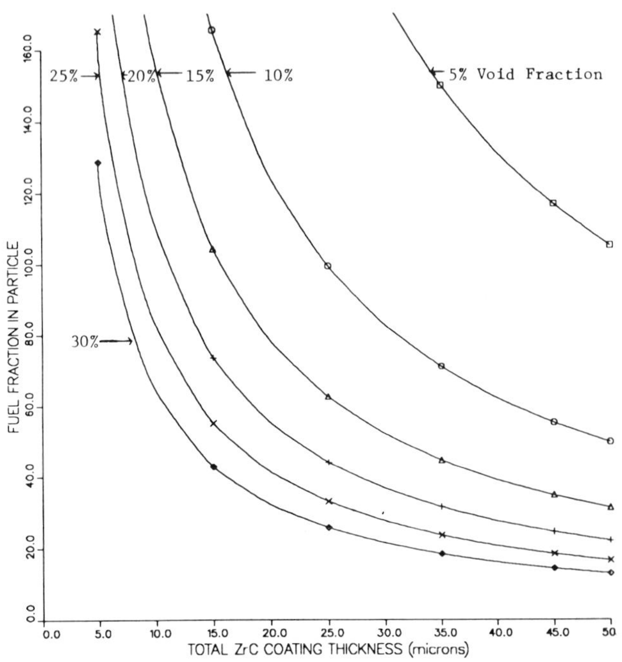

Figure 2 ZrC Maximum Stress verses Coating Thickness for a 1-MWt Reactor with 60% Enrichment.

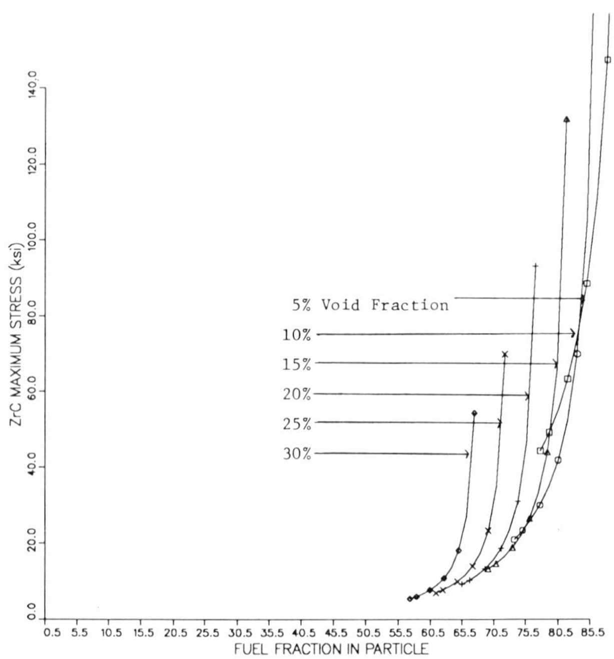

Figure 3 ZrC Maximum Stress versus Fuel Fraction per Particle for a 1-MWt Reactor with 60% Enrichment.

3. The maximum fuel centerline temperature is 1773 K.

The fuel designs are given in Table 1. The amount of fuel per particle, void fraction and ZrC coating thickness were optimized to obtain the maximum amount of fuel per particle. The void fraction within the fuel was incremented from 5% to 30%, in steps of 5%. In further studies, the void fraction should be incremented in smaller steps to ensure uniform results as the enrichment is increased.

Lifetime Calculations

Methodology

The reactor core consists of a bed of particulate fuel which is packed to 60% of the total volume. The core vessel was not included in the model. The helium coolant was included in the analysis, although its effects are relatively small.

Two reflector materials were chosen for this study, beryllium and graphite. The beryllium reflector is the optimum choice for space reactors due to its low material density, high melting point, and good neutron reflection characteristics. However, it may be limited due to the potential health risk of material dispersion in the case of launch abort or re-entry. When considering large systems this may be a key factor.

Graphite is a less effective reflector than beryllium. Using graphite will require a thicker reflector, which will result in a mass penalty, not primarily from the increased mass of the reflector, but rather from the increase in the radius of the radiation shield. Graphite does not present the biological hazard of beryllium and is a conservative alternative.

In the model the reflector is divided into three regions of reflector and poison (Figure 4). The reflector's available reactivity control is directly proportional to its thickness. The optimum reflector thickness can be determined by increasing this thickness until there is only a slight increase in the reactivity effect or until the size of the reflector exceeds system limitations. A reflector thickness of 15 cm was used for the beryllium reflector and 20 cm for the graphite reflector. Graphite is not as good of a neutron reflector as beryllium; thus a thicker reflector is required.

The core radius was determined for each case with the B_4C rotated toward the reactor and k_{eff} set at 0.95. The initial value of k_{eff} was set at this value to insure enough safety shutdown margin at the beginning of life. If a less conservative value of k_{eff} were chosen in establishing the reactor size, the lifetime of the system could be extended. This is due to the increase in the reactor size, which thus increases the fuel inventor. FEMP1D was used for all criticality calculations re-

Table 1 *Summary of the Reactor Fuel Design.*

		Power = 1 MWt			
Enrichment (%)	45	60	75	93	97
Fuel Diameter (μm)	857.6	859.0	873.0	888.0	890.0
%UC_2	81	81	79	77	76
%ZrC	8	9	11	13	9
%Void	9	9	9	9	13
%PyC	2	2	2	2	2
		Power = 10 MWt			
Enrichment (%)	45	60	75	93	97
Fuel Diameter (μm)	880.0	904.0	894.0	886.0	892.0
%UC_2	73	70	67	64	59
%ZrC	12	16	14	13	14
%Void	13	12	17	21	25
%Pyc	2	2	2	2	2
		Power = 100 MWt			
Enrichment (%)	45	60	75	93	97
Fuel Diameter (μm)	892.0	894.0	890.6	910.0	915.0
%UC_2	68	63	59	57	55
%ZrC	14	14	14	17	14
%Void	17	21	25	25	29
%PyC	2	2	2	2	2

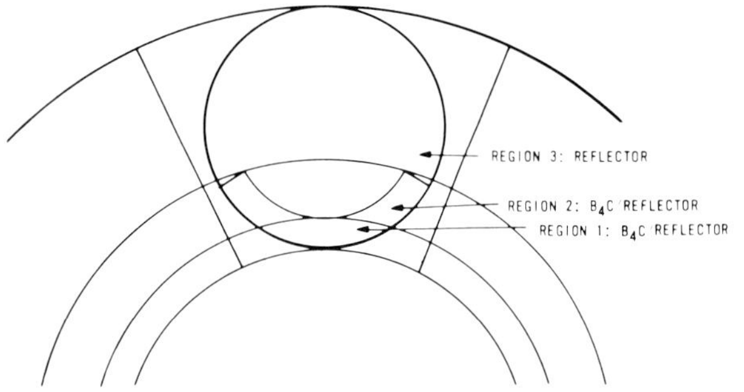

Figure 4 *Reflector Design Model.*

ported here, coupled with a 24-group neutron cross-section set.

Once the reactor core was sized, the total mass of each core material was determined and input into the ORIGEN code. ORIGEN, was used to obtain the ^{235}U burnup and the ^{239}Pu, ^{135}Xe, and ^{149}Sm production as a function of operating power and time. Obtaining the plutonium production is important when determining the effects of in-core breeding of fuel. The atoms of ^{235}U burned were replaced by molybdenum and barium, to model the production of the fission products. Molybdenum and barium are two of the primary products from the fission of ^{235}U.

Using the ORIGEN output, k_{eff} as a function of time was determined by placing the time dependent number densities into FEMP1D and calculating a new value for k_{eff}.

A set of curves were then obtained of k_{eff} versus time for each enrichment at a specified power level. The lifetime was obtained by extrapolating to the point where the curves fell below a k_{eff} of 0.99. Using these data, enrichment versus lifetime curves were obtained.

Examination of the results will help determine whether burnup and reactivity control will be major design issues for each power level and enrichment. These results may be quantitatively applicable to systems other than this particular reactor design.

Results

The core diameter is nearly the same for both reactors reflected by either beryllium or graphite for the same enrichment and power levels. One can conclude that the size of the reactor is not strongly dependent upon reflector material once the reflector thickness is past its saturation value. Rather, it is dependent upon the reactor geometry and size.

The results of the 1-MWt study showed that for both the beryllium and graphite reflected cases, neither the loss or reactivity control nor the burnup are the system limiters. Rather, both systems are criticality limited. The core size and lifetime are dominated by the minimum amount of fuel required to make the system critical. In these cases, the reactor core is large, so that the overall burnup is very low. The burnup is less than 1.6% at 20 years of operation at 1 MWt.

The results of the graphite system are very similar to those obtained for the beryllium system, with the exception that the beginning of life control margin is much less. For the graphite reflected, 97% enriched case, the reactor is barely critical at the beginning of life (BOL) and is subcritical in a short amount of time. To boost the highly enriched cases, which are nearly sub-critical at BOL, the initial shutdown margin can be reduced. This will increase the reactor size and the amount of reactivity control over the core lifetime.

Due to the relatively long lifetime obtained for the beryllium reflected system, incore breeding of ^{239}Pu at the lower enrichment was not an issue because only 1 kg of ^{239}Pu was generated over 16 years of operation for the 45%, enriched fuel.

The relative lifetime versus enrichment results show that lifetime varies inversely with the initial enrichment. Intuition would lead one to believe that the cases with higher enriched fuel would have a correspondently higher k_{eff} where in fact the opposite was true. There are a number of factors which contribute to this overall trend; first, the fuel design is optimized for each enrichment and power level, therefore a particle which has high enriched uranium will also require more void space and a thicker coating thereby decreasing the overall percentage of fuel in each particle and secondly the core size is based on an initial value for k_{eff} or 0.95 with the poison rotated fully towards the core. The size of the reactor core decreases as the enrichment increases. When the drums are then rotated outward the beginning of life k_{eff} for the higher enriched cases are lower relative to their smaller core size and therefore have less reactivity swing over the entire lifetime. It is important to understand the design philosophy behind the final results.

The size of the reactor may become the main design feature that will have to be optimized rather than the enrichment level. The smallest reactor system can be obtained when using the most highly enriched ^{235}U, but this also results in the shortest lifetime. An alternative means of optimization may include varying the reactor length to diameter ratio as a function of the reflector thickness.

At 10 MWt the beryllium reflected case is less dependent upon enrichment over time than is the graphite reflected case. This can be attributed to the large beginning of life control margin in all of the beryllium cases. The results show that for the beryllium reflector, the lifetime extends past 20 years for all enrichments and is not limited by either burnup or reactivity control. Again, the limitation may be system related, external to the reactor core, but it is not due to the loss of reactivity control. The graphite case was limited by loss of reactivity control by the external control elements. The maximum predicted life is 11.5 years using a k_{eff} of 0.99 as marking end of life. The lifetime results versus enrichment are shown in Figure 5. The corresponding burnup is less then 8%, this is well below the initial design value of 10%, which confirms that burnup is not a major factor for this design. The particulate fuel is very attractive in this area because of its tolerance to very high fuel burnup levels. Another advantage of the particulate fuel is its ability to withstand failure of a percentage of the fuel spheres without causing failure of the reactor system.

The results of the 100 MWt study are show in Figures 6 and 7. For both the beryllium and graphite reflected cases, loss of reactivity control is the overriding lifetime limiter. The maximum design burnup of 25% is not exceeded prior to system subcriticality due to loss of reactivity.

As a means of refining this study, the fuel needs to be optimized with respect to the maximum amount of burnup to eliminate over-design of the particulate fuel. Both the amount of void space and the coating thickness are determined by the predicted burnup level. Optimization of the fuel versus lifetime is an iterative process between obtaining lifetime results, comparing them to the burnup levels and redesigning the fuel with adjusted burnup levels. This procedure would take several iterations but would provide more precise fuel design information.

The effect of breeding in the lower enrichments was clear for both cases at 100 MWt. The change in k_{eff} overtime was less for the 45% enriched cases than for

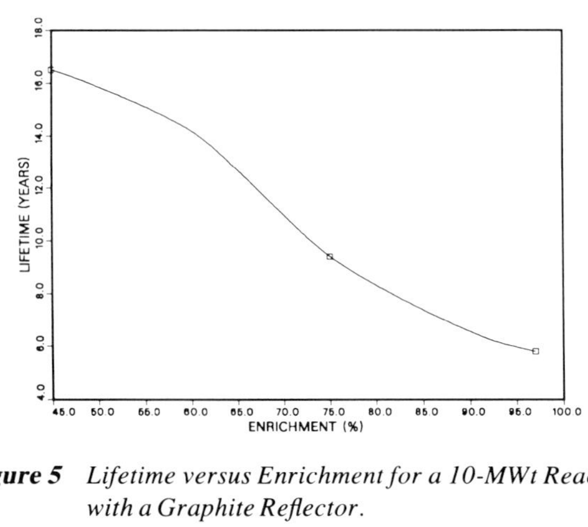

Figure 5 *Lifetime versus Enrichment for a 10-MWt Reactor with a Graphite Reflector.*

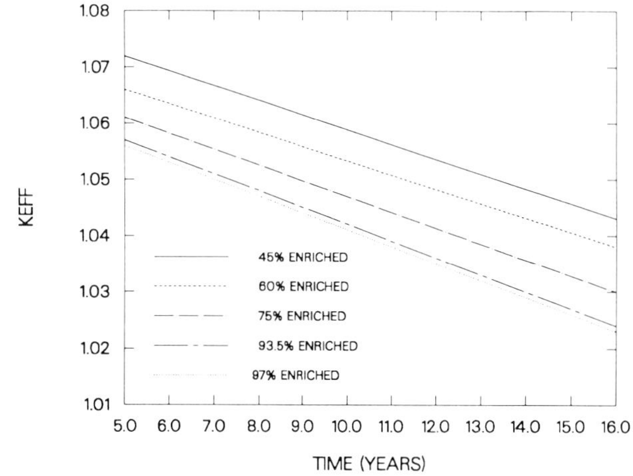

Figure 6 *Keff versus Time for a 10-MWt Reactor with a Beryllium Reflector.*

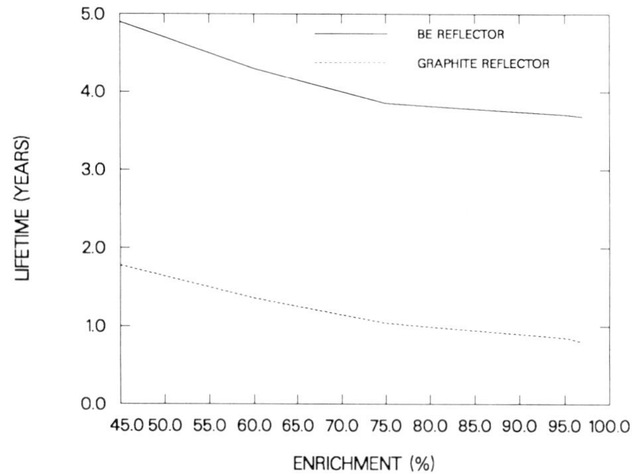

Figure 7 *Lifetime versus Enrichment for a 100-MWt Reactor with either a Beryllium or a Graphite Reflector.*

the other enrichments, due to the formation of ^{239}Pu from ^{238}U. This may play an important role in extending the lifetime of the reactor system. The maximum lifetime obtained for the beryllium reflector, 45% enriched case, is 4.5 years compared to 3.3 years for the 97% enriched case.

The graphite reflected case exemplifies the effect of decreased reactivity excess at the beginning of life and a drastic decrease of the maximum reactor lifetime. With the graphite reflector, the maximum lifetime is reduced to 1.79 years at 45% enriched and 0.8 years at 97% enriched. A beryllium/graphite zoned reflector may be the needed compromise between ensuring safety and maximizing reactivity control.

Conclusions and Recommendations

In conclusion, it is clear that the limitations on lifetime are directly related to the power level, enrichment and burnup of the reactor fuel. Each limitation should be carefully examined to ensure that the optimum system can be designed for lifetime, size, and efficiency requirements.

The results of this work show that the beryllium reflector is more effective than the graphite reflector by as much as 6% delta-k at the beginning of life. This greatly decreases the maximum lifetime that can be obtained for the graphite cases.

The effect of enrichment is not as significant for the beryllium reflector as it is for the graphite reflector. The lifetime versus enrichment results are clear for the graphite reflected cases, because the difference in k_{eff} is larger between enrichments.

In both the 1 and 10 MWt cases, criticality is the main design driver in determining the reactor size. At such low power levels, the over-riding design concern may no longer be enrichment or lifetime, but rather optimization of the reactor size and mass.

At 100 MWt the lifetime was limited by the loss of excess reactivity for control of the reactor. With the maximum lifetime being 4.5 years, alternate means will be needed to obtain desired lifetimes of 7 to 10 years. Achieving longer lifetimes at high power levels may be a primary goal in meeting space power requirements. Since reactivity is established as the primary factor that limits lifetime, alternate means of extending lifetime need to be addressed.

Three methods of extending the reactor lifetime are incore breeding of fissile material (that is, ^{233}U or ^{239}U), incore control elements and burnable poisons. In the 100 MWt case, almost 33 kg of ^{239}Pu were produced after five years which represents 10% of the total fissile fuel inventor within the core. The relative amount of power produced by ^{239}Pu versus ^{235}U over time needs to be examined. In this regard there must be a tradeoff between safety and safeguards. Re-entry limitations may prohibit in-core breeding of plutonium on such a large scale.

Burnable poisons may present a less hazardous approach, but further study is needed. Note that burnable poisons were applied on the inside of the fuel cladding for the SNAP 2, 8, and 10 reactors to extend their lifetimes.

For each case, several parameters can be modified to extend lifetime, such as increasing k_{eff} or decreasing the relative burnup; but each of these will depend upon the actual design requirements. If lifetime is the primary design criteria and size in inconsequential, the reflector thickness can be increased and burnable poisons may be used. In the lower power cases, lifetime may not be limited by loss of reactivity or burnup, and size could be the primary driver.

Note that the reactor was not designed to maximize heat transfer, but it was designed with criticality limitations in mind. A set centerline temperature of 1773 K was used and was considered constant across the core. Design tradeoffs would be needed to optimize the reactor with respect to both the heat transport requirements and criticality limitations.

Acknowledgments

The continued encouragement and useful discussion provided by Dr. James H. Lee, Jr. of Sandia National Laboratories is greatly appreciated. The current affiliation for Dr. Woodall is the Idaho National Engineering Laboratory. The writing of this paper was funded by the Air Force Weapons Laboratory Nuclear Power Section.

References

Bartine, D. E. et al. (1982) *Characterization of an HTGR for Deep Space Applications*, ORNL-GCR-82-15, Oak Ridge National Laboratory, Oak Ridge, TN, September 1982.

Bennett, D. D. (1979) *Sandia ORIGEN User's Manual*, NUREG/CR-0987, Sandia National Laboratories, Albuquerque, NM, October 1979.

Fukuda, K., K. Ikawa, and K. Iwamoto (1981) "A Study of the Reaction Between ZrC and Some Metal Oxides," *J. Nucl. Mat.*, 98:18-26.

GA Technologies (1983) *SP-100 Coated Particle Fuel Development, Phase I, Final Report*, GA-A16992, GA Technologies, San Diego, CA, March 1983.

Gulden, T. D. and H. Nickel (1977) "Coated Particle Fuels," *Nucl. Tech*, 35:206-203.

Lazo, M. S., V. S. Dean, and P. J. McDaniel (1985) *A User's Manual to FEMP1 D-A Finite Element Multigroup, PN, One Dimensional Transport Code*, Chemical and Nuclear Engineering Department, University of New Mexico, Albuquerque, NM, January 1985.

Ogawa, T. and K. Ikawa (1981) "Crushing Strengths of SiC-TRESO and ZrC Coated fuel particles," *J. Nucl. Mat.*, 98:18-26.

Price, R. J. (1977) "Properties of Silicon Carbide for Nuclear Fuel Particle Coatings," *Nucl. Tech.*, 35:56-60.

Reynolds, G. H., J. C. Javier, and J. L. Kaae (1976) "Irradiation Behavior of Experimental Fuel Particles Containing Chemically desposited Vapor Zirconium Carbide Coatings," *J. Nucl. Mat.*, 62:9-16.

Shamasunder, B. I. et al. (1982) *HTGR Fuel Performance Basis*, GA-A16767, GA Technologies, San Diego, CA, May 1982.

Taub, J. M. (1975) *Review of Fuel Element Development for Nuclear Rocket Engines*, LA-5931, Los Alamos National Laboratory, Los Alamos, NM, March 1985.

Wolf, L., G. Ballensiefen, and W. Frohling (1975) "Fuel Elements of the High Temperature Pebble Bed Reactor," *Nucl. Eng. and Des.*, 34:93-108.

Part 14

RADIOISOTOPE GENERATORS

Space Nuclear Power Systems 1986
Edited by M. S. El-Genk and M. D. Hoover

Chapter 42

Power Performance of the General-Purpose Heat Source Radioisotope Thermoelectric Generator

Gary L. Bennett, James J. Lombardo, and B. J. Rock

Office of Nuclear Energy
U.S. Department of Energy
Washington, D. C. 20545

The General-Purpose Heat Source Radioisotope Thermoelectric Generator (GPHS-RTG) has been developed to provide power for the Galileo mission to Jupiter and the Ulysses mission to study the polar regions of the Sun. The results of power performance measurements and analyses on five nuclear-heated generators and one electrically heated generator are presented. The analyses are based on a computer model for power predictions that was developed using results obtained from previous silicon-germanium programs. This model includes the effects of radioisotope fuel decay, dopant precipitation, thermal conductivity changes, material sublimation, and carbon monoxide effects. Based on extensive tests and analyses it is concluded that the GPHS-RTGs will meet or exceed their power performance requirements.

Introduction

The General-Purpose Heat Source Radioisotope Thermoelectric Generator (GPHS-RTG) has been developed under the sponsorship of the Department of Energy (DOE) to provide electrical power for the National Aeronautics and Space Administration (NASA) Galileo mission to Jupiter (see Figure 1) and the joint NASA/European Space Agency (ESA) Ulysses mission to study the polar regions of the Sun (see Figure 2). Five nuclear-heated generators and one electrically heated generator have been built and tested, establishing the design concept and meeting the specification requirements. This paper provides a summary of the power performance of the GPHS-RTGs. Information on the other performance attributes are supplied by Bennett (1986).

GPHS-RTG Description

The GPHS-RTG design is built upon the successful technology used in the Multi-Hundred Watt RTGs (MHW-RTGs) flown on the two NASA Voyager spacecraft and two U.S. Air Force (USAF) communications satellites known as Lincoln Experimental Satellites 8 and 9 (LES 8/9). The GPHS-RTG converts about 4410 Wt from the nuclear heat source into at least 285 We at beginning of mission (BOM). The GPHS-RTG case allows for attachment to the spacecraft and for tandem attachment of additional RTGs. The GPHS-RTG is configured for space shuttle launches and is designed to withstand the launch and ascent dynamic environment.

The GPHS-RTG consists of two major components: the General-Purpose Heat Source (GPHS) and the Converter. A conceptual drawing of the GPHS-RTG is presented in Figure 3, which shows a cutaway portion of the converter to illustrate its internal construction and the position of the 18 GPHS modules comprising the radioisotope heat source.

The converter is composed of a thermopile that converts the radioisotope-generated heat into electrical power; support assemblies for the GPHS; and a housing that encloses the generator and acts as a heat rejection radiator. The thermopile consists of 572 silicon germanium (SiGe) thermoelectric elements called ''unicouples'' connected in a series-parallel network. The unicouple, shown schematically in Figure 4, is identical in design to that used in the generators flown on LES 8/9 launched in 1976 and Voyagers 1 and 2 launched in 1977 (Bennett 1981, Cockfield 1980, and Cockfield 1981).

The unicouple consists of an 85 weight percent ($\sim$95 atom percent) silicon in a silicon molybdenum alloy (SiMo) electrical connector and heat collector, or hot shoe, to which the N and P couple legs are bonded. The N and P legs are equal in size, 2.74 mm $\times$ 6.50 mm in cross section, with a total length of 20.3 mm. Couple height is 31.1 mm and the hot

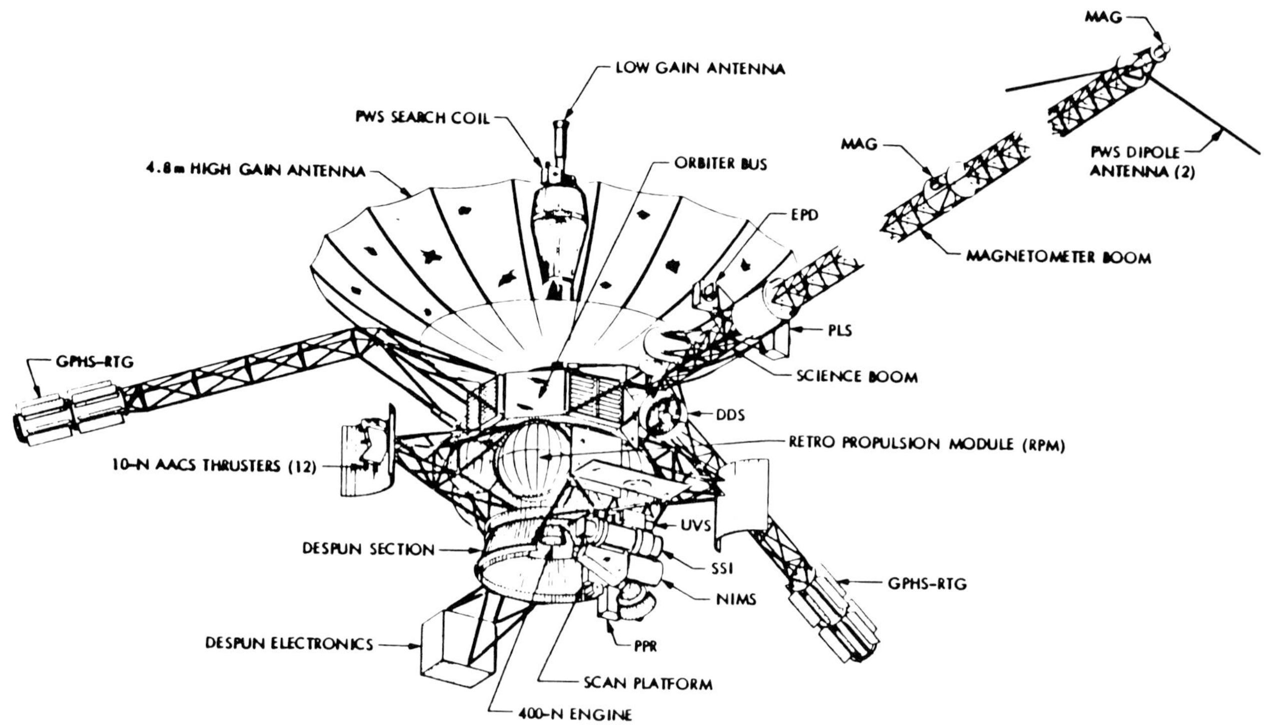

Figure 1 *Galileo Spacecraft Orbiter with the two GPHS-RTGs shown Mounted on the two Booms. (From NASA/JPL.)*

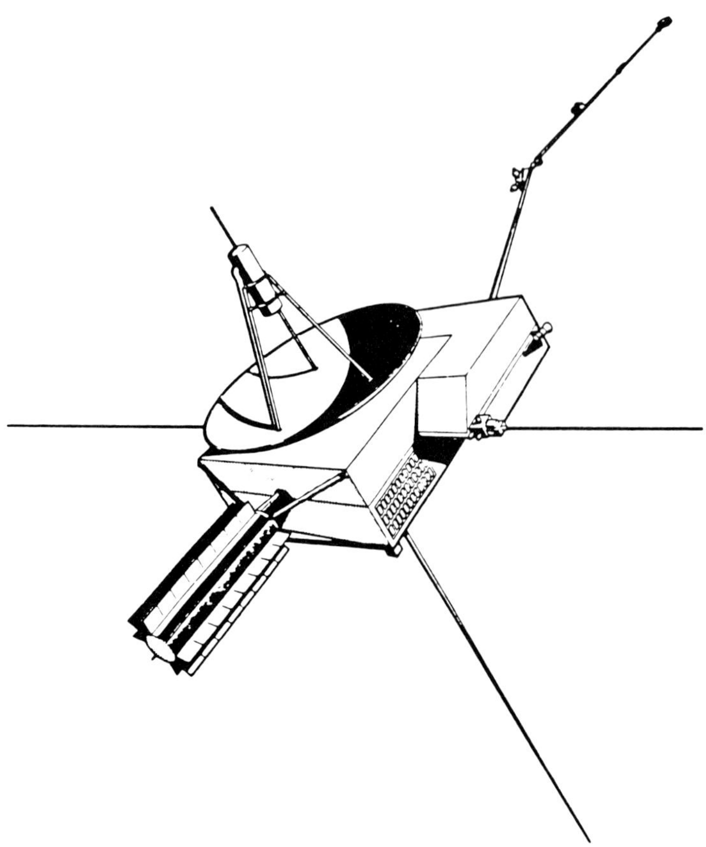

Figure 2 *Ulysses Spacecraft with the General-Purpose Heat Source RTG Mounted on the Side. (From European Space Agency.)*

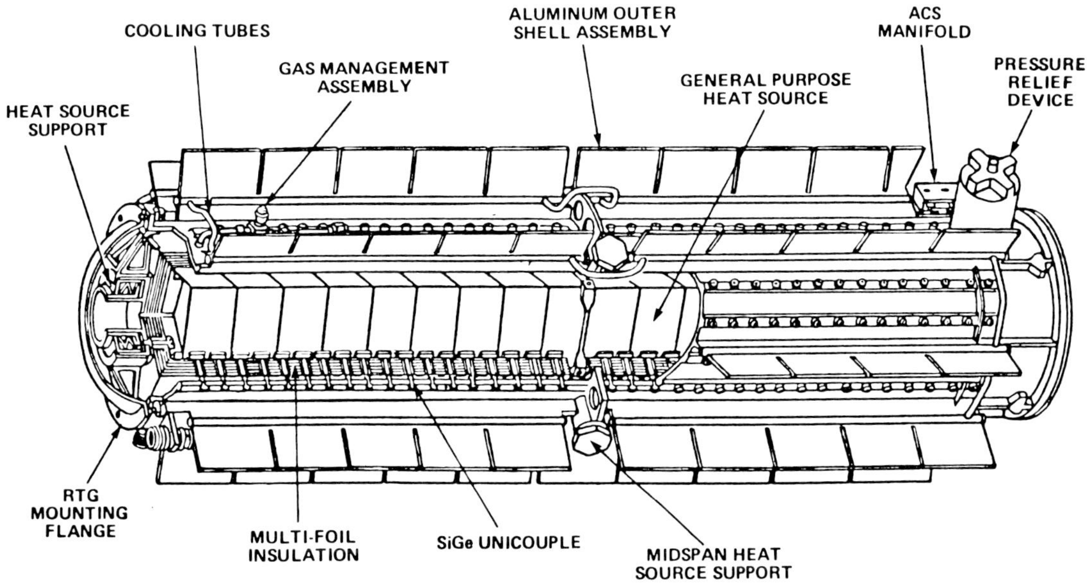

Figure 3 *Cutaway of the General-Purpose Heat Source Radioisotope Thermoelectric Generator (GPHS-RTG).*

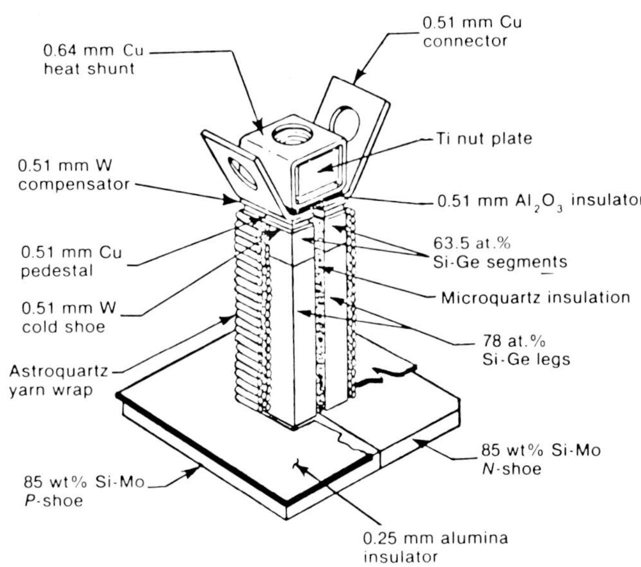

Figure 4 *Silicon-Germanium Thermoelectric Element (Unicouple) used in the GPHS-RTG.*

shoe measures 22.9 mm × 22.9 mm and is 1.9 mm thick. Two compositions of SiGe are used in the legs: 78 atom percent silicon for most of the length and a short (3.18 mm) 63.5 atom percent section at the cold end. The lower content segment is used to provide improved matching for thermal expansion of the bonded parts. The two SiGe legs and their corresponding sections of the hot shoe are doped to provide thermoelectric polarity; the N-type material is doped with phosphorus and the P-type with boron. The dopant level is $\sim 10^{20}$ carriers per cm^3 ($\sim$ 0.1 atom percent). The silicon alloy couple is bonded to a cold stack assembly of tungsten, copper, and alumina parts which separate the electrical and thermal currents. The thermal current crosses the alumina insulator and passes through the radiator attachment into the outer shell. The electric current between couples is shunted through separate copper straps bonded to each leg assembly. These straps are riveted together in the space between the inside of the outer case and the outside of the insulation system to form the thermopile electrical circuit. A two string, series-parallel, electric wiring circuit is used. This circuit permits continued operation in the event a unicouple fails in either the open or short circuit mode. The circuit loops are also arranged to minimize the net magnetic field of the generator. The cold stack is designed to flex laterally to ensure unicouple load sharing from insulation loads and thermal deformations. These couples are supported in a cantilever fashion from the aluminum outer case and they in turn support the insulation packet. The unicouples are electrically insulated and mechanically cushioned from the multifoil insulation by three layers of astroquartz (SiO_2) yarn (nominal diameter of 0.76 mm) wound tightly around the two SiGe legs of each couple and by an alumina wafer beneath the hot shoe. A silicon nitride (Si_3N_4) coating approximately 1μm thick is applied to the unicouple legs and hot shoes to retard silicon sublimation. The unicouples operate at a nominal 1273 K/573 K hot junction/cold junction in a vacuum environment. (The corresponding nominal hot shoe temperature is 1308 K (Kelly 1973 and 1982).)

The thermopile insulation consists of an insulation canister and end caps composed of 60 layers of 7.6μm molybdenum foil with astroquartz cloth separators and a molybdenum frame assembled in the shape of an octagonal prism with joints designed to accommodate thermal expansion. The unicouples penetrate the insulation system and are bolted to a cylindrical housing of type 2219 aluminum, which is the primary structural element of the converter. The housing diameter is 218.4 mm with 101.6 mm wide heat rejection fins bringing the radial envelope to 421.6 mm. The axial envelope, including end domes, is 1140.5 mm. The nominal housing thickness is 1.5 mm with stiffening provided by eight axial ribs and three circumferential rings. The housing is coated with a 50.8μm thick electrically conductive silicone paint to provide an emissivity of 0.9. To limit the energy radiated from the converter surface to the shuttle bay and sensitive spacecraft instruments, an active cooling system (ACS) consisting of tubular passages near the base of each fin permits water circulation to remove approximately 3500 Wt (Cockfield 1980, Bennett 1981, Cockfield 1981, and Kelly 1982).

There are two principal modes of operation for the GPHS-RTG: air and vaccuum. During air operation, the RTG is filled with an inert gas (normally argon for testing and storage and xenon at launch) to protect the molybdenum and graphitic components from oxidation. A gas management system consisting of a pressure relief device (PRD) and a gas management valve (GMV) is used in conjunction with the pressure tight converter case to maintain the desired internal environment of the generator. This system maintains the inert gas environment (nominal pressure 172 kPa absolute) to allow partial power operation on the launch pad. (The RTGs are to remain positively pressurized (greater than 3.4 kPa gauge) for 30 days on-pad after being initially pressurized to 172 kPa absolute. To satisfy this requirement, the pressure drop in the RTGs must be less than 3.2 kPa gauge in 12 hours after the initial 172-kPa pressurization.) Full power operation in space is achieved after venting the inert gas through the PRD. The GMV is a manually actuated valve required for pressure maintenance of the generator during ground storage. During the thermal vacuum tests of the flight RTGs, the GMV was used to simulate the effects of venting of the cover gas into space.

The thermal energy provided to the converter comes from the GPHS, which consists of a stacked column of 18 individual modules each providing about 245 Wt from the decay of encapsulated plutonium-238 oxide, which has a half-life of 87.8 years. As a result of this half-life, the reduction of thermal power is approximately 0.8 percent per year (Schock 1980).

The average mass of a flight RTG is 55.90 kilograms.

Electrical Power Requirements

The GPHS-RTG has been designed to satisfy the electrical power requirements of both the Galileo and Ulysses spacecraft. For the Galileo spacecraft the RTG must provide DC power at an output voltage of $30 + 0.7/-0.5$ V. The BOM power is to be 285 We to 300 We. The end-of-mission (EOM) is defined to be 4.2 years (36,800 h) after spacecraft launch for Galileo at which time the RTG is to produce at least 255 We. Upon cessation of cooling for both Galileo and Ulysses missions, the RTG must achieve 95% of full BOM power within 8 hours after venting the xenon cover gas through the PRD. On the launch pad with the xenon cover gas the RTG must provide a launch power of 200 We to 240 We for both Galileo and Ulysses missions. For the more conductive argon cover gas usually used in the ground test phase this translates into a power output of at least 130 We during air performance tests of a generator filled with argon.

For the Ulysses spacecraft the RTG must provide DC BOM power of 285 We to 310 We at an output voltage of 28 +0.7/ − 0.5 V. The EOM was defined to be 4.7 years (41,200 h) after spacecraft launch for the Ulysses mission at which time the RTG is to produce at least 250 We.

The RTG power output is defined to be the power at the RTG power connector pins. Because the test measurements were made from voltage taps located in the test power cable a correction had to be applied to account for connector resistance. From measurements made on the more extensively instrumented Engineering Unit a connector contact resistance of 0.020 ohm was derived. Thus, to determine the RTG power output from test measurements the following correction must be applied

$$P = (0.020)\ I^2 \text{ watts}$$

where I = measured RTG output current. Typically, for thermal vacuum conditions where the current is approximately 10 A, the power correction is about 2 We, while for an argon-filled operation where the current is about 5 A, the correction is about 0.5 We. For comparison purposes the electrical power is usually normalized to 4410 Wt by using a computed adjustment of 0.12 We/Wt. With this normalization, acceptance power was defined as 292 We (minimum) at 30 V DC for Galileo and 287 We (minimum) at 28 V DC for Ulysses (GE 1985a).

In addition to monitoring power during the test phase, several other electrical parameters were measured. In order to ascertain the integrity of the electrical insulation system, the insulation (shunt) resistance between the thermoelectric circuit and the outer case was measured. The requirement was that this resistance be equal to or greater than 1000 ohms.

The change in the internal circuit resistance from one functional test to another was measured. After correcting for temperature and aging effects the requirement was that this resistance change should be less than 0.008 ohm. Satisfaction of this requirement provides assurance of the integrity of the unicouples since one broken unicouple will result in an increase in resistance of 0.008 ohm (GE 1985a).

Test Program

The GPHS-RTG test program consisted of several phases to verify that the power requirements had been met. Six 18-couple modules were put on long-term testing at General Electric (GE) and the Jet Propulsion Laboratory (JPL). These modules provided long-term life test data, including some under accelerated (high temperature) conditions.

Next a nonnuclear Engineering Unit was built and tested to prove out the overall design including power requirements. The Engineering Unit was heated by an electric heat source (EHS) consisting of a serpentine graphite heating element electrically isolated from its outer shell by alumina spacers and having an inner molybdenum rod to provide the proper mass simulation. The EHS was supported at the ends and midspan and delivered 4400 Wt by radiation to the thermopile. During the life testing of the Engineering Unit the EHS was adjusted to follow the power decay of plutonium-238 (Kelly 1982).

The first RTG built was the Qualification Unit which was put through a series of flight acceptance and type acceptance tests (Ambrose 1985a). The Qualification Unit is currently on life test in a simulated space environment at the Mound Plant where it has accumulated 9,850 hours as of 5 January 1986 (Deaton 1986).

Four flight RTGs were assembled and tested by the DOE complex for NASA (Ambrose 1985b, 1985c, 1985d, and 1985e). Of these four, two (F1 and F5) are to be flown on the Galileo spacecraft, one (F3) is to be flown on the Ulysses spacecraft and one (F4) is a common spare for both missions. Both the Qualification Unit and F4 were tested under conditions simulating Galileo (30 V) and Ulysses (28 V).

Figure 5 shows the Qualification Unit and flight RTG configurations. (The Engineering Unit configuration was similar to that shown for the Qualification Unit.) The principal difference between the Qualification Unit and a flight RTG is the spool piece assembly attached to the former which permits extensive internal instrumentation for diagnostic purposes. In the case of the Engineering Unit the spool piece also allowed connection of electrical power leads for the EHS.

Figure 6 shows the overall RTG performance test sequence. The converters were designed and built by GE and shipped to the Mound Plant for fueling and testing as RTGs. Upon completion of fueling an initial power check was obtained by evacuating the bell jar in the Mound Plant's Inert Atmosphere Assembly Chamber (IAAC).

Functional tests were run before and after vibration to establish the performance baseline and determine if

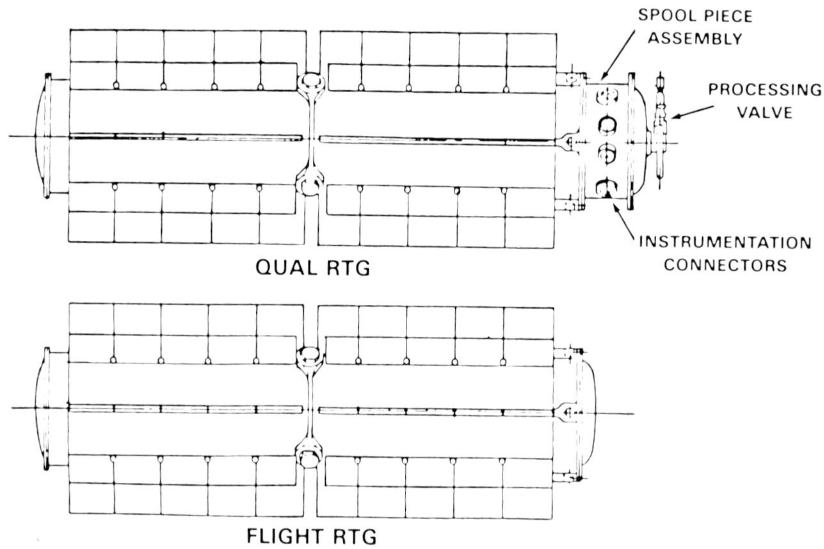

Figure 5 *Comparison of Qualification and Flight RTG Configurations for the GPHS-RTG.*

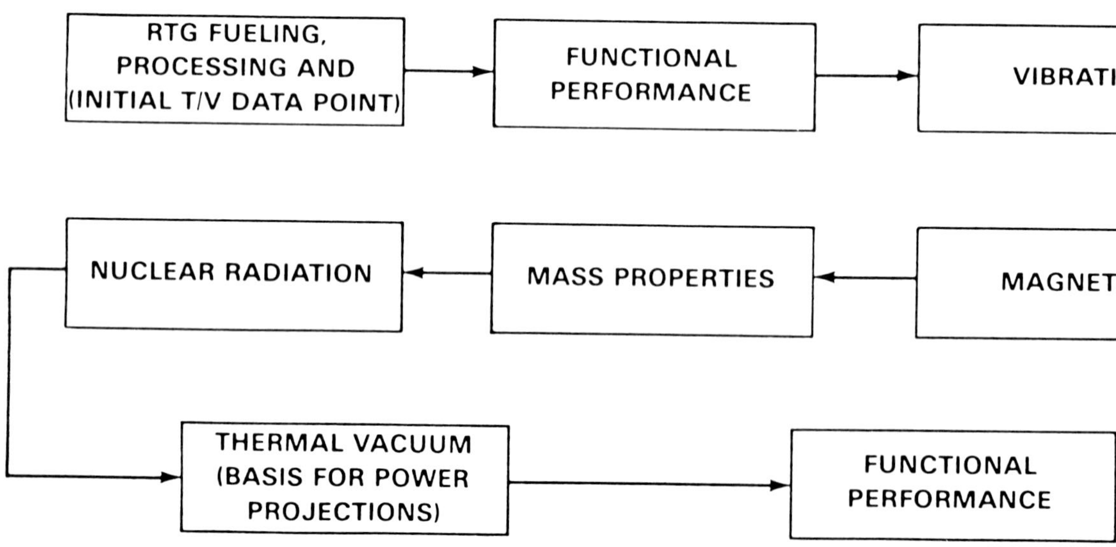

Figure 6 *Performance Sequence for GPHS-RTG Assembly and Testing.*

any changes occurred as a result of the vibration tests. The post-vibration functional tests also provided a performance baseline before thermal vacuum testing. For Galileo the RTGs (F1, F4, and F5) had to provide a minimum of 130 We tested in air and filled with argon. For Ulysses, the RTGs (F3 and F4) had to provide 140 We under the same conditions. From tests run on the Engineering Unit this requirement ensures that there will be 200 We of prelaunch power available when the RTGs are filled with xenon after accounting for degradation losses caused by the decay of the isotope fuel and precipitation of the boron and phosphorus dopants following 25,000 hours of storage. (The launch pad power can also be calculated from the power measured in an evacuated RTG. From tests conducted with the Engineering Unit, the power output when xenon-filled is 0.77 of the power output in vacuum operation. The corresponding factor for the Qualification Unit was 0.764 and this was applied to the measurements made on the flight RTGs (Ambrose 1985b and 1985c).) The functional tests also included measurements of electrical isolation (''insulation resistance''), internal resistance, and pressure decay.

The thermal vacuum test provided the basis for the

power projections, both BOM and EOM, for each RTG. Each flight RTG provided electrical power under simulated space conditions from 6 hours to more than 40 hours in the thermal vacuum chamber at a pressure of 0.1 mPa or less and an average sink temperature of 308.9 K. Following the thermal vacuum test, each RTG was backfilled with argon for a six-hour performance and pressure decay test, which provides strong confirmation of the performance predictions.

As of 5 January 1986, more than 903,525 hours have been accumulated in fueled SiGe RTGs, which provides strong confirmation of the performance predictions.

Test Results

Once assembled and before shipment to Mound, each converter was tested as an electric (electrically heated) thermoelectric generator (ETG). Table 1 shows the flight GPHS-ETG acceptance performance. This demonstrates that each converter will meet the performance requirements.

Table 2 shows the thermal vacuum performance for each of the four flight RTGs. The spare RTG (F4) was tested twice, once under Galileo load conditions (30 V) and once under Ulysses load conditions (28 V). The thermal inventory of the GPHS is shown for the indicated date of completion of the thermal vacuum test. As discussed previously, three power outputs are shown: the measured power, the power corrected for losses in the connector resistance, and the power normalized to 4410 Wt. All four flight RTGs met the performance requirements.

The RTG air performance comparisons are shown in Table 3. Each RTG met the power requirement for an argon fill thus ensuring proper on-pad power. The internal resistance change showed no evidence of broken unicouples and the RTGs demonstrated their ability to maintain the desired pressure from the cover gas for more than 30 days unattended.

Power Projections

Using the test results in Table 2, projections can be made of the power performance of each RTG. The reliability of these projections is based on estimates of the uncertainty in the factors that affect BOM power and uncertainties in the degradation model that predicts power output as a function of time.

Table 1 *Comparison of Performance and Requirements for the GPHS-ETGs.*

Parameter	Requirement	F1	F3	F4	F5
BOL Power (W)[a]	≥293	294.6	294.4	293.8	295.4
Voltage (V)	30	30	30	30	30
Shunt Resistance (ohms)	>1000	2000	2100	3100	5000
Power in Argon (W)	>130	151.8	156.0	154.7	158.7
Mass (kg)	≤30.39	30.05	29.95	30.09	29.90
Storage Life without Repressurization (days)	≥30	300	1000	1000	300

[a]Normalized to 4400 Wt.

Table 2 *Comparison of Thermal Vacuum Performance and Requirements for the Flight GPHS-RTGs. (From GE 1985a.)*

Parameter	Requirement	F1	F3	F4	F4	F5
Thermal Inventory (W)		4460.1	4479.1	4435.2	4435.1	4443.5
Date		12/14/84	8/15/85	7/20/85	7/21/85	5/22/85
RTG Power Output (W)		306.1	303.1[a]	301.4	302.8	305.2
as measured						
corrected		308.2	305.4[a]	303.7	304.8	307.3
normalized to 4410 W	>292	302.0	279.1[a]	300.7	301.8	303.3
Load Voltage (V)	30	30.02	28.02	28.04	29.97	30.03
Open Circuit Voltage (V)		52.42	51.75	51.37	51.94	52.31
Internal Resistance (ohms)		2.197	2.194	2.169	2.174	2.191
Shunt Resistance (kohms)	>1.0	2.2	2.1	3.4	3.4	5.6
RTD Temperature (°C)	<260	247	247	246	246	244

[a]Power Not Stabilized

Beginning-of-Mission (BOM) Power

To determine the BOM power the corrected powers shown in Table 2 must be reduced to account for aging while the RTG is in storage and increased to account for the restrictive venting characteristics of the gas management system used in thermal vacuum testing. Figure 7 shows the process used to project the BOM power and the power adjustments made for each RTG from the thermal vacuum power measurements to the estimated BOM power.

The first adjustment was a correction for the accuracy of the thermal vacuum test power measurements. The accuracy of these measurements was estimated to be ± 0.2%. Next the power loss resulting from the decay of the radioisotope fuel during storage was calculated. As noted earlier, plutonium-238 decays at a rate of approximately 0.8% per year and the associated electrical power loss, based on computer models, is 0.12 We/Wt. The third adjustment was to reduce the power to account for precipitation of the boron and phosphorus dopants during storage. This power loss can be estimated from Figure 8 which was derived from previous generator tests (MHW-RTG Test Bed Converter No. 4) and which correlates with the data obtained from the Voyager RTGs. An uncertainty of ± 20% was applied to the nominal value taken from Figure 8. The final adjustment was for GMV venting during the RTG thermal vacuum tests. In space the RTGs will be vented through the PRD which is less restrictive than the GMV so that there is less build-up of the gas pressure (primarily helium from the alpha decay of the fuel). The experience with the MHW-RTGs showed a 0.6% to 1% reduction in power output when venting through the GMV. On this basis it would be expected that the GPHS-RTG power would be reduced by 1.8 We to 3 We when venting through the GMV. Data obtained from the Qualification Unit indicated a GMV effect of 3.8 We but because of complications in the measurements this higher correction was not used (GE 1985a and Ambrose 1985b).

The launch power, which is that power obtained when the RTG is filled with xenon, was obtained by multiplying the BOM power by the previously discussed factor of 0.764. Some reduction in launch power can also occur as a result of the helium buildup which occurs after the generator has been filled with xenon. The magnitude of this reduction depends upon the time between the xenon gas exchange and launch which, for this analysis, was assumed to be 30 days. Measurements of the on-pad power of the LES 8/9 MHW-RTGs showed that this reduction in launch power will be 1.3% (Ambrose 1985c).

The results of these power projections and performance measurements are shown in Table 4. In each case it can be seen that the four flight RTGs meet their performance requirements.

Degradation Model

A degradation model has been developed to predict the EOM power based on the BOM power ranges described in the preceding section (Loffreda 1973, Kelly 1973, and also see Shields 1981 and Chmielewski 1982). This degradation model, known as TEG 3, was originally developed for the Apollo SNAP-27 RTGs and was modified for the MHW-RTGs used on the Voyager 1/2 and LES 8/9 missions and was then adapted to the GPHS-RTGs. TEG 3 has the capability to include the following effects (GE 1985a):

- Seebeck voltage changes in the N and P legs,
- Resistivity changes in the N and P legs,
- Thermal conductivity changes in the N and P legs,
- Junction contact resistivities,
- Hot and cold shoe resistances,
- Insulation thermal conductance change with time,
- Heat source support losses,
- Radiator surface emissivity,
- Heat sink temperature variation,
- Radioisotope fuel decay, and
- Sublimation effects on the geometry of the thermoelectric legs and hot shoe.

The following subsections discuss the principal mechanisms which affect the RTG output power at EOM.

Table 3 *Comparison of the Air Performance of the Flight GPHS-RTGs.*

Parameter	Requirement	F1	F3	F4	F5
Power in Argon (W)	>130	166	178	174	168
Internal Resistance Change (ohms)	<.008	<.008	<.008	<.008	<.008
Pressure Decay Life (days)	>30	>300	>300	>300	>300

	POWER VARIATION — WATTS			
	F1	F3	F4	F5
THERMAL VACUUM TEST POWER (MEASUREMENT ACCURACY)	±0.6	±0.6	±0.6	±0.6
FUEL DECAY EFFECT DURING STORAGE	−5.8	−3.0	−3.2	−3.9
POWER LOSS FROM DOPANT PRECIPITATION DURING STORAGE	−2.4 TO −3.6	−1.3 TO −1.9	−1.4 TO −2.0	−1.8 TO −2.6
GMV EFFECT	+1.8 TO +3.0	+1.8 TO +3.0	+1.8 TO +3.0	+1.8 TO +3.0
BOM POWER PROJECTION				

Figure 7 *Basis for Projecting the Beginning-of-Mission (BOM) Power.*

Radioisotope Fuel Decay

This mechanism, which decreases both the output power and hot junction temperature, has been described in previous sections and can be predicted with a high degree of accuracy. Radioisotope fuel decay is by far the largest single contributor to the total generator output power degradation (Stapfer 1976).

Dopant Precipitation

Dopant precipitation increases both the resistivity and Seebeck coefficient causing a net reduction in power output. The magnitude of the dopant precipitation is a function of the previous time-temperature history of the thermoelectric elements as well as of the initial dopant level. The dopant is initially added to the solubility limit at the higher temperatures associated with operation. As a result when the unicouple is at lower temperatures the solubility limit is exceeded and precipitation takes place resulting in a reduction in charge carriers and an increase in resistivity and Seebeck coefficient. This precipitation proceeds most rapidly in the N-type material in the range from 623 K to 923 K and in the P material in the range 973 K to 1273 K. Because of these dependencies on time and temperature the exact amount of dopant precipitation at any time is not known exactly. A concept called "effective age" (T_o) has been used to determine the RTG performance with respect to a data base of measurements made on SiGe N and P type elements by Syncal Corporation for JPL. In this approach, the lower the effective age the more rapidly the precipitation process is occurring (GE 1985a, Stapfer 1976, Kelly 1975, Raag 1976, and Kelly 1973).

Thermal Conductivity Changes

The thermal conductivity of the SiGe alloys decreases slightly with time, primarily as a result of a homogenization of the alloys at temperatures above 973 K. The effect of a decreasing conductivity will be a tendency for the hot junction temperature to rise although analyses show that a thermal runaway will not occur even for significant assumed changes in the leg thermal conductivity (Kelly 1973). The extent of the thermal conductivity reduction is dependent on the initial alloy inhomogeneity, including the dopant precipitation. As in the case of dopant precipitation the thermal conductivity will depend upon the previous time-temperature history of the alloys which means

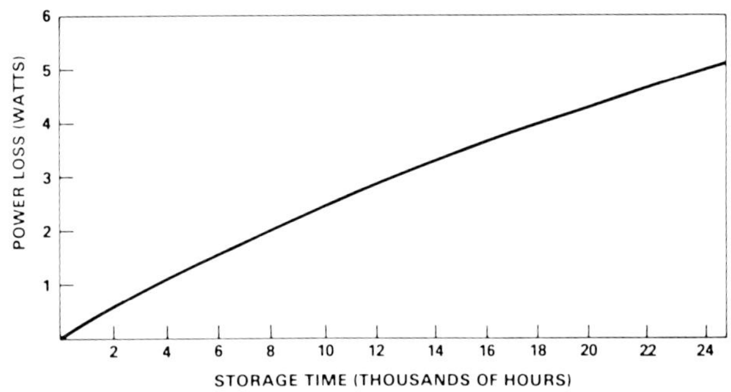

Figure 8 *Storage Power Loss due to Dopant Precipitation in Silicon-Germanium Unicouples. (From GE 1985a.)*

that there will be an "effective age" (k-function) with respect to the SiGe alloy thermal conductivity data base. This effective age (k-function) will be different from the effective age (T_o) for dopant precipitation effects. This difference will be seen in the discussion of the power performance of the generators used on LES 8/9 and Voyager 1/2.

Material Sublimation

At the operating temperatures of the unicouples the potential exists for sublimation of the alloy materials which can lead to a reduction in power output. The Si_3N_4 coating applied to each unicouple retards this sublimation.

One result of sublimation of the thermoelectric legs is a reduction in the cross-sectional area of the legs which in turn results in an increase in the thermal and electrical resistance of the legs and a consequent reduction in power output. The increase in thermal resistance also increases the hot junction temperature which increases the sublimation rate (GE 1985a).

The sublimation products, if allowed to disperse and condense within the foil layers of the insulation system, can increase the thermal conductivity of the multifoil insulation which would decrease the hot junction temperature and reduce the RTG power output (GE 1985a and Stapfer 1976).

The sublimation products can also diffuse along the unicouple legs, through the astroquartz yarn, and form a porous solid between the unicouple legs and the adjacent multifoil insulation. The silicon can react with the astroquartz yarns to produce SiO which dissociates into SiO_2 and Si. The electrically conductive silicon can establish a shunting path between the unicouples and the foil which tends to dissipate energy within the thermopile rather than providing the energy as useful electricity (GE 1985a and Stapfer 1976).

Sublimation can also affect the geometry of the unicouple hot shoes causing changes in both the thermal and electrical performance of the thermopile (GE 1985a).

As noted previously the Si_3N_4 coating as well as the alumina plate pressed against the hot shoe act to retard the sublimation effects.

Carbon Monoxide Effects

It has been established by experiments that the presence of carbon monoxide (CO) will reduce the longevity of the Si_3N_4 coating which is used on the unicouples to suppress the silicon sublimation. The amount of CO which is present in the GPHS-RTG is derived from three sources:

1. The residual gases (including oxygen, water vapor, CO_2, and CO) which can evolve from the heat source graphitic components (these components are preoutgassed to reduce this release to a minimum);
2. CO which is generated by the reaction of free silicon with SiO_2 (astroquartz) which in turn releases SiO which reacts with the graphitic components of the heat source; and
3. Chemical reaction between oxygen transported from the PuO_2 fuel and the carbon of the graphite heat source components (the fuel is kept substoichiometric to reduce this release to a minimum) (GE 1985a and Stapfer 1976).

Table 4 *Projected Flight GPHS-RTG Performance (based on a Beginning-of-Mission of 1 May 1986). (From GE-1985a.)*

	Requirement	F1	F3	F4	F5
Launch Power (W)	200–240	230.5	226.8	226.1	227.1
Time to Achieve 95% Full Vacuum Power	8 h after Venting	3.7 h	3.7 h	3.7 h	3.7 h
BOM (W)					
Galileo	285–300	300.0–303.6		300.8–303.8	302.0–305.2
Ulysses	285–310		302.7–306.7	299.7–302.7	
EOM (W)					
Galileo	255 Min	268.2–276.6		268.9–276.8	270.0–278.0
Ulysses	250 Min		264.9–274.1	262.2–270.5	

Power Output as a function of Time and EOM Power

An extensive data base on the behavior of SiGe unicouples under actual space operating conditions exists as a result of the power measurements made on the Voyager 1/2 and LES 8/9 RTGs (Bennett 1983 and GE 1985a). This data base includes the effects of the mechanisms described in the previous section since the MHW-RTG unicouple is identical to the GPHS-RTG unicouple and both RTGs have a similar molybdenum foil/astroquartz insulation system. Thus, the degradation model parameters that fit the flight data provide the most reliable prediction for the flight performance of the GPHS-RTGs to be used on the Galileo and Ulysses spacecraft (GE 1985a).

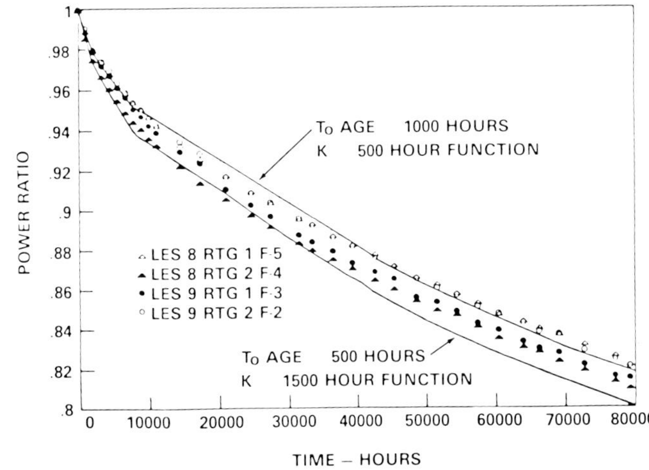

Figure 9 *Power Ratios for the LES 8 and LES 9 Multi-Hundred Watt RTGs as a function of Time. (From GE 1985a.)*

Figure 9 shows the ratio of power output to initial power as a function of time for the LES 8/9 RTGs as determined from flight telemetry data. The upper model curve was based on a 1,000-hour effective age for dopant precipitation effects and a 500-hour effective age for thermal conductivity changes. The lower model curve was based on a 500-hour effective age for dopant precipitation effects and a 1,500-hour effective age for thermal conductivity changes. It should be noted that these curves are not unique solutions in that other combinations of dopant precipitation age and thermal conductivity age could be found to fit the telemetry data; however, the combinations selected provide a good fit to the telemetry data band in the times of interest for the Galileo and Ulysses missions. The computer model was run at the LES 8/9 conditions of 26 V and a constant average daily sink temperature. (In actuality, the daily rotation of the satellites with respect to the Sun causes a variation in the received solar flux which causes temperature and output power changes.) The selected age parameters bracket the data well beyond the 36,800 hours of the Galileo mission and the 41,200 hours of the Ulysses mission. Beyond these times the telemetry data slope is slightly less steep than the model prediction (GE 1985a).

Figure 10 shows the Voyager 1/2 telemetry data power ratios and the model predictions using the same effective ages shown in Figure 9. Again, the model provides a good fit to the telemetry data band in the times of interest for the Galileo and Ulysses missions. As before, beyond these times the telemetry data slope is slightly less steep than the model prediction. Differences between the LES 8/9 and Voyager 1/2 data are caused by differences in the operating voltage level (the LES 8/9 RTGs operate at 26 V while the

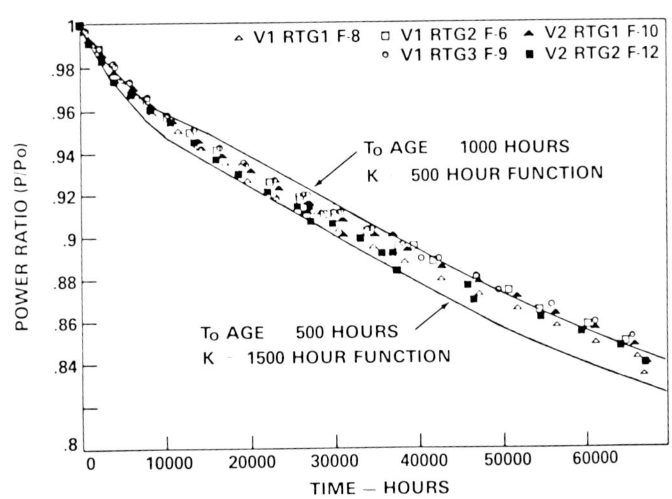

Figure 10 *Power Ratios for the Voyager 1 and Voyager 2 Multi-Hundred Watt RTGs as a function of Time. (From GE 1985a.)*

Voyager 1/2 RTGs operate at 30 V) and in the heat sink temperature (Voyagers 1 and 2 are moving away from the Sun) (GE 1985a).

Additional performance data are shown in Figure 11 for the GPHS-RTG Engineering Unit and in Figure 12 for the Qualification Unit. As of 5 January 1986, over 13,000 hours have been accumulated on the Engineering Unit and 9,850 hours on the Qualification Unit. Since the Qualification Unit did not have the long-term storage that the Voyager 1/2 and LES 8/9 RTGs experienced, a slightly lower effective dopant precipitation age effect is expected. The Qualification Unit did experience a 300-hour thermal vacuum test prior to life testing, which would cause similar dopant precipitation and leg thermal conductivity changes (GE 1985a). The Qualification Unit power is within 0.2% of the model prediction for a 500-hour effective age for dopant precipitation effects and a 1,500-hour effective age for thermal conductivity changes. This minor deviation may be caused by an accidental air leakage that occurred during the testing phase.

The current-voltage-power (I-V-P) characteristics of the Qualification Unit are shown in Figure 13. Note that the 28 V and 30 V operating levels are near the peak power points (GE 1986).

Using the same effective age parameters that bracket the Voyager 1/2 and LES 8/9 telemetry data, the power ratios shown in Figures 14 and 15 were developed for the Galileo and Ulysses missions respec-

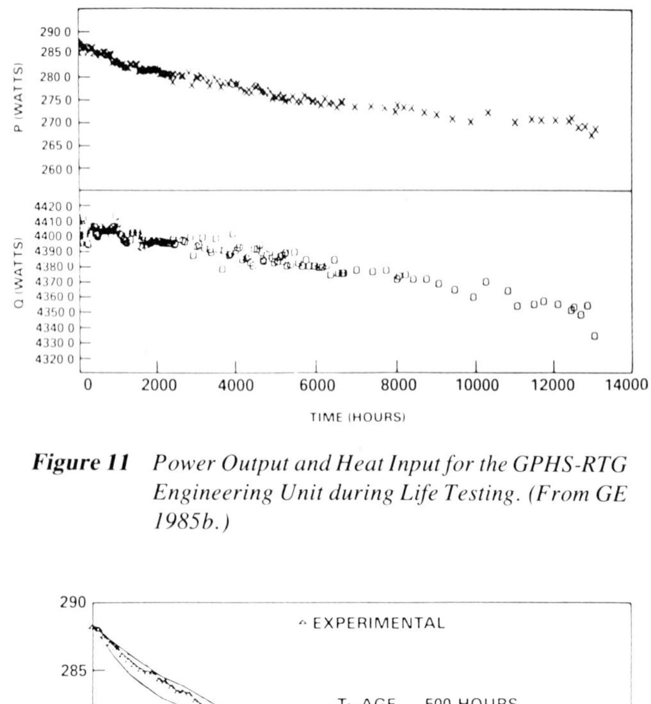

Figure 11 *Power Output and Heat Input for the GPHS-RTG Engineering Unit during Life Testing. (From GE 1985b.)*

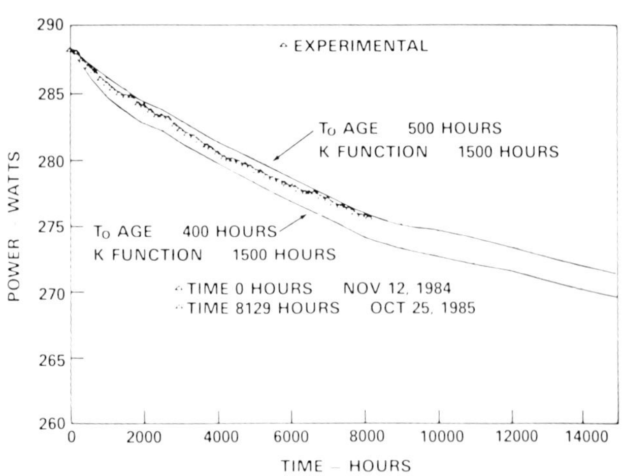

Figure 12 *Power Output of the GPHS-RTG Qualification Unit during Life Testing. (From Deaton 1986.)*

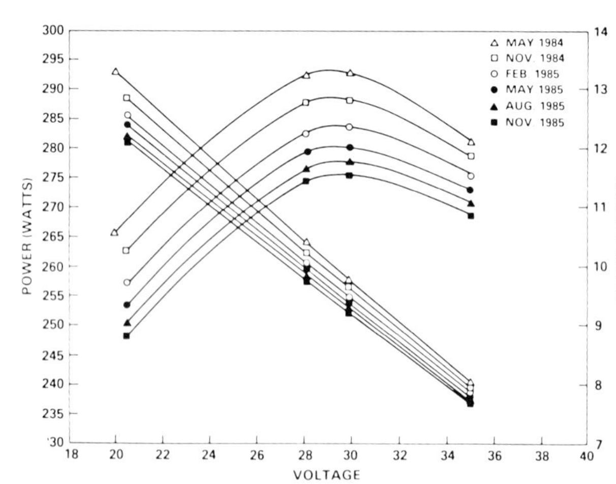

Figure 13 *Map of the Current, Voltage, and Power(I-V-P) for the GPHS-RTG Qualification Unit at Different Times in the Life Test. (From GE 1986.)*

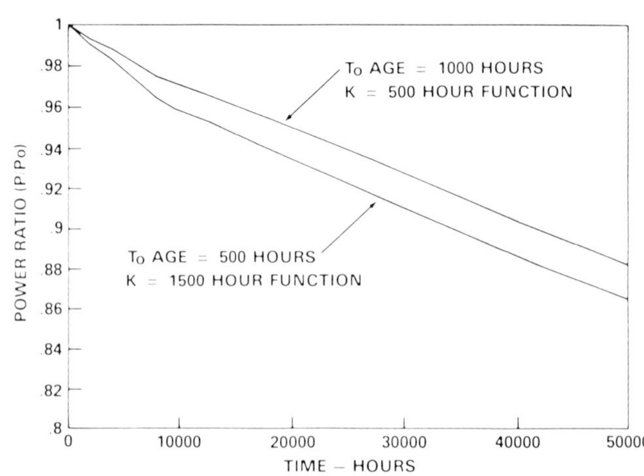

Figure 14 *Projected Power Ratios for the Galileo GPHS-RTGs based on Two Different Combinations of Effective Ages for Dopant Precipitation Effects and Thermal Conductivity Changes. (From GE 1985a.)*

tively. Differences between the two sets of power ratios are caused by differences in the operating voltage level (the Galileo RTGs operate at 30 V and the Ulysses RTG operates at 28 V) and differences in the sink temperature (see Figure 16) (GE 1985a).

These power ratios and the previously described BOM power ranges were used to predict the EOM power for each of the flight RTGs for the Galileo and Ulysses missions as shown in Table 4. The lower value for each RTG is based on the lower estimate of BOM power and the lower power ratio curve. The higher value is based on the higher estimate of BOM power and the upper power ratio curve. The Galileo RTGs (including the backup F4) are projected to exceed the minimum power requirement of 255 We by 13.2 We to 15 We. The Ulysses RTG (including the backup F4) is projected to exceed the minimum power requirement of 250 We by 12.2 We to 14.9 We (GE 1985a).

Conclusion

Based on an extensive ground testing program and the correlation between the predicted and actual performance of the previous SiGe RTGs and test modules over extended times, it is concluded that the GPHS-RTGs to be launched aboard the Galileo and Ulysses spacecraft will meet or exceed the power performance requirements of the two missions.

Acknowledgments

The authors acknowledge with thanks the contributions made by members of the staffs of the DOE Division of Special Applications, General Electric Company, Mound Plant, Savannah River Plant and Laboratory, Los Alamos National Laboratory, Oak Ridge National Laboratory, Sandia National Laboratories, Hanford Engineering Development Laboratory, Ames Laboratory, the associated Department of Energy Operations Offices, Fairchild Industries, NUS Corporation, the Applied Physics Laboratory, and Battelle Columbus Laboratories. Special thanks are due to G. R. Ambrose, R. D. Cockfield, and C. E. Kelly of General Electric Company and R. L. Deaton of Mound for data collection and analyses.

References

Ambrose, G. R., R. D. Cockfield, and C. E. Kelly (1985a) *Qualification Unit RTG Engineering Report*, GESP 7185, Rev "C", General Electric Company, King of Prussia, PA, 21 June 1985.

Ambrose, G. R., R. D. Cockfield, C. E. Kelly (1985b) *F1 Flight Unit RTG Engineering Report*, GESP 7192, Rev. A, General Electric Co., King of Prussia, PA, 18 June 1985.

Ambrose, G. R., R. D. Cockfield, and C. E. Kelly (1985c) *F3 Flight Unit RTG Engineering Report*, GESP 7202, Rev. A, General Electric Co., King of Prussia, PA, 14 October 1985.

Ambrose, G. R., R. D. Cockfield, and C. E. Kelly (1985d) *F4 Flight Unit RTG Engineering Report*, GESP 7199, Rev. A, General Electric Co., King of Prussia, PA, 20 September 1985.

Ambrose G. R., R. D. Cockfield, and C. E. Kelly (1985e) *F5 Flight Unit RTG Engineering Report*, GESP 7193, General Electric Co., King of Prussia, PA, 19 July 1985.

Bennett, G. L., J. J. Lombardo, and B. J. Rock (1981) "Development and Use of Nuclear Power Sources for Space Applications," *The Journal of the Astronautical Sciences*, XXIX (4): 321–342.

Bennett, G. L., J. J. Lombardo, and B. J. Rock (1983) "U.S. Radioisotope Thermoelectric Generator Space Operating Experience (June 1961 - December 1982)" in *Proceedings of the 18th Intersociety Energy Conversion Engineering Conference*, held in Orlando, FL, 21–26 August 1983, p. 1044.

Bennett, G. L. et al. (1986) "The General-Purpose Heat Source Radioisotope Thermoelectric Generator: Power for the Galileo and Ulysses Missions," in *Proceedings of the 21st Intersociety Energy Conversion Engineering Conference*, held in San Diego, CA, 25–29 August 1986.

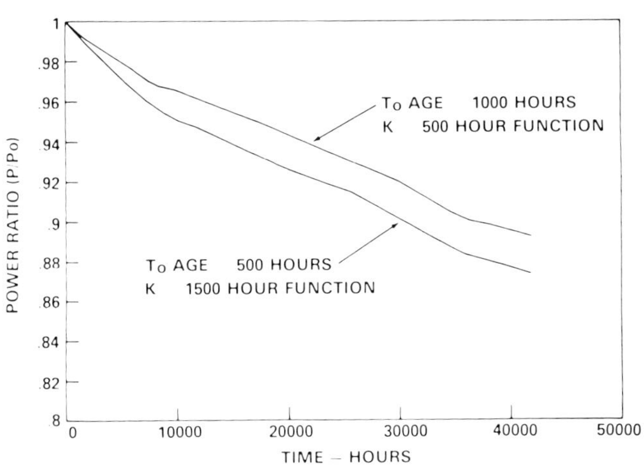

Figure 15 *Projected Power Ratios for the Ulyssess GPHS-RTGs based on Two Different Combinations of Effective Ages for Dopant Precipitation Effects and Thermal Conductivity Changes. (From GE 1985a.)*

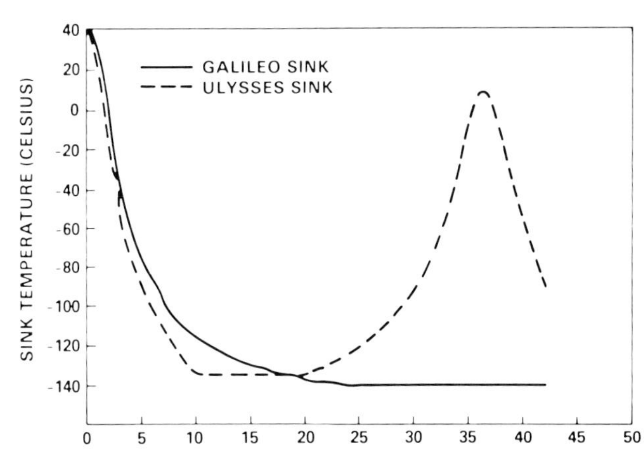

Figure 16 *Calculated Sink Temperatures for the Galileo and Ulysses GPHS-RTGs as a function of Mission Time. (From GE 1985a.)*

Chmielewski, A. B. and V. Shields (1982) "Computer Modeling of Thermoelectric Generator Performance," in *Proceedings of the 17th Intersociety Energy Conversion Engineering Conference*, held in Los Angeles, CA, 8–12 August 1982, p. 1387.

Cockfield, R. D., R. F. Hartman, and C. E. Kelly (1980) "RTG Power Sources for the International Solar Polar Mission," in *Proceedings of the 15th Intersociety Energy Conversion Engineering Conference*, held in Seattle, WA, 18–22 August 1980, p. 1043.

Cockfield, R. D. (1981) "Engineering Development Testing of the GPHS-RTG Converter," in *Proc. of the 16th Intersociety Energy Conversion Engineering Conference*, held in Atlanta, GA, 9–14 August 1981, p. 321.

Deaton, R. L. (1986) *Letter to C. E. Kelly*, General Electric Co., 7 January 1986.

General Electric (1985a) *Final Reliability Assessment for the Galileo/Ulysses General Purpose Heat Source Radioisotope Thermoelectric Generator (GPHS-RTG)*, to be published as GESP XXXX, General Electric Co., King of Prussia, PA.

General Electric (1985b) *Silicon Germanium (SiGe) Radioisotope Thermoelectric Generator (RTG) Program for Space Missions, Sixty-Eighth Technical Progress Report, Period from 1 April 1985 through 30 April 1985*, General Electric Co., King of Prussia, PA, 29 May 1985.

General Electric (1986) *Silicon Germanium (SiGe) Radioisotope Thermoelectric Generator (RTG) Program for Space Missions, Seventy-Fifth Technical Progress Report, Period from 1 November 1985 through 30 November 1985*, General Electric Co., King of Prussia, PA, 6 January 1986.

Kelly, C. E., J. A. Loffreda, and R. Lorentzen (1973) "Performance Models for the MHW Converter," in *8th Intersociety Energy Conversion Engineering Conference Proceedings Addendum*, held in Philadelphia, PA, 13–16 August 1973, p. 37.

Kelly, C. E. (1975) "The MHW Converter (RTG)", in *Record of the Tenth Intersociety Energy Conversion Engineering Conference*, held in Newark, DE, 18–22 August 1975, p. 880.

Kelly, C. E. and G. R. Ambrose (1982) "Testing of the GPHS Electrically Heated Thermoelectric Converter" in *Proc. 17th Intersociety Energy Conversion Engineering Conference*, held in Los Angeles, CA, 8–12 August 1982, p. 1382.

Loffreda, J. A. (1973) *TEG 2: A Modified Version of the LOLLIPOPS Computer Program*, Internal Memorandum PIR ISAO MHW 4663, General Electric Co., King of Prussia, PA, 4 October 1973.

Raag, Valvo (1976) "The Time and Temperature Behavior of the Thermoelectric Properties of 78 a/o Si - 22 a/o Ge Alloy," in *Eleventh Intersociety Energy Conversion Engineering Conference Proceedings*, held in State Line, NV, 12–17 September 1976, p. 1527.

Schock, A. (1980) "Design Evolution and Verification of the 'General-Purpose Heat Source'," in *Proc. of the 15th Intersociety Energy Conversion Engineering Conference*, held in Seattle, WA, 18–22 August 1980, p. 1032.

Shields, V. (1981) "Long Term Behavior of Silicon Germanium Thermoelectric Generators," in *Proc. 16th Intersociety Energy Conversion Engineering Conference*, held in Atlanta, GA, 9–14 August 1981, p. 315.

Stapfer, G. and V. C. Truscello (1976) "The Long-Term Performance Degradation of a Radioisotope Thermoelectric Generator Using Silicon Germanium," in *Eleventh Intersociety Energy Conversion Engineering Conference Proceedings*, held in State Line, NV, 12–17 September 1976, p. 1533.

Space Nuclear Power Systems 1986
Edited by M. S. El-Genk and M. D. Hoover

Chapter 43

Technology and Hardware Status of Advanced Modular Radioisotope Generator Development

Robert F. Hartman, Paul D. Gorsuch, and Jerry R. Peterson

General Electric Company
Space Systems Division
Valley Forge, PA 19101

The Modular RTG Program was initiated in October 1983. The program is sponsored by the U. S. Department of Energy and is directed toward improving current state-of-the-art radioisotope thermoelectric generator technology. This paper presents the program's current hardware and technology status. New technology has been developed to reproducibly fabricate thermoelectric multicouple devices. Construction of all major components, with the exception of the multicouple, needed to assemble a Ground Demonstration System is nearing completion. The Ground Demonstration System contains 48 multicouple devices, produces 96 watts electric, and is expected to confirm a specific power greater than 6.6 W/kg. The Ground Demonstration System will establish the data base needed to verify performance and long-life predictions for a MOD-RTG flight program.

Introduction

Radioisotope Thermoelectric Generators (RTGs) have been employed as nuclear-electric power sources for a variety of space missions since the 1961 launch of the RTG-powered Transit-4A spacecraft. SNAP-27 RTGs were used to power the experiments left on the moon by the Apollo astronauts.

Multi-Hundred Watt (MHW) RTGs were employed for the Lincoln Experiment Satellite (LES 8/9) communication spacecraft launched in 1976 and the Voyager 1 and 2 outer planet scientific missions, launched in 1977. On these last two missions alone, nearly 300 million thermocouple hours of operation in space have been achieved to date. Current technology General Purpose Heat Source (GPHS) RTGs have been built and tested for use on the upcoming Galileo mission to Jupiter and the Ulysses mission to the sun's polar region.

The U. S. Department of Energy has sponsored a program of RTG technology improvement, resulting in a pattern of exceptional reliability and long life on the missions flown to date, as well as continuous improvements in performance and safety. Voyager and LES satellites are still operational after more than 9 years in space. The performance levels attained have improved from about 2.2 W/kg for the earliest units to 5.3 W/kg for the current technology GPHS-RTG.

The Modular Radioisotope Thermoelectric Generator (MOD-RTG) is under development for future space missions. The MOD-RTG, targeted to achieve performance greater than 7.7 W/kg, embodies improved thermoelectric materials and a "multicouple" thermoelectric converter that incorporates 20 thermoelectric conversion elements in a single device. The multicouple approach allows design modularity as well as significant performance improvement.

This paper presents the hardware and technology status achieved by the ongoing MOD-RTG development program. Multicouple and materials technology has progressed to the point where multicouples are in production and fabrication of a Ground Demonstration System (GDS) converter is under way. The GDS, which will produce 96 We and contains 48 multicouples, is expected to confirm specific power capability greater than 6.6 w/kg and establish readiness for production of flight units for early 1990s space missions.

Ground Demonstration System Overview

The MOD-RTG Ground Demonstration System (GDS) program was initiated in October 1983. Its objective is to develop, fabricate and test an electrically heated, modular GDS that will project a specific power of greater than 7.7 W/kg for an equivalent

MOD-RTG flight system. The MOD-RTG flight design and GDS detail design have been described by Hartman et al. (1985). The overall size and power level of the GDS are designed to demonstrate the modularity of the MOD-RTG design; that is, scalability of power level. The GDS is designed to be a reliable power system with a minimum five-year operating life. In addition, it is designed to meet the NASA Shuttle launch environments.

The GDS will undergo performance and dynamic testing at General Electric Company. Subsequently, it will be delivered to Fairchild Industries, in Germantown, MD, where it will be placed on long-life test.

The GDS is shown in Figure 1 and its design attributes summarized in Table 1. It is designed to contain the equivalent of six stacked GPHS heat source modules. In effect, it is approximately a one-third model of the reference MOD-RTG flight design. It incorporates the key design features and its components are flight prototypical. The major differences between the GDS and the flight design are

- The GDS is electrically operated,
- The GDS is heavily instrumented and requires a special instrumentation end dome, and
- The GDS does not include a Pressure Relief Device (PRD) to vent the converter after launch.

The GDS utilizes a thermoelectric multicouple (Figure 2) to convert the thermal energy from the heat source to usable electrical energy. A total of 48 multicouples are interconnected in a series-parallel network for redundancy. The multicouples penetrate a molybdenum multifoil insulation system that surrounds the electric heat source. All are contained within a finned aluminum outer shell. The outer shell contains external coolant tubes, required for the flight

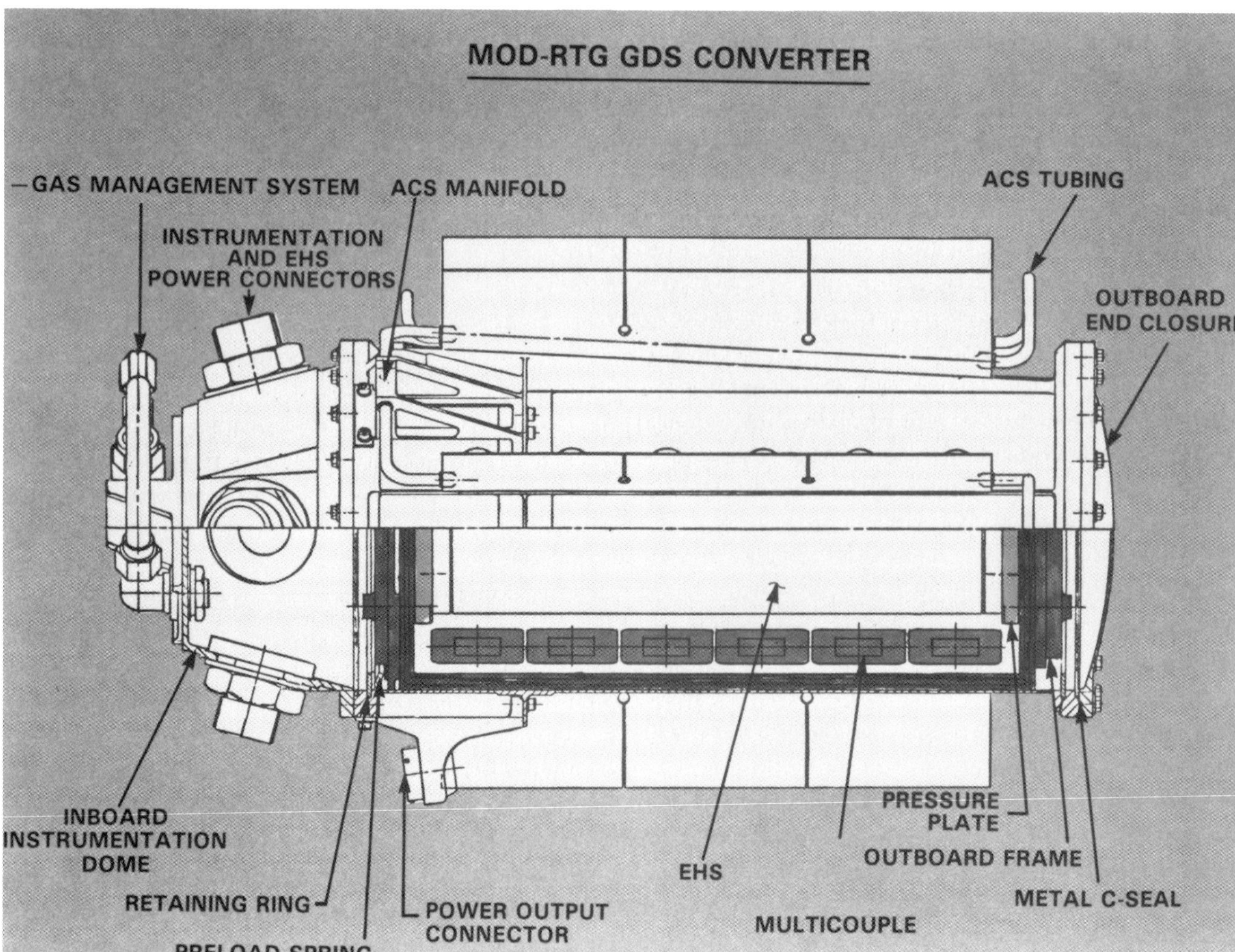

Figure 1 *Ground Demonstration System Converter Design.*

design, to make it thermally compatible with Shuttle launch requirements.

Ground Demonstration System Technology Status

The GDS design utilizes current state-of-the-art thermoelectric and structural materials. However, new technology was required to reproducibly fabricate the multicouple device and achieve the required performance. Manufacturing process development has been completed and manufacturing specifications have been prepared for production of the GDS multicouples. The following paragraphs summarize the multicouple design and current technology status of the multicouple device.

The multicouple design (Figure 3) consists of a graphite heat collector, 20 thermoelectric couples connected electrically in series, a graphite cold cushion and a mounting stud for attachment to the GDS outer shell. A hybrid combination of thermoelectric materials; namely, silicon-germanium gallium phosphide type N and silicon-germanium type P, were selected for the multicouple device. Analysis of current thermoelectric property data showed the performance of the hybrid design to be 13% greater than a design us-

Table 1 *GDS Design Attributes and Predicted Performance.*

Number of Equivalent GPHS Modules	6
Number of Multicouples	48
Length	0.44 m
Overall Diameter	0.33 m
Operating Life	5 yr
Optimum Load Voltage	28.6 V
Open Circuit Voltage	52.0 V
Power Output	96 We
Cold/Hot Junction Temperature	573 K / 1273 K
Converter Efficiency	6.2%

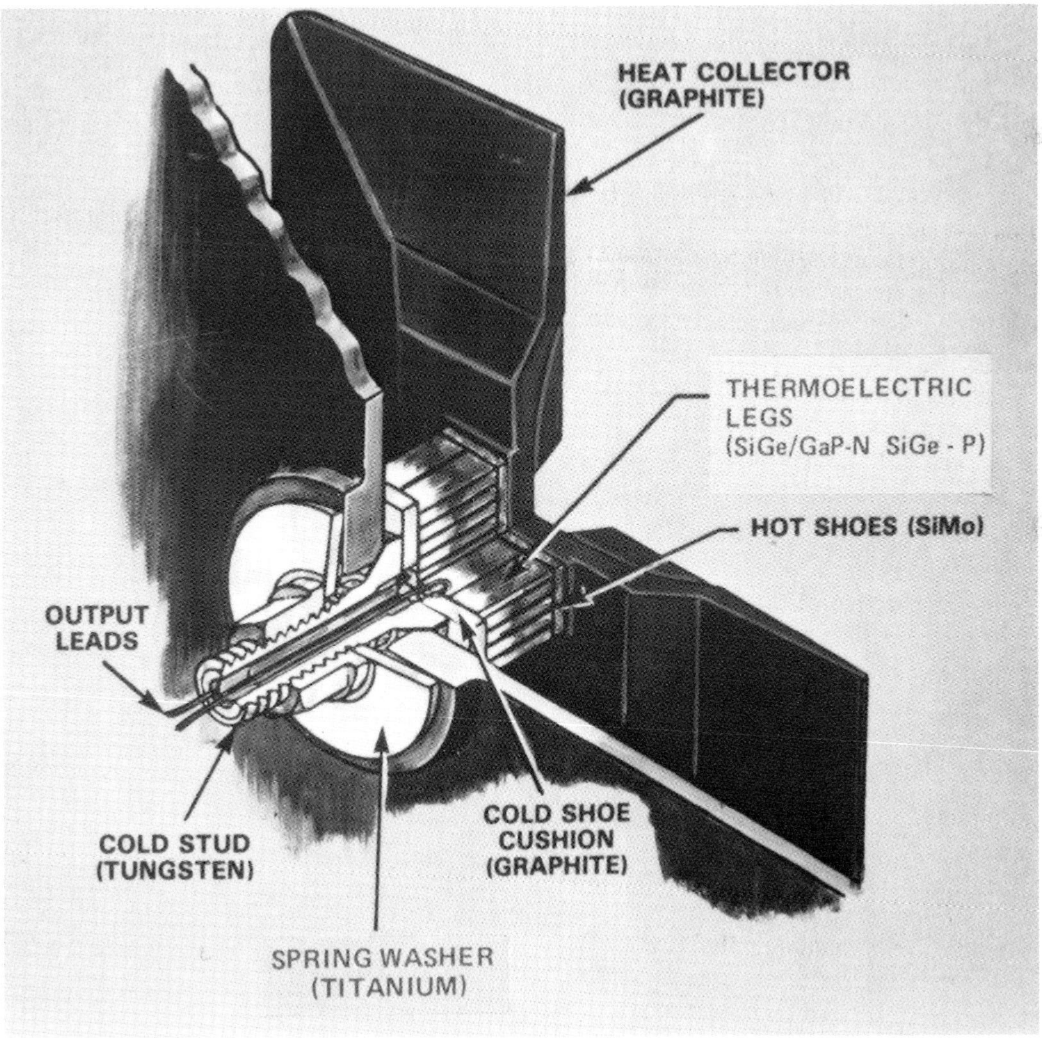

Figure 2 *Exploded View of Multicouple.*

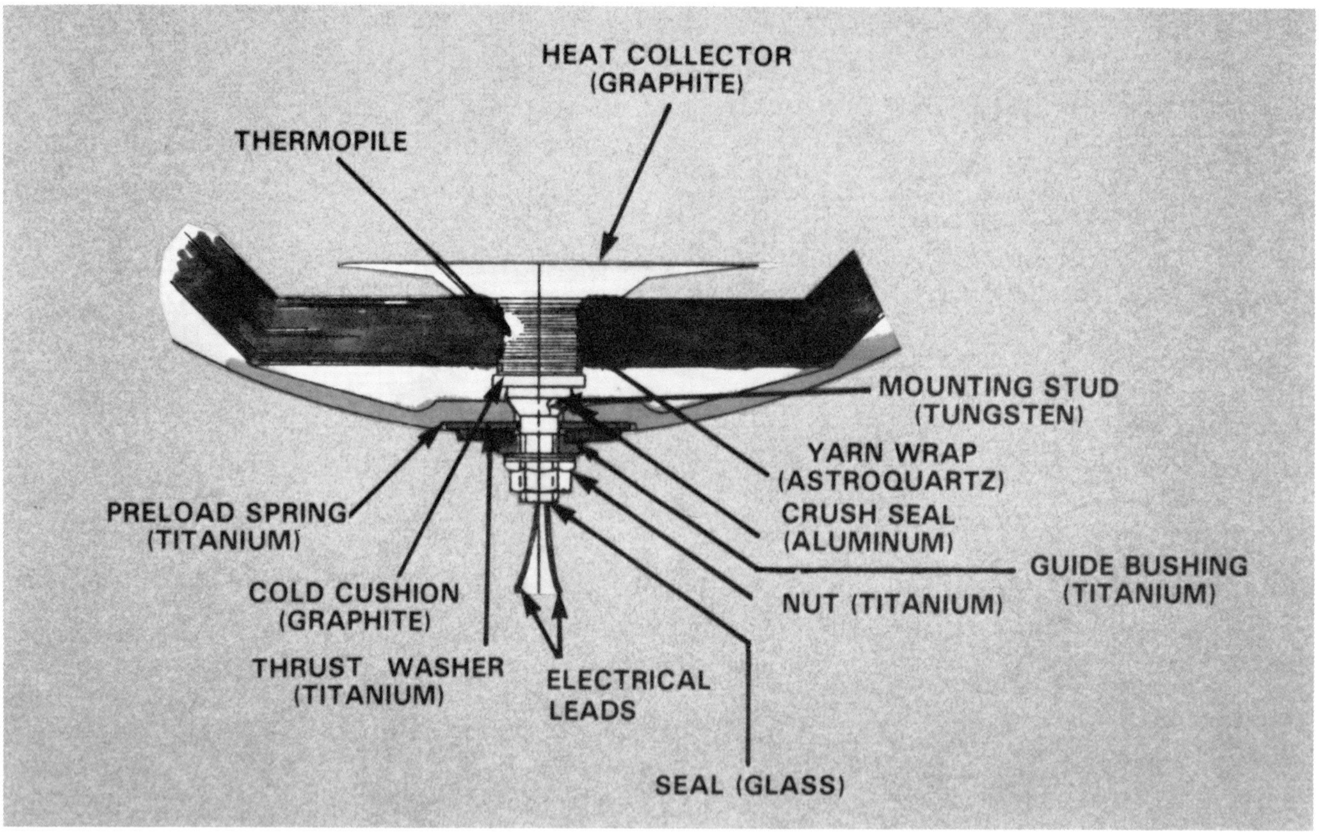

Figure 3 *Multicouple Cross Section.*

ing silicon-germanium gallium phosphide type N and P materials.

The heat collector maintains the heat source at 1373 K, with the thermoelectric hot junction at 1273 K. The heat collector is bonded to the thermoelements with an alumino-silicate high-temperature glass. This bond is critical to the performance of the multicouple because it provides electric isolation. The bond must minimize temperature drop and must have adequate structural integrity to withstand loading due to thermal stresses and dynamic environments anticipated for a Shuttle launch. The bond is controlled to a thickness of $50\mu m \pm 20/5\mu m$ ($0.0020 \pm 0.0008/0.0002$ inches). Each bond is mechanically proof-load tested to ensure adequate structural integrity.

Hot pressed compacts of SiGe/GaP type N and SiGe type P material are wafered and sliced to the final thermoelement thickness. N and P slices, screen printed with an alumino-silicate glass, are bonded to form an N/P sandwich (Figure 4). Silicon molybdenum hot shoes, screen printed with germanium ink, are bonded to the N/P sandwich to provide high temperature contacts (Figure 5). Five hot shoe bond su-

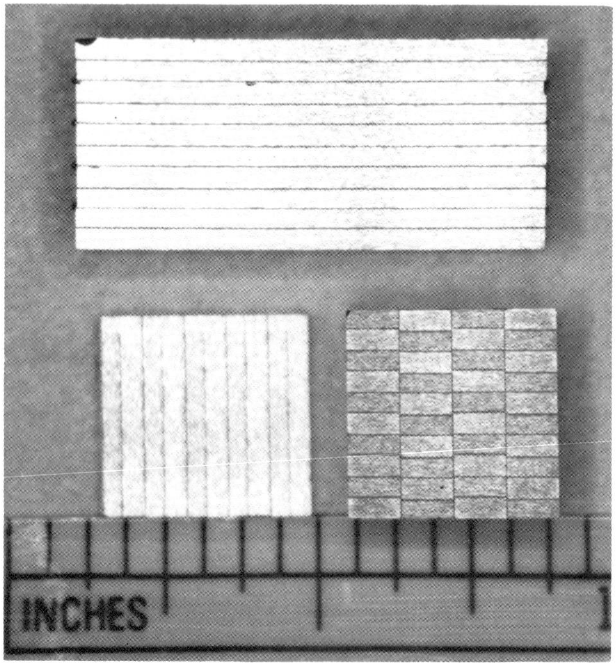

Figure 4 *Multicouple N/P Sandwich.*

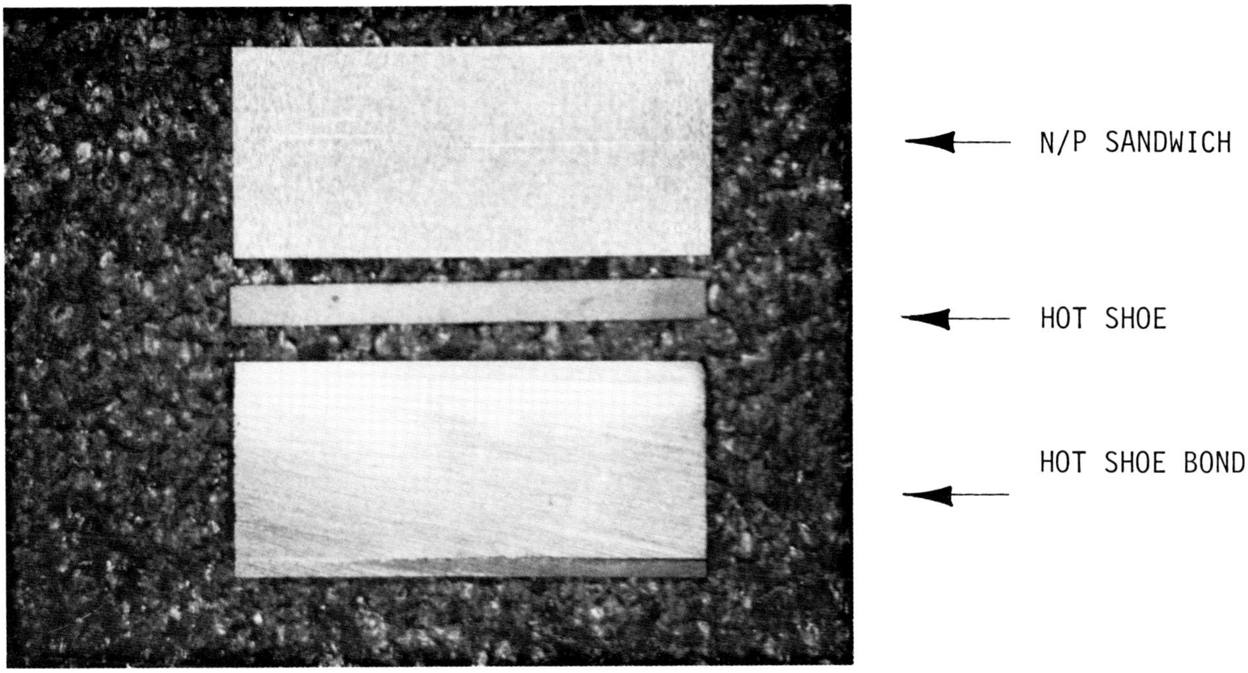

Figure 5 *Multicouple Hot Shoe Bond.*

bassemblies, each separated with high temperature glass, are stacked and bonded to form a 5-sandwich stack (Figure 6). These are sliced to form 5-couple slices (Figure 6), which are subsequently arranged in a 4 by 10 array and bonded with high temperature glass to form a thermopile (Figure 6). Silver plated tungsten contacts are bonded to the low temperature end of the thermopile to complete the contacted thermopile. Molybdenum wire is brazed to the centrally located thermoelements to provide the power leads. The outer surfaces of the thermopile are coated with glass to suppress sublimation of the thermoelectric materials. The graphite cold cushion is bonded to the thermoelements using a low temperature solder glass to maintain electrical isolation. It, in turn, is brazed to the tungsten mounting stud. The power leads exit through the tungsten mounting stud. Glass headers isolate the lead wire from the mounting stud and provide a hermetic seal. A low melting temperature glass has been developed for this purpose.

The multicouples are installed into the outer shell using a titanium disc spring and nut. A disc spring is required to compensate for differential thermal expansions between multicouple and shell while maintaining sufficient load for heat transfer. A conical seal is used to hermetically seal the multicouple penetration through the shell. Astroquartz yarn surrounds the

Figure 6 *Multicouple Sub-Assemblies: 5 Sandwich Stack (above), 5 Couple Slice (left), and Hybrid Thermopile (right).*

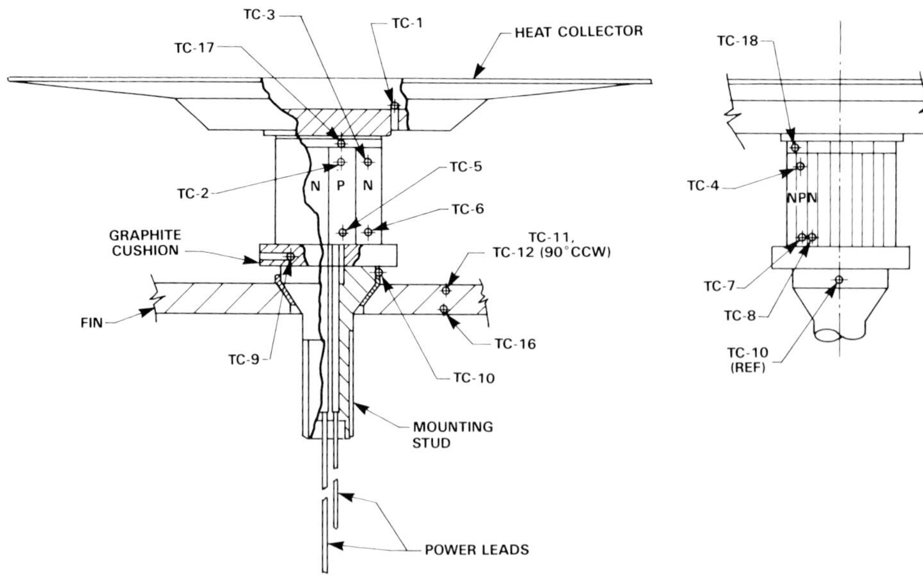

Figure 7 *Schematic Diagram of the Instrumented Multicouple.*

multicouple thermopile to provide electrical isolation with the multifoil insulation and to provide compliance during dynamic loading.

Several hybrid multicouples were fabricated for design verification performance testing. As shown in Figure 7, these multicouples, designated M/C 107 and 108, were uniquely equipped with diagnostic instrumentation to permit in-depth analysis of multicouple performance. Thermocouples, 76-μm (0.003-inch) in diameter, were mounted in the hot shoes, thermoelectric legs and cold cushion to permit measurement of temperature drops across critical bond interfaces. These leads also served as voltage taps to measure individual leg resistances and in-situ thermoelectric properties (Seebeck and resistivity). Both multicouples were operated in-gradient for more than 1500 hours.

The normalized power output as a function of operating time is shown in Figure 8. Initial measured performance was approximately 9% below the predicted performance. Analysis of the instrumentation data indicates that the temperature drops through the glass interfaces account for approximately 3% of the shortfall. The low temperature glass interface between thermopile and graphite cold cushion appears to be the principal contributor. The remaining power loss appears to be a variation in the electrical properties of the thermoelectric materials. Testing also indicates improved long term stability of the hybrid device compared to the non-hybrid configuration. An average degradation of 5.2% after 1000 hours was measured for the hybrid multicouples compared to an average degradation of 9.2% for non-hybrid multicouples. GE has delivered eight multicouples from the GDS production lot to Fairchild Industries for long life testing. Testing is scheduled to begin the first quarter of calendar 1986.

Ground Demonstration System Hardware Status

The GDS is scheduled to be assembled, instrumented and processed at GE. Construction of all the major components, with the exception of the multicouple, is nearing completion. The following para-

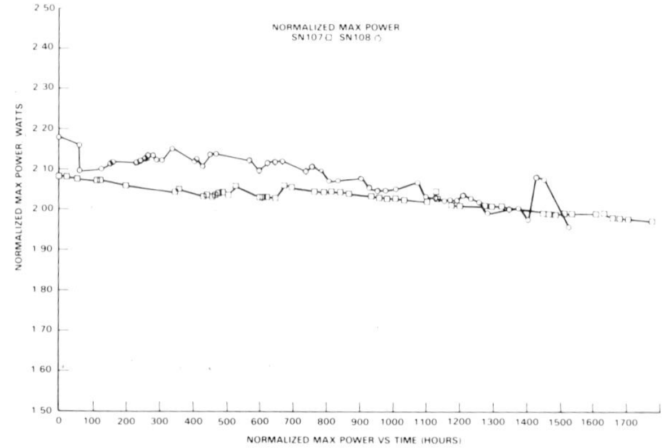

Figure 8 *Multicouple Normalized Maximum Power Output (Watts) versus Time (Hours).*

graphs describe the status of each of the major components.

The GDS will be operated with an electric heat source (EHS) that duplicates the thermal inventory, size and mass of six stacked GPHS heat source modules. The design consists of a graphite heating element separated by silicon nitride insulator supports. Tungsten end caps are used to duplicate the required mass and a graphite shell simulates the GPHS module aeroshell. The EHS hardware has been fabricated and is ready for use in the GDS. The hardware is shown in Figure 9.

A variety of manufacturing tooling is required to fabricate the molybdenum multifoil insulation system. The tooling required consists of more than 125 trim templates, punch and dies, and lay-up and transfer fixtures. All tooling has been manufactured (see Figure 10) and fabrication of the multifoil insulation is under way. The multifoil insulation system design requires that zirconia powder be electrostatically sprayed onto the side of each of the 60 layers of foil. A facility has been constructed to perform this function and the process parameters have been developed.

Multicouple construction is in progress; a total of 48 multicouples are being fabricated for the GDS. All mounting hardware required to install the multicouples into the GDS have been fabricated and are shown in Figure 11.

Hardware for the GDS heat source support system has been fabricated and is shown in Figure 12. Titanium components (preload frames, disc spring, and retaining ring) are utilized to minimize system mass. Graphite pressure plates are used to interface with the end caps of the heat source.

The GDS outer shell is a single-piece construction manufactured from 2219 aluminum alloy. Radiating fins are electron beam welded to the outer shell. Cooling channels are machined integral into the fin and are interconnected for auxiliary cooling by tubing and manifolds. Fabrication of the outer shell, radiating fins, and other components, is complete (see Figure 13). Welding operations are in progress to complete the shell/fin assembly.

End domes are used to seal the ends of the GDS converter. The inboard dome contains the gas management system required to pressurize the GDS, and electrical connectors to provide heater input power and to interconnect internal instrumentation. All hardware has been fabricated and is shown in Figure 14.

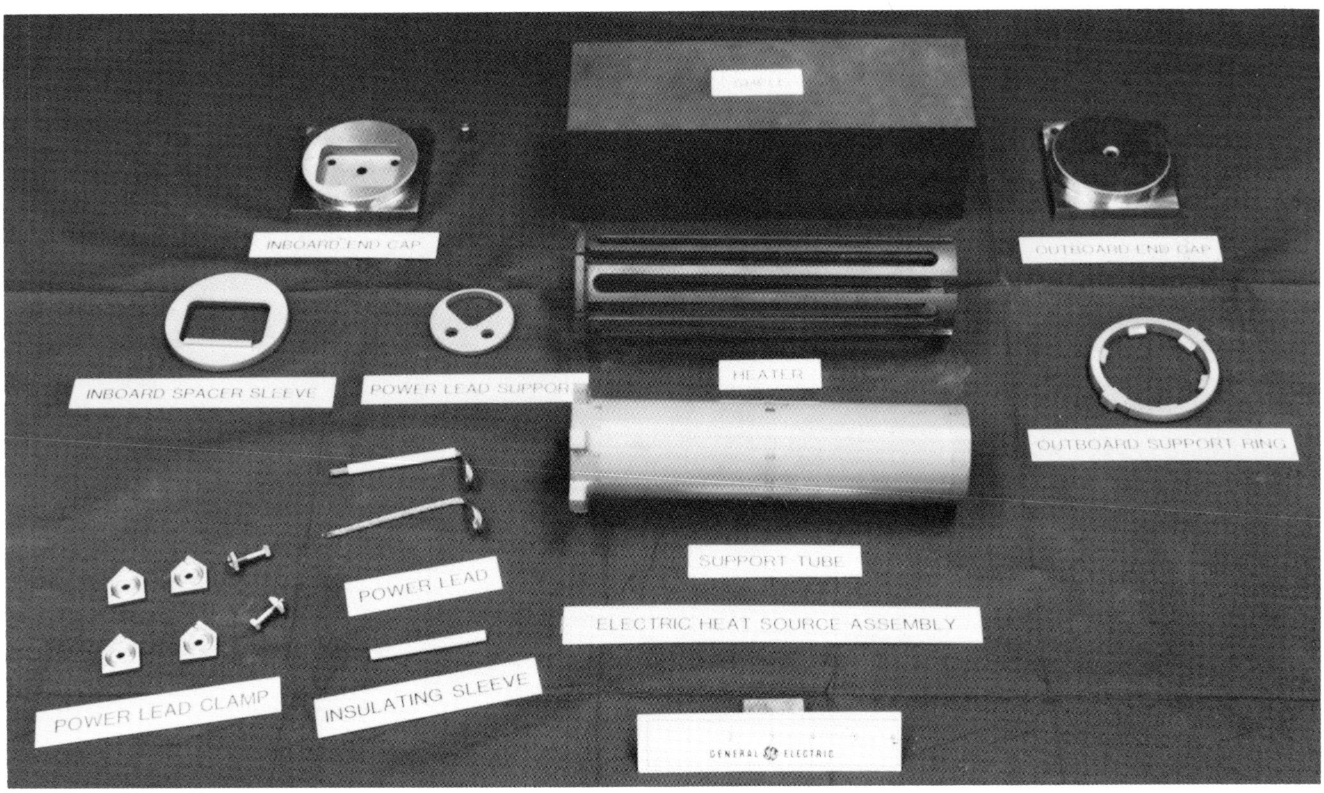

Figure 9 *GDS Electric Heat Source Components.*

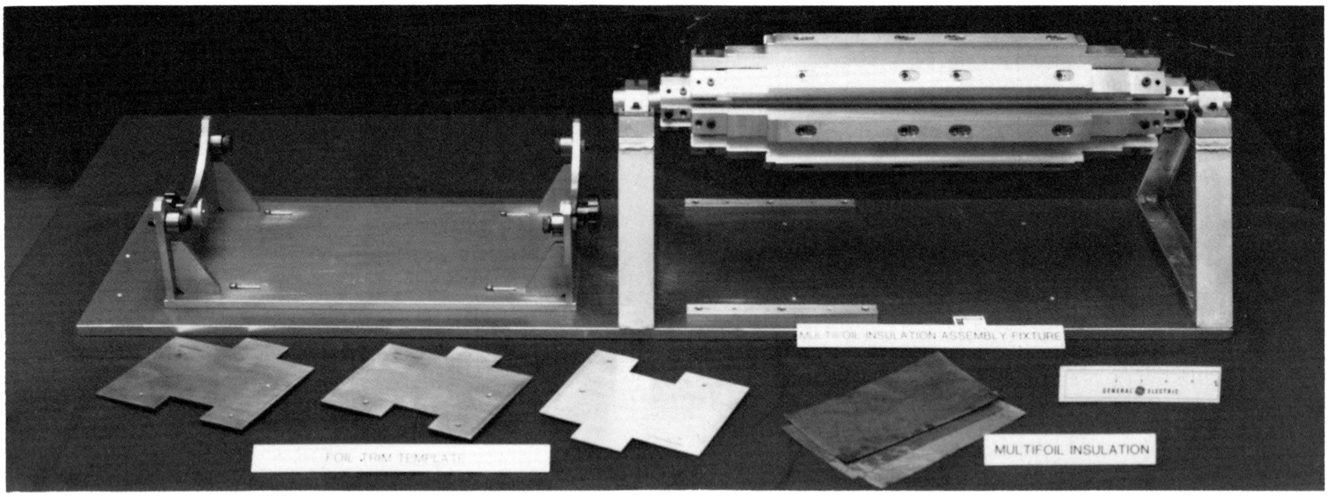

Figure 10 *GDS Multifoil Insulation Manufacturing Fixtures.*

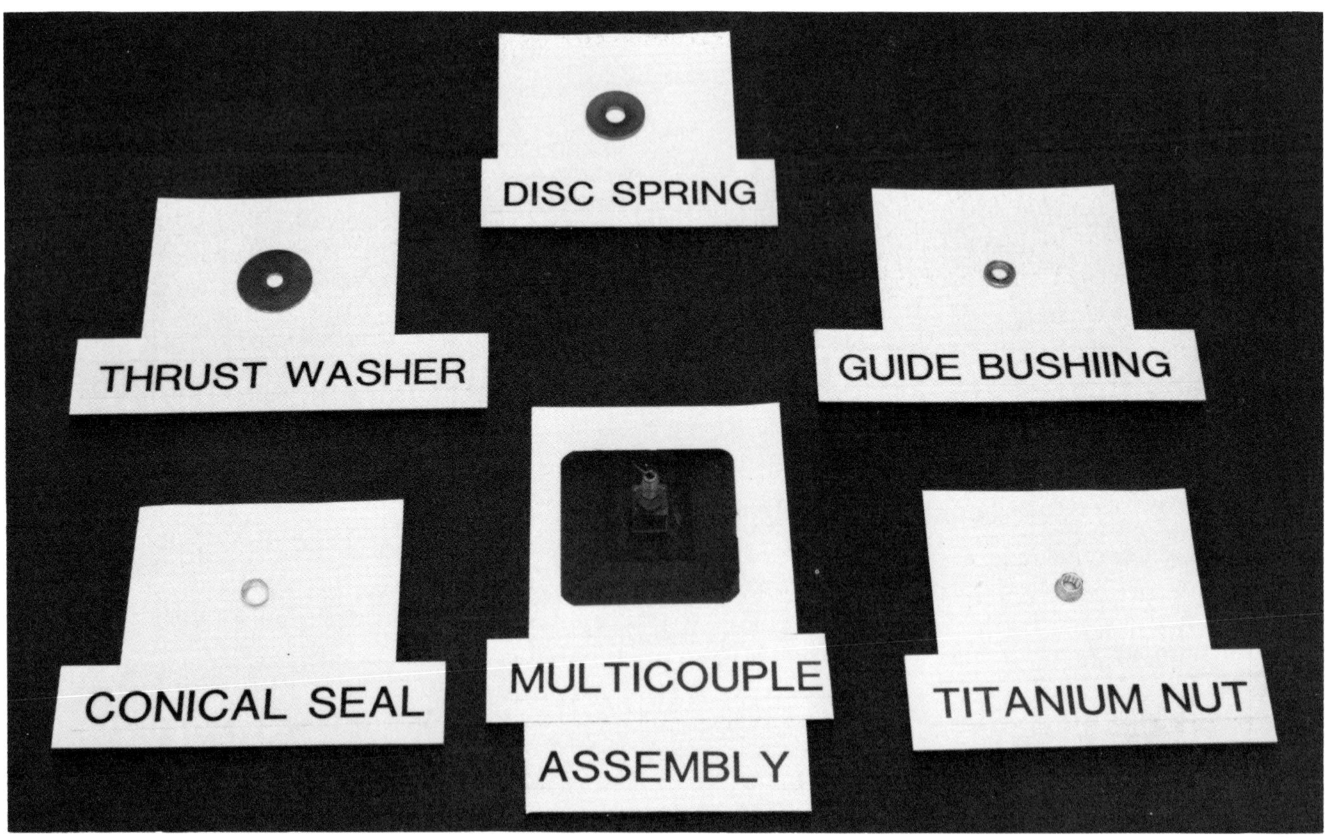

Figure 11 *GDS Multicouple and Mounting Hardware.*

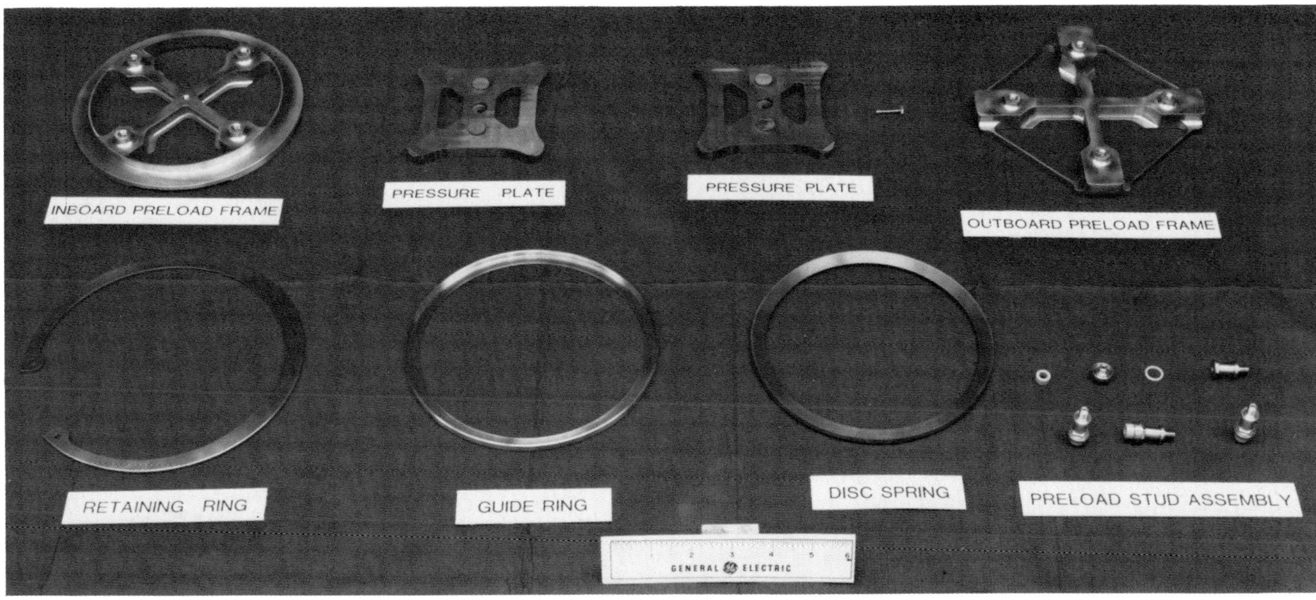

Figure 12 *GDS Heat Source Support Hardware.*

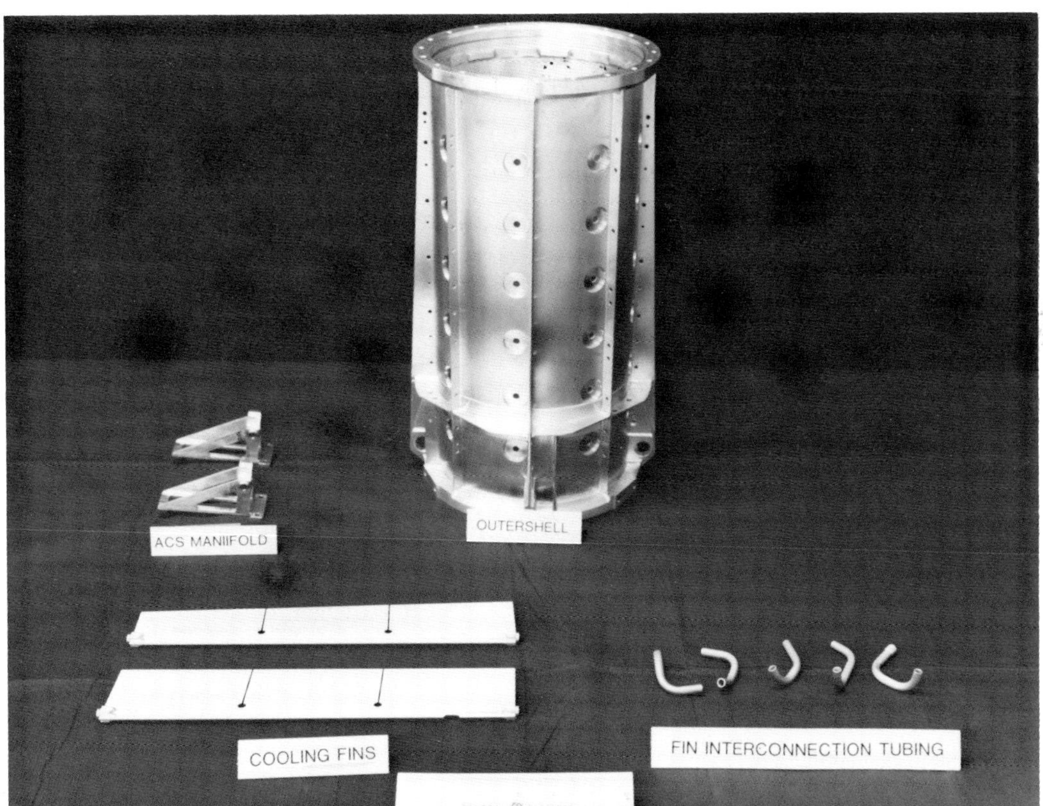

Figure 13 *GDS Outer Shell, Cooling Fins and Interconnection Hardware.*

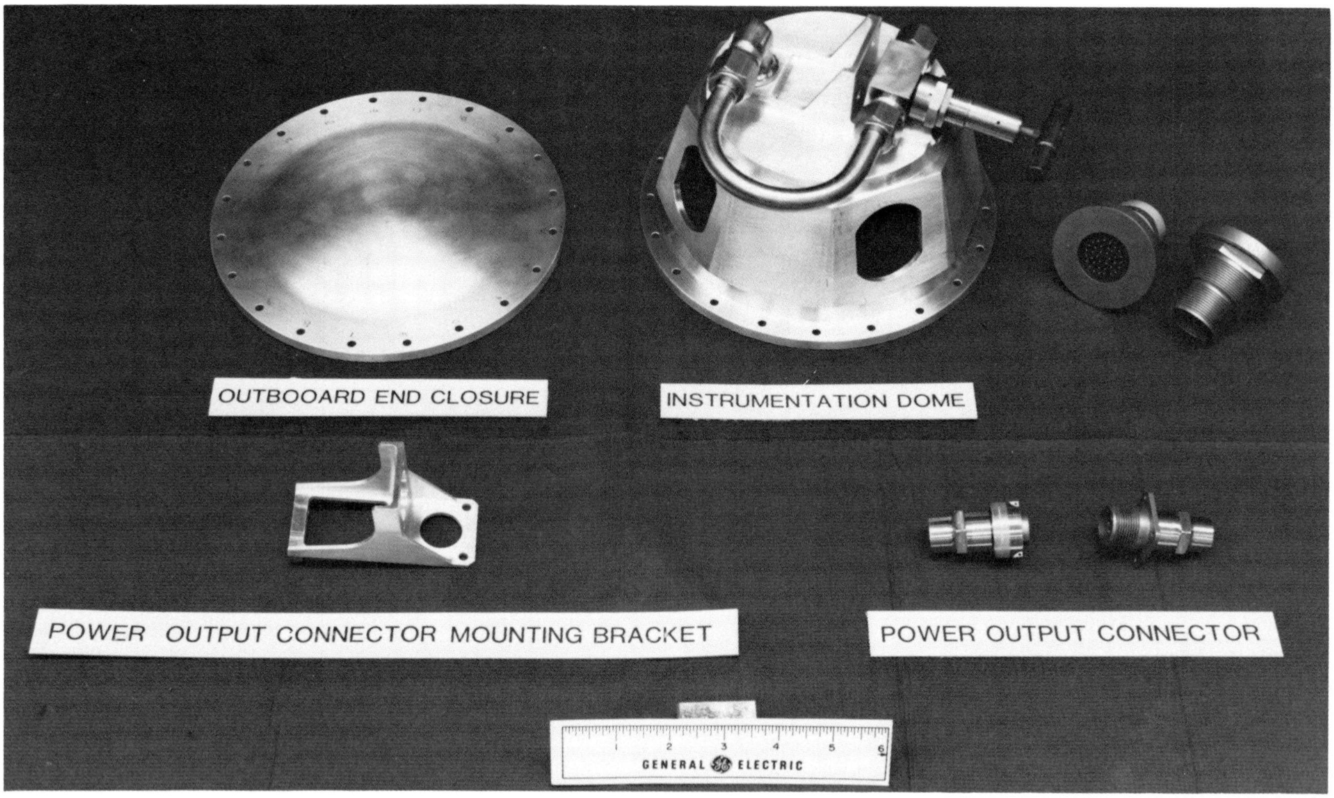

Figure 14 *GDS End Domes and Connector Hardware.*

Conclusion

The Ground Demonstration System is the precursor for a MOD-RTG flight program. Critical new technology has been developed. Manufacturing drawings, specifications and tooling are in place, and most of the required hardware has been built. Fabrication and testing will establish the data base needed to verify performance and long life predictions for a MOD-RTG flight program.

Acknowledgments

The authors gratefully acknowledge the contributions and assistance provided by the MOD-RTG/GDS team members at General Electric Company.

References

Hartman, R. F., E. Kelly, and A. Kull (1985) ''Modular Radioisotope Thermoelectric Generator Ground Demonstration System,'' in *Proc. Twentieth Intersociety Energy Conversion Engineering Conference*, held in Miami, FL, August 1985.

Space Nuclear Power Systems 1986
Edited by M. S. El-Genk and M. D. Hoover

Chapter 44

Long-Life, Highly Survivable Rankine Cycle Dynamic Isotope Power Systems

R.E. Niggemann and P.D. Lacey

Sundstrand Advanced Technology Group
4747 Harrison Avenue
Rockford, IL 61125

The requirements of high reliability, long life and high survivability are all critical parameters that must be considered when designing a satellite power system today. High survivability, coupled with today's higher satellite power requirements, drives the selection of the satellite power system away from conventional solar arrays. Using radioisotope fuel as the heat source and a turbogenerator for the electric power conversion, high survivability can be inherently achieved through integrated satellite design. High reliability of the turbogenerator dynamic isotope power system (DIPS) can be accomplished through component and subsystem redundancy. This paper will address methods of attaining high reliability in a dynamic power source in addition to discussing how high survivability can be achieved through design.

Introduction

The characteristics of the long-life, highly survivable DIPS using organic Rankine cycle (ORC) power conversion systems are presented along with a description of the approaches to enhance survivability. While the data describe systems in the 1- to 10-kWe range, larger systems could be built if required.

Integration of DIPS into the spacecraft and its impact on the overall spacecraft is also discussed.

Dips Configuration

To achieve long life and reliability goals, the system must include redundancy. Sufficient fuel must be included so that at the end of life, energy from the isotope is sufficient to produce rated power. Both of these requirements result in increased system mass compared to systems with shorter lifetimes.

The major elements that make up the system are

- Heat source assembly (HSA) consisting of the isotope heat source and vaporizer;
- Power conversion unit (PCU), which consists of three major elements:
  - —Combined rotating unit (CRU): turbine pump, alternator,
  - —Rotating fluid management device (RFMD), and
  - —Regenerator;
- Condenser; and
- Radiator.

The high reliability necessary for long life is achieved by providing redundancy with two completely independent ORC systems. Figure 1 is a block diagram of such a system. Redundancy in the vaporizer, PCU, and condenser ensure a highly reliable system. Because of their inherent high reliability, redundancies in the HSA and heat pipe radiator are not required. The heat pipe radiator may be oversized to allow for failure of some heat pipe elements while maintaining rated power output. The two fluid loops are completely independent.

Figure 2 summarizes a reliability assessment for this configuration. An overall reliability in excess of 0.99 can be achieved for a system with a 10-year life.

Several working fluids could be considered for this application. The choice will depend on several factors, with the key criteria being power level and optimization.

Figure 3 shows the comparison between 5-kWe systems using toluene and Dowtherm A® working fluids. (Dowtherm A is a Registered Trademark of Dow Chemical Company.) The curves indicate that different ranking of optimization criteria will affect the selection of the optimum working fluid. As power levels

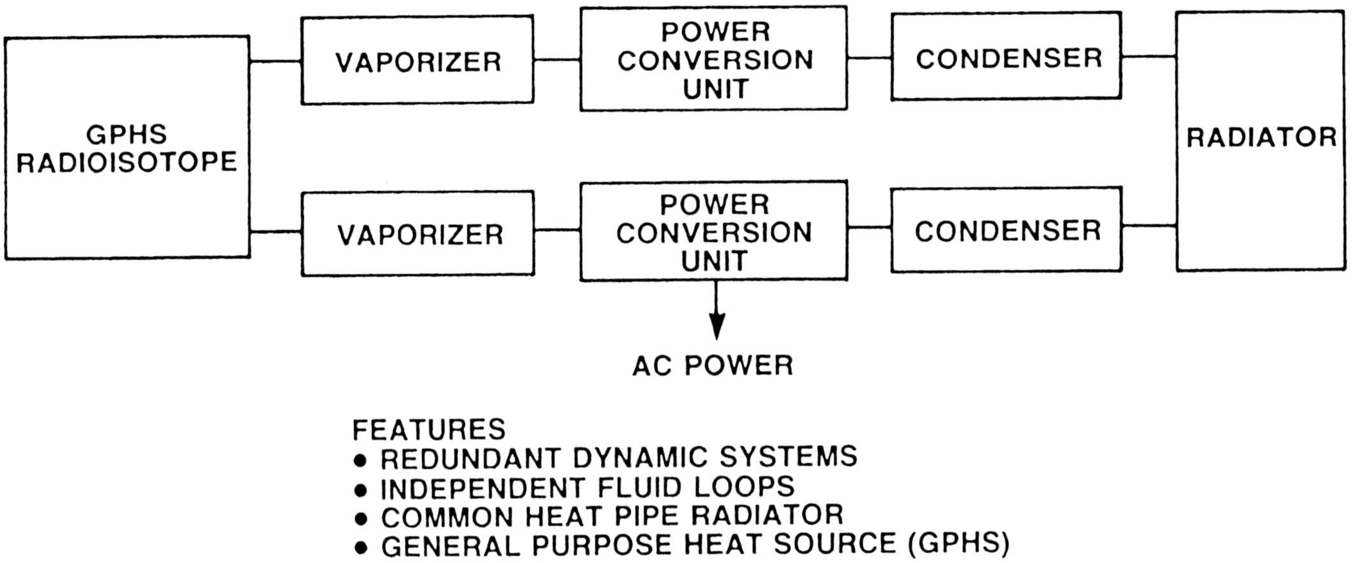

Figure 1 *Block Diagram of the Organic Rankine Cycle Dynamic Isotope Power System.*

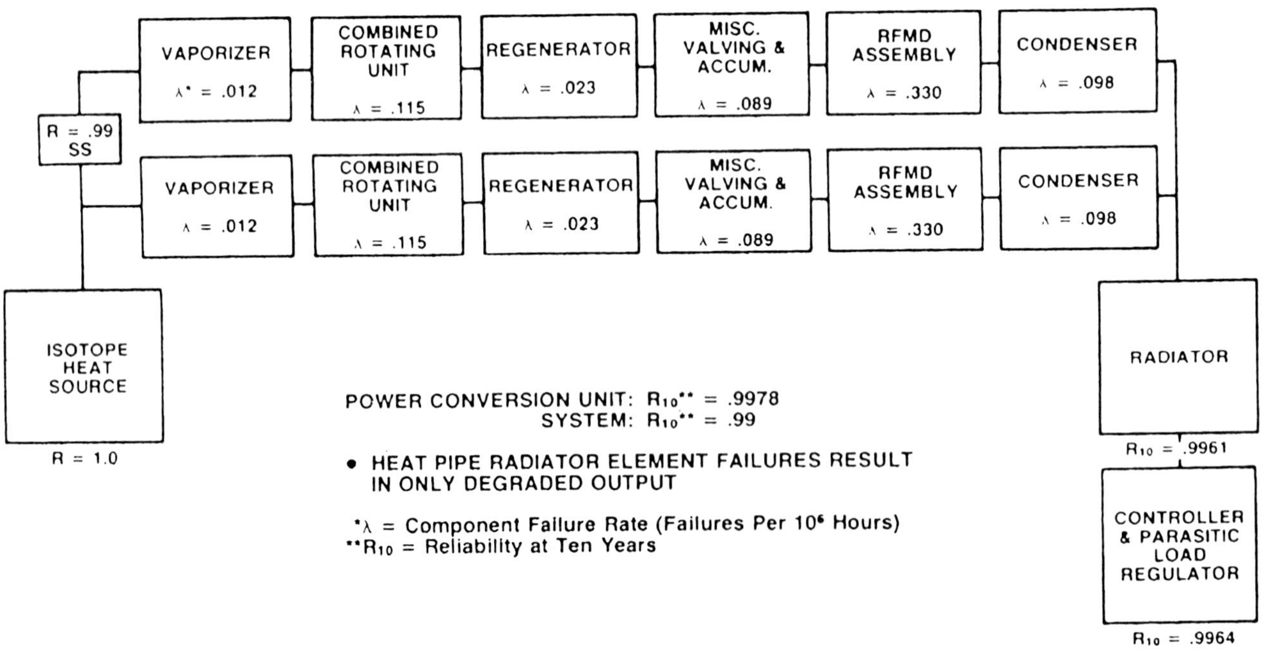

Figure 2 *Dynamic Isotope Power System Reliability Block Diagram.*

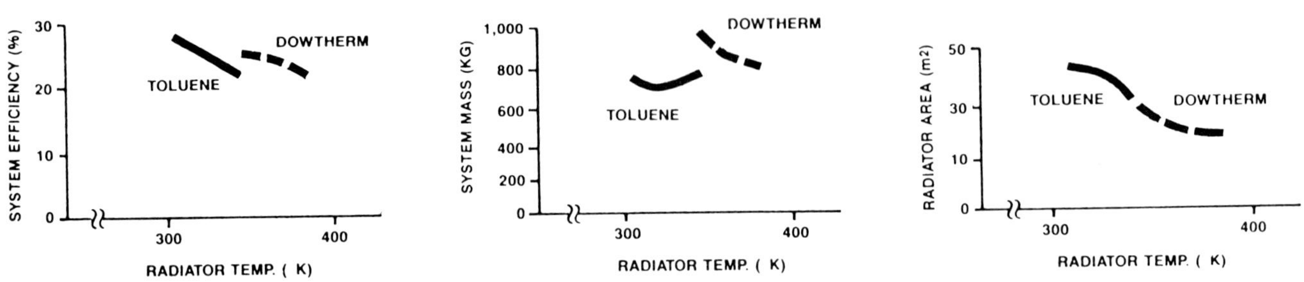

Figure 3 · *5-kWe System Fluid Properties Comparison.*

increase, toluene systems become more attractive; while at low power ratings, Dowtherm A will most likely be the optimum choice.

Operating Characteristics

The operating state points and system schematic for a 5-kWe DIPS are shown in Figure 4. The overall system characteristics are

- ORC Efficiency: 22.5%,
- Radiator Area: 46 m^2, and
- System Mass: 710 kg.

The parameters shown are for a system with optimization weighting factors of 70% system efficiency and 30% system mass. Specific mass and specific radiator area, as a function of power level, are shown in Figures 5 and 6, respectively.

The overall efficiency of the system can be significantly enhanced by incorporating thermoelectric (TE) generators in the system. The characteristics of the TE generators result in a good match between the operating temperature requirements of the general purpose heat source (GPHS) and the ORC. Figure 7 is a block diagram of this arrangement.

Comparisons of system mass and radiator area between TE topped and untopped systems for both toluene and Dowtherm A working fluids are shown in Figures 8 and 9, respectively. Figure 10 is a comparison of system efficiency between TE topped and untopped systems using Dowtherm A and toluene as working fluids. Addition of TE topping increases efficiency by 15% to 20% over the untopped system.

Performance data for the TE portion of the system, and the selection of TE cold junction temperature and ORC turbine inlet temperature, are based on preliminary data for the modular radioisotope thermoelectric generator (MOD RTG), which the Department of Energy is currently developing.

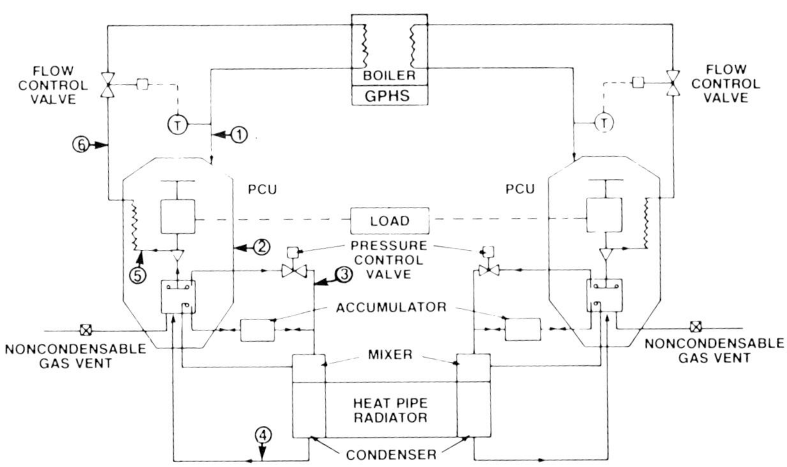

STATE POINTS

	PRESSURE		TEMPERATURE	
	Bars	(PSIA)	K	(°F)
1	40.90	(591)	671.9	(750)
2	0.32	(4.64)	565.2	(558)
3	0.27	(3.90)	370.8	(208)
4	0.20	(2.96)	334.1	(142)
5	48.79	(705)	360.8	(190)
6	45.33	(655)	512.4	(463)

Figure 4 *Schematic of the 5-kWe Dual-Channel Dynamic Isotope Power System (Toluene - One System Operating).*

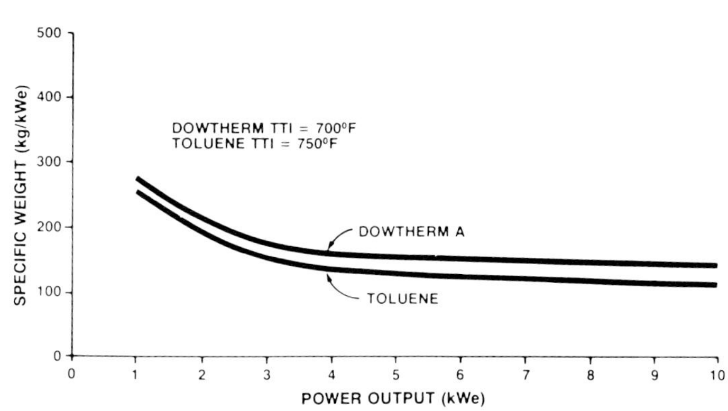

Figure 5 *Specific Weight Comparison Between Toluene and Dowtherm A Working Fluid Systems.*

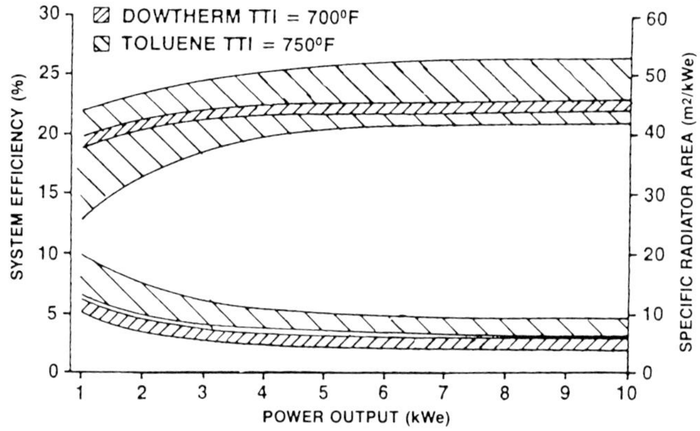

Figure 6 *System Efficiency and Specific Radiator Area Comparison between Toluene and Dowtherm A Working Fluid Systems.*

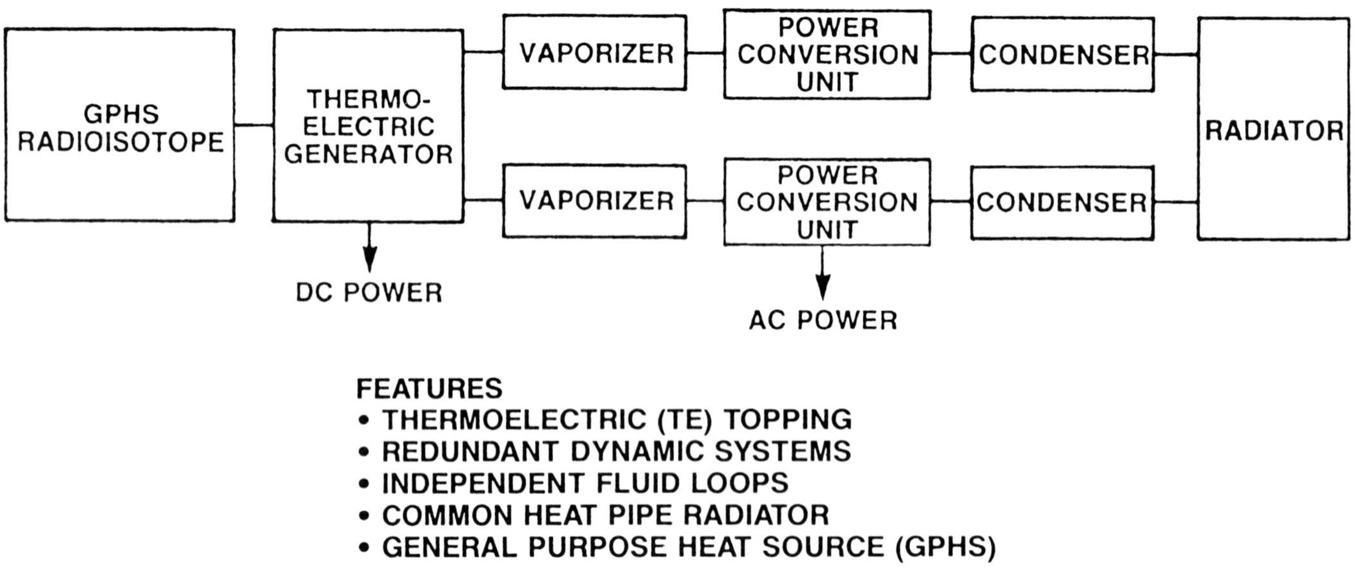

Figure 7 *Block Diagram of the Organic Rankine Cycle Dynamic Isotope Power System (Thermoelectric Topped).*

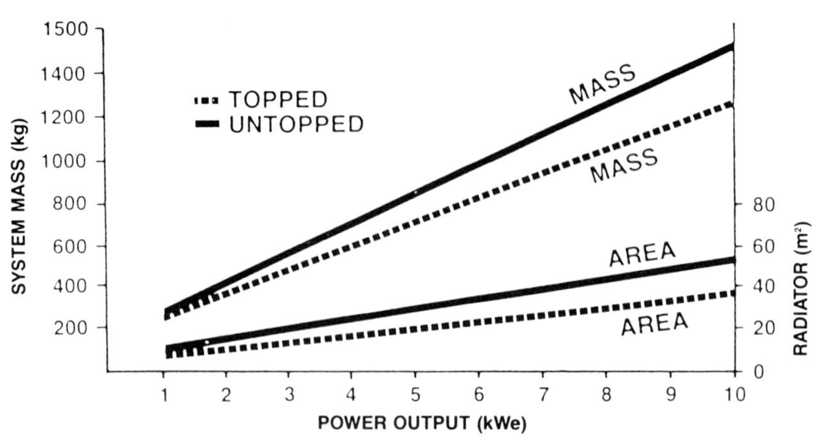

Figure 8 *System Mass/Radiator Area Comparison (Thermoelectric Topped and Untopped Toluene Systems).*

Spacecraft Configuration and Integration

The DIPS has inherent characteristics that make it highly survivable, especially when compared to conventional satellite power sources. By considering the DIPS as an integral element of the spacecraft, it can be configured to enhance overall spacecraft survivability. Figure 11 shows how the DIPS would be integrated into a spacecraft using the radiator as the outer surface of the spacecraft.

This configuration provides several benefits: it shields the entire area covered by the radiator from laser threats, helps provide electromagnetic shielding of electronic components inside the radiator, and provides shielding from space debris.

With proper radiator design, the DIPS can continue to operate at rated power while being illuminated by a laser. The basic approach is to use heat pipes that act as thermal diodes, thus decoupling the working fluid loop from any radiator heat pipes that are being illuminated by a high energy laser. Full power can be maintained by oversizing the radiator so that design point conditions can be fully maintained with only one half of the radiator active. To maintain full power capability under threat conditions, a mass penalty must be accepted. Advanced radiator materials, such as carbon, will be required to minimize this penalty.

The radiation level produced by DIPS is small in comparison to the reactor system and will require only minimal shielding of sensitive electronics. Figure 12 shows the upper bound of the total dose rate expected

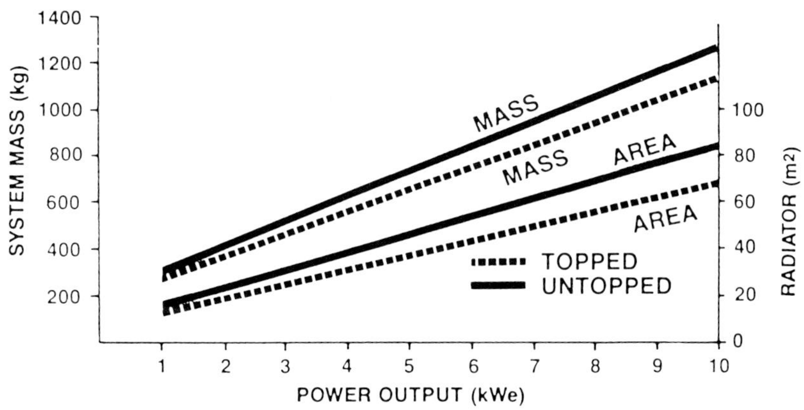

Figure 9 *System Mass/Radiator Area Comparison (Thermoelectric Topped and Untopped Toluene Systems).*

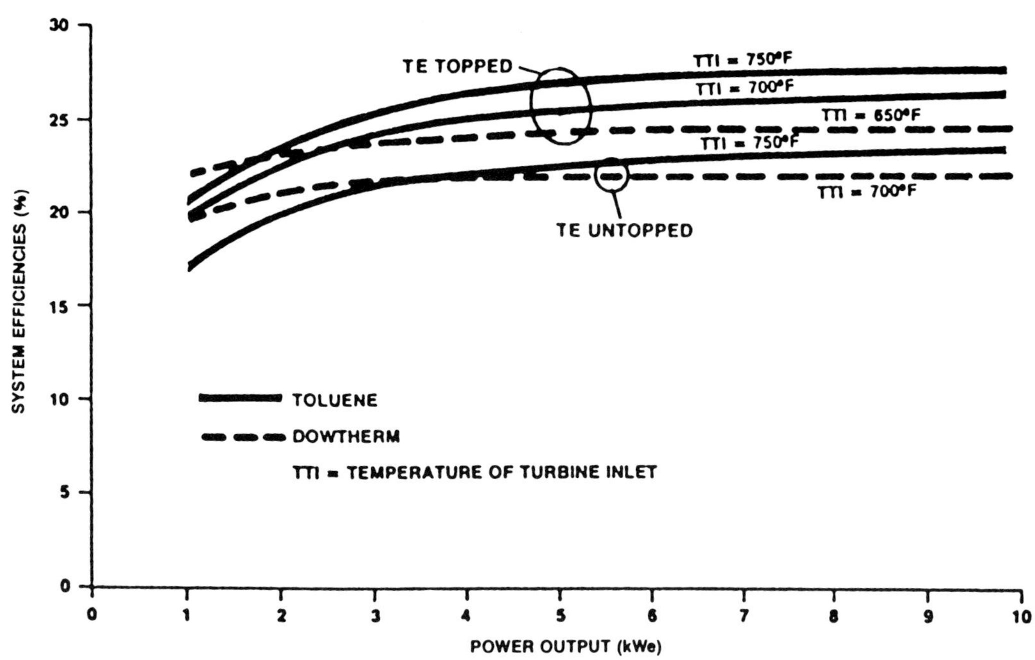

Figure 10 *Thermoelectric Topped and Untopped System Efficiency.*

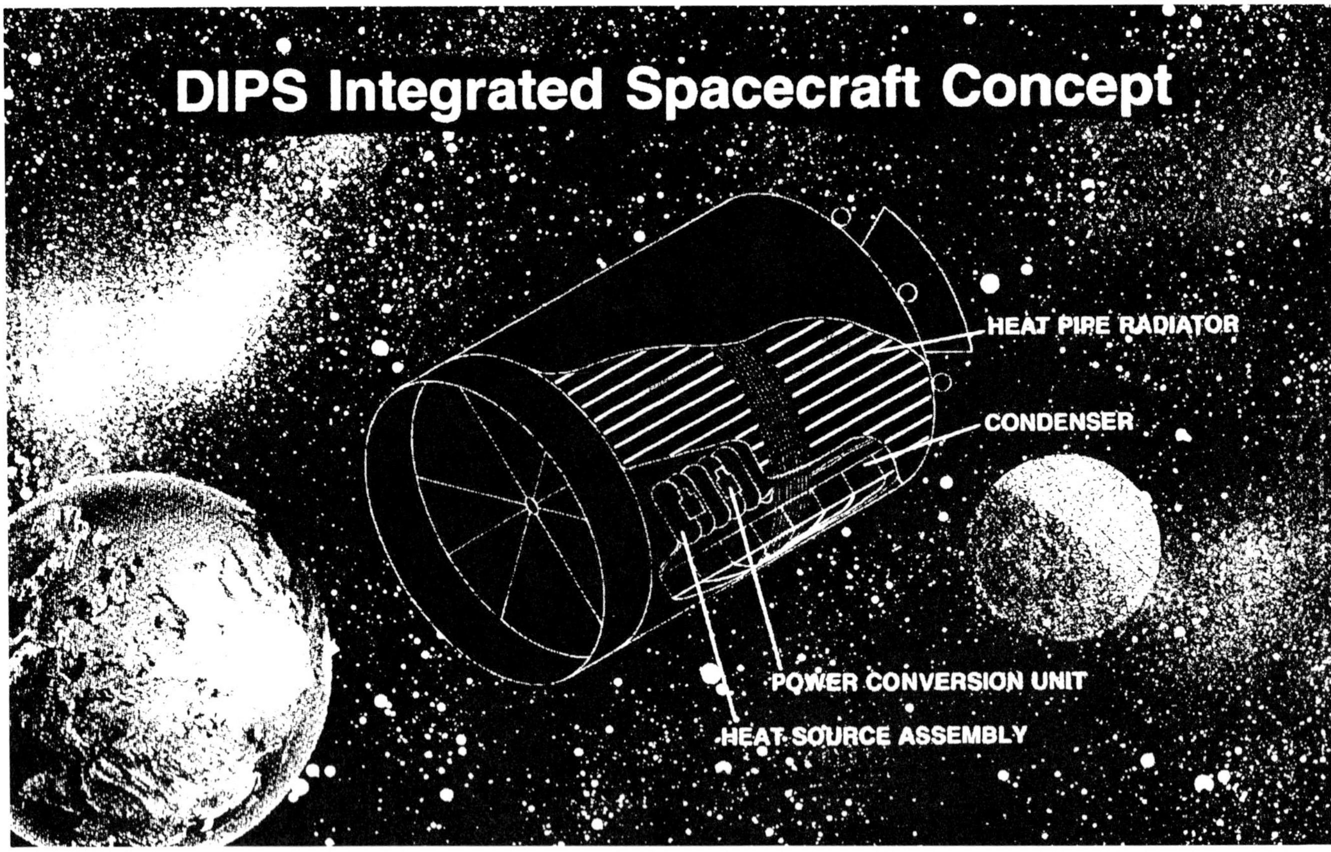

Figure 11 *Dynamic Isotope Power System Integrated Spacecraft Concept.*

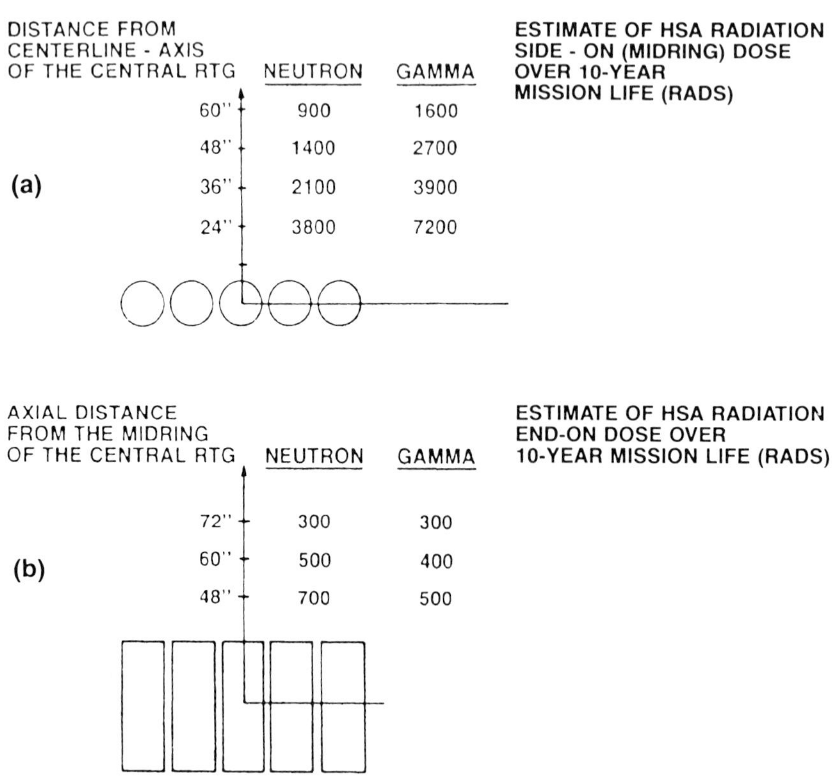

Figure 12 *Estimated 5-kWe DIPS Radiation Environment at the (a) Side-on and (b) End-on Orientations.*

for a 10-year mission. The numbers presented assume no shielding from other components.

By using the DIPS working fluid as a coolant for electronics and other equipment, an acceptable temperature level can be maintained inside the satellite.

Conclusion

A DIPS employing ORC power conversion technology can be designed to meet long-life, high-reliability mission requirements by including redundancy in the design. By taking advantage of DIPS's inherent high survivability, spacecraft survivability can be enhanced.

Acknowledgments

This research was done under support from the Sundstrand Advanced Technology Group.

References

Bland, T. J., R. E. Niggemann, and P. W. Wren (1983) "Organic Rankine Cycle Power Conversion Systems for Space Applications," Paper Number 839162, IECEC Meeting, held in Orlando, FL, 21–26 August 1983.

Bland, T. J., R. S. Downing, and D. P. Rogers (1984) "A Two-Phase Thermal Management System for the Space Station," Paper Number SAE 841525, Aerospace Congress and Exposition, held in Long Beach, Ca, 15–18 October 1984.

Bland, T. J., M. P. Ciaccio, P. D. Lacey, and J. T. Eliason (1985) "Recent Advances in the Design of Organic Rankine Cycle Power Conversion for Space Power Systems," IECEC Meeting, held in Miami, FL, 18–23 August 1985.

Space Nuclear Power Systems 1986
Edited by M. S. El-Genk and M. D. Hoover

Chapter 45

Brayton Dynamic Isotope Power Systems Update

Keith A. Davis and Anthony Pietsch
Garrett Fluid Systems Company
1300 W. Warner Road
P.O. Box 22200
Tempe, AZ 85282

Roger D. Casagrande
General Electric Company
Space Systems Division
P.O. Box 8555 29B12-KB
Philadelphia, PA 19101

Brayton dynamic power systems are uniquely suited for space applications. They are compact and highly efficient; offer inherent reliability through the use of only one moving part, a rotor mounted on gas bearings; and utilize a single-phase and inert working fluid. Brayton power systems for space applications have been under development since 1962. The first Brayton Cycle Demonstrator (BCD) hardware, funded by the Garrett Corporation in 1962, provided the proof-of-principle demonstration. Subsequently, the NASA, USAF, USN, DARPA, AEC, and DoE have been sponsors of this work. The foil gas bearing version of the Brayton Rotating Unit (BRU), designated the BRU-F, was developed under NASA LeRC sponsorship and was the forerunner of the Mini-BRU. The Mini-BRU, developed under a NASA/DoE joint venture program, became the Brayton Isotope Power System (BIPS) in a program (Brayton Isotope Power System phase I) completed by Garrett AiResearch in 1978. This paper describes the salient features of the BIPS as a Dynamic Isotope Power System (DIPS), summarizes the development work to date, establishes the maturity of the design, provides an update on materials technology, and reviews systems integration considerations.

Introduction

The Dynamic Isotope Power System (DIPS) development in the 1970s was motivated by several requirements for space missions. These included a need to reduce spacecraft complexity and mass, capability to operate in arbitrary orbits with arbitrary attitudes, low detectability, hardenability, maneuverability, and growth capability. Today space missions are more ambitious, requiring larger and more complex payloads and longer missions. Longer missions with durations of 10 to 15 years, requirements for uninterrupted power, and various threats to the spacecraft have added new emphasis to DIPS reliability.

The Brayton Isotope Power System (BIPS) has many advantages that make it uniquely suited for use in space:

- Single-phase, inert working fluid that is not affected by zero gravity or freezing;
- Space environment testing that can be accomplished entirely on earth;
- High cycle efficiency, minimum isotope power requirements;
- A high degree of design flexibility, which facilitates spacecraft integration;
- A single machine design that can operate at different power levels by varying closed-loop pressure levels;
- System growth potential due to the BIPS flexibility to operate over a wide range of turbine inlet temperatures (TIT) and pressure levels;
- Easily accomplished starting and restarting in space;
- Redundant units that can be operated with each at 50 percent load or with one at full power and the other on standby;
- Long-life features including minimum temperature cycling, constant-speed operation, and compliant foil gas bearings;
- Design that is highly survivable against external threats; and
- Reliability resulting from the utilization of only one moving part.

Key technology has continued to develop under Garrett-funded programs since the conclusion of the BIPS contract in 1978. Improved system computer modeling, a mature foil gas bearing technology, and major developments of superalloys for open-cycle gas turbines are available for the next generation of DIPS.

Development Background

Virtually all closed-Brayton-cycle (CBC) technology developed in this country was an outgrowth of prime contracts awarded to Garrett or internally-funded by Garrett for the study, design, fabrication, and testing of CBC systems and components.

Among the significant milestones achieved in CBC power systems development are the Brayton Cycle Demonstrator (BCD), the Brayton Rotating Unit (BRU), and the Brayton Isotope Power System (BIPS).

The 3-kWe BCD, built in 1962, was the first small gas turbine to successfully demonstrate the closed-cycle concept in the United States. The initial program, consisting of rotating unit design, fabrication and open-loop testing, fabrication of the heat-transfer equipment, design and fabrication of the test equipment, assembly of the system, and system test, was accomplished in less than one year on a Garrett-funded program. Systems tests included vacuum chamber testing in 1962 at Garrett and in 1966 at the NASA Johnson Space Center for familiarization training for NASA personnel. Through the years, the BCD has been upgraded to serve as a test bed for aerodynamic components, gas bearing systems, and control systems.

As successor to the BCD, the BRU produces from 2.25 kWe to 10.5 kWe of 1200-Hz power and incorporates solid-geometry gas journal and thrust bearings. Of the four units built for NASA, one has accumulated more than 41,000 hours and a second 11,000 hours of successful operation. The advanced state of CBC technology is evidenced by the fact that the BRUs were designed, built, and shipped without the benefit of extensive development testing at Garrett. The BRU program has been a dramatic demonstration of design skill, product confidence, and maturity of the technology. An updated version, designated the BRU-F, produces 15 kWe and utilizes a hydrodynamic, compliant foil gas bearing system.

The BIPS program represented a joint venture by NASA LeRC and the U.S. DoE to develop the nuclear-isotope-powered, dynamic power conversion system for space. Garrett turbocompressor/alternator, recuperator, radiator, controls, ducting, and insulation were to be combined with two multi-hundred-watt isotope, heat-source assemblies to make a 1300-watt electrical system adaptable to a variety of space applications. Hastelloy-X was selected for use as the material for the key high-temperature components because of its extensive industrial history and data base. The initial ground-test system utilized electrical resistance heaters for the heat source and a gas-to-liquid heat exchanger for heat rejection. These components, combined with the mini-BRU turbocompressor, mini-Brayton recuperator, control system, ducting, multi-foil insulation, and instrumentation made up a workhorse loop capable of generating 1200 watts with a 1033 K turbine inlet temperature (TIT). This workhorse loop completed more than 1,000 hours of testing with no system problems in a vacuum chamber that provided a 10^{-5}-torr range, simulated space environment. Efficiencies up to 25 percent were demonstrated with this small system.

System Features

Cycle Description

Figure 1 presents a flow diagram of the BIPS. Figure 2 illustrates the Brayton thermodynamic cycle as shown on a temperature-versus-entropy diagram. Recuperation is a key cycle modifier for achieving high conversion efficiency. As shown in the referenced figures, energy from an isotope heat source is used to heat the high-pressure gaseous working fluid to the maximum cycle temperature. The high-pressure gas is expanded through a turbine to produce useful output power and drive the compressor. The turbine dis-

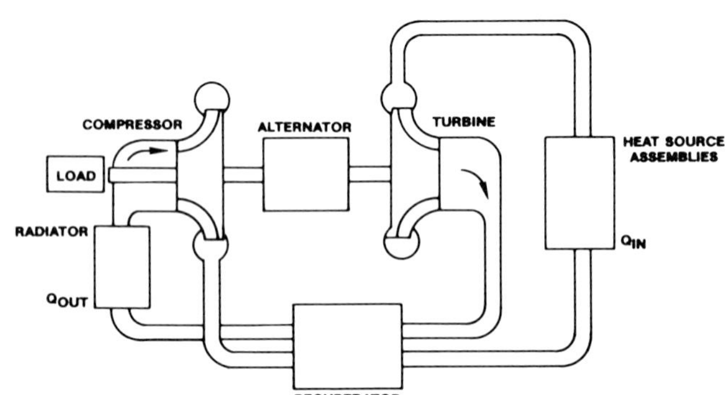

Figure 1 *Brayton Dynamic Isotope Power Systems (BIPS) Flow Diagram.*

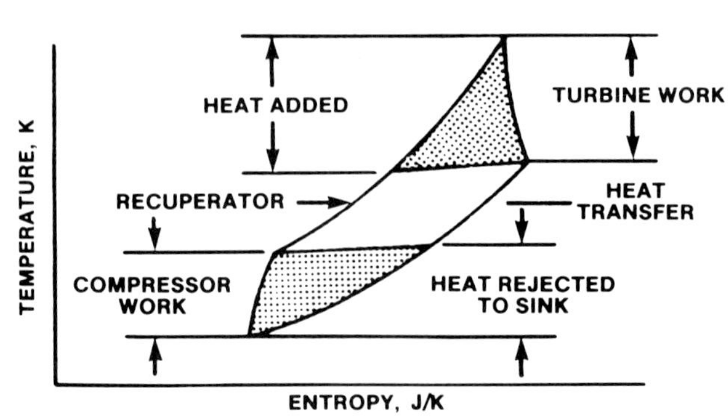

Figure 2 *Brayton Thermodynamic Cycle.*

charge gas is cooled, first in the recuperator/heat exchanger and then the radiator, before being introduced into the compressor. The low-pressure gas is compressed to the highest cycle pressure. The gas is then heated at essentially constant pressure in the recuperator before being directed to the heat source assembly (HSA) heat exchanger, thus completing the closed loop. The recuperator component recovers a significant portion of the turbine waste heat, thereby reducing the amount of heat that must be added in the heat source assembly.

Component Description

For space applications, the rotating group, recuperator, cooler/radiator, and heat source are generally separate subsystems connected by ducting.

Rotating Group

Figure 3 shows the cross section of a typical BIPS rotating unit. The turbine and compressor are both radial wheels and are located on the same shaft with the alternator rotor. The turbine-alternator-compressor assembly is supported by gas-lubricated, compliant foil bearings.

Key contributors to the typical high cycle efficiency of the BIPS are the high component efficiencies of the compressor (83 to 86 percent) and turbine (87 to 90 percent). These high component efficiencies are a result of the low pressure ratios and low tip speeds associated with CBC engines.

Foil Bearings

The use of foil gas bearings provides a long-life, low-power-loss, bearing system that requires no seals. Hydrodynamic capability of the foil bearings for starting, operation, and stopping permits the design of a bearing system that does not require externally supplied gas for cooling. Typical foil journal and thrust bearings are shown in Figure 4.

Foil gas bearings have been used extensively in commercial and military aircraft. Foil bearing technology today is a fully mature technology. This status is updated in the following Key Technology section.

The foil gas bearing concept, proprietary to Garrett, is illustrated in Figure 5. Curved, flexible foils are first inserted into the journal-bearing housing; a slot in the housing retains a tang on each foil and, thereby, positions the foil. Inserting the shaft into the housing requires deflecting and further curving the foils, so that each foil contacts both the shaft and an adjacent foil and provides a resilient preload on the bearing. Shock and vibration during the nonoperating launch phase are easily absorbed by the resilient mounting without Brinneling as in a conventional solid bearing. In the operational mode, viscous drag between the shaft and gas in the bearing cavities gen-

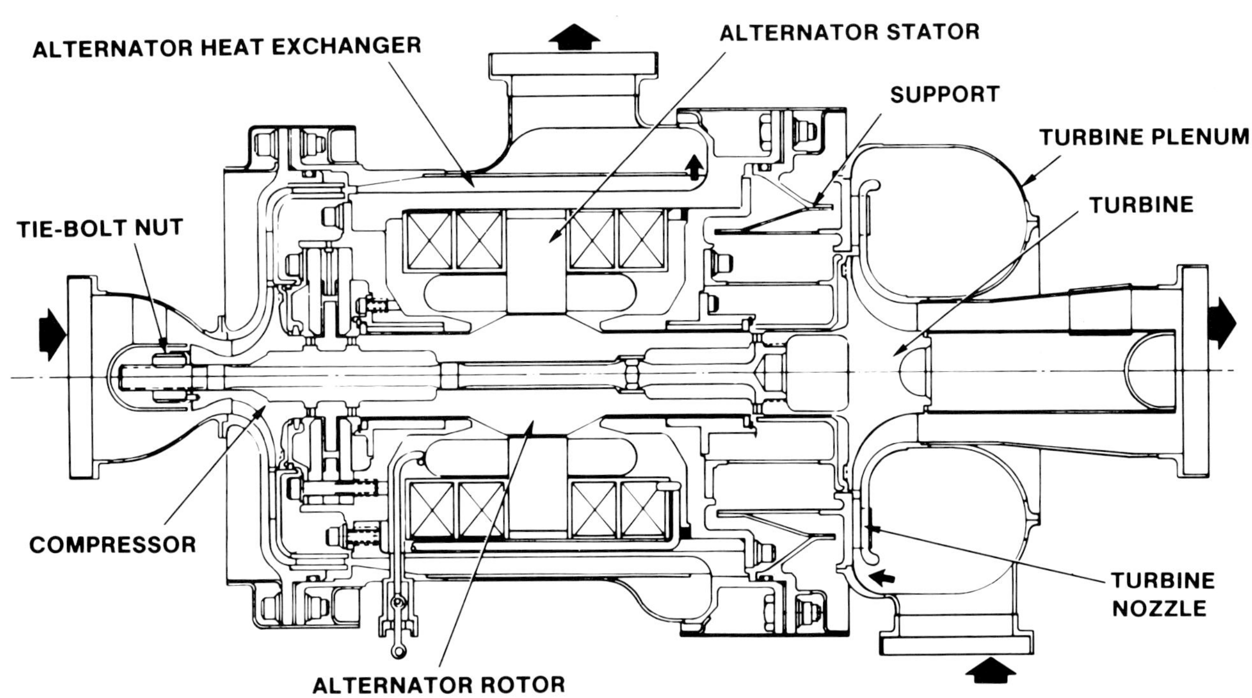

Figure 3 *Typical BIPS Rotating Unit.*

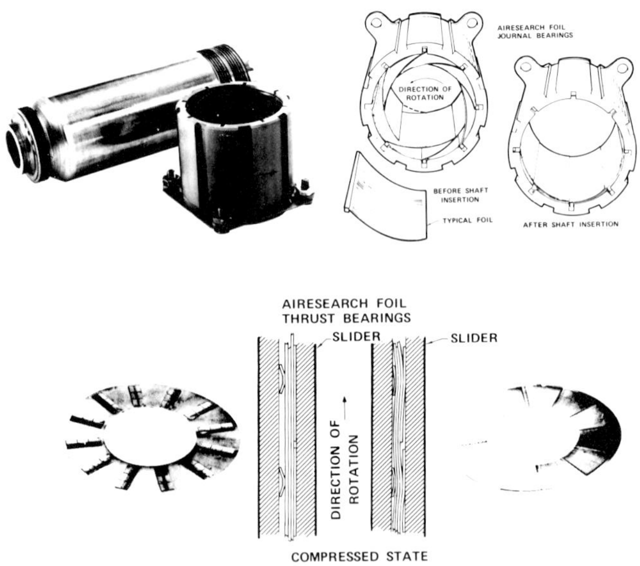

Figure 4 *Foil Gas Bearings.*

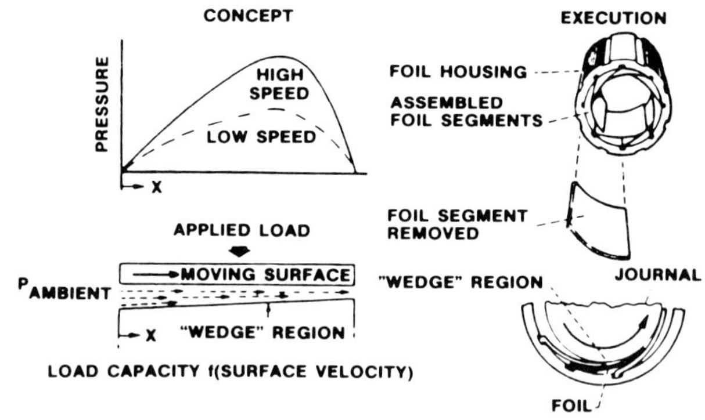

Figure 5 *Foil Gas Bearings Concept.*

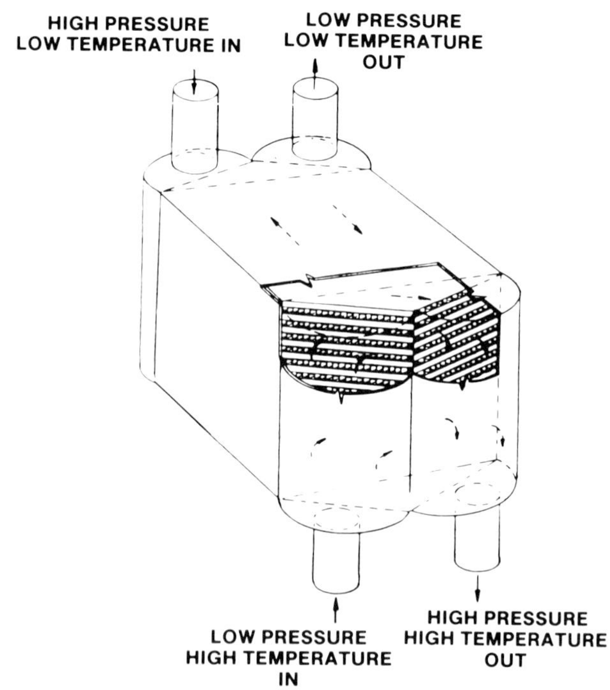

Figure 6 *Fin Counterflow Recuperator.*

erates a gas-pressure field in the bearings which supports the rotating group. During normal operation, the foil rides on a film of gas and does not touch the shaft: bearing operation is hydrodynamic. Differential thermal deformation of shaft and housing are accommodated by flexing of the foils, the clearance between foil and shaft remaining nearly constant. Foils are treated with a low-friction, solid coating to prevent wear during starting and stopping. Space power systems, using foil gas bearings, experience a very small number of start/stop cycles compared to bearings used in typical commercial or military applications.

Recuperator

Typically, the turbine exhaust is ducted to a high-effectiveness, plate/offset-fin, pure counter-flow recuperator as shown in Figure 6.

The recuperator is a stacked assembly of thin, alternating, low-pressure and high-pressure, counter-flow panels through which the cycle gas is passed. Low-pressure and high-pressure gas enters and leaves the recuperator by manifolding the corresponding pressure panels to four ducts.

The recuperator is a mature technology used in the NASA BRU family, the NASA/DoE BIPS, and many commercial applications. Typical recuperator construction is shown in Figure 7.

Radiator

Cycle waste heat is transferred from the working fluid in the radiator. Several radiator designs are possible. Cycle working fluid can be routed directly through a gas radiator as in the original BIPS. A separate radiator loop interfacing with the working fluid at a heat exchanger (cooler) can also be used. This separate loop can then be connected to either a pumped liquid or heatpipe radiator.

Heat Source Assembly

The General Purpose Heat Source (GPHS) transfers energy to the heat source heat exchanger (HSHX) by radiation. The GPHS is composed of individual

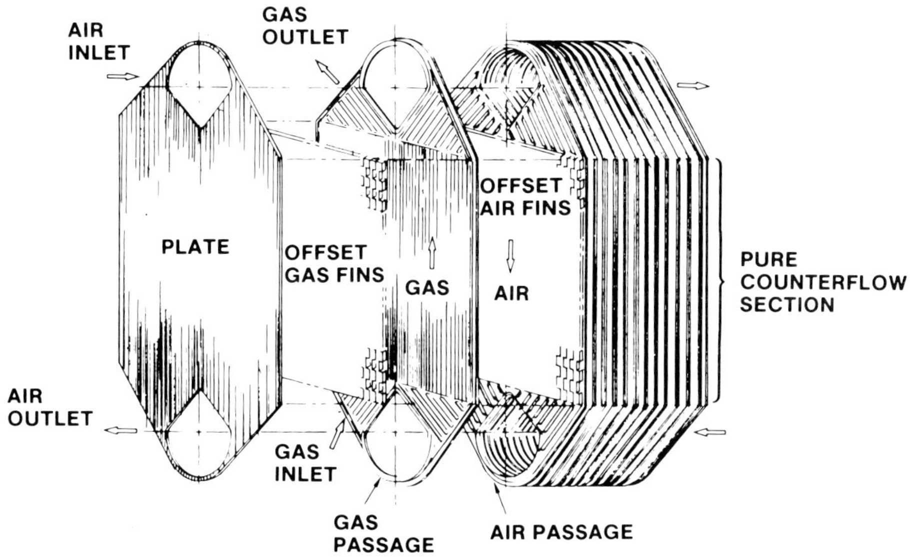

Figure 7 *Recuperator Construction.*

250-watt isotope cartridges. The cartridge consists of an aeroshell, two graphite impact shells (GIS), and four fuel clads (FC). Each FC consists of a fuel pellet of ^{238}Pu in the form of the oxide PuO_2 encased in an iridium shell that contains the fuel. Two of the FCs, separated by a graphite floating membrane, are then encased in a GIS to provide the primary resistance to mechanical impact loads. A graphite aeroshell, which serves as the cartridge's primary structural element and ablator, provides protection for the two GIS. Each GIS is separated from the aeroshell by a graphite insulator. Figure 8 shows elements of each cartridge and the stacking arrangement of cartridges. The HSA heat exchanger features two helical-wound and joined tubes, as shown in Figure 9, to provide the independent gas paths.

The containment of the plutonium fuel is ensured to the maximum extent possible during all mission phases including ground handling, transportation, launch, ascent, temporary orbit, and during unplanned events such as reentry, impact, and post-impact situations. The relative motion of the GPHS

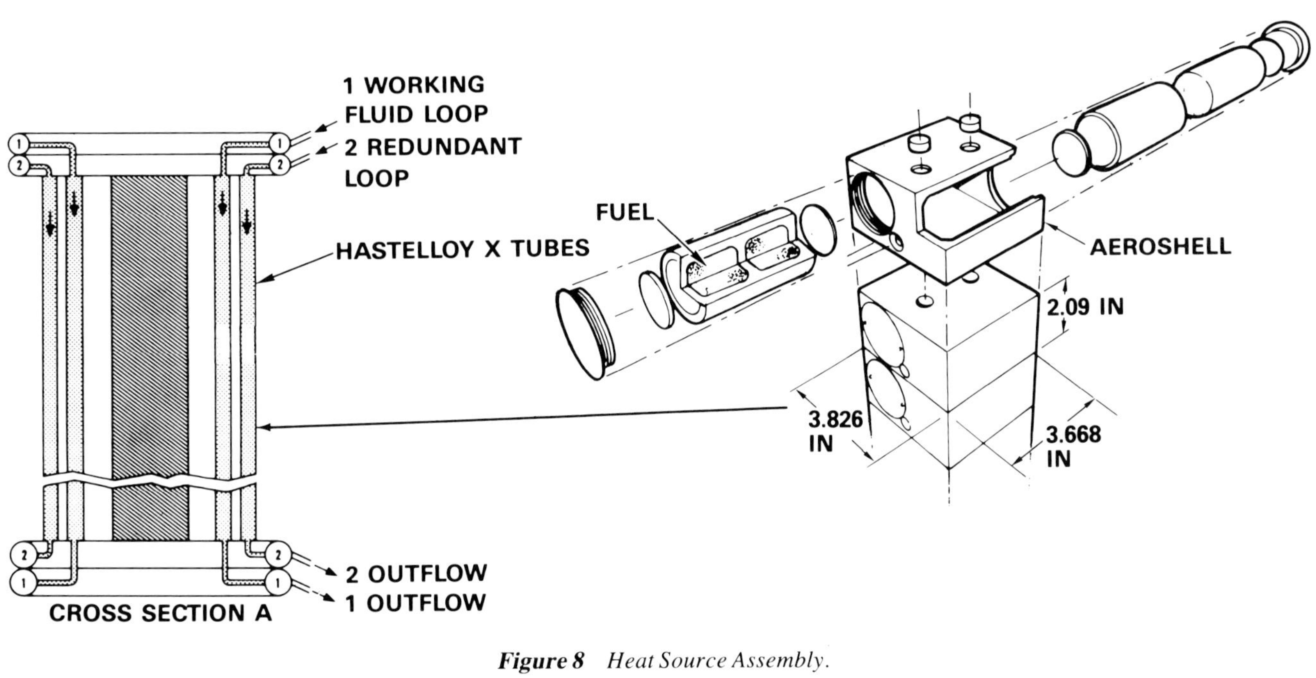

Figure 8 *Heat Source Assembly.*

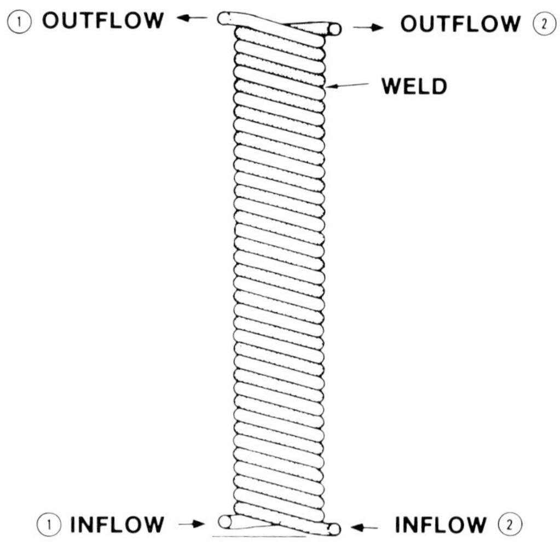

Figure 9 *Heat Source Heat Exchanger.*

cartridges is restricted by the heat source support system, which provides axial compression to prevent separation of the modules and locking members that minimize any relative lateral motion of the individual GPHS cartridges.

Each cartridge has a power output of approximately 250 watts. The reduction of thermal power is approximately 0.8 percent per year due to the half life of the ^{238}Pu. The required power level is obtained by providing multiple cartridges for the system.

SYSTEMS CONCEPTS

Redundant Systems

The ambitious space missions, increased minimum mission power level requirements, and uninterrupted power mandate reliability as a key driver for the DIPS development program. A redundant power-conversion system concept will meet these program requirements. Figure 10 illustrates the system. Each loop, with an independent gas path through the HSA and radiator, has an independent start and restart capability.

The redundant system can operate by two different approaches. As previously noted, the BIPS can be started in space. The single-phase gas working fluid and low-friction or drag characteristic of the foil bearings make this possible. Therefore, the first redundant operating approach would have one BIPS operating at the required power level with the other unit shut down on standby.

The second approach would be to operate the two units in parallel at 50 percent of the required power level. Both units would operate at reduced pressure, but at rated speed and TIT.

Should one power conversion loop shut down, a signal would cause the release of additional gas into the surviving loop. This action would raise the working pressure in the loop so that the single unit would continue to meet the full mission power requirement. Uninterrupted power is assured.

The gas storage system would be based upon an existing Garrett product line utilized for missile systems. The pressure vessel includes a pyrotechnic- or squib-type valve. Current applications require helium or helium-neon storage for pressure as high as 10,000 psia with a 10-year shelf or service life.

An isometric view of the redundant system with the radiator removed to more clearly illustrate the two units is shown in Figure 11. Four HSAs are manifolded in parallel to provide the heat source.

System Flexibility

BIPS system flexibility should be considered in two forms, flexibility in the design point and flexibility in operation. Operational flexibility is exhibited by excellent off-design performance, as shown in the BRU performance of Figure 12.

Design flexibility is exhibited by the fact that the design driver may be maximum cycle efficiency, minimum system mass and/or volume, minimum heat-rejection system size, or a desirable combination of these. The major independent cycle parameters selected during the optimization process are

- Recuperator Effectiveness (E_r),
- Compressor and Turbine Inlet Temperature (CIT and TIT),
- Compressor pressure ratio (R_c),
- Pressure loss parameter (β),
- Molecular weight of working fluid (He-Xe mixture),
- BRU shaft speed (N), and
- Compressor-specific speed (N_s).

As noted above, changing the system pressure allows one machine to operate at different power levels. The pressure vessel storage arrangement discussed in the systems concept section provides automatic or remote control, with a simple, high-reliability system to increase the power level.

The flexibility of the BIPS design permits system trades to be made to optimize size, mass, cost, redundancy, and reliability. An example is the case where dual BIPS are used in a system, each operating at 50 percent power. Each of the redundant loops is designed to carry mission power independently through

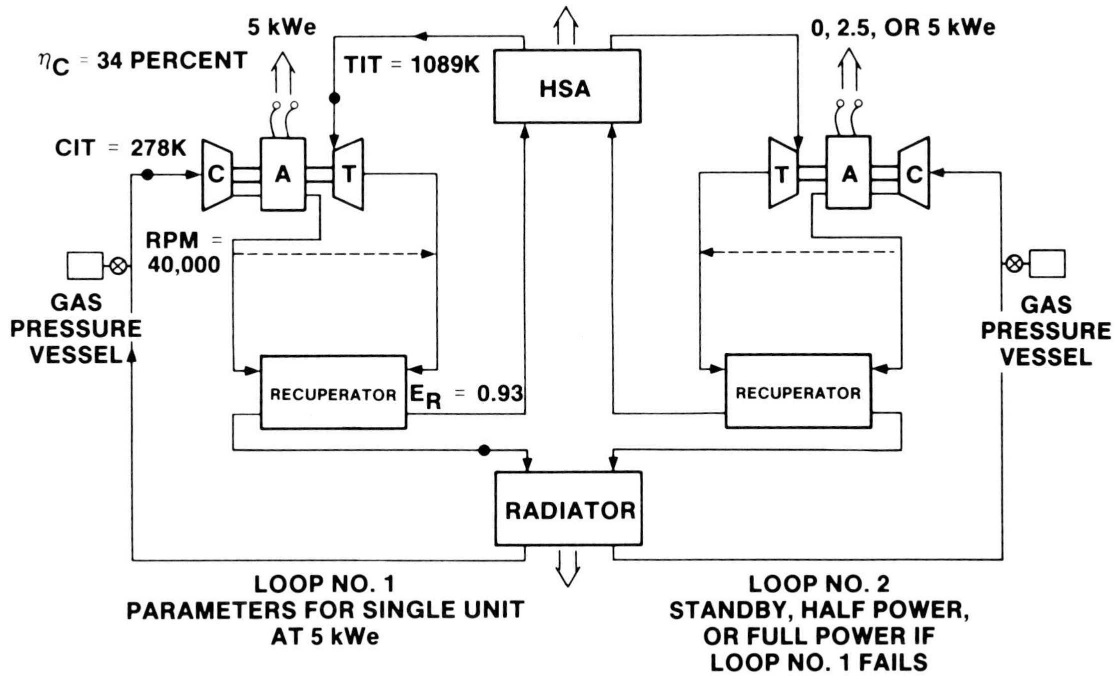

Figure 10 *Redundant BIPS System Schematic.*

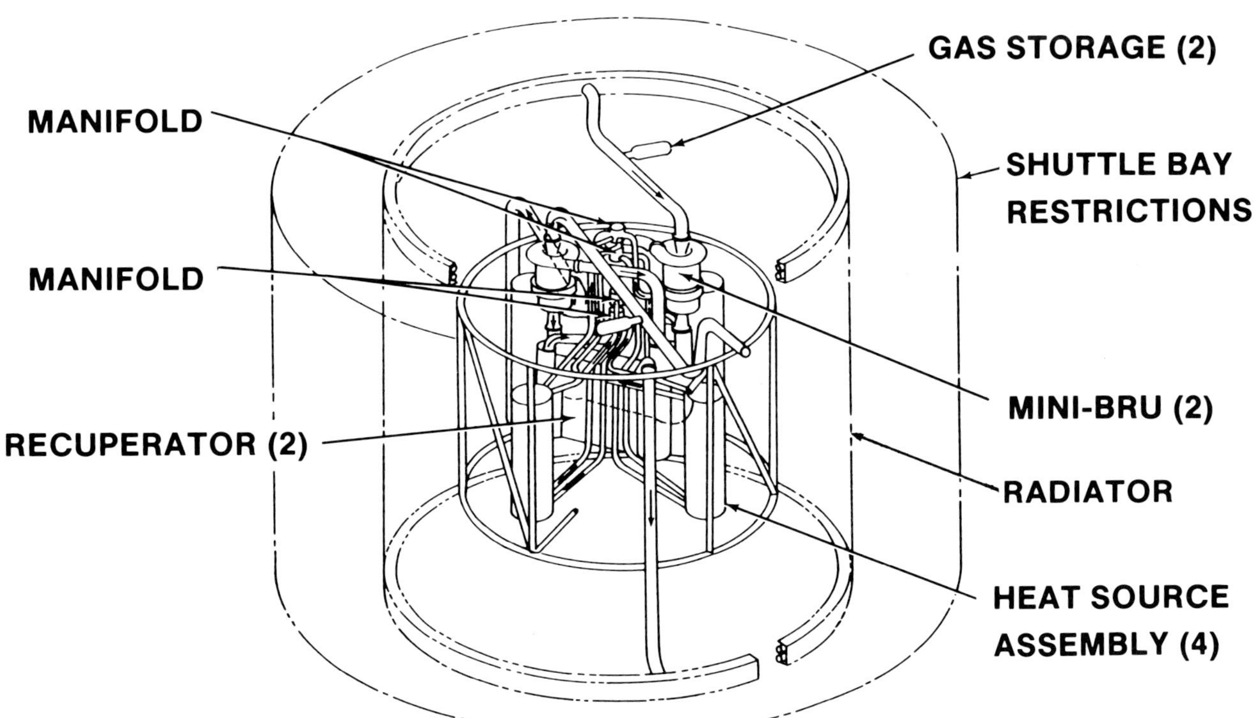

Figure 11 *Redundant BIPS System Isometric.*

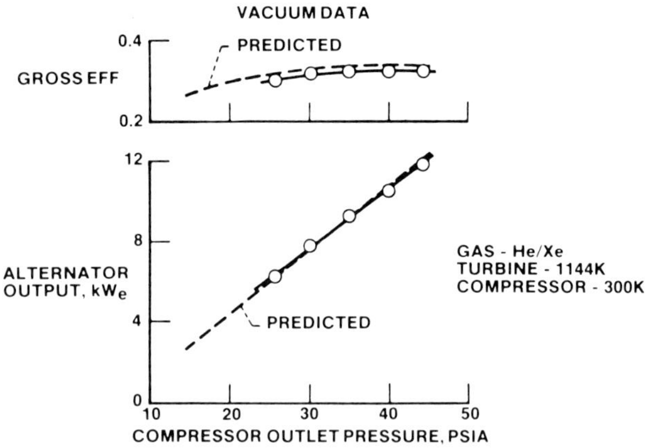

Figure 12 *Off-Design Performance of NASA Engine "B".*

the optimization process noted above. The system is designed to run at constant speed and TIT. The stage efficiencies of the radial compressor and turbine are relatively flat over the necessary operating range. The windage loss in the alternator is reduced with the lowered pressure of 50 percent power. The net cycle efficiency of the redundant system will be reduced approximately 2 percent compared with the efficiency of the single 5-kWe system.

Performance

A preliminary flow schematic for a redundant 5-kWe BIPS is shown in Figure 10. Preliminary performance curves are shown in Figures 13 and 14. The schematic reflects a cycle efficiency of 32.5 percent as a redundant system. The efficiency of a single conversion loop delivering 5 kWe would be 34 percent. It should be noted that the 15.9 kWt includes the beginning-of-life (BOL) power level to ensure end-of-life (EOL) performance for a 10-year service. This is based upon a 7.75-percent decay in HSA output.

Figures 13 and 14 reflect a range of TIT from 1033 K to 1145 K. Excellent cycle efficiency can be achieved with conservative turbine and HSA operating temperatures. This conservatism coupled with state-of-the-art superalloys is consistent with the 10- to 15-year service life as will be discussed in the materials section.

Design Optimization Methods

Maturity of the Foil Bearing System

Development of foil bearings has spanned more than two decades with initial production beginning in the mid-1960s. Foil bearings were used in military fighter aircraft cooling turbines, nitrogen compressors, and C-5A tire-pumping compressors. High production of foil gas bearings began in 1970 with their use in the DC-10 aircraft environmental control system (ECS) cooling turbine. The original hand fabrication for development and limited tooling in early production were replaced by a dedicated production facility in 1973. Production designs, hard tooling, automated fabrication, material/coating improvements, volume production, and inspection tooling/methods complete the maturity of foil bearing production.

Current production is primarily divided between airborne cooling turbines and industrial turboexpanders. To date, more than 89 million hours of service have been accumulated on 5000 delivered systems. Although much of this experience has been achieved

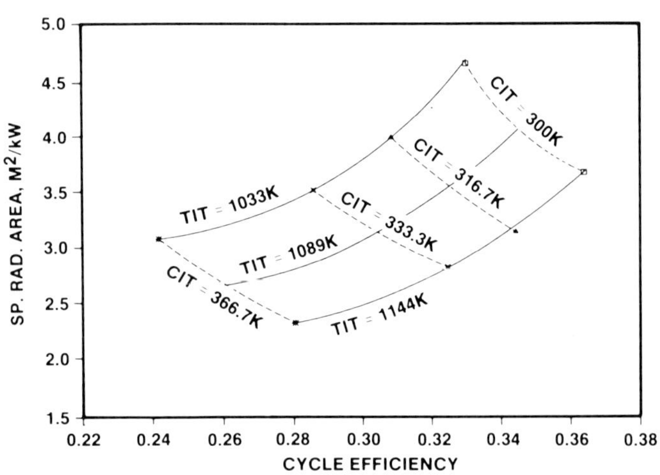

Figure 13 *Minimum Specific Radiator Area versus Cycle Efficiency.*

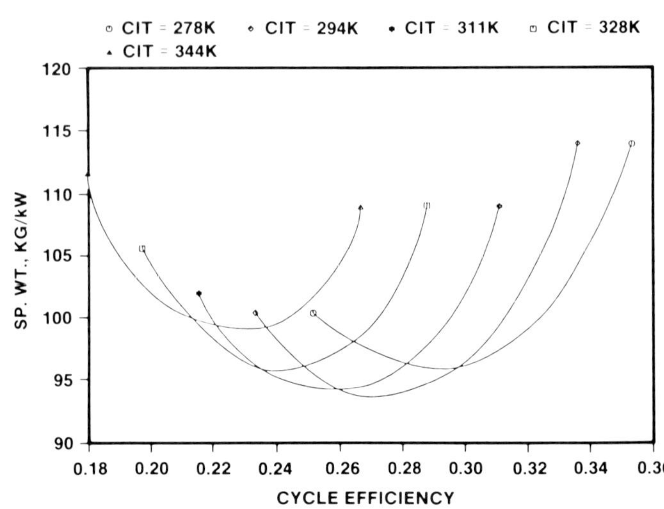

Figure 14 *Minimum Specific Weight versus Cycle Efficiency for TIT = 1089 K.*

on turbomachinery which operates at low turbine inlet temperatures relative to the CBC, the actual bearing temperatures are equal to or greater than those encountered on the Garrett CBC systems. Therefore, the data obtained from the 89 million hours of service is useful in predicting reliability for space applications. Annual production rates now exceed 1000 units. Garrett has also been developing high-temperature foil bearings for open-cycle gas turbines. The Advanced Gas Turbine (AGT) research program for DoE, initiated in 1979, is continuing. Ceramic turbines will operate at TITs up to 1644 K. Foil bearing temperatures will range between 811 K and 922 K.

Reliability has become the hallmark of the foil bearing systems. The achieved mean time between unscheduled removals for several systems is shown in Table 1. No random causes for removal have been identified to date. Removals were related to three causes: contamination, cooling air blockage, and wear due to 10,000 or more start cycles. Contamination, primarily dirt or oil, was introduced into the bearings from the external air supply. Cooling air blockage was related to inlet door malfunction. Air cycle ECS typically experience a high number of starts, as many as one start per operating hour. Since the foils are not in contact with the shaft at speed, starts are the primary wear source. Service life limits are, therefore, measured in thousands of starts.

The BIPS design and environment inherently resolve each of the problems historically associated with the premature removal of open-cycle units. The BIPS closed loop is filled with a clean, inert gas with no source of dirt and oil contamination. Bearings are supplied with cool, clean gas from the system loop. The number of start/stop cycles predicted for the BIPS units over the 10- to 15-year service life is less than 200 compared to the 10,000 start service life of open-cycle gas bearings.

Table 1 *Reliability of Foil Gas Bearings Production Air Cycle Equipment.*

APPLICATION	SERVICE (million hours)	MTBUR[a] (hours)
DC-10 Aircraft Environmental Control System Cooling Turbines	> 85	71,000
General Aviation Cooling Turbines	1.9	> 66,000
B-757/b-767 Aircraft Cooling Turbines	> 2.1	64,000

[a]Mean Time Between Unscheduled Removals.

Therefore, the premature removal causes identified in more than 89 million fleet operating hours to date can be justifiably audited out for the BIPS application. Since no inherent bearing failures relevant to closed-Brayton-cycle (CBC) systems have occurred in 87 million hours, the failure rate is conservatively assessed at less than 0.2×10^{-6} failures per hour.

The CBC offers a high degree of flexibility in addition to high efficiency. Parametric design and optimization of the CBC are conducted with the aid of a unique series of computer programs for design, off-design, and transient analyses. The accuracy of these computer programs has been proven through correlation with component and system test data. These programs provide the capability to explore design questions in detail. The approach to design-point selection is discussed below.

The basic design program matches empirical turbine and compressor performance data to the thermodynamics of the Brayton cycle and defines the complete state points throughout the closed loop. Component and duct pressure drops throughout the cycle are considered in the analysis as well as bearing losses, cooling flows, and the variation in working fluid transport properties with temperature. Various working fluids can also be investigated if system optimization requires different mass flow, heat transfer, size, and mass.

The second part of the design program includes design routines, utilizing empirically determined heat-transfer and pressure-drop data for the plate-fin, pure counter-flow cooler configurations that employ finned tube or plate-fin constructions. The program also includes a design program for a gas, liquid, or heatpipe space radiator. Figure 15 schematically shows the computer program steps.

The last portion of the design program consists of a summary routine that compiles selected data from the first two parts. It then prints the summary information and automatically plots the combined system mass and volume versus the cycle efficiency.

The design program is used in two operating modes. The first mode allows simultaneous evaluation of the major system parameters, and the results can be plotted as shown in Figure 16. These "shotgun" plots are generated by varying the major design parameters over wide ranges. The result is the definition of the minimum system mass (or minimum volume) line as a function of the cycle efficiency. It is common to evaluate hundreds or even thousands of complete systems during the "shotgun" performance evaluations.

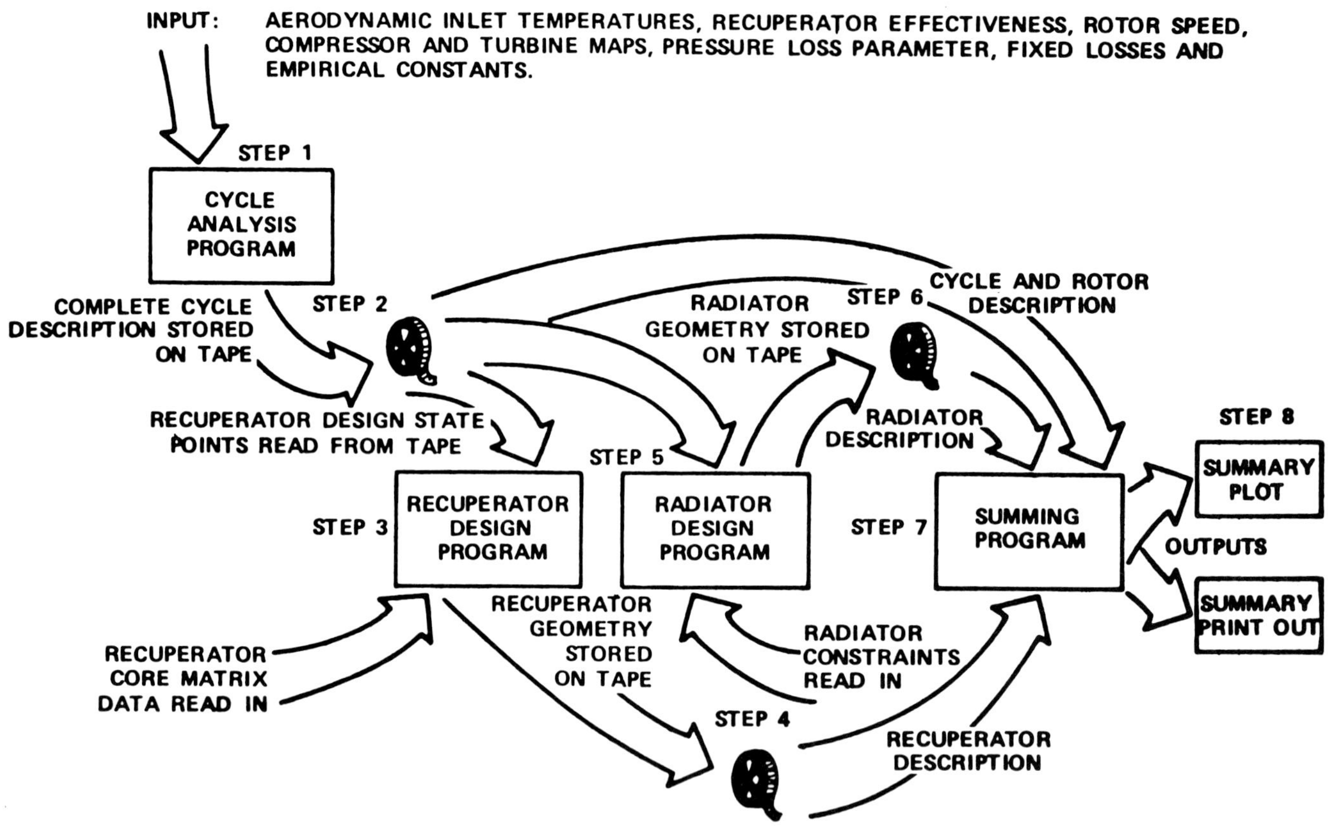

Figure 15 *Power System Design Methodology.*

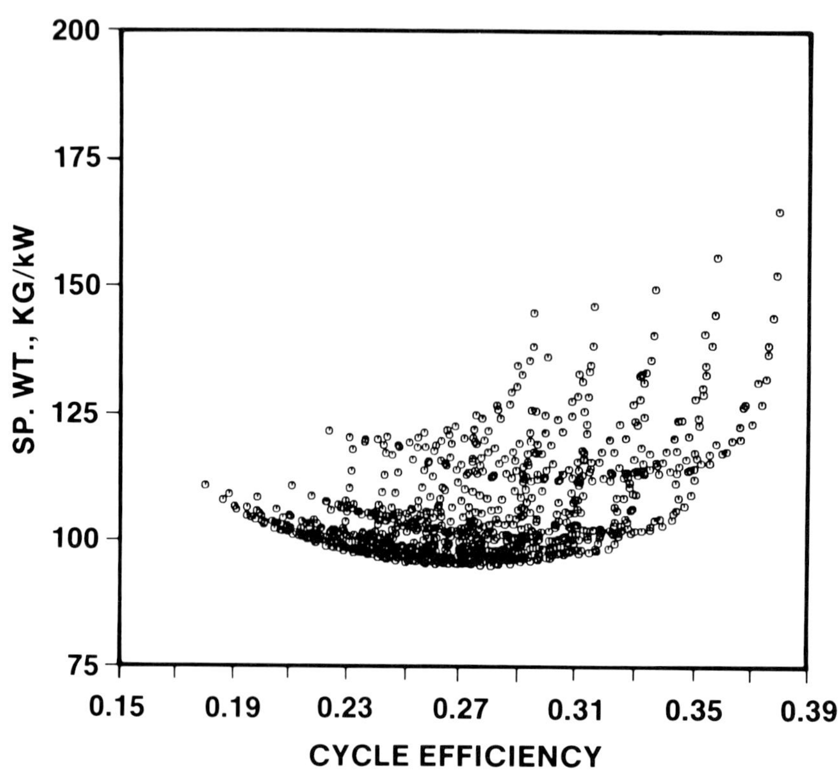

Figure 16 *Shotgun Plot for Specific Weight versus Cycle Efficiency for TIT = 1089 K.*

The second operating mode is defined as the "single-value parametric study." Several conversion system designs along the minimum line shown in Figure 16 are selected for in-depth review. Each of the major parameters used to define the design is then systematically varied over a small range and the results are plotted. This analysis is used for fine tuning the systems that were selected during the shotgun analysis, as shown in Figure 17. The final step is to obtain detailed computer output that precisely lists: all of the temperature and pressure state points around the cycle; the component sizes, efficiencies, masses, windage losses, heat-transfer core geometries, surface finish requirements, and other data. From these data, layouts can be made to satisfy packaging constraints. The final result is the selection of the best design to meet all of the requirements.

Materials

The BIPS of 1976 to 1978 was designed to produce 1.37 kWe with a TIT of 1144 K. The service life was 100,000 hours and 1000 start cycles. The turbine rotor and nozzle were cast in IN713C. The rotor design has a burst margin of 3.53, a minimum 0.1 percent creep life of 7.43 million hours, and a low-cycle-fatigue life greater than one million start cycles.

Today a turbine rotor can be designed for a 5-kWe power system with this same conservatism. The standard production material today is MAR-M-247 (fine grain) superalloy. The creep response of this alloy, the key property for the BRU turbine operating at constant speed and TIT, is a conservative five times better than that of IN713C in the range to 1144 K. The nozzle will be cast in the same MAR-M-247 alloy.

In the 1970s, the BIPS turbine scroll, exhaust diffuser duct, and HSHX were fabricated of Hastelloy-X and the shroud of Hastelloy-C. Today these alloys would be replaced with Inconel 617 and Hastelloy-S, respectively. Inconel 617 may also be used for the shroud for commonality although operating temperatures do not require it. The gains are in long-term thermal stability and ductility.

The above material considerations are key to the BRU and provide an example of improvements expected throughout the BIPS by applying state-of-the-art and well-characterized materials.

Spacecraft Integration

Approaches Considered

The most important issue confronting the spacecraft designer is how the DIPS can be configured to of-

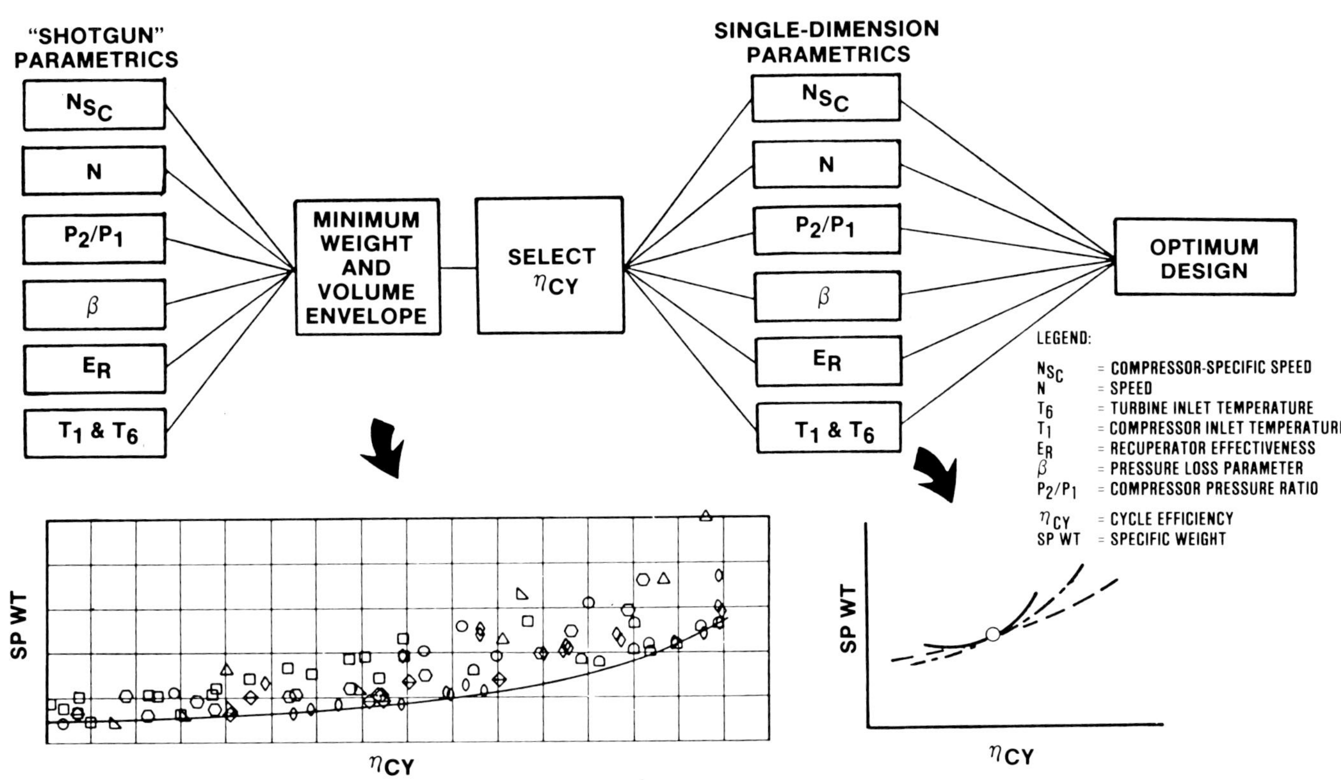

Figure 17 *Analytical Tools—Logic Network.*

fer the best compromise in physical location, effect on the center of gravity, requirements imposed by an STS launch, and requirements of the spacecraft payload and support systems. Three configurations have been investigated using a surveillance-type spacecraft model:

1. A separate DIPS module,
2. A hybrid package in which the radiator becomes part of the spacecraft structure while the payload is kept separate from the DIPS in terms of thermal control, and
3. A completely integrated DIPS/spacecraft in which the spacecraft components are tied to the DIPS thermal control system as in the Nuclear Integrated Multimission Spacecraft (NIMS) study conducted in 1979.

The DIPS module approach shown in Figure 18 offers a self-contained power source that includes all of the DIPS components in a complete package. The radiator shell would be designed to incorporate structural attachments providing a simple mechanical attachment/separation interface to the spacecraft. The mission payload is rigidly attached to the DIPS module forward end and is thermally independent and isolated. Thermal control of the DIPS would also be self-contained with the radiator optimized for the power system considerations alone. An electrical umbilical connection would supply the spacecraft with power and a second umbilical connection would be used for telemetry and control interfaces.

The hybrid approach shown in Figure 19 integrates the radiator with the spacecraft to allow the DIPS volume to be shared with the spacecraft components. Consideration for this arrangement is to provide the potential for a more compact spacecraft package. The DIPS radiator covers most of the spacecraft surface, but an axial portion of the surface area is kept free to provide mounting panels for spacecraft components. A single area is shown for passive component thermal control, which can be kept oriented away from the sun for temperature control. This approach still provides separation between the spacecraft and the DIPS from a thermal control point of view, but the structures are integrated.

The totally integrated system shown in Figure 20 is the embodiment of the NIMS approach. In this configuration, the radiator becomes the total outer shell of the spacecraft, and the components are mounted internally and tied thermally to the radiator as well. This approach is very attractive for spacecraft in the kilowatt power range where the radiator length requirement is not significantly different from that dictated by the overall spacecraft arrangement. As the power level increases, the total radiator area becomes dominant, and minimizing the radiator size may be necessary. Coupling the thermal control of the components to the DIPS system requires radiator temperatures dictated by operating temperature limits of the components. The Brayton system has a very flexible temperature requirement, a broader range than electronics experience. Therefore, a larger radiator area will most likely be required to satisfy the integrated approach.

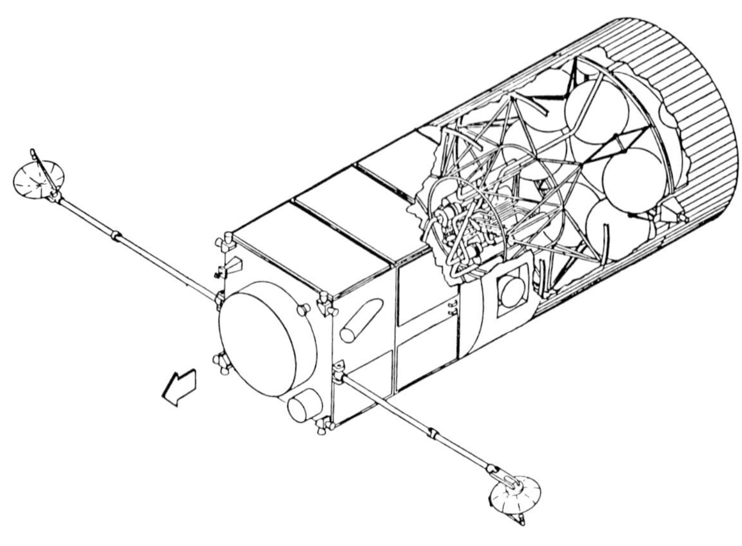

Figure 18 *Modular DIPS Configuration.*

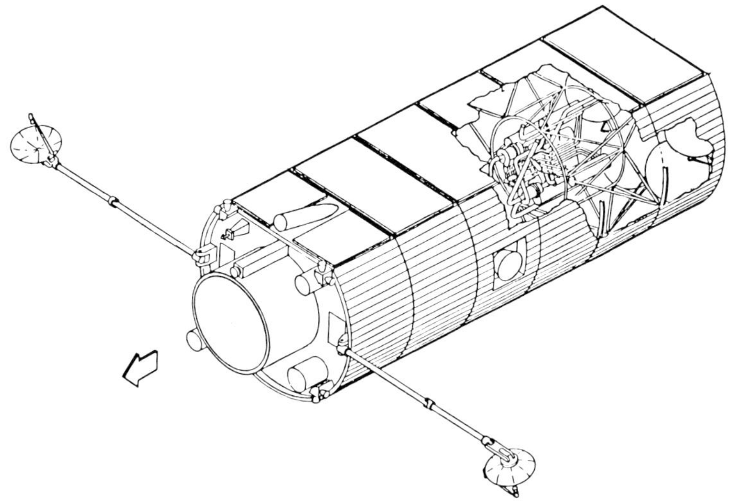

Figure 19 *Hybrid DIPS Configuration.*

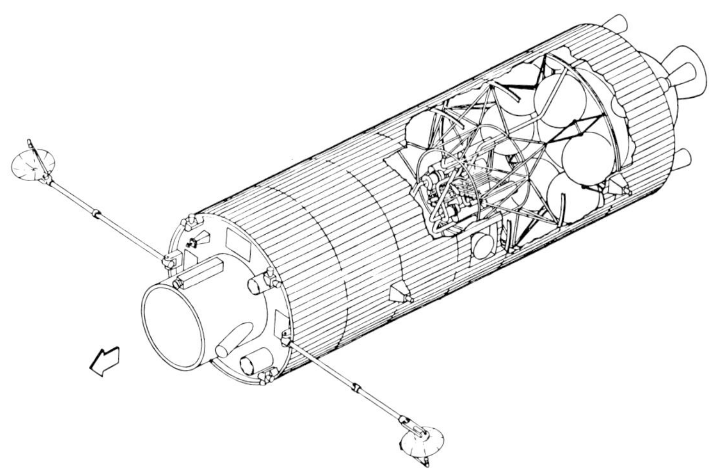

Figure 20 *Integrated DIPS Configuration (NIMS).*

Evaluation

The DIPS module approach is the simplest system to implement. The separate system provides attachment points and acts as the load path between the spacecraft payload section and the propulsion stage. The power system can be designed for universal use and developed totally independent of spacecraft. The resulting system will not be optimally packaged as a spacecraft since the radiator will provide for a much larger internal volume than would be necessary for the Brayton conversion system.

The hybrid system was configured as a means to improve the packaging efficiency. The volume within the DIPS is used by the spacecraft components to minimize the overall spacecraft length. Power system independence is reduced because the radiator structure now serves as part of the spacecraft payload structure as well. However, thermal independence is maintained. The provision made for spacecraft components to be mounted on the outer panels allows the use of passive thermal control techniques. The component panel surface could be constantly positioned away from the sun if there are critical components requiring special temperature-control considerations. This independence allows for separate development and testing of the DIPS components and the spacecraft payload components.

The integrated system (NIMS) had the advantage of being the most robust in terms of hardening capability. The totally enclosing radiator offers protection from threats for all internally mounted components. As was shown in the NIMS study, the increased radiator size allows the system to not only survive, but to continue operating at full power continuously during a threat encounter. This system requires the highest degree of integration between the power system developer and the spacecraft developer. It will also be the heaviest system because of the increased radiator area requirement and the corresponding enhanced capability of the system.

Summary

This paper has shown the status of technology for the closed-Brayton version of a dynamic isotope power system. Preliminary cycle data have been presented for typical 5-kWe, 10- to 15-year space missions of the future. System efficiencies in excess of 30 percent are achievable, even in a redundant configuration. System mass can be traded for higher efficiency. Nominal system mass is 518 kg at 34 percent cycle efficiency. Radiator area can be traded for efficiency and mass. Nominal radiator area is 21.0 m^2 at 34 percent cycle efficiency. The redundant system offers high mission reliability and survivability. The high system efficiency allows fuel loading to be minimized, thereby saving fuel mass and cost.

Acknowledgments

Brayton power system work was performed by The Garrett Corporation under several sponsorships. The BIPS effort was performed under NASA- and DoE-sponsorship. Continuing technology research has been internally funded by The Garrett Corporation.

References

AiResearch (1978) *Brayton Isotope Power System Phase I Final Report, 31–2919*, prepared for the U.S. Department of Energy by AiResearch Manufacturing Company of Arizona, Phoenix, AZ, July 1978.

LIST OF PEER REVIEWERS

(We gratefully acknowledge the following individual technical reviewers, along with those whose names we do not have, for their contributions to the peer review of the manuscripts published in this book.)

Christopher C. Alexion
Westinghouse Electric Corporation

S. G. Bankoff
Northwestern University

W. J. Barnett
U. S. Department of Energy

Everet Beckner
Sandia National Laboratories

Frederick R. Best
Texas A&M University

Richard Bohl
Los Alamos National Laboratory

Jack L. Boone
University of Missouri-Rolla

Michael Brasher
Air Force Space Technology Center

David Buden
Science Applications International Corporation

Donald E. Carlson
Los Alamos National Laboratory

John W. H. Chi
Westinghouse Electric Corporation

Wen Chiu
General Electric Company

Lee Danielson
Thermo Electron Corporation

Dale Dutt
Hanford Engineering Development Laboratory

William L. Frankhauser
System Planning Corporation

Allen E. Asselmeier
Sandia National Laboratories

William J. Barattino
United States Air Force

George A. Beale
Air Force Rocket Propulsion Laboratory

David Bents
NASA Lewis Research Center

Harvey S. Bloomfield
NASA Lewis Research Center

James A. Bond
General Electric Company

Frank L. Bouquet
Jet Propulsion Laboratory

Alfred O. Britting
Aerospace Corporation

John J. Buksa
Science Applications International Corporation

Roger D. Casagrande
General Electric Company

Donald Chick
EG&G Idaho, Inc.

Benjoy Choudhway
General Electric Company

William R. Determan
Rockwell International

Robert English
NASA Lewis Research Center

Robert Frisbee
Jet Propulsion Laboratory

T. Fujita
Jet Propulsion Laboratory

Y. Gohar
Argonne National Laboratory

Fahamy Haggag
Battelle Columbus Laboratory

Richard B. Harty
Rockwell International

R. L. Heestand
Oak Ridge National Laboratory

M. M. Hui
General Electric Company

Ross M. Jones
Jet Propulsion Laboratory

Albert Juhasz
NASA Lewis Research Center

John J. Kirkpatrick
Babcock and Wilcox

James A. Lake
Idaho National Engineering Laboratory

James H. Lee, Jr.
Sandia National Laboratories

Patrick Magee
General Electric Company

Albert Marshall
Sandia National Laboratories

Arthur T. Mattick
University of Washington

Raymond A. Meyer
General Electric Company

Robert C. Nelson
United States Air Force

Alexander Parlos
University of New Mexico

William A. Ranken
Los Alamos National Laboratory

John Rice
United States Air Force

P. Roy
General Electric Company

Mohamed E. Sawan
University of Wisconsin

David Furgal
Sandia National Laboratories

Robert E. Gold
Westinghouse Electric Corporation

Maurice H. Hanes
Westinghouse Electric Corporation

E. A. Harvego
Idaho National Engineering Laboratory

James A. Horak
Oak Ridge National Laboratory

Michael Jacox
United States Air Force

Wayland T. Joyner
Hampden-Sydney College

Michael Kangilaski
General Electric Company

Prabhat Kumar
Boyertown, PA

Clarence Lee
Sandia National Laboratories

Zdzislaw Maczenski
University of Missouri-Rolla

Ram Manri
Jet Propulsion Laboratory

Peter K. Mast
Science Applications International Corporation

John Metzger
Los Alamos National Laboratory

A. Monteque
General Electric Company

E. E. Olich
General Electric Company

Robert Protsik
General Electric Company

Richard Rice
Idaho National Engineering Laboratory

John H. Rosenfeld
Thermacore, Inc.

Brian Rushton
Idaho National Engineering Laboratory

Michael Schuller
United States Air Force

Jong T. Seo
University of New Mexico

E. A. Skrabek
Fairchild Space Company

John A. Spitznagel
Westinghouse Electric Corporation

J. D. Stephen
General Electric Company

James Kelly Thomas
Texas A&M University

Cronin B. Vining
General Electric Company

Terry Wallace
Los Alamos National Laboratory

F. William Wiffen
Oak Ridge National Laboratory

R. Yahalom
General Electric Company

Joel C. Sercel
Jet Propulsion Laboratory

Ronald J. Sovie
NASA Lewis Research Center

Gerry Stapfer
Jet Propulsion Laboratory

James R. Stone
NASA Lewis Research Center

Frank Thome
Sandia National Laboratories

J. Wadworth
Lockheed Palo Alto Research Lab

Joe Wetch
Space Power, Inc.

C. H. Wu
University of Missouri-Rolla

AUTHOR INDEX

(Bold chapter numbers indicate senior authorship.)

KEY WORD LIST

(Numbers indicate chapters covering these subjects.)